MONOCLONAL ANTIBODY AND PEPTIDE-TARGETED RADIOTHERAPY OF CANCER

MONOCLONAL ANTIBODY AND PEPTIDE-TARGETED RADIOTHERAPY OF CANCER

Edited by

Raymond M. Reilly

WILEY

A JOHN WILEY & SONS, INC., PUBLICATION

Published by John Wiley & Sons, Inc., Hoboken, New Jersey
Published simultaneously in Canada

For general information on our other products and services or for technical support, please contact our Customer Care Department within the United States at (800) 762-2974, outside the United States at (317) 572-3993 or fax (317) 572-4002.

Wiley also publishes its books in a variety of electronic formats. Some content that appears in print may not be available in electronic formats. For more information about Wiley products, visit our web site at www.wiley.com.

Library of Congress Cataloging-in-Publication Data:

Monoclonal antibody and peptide-targeted radiotherapy of cancer / edited by Raymond M. Reilly.
    p. ; cm.
  Includes bibliographical references and index.
  ISBN 978-0-470-24372-5 (cloth)
  1. Cancer–Radioimmunotherapy. 2. Monoclonal antibodies–Therapeutic use.
3. Tumor proteins. I. Reilly, Raymond.
  [DNLM: 1. Neoplasms–radiotherapy. 2. Antibodies, Monoclonal–therapeutic use. 3. Peptides–therapeutic use. 4. Radioimmunotherapy–methods. 5. Receptors, Peptide–therapeutic use. QZ 269 M751 2010]
  RC271.R26M66 2010
  616.99′40642–dc22   2009045879

Printed in Singapore

10 9 8 7 6 5 4 3 2 1

To my mother, who always used to ask me, "What is a monoclonal antibody?" and, in another life would have been a wonderful scientist with her inborn fascination with medical discovery and knowledge.

CONTENTS

7. Radioimmunotherapy of Acute Myeloid Leukemia   **219**

Todd L. Rosenblat and Joseph G. Jurcic

In June 2009 at the 56th annual meeting of the Society of Nuclear Medicine in Toronto, the "Image of the Year" was selected by Dr. Henry N. Wagner Jr. from Johns Hopkins University [Figure 1(1)]. This image illustrated the high sensitivity of positron emission tomography (PET) with ^{18}F-2-fluorodeoxyglucose (^{18}F-FDG) to reveal complete responses as early as 3 months post-treatment with ^{90}Y-ibritumomab tiuxetan (Zevalin) or ^{131}I-tositumomab (Bexxar) in patients with non-Hodgkin's lymphoma (NHL) (2). These two radioimmunotherapeutics are the first to be approved by regulatory authorities for treating cancer. By highlighting this image, Dr. Wagner not only recognized the great advances that have been made over the past three decades in radioimmunotherapy (RIT) of NHL (3) but also pointed the way toward how this approach could be combined with achievements in imaging (4) to help further advance the field of molecularly targeted radiotherapy.

There remain many challenges to be overcome, however, particularly to extend the impressive results seen in NHL to RIT of the more prevalent solid tumors (3). RIT and peptide-directed radiotherapy (PDRT) of solid tumors have been restricted by low tumor uptake, dose-limiting toxicity to normal tissues including the bone marrow, and an intrinsically greater radioresistance (3). Nonetheless, the success of RIT of NHL has proven that this approach is scientifically sound, translatable to clinical practice, and feasible. Moreover, there has recently been progress in the treatment of solid tumors with targeted radiotherapeutics, particularly using innovative pretargeting techniques and in the setting of minimal residual disease (3).

My goal in assembling this book was to provide a single resource that would constitute an expert discussion of the diverse aspects of the field of monoclonal antibody and peptide-targeted radiotherapy of cancer. The chapters cover a wide range of topics including the optimization of design of biomolecules and their radiochemistry, cell and animal models for preclinical evaluation, important discoveries from key clinical trials of their effectiveness for the treatment of malignancies, an understanding of their radiation biology and dosimetry, considerations in their regulatory approval, and health economics issues that need to be appreciated to ultimately see their widespread use in clinical oncology. New emerging areas such as the role of molecular imaging in evaluating the response and resistance to targeted radiotherapy, a discussion of the bystander effect that may enhance its effectiveness, and the potential of combining cytolytic virus therapy with targeted radiotherapy have also been included.

Many of the chapters were authored by internationally renowned experts who have made seminal discoveries in the field and by others who are leaders in areas that will be important to its future. I am grateful to all authors for their excellent contributions and

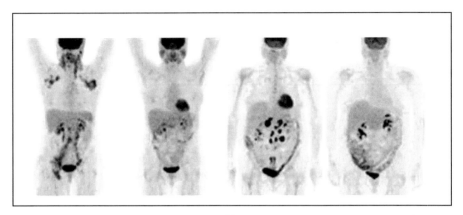

FIGURE 1 Whole-body PET scans using ^{18}F-2-fluoro-deoxyglucose demonstrating complete response in two patients receiving ^{131}I-tositumomab (Bexxar; left two images showing pre- and post-treatment) or ^{90}Y-ibritumomab tiuxetan (Zevalin; right two images showing pre- and post-treatment). (Reprinted with permission from Reference 1.)

thank them all for their patience as this book emerged. I am also indebted to my wife, Anita who tolerated the workload and spared some of the precious time that we have to spend together to accomplish this task. I believe that the book not only celebrates the substantial achievements of mAb and peptide-targeted radiotherapy of cancer but also acknowledges its limitations and failures—as Henry Ford said, "Failure is simply an opportunity to begin again, this time more intelligently." A great deal has certainly been learned, approaches are now more informed and elegant, and it is expected that this new knowledge will build on the pioneering discoveries in targeted radiotherapy of NHL that have proven so successful as aptly presented in Dr. Wagner's selection of the Image of the Year. I hope that this book will provide the impetus for discussion, encourage continued contributions to the advancement of the field, and stimulate the imagination of those who would aspire to set its future.

RAYMOND M. REILLY

Toronto, Ontario, Canada
January 2010

REFERENCES TO THE PREFACE

1. Anonymous . International interest focuses on SNM annual meeting. *J Nucl Med*. 2009; **50**: 16N–18N.
2. Iagaru A, Mittra E, Goris M. ^{131}I-tositumomab (Bexxar) vs. ^{90}Y-ibritumomab (Zevalin) therapy of low grade refractory/relapsed non-Hodgkin lymphoma. *J Nucl Med*. 2009; **50** (Suppl. 2): 12P (abstract no. 47).
3. Goldenberg DM, Sharkey RM. Advances in cancer therapy with radiolabeled monoclonal antibodies. *Quarterly J Nucl Med*. 2006; **50**: 248–264.
4. McLarty K, Reilly RM. Molecular imaging as a tool for targeted and personalized cancer therapy. *Clin Pharmacol Ther*. 2007; **81**: 420–424.

CONTRIBUTORS

Norbert Avril, Queen Mary University of London, London, United Kingdom.

Darell D. Bigner, Departments of Radiology, Surgery, and Pathology and the Preston Robert Tisch Brain Tumor Center, Duke University, Durham, North Carolina.

Lisa Bodei, Departments of Nuclear Medicine and Internal Medicine, Erasmus University, Rotterdam, The Netherlands.

Ann F. Chambers, Department of Oncology, University of Western Ontario and London Regional Cancer Program, London, Ontario, Canada.

Martijn van Essen, Departments of Nuclear Medicine and Internal Medicine, Erasmus University, Rotterdam, The Netherlands.

John B. Fiveash, Department of Radiation Oncology, University of Alabama, Birmingham, Alabama.

David M. Goldenberg, Garden State Cancer Center, Center for Molecular Medicine and Immunology, Belleville, New Jersey.

Marion de Jong, Departments of Nuclear Medicine and Internal Medicine, Erasmus University, Rotterdam, The Netherlands.

Joseph G. Jurcic, Leukemia Service, Department of Medicine, Memorial Sloan Kettering Hospital, New York.

Wouter W. de Herder, Departments of Nuclear Medicine and Internal Medicine, Erasmus University, Rotterdam, The Netherlands.

Jeffrey S. Hoch, Pharmacoeconomics Research Unit, Cancer Care Ontario and the Department of Health Policy, Management and Evaluation, University of Toronto; Centre for Research on Inner City Health, The Keenan Research Centre, Li Ka Shing Knowledge Institute, St. Michael's Hospital, Toronto, Ontario, Canada.

Boen L. R. Kam, Departments of Nuclear Medicine and Internal Medicine, Erasmus University, Rotterdam, The Netherlands.

Amin Kassis, Department of Radiology, Harvard Medical School, Harvard University, Boston, Massachusetts.

Eric P. Krenning, Departments of Nuclear Medicine and Internal Medicine, Erasmus University, Rotterdam, The Netherlands.

Dik J. Kwekkeboom, Departments of Nuclear Medicine and Internal Medicine, Erasmus University, Rotterdam, The Netherlands.

John Lewis, Department of Oncology, University of Western Ontario and London Regional Cancer Program, London, Ontario, Canada.

Leonard G. Luyt, Departments of Oncology, Chemistry, and Medical Imaging, University of Western Ontario and London Regional Cancer Program, London, Ontario, Canada.

Judith Andrea McCart, Institute of Medical Sciences and Department of Surgery, University of Toronto and Toronto General Research Institute, University Health Network, Toronto, Ontario, Canada.

Diane E. Milenic, National Cancer Institute, Bethesda, Maryland.

Carmel Mothersill, Department of Medical Physics and Applied Radiation Sciences, McMaster University, Hamilton, Ontario, Canada.

David Murray, Department of Oncology, Division of Experimental Oncology, University of Alberta, Edmonton, Alberta, Canada.

Kathryn Ottolino-Perry, Institute of Medical Sciences, University of Toronto and Toronto General Research Institute, University Health Network, Toronto, Ontario, Canada.

David A. Reardon, Departments of Radiology, Surgery, and Pathology and the Preston Robert Tisch Brain Tumor Center, Duke University, Durham, North Carolina.

Raymond M. Reilly, Leslie Dan Faculty of Pharmacy, Departments of Pharmaceutical Sciences and Medical Imaging, University of Toronto and the Toronto General Research Institute University Health Network, Toronto, Ontario, Canada.

Todd L. Rosenblat, Leukemia Service, Department of Medicine, Memorial Sloan Kettering Hospital, New York.

Colin Seymour, Department of Medical Physics and Applied Radiation, McMaster University, Hamilton, Ontario, Canada.

Robert M. Sharkey, Garden State Cancer Center, Center for Molecular Medicine and Immunology, Belleville, New Jersey.

Sui Shen, Department of Radiation Oncology, University of Alabama, Birmingham, Alabama.

Connie J. Sykes, E.G.A. Biosciences Inc., Edmonton, Alberta, Canada.

Thomas R. Sykes, E.G.A. Biosciences Inc. and Division of Oncologic Imaging, Department of Oncology, Faculty of Medicine and Dentistry, University of Alberta, Edmonton, Alberta, Canada.

Eva A. Turley, Department of Oncology, University of Western Ontario and London Regional Cancer Program, London, Ontario, Canada.

Roelf Valkema, Departments of Nuclear Medicine and Internal Medicine, Erasmus University, Rotterdam, The Netherlands.

Michael Weinfeld, Department of Oncology, Division of Experimental Oncology, University of Alberta, Edmonton, Alberta, Canada.

Thomas E. Witzig, Division of Internal Medicine and Hematology, Mayo Clinic and Mayo Foundation, Rochester, Minnesota.

Michael R. Zalutsky, Departments of Radiology, Surgery, and Pathology and the Preston Robert Tisch Brain Tumor Center, Duke University, Durham, North Carolina.

CHAPTER 1

Antibody Engineering: Optimizing the Delivery Vehicle

DIANE E. MILENIC

1.1 INTRODUCTION

The progression of monoclonal antibodies (MAbs) for radioimmunotherapy (RIT) has been driven by the need to solve a series of problems. As variants of antibodies have been developed and evaluated in preclinical studies, opportunities and limitations have become evident. Recent advances in DNA technology have led to the ability to tailor and manipulate the immunoglobulin (Ig) molecule for specific functions and *in vivo* properties. This chapter discusses the use of monoclonal antibodies for radiotherapy with an emphasis on the problems that have been encountered and the subsequent solutions.

The exploration of monoclonal antibodies as vehicles for the delivery of radio-nuclides for therapy has been ongoing for almost 50 years (1). In 1948, Pressman and Keighley reported the first *in vivo* use of a radiolabeled antibody for imaging (2). Ten years later, the first report of radiolabeled tumor-specific antibodies was utilized for radioimmunodiagnosis, and in 1960, radiolabeled antibodies were used to selectively deliver a therapeutic dose of radiation to tumor tissue (1, 3). Even at these early stages, investigators were quick to realize the obstacles associated with utilizing antibodies for radioimmunotherapy. Radiation doses delivered to tumors in patients were too low to have significant effects on tumor growth, and the prolonged retention of the radiolabeled antibodies in the blood led to toxicity complications (4). The inherent heterogeneity in specificity and affinity of polyclonal antibodies resulted in *in vivo* variability. The advent of hybridoma technology and the ability to generate mono-specific, monoclonal antibodies produced a resurgence in the use of antibodies as "magic bullets" (5, 6). In the 1980s, the literature exploded with reports of radiolabeled MAbs being evaluated in the clinical setting, initially in radioimmunodiagnostic applications, confirming that MAbs against tumor-associated antigens could target

Monoclonal Antibody and Peptide-Targeted Radiotherapy of Cancer, Edited by Raymond M. Reilly
Copyright © 2010 John Wiley & Sons, Inc.

tumors in patients. Subsequently, RIT clinical trials were initiated to deliver systemi-cally administered radiation to tumors with a specificity that would spare normal tissues from damage (7). This optimistic viewpoint was quickly tempered by the realization of the obstacles inherent to the use of a biological reagent, especially one of xenogeneic origin.

The preclinical and clinical RIT trials exposed the major constraints to the successful clinical use of radiolabeled MAbs: (i) development of human anti-murine immunoglobulin antibodies (HAMA); (ii) inadequate (low) therapeutic levels of radiation doses delivered to tumor lesions; (iii) slow clearance of the radiolabeled MAbs (radioimmunoconjugates) from the blood compartment; (iv) low MAb affinity and avidity; (v) trafficking to, or targeting of, the radioimmunoconjugates to normal organs; and (vi) insufficient penetration of tumor tissue (8, 9). In addition, there were toxicities associated with conjugated radionuclides when the radioimmunoconju-gates were metabolized or when the radionuclide dissociated from the immunocon-jugate (9). With these problems in mind, a primary focus has been to optimize RIT by manipulating the MAb molecule. As technology permitted, this was initially accom-plished with chemical or biochemical techniques to generate a variety of immuno-globulin forms but is now predominated by genetic engineering.

1.2 INTACT MURINE MONOCLONAL ANTIBODIES

In May 2008, a perspective on MAbs by Reichert and Valge-Archer (10) reported that in the periods 1980–1989, 1990–1999, and 2000–2005, 37, 25, and 8 murine MAbs, respectively, were evaluated in the clinic as cancer therapeutics. During this entire 25-year period, radiolabeled MAbs comprised 33% of the murine MAbs (10). To date, only two radiolabeled murine (mu) MAbs, both targeting CD20, have received FDA approval. Zevalin, ^{90}Y-rituxan (ibritumomab-tiuxetan), was approved in 2002 and is indicated for relapsed or refractory low-grade follicular transformed non-Hodgkin's lymphoma (NHL). The overall response rate of patients is reported to be 80%; 46% for those with rituximab refractory disease (11). Bexxar (^{131}I-tositumomab) was ap-proved in 2003 for the treatment of non-Hodgkin's B-cell lymphoma in rituximab refractory patients (see Chapter 6). Objective responses following ^{131}I-tositumomab therapy have ranged from 54% to 71% in patients who have undergone previous therapies while for newly diagnosed patients the response rates are 97% with 63% of those experiencing a complete response (12).

In clinical trials using muMAbs for RIT of solid tumors, approximately 73% (ranging from 16% to 100%) of the patients developed HAMA following a single infusion of MAb (13). In contrast, only about 42% of the patients in RIT trials for treatment of hematologic malignancies develop HAMA. When multiple doses of a radioimmunoconjugate have been administered, the amount of MAb that effectively targets tumor tissue is usually compromised after the second administration (13). In general, the human antibody response, especially at earlier time points, is directed against the Fc portion of the MAb molecule (Fig. 1.1). With the passage of time and particularly after repeated infusions, the specificity of the human antibody response

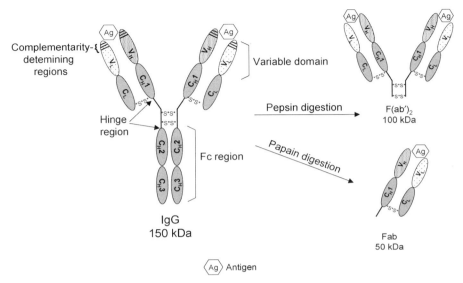

FIGURE 1.1 Schematic of an immunoglobulin structure. Enzymatic digestion of the intact IgG molecule yields F(ab')$_2$ and Fab fragments.

matures and becomes increasingly specific for the variable region of the MAb (13). In some instances, anti-variable region antibodies develop after a single infusion of the MAb (13, 14). This response has the potential of directly inhibiting the ability of the injected MAb from interacting with the targeted tumor (14). As with any therapeutic regimen, for RIT to be effective, multiple treatment cycles will be necessary. Immunomodulatory drugs such as deoxyspergualin, cyclosporin A, or cyclophosphamide have been evaluated as a means of minimizing or suppressing a patient's immune response during RIT (15).

To address these challenges of MAb-directed therapy, several strategies have been employed that center around modifying the MAb molecule. These alterations include reduction in the size of the MAb molecule, deglycosylation, or the addition of side groups. Reduction in size of the MAb molecule has been accomplished through methods such as enzymatic cleavage or genetic engineering (16–18). Digestion of an antibody with pepsin removes the Fc region of the heavy chain on the carboxyl terminus of cysteamine producing F(ab')$_2$ fragments that retain two antigen binding sites and have a molecular weight of ~100 kDa (Fig. 1.1). Fab fragments are generated by digestion with papain, an enzyme with a specificity for the amino group of cysteines. In this case, the disulfide bridges between the heavy chains are removed with the Fc region, which results in a molecule ($M_r \sim 50$ kDa) with one antigen binding site. Fab' fragments are produced through reduction and alkylation of F(ab')$_2$, which also yields a MAb molecule with a single antigen binding site and an M_r of ~50 kDa (16–18). Comparisons of intact MAbs and F(ab')$_2$ fragments (Fig. 1.1) in RIT clinical trials have demonstrated that the F(ab')$_2$ fragments do have a shorter serum half-life than intact MAbs. Patient antibody responses against F(ab')$_2$ fragments appear to occur with lower frequency after a single administration of the radio-immunoconjugate. Furthermore, some objective responses to treatment with a radi-olabeled $F(ab')_2$ fragment have been observed (19, 20). Autoradiographic studies of radiolabeled MAbs administered to athymic mice bearing human tumor xenografts have illustrated the ability of Fab' and $F(ab')_2$ fragments to penetrate tumor tissue with greater efficiency than intact MAbs (20, 21). The pharmacokinetics of Fab or Fab' fragments is even more rapid than $F(ab')_2$ fragments ($t_{1/2}\alpha \sim 10$ min, $t_{1/2}\beta \sim 1.5$ h for Fab' fragments versus $t_{1/2}\alpha \sim 30$ min, $t_{1/2}\beta \sim 12$ h for $F(ab')_2$ fragments) (22). In general, Fab and Fab' fragments have proven to be less immunogenic than intact MAbs (23). Their greatest disadvantage for RIT applications is their high and persistent renal localization, which appears to be a function of molecular size (22), which greatly increases the risk for renal toxicity. The degree to which the radiolabel is retained in the kidneys depends on the radionuclide and the radiolabeling chemistry (see Chapter 2). Radioiodinated MAbs are rapidly dehalogenated and the radioiodine excreted via the kidneys or into the stomach and intestines. Free radioiodine is trapped in the thyroid gland if there is inadequate blocking with stable iodine. Chelated radiometallonu-clides, that is, ^{111}In, ^{90}Y, and ^{177}Lu, are not as readily eliminated from normal tissues when the radioimmunoconjugate is metabolized (24). The retention of radiometals in the kidneys is due to the reabsorption of antibody fragments after their glomerular filtration followed by degradation of the radioimmunoconjugates with trapping of radioactive metabolites within the renal tubular cells (22, 24, 25). Although they are readily eliminated from the body, radioiodines may also pose a concern for toxicity to renal tissue, depending on the dose of radioactivity administered. An effective means of enhancing renal excretion of the radioimmunoconjugates is the blocking of its readsorption from the luminal fluid in the proximal tubules by administering basic amino acids such as lysine or arginine, prior to or with the radiolabeled MAb fragment (26, 27).

Fragments of MAb that retain immunoreactivity, however, are often difficult to generate (22). As mentioned, they are prepared by proteolytic digestion of intact MAb using enzymes, a procedure that must be optimized for each MAb and usually requires threefold or more MAbs to obtain the final desired quantity of the fragment. The process is inefficient and costly when producing the amounts necessitated by a RIT clinical trial.

1.3 RECOMBINANT IMMUNOGLOBULIN MOLECULES

Antibodies consist of four polypeptide chains, two heavy and two light chains, connected by disulfide bonds; the heavy chains are glycosylated (Fig. 1.1). Several criteria must be met to generate and produce genetically engineered antibodies. First, a host cell is needed that would produce and secrete a properly assembled functional antibody molecule with the appropriate carbohydrate side chains. Second, the DNA must be introduced into the recipient cell in an efficient manner. Finally, expression vectors must be available that permit the expression of the introduced genes as well as the isolation of the cells expressing the introduced antibody genes (28). The vectors require a plasmid origin for replication, a gene encoding a selectable biochemical phenotype in bacteria and a gene encoding a selectable marker in eukaryotic cells. The creation of recombinant immunoglobulin molecules also requires the transfection of the host cell with two expression vectors, one containing the gene for the heavy chain and the other containing the gene that encodes the light chain.

1.3.1   Chimeric Monoclonal Antibodies

Chimeric MAbs are constructed by ligating the gene encoding the variable region of a murine MAb to the gene encoding the constant region of a human Ig (Fig. 1.2). There are a variety of vectors available into which the murine and human Ig gene sequences can be inserted. In turn, there are a number of expression systems, prokaryotic and eukaryotic, into which the recombinant Ig genes can be introduced and the protein expressed (28, 29). The ability to tailor a MAb of a particular specificity for a certain function broadens the horizon for MAb-directed therapies.

The first clinical trial involving a recombinant/chimeric MAb employed MAb 17-1A, which recognizes the 40 kDa glycoprotein designated epithelial-specific cell adhesion molecule (EpCAM) (30–32). The variable region of MAb 17-1A was fused with a human $IgG_{1\kappa}$ sequence. Ten patients with metastatic colon carcinoma were given injections of the chimeric (ch) 17-1A. Only one of the patients who received multiple injections developed a low titer antibody response against ch17-1A that was directed against the variable region of the chMAb and not against the human constant domains. In addition, the pharmacokinetics of the ch17-1A was slower than the original murine MAb by sixfold.

Several chMAbs have since been constructed, characterized in preclinical *in vitro* and *in vivo* studies, and have been evaluated in RIT clinical trials. Direct comparisons

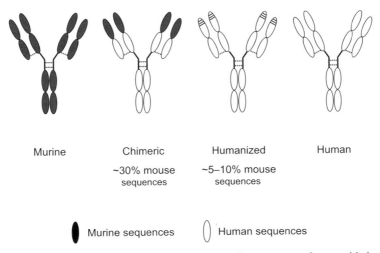

Murine   Chimeric   Humanized   Human

~30% mouse   ~5–10% mouse
sequences   sequences

● Murine sequences   ◖ Human sequences

FIGURE 1.2   The "humanization" of the murine IgG to generate forms with increasing percentages of human sequences.

of the chimeric and murine forms of a MAb (B72.3) determined that the chimeric form was quantitatively superior in tumor targeting (33). This enhanced tumor targeting of the chMAb was attributed to its longer plasma half-life, approximately 4.7-fold longer than that of muB72.3. Unfortunately, chMAbs have also proven to be immunogenic in patients. Evidence suggests that the degree of immunogenicity may be dependent on the particular MAb. The murine MAb 17-1A elicited antibody responses in 77% of the patients, while the chimeric 17-1A evoked a humoral response in only 5–10% of the patients. In contrast, chB72.3 evoked an antibody response in patients with at least the same frequency as muMAb B72.3 (13). Minimal antibody responses have been reported for patients receiving rituximab, a chimeric anti-CD20 MAb used for non-Hodgkin's B-cell lymphoma; this may be attributed in part to the impaired immune status of these patients (11, 34). The antibody responses appear not as robust as the HAMA responses, and in some cases, more than one dose of chMAb may be administered before an antibody response against the chMAb is elicited. Further humanization of the murine MAb has been accomplished by grafting the complementarity-determining regions (CDRs) of the murine MAb into the variable light (V_L) and the variable heavy (V_H) frameworks of a human MAb (Fig. 1.2) (35).

1.3.2 Humanized Monoclonal Antibodies

1.3.2.1 CDR Grafting X-ray crystallographic studies have shown that the contact of antibodies with antigen is through amino acid residues within the complementarity-determining regions (36). Some of the surrounding framework amino acid residues are also involved in interactions with the cognate antigen (36, 37). It is crucial to maintain the CDRs as well as the interactions of the CDRs with each other and the rest of the variable domains if the binding specificity of the MAb is to be preserved. The proper configuration, or conformation, of the binding site requires retention of crucial framework residues, which include those involved in V_H and V_L associations and those that affect the overall domain and combining site structure (36). The necessary framework residues can be identified through high-resolution X-ray crystallographic studies; otherwise, molecular modeling based on the structure of related molecules or the ligand binding properties of site-specific mutants can facilitate identification of required amino acids for correct conformational positioning of CDRs for antigen binding. It is estimated that chimeric antibodies contain ~30% murine sequences; CDR grafting would reduce the nonhuman content to 5–10% (38). Selecting the appropriate human acceptor template for the CDRs is another crucial element for the successful humanization of a murine MAb. The strategy is usually to choose a human template with the greatest sequence homology to the murine MAb being grafted (39). The polymerase chain reaction (PCR) technology has enabled investigators to graft the entirety of CDRs along with the necessary framework residues from a murine MAb into the human frameworks of human Igs (40).

Humanized (hu) MAbs have progressed through evaluation in animal models and into clinical trials. Trastuzumab (Herceptin®, Genentech) that targets HER2 was the first humanized MAb to gain FDA approval (1998) for the treatment of HER2-positive metastatic breast cancer. Three other humanized MAbs have since been approved for the treatment of cancer patients. Two of these, bevacizumab (Avastin, Genentech) and alemtuzumab (Campath-1H, Berlex), are administered as naked MAbs and one, Mylotarg (gemtuzumab ozogamicin, Wyeth), is conjugated with the toxin calicheamicin. The naming of drugs is a consensus between the United States Adopted Names system, the inventor/discoverer of the drug, and the FDA. The American Medical Association established the guidelines for assigning generic names of MAb drugs. The foundation to the designation is the suffix "MAb" for monoclonal antibody. Letters, or infixes, before the suffix denote the source, that is, o for mouse, xi for chimeric, zu for humanized, and u for human.

The transition to humanized MAbs proved that empirical evaluation of each MAb was required. CAMPATH-1H, an anti-human CD52 MAb, was found to have a lower affinity than the original rat MAb (41). This was remedied when two amino acids in the V_H framework were mutated back to the original rat MAb sequence. In general, preclinical studies have demonstrated that the CDR-grafted MAbs retain the ability to react with their tumor antigen. In some instances, the huMAbs have had higher affinities than the original murine MAb. HuM195, an anti-CD33 MAb, was found to have a 3–8.6-fold increase in affinity and avidity (42). Other huMAbs, that is, MN-14, an anti-CEA MAb, also proved to have improved tumor targeting over the murine MAb (43). In contrast, the CDR grafting of other MAbs has yielded Ig molecules with decreased antigen affinity. For example, huCC49 has been found to have a two- to threefold lower relative affinity compared to the murine CC49 MAb (39).

Perhaps more interesting is the plasma pharmacokinetic data collected from clinical trials with some of the huMAbs. In general, one would anticipate that a huMAb injected into patients would have a longer residence time in the blood than a xenogeneic muMAb. This was in fact true for some huMAbs. The plasma half-life of huBrE-3, a MAb that targets breast epithelial mucin, was twofold greater than that for the murine BrE-3, 114.2 ± 39.2 and 56 ± 25.4 h, respectively (44). This prolonged retention of the radioimmunoconjugate in the blood may result in increased myelotoxicity. However, if the huMAb has reduced immunogenicity, then multiple cycles of radioimmunoconjugate at lower radiation doses (dose fractionation) would be possible and still result in effective RIT. On the other hand, the plasma pharmacokinetics of two other huMAbs (MN-14 and M195) proved to be similar to their parental murine forms (43, 45). This latter phenomenon may be reflective of the MAb interacting with antigen and/or tumor cells present in the blood.

As mentioned previously, huMAbs possess a murine sequence content of 5–10% and this amount of xenogenic sequence has proven to be sufficiently immunogenic in patients to elicit humoral responses. The huMAbs of anti-TAC and anti-CD18 were evaluated in subhuman primates and found to be immunogenic with anti-idiotypic antibody responses detected (46, 47). Humanized anti-TNFα, when administered at doses of 1, 2, 5, and 10 mg/kg, elicited antibody (IgM) responses in normal human volunteers (48). Antibody responses have also been detected in patients receiving weekly 2–4 mg/kg (i.v.) of trastuzumab (49). In general, the protein amounts of radiolabeled MAbs that are injected into patients are lower; the immune responses directed against each of these huMAbs may not be relevant to the use of radiolabeled MAbs. No evidence of a human anti-human antibody (HAHA) response in patients receiving huJ591, radiolabeled with ^{131}I, ^{90}Y, or ^{177}Lu, has been detected nor has a response been detected in patients receiving as many as three injections of ^{131}I-huMN-14 (45, 50, 51).

Studies have been conducted to characterize the immune response against two MAbs in greater detail. Schneider et al. identified specific CDRs in the huMAb anti-Tac that were recognized by antibodies in the sera of cynomolgus monkeys that had received these antibodies (46). The majority of the antibody response was found to be directed against the heavy-chain CDRs 1 and 2 as well as the light-chain CDR3 of the humanized anti-Tac. No detectable response was directed solely against any one CDR or to the modified framework of the human variable regions. A similar study was performed using the serum of a patient who had received ^{177}Lu-labeled muMAb CC49 for RIT (52). In this particular study, the patient's antibody response was determined to be directed toward the heavy-chain CDR2 and the light-chain CDRs 1 and 3 (53). It was also found that these same CDRs were required for antigen binding. The information from such studies led to the development of huMAb using SDR (specificity-determining residue) and abbreviated CDR grafting, with the objective of creating a minimally immunogenic Ig molecule that retained optimal antigen binding and affinity.

1.3.2.2 *SDR Grafting* The specificity-determining residues comprise only 20–33% of the CDR residues; therefore, the CDRs could be humanized by up to 80% when only the SDRs are grafted (54). The process requires identification of SDR and non-SDR residues within the CDR. When a crystal structure of the antibody–antigen complex is available, SDR/non-SDR residues are readily identified. Lacking the crystal structure, the indispensability of SDR residues can be tested through genetic engineering. Based on known antibody–antigen complexes, it appears that there is little variation in the regions that contain SDRs; antibodies with unknown structures will likely have SDR residues in the same positions. Therefore, only a few variants are required to identify those SDR residues that are required for antigen binding and the non-SDRs can then be replaced with corresponding human residues (54). The muMAb COL-1, which reacts with carcinoembryonic antigen (CEA), was humanized by SDR grafting while huCC49 was subjected to further refinement (14, 55). Variants of both MAbs were generated using a baculovirus expression system and tested *in vitro* for antigen binding. One variant of HuCC49 exhibited superior binding and tumor targeting properties compared to the original huCC49. As with the grafting of whole CDRs, it is crucial in SDR grafting that an appropriate human framework is chosen as well as retaining the framework residues that are needed for maintaining the conformation of the antigen binding site. The evaluation of such Ig molecules in clinical trials will determine if the objective of minimizing immunogenicity has been achieved. The affinities of the CDR- or SDR-grafted MAbs can be further manipulated with methods such as *in vitro* affinity maturation using phage display techniques (56).

1.3.2.3 *Abbreviated CDR Grafting* To further reduce the number of murine residues of the huMAb, grafting of only the SDRs into the human Ig framework, coined as "abbreviated" CDR grafting, has been proposed (54). Engineering huCOL-1 in this fashion resulted in a 2.7-fold decrease in affinity compared to CDR-grafted huCOL-1 and a 4.3–5-fold lower affinity compared to muCOL-1 (57). Unfortunately, these humanized forms were not evaluated *in vivo* for tumor targeting, but the trend in decreasing affinity provides an argument for retaining residues from the murine framework to maintain the binding site conformation of the MAb.

One fact is clear, the insertion of mouse sequences into human sequences and the further replacement of sequences in the Ig to alter properties and reduce immunogenicity of the MAb is laborious, requiring remodeling and engineering MAb by MAb. After all of these manipulations, mouse sequences remain and even though each step has reduced the immunogenicity, the molecules still elicit an antibody response in patients.

1.3.3 Human Monoclonal Antibodies

Human MAbs against tumor-associated antigens are believed to be the ideal agent for clinical applications. One of the main obstacles to administering a xenogeneic protein, immunogenicity, would be absent, or minimal, if a syngeneic antibody was available. The biological characteristics (metabolism and pharmacokinetics) (58) would, however, differ appreciably from muMAbs. Human MAbs have been generated that are reactive with antigens present in human tumors by fusing lymphocytes (myeloma cells) from cancer patients, thus creating human–human hybridomas. However, very few have demonstrated the necessary specificity or affinity to merit their use in clinical trials (59, 60). Inherent human tolerance to human antigens along with the reality that human subjects will not undergo the immunization regimen required to generate antibodies has understandably limited the possibilities. Recombinant DNA technology has hence provided the tools to create completely human MAbs. In the early 1980s, the race began to create a transgenic mouse for human Ig that possessed the heavy- and light-chain repertoires that would be capable of generating a secondary immune response that would result in high-affinity antibodies. Strategies taken to accomplish this utilized homologous recombination in mouse embryonic cells to disrupt the endogeneous heavy- and light-chain genes. Construction and introduction of human unrearranged gene segment sequences is where strategies differ. One method used fusions of yeast protoplasts to deliver yeast artificial chromosome (YAC)-based minilocus transgenes into mouse embryonic cells (61). A second method used pronuclear microinjection of reconstructed minilocus transgenes into the mouse cell (62). The numbers of heavy-chain variable (V), diversity (D), and joining (J) segments varied in each of these transgene reconstructions and were not the whole repertoire. However, each could be shown to undergo VDJ joining and class switching in the transgenic mice. In both studies reported, the mouse heavy-chain genes were inactivated, the light-chain genes were not, and expression of a functional mouse light chain was observed. Further analysis determined that the resulting subpopulation of B cells did not interfere with the isolation of hybridoma cell lines that secreted fully human MAbs that were reactive with the immunizing antigen. Subsequently, transgenic mice have been created that express complete human heavy- and light-chain repertoires with high-affinity MAb isolated (63, 64).

There is also the *in vitro* approach to generate human MAbs using phage display. Methods were developed for cloning expressed Ig variable region cDNA repertoires to create phage display libraries of antibody variable fragments. Sequences can be selected based on the desired properties and then enriched (65). The libraries are restricted to the donor's exposure to antigen, which dictates whether early or mature B-cell response Igs are present for selection. The technology has been further refined since the first description of generating antibody variable domains with affinity maturation (66).

The first fully human MAb that gained FDA approval in 2002 was created using the phage display platform; adalimumab, an anti-TNFα antibody, was approved for the treatment of inflammatory diseases (67). Panitumumab (Amgen and Abgenix), approved in 2006 for the treatment of patients with EGFR-expressing metastatic colorectal cancer, is the first commercially available human MAb generated using transgenic mice (68, 69). The human MAbs have been well tolerated and to date appear to be less immunogenic than the chimeric MAbs (69). High-affinity human MAbs have been generated with specificities for numerous antigens that include cytokines, growth factors, CD antigens, and nuclear factor receptors using both phage display and transgenic technologies (67). A survey of the literature suggests that the latter approach though may be the favored route to obtaining human MAbs. In a recent report tabulating selected human MAbs that are in clinical development, 45 are from transgenic mice while 16 are from phage display libraries (67). Thirty-five of these human MAbs were developed as cancer therapeutics with 28 derived from transgenic mice. The favoring of the transgenic mouse platform most likely is a reflection of the processes involved in moving from discovery to the clinical setting. In general, the MAbs that are initially identified when generated from a transgenic mouse will go into production and development without the need for further manipulation. In contrast, it appears that the phage display-generated MAbs have consistently required additional tweaking such as affinity maturation.

1.4 NANOBODIES

Nanobodies are the smallest antigen binding regions or fragments of naturally occurring heavy-chain antibodies (HCAbs). Lacking a light chain, these fully functional HCAbs were identified as part of the humoral response in camels, dromedaries, and llamas (Camelidae) (70). HCAbs have also been identified in wobbegong and nurse shark (71). The structure of the HCAbs consists of a single variable domain (VHH), a hinge region, and two constant domains, C_H2 and C_H3 (Fig. 1.3). The VHH region contains three CDRs for antigen binding. The HCAbs lack the C_H1 domain, which is actually contained in the genome, but is spliced out during mRNA processing. This absence would explain the lack of light chain since it is the C_H1 domain that interacts with the C_L domain of the light chain. The CDRs of HCAbs appear to be structurally larger, those from the dromedary contain 16–18 amino acids (a.a.), compared to human (12 a.a.) or mouse (~9 a.a.) CDRs. This structural difference might serve as a means of providing a larger repertoire to the organism

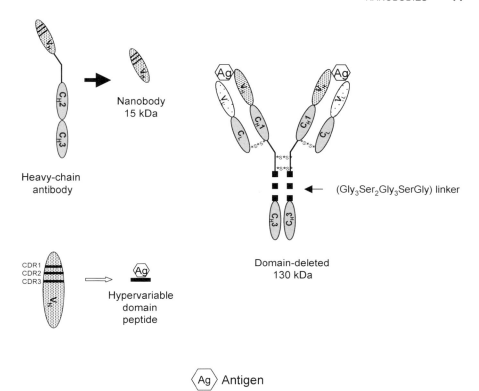

Nanobody
15 kDa

Heavy-chain
antibody

(Gly₃Ser₂Gly₃SerGly) linker

CDR1
CDR2
CDR3

Hypervariable
domain
peptide

Domain-deleted
130 kDa

⟨Ag⟩ Antigen

FIGURE 1.3 Illustration of a heavy-chain antibody, nanobody, domain-deleted MAb, and a hypervariable domain peptide.

since the V_L region and three CDRs are missing. CDR3 appears to be more exposed and the antigen binding site of the VHH also has protruding loops. This has the affect of increasing the surface of the HCAb paratope, making it as large as conventional antibodies.

The single domain of nanobodies (Fig. 1.3) simplifies the cloning, expression, and selection of antigen-specific molecules. Only one set of primers is needed and the HCAb has undergone affinity maturation *in vivo*; therefore, the library is relatively small ($10^6–10^7$ nanobody genes) from which high-affinity nanobodies are isolated (72). The nanobodies are soluble, nonaggregating proteins with an M_r of 15 kDa and can easily be produced in bacterial or eukaryotic systems. High-level nanobody production has been noted in a variety of expression systems (73). Nanobodies have also been shown to have high thermal and conformational stability. The melting points of the nanobodies range from 60 to 78 °C; following incubations at 90 °C, they have regained antigen binding/specificity (71). With such properties, nanobodies may prove to have a long shelf-life and may be manipulated under conditions not acceptable for other antibody forms such as radiolabeling at higher temperatures to obtain higher labeling efficiencies.

Radiolabeled nanobodies have been shown to efficiently target tumor xenografts in mice by microSPECT/CT and biodistribution studies. For the former, images of tumor xenografts were obtained 1 h after i.v. injection of anti-EGFR nanobodies labeled with ^{99m}Tc (74). For the latter, Balb/c mice bearing syngeneic tumors were injected i.v. with ^{125}I-labeled nanobodies that react with lysozyme. Tumor targeting was demonstrated at 2 and 8 h post-injection. More importantly, this study was conducted in immuno-competent mice; no antibody or T-cell responses were detected against the nanobody, suggesting that the nanobodies may not be immunogenic or their immunogenicity is very low, at least in this host (75). The single domain nature of nanobodies along with their physical properties makes them particularly interesting and appealing as a delivery vehicle for radionuclides or any other desired payload.

1.5 DOMAIN-DELETED MONOCLONAL ANTIBODIES

The recent advances in molecular cloning that led to the CDR grafting of MAbs have also led to modifying the domains of MAbs to alter their biological properties, that is, pharmacokinetics, with the objective of optimizing their therapeutic potential. Gillies and Wesolowski were the first to construct a F(ab')$_2$ fragment using genetic engineering techniques (76). They were unable to generate a bivalent molecule, nor were the resulting molecules reactive with antigen. In the pursuit of determining what portions of the Ig molecule were required for antigen binding, a construct with a C_H2 domain deletion was generated (Fig. 1.3) (77). This new MAb form, in this case a construct from chimeric MAb 14.18, which recognizes the ganglioside GD2, demonstrated a significantly faster elimination from the plasma compared to the intact and aglyco-sylated form of the same MAb. The pharmacokinetics was found to be similar to human IgG F(ab')$_2$ fragments. Maximal tumor targeting with radiolabeled ch18.14ΔC_H2 occurred at 12–16 h versus 96 h postinjection for the intact ch18.14. In addition, the domain-deleted variant did not exhibit the renal uptake of radioactivity that is usually associated with radiolabeled F(ab')$_2$ fragments (78). A similar C_H2 domain-deleted chimeric antibody of MAb B72.3 was reported by Slavin-Chiorini et al. (79). The chB72.3ΔC_H2 differed from the ch18.14ΔC_H2 in that a 10-amino acid linker (gly$_3$-ser$_2$-gly$_3$-ser-gly) was inserted in place of the deleted C_H2 domain, which provided stability to the molecule. Domain-deleted mutants have subsequently been produced of chimeric and CDR-grafted humanized forms of MAb CC49 that have been analyzed in preclinical *in vitro* and *in vivo* studies (80, 81). A C_H1 domain-deleted mutant of chCC49 was also produced and was compared to chCC49 and the chCC49ΔC_H2 Ig forms (81). The chCC49ΔC_H1 exhibited pharmacokinetics and tumor localization that were similar to those of the intact chCC49. In contrast, the pharmacokinetics of the chCC49ΔC_H2 was significantly faster in nontumor bearing athymic mice and rhesus monkeys than chCC49. Tumor targeting was also more efficient and occurred within an earlier time frame than that of chCC49. When labeled with a radiometal, that is, ^{177}Lu, the pharmacokinetics exhibited a profile similar to ^{131}I-chCC49ΔC_H2. The domain-deleted huCC49 has demonstrated these same desirable characteristics (80). The huCC49ΔC_H2 has shown efficacy in animal models for the treatment of peritoneal tumor deposits and subcutaneous tumors when radiolabeled with ^{177}Lu and ^{213}Bi, respectively (82, 83).

Two clinical trials with HuCC49ΔC_H2 have been conducted (84, 85). The first was a small pilot study of four colorectal cancer patients receiving 10 mCi (370 MBq) of ^{131}I-huCC49ΔC_H2 (84). Pharmacokinetics, biodistribution, dosimetry, and immune responses were evaluated. Targeting of metastatic disease was observed in all patients with no toxicities reported. The mean plasma elimination half-life was 20 $\pm$ 3 h with a mean residence time of 29 $\pm$ 2 h; this was a faster elimination rate than murine CC49. One of the patients appeared to develop a detectable antibody response at 6 weeks. The second trial enrolled 21 patients with recurrent and metastatic colorectal cancer (85). In this trial in which patients were administered 2 mCi (74 MBq) of ^{125}I-huCC49ΔC_H2, the pharmacokinetics was found to be similar to murine CC49, tumors were detectable, and no antibody response to the injected huCC49ΔC_H2 was detected. Overall, the investigators performing the trials reported that the huCC49ΔC_H2 was well tolerated.

Production of chCC49ΔC_H2, huCC49ΔC_H2, and chB72.3ΔC_H2 was found to result in what initially appeared to be impurities by SDS-polyacrylamide gel electrophoresis. The impurities were subsequently identified as isoforms of the domain-deleted molecules. Form A was proposed to contain the appropriate interchain disulfide between the two heavy chains. Form B was thought to be a result of the heavy chains associating through noncovalent interactions of the C_H3 domains. Instead of the disulfides forming interchain bonds, form B contained intrachain bonds. To favor production of form A with the interchain disulfide bonds, investigators at Biogen Idec modified the hinge region linker sequence to stabilize the hinge region and thus favor the correct disulfide bond formation (86). Insertion of a cysteine-rich 15-amino acid IgG$_3$ hinge motif along with an additional alanine and proline resulted in a product that was 98% form A isoform, with little or no form B detected. This alteration of the hinge region, unfortunately, resulted in a 1.4-fold decrease in the affinity of the huCC49ΔC_H2 (86).

1.6 HYPERVARIABLE DOMAIN REGION PEPTIDES

It is the CDRs in the variable domains that interact with the antigen epitope (36). As previously mentioned, this interaction depends on the tertiary structure of the antigen combining site. However, in some instances, the CDR sequence that interacts with the antigen may be linear. As a result, hypervariable (HV) domain region peptides, or molecular recognition units (MRUs), may be identified and synthesized that can target and bind to tumor-associated antigens (Fig. 1.3). A peptide, designated αM2, was synthesized based on the heavy-chain CDR3 and some framework residues of the anti-MUC-1 antibody, ASM2 (87). Analysis of the αM2 peptide determined that it adopted the β-strand conformation of the antigen binding structure of the intact MAb. Radiolabeled peptide was able to bind antigen, albeit at a lower affinity. Studies with the αM2 peptide progressed to a pilot clinical study. Twenty-six women with primary, recurrent, or metastatic breast cancer were injected with ^{99m}Tc-labeled αM2 (87).

Optimal radioimmunodetection occurred by 3 h postinjection and 77% of known lesions were visualized.

Reducing the antigen–antibody interaction to a single CDR may increase the potential for cross-reactivity with other antigens. Furthermore, the use of HV peptides is limited to those antibody–antigen interactions that do not rely on spatial conformations. Either of these obstacles, however, may not be insurmountable. Through molecular modeling and other sophisticated techniques, peptides may be designed with improved binding properties. Higher affinity binding HV peptides have been obtained by either dimerizing or constraining the conformation of the HV peptide by introducing cysteine residues that result in looping of the peptide and restricting the conformations it could assume as a linear peptide (88). It is also conceivable that HV peptides will be designed and synthesized with chelates or additional amino acids to facilitate labeling with radionuclides and/or to increase the specific activity of the radioimmunoconjugate for RIT procedures. The rapid pharmacokinetics in conjunction with the ease of synthesizing and producing peptides are desirable characteristics for radiopharmaceutical development.

1.7 FV FRAGMENTS

In 1988, single variable domains of a mouse Ig were expressed in *Escherichia coli* and shown to be functional (89, 90). Again, where enzymatic methods were limited and greatly variable in reproducibility, recombinant technology has greatly facilitated the generation and production of this antibody form. Fv fragments (Fig. 1.4) consist of the V_L and V_H domains of the antibody molecule. The associations between the domains are weak noncovalent interactions due to the lack of the C_H1 and C_L domains (91–93). In many cases the V_L–V_H associations were found to be reproducible and resulted in a functional binding site (92, 94). Unfortunately, the Fv fragments dissociated at low protein concentrations and were found to be unstable at physiological temperatures. The strategies taken to obtain stable Fv fragments were (i) chemical cross-linking, (ii) engineering of disulfide bonds into the molecule, or (iii) the insertion of a peptide linker between the V_L and V_H domains (Fig. 1.4). The employment of the peptide linker has been the most favored strategy and the resulting Ig form has been designated as scFv (single-chain Fv). The scFv is constructed by connecting the V_L and V_H genes with an oligonucleotide that encodes 15–25 amino acids and is expressed as a single polypeptide chain (89, 95). The variable domains can be assembled in either order (V_H–V_L or V_L–V_H) with examples in the literature of one orientation proving superior to the other (96, 97). The most common linker is (Gly_4-Ser)$_3$; linkers of 18-amino acid residues, or more, have been found to favor folding of the scFv resulting in a monomer form (98).

The scFv form has demonstrated several advantages over intact Ig MAbs, F(ab')$_2$, and Fab' fragments. The pharmacokinetics of elimination of the scFv and clearance from normal tissues is appreciably faster (22, 99). In therapeutic applications, this would translate to a reduction in radiation exposure to normal tissues. It has also been shown that scFvs penetrate tumor tissue more rapidly, farther, and with a greater

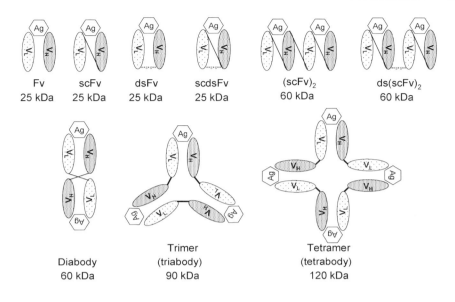

FIGURE 1.4 Schematic diagram of the various Fv forms.

degree of homogeneity (21). With these properties, scFvs have the potential of delivering a radiation dose more homogeneously throughout a tumor lesion (100). However, the low %ID/g may not permit the delivery of an adequate therapeutic dose of radioactivity if it is not matched with a radionuclide of an appropriate half-life (101).

Since scFv molecules have such rapid pharmacokinetics, the overall amount that accumulates in the tumor is low. The tumor uptake of two scFvs, both labeled with ^{125}I and evaluated in the same tumor model, ranged from 0.3 to 3.4%ID/g at 24 h postinjection with tumor-to-blood ratios ranging from 3.8 to 26.5 (22, 102). The tumor %ID/g at 24 h for the scFv of CC49 was 12.5-fold lower than that obtained with the intact murine CC49 MAb (22). Due to the low percentage of the radioimmunoconjugates in the blood, the scFv was actually a more attractive candidate for RIT especially when labeled with a short-lived radionuclide such as one of the α-emitters that are very potent even at low amounts of radioactivity delivered to tumors (100).

In general, the scFv has a diminished affinity, which is related to the loss of the second antigen binding site (bivalency) that would stabilize the antigen–antibody interaction (22). In one instance, an scFv was reported to have an affinity constant comparable to the parental IgG (99). Adams et al. (103) have been able to enhance the affinity of the scFv of C6.5 (which recognizes HER2/*neu*) through site-directed mutagenesis. Using a phage display library, C6.5 scFv mutants were isolated that varied 320-fold in their affinity compared to the nonmutated C6.5 scFv. The mutant with the highest affinity differed in only three amino acids in the V_L CDR3. This high-affinity mutant scFv showed a twofold increase in the ability to target tumor.

The mutant also demonstrated an improvement in the radiolocalization indices (tumor-to-normal organ ratio). More elegant studies by the same group have evaluated the effect of affinity on scFv tumor targeting, intratumoral diffusion, and tumor retention in greater detail (104). Variants of an anti-HER2/*neu* scFv were produced with affinities from 10^{-7} to 10^{-11} M. Biodistribution studies comparing these scFvs radiolabeled with ^{125}I revealed that tumor uptake/retention did not significantly increase beyond an affinity of 10^{-9} M and the differences in the pharmacokinetics of the scFvs were not a function of renal clearance. Immunohistochemical analysis of tumor xenografts following injection of the scFv affinity variants indicated that the scFvs with the lower affinity distributed diffusely throughout the tumor. Meanwhile, the scFvs with the higher affinities were localized primarily to the perivascular regions of the tumors. The implication is that antibody-based molecules with high affinities are restricted in their ability to penetrate tumors, which is yet another consideration in designing a targeting agent.

The ability of scFv to target tumor efficiently and to clear from normal tissues has been investigated in the clinical setting. Single-photon emission computed tomography (SPECT) and whole-body planar imaging were performed with CC49 scFv, radiolabeled with ^{123}I (105). Tumors were visualized; uptake of the radioimmunoconjugate by tumor tissue was determined directly from biopsy samples. The scFv was also determined to have a biphasic clearance with a distribution half-life ($T_{1/2}\alpha$) of 30 min and elimination half-life ($T_{1/2}\beta$) of 10.5 h. The patients did not receive any treatment to inhibit renal sequestration of the radiolabeled scFv; thus, significant uptake was evident in the kidneys. Similar results were also reported with an anti-CEA scFv radiolabeled with ^{123}I (106). More recently, promising results were reported for an anti-CEA scFv (MFE-23) for application in radioimmunoguided surgery (107). In this particular study, the $T_{1/2}\beta$ of the MFE-23 scFv was twofold greater than that reported previously.

An alternative to the peptide linker for stabilizing the orientation of the V_H and V_L domains has been the use of disulfide bridging introduced through cysteine residues in the sFv. This strategy has proven quite effective; disulfide sFvs (dsFvs) have been produced with reactivity against the IL-2 receptor, LYM-1, Lewis Y-related carbohydrate, CD19, and p185^{HER2} (108–112). As with the sFvs, the dsFvs have shown good tumor targeting, with rapid pharmacokinetics and excellent tumor-to-normal tissue ratios.

The combination of the peptide linker with disulfide bridging has also been explored as a means of providing stability and proper binding site configuration. The purification yield of this form, a single-chain dsFv (scdsFv), is appreciably better than that of the scFv form, with less aggregation occurring in the final product (113). In addition, the *in vitro* and *in vivo* properties of the scdsFv have proven to be equivalent to the scFv form (113, 114).

1.7.1 Multimeric Fv Forms

A variety of multimeric Fv forms have been created and assessed with the goal of improving the affinity of Fvs with desired pharmacokinetic properties. These multimeric forms include Fv dimers (diabodies), Fv trimers (triabodies), Fv tetramers, and minibodies. In addition, multimeric Fvs have been created with mono- and bispecificity. Diabodies (55 kDa) are noncovalent dimers of scFv fragments that are formed using short peptide linkers (3–12 amino acids) that promote cross-pairing or association of the V_H and V_L domains of two polypeptides (Fig 1.4) (115, 116). A scFv dimer of MAb T84.66, labeled with ^{125}I, was reported to have a three- to fivefold greater uptake in tumor than its scFv monomer counterpart with reduced radioactivity uptake observed in the kidneys (102). When labeled with the radiometal, ^{111}In, the T84.66 diabody again demonstrated good tumor targeting; however, renal and hepatic accumulation of radioactivity remained a problem (117). Diabodies have also been generated for the MAb CC49 scFv and the anti-HER2/neu scFv (C6.5) that were also found to have improved tumor targeting over their scFv monomer form (22, 118, 119). The C6.5 diabody was reported to have an increase in affinity of 40-fold over the C6.5 scFv. In contrast to the aforementioned scFv dimers, a diabody of the anti-MUC1 C595 MAb, created by replacing the $(Gly_4\text{-}Ser)_3$ with $(Gly_6\text{-}Ser)$, displayed binding reactivity to MUC-1 that was similar to the intact MAb C595 (120). Similar findings were reported for a dimer of the anti-HER2/*neu* scFv that was prepared in a comparable manner (118).

The use of diabodies for targeted radiation therapy has been pursued using the α-emitting radionuclides, [213]Bi and [211]At (121, 122). The rapid pharmacokinetics of a diabody presents itself as a good match with the half-lives of these radionuclides ([213]Bi at 45.6 min and [211]At at 7.2 h). In a therapy study treating subcutaneous (s.c.) ovarian carcinoma xenografts with the [213]Bi-C6.5 diabody, acceptable toxicity occurred at the lowest dose administered (121). To minimize renal exposure to the [213]Bi, mice were pretreated with D-lysine. Unfortunately, the therapeutic effect was found to be nonspecific, leading the investigators to conclude that the half-life of the [213]Bi was too short to be effectively paired with a systemically administered diabody. A subsequent study pairing [211]At ($t_{1/2} = 7.2$ h) and the C6.5 diabody proved more successful in the therapy of HER2-positive breast cancer tumor xenografts (122). A single i.v. injection of [211]At-C6.5 diabody resulted in durable complete responses in 60% of the mice; the remaining mice experienced a significant delay in tumor growth. The C6.5 diabody has also been shown to be an effective vehicle for targeting the β-emitter, [90]Y (123). Growth inhibition of breast and ovarian cancer xenografts in mice was reported; the maximum tolerated dose appeared dependent on the tumor model. Renal toxicity of the ^{90}Y-C6.5 was evaluated in nontumor bearing mice with mixed results. One mouse showed no overt signs of renal damage, another demonstrated early-stage renal disease, while a third had severe kidney damage. Renal toxicity was also evaluated in the [211]At study. After 1 year, histopathologic examination of the kidneys revealed that two of the three mice exhibited regions of fibrosis amidst healthy tissue (122). This renal damage was modest compared to that observed in the mice that received the ^{90}Y-C6.5, providing an argument for the pursuit of the halogen-based radioisotopes for therapeutic applications with diabodies as the targeting vector.

Noncovalent trimers (triabodies) and tetramers (tetrabodies) have also been produced and evaluated. Initial studies with scFvs had made it apparent that several factors such as the length of linker sequences connecting the V_H and V_L domains, the linker composition, as well as concentration of the molecules influence the formation of multimeric Fv molecules (97, 116, 124). Several multimers of the anti-Lewis Y antigen MAb, HuS193, were created by directly ligating the V_H and V_L domains by either inserting one or two amino acid residues or removing one or two amino acids (97). The addition of residues favored the formation of dimers while the direct linkage of the domains, or the removal of amino acids, led to the trimeric and tetrameric forms. The stability of the multimers was directly related to the apparent affinity of the domains for each other. This study also illustrates the difficulty in maintaining a homogeneous product. Over a 2-week period, the investigators found that a solution of a given multimer would contain other sizes of multimers suggesting the formation of multimers is a dynamic process. The engineering of "knobs into holes", that is, the introduction of amphipathic helices or leucine zipper motifs into the scFv molecule, is one of the measures being taken to enhance the formation of the noncovalent multimeric forms (125–127).

An alternative to the noncovalent multimeric scFv forms are those that are covalently associated. Difficulties of diabodies arise from their inherent compactness and inflexibility. The two binding sites in this Ig form are in an opposing orientation to each other, which may restrict interactions with antigen at both combining sites. The introduction of a peptide linker between the two scFv chains, tethering the V_H domain of one chain with the V_L domain of the other chain, not only lends stability to the molecule but also maintains the binding sites in an appropriate configuration for interacting with antigen (Fig. 1.4). Recently, Beresford et al. (119) compared two dimeric scFv forms (covalent and noncovalent) of MAb CC49. The covalent form was found to target tumor and had pharmacokinetics similar to the noncovalent form. The two forms differed in their tumor-to-normal tissue ratios, with the covalent form yielding superior ratios. This dimeric CC49 scFv utilized a helical linker, consisting of 25 amino acids, connecting the V_H and V_L domains of each peptide chain as well as the two chains (128). In the same study, the investigators also modified the charge of the sFv and evaluated the effect on renal retention of the scFv. Negatively charged amino acids were added to the carboxyl terminus of the CC49 V_H by including oligonucleotide sequences in a polymerase chain reaction amplification. Interestingly, decreasing the isoelectric point of the scFv molecule from pH 8.1 to pH 5.1 did not significantly affect the accretion of the scFv in the kidney; it is believed that cationic amino acids promote renal uptake.

Studies have progressed with the dimeric CC49 scFv with the creation and evaluation of a tetravalent molecule designated $[sc(Fv)_2]_2$ (Fig. 1.4) (129, 130). This molecule consists of CC49 scFv dimers that are noncovalently associated through interactions between opposing V_L and V_H domains of each dimer. When compared to the CC49 scFv dimer and the original CC49 MAb, all radioiodinated, the tetravalent molecule was comparable to the IgG in its affinity as well as in overall tumor uptake. Where the CC49 $[sc(Fv)_2]_2$ differed was in its clearance from the blood. The tetravalent molecule had a residence time that was twofold longer than the dimer, but was twofold shorter than the IgG. The same pharmacokinetic behavior was observed when the CC49 $[sc(Fv)_2]_2$ was labeled with ^{177}Lu. More surprising was the fact that high renal uptake of the radioimmunoconjugates occurred that could not be abrogated with pretreatment of the mice with D-lysine (131). This *in vivo* behavior is inconsistent with a molecule with a molecular weight of $\sim$120 kDa, since the threshold for renal filtration of proteins is <50 kDa. In theory, the tetravalent form should not be subjected to first-pass renal clearance. The renal uptake and faster pharmacokinetics of the tetravalent molecule may be a result of its dissociation into its two dimer components. In a subsequent study, both the divalent and the tetravalent molecules were found to form higher molecular weight species in the serum but with time, lower molecular weight species appeared, suggesting degradation, thus providing an explanation for the results of the earlier study (130).

The introduction of cysteine residues in the C-terminus of the V_H and V_L domains has also been employed to create multimeric forms via disulfide bond formation. A divalent scFv of the anti-HER2/*neu* MAb 741F8 was prepared in this manner by Adams et al. (132). Compared to the monomer scFv, an improvement in tumor targeting was observed that was attributed to the increased avidity of the divalent 741F8 scFv molecule.

Adams et al. (133) have provided proof that improved tumor uptake is a function of the valency rather than due to a longer retention time in the blood of the larger molecules. Cysteinyl residues (Ser-Gly$_4$-Cys) were introduced at the COOH-terminal of an anti-HER2/*neu* scFv and an anti-digoxin scFv creating an scFv' of each. Monospecific (scFv')$_2$ with each specificity and a bispecific (scFv')$_2$ consisting of anti-HER2/*neu* and anti-digoxin scFv' were constructed and compared *in vivo*. The monomer scFv of each was also included in their comparison. The homodimer of the anti-HER2/*neu* resulted in tumor uptake that was threefold higher than the heterodimer at 24 h postinjection while the blood levels of all three (scFv')$_2$ molecules were similar.

1.8 MINIBODIES

Another route to provide a multivalent fragment with a molecular weight greater than the 60 kDa molecular cutoff for renal elimination is to reintroduce Ig domains that homodimerize. One such molecule is the minibody that is constructed by ligating the gene encoding the scFv to the human IgG$_1$ C$_H$3 domain. Dimerization of two polypeptide chains occurs spontaneously as a result of interactions between the two C$_H$3 domains. The minibody resembles a F(ab')$_2$ antibody fragment in size ($M_r \sim 80$ kDa for the minibody versus $\sim$100 kDa for F(ab')$_2$) and bivalency. Two forms of this novel engineered Ig have been generated from MAb T84.66, an anti-CEA MAb, and evaluated (134). One, designated LD minibody, contains a two-amino acid peptide between the V_H and C$_H$3 domains on each of the chains. The other minibody form, termed Flex minibody, contains the human IgG$_1$ hinge region and a 10-amino acid peptide linker to the C$_H$3 domain (Fig. 1.5). The Flex minibody has the advantage of covalent linkage via the formation of disulfide bonds in the hinge region. *In vitro* analysis of the two T84.66 minibody forms indicated that both molecules maintained binding affinity. The Flex minibody was very slightly better than the LD minibody with affinity constants (K_a) of 2.7×10^9 and $2.0 \times 10^9 \, M^{-1}$, respectively (134). *In vivo*

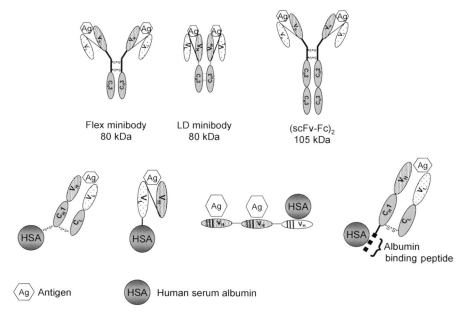

FIGURE 1.5   Schematic representation of different minibody formats and modification strategies.

studies revealed excellent tumor targeting of both molecules, as well as stability in serum. The Flex minibody demonstrated an overall higher %ID/g in the tumor than did the LD minibody and as such was taken forward for further characterization. When labeled with ^{111}In, the minibody maintained good tumor uptake and, in contrast to the diabody, renal uptake was lower. Unfortunately, high hepatic uptake occurs that is comparable to or higher than that observed in the tumor (117). The observed uptake may be antigen driven since a lower percentage of the radioactivity accumulated in the liver in nontumor bearing mice. The hepatic uptake may restrict the use of minibodies to radioiodine labeling or perhaps a short-lived α-emitter such as [211]At for therapeutic applications. In fact, the T84.66 minibody has progressed from preclinical studies to a pilot clinical trial labeled with [124]I and ^{123}I, respectively (135, 136). In the clinical trial, 10 colorectal patients were injected with ^{123}I-labeled T84.66 minibody and imaged over 2–3 days. The mean serum residence time of the minibody was 29.8 h, 4.3–5.6-fold longer than that observed in the mouse model (117, 134). Tumor imaging was observed in seven patients. The findings of the minibody scans were compared to CT (computed tomography) scans and confirmed by surgery. Three patients had lesions reported as false negative by CT that were detected (positive) by minibody nuclear scans; one false positive by CT was found as a true negative with the minibody scans. Three patients underwent PET imaging, the results of which agreed with the minibody scans. These findings suggest that the minibody has potential for targeted radiotherapy applications.

A minibody has since been engineered with specificity for HER2/neu with the purpose of expanding the repertoire of these molecules (137). The pharmacokinetics was found to be similar to the T84.66 minibody; however, the tumor uptake was lower, which may simply be due to differences in the animal model and the molecule being targeted.

1.9 SELECTIVE HIGH AFFINITY LIGANDS

Monoclonal antibodies generated against "tumor antigens" have in turn been useful in the identification, isolation, and purification of the antigens themselves. In some instances, identification of a MAb epitope has been reported (138). As indicated previously, the hypervariable domain of the heavy and light chains of several MAbs has been sequenced (139). Information such as this has allowed investigators to design and synthesize small molecules that bind to cavities within epitopic regions. These antibody mimics, designated selective high-affinity ligands or SHALs, are attractive targeting vehicles of radionuclides for tumor imaging or radiotherapy (140–143). With the objective of developing a more effective therapeutic for non-Hodgkin's lymphoma patients, SHALs have been synthesized and tested for efficacy in animal models and reactivity with patient tumor biopsy samples. The first-generation molecules were bidentate, consisting of two ligands linked by polyethylene glycol (PEG) and having a molecular weight <3 kDa (143). In order to improve affinity, a dimer was also synthesized. The SHALs demonstrated selective or specific binding for the target antigen (HLA-DR10) with nM affinities. Unlike the parental MAb (Lym-1), they did not inhibit cell growth or induce the death of lymphoma cells. It was also realized that although the dimeric SHAL was designed to dock into two cavities, that was not the case; it was concluded that the linker was too short to permit bivalent binding. SHALs are readily conjugated with bifunctional chelates such as DOTA and labeled with radiometals efficiently. High radiochemical yields and high specific activities, that is, 2.1–5.3 MBq/μg, can be obtained (141). Several SHALs were compared *in vivo*, and as expected a very rapid clearance from the plasma compartment was observed. Surprisingly, considering the small size of these molecules, low renal uptake was observed, much lower than that reported for scFvs (141). With the feasibility of the SHALs confirmed, the focus of the work with these molecules will be on synthesizing SHALs with higher affinities, optimizing the linkers for dimeric forms and creating multidentate forms using additional SHALs for other antigen cavities (140). The synthetic nature of SHALs should make these molecules attractive to pharmaceutical companies at the level of production and from a regulatory viewpoint. Their rapid clearance from the blood, the ease of radiolabeling, and the high specific activities that can be attained certainly make them attractive candidates for targeted radiotherapy as well as imaging.

1.10 AFFIBODIES

By now it is apparent that evolving techniques in molecular biology have led to targeting agents of ever decreasing size. This section entails describing small nonantibody affinity ligands (M_r 8.7 kDa) designated affibodies. These molecules are based on a 58-amino acid domain derived from staphylococcal protein A that binds Ig (144). The Z domain is a cysteine-free three-helix bundle that has been used as a platform to generate a number of affibodies (144). The first cancer target for which affibodies were generated and evaluated was HER2 (145). Phage display technology was used to isolate affibody ligands specific for the extracellular domain of HER2 (HER2-ECD). Four rounds of selection resulted in the sequencing of 49 colonies. Analysis of the sequencing indicated that there were actually seven colonies represented, one of which appeared 33 times; the dominance of this sequence is an indicator of the high binding affinity of this particular affibody. Four affibodies were chosen for production, purification, and further characterization. All bound to HER2-ECD (extracellular domain) with affinities in the nanomolar range; one of the two that were radiolabeled with ^{125}I demonstrated binding with HER2-positive cells (145). The small size, stability, and high affinity of affibodies render the molecules attractive candidates for tumor targeting applications. As a monovalent binding agent, the dissociation rate of the affibody from the target molecule was too rapid. Thus, studies with affibodies have progressed with the goal of improving affinity, which has been achieved through the head-to-tail dimerization of two affibodies resulting in a molecule with an M_r of $\sim$15.6 kDa (146) The bivalency resulted in a product with an affinity that was improved by $\sim$17-fold, which approximates that of trastuzumab. The *in vivo* potential of the bivalent affibody [($Z_{HER2:4}$)$_2$] was demonstrated by biodistribution and imaging studies (147). Radiolabeled with ^{125}I, tumor uptake was observed with minimal uptake in normal organs. Renal uptake was also observed but the values were still within acceptable levels. A second generation of affibodies has been since produced with affinity constants in the picomolar range (148). This was achieved through affinity maturation, a process in which random mutagenesis of select amino acids is introduced and through several rounds of screening molecules with higher affinity are selected. The *in vivo* tumor targeting was several times higher than the earlier affibody studied and clear visualization was obtained with γ-scintigraphy 1 h after injection (149). There has also been success at producing a completely synthetic affibody to which a chelating agent for complexing with radiometals was added (148). When this higher affinity affibody was radiolabeled with ^{111}In and administered to tumor bearing mice for evaluation, tumor targeting was comparable to that reported earlier. However, and not surprising, when a radiometal is used to label a molecule with $M_r < 60$ kDa, the radioactivity in the kidneys was appreciably higher (24, 150). However, this is not viewed as an insurmountable setback for affibodies in the arena of targeted radiation therapy. An affibody labeled with a radiometal would have potential in locoregional administration for the treatment of cancers confined to anatomical regions, for example, ovarian cancer, urinary bladder cancer, or primary brain tumors. Affibodies, with their very rapid pharmacokinetics and ability to penetrate tumors, may also have potential as delivery vehicles for short-lived radionuclides such as ^{212}Bi ($t_{1/2} = 60.6$ min), ^{213}Bi ($t_{1/2} = 45.6$ min), ^{211}At ($t_{1/2} = 7.2$ h), or even ^{212}Pb ($t_{1/2} = 10.6$ h). The ability to introduce residues for site-specific modifications, that is, conjugation with a bifunctional chelate, for complexing radiometals provides a well-defined radiopharmaceutical that is homogeneous (151).

1.11 OTHER STRATEGIES

Rapid clearance from the blood along with tumor penetration has been achieved with molecules such as the scFvs, Fabs, diabodies, and other MAb-based forms. Those properties though, as indicated throughout this chapter, have required compromises and have come at a cost. As indicated when discussing each of these molecules, reducing the size of a protein to less than 60 kDa results in their elimination through renal filtration, and when conjugated with a radionuclide, there is a concomitant increase in the radiation doses to the kidneys. Aside from increasing the size/valency of the engineered antibody fragments by tethering fragments together, a number of strategies have been taken to increase the size and/or circulation time of smaller MAb variants. The approaches have been to modify fragments, either chemically or by genetic engineering with, for example, albumin, the Fc domain of antibodies, or polyethylene glycol.

1.11.1 Fc Domain and the Neonatal Fc Receptor

The Fc domain of an Ig is responsible for its retention in the blood and as such offers the chance to manipulate the pharmacokinetics of an antibody. Interactions of the Fc domain with the neonatal Fc receptor (FcRn) protect Igs (IgGs) from degradation by lysosomes, such that the FcRn is attributed with maintaining IgG levels in the blood (58). Igs are taken up by vascular endothelial cells by endocytosis and transported to early endosomes where the FcRn is also located. Under slightly acidic conditions (pH < 6.5), IgG will bind to the FcRn with high affinity; low-affinity binding occurs under neutral conditions (152). IgGs that are bound to FcRn will be transported back to the circulation or transported across the endothelial cell and released into the interstitial fluids. The neutral pH of the fluids in either location triggers the dissociation of the IgG from FcRn. Three amino acids that are conformationally in close proximity, an Ile and a His in the C_H2 domain and a His in the C_H3 domain, are required for IgG binding with the FcRn. Mutations of any of these three residues results in an abbreviated serum half-life of the IgG (153). Conversely, residue mutations near the FcRn binding site can increase the binding affinity for the receptor, resulting in an IgG with a longer retention time in the serum (154). These studies provided the basis for Kenanova et al. (155) to engineer a single-chain Fv-Fc fragment and five variants with the objective of modulating the serum half-life of an scFv and tailoring molecules for specific applications (Fig. 1.5). Biodistribution studies in mice proved that tumor targeting molecules could be produced that varied in serum half-lives ranging from 8 h to 12 days (155). Tumor targeting was visualized with the anti-CEA scFv-Fc variants using a microPET scanner. Images with high signal-to-noise background were obtained ∼18 h post-injection of the [124]I-labeled scFv-Fc variants; the variant with the fastest serum pharmacokinetics produced the earliest image. The pH sensitivity of the Fc–FcRn interaction provided the rationale for the mutations that were inserted into the Fc domain. For example, at an acidic pH, the His residue is positively charged; an Arg was substituted to maintain this positive charge. What was of further consideration was that Arg would also be positively charged in the more neutral milieu of the serum, interstitial fluids, or endosomes, thus discouraging dissociation of the scFv-Fc from the FcRn (155). A His substituted with a glutamine resulted in the variant having the shortest residence time in the serum; Gln remains uncharged in acidic or neutral backgrounds. When both histidines in the FcRn binding site of the Fc were modified, the resulting variant demonstrated the greatest reduction in serum retention of all of the variants.

Three of the anti-CEA scFv-Fc variants underwent further evaluation to assess their potential in radioimmunotherapy regimens (156). Not surprising, *in vivo* studies indicate that there is an inverse relationship between residence time in the blood and tumor uptake of the radiolabeled scFv-Fc. However, even the variant with the fastest clearance from the blood demonstrated good tumor targeting. Renal accretion of radioactivity is not an issue with these molecules since they were engineered to exceed the molecular weight threshold for clearance through the kidneys. Hepatic uptake was observed when the scFv-Fc variants were labeled with ^{111}In; this was comparable with that observed for a minibody of the same MAb (T84.66) (117). Due to this hepatic uptake in three mouse models, the investigators argued that if the fastest clearing variant (both histidines substituted) were labeled with ^{90}Y, dosimetry estimates indicated that the liver would receive a similar radiation absorbed dose as the tumor (156). The scFv-Fc would be a more appropriate agent for targeted radiation therapy if radiolabeled with ^{131}I, due to the catabolism and elimination of radioiodine and the absence of its sequestration in the liver. It is also suggested that this particular molecule would have applications in multistep approaches that employ bispecific forms for pretargeting.

The data gathered from biodistribution studies in which mice were coinjected with anti-CEA scFv-Fc constructs labeled with ^{111}In and ^{125}I have allowed investigators to develop a physiologically based pharmacokinetic model that can simulate the *in vivo* behavior of an scFv-Fc labeled with either ^{111}In or radioiodine. This model will prove quite useful in designing antibody-based therapeutics as well as assist in establishing dosing schedules for patients in the clinic by translating preclinical animal data to humans (157).

Production of recombinant MAb forms does not always go smoothly or as planned. The scFv-Fc molecules can self-assemble into multimeric forms. An anti-CD20 scFv-Fc was generated for evaluation as an immunotherapeutic and/or radioimmunotherapeutic agent. The purified product contained not only the expected monomeric form of 104 kDa but also discrete multiple species of incrementally higher molecular weights based on the size of the single unit (158). This particular scFv-Fc demonstrated a propensity for multimerization that appeared to be driven by intermolecular pairing of variable regions of an scFv-Fc. Systematically, this group demonstrated that in this particular case, it was not the linker they had chosen nor was it interactions due to the hinge region of the molecule. The investigators postulated that during production of an anti-CD20 scFv-Fc, the variable domains of the nascent polypeptides are juxtaposed as they are synthesized with the domains of another strand. Further investigation of suspect residues that interact at the V_L–V_H interface or in the framework that affect folding or to identify residues that become available for chain–chain interactions is warranted. Elucidation of such factors would provide critical information for the design and production of future scFv-Fc molecules as well as other recombinant MAb base molecules.

1.11.2 PEGylation

Conjugation of proteins with polyethylene glycol has been shown to minimize or abrogate the immunogenicity of a protein and to increase its residence time in the circulation (159–162). In addition, PEGylation of the smaller scFv provides yet another method of increasing their size to avoid renal accumulation of radioactivity when conjugated with radiometals.

Random conjugation of scFvs with PEG has been demonstrated to be effective. PEGylation of CC49 scFv with increasing sizes of PEG (molecular masses ranging from 2000 to 20,000 Da) yielded conjugates with corresponding increasing half-lives in the serum (163). The caveat to this PEGylation was that careful adjustments to the reaction were required to ensure that the final product had a PEG:scFv ratio no greater than 2; loss of immunoreactivity was noted at higher ratios. These studies also illustrated the requirement for careful consideration of the chemistry employed for the PEGylation, that is, employing carboxylic or amine moieties. Immunoreactivity of the scFv was influenced by the chemical route chosen. Interestingly, it was determined that the length of the PEG polymer was more effective in prolonging the serum half-life of the conjugate than did an increase in the total mass of PEG.

In another study, an anti-CEA diabody was coupled with PEG, conjugated with cysteinyl-DOTA and labeled with [111]In (164). With an apparent molecular weight of 75,000 Da, the radiolabeled PEG-diabody retained immunoreactivity. Compared to the unmodified diabody, tumor uptake was greatly improved at early time points and was retained at higher levels for a longer period. Renal uptake was reduced by ~2.5-fold but was still high with ~50%ID/g at 24 h. An increase in hepatic uptake was also observed. The PEG was effective in improving the tumor uptake, a function of the longer circulation time; however, the renal uptake still remained an obstacle to using this MAb form for RIT, restricting its use as a delivery vehicle for the radioiodines or α-particle emitters.

As discussed, the random conjugation of a molecule with PEG can affect its bioactivity, which therefore requires testing. Efficiency of these reactions will vary from batch-to-batch as will the number of PEG molecules that will be attached per protein molecule. The conjugation invariably results in a heterogeneous product that is difficult to characterize and therefore problematic to standardize for a potential pharmaceutical product (165–167). A solution to these obstacles is the insertion of amino acids for the purpose of performing site-specific conjugations. Recent work has taken the approach of introducing unpaired cysteine residues in the scFv construct to provide a specific site for PEGylation with the goal of preserving the immunoreactivity of the di-scFv (168). First, a systematic investigation of the appropriate length of the peptide linker determined that a 20-amino acid linker resulted in the best production yield of an anti-MUC-1 di-scFv. Again, the G_4S repeat was chosen based on the flexibility and hydrophilicity it provides to the molecule along with its low immunogenicity (169). A comparison of an anti-MUC-1 di-scFv with a cysteine introduced at five locations within the molecule led to the identification of a form that retained immunoreactivity. Although all five versions were reactive with the cognate antigen, some differences were discernible. However, the cysteine location seemed to have a greater effect on the efficiency of the PEGylation. The highest PEGylation efficiency occurred when the available cysteine was located in the inter-scFv linker. It was also reported that PEG of various sizes (up to 40 kDa) could be conjugated to the di-scFv without loss of immunoreactivity.

1.11.3 Albumin Binding

Human serum albumin (HSA), the most abundant protein in the blood plasma, serves as a transporter and scavenger (i.e., fatty acids, bilirubin, anions) for an assortment of molecules. Binding with albumin (i) increases the solubility of a molecule in the plasma, (ii) provides protection from oxidation/degradation, and (iii) lowers the toxicity of a molecule (170). HSA interacts reversibly with a wide spectrum of drugs with varying affinity; valproate, warfarin, ibuprofen, and indomethacin are some examples. More importantly, with a 19-day half-life, albumin can significantly prolong the circulating time of a drug. In fact, binding with albumin has become a common strategy for altering the pharmacokinetics of small molecular weight drugs such as analogues of insulin (171). Associating albumin with immunoglobulin has been evaluated using three tactics: chemical conjugation, genetic fusion, and creation of a bifunctional molecule with albumin and antigen binding capabilities (Fig. 1.5) (172). All three methods are successful in generating molecules that retain reactivity with antigen and also have an extended residence time in the plasma. A chemical conjugation of a Fab' with HSA exploits the exposed cysteine in the former with the free cysteine of the latter. The proteins are subjected to reduction with reagents, combined, and then the reducing agents removed to allow disulfide bond formation to occur between the two proteins. Such a conjugate was created with an anti-TNF Fab fragment and HSA. Compared to the Fab' alone, the addition of the HSA increased the $t_{1/2}\alpha$ plasma pharmacokinetics from 1 to 4.6 h while the $t_{1/2}\beta$ increased from 31.4 to 39.6 h. The conjugation, however, is not very efficient (a 5% yield was reported) and even when optimized resulted in a heterogeneous mixture requiring several purification steps and a product that would not contain a uniform ratio of Fab to HSA (172). Fusion proteins consisting of an antibody fragment and HSA eliminate the difficulties just mentioned and provide a uniform product that can be well characterized. The scFv of an antibody is either fused directly to the N-terminus of HSA, or linkers such as $(Gly)_4$-Ser repeats are inserted between the two as spacers, providing some flexibility to the fusion protein and avoiding steric hindrance. The *in vivo* evaluation of these fusion proteins has indicated that in a biphasic analysis, their plasma clearance is similar to that of HSA and was found to have a considerably longer retention time in the plasma than that reported for an scFv (22, 172, 173). As mentioned previously, unacceptably high accretion in the kidneys and liver is observed when sFv fragments, diabodies, and minibodies are labeled with radiometals (24, 117). High renal and hepatic accumulation of radioactivity has not been evident with an ^{111}In-labeled fusion protein consisting of an scFv and HSA (173).

A bifunctional molecule, produced either by chemical methods or through genetic engineering, that would bind a target antigen and albumin represents another means of recruiting albumin. In one instance, the Fab′ of an anti-rat serum albumin (RSA) antibody was cross-linked with the Fab′ of an anti-TNF antibody (172). The resulting F(ab′)$_2$ demonstrated reactivity with both TNF and RSA. More importantly, the bispecific F(ab′)$_2$ had a 6.6- and 2.2-fold increase in the $t_{1/2}\beta$ plasma clearance phase compared to a control F(ab′)$_2$ and RSA, respectively. This approach has been taken one step further with the creation of multivalent fusion proteins constructed from nanobodies. One such fusion protein contained two nanobodies with specificity for EGFR and a third was reactive with albumin (174). The bispecific nanobody was radiolabeled with ^{177}Lu and evaluated in a tumor bearing mouse model. Excellent tumor targeting, comparable to cetuximab, was achieved while modest uptake of radioactivity was observed in the kidneys. The fusion protein was also shown by immunohistochemical analysis to penetrate tumor tissue.

Another interesting approach borrows from the fact that albumin contains several domains that bind to small molecules and metal ions (171). Using phage display techniques, a library of high-affinity albumin binding peptides was developed (175). Peptides that bound rat, human, and rabbit albumin were synthesized and characterized. Based on favorable pharmacokinetics and stability, two peptides were chosen and recombinantly fused with a Fab fragment. In an elegant set of studies, these investigators demonstrated high-affinity binding of the Fab fusion protein with albumin as well as binding with the cognate antigen. The half-life of the molecule was increased ~40-fold. Having demonstrated proof of concept, this tactic was then translated to trastuzumab (Herceptin), the anti-HER2 MAb currently used for treatment of metastatic breast cancer patients (176, 177).

Several variants of a trastuzumab Fab with the albumin binding peptide (AB.Fab) were generated and evaluated for albumin and HER2 binding. Variations consisted of introducing a peptide linker (Gly$_3$Ser) or altering the length of the albumin binding peptide (177). The variant with the linker sequence exhibited a higher affinity for albumin, maintained its reactivity with HER2, and had a prolonged retention in the plasma. Compared to the Herceptin Fab, the $t_{1/2}\beta$ in mice, rats, and rabbits was extended by 15.4-, 21.0-, and 11.5-fold, respectively (176, 177). Tumor targeting was visualized and quantitated by SPECT and CT (176). The AB.Fab resulted in a more rapid uptake in the tumor, which at 48 h was comparable to intact trastuzumab IgG and ~4-fold greater than the Fab fragment. The AB.Fab cleared from the blood rapidly and did not appear to accumulate in the kidneys. Greater uniformity of distribution through tumor tissue was evident by histological analysis of HER2-positive tumor xenografts. These properties of the AB.Fab indicate that these fusion proteins may be effective vehicles for the delivery of therapeutic doses of radiation in RIT applications. The ability to select a peptide that has relevance for humans and yet be characterized in another species can facilitate its translation to clinical trials with cancer patients. These studies illustrate the exquisite fine-tuning that can be performed to obtain a targeting molecule of the desired properties.

1.12 CONCLUDING REMARKS

This chapter has presented an overview of various forms of MAb and their application in targeted radiotherapy of cancer. RIT still offers the potential of delivering a high radiation absorbed dose that is tumor selective with modest to tolerable, and in some cases minimal, damage to normal tissues. The clinical RIT trials with murine monoclonal antibodies identified a number of obstacles to effective treatment, which included inadequate delivery of a therapeutic dose to and throughout the tumor, bone marrow toxicity, and the development of human anti-mouse antibodies. Genetic engineering has greatly facilitated our ability to address these challenges. The antigen binding, specificity, affinity, and *in vivo* characteristics of a MAb can be optimized. Recombinant technology has also shown us the compromises of antibody-based therapy of cancer. There are a myriad of MAb forms and treatment strategies to evaluate. The availability of targeted radiation therapy provides a viable alternative for the treatment of cancer patients.

REFERENCES

1. Bale WF, Spar IL, Goodland RL. Experimental radiation therapy of tumors with I^{131}-carrying antibodies to fibrin. *Cancer Res* 1960;20:1488–1494.
2. Pressman D. The development and use of radiolabeled antitumor antibodies. *Cancer Res* 1980;40:2960–2964.
3. Hiramoto R, Yagi Y, Pressman D. *In vivo* fixation of antibodies in the adrenal. *Proc Soc Exp Biol Med* 1958;98:870–874.
4. McCardle RJ, Harper PV, Spar IL, Bale WF, Andros G, Jiminez F. Studies with iodine-131-labeled antibody to human fibrinogen for diagnosis and therapy of tumors. *J Nucl Med* 1966;7:837–847.
5. Ehrlich P, Herta C, Shigas K. Ueber einige verwendungen der naphtochinosuflsaure. *Ztschr f Physiol Chem* 1904;61:379–392.
6. Kohler G, Milstein C. Continuous cultures of fused cells secreting antibody of predefined specificity. *Nature* 1975;256:495–497.
7. Knox SJ, Meredith RF. Clinical radioimmunotherapy. *Semin Radiat Oncol* 2000;10:73–93.
8. Jurcic JG, Scheinberg DA. Radioimmunotherapy of hematological cancer: problems and progress. *Clin Cancer Res* 1995;1:1439–1446.
9. Schlom J. Monoclonal antibodies: they're more and less than you think. In: Broder S, editor. *Molecular Foundations of Oncology*. Williams and Wilkins, Baltimore, MD, 1990.
10. Reichert JM, Wenger JB. Development trends for new cancer therapeutics and vaccines. *Drug Discov Today* 2008;13:30–37.

11. Witzig TE. The use of ibritumomab tiuxetan radioimmunotherapy for patients with relapsed B-cell non-Hodgkin's lymphoma. *Semin Oncol* 2000;27:74–78.

12. Vose JM. Bexxar: novel radioimmunotherapy for the treatment of low-grade and transformed low-grade non-Hodgkin's lymphoma. *Oncologist* 2004;9:160–172.

13. Khazaeli MB, Conry RM, LoBuglio AF. Human immune response to monoclonal antibodies. *J Immunother* 1994;15:42–52.

14. Tamura M, Milenic DE, Iwahashi M, Padlan E, Schlom J, Kashmiri SV. Structural correlates of an anticarcinoma antibody: identification of specificity-determining residues (SDRs) and development of a minimally immunogenic antibody variant by retention of SDRs only. *J Immunol* 2000;164:1432–1441.

15. Blanco I, Kawatsu R, Harrison K, et al. Antiidiotypic response against murine monoclonal antibodies reactive with tumor-associated antigen TAG-72. *J Clin Immunol* 1997;17:96–106.

16. Milenic DE, Esteban JM, Colcher D. Comparison of methods for the generation of immunoreactive fragments of a monoclonal antibody (B72.3) reactive with human carcinomas. *J Immunol Methods* 1989;120:71–83.

17. Parham P, Androlewicz MJ, Brodsky FM, Holmes NJ, Ways JP. Monoclonal antibodies: purification, fragmentation and application to structural and functional studies of class I MHC antigens. *J Immunol Methods* 1982;53:133–173.

18. Waller M, Curry N, Mallory J. Immunochemical and serological studies of enzymatically fractionated human IgG globulins. I. Hydrolysis with pepsin, papain, ficin and bromelin. *Immunochemistry* 1968;5:577–583.

19. Juweid ME, Hajjar G, Swayne LC, et al. Phase I/II trial of ^{131}I-MN-14F(ab)$_2$ anti-carcinoembryonic antigen monoclonal antibody in the treatment of patients with metastatic medullary thyroid carcinoma. *Cancer* 1999;85:1828–1842.

20. Lane DM, Eagle KF, Begent RH, et al. Radioimmunotherapy of metastatic colorectal tumours with iodine-131-labelled antibody to carcinoembryonic antigen: phase I/II study with comparative biodistribution of intact and F(ab')$_2$ antibodies. *Br J Cancer* 1994;70:521–525.

21. Yokota T, Milenic DE, Whitlow M, Schlom J. Rapid tumor penetration of a single-chain Fv and comparison with other immunoglobulin forms. *Cancer Res* 1992;52:3402–3408.

22. Milenic DE, Yokota T, Filpula DR, et al. Construction, binding properties, metabolism, and tumor targeting of a single-chain Fv derived from the pancarcinoma monoclonal antibody CC49. *Cancer Res* 1991;51:6363–6371.

23. Carrasquillo JA, Krohn KA, Beaumier P, et al. Diagnosis of and therapy for solid tumors with radiolabeled antibodies and immune fragments. *Cancer Treat Rep* 1984;68:317–328.

24. Schott ME, Milenic DE, Yokota T, et al. Differential metabolic patterns of iodinated versus radiometal chelated anticarcinoma single-chain Fv molecules. *Cancer Res* 1992;52:6413–6417.

25. Yokota T, Milenic DE, Whitlow M, Wood JF, Hubert SL, Schlom J. Microautoradiographic analysis of the normal organ distribution of radioiodinated single-chain Fv and other immunoglobulin forms. *Cancer Res* 1993;53:3776–3783.

26. Behr TM, Sharkey RM, Juweid ME, et al. Reduction of the renal uptake of radiolabeled monoclonal antibody fragments by cationic amino acids and their derivatives. *Cancer Res* 1995;55:3825–3834.

27. DePalatis LR, Frazier KA, Cheng RC, Kotite NJ. Lysine reduces renal accumulation of radioactivity associated with injection of the [^{177}Lu]alpha-[2-(4-aminophenyl) ethyl]-1,4,7,10-tetraaza-cyclodecane-1,4,7,10-tetraacetic acid-CC49 Fab radioimmunoconjugate. *Cancer Res* 1995;55:5288–5295.

28. Morrison S, Schlom J. Recombinant chimeric monoclonal antibodies. In: DeVita V, Rosenberg S, editors. *Important Advances in Oncology.* JB Lippincott, Philadelphia, PA, 1990, pp. 3–18.

29. Hiatt A, Cafferkey R, Bowdish K. Production of antibodies in transgenic plants. *Nature* 1989;342:76–78.

30. LoBuglio AF, Wheeler RH, Trang J, et al. Mouse/human chimeric monoclonal antibody in man: kinetics and immune response. *Proc Natl Acad Sci USA* 1989;86:4220–4224.

31. Baeuerle PA, Gires O. EpCAM (CD326) finding its role in cancer. *Br J Cancer* 2007;96:417–423.

32. Litvinov SV, Bakker HA, Gourevitch MM, Velders MP, Warnaar SO. Evidence for a role of the epithelial glycoprotein 40 (Ep-CAM) in epithelial cell–cell adhesion. *Cell Adhes Commun* 1994;2:417–428.

33. Meredith RF, Khazaeli MB, Grizzle WE, et al. Direct localization comparison of murine and chimeric B72.3 antibodies in patients with colon cancer. *Hum Antibodies Hybridomas* 1993;4:190–197.

34. Maloney DG, Grillo-Lopez AJ, White CA, et al. IDEC-C2B8 (Rituximab) anti-CD20 monoclonal antibody therapy in patients with relapsed low-grade non-Hodgkin's lymphoma. *Blood* 1997;90:2188–2195.

35. Jones PT, Dear PH, Foote J, Neuberger MS, Winter G. Replacing the complementarity-determining regions in a human antibody with those from a mouse. *Nature* 1986;321:522–525.

36. Padlan EA. Anatomy of the antibody molecule. *Mol Immunol* 1994;31:169–217.

37. Amit AG, Mariuzza RA, Phillips SE, Poljak RJ. Three-dimensional structure of an antigen–antibody complex at 2.8 A resolution. *Science* 1986;233:747–753.

38. Kashmiri SV, De Pascalis R, Gonzales NR. Developing a minimally immunogenic humanized antibody by SDR grafting. *Methods Mol Biol* 2004;248:361–376.

39. Kashmiri SV, Shu L, Padlan EA, Milenic DE, Schlom J, Hand PH. Generation, characterization, and *in vivo* studies of humanized anticarcinoma antibody CC49. *Hybridoma* 1995;14:461–473.

40. Winter G, Harris WJ. Humanized antibodies. *Immunol Today* 1993;14:243–246.

41. Riechmann L, Clark M, Waldmann H, Winter G. Reshaping human antibodies for therapy. *Nature* 1988;332:323–327.

42. Caron PC, Schwartz MA, Co MS, et al. Murine and humanized constructs of monoclonal antibody M195 (anti-CD33) for the therapy of acute myelogenous leukemia. *Cancer* 1994;73:1049–1056.

43. Sharkey RM, Juweid M, Shevitz J, et al. Evaluation of a complementarity-determining region-grafted (humanized) anti-carcinoembryonic antigen monoclonal antibody in preclinical and clinical studies. *Cancer Res* 1995;55:5935s–5945s.

44. Kramer EL, Liebes L, Wasserheit C, et al. Initial clinical evaluation of radiolabeled MX-DTPA humanized BrE-3 antibody in patients with advanced breast cancer. *Clin Cancer Res* 1998;4:1679–1688.

45. Caron PC, Jurcic JG, Scott AM, et al. A phase 1B trial of humanized monoclonal antibody M195 (anti-CD33) in myeloid leukemia: specific targeting without immunogenicity. *Blood* 1994;83:1760–1768.

46. Schneider WP, Glaser SM, Kondas JA, Hakimi J. The anti-idiotypic response by cynomolgus monkeys to humanized anti-Tac is primarily directed to complementarity-determining regions H1, H2, and L3. *J Immunol* 1993;150:3086–3090.

47. Singer II, Kawka DW, DeMartino JA, et al. Optimal humanization of 1B4, an anti-CD18 murine monoclonal antibody, is achieved by correct choice of human V-region framework sequences. *J Immunol* 1993;150:2844–2857.

48. Stephens S, Emtage S, Vetterlein O, et al. Comprehensive pharmacokinetics of a humanized antibody and analysis of residual anti-idiotypic responses. *Immunology* 1995;85:668–674.

49. Pendley C, Schantz A, Wagner C. Immunogenicity of therapeutic monoclonal antibodies. *Curr Opin Mol Ther* 2003;5:172–179.

50. Nanus DM, Milowsky MI, Kostakoglu L, et al. Clinical use of monoclonal antibody HuJ591 therapy: targeting prostate specific membrane antigen. *J Urol* 2003;170: S84–S88; discussion S88–S89.

51. Smith-Jones PM. Radioimmunotherapy of prostate cancer. *Q J Nucl Med Mol Imaging* 2004;48:297–304.

52. Mulligan T, Carrasquillo JA, Chung Y, et al. Phase I study of intravenous Lu-labeled CC49 murine monoclonal antibody in patients with advanced adenocarcinoma. *Clin Cancer Res* 1995;1:1447–1454.

53. Iwahashi M, Milenic DE, Padlan EA, Bei R, Schlom J, Kashmiri SV. CDR substitutions of a humanized monoclonal antibody (CC49): contributions of individual CDRs to antigen binding and immunogenicity. *Mol Immunol* 1999;36:1079–1091.

54. Padlan EA, Abergel C, Tipper JP. Identification of specificity-determining residues in antibodies. *FASEB J* 1995;9:133–139.

55. Kashmiri SV, Iwahashi M, Tamura M, Padlan EA, Milenic DE, Schlom J. Development of a minimally immunogenic variant of humanized anti-carcinoma monoclonal antibody CC49. *Crit Rev Oncol Hematol* 2001;38:3–16.

56. De Pascalis R, Gonzales NR, Padlan EA, et al. *In vitro* affinity maturation of a specificity-determining region-grafted humanized anticarcinoma antibody: isolation and characterization of minimally immunogenic high-affinity variants. *Clin Cancer Res* 2003;9:5521–5531.

57. De Pascalis R, Iwahashi M, Tamura M, et al. Grafting of "abbreviated" complementarity-determining regions containing specificity-determining residues essential for ligand contact to engineer a less immunogenic humanized monoclonal antibody. *J Immunol* 2002;169:3076–3084.

58. Waldmann TA, Strober W. Metabolism of immunoglobulins. *Prog Allergy* 1969;13:1–110.

59. Shuke N, Steis R, McCabe R. Pharmacokinetics of two human IgM monoclonal antibodies (16.88 and 28A32). *J Nucl Med* 1989;30:909.

60. Cole SP, Vreeken EH, Roder JC. Antibody production by human X human hybridomas in serum-free medium. *J Immunol Methods* 1985;78:271–278.

61. Green LL, Hardy MC, Maynard-Currie CE, et al. Antigen-specific human monoclonal antibodies from mice engineered with human Ig heavy and light chain YACs. *Nat Genet* 1994;7:13–21.

62. Lonberg N, Taylor LD, Harding FA, et al. Antigen-specific human antibodies from mice comprising four distinct genetic modifications. *Nature* 1994;368:856–859.

63. Lonberg N. Human antibodies from transgenic animals. *Nat Biotechnol* 2005; 23:1117–1125.

64. Tomizuka K, Shinohara T, Yoshida H, et al. Double trans-chromosomic mice: maintenance of two individual human chromosome fragments containing Ig heavy and kappa loci and expression of fully human antibodies. *Proc Natl Acad Sci USA* 2000;97:722–727.

65. McCafferty J, Griffiths AD, Winter G, Chiswell DJ. Phage antibodies: filamentous phage displaying antibody variable domains. *Nature* 1990;348:552–554.

66. Rajpal A, Beyaz N, Haber L, et al. A general method for greatly improving the affinity of antibodies by using combinatorial libraries. *Proc Natl Acad Sci USA* 2005; 102:8466–8471.

67. Lonberg N. Fully human antibodies from transgenic mouse and phage display platforms. *Curr Opin Immunol* 2008.

68. Foon KA, Yang XD, Weiner LM, et al. Preclinical and clinical evaluations of ABX-EGF, a fully human anti-epidermal growth factor receptor antibody. *Int J Radiat Oncol Biol Phys* 2004;58:984–990.

69. Rowinsky EK, Schwartz GH, Gollob JA, et al. Safety, pharmacokinetics, and activity of ABX-EGF, a fully human anti-epidermal growth factor receptor monoclonal antibody in patients with metastatic renal cell cancer. *J Clin Oncol* 2004;22:3003–3015.

70. Hamers-Casterman C, Atarhouch T, Muyldermans S, et al. Naturally occurring antibodies devoid of light chains. *Nature* 1993;363:446–448.

71. Dumoulin M, Conrath K, Van Meirhaeghe A, et al. Single-domain antibody fragments with high conformational stability. *Protein Sci* 2002;11:500–515.

72. Revets H, De Baetselier P, Muyldermans S. Nanobodies as novel agents for cancer therapy. *Expert Opin Biol Ther* 2005;5:111–124.

73. Joosten V, Lokman C, Van Den Hondel CA, Punt PJ. The production of antibody fragments and antibody fusion proteins by yeasts and filamentous fungi. *Microb Cell Fact* 2003;2:1.

74. Gainkam LO, Huang L, Caveliers V, et al. Comparison of the biodistribution and tumor targeting of two ^{99m}Tc-labeled anti-EGFR nanobodies in mice, using pinhole SPECT/micro-CT. *J Nucl Med* 2008;49:788–795.

75. Cortez-Retamozo V, Lauwereys M, Hassanzadeh GhG, et al. Efficient tumor targeting by single-domain antibody fragments of camels. *Int J Cancer* 2002;98:456–462.

76. Gillies SD, Wesolowski JS. Antigen binding and biological activities of engineered mutant chimeric antibodies with human tumor specificities. *Hum Antibodies Hybridomas* 1990;1:47–54.

77. Dorai H, Mueller BM, Reisfeld RA, Gillies SD. Aglycosylated chimeric mouse/human IgG1 antibody retains some effector function. *Hybridoma* 1991;10:211–217.

78. Mueller BM, Romerdahl CA, Gillies SD, Reisfeld RA. Enhancement of antibody-dependent cytotoxicity with a chimeric anti-GD2 antibody. *J Immunol* 1990; 144:1382–1386.

79. Slavin-Chiorini DC, Horan Hand PH, Kashmiri SV, Calvo B, Zaremba S, Schlom J. Biologic properties of a C_H2 domain-deleted recombinant immunoglobulin. *Int J Cancer* 1993;53:97–103.

80. Slavin-Chiorini DC, Kashmiri SV, Lee HS, et al. A CDR-grafted (humanized) domain-deleted antitumor antibody. *Cancer Biother Radiopharm* 1997;12:305–316.

81. Slavin-Chiorini DC, Kashmiri SV, Schlom J, et al. Biological properties of chimeric domain-deleted anticarcinoma immunoglobulins. *Cancer Res* 1995;55:5957s–5967s.

82. Milenic D, Garmestani K, Dadachova E, et al. Radioimmunotherapy of human colon carcinoma xenografts using a ^{213}Bi-labeled domain-deleted humanized monoclonal antibody. *Cancer Biother Radiopharm* 2004;19:135–147.

83. Rogers BE, Roberson PL, Shen S, et al. Intraperitoneal radioimmunotherapy with a humanized anti-TAG-72 (CC49) antibody with a deleted C_H2 region. *Cancer Biother Radiopharm* 2005;20:502–513.

84. Forero A, Meredith RF, Khazaeli MB, et al. A novel monoclonal antibody design for radioimmunotherapy. *Cancer Biother Radiopharm* 2003;18:751–759.

85. Xiao J, Horst S, Hinkle G, et al. Pharmacokinetics and clinical evaluation of ^{125}I-radiolabeled humanized CC49 monoclonal antibody (HuCC49ΔC_H2) in recurrent and metastatic colorectal cancer patients. *Cancer Biother Radiopharm* 2005;20:16–26.

86. Glaser SM, Hughes IE, Hopp JR, Hathaway K, Perret D, Reff ME. Novel antibody hinge regions for efficient production of C_H2 domain-deleted antibodies. *J Biol Chem* 2005;280:41494–41503.

87. Sivolapenko GB, Douli V, Pectasides D, et al. Breast cancer imaging with radiolabelled peptide from complementarity-determining region of antitumour antibody. *Lancet* 1995;346:1662–1666.

88. Williams WV, Kieber-Emmons T, VonFeldt J, Greene MI, Weiner DB. Design of bioactive peptides based on antibody hypervariable region structures. Development of conformationally constrained and dimeric peptides with enhanced affinity. *J Biol Chem* 1991;266:5182–5190.

89. Bird RE, Hardman KD, Jacobson JW, et al. Single-chain antigen-binding proteins. *Science* 1988;242:423–426.

90. Skerra A, Pluckthun A. Assembly of a functional immunoglobulin Fv fragment in *Escherichia coli*. *Science* 1988;240:1038–1041.

91. Azuma T, Hamaguchi K, Migita S. Interactions between immunoglobulin polypeptide chains. *J Biochem* 1974;76:685–693.

92. Glockshuber R, Malia M, Pfitzinger I, Pluckthun A. A comparison of strategies to stabilize immunoglobulin Fv-fragments. *Biochemistry* 1990;29:1362–1367.

93. Horne C, Klein M, Polidoulis I, Dorrington KJ. Noncovalent association of heavy and light chains of human immunoglobulins. III. Specific interactions between V_H and V_L. *J Immunol* 1982;129:660–664.

94. Ward ES, Gussow D, Griffiths AD, Jones PT, Winter G. Binding activities of a repertoire of single immunoglobulin variable domains secreted from *Escherichia coli*. *Nature* 1989;341:544–546.

95. Huston JS, Levinson D, Mudgett-Hunter M, et al. Protein engineering of antibody binding sites: recovery of specific activity in an anti-digoxin single-chain Fv analogue produced in *Escherichia coli*. *Proc Natl Acad Sci USA* 1988;85:5879–5883.

96. Desplancq D, King DJ, Lawson AD, Mountain A. Multimerization behaviour of single chain Fv variants for the tumour-binding antibody B72.3. *Protein Eng* 1994; 7:1027–1033.

97. Power BE, Doughty L, Shapira DR, et al. Noncovalent scFv multimers of tumor-targeting anti-Lewis(y) hu3S193 humanized antibody. *Protein Sci* 2003;12:734–747.

98. Wu AM. Engineering multivalent antibody fragments for *in vivo* targeting. *Methods Mol Biol* 2004;248:209–225.

99. Colcher D, Bird R, Roselli M, et al. *In vivo* tumor targeting of a recombinant single-chain antigen-binding protein. *J Natl Cancer Inst* 1990;82:1191–1197.

100. McDevitt MR, Sgouros G, Finn RD, et al. Radioimmunotherapy with alpha-emitting nuclides. *Eur J Nucl Med* 1998;25:1341–1351.

101. Milenic DE, Brechbiel MW. Targeting of radio-isotopes for cancer therapy. *Cancer Biol Ther* 2004;3:361–370.

102. Wu AM, Chen W, Raubitschek A, et al. Tumor localization of anti-CEA single-chain Fvs: improved targeting by non-covalent dimers. *Immunotechnology* 1996;2:21–36.

103. Adams GP, Schier R, Marshall K, et al. Increased affinity leads to improved selective tumor delivery of single-chain Fv antibodies. *Cancer Res* 1998;58:485–490.

104. Adams GP, Schier R, McCall AM, et al. High affinity restricts the localization and tumor penetration of single-chain fv antibody molecules. *Cancer Res* 2001;61:4750–4755.

105. Larson SM, El-Shirbiny AM, Divgi CR, et al. Single chain antigen binding protein (sFv CC49): first human studies in colorectal carcinoma metastatic to liver. *Cancer* 1997;80:2458–2468.

106. Begent RH, Verhaar MJ, Chester KA, et al. Clinical evidence of efficient tumor targeting based on single-chain Fv antibody selected from a combinatorial library. *Nat Med* 1996;2:979–984.

107. Mayer A, Tsiompanou E, O'Malley D, et al. Radioimmunoguided surgery in colorectal cancer using a genetically engineered anti-CEA single-chain Fv antibody. *Clin Cancer Res* 2000;6:1711–1719.

108. Almog O, Benhar I, Vasmatzis G, et al. Crystal structure of the disulfide-stabilized Fv fragment of anticancer antibody B1: conformational influence of an engineered disulfide bond. *Proteins* 1998;31:128–138.

109. Bin Song K, Won M, Meares CF. Expression of recombinant Lym-1 single-chain Fv in *Escherichia coli*. *Biotechnol Appl Biochem* 1998;28 (Pt 2): 163–167.

110. Li Q, Hudson W, Wang D, Berven E, Uckun FM, Kersey JH. Pharmacokinetics and biodistribution of radioimmunoconjugates of anti-CD19 antibody and single-chain Fv for treatment of human B-cell malignancy. *Cancer Immunol Immunother* 1998;47:121–130.

111. Rodrigues ML, Presta LG, Kotts CE, et al. Development of a humanized disulfide-stabilized anti-p185HER2 Fv-beta-lactamase fusion protein for activation of a cephalosporin doxorubicin prodrug. *Cancer Res* 1995;55:63–70.

112. Webber KO, Kreitman RJ, Pastan I. Rapid and specific uptake of anti-Tac disulfide-stabilized Fv by interleukin-2 receptor-bearing tumors. *Cancer Res* 1995;55:318–323.

113. Rajagopal V, Pastan I, Kreitman RJ. A form of anti-Tac(Fv) which is both single-chain and disulfide stabilized: comparison with its single-chain and disulfide-stabilized homologs. *Protein Eng* 1997;10:1453–1459.

114. Kobayashi H, Han ES, Kim IS, et al. Similarities in the biodistribution of iodine-labeled anti-Tac single-chain disulfide-stabilized Fv fragment and anti-Tac disulfide-stabilized Fv fragment. *Nucl Med Biol* 1998;25:387–393.

115. Holliger P, Prospero T, Winter G. "Diabodies": small bivalent and bispecific antibody fragments. *Proc Natl Acad Sci USA* 1993;90:6444–6448.

116. Whitlow M, Filpula D, Rollence ML, Feng SL, Wood JF. Multivalent Fvs: characterization of single-chain Fv oligomers and preparation of a bispecific Fv. *Protein Eng* 1994;7:1017–1026.

117. Yazaki PJ, Wu AM, Tsai SW, et al. Tumor targeting of radiometal labeled anti-CEA recombinant T84.66 diabody and t84. 66 minibody: comparison to radioiodinated fragments. *Bioconjug Chem* 2001;12:220–228.

118. Adams GP, Schier R, McCall AM, et al. Prolonged *in vivo* tumour retention of a human diabody targeting the extracellular domain of human HER2/neu. *Br J Cancer* 1998;77:1405–1412.

119. Beresford GW, Pavlinkova G, Booth BJ, Batra SK, Colcher D. Binding characteristics and tumor targeting of a covalently linked divalent CC49 single-chain antibody. *Int J Cancer* 1999;81:911–917.

120. Denton G, Brady K, Lo BK, et al. Production and characterization of an anti-(MUC1 mucin) recombinant diabody. *Cancer Immunol Immunother* 1999;48:29–38.

121. Adams GP, Shaller CC, Chappell LL, et al. Delivery of the alpha-emitting radioisotope bismuth-213 to solid tumors via single-chain Fv and diabody molecules. *Nucl Med Biol* 2000;27:339–346.

122. Robinson MK, Shaller C, Garmestani K, et al. Effective treatment of established human breast tumor xenografts in immunodeficient mice with a single dose of the alpha-emitting radioisotope astatine-211 conjugated to anti-HER2/neu diabodies. *Clin Cancer Res* 2008;14:875–882.

123. Adams GP, Shaller CC, Dadachova E, et al. A single treatment of yttrium-90-labeled CHX-A″-C6.5 diabody inhibits the growth of established human tumor xenografts in immunodeficient mice. *Cancer Res* 2004;64:6200–6206.

124. Whitlow M, Bell BA, Feng SL, et al. An improved linker for single-chain Fv with reduced aggregation and enhanced proteolytic stability. *Protein Eng* 1993;6:989–995.

125. Kostelny SA, Cole MS, Tso JY. Formation of a bispecific antibody by the use of leucine zippers. *J Immunol* 1992;148:1547–1553.

126. Pack P, Pluckthun A. Miniantibodies: use of amphipathic helices to produce functional, flexibly linked dimeric FV fragments with high avidity in *Escherichia coli*. *Biochemistry* 1992;31:1579–1584.

127. Ridgway JB, Presta LG, Carter P. 'Knobs-into-holes' engineering of antibody CH3 domains for heavy chain heterodimerization. *Protein Eng* 1996;9:617–621.

128. Pavlinkova G, Beresford G, Booth BJ, Batra SK, Colcher D. Charge-modified single chain antibody constructs of monoclonal antibody CC49: generation, characterization, pharmacokinetics, and biodistribution analysis. *Nucl Med Biol* 1999;26:27–34.

129. Goel A, Colcher D, Baranowska-Kortylewicz J, et al. Genetically engineered tetravalent single-chain Fv of the pancarcinoma monoclonal antibody CC49: improved biodistribution and potential for therapeutic application. *Cancer Res* 2000;60:6964–6971.

130. Wittel UA, Jain M, Goel A, Chauhan SC, Colcher D, Batra SK. The *in vivo* characteristics of genetically engineered divalent and tetravalent single-chain antibody constructs. *Nucl Med Biol* 2005;32:157–164.

131. Chauhan SC, Jain M, Moore ED, et al. Pharmacokinetics and biodistribution of ^{177}Lu-labeled multivalent single-chain Fv construct of the pancarcinoma monoclonal antibody CC49. *Eur J Nucl Med Mol Imaging* 2005;32:264–273.

132. Adams GP, McCartney JE, Tai MS, et al. Highly specific *in vivo* tumor targeting by monovalent and divalent forms of 741F8 anti-c-erbB-2 single-chain Fv. *Cancer Res* 1993;53:4026–4034.

133. Adams GP, Tai MS, McCartney JE, et al. Avidity-mediated enhancement of *in vivo* tumor targeting by single-chain Fv dimers. *Clin Cancer Res* 2006;12:1599–1605.

134. Hu S, Shively L, Raubitschek A, et al. Minibody: A novel engineered anti-carcinoembryonic antigen antibody fragment (single-chain Fv-C_H3) which exhibits rapid, high-level targeting of xenografts. *Cancer Res* 1996;56:3055–3061.

135. Sundaresan G, Yazaki PJ, Shively JE, et al. [124]I-labeled engineered anti-CEA minibodies and diabodies allow high-contrast, antigen-specific small-animal PET imaging of xenografts in athymic mice. *J Nucl Med* 2003;44:1962–1969.

136. Wong JY, Chu DZ, Williams LE, et al. Pilot trial evaluating an ^{123}I-labeled 80-kilodalton engineered anticarcinoembryonic antigen antibody fragment (cT84.66 minibody) in patients with colorectal cancer. *Clin Cancer Res* 2004;10:5014–5021.

137. Olafsen T, Tan GJ, Cheung CW, et al. Characterization of engineered anti-p185HER-2 (scFv-C_H3)$_2$ antibody fragments (minibodies) for tumor targeting. *Protein Eng Des Sel* 2004;17:315–323.

138. Rose LM, Deng CT, Scott SL, et al. Critical Lym-1 binding residues on polymorphic HLA-DR molecules. *Mol Immunol* 1999;36:789–797.

139. Shi XB, Gumerlock PH, Kroger L, DeNardo GL, DeNardo SJ. Efficient recombination of Lym-1 scFv gene using multiple doubly-restricted DNA fragments. *Cancer Biother Radiopharm* 1999;14:139–143.

140. Balhorn R, Hok S, Burke PA, et al. Selective high-affinity ligand antibody mimics for cancer diagnosis and therapy: initial application to lymphoma/leukemia. *Clin Cancer Res* 2007;13:5621s–5628s.

141. DeNardo GL, Natarajan A, Hok S, et al. Pharmacokinetic characterization in xenografted mice of a series of first-generation mimics for HLA-DR antibody, Lym-1, as carrier molecules to image and treat lymphoma. *J Nucl Med* 2007;48:1338–1347.

142. Hok S, Natarajan A, Balhorn R, DeNardo SJ, DeNardo GL, Perkins J. Synthesis and radiolabeling of selective high-affinity ligands designed to target non-Hodgkin's lymphoma and leukemia. *Bioconjug Chem* 2007;18:912–921.

143. West J, Perkins J, Hok S, et al. Direct antilymphoma activity of novel, first-generation "antibody mimics" that bind HLA-DR10-positive non-Hodgkin's lymphoma cells. *Cancer Biother Radiopharm* 2006;21:645–654.

144. Nilsson B, Moks T, Jansson B, et al. A synthetic IgG-binding domain based on staphylococcal protein A. *Protein Eng* 1987;1:107–113.

145. Wikman M, Steffen AC, Gunneriusson E, et al. Selection and characterization of HER2/neu-binding affibody ligands. *Protein Eng Des Sel* 2004;17:455–462.

146. Steffen AC, Wikman M, Tolmachev V, et al. *In vitro* characterization of a bivalent anti-HER-2 affibody with potential for radionuclide-based diagnostics. *Cancer Biother Radiopharm* 2005;20:239–248.

147. Steffen AC, Orlova A, Wikman M, et al. Affibody-mediated tumour targeting of HER-2 expressing xenografts in mice. *Eur J Nucl Med Mol Imaging* 2006;33:631–638.

148. Orlova A, Tolmachev V, Pehrson R, et al. Synthetic affibody molecules: a novel class of affinity ligands for molecular imaging of HER2-expressing malignant tumors. *Cancer Res* 2007;67:2178–2186.

149. Orlova A, Magnusson M, Eriksson TL, et al. Tumor imaging using a picomolar affinity HER2 binding affibody molecule. *Cancer Res* 2006;66:4339–4348.

150. Orlova A, Rosik D, Sandstrom M, Lundqvist H, Einarsson L, Tolmachev V. Evaluation of [$^{111/114m}$In]CHX-A″-DTPA-ZHER2:342, an affibody ligand conjugate for targeting of HER2-expressing malignant tumors. *Q J Nucl Med Mol Imaging* 2007; 51:314–323.

151. Tolmachev V, Xu H, Wallberg H, et al. Evaluation of a maleimido derivative of CHX-A″ DTPA for site-specific labeling of affibody molecules. *Bioconjug Chem* 2008.

152. Rodewald R. pH-dependent binding of immunoglobulins to intestinal cells of the neonatal rat. *J Cell Biol* 1976;71:666–669.

153. Kim JK, Firan M, Radu CG, Kim CH, Ghetie V, Ward ES. Mapping the site on human IgG for binding of the MHC class I-related receptor. *FcRn. Eur J Immunol* 1999; 29:2819–2825.

154. Ghetie V, Popov S, Borvak J, et al. Increasing the serum persistence of an IgG fragment by random mutagenesis. *Nat Biotechnol* 1997;15:637–640.

155. Kenanova V, Olafsen T, Crow DM, et al. Tailoring the pharmacokinetics and positron emission tomography imaging properties of anti-carcinoembryonic antigen single-chain Fv-Fc antibody fragments. *Cancer Res* 2005;65:622–631.

156. Kenanova V, Olafsen T, Williams LE, et al. Radioiodinated versus radiometal-labeled anti-carcinoembryonic antigen single-chain Fv-Fc antibody fragments: optimal pharmacokinetics for therapy. *Cancer Res* 2007;67:718–726.

157. Ferl GZ, Kenanova V, Wu AM, DiStefano JJ 3rd. A two-tiered physiologically based model for dually labeled single-chain Fv-Fc antibody fragments. *Mol Cancer Ther* 2006;5:1550–1558.

158. Wu AM, Tan GJ, Sherman MA, et al. Multimerization of a chimeric anti-CD20 single-chain Fv-Fc fusion protein is mediated through variable domain exchange. *Protein Eng* 2001;14:1025–1033.

159. Abuchowski A, McCoy JR, Palczuk NC, van Es T, Davis FF. Effect of covalent attachment of polyethylene glycol on immunogenicity and circulating life of bovine liver catalase. *J Biol Chem* 1977;252:3582–3586.

160. Bailon P, Palleroni A, Schaffer CA, et al. Rational design of a potent, long-lasting form of interferon: a 40 kDa branched polyethylene glycol-conjugated interferon alpha-2a for the treatment of hepatitis C. *Bioconjug Chem* 2001;12:195–202.

161. Wilkinson I, Jackson CJ, Lang GM, Holford-Strevens V, Sehon AH. Tolerogenic polyethylene glycol derivatives of xenogeneic monoclonal immunoglobulins. *Immunol Lett* 1987;15:17–22.

162. Takashina K, Kitamura K, Yamaguchi T, Noguchi A, Tsurumi H, Takahashi T. Comparative pharmacokinetic properties of murine monoclonal antibody A7 modified with neocarzinostatin, dextran and polyethylene glycol. *Jpn J Cancer Res* 1991; 82:1145–1150.

163. Lee LS, Conover C, Shi C, Whitlow M, Filpula D. Prolonged circulating lives of single-chain Fv proteins conjugated with polyethylene glycol: a comparison of conjugation chemistries and compounds. *Bioconjug Chem* 1999;10:973–981.

164. Li L, Yazaki PJ, Anderson AL, et al. Improved biodistribution and radioimmunoimaging with poly(ethylene glycol)-DOTA-conjugated anti-CEA diabody. *Bioconjug Chem* 2006;17:68–76.

165. Molineux G. Pegylation: engineering improved pharmaceuticals for enhanced therapy. *Cancer Treat Rev* 2002;28 Suppl A: 13–16.

166. Natarajan A, Xiong C, Albrecht H, DeNardo GL, DeNardo SJ. Characterization of site-specific ScFv PEGylation for tumor-targeting pharmaceuticals. *Bioconjug Chem* 2005;16:113–121.

167. Yang K, Basu A, Wang M, et al. Tailoring structure–function and pharmacokinetic properties of single-chain Fv proteins by site-specific PEGylation. *Protein Eng* 2003;16:761–770.

168. Xiong CY, Natarajan A, Shi XB, Denardo GL, Denardo SJ. Development of tumor targeting anti-MUC-1 multimer: effects of di-scFv unpaired cysteine location on PE-Gylation and tumor binding. *Protein Eng Des Sel* 2006;19:359–367.

169. Albrecht H, Denardo GL, Denardo SJ. Monospecific bivalent scFv-SH: effects of linker length and location of an engineered cysteine on production, antigen binding activity and free SH accessibility. *J Immunol Methods* 2006;310:100–116.

170. Kragh-Hansen U, Chuang VT, Otagiri M. Practical aspects of the ligand-binding and enzymatic properties of human serum albumin. *Biol Pharm Bull* 2002;25:695–704.

171. Chuang VT, Kragh-Hansen U, Otagiri M. Pharmaceutical strategies utilizing recombinant human serum albumin. *Pharm Res* 2002;19:569–577.

172. Smith BJ, Popplewell A, Athwal D, et al. Prolonged *in vivo* residence times of antibody fragments associated with albumin. *Bioconjug Chem* 2001;12:750–756.

173. Yazaki PJ, Kassa T, Cheung CW, et al. Biodistribution and tumor imaging of an anti-CEA single-chain antibody–albumin fusion protein. *Nucl Med Biol* 2008;35:151–158.

174. Tijink BM, Laeremans T, Budde M, et al. Improved tumor targeting of anti-epidermal growth factor receptor Nanobodies through albumin binding: taking advantage of modular Nanobody technology. *Mol Cancer Ther* 2008;7:2288–2297.

175. Dennis MS, Zhang M, Meng YG, et al. Albumin binding as a general strategy for improving the pharmacokinetics of proteins. *J Biol Chem* 2002;277:35035–35043.

176. Dennis MS, Jin H, Dugger D, et al. Imaging tumors with an albumin-binding Fab, a novel tumor-targeting agent. *Cancer Res* 2007;67:254–261.

177. Nguyen A, Reyes AE, 2nd Zhang M, et al. The pharmacokinetics of an albumin-binding Fab (AB.Fab) can be modulated as a function of affinity for albumin. *Protein Eng Des Sel* 2006;19:291–297.

CHAPTER 2

The Radiochemistry of Monoclonal Antibodies and Peptides*

RAYMOND M. REILLY

2.1 INTRODUCTION

Cancer is a major health problem worldwide. It is estimated that there are more than 10 million new cases of cancer diagnosed annually around the globe and more than 4 million individuals die each year from the disease (1). Intense research over the past few decades has led to important discoveries by cancer biologists that are just now stimulating the development of new potentially more effective and safer biologically targeted therapies for the disease. One promising strategy for treating malignancies that exploits their biological phenotype is targeted *in situ* radiotherapy (2). In this approach, monoclonal antibodies (mAbs) that recognize tumor-associated antigens or peptide ligands that specifically bind to cell-surface growth factor receptors, are used as targeting vehicles to selectively deliver radionuclides to cancer cells for *in situ* radiation therapy. These two approaches are known as radioimmunotherapy (RIT) or peptide-directed radiotherapy (PDRT), respectively. An extension of RIT or PDRT is to employ radiolabeled mAbs or peptides for imaging metastatic deposits for detection or to noninvasively characterize their phenotype *in situ*, known as molecular imaging (3). Phenotypic characterization of tumors will be critical in the future to be able to appropriately select patients for treatment with new biologically targeted anticancer agents, including targeted radiotherapeutics. For example, it is known that breast cancers that exhibit high levels of amplification of the human epidermal growth factor receptor-2 (HER2/neu) gene respond best to treatment with the humanized anti-HER2/neu mAb, trastuzumab (Herceptin; Roche Pharmaceuticals) (4). HER2/neu gene amplification is currently evaluated in a primary breast cancer biopsy by

---

*This chapter is updated and reprinted with permission from Gad SC, editor. *Handbook of Pharmaceutical Technology*, John Wiley & Sons 2007: Chapter 6.7: Reilly RM. The radiopharmaceutical science of monoclonal antibodies and peptides for imaging and targeted *in situ* radiotherapy of malignancies.

---

immunohistochemical staining for the HER2/neu protein or by fluorescence *in situ* hybridization (FISH) for the gene copy number. A recent report, however, suggests that molecular imaging of HER2/neu expression in breast tumors using indium-111 (^{111}In)-labeled trastuzumab may reliably predict which patients respond to Herceptin as well as identify those who may be at risk for toxicity from the drug (5). Moreover, imaging studies using radiolabeled forms of biopharmaceuticals could allow a noninvasive *in situ* assessment of their pharmacokinetic properties and in particular, their tumor and normal tissue uptake and elimination. This may provide insight into their effectiveness as antitumor agents as well as their potential sites of normal organ toxicity. There have been many comprehensive reviews on imaging and targeted *in situ* radiotherapy of malignancies with radiolabeled mAbs and peptides (2, 3, 6). In this chapter, the radiochemical science that provides the foundation for these biomolecules for targeting radionuclides to tumors is discussed.

2.2   TUMOR AND NORMAL TISSUE UPTAKE OF MONOCLONAL ANTIBODIES AND PEPTIDES

The tumor and normal tissue delivery properties of mAbs and their fragments and peptides must to be considered when designing a radiopharmaceutical for imaging or targeted radiotherapy of cancer. Intact IgG mAbs ($M_r = 150$ kDa) (Fig. 2.1) (also see Chapter 1) are macromolecules that are cleared slowly from the blood with an elimination half-life of 2–3 days for murine forms and 4 days for chimeric and humanized mAbs (7). This slow elimination provides multiple passes through a tumor at which extravasation and interaction with tumor cells may occur. Thus, the tumor accumulation (percent injected dose per gram (% i.d./g)) of radiolabeled intact IgG mAbs is much greater than that of smaller and more rapidly cleared antibody fragments (e.g., Fab or scFv) (Fig. 2.1) or that of small peptides. This property renders intact IgG mAbs more attractive for RIT, since it maximizes the amount of radioactivity that is delivered to the tumor and thus, the radiation-absorbed dose delivered. However, the macromolecular properties of intact IgG mAbs severely restrict their tumor penetration. It has been shown that radiolabeled mAbs remain close to tumor blood vessels, whereas Fab and single-chain Fv (scFv) fragments penetrate more deeply into

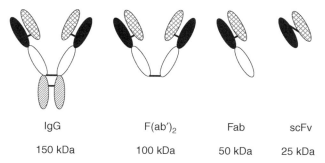

IgG	F(ab')$_2$	Fab	scFv
150 kDa	100 kDa	50 kDa	25 kDa

FIGURE 2.1   Structures and corresponding molecular weights of various antibody forms used for imaging or targeted radiotherapy of cancer.

tumors (8). This differential depth of penetration for intact IgG mAbs compared to their fragments may be related in part to antigen binding affinity and avidity, since it has been hypothesized that there is a "binding-site barrier" that restricts the penetration of antibodies into tumors as a consequence of interaction with antigens on cancer cells proximal to the blood vessels (9, 10). Antibody fragments (Fab and scFv) are monovalent, a property that diminishes their avidity and affinity compared to divalent IgG mAbs. The poor tumor penetration properties of intact IgG mAbs produce heterogeneous distribution of radioactivity within a tumor. Moreover, the Fc-domain of intact mAbs is recognized by asialoglycoprotein receptors on hepatocytes, which promotes liver accumulation of radioactivity (7). Retention of radioactivity in the liver interferes with detection of hepatic metastases, especially when radiometal-labeled mAbs and peptides are used for imaging. This is less problematic for radioiodinated biomolecules due to their intracellular catabolism and release of radioiodine from hepatocytes.

For tumor imaging, antibody fragments ($F(ab')_2$, Fab, or scFv) or peptides (see Chapter 3) are the most useful because they are rapidly eliminated from the blood and most normal tissues (except the kidneys), which minimizes circulating background radioactivity and yields high tumor/blood (T/B) and tumor/normal tissue (T/NT) ratios that occur at early times after injection. The major challenge with radiolabeled mAb fragments and peptides is their high accumulation in the kidneys. Kidney uptake is thought to be due to glomerular filtration of the fragments or peptides followed by charge interactions between cationic amino acid residues (e.g., lysine and arginine) in the proteins and the negatively charged renal tubular cell membrane (11). High renal uptake of the somatostatin octapeptide analogue, DOTATOC labeled with the β-emitter, ^{90}Y is associated with serious kidney toxicity in patients when used for PDRT of neuroendocrine malignancies (see Chapter 4) (2). However, renal toxicity can be avoided by coadministering intravenous solutions of lysine or arginine that competitively inhibit the interaction between renal tubular cells and radiolabeled mAb fragments or peptides (11, 12). Furthermore, renal toxicity is not associated with the use of these radiopharmaceuticals for tumor imaging, since they are administered at much lower doses and labeled with low linear energy transfer (LET) γ-emitting radionuclides.

2.3   SELECTION OF A RADIONUCLIDE FOR TUMOR IMAGING

Radionuclides suitable for labeling mAbs or peptides for tumor imaging may be single γ-photon emitters (Table 2.1) or positron emitters (Table 2.2). Images using single γ-photons are usually acquired in 3-dimensional mode, known as single photon emission computerized tomography (SPECT) (Fig. 2.2). The single γ-photons are detected by thallium-doped sodium iodide [NaI(Tl)] scintillation crystals that are housed in two opposing heads of a γ-camera. The heads are rotated 180° around the subject to obtain a series of images. Images of the body, organ or tissue of interest are reconstructed by back-extrapolation of the lines of detection acquired by each camera head. This technique allows the reconstructed images to be "sliced" to visualize a single image plane separated from any overlying or underlying interfering planes. Optimal γ-energies for SPECT imaging are 100–300 keV. The corresponding imaging technique

TABLE 2.1   Radionuclides Suitable for Labeling Biomolecules for SPECT Imaging of Tumors

Radionuclide	Production Method	$E\gamma$ (keV) (Abundance)	$T_{1/2}$phys	Labeling Methods
^{99m}Tc	^{99}Mo/^{99m}Tc generator	140.5 (98.9%)	6.0 h	Binding to thiols; chelation by tetradentate complexes; HYNIC; interaction of carbonyl complex with histidine residues
^{111}In	^{112}Cd($p,2n$)^{111}In	171.3 (90.2%), 245.4 (94.0%)	2.8 days	Chelation by DTPA, SCN-Bz-DTPA, or DOTA
^{67}Ga	^{68}Zn($p,2n$)^{67}Ga	93.3 (35.7%), 184.6 (19.7%), 300.2 (16.0%)	3.3 days	Chelation by DFO or DOTA
^{123}I	^{124}Xe($p,2n$)^{123}Cs → ^{123}Xe → ^{123}I; ^{124}Xe(p,pn) ^{123}Xe → ^{123}I	159 (83.4%)	13.2 h	Direct radioiodination with chloramine-T or Iodogen; indirect conjugation using ATE, SIPC, SGMIB; indirect conjugation using carbohydrate adducts such as TCB or dextran
^{131}I	^{130}Te(n,γ) ^{131}Te → ^{131}I	364 keV (81.2%)	8.0 days	Same as for ^{123}I

for positron emitters is called positron emission tomography (PET). Positrons are β^+-particles that are emitted when a proton is converted to a neutron in the decay scheme of certain radionuclides to stable elements. The positrons travel a distance (0.7–5 mm) that is directly proportional to their energy before they are annihilated through interaction with an electron in tissues. Positron annihilation creates two 511 keV γ-photons that travel out at approximately 180° from one another at the site of annihilation. PET relies on the simultaneous detection of these two γ-photons within a narrow time window ("time of flight") by a ring of lutetium silicate (LSO) or bismuth subgerminate (BGO) scintillation crystals surrounding the subject (Fig. 2.2). Back-extrapolation of the lines of detection

TABLE 2.2 Radionuclides Suitable for Labeling Biomolecules for PET Imaging of Tumors

Radionuclide	Production Method	β^+-Energy (MeV) (Abundence)	Intrinsic Spatial Resolution (mm)	$T_{1/2}$phys	Labeling Methods
^{124}I	^{124}Te$(p,n)^{124}$I or ^{124}Te$(d,2n)^{124}$I	0.8–2.1 (23%)	2.3	4.2 days	Direct radioiodination with chloramine-T or Iodogen
^{76}Br	^{76}Se$(p,n)^{76}$Br	3.4 (54%)	5.0	16.2 h	Direct radiobromination with chloramine-T; indirect conjugation using SpBrB, HPEM, brom-3-pyridine-carboxylate, *closo*-dodecaborate, *nido*-undecaborate
^{18}F	$^{18}O(p,n)^{18}$F	0.6 (100%)	0.7	110 min	Indirect conjugation using SFBS, SFB, NPFP, FB-CHO, AFP
^{94m}Tc	^{94}Mo$(p,n)^{94m}$Tc	2.5 (72%)	3.3	52 mins	N$_4$-tetradentate chelator
^{110m}In	^{110}Cd$(p,n)^{110m}$In	2.3 (62%)	3.0	69 min	Chelation by DTPA
^{86}Y	^{86}Sr$(p,n)^{86}$Y	1.2–1.5 (33%)	1.8	14.7 h	Chelation by DOTA
^{68}Ga	^{68}Ge/^{68}Ga generator	1.9 (89%)	2.4	68 min	Chelation by DOTA
^{64}Cu	^{64}Ni$(p,n)^{64}$Cu	0.7 (19%)	0.7	12.7 h	Chelation by DOTA, BAT, or TETA

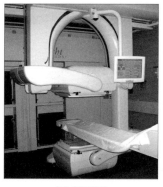 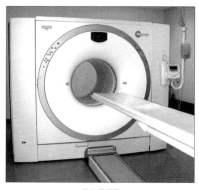

(a) SPECT   (b) PET

FIGURE 2.2   Clinical imaging systems used in nuclear medicine to visualize the distribution of radiopharmaceuticals labeled with single photon-emitting radionuclides or with positron-emitting radionuclides. (a) A gamma camera used for SPECT. (b) A positron-emission tomograph used for PET scanning.

allows accurate identification of the site of positron annihilation and yields high spatial resolution images. However, it is important to recognize that the spatial resolution of PET is impacted by the finite distance that the positron travels before its annihilation. This distance ranges from 0.7 mm for the low energy, 0.6 MeV β^+-particles of ^{18}F to 5.3 mm for the higher energy 1.7 MeV β^+-particles emitted by ^{76}Br (Table 2.2) (13).

It is important to consider the pharmacokinetic properties of the targeting vehicle in selecting an appropriate radionuclide for tumor imaging. Intact IgG mAbs require several days to reach maximum tumor uptake and to be cleared from the blood and normal tissues. Therefore, single γ-photon emitters such as ^{111}In or ^{131}I with a half-life of 2.8 or 8 days, respectively (Table 2.1) or longer lived positron emitters such as ^{124}I (half-life of 4 days) (Table 2.2) are most suitable for labeling these vehicles for tumor imaging. Antibody fragments such as Fab or scFv are cleared more rapidly from the blood and most normal tissues (except the kidneys), yielding high T/B and T/NT ratios within 24 h. Therefore, these vehicles may be labeled with short to intermediate half-life single γ-photon emitters such as ^{99m}Tc, ^{123}I, or ^{67}Ga for SPECT (Table 2.1) or with ^{64}Cu or ^{86}Y for PET (Table 2.2). The shorter half-life positron emitters such as ^{18}F and ^{68}Ga are reserved for labeling peptides and scFv fragments of mAbs that exhibit very rapid tumor uptake and elimination from the blood and normal tissues. Ultrashort-lived positron emitters, such as ^{15}O, ^{13}N, or ^{11}C (half-lives of 2, 10, or 20 min, respectively) that have been used for labeling small molecules for PET, are not feasible for imaging using mAbs or peptides.

2.4 SELECTION OF A RADIONUCLIDE FOR TARGETED RADIOTHERAPY

Radionuclides suitable for targeted *in situ* radiotherapy of tumors (Table 2.3) emit either α-particles, β-particles, or Auger and conversion electrons (2). The important differences between these different forms of radiation are their range in tissues and

TABLE 2.3 Radionuclides Suitable for Labeling Biomolecules for Targeted *In Situ* Radiotherapy of Tumors

Radionuclide	Production Method	Particulate Emissions (Energy)	Maximum Range in Tissues	$T_{1/2}$phys	Labeling Methods
^{125}I	^{124}Xe$(n,\gamma)^{125}$Xe $\rightarrow$ ^{125}I	Auger electrons (<30 keV)	$<10\,\mu$m	59.4 days	Direct radioiodination with chloramine-T or Iodogen; indirect conjugation using ATE, SIPC, SGMIB
^{123}I	^{124}Xe$(p,2n)^{123}$Cs $\rightarrow$ ^{123}Xe $\rightarrow$ ^{123}I	Auger electrons	$<10\,\mu$m	13.2 h	Same as for ^{125}I
^{131}I	Neutron irradiation of ^{130}Te	β-Particles (0.6 MeV)	2 mm	8.0 days	Same as for ^{125}I
^{211}At	^{209}Bi$(\alpha,2n)^{211}$At	α-Particles (5.9–7.4 MeV)	50–100 μm	7.2 h	Indirect conjugation with ATE, SAB, SAPC
^{186}Re	^{185}Re$(n,\gamma)^{186}$Re	β-Particles (1.1 MeV)	3 mm	3.7 days	Binding to thiols; chelation by tetradentate complexes; HYNIC; interaction of carbonyl complex with histidine residues; trisuccin
^{188}Re	^{188}W/^{188}Re generator	β-Particles (1.1 MeV)	8 mm	3.7 days	Same as for ^{186}Re
^{90}Y	^{90}Sr/^{90}Y generator	β-Particles (2.3 MeV)	12 mm	2.7 days	Chelation by DOTA
^{67}Ga	^{68}Zn$(p,2n)^{67}$Ga	Auger electrons (<30 keV)	$<10\,\mu$m	3.3 days	Chelation by DFO and DOTA
^{64}Cu	^{68}Zn$(p,\alpha n)^{64}$Cu	β-Particles (0.6 MeV)	2 mm	12.7 h	Chelation by DOTA, BAT, or TETA
^{67}Cu	^{nat}Zn$(p,2p)^{67}$Cu or ^{68}Zn$(p,2p)^{67}$Cu	β-Particles (0.4–0.6 MeV)	2 mm	2.6 days	Same as for ^{64}Cu
^{177}Lu	^{176}Yb$(n,\gamma)\rightarrow$ ^{177}Yb $\rightarrow$ ^{177}Lu	β-Particles (0.5 MeV)	2 mm	6.6 days	Chelation by DOTA or CHX-DTPA
^{213}Bi	^{225}Ac/^{213}Bi generator	α-Particles (8 MeV)	50–100 μm	46 min	Chelation by DOTA or CHX-DTPA
^{225}Ac	^{233}U $\rightarrow$ ^{225}Ac	α-Particles (several daughter radionuclides with different energies)	50–100 μm	10 days	Chelation by DOTA

their LET. α-Particles consist of two protons and two neutrons and carry a 2^+ charge. These particles have the highest LET (100 keV/μm), are densely ionizing, and travel 50–100 μm (5–10 cell diameters) in tissues. α-Emitters such as ^{211}At, ^{212}Bi, or ^{225}Ac are most useful for eradicating small clusters of cancer cells or micrometastases. β-Particles are high-energy electrons that have a 1^- charge and travel 2–12 mm in tissues (200–1200 cell diameters). β-Particles deposit most of their energy at the end of their track length in tissues. However, for comparison with α-particles, the average LET of the β-particles emitted by ^{131}I ($E\beta = 0.6$ MeV) over their 2 mm track length is 0.3 keV/μm. The average LET of the β-particles emitted by ^{90}Y ($E\beta = 2.3$ MeV) over their 10–12 mm track length is 0.2 keV/μm. Due to the long range of the β-particles emitted by ^{131}I, ^{90}Y, ^{186}Re/^{188}Re, or ^{175}Lu conjugated to mAbs or peptides that are targeted to tumor cells, it is possible to irradiate and kill more distant nontargeted tumor cells ("cross-fire" effect). This is advantageous for larger lesions (i.e., 2–10 mm in diameter) in which there is likely to be incomplete targeting of tumor cells. Inadequate targeting could allow some cells to survive. However, the "cross-fire" effect of the β-particles also contributes to dose-limiting bone marrow toxicity in RIT, due to nonspecific irradiation of hematopoietic stem cells by circulating radioactivity perfusing the bone marrow (2).

Auger electrons are very low-energy electrons emitted by radionuclides that decay by electron capture (EC) (see Chapter 9). In EC, a proton in the nucleus captures an electron from an inner orbital shell, creating a vacancy in the shell. This vacancy is filled by the decay of an electron from a higher shell. The excess energy released is transferred to an outer orbital electron, which is then ejected from the atom, creating a 2^+ atomic species. Because of their very low energy (<30 keV), Auger electrons travel only a few nanometers to at most a few micrometers in tissues (less than one cell diameter). Their LET approaches that of α-emitters (100 keV/μm). However, an antibody or peptide carrying an Auger electron-emitting radionuclide such as ^{125}I, ^{123}I, ^{111}In, or ^{67}Ga must be internalized and ideally, translocated to the nucleus for the electrons to be most damaging DNA. Biomolecules that recognize peptide growth factor receptors (e.g., epidermal growth factor receptors (EGFRs) or somatostatin receptors (SMSRs)) are frequently internalized into cells and in some cases translocated to the nucleus, which allows them to be employed for targeted Auger electron radiotherapy of malignancies. The advantage of Auger electron-emitting radionuclides is that it is possible to restrict killing to cells that specifically bind and internalize radiolabeled mAbs or peptides. There is no "cross-fire" effect from Auger electron-emitting radionuclides, which should obviate any major nonspecific radiotoxicity to bone marrow stem cells, particularly if these cells do not express the target epitopes/receptors. On the other hand, in contrast to the more energetic β-emitters, the lack of a "cross-fire" effect from Auger electron-emitters does not permit killing of nontargeted cancer cells, although a "bystander effect" and a more local "cross-dose" effect has been reported (see Chapter 9) (14). Auger electron-emitting radionuclides are therefore most useful for treating small tumor deposits or micrometastases for which delivery of radiolabeled biomolecules is more homogeneous.

2.5 LABELING ANTIBODIES AND PEPTIDES WITH RADIOHALOGENS

2.5.1 Iodine Radionuclides

Radioiodination is one of the simplest ways to radiolabel a biomolecule. Several radionuclides of iodine are available for SPECT or PET (^{123}I, ^{124}I, and ^{131}I) (Tables 2.1 and 2.2) or for targeted radiotherapy of cancer (^{125}I and ^{131}I) (Table 2.3). Iodine is present in a valence state of 1^- in the alkaline solution in which it is supplied. It requires oxidation to a 1^+ valence state for electrophilic substitution into tyrosine amino acids in antibodies and peptides (15). Reaction of radioiodine with histidine, cysteine, methionine, phenylalanine, and tryptophan residues is possible, but less likely. The most commonly used oxidizing agents are chloramine-T (N-chloro-4-methylbenzene sulfonamide) (16) and Iodogen® (1,3,4,6-tetrachloro-3α, 6α-diphenylglycouril; Pierce) (17). Chloramine-T provides higher radiolabeling yields than Iodogen (70–90% versus 40–60%) but since it is a stronger oxidizing agent, it may damage mAbs, especially if the conditions are not carefully controlled. A typical chloramine-T-mediated radioiodination involves diluting the antibody or peptide in a slightly alkaline buffer and incubating it with radioiodine and 10–20 µg of chloramine-T in a glass tube for 30–60 s at room temperature (18). The radioiodination reaction is stopped by adding 20–40 µg of the reducing agent, sodium metabisulfite. Radioiodinated antibodies and peptides can be purified from free radioiodide by size-exclusion chromatography (SEC); alternatively peptides may be purified by reversed-phase chromatography. Iodogen is a water-insoluble oxidizing agent which is dissolved in chloroform; 10–20 µg are then aliquoted into a glass tube and the chloroform is evaporated using a gentle stream of nitrogen to leave a coating of Iodogen on the inside surface of the tube. The antibody or peptide contained in a suitable buffer and radioiodide are then added to the tube and incubated for 1–2 min at room temperature. The reaction is stopped simply by transferring the radioiodination mixture to a chromatography column for purification, leaving the water-insoluble Iodogen remaining in the tube. Iodogen is a more gentle oxidizing agent than chloramine-T and is preferred for radioiodinating mAbs and their fragments.

The main challenge for radioiodination of mAbs and peptides is their instability *in vivo* to proteolysis, deiodination and loss of radioactivity from tumor cells. It is widely recognized that radioiodinated antibodies and peptides are proteolytically degraded in cells to radioiodotyrosine that is efficiently exported from the cells by membrane amino acid transporters. Released radioiodotyrosine is deiodinated by deiodinases found in tissues and the free radioiodine redistributes and accumulates in organs with sodium iodide symporter expression, particularly the thyroid, stomach and salivary glands. For tumor imaging applications, these catabolic processes diminish the tumor signal and increase normal tissue uptake of radioactivity. Moreover, in radiotherapeutic applications, these processes diminish the radiation absorbed dose to the tumor and increase the dose to normal tissues, thus narrowing the therapeutic index. To address this issue, several new radioiodination methods have been developed that retain the radioactive catabolites within cells; these methods are known as "residualizing" techniques. Zalutsky et al. (19) first reported a method for

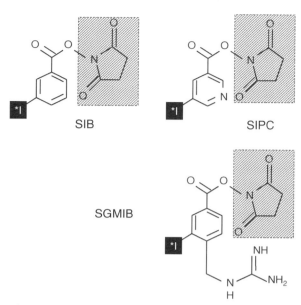

FIGURE 2.3 Precursors used for residualizing radioiodination of monoclonal antibodies and peptides. Chemically reactive groups on the precursors are shaded, and the site for radioiodination of the precursor is indicated in black and with an asterisk. SIB: *N*-succinimidyl-3-iodobenzoate; SIPC: *N*-succinimidyl-5-iodo-3-pyridinecarboxylate; SGMIB: *N*-succinimidyl-4-guanidinomethyl-3-iodobenzoate.

residualizing radioiodine in cells in 1987. His group synthesized the *N*-succinimidyl ester of 3-(tri-*n*-butylstannyl) benzoate (ATE) (Fig. 2.3), a precursor that could be radioiodinated in anhydrous chloroform by iododestannylation using anhydrous *t*-butylhydroperoxide (TBHP) as an oxidant. Radioiodinated ATE, termed *N*-succinimidyl-3-iodobenzoate (SIB), was purified on a silica gel Sep-Pak® (Waters) and conjugated to the N-terminus or ε-amino groups of lysine residues on IgGs through its reactive *N*-succinimidyl ester side chain. In this method, radioiodine is situated at a position on the aromatic ring of ATE that is not ortho to a phenolic hydroxyl group (as in the case of radioiodotyrosine); this renders the molecule resistant to *in vivo* deiodination. In mice, thyroid uptake of radioactivity was substantially decreased compared to IgGs labeled with radioiodine using Iodogen. A paired-label experiment comparing the tumor and normal tissue uptake of an F(ab')$_2$ fragment of the OC125 mAb labeled with ^{125}I using ATE or ^{131}I using Iodogen in mice implanted with OVCAR-3 ovarian carcinoma xenografts, revealed that thyroid uptake of radioiodine was reduced 100-fold using ATE (20). There was also more rapid elimination of radioactivity from normal tissues, and at 96 h postinjection (p.i.), T/NT ratios were fourfold higher. Similar promising results were observed in mice implanted subcutaneously (s.c.) with D-54 MG glioblastoma xenografts administered antite-nascin mAb 81C6 labeled with ^{125}I using ATE compared to labeling with ^{131}I using Iodogen (21). Zalutsky et al. (22) extended this residualizing strategy with an alternative radioiodination agent, N-succinimidyl 5-[^{131}I]iodo-3-pyridinecarboxylate (SIPC) (Fig. 2.3). SIPC was used to radioiodinate 81C6 mAb IgG and the F(ab$'$)$_2$ fragment of antimelanoma mAb Mel-14. It was hypothesized that SIPC, an iodopyridine would be more dissimilar than SIB to iodotyrosine, and thus more resistant *in vivo* to deiodination. The normal tissue distribution in mice of the 81C6 mAb and Mel-14 F(ab$'$)$_2$ fragment radioiodinated using SIPC or ATE were similar and there was very low thyroid uptake of radioactivity (<0.2–0.3% i.d. at 7 days p.i.) using either reagent. The tumor-specific anti-EGFRvIII mAb L8A4, which internalizes into receptor-positive cells and is highly susceptible to deiodination, was radioiodinated using SIPC resulting in improved tumor cell retention of radioactivity *in vitro* and increased accumulation in EGFRvIII-positive tumor xenografts in mice providing enhanced T/NT ratios in comparison to L8A4 antibodies radioiodinated using Iodogen (23, 24). Radioiodination using SIPC has also been applied to two 13-mer peptides: α-melanocyte-stimulating hormone (α-MSH) and its analogue, [Nle^4, D-Phe^7]-α-MSH, resulting in preservation of receptor binding properties *in vitro* and stability against deiodination *in vivo* in mice (25). However, conjugation of the peptides with SIPC increased their hydrophobicity. The catabolite of mAbs and peptides radioiodinated with SIPC was found to be lysine-iodobenzoic acid (Lys-IBA).

   One strategy that has been investigated to further enhance the retention of radioiodine in cells is to use labeling techniques that generate a charged catabolite following intracellular proteolysis, which cannot traverse the lysosomal membrane and is thus resistant to exocytosis. Zalutsky et al. (26) synthesized N-succinimidyl 4-guanidinomethyl-3-[^{131}I]iodobenzoate (SGMIB), a radioiodinating reagent that was expected to generate a positively charged catabolite at the acidic pH in lysosomes. They used this reagent to radioiodinate anti-EGFRvIII mAb L8A4. There was three- to fourfold greater retention of radioactivity in receptor-positive U87MG glioblastoma cells when these antibodies were radioiodinated using SGMIB than when Iodogen was used. Analysis confirmed that the final catabolites were cationic. SGMIB has advantages over SIPC as a residualizing radioiodination reagent, in that a twofold improvement in tumor retention of radioactivity was observed for L8A4 mAbs in mice implanted s.c. with D-256 glioblastoma xenografts expressing EGFRvIII. Again, similar to the SIPC reagent, thyroid radioactivity was very low (<0.35% i.d.) (27). An analogous approach was reported by Shankar et al. (28) who used N-succinimidyl 3-[^{131}I]iodo-4-phosphonomethylbenzoate ([^{131}I]SIPMB) (Fig. 2.3) to radioiodinate mAb L8A4; this reagent generates an anionic catabolite that is retained within cells.

   A different strategy for residualizing radioiodine in cells uses radioiodinated diethylenetriaminepentaacetic acid (DTPA)-containing peptides composed of one or more D-amino acids including D-tyrosine that are conjugated through a maleimide functional group to chemically reduced IgG mAbs (29). Two such peptides R-Gly-D-Tyr-D-Lys[1-(p-thiocarbonylaminobenzyl)DTPA], termed IMP-R1, and [R-D-Ala-D-Tyr-D-Tyr-D-Lys]$_2$(CA-DTPA), termed IMP-R2, were described by Govindan et al. (29). The BOC-protected peptides were radioiodinated at the D-tyrosine amino acid using chloramine-T and derivatized at their N-terminus with sulfo-SMCC to introduce maleimide groups. Following deprotection, the maleimide-derivatized and radioiodinated peptides were conjugated to thiol groups generated by dithiothreitol (DTT)-reduction of some disulfide linkages on the anti-CD20 mAb LL2 or the anti-EGP-1 mAb RS7. The premise of including DTPA in the peptides was that since it had been previously shown that mAbs labeled with ^{111}In through introduction of a DTPA metal chelator were catabolized to ^{111}In-DTPA-lysine which was retained within the cells (30), it would be expected that radioiodinated peptides containing this group would be trapped. The DTPA groups in the peptides are not used for radiolabeling. D-Amino acids were included since these are more resistant to proteolysis than L-amino acids and in particular D-iodotyrosine is more resistant to *in vivo* deiodination than L-tyrosine (31). The immunoreactivity of the RS7 antibody was maintained using this labeling strategy (radioiodinated LL2 was not tested for immunoreactivity) and both antibodies exhibited greater retention of radioactivity in tumor cells *in vitro* than antibodies radioiodinated using chloramine-T. A series of radioiodinated DTPA-D-amino acid peptides (IMP-R1 to IMP-R8) that differed in their hydrophobicity and charge was synthesized and conjugated to DTT-reduced RS7 mAb (32). These peptides varied in their DTPA content and extent of maleimide derivatization. Increasing the maleimide substitution in the peptides from one to two functional groups per molecule increased the conjugation efficiency with the R27 mAb from 30% to >80%. In mice implanted s.c. with Calu-3 lung carcinoma xenografts, tumor radioactivity was enhanced for mAb RS7 radioiodinated using any of the DTPA-D-amino acid peptides compared to chloramine-T radioiodinated mAbs. However, normal tissue uptake was highest for the most hydrophobic peptides. IMP-R4: MCC-Lys(MCC)-Lys(1-((*p*-CSNH)benzyl)DTPA)-D-Tyr-D-Lys(3-((*p*-CSNH)-benzyl) DTPA)-OH, where MCC represents the maleimide groups, provided the greatest retention of radioiodine in tumors and lowest normal tissue accumulation. This approach has been optimized using a one-vial kit labeling procedure under good manufacturing practices (GMP) conditions to yield at least 100 mCi of highly pure (>95%) ^{131}I-labeled humanized anti-CEA mAb hMN-14 that exhibits preserved immunoreactivity (>95%) for targeted radiotherapy of malignancies (33). A similar strategy was recently described by Foulon et al. (34). This group used a polycationic peptide composed of D-amino acids: D-Lys-D-Arg-D-Tyr-D-Arg-D-Arg to radioiodinate anti-EGFRvIII mAb L8A4. These peptides were first radioiodinated using Iodogen and then conjugated in 60% overall yield via a maleimide group to mAb L8A4, which was thiolated using 2-iminothiolane. Paired-label experiments in mice implanted s.c. with U87 glioblastoma xenografts expressing EGFRvIII and administered mAb L8A4 labeled with ^{125}I using this approach showed up to a fivefold higher tumor accumulation compared to L8A4 labeled directly with ^{131}I using Iodogen. However, use of the peptides for radioiodinating L8A4 increased accumulation in the kidneys, perhaps because of binding of the positively charged catabolites to renal tubular cells (34).

Radioiodine may also be residualized in cells by substitution onto an aromatic residue that is linked to a carbohydrate moiety attached to the biomolecule (35). A sevenfold increase in tumor retention of radioactivity was observed in mice implanted with Calu-3 lung carcinoma xenografts at 7 days p.i. of radioiodinated mAb RS7 using a dilactitol-tyramine (DLT) carbohydrate adduct compared to chloramine-T radio-iodination (38.0 versus 5.5% i.d./g, respectively) (36). Similar results were achieved using tyramine cellobiose (TCB)-radioiodinated anti-EGFRvIII mAb L8A4 in mice bearing receptor-positive tumor xenografts (37). However, a limitation of this approach appears to be the higher liver and spleen uptake of radioactivity of mAbs radioiodinated using TCB (38). Epidermal growth factor (EGF), a 53-amino acid peptide ligand for the EGFR present on many epithelial malignancies has been labeled using an analogous approach by radioiodinating a tyrosine-modified dextran and then conjugating this dextran molecule to EGF (39). Glioblastoma cells expressing EGFR showed significantly greater retention of radioactivity when incubated with dextran-EGF in which the radioiodine was present on the dextran moiety compared to EGF or dextran-EGF where the radioiodine was substituted into tyrosine amino acids on the EGF molecule itself.

Since the advent of high-resolution PET/CT (40) and small animal PET tomographs (41) and building on the success of ^{18}F-fluorodeoxyglucose (^{18}F-FDG) for PET of malignancies (42), there has been a growing interest in labeling mAbs and peptides with positron emitters. ^{124}I is a positron emitter with a sufficiently long half-life of 4.2 days that is feasible for labeling mAbs and peptides for PET. Nevertheless, ^{124}I has limitations compared to ^{18}F. These include its lower abundance of positron decay compared to ^{18}F (23% versus 100%); the higher β^+-energies for ^{124}I compared to ^{18}F (0.8–2.1 MeV versus 0.6 MeV) which result in poorer spatial resolution (2.3 mm versus 0.7 mm); and the emission of high-energy γ-photons by ^{124}I that degrade the PET image and contribute along with the higher positron energy and longer half-life to higher radiation-absorbed doses (13). Nevertheless, intact IgG mAbs (43–46), genetically engineered antibody fragments (e.g., minibodies and diabodies) (47, 48) and peptides (49) have been labeled with ^{124}I, in most cases using chloramine-T or Iodogen. It is important to add a small amount of ascorbic acid to the formulation after labeling biomolecules with ^{124}I in order to protect them from radiolytic decomposition caused by the high-energy positrons (46). ^{124}I-labeled antibodies have been successfully used for PET imaging of receptor/antigen-positive tumors in mouse xenograft models (43, 45–48) and in one study in a child with neuroblastoma (44). One advantage of PET using ^{124}I-labeled antibodies, is that it allows accurate quantification of radioactivity uptake in tumors and normal tissues. This could be useful for predicting the radiation-absorbed doses for subsequent RIT using the corresponding ^{131}I-labeled antibodies (47, 50).

2.5.2 Bromine Radionuclides

While there has been interest in labeling mAbs and peptides with ^{124}I for PET, another attractive positron emitter is bromine-76 (^{76}Br). ^{76}Br decays with a half-life of 16.2 h, emitting positrons in 54% abundance with energy of 3.4 MeV. The almost sixfold higher positron energy of ^{76}Br compared to that of ^{18}F (0.6 MeV) provides poorer spatial resolution (>5 mm versus 0.7 mm, respectively) and a higher radiation-absorbed dose (13). Nevertheless, ^{76}Br is more practical for labeling biomolecules for PET than ^{124}I due to its greater abundance of positron emissions, and because it can be produced using low-energy biomedical cyclotrons by the ^{76}Se$(p,n)^{76}$Br reaction (51). Chloramine-T was initially studied for radiobromination of mAbs and peptides using [76]Br. The weaker oxidizing agent, Iodogen does not appear to be useful for [76]Br labeling because of the greater resistance to oxidation compared with radioiodine (52). The colon cancer mAbs A33, 3S193, and 38S1 as well as EGF have been labeled with [76]Br using chloramine-T in yields of 63–77% and with good preservation of antigen/receptor binding characteristics (53). Nevertheless, in one study, the immunoreactivity of [76]Br-mAb 38S1 was diminished to 45% and only radiobromination using bromoperoxidase was found to generate radiolabeled antibodies with preserved immunoreactivity (52). Human colon cancer xenografts implanted into nude rats have been visualized by PET using [76]Br-mAb 38S1 (54, 55).

Despite the ability to directly radiobrominate some antibodies using chloramine-T, many investigators obtained inconsistent and low yields and as mentioned, in some cases, decreased immunoreactivity using this approach. Therefore, newer strategies were explored for labeling mAbs with [76]Br using N-succinimidyl *para*-[[76]Br]bromobenzoate ([76]Br-SPBrB) (Fig. 2.4) (56). [76]Br-SPBrB was generated by bromodestannylation of N-succinimidyl-*para*-tri-*n*-butylstannylbenzoate (SPMB) using chloramine-T (56, 57). The resulting [76]Br-SPBrB was purified by HPLC and conjugated through the activated N-succinimidyl moiety to ε-amino groups of lysines on the antibodies. Using this approach, mAb 38S1 was labeled with [76]Br in 49% yield and with good preservation of immunoreactivity (69–76%) (56). Building on the strategy described by Zalutsky et al. (26) to radioiodinate antibodies using reagents

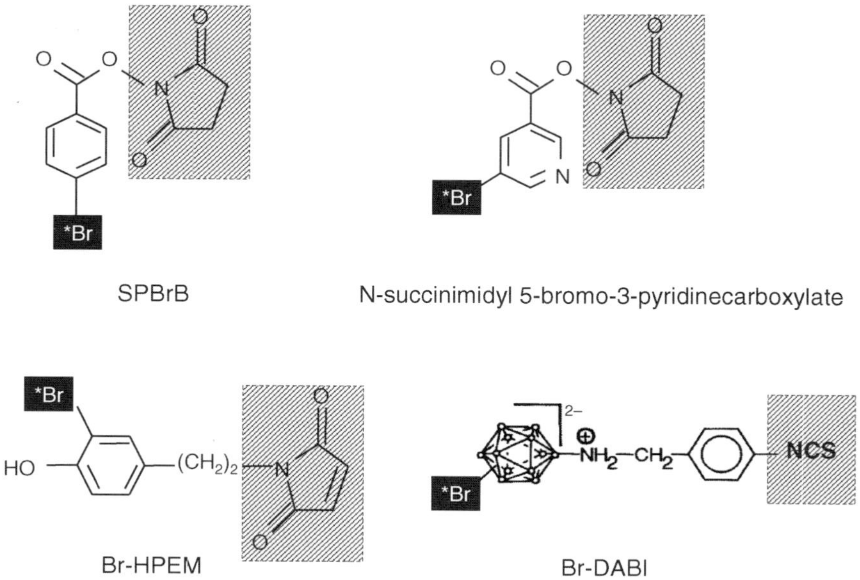

SPBrB   N-succinimidyl 5-bromo-3-pyridinecarboxylate

Br-HPEM   Br-DABI

FIGURE 2.4 Precursors used for residualizing radiobromination of monoclonal antibodies and peptides. Chemically reactive groups on the precursors are shaded, and the site for radiobromination of the precursor is indicated in black and with an asterisk. SPBrB: N-succinimidyl-*para*-bromobenzoate; Br-HPEM: bromo-((4-hydroxyphenyl)ethyl) maleimide; Br-DABI: bromo-(4-isothiocyanatobenzyl-ammonio)-bromo-decahydro-closo-dodecaborate.

(e.g., SGMIB) that yield positively charged catabolites that are retained in cells, Mume et al. (58) synthesized N-succinimidyl 5-[^{76}Br]bromo-3-pyridinecarboxylate (Fig. 2.4) by bromodestannylation of N-succinimidyl-5-(tributylstannyl)3-pyridine-carboxylate and conjugated it to the HER-2/neu mAb trastuzumab (Herceptin). The labeling yield was 45% but after purification, the immunoreactivity with HER2/neu-positive SKOV-3 ovarian carcinoma cells was >75%. However, this method of radiobromination did not improve cellular retention of radioactivity compared to trastuzumab labeled with ^{76}Br-SPBrB. A site-specific radiobromination technique was reported for labeling the affibody molecule $(Z_{HER2-4})_2$-Cys with ^{76}Br at a cysteine residue (59). $(Z_{HER2-4})_2$-Cys affibodies are small recombinant proteins ($M_r = 7$ kDa (monomer) or 15 kDa (dimer)) that bind with high specificity to HER2/neu receptors, similar to antibody fragments but produced by phage display techniques (see Chapter 1) (60). A bifunctional reagent, ((4-hydroxy-phenyl)ethyl)maleimide (HPEM) (Fig. 2.4) reactive with thiols was synthesized and radiobrominated with ^{76}Br using chloramine-T. The radiobrominated HPEM was conjugated through its maleimide group to the free thiol on the cysteine residue of the $(Z_{HER2-4})_2$-Cys affibodies. ^{76}Br-$(Z_{HER2-4})_2$-Cys exhibited preserved binding to HER2/neu-positive SKOV-3 cells *in vitro* and achieved high tumor uptake (5% i.d./g) and T/NT ratios (up to 31:1) at 4 h p.i. *in vivo* in mice implanted with SKOV-3 tumor xenografts.

Finally, a totally new approach to labeling antibodies with ^{76}Br employed derivatives of polyhedral boron clusters such as *closo*-dodecaborate (2^-) or *nido*-undecaborate (1^-) anions (Fig. 2.4) (61, 62). These structures form strong boron–bromine bonds and were labeled with ^{76}Br using chloramine-T. Once labeled with ^{76}Br, they were conjugated through their benzylisothiocyanato side chain to ε-amino groups of lysine residues on mAbs. The idea was that, since ^{76}Br is attached through a bromine–boron bond to a charged molecular structure that is completely foreign to the body, enzymatic debromination and exocytosis of radioactivity from cells would be substantially reduced. Bruskin et al. (61) labeled trastuzumab (Herceptin) with ^{76}Br in >80% yield using this approach resulting in a preparation that was immunoreactive with HER2/neu-positive SKBR-3 human breast cancer cells. Trastuzumab has also been labeled with ^{76}Br using the related *nido*-undecaborate (1^-) anion (62).

2.5.3 Fluorine Radionuclides

Fluorine-18 (^{18}F) is the most widely used radionuclide for PET, usually in the form of ^{18}F-FDG (42). Due to its relatively low positron energy (0.6 MeV), ^{18}F provides excellent spatial resolution (0.7 mm); it also is associated with lower radiation-absorbed doses compared to ^{124}I or ^{76}Br (13). There has been considerable interest in labeling mAb fragments and peptides with ^{18}F (63). The pharmacokinetics of intact IgG mAbs, in particular their slow elimination from the blood, do not lend themselves well to the use of ^{18}F, due to its short half-life of 110 min. In contrast, antibody fragments (e.g., F(ab')$_2$, Fab, and scFv) and peptides are accumulated rapidly in tumors and eliminated quickly from the blood and most normal tissues. Antibody fragments and peptides labeled with ^{18}F could therefore be useful for PET. The principal challenge in labeling these molecules with ^{18}F is that nucleophilic

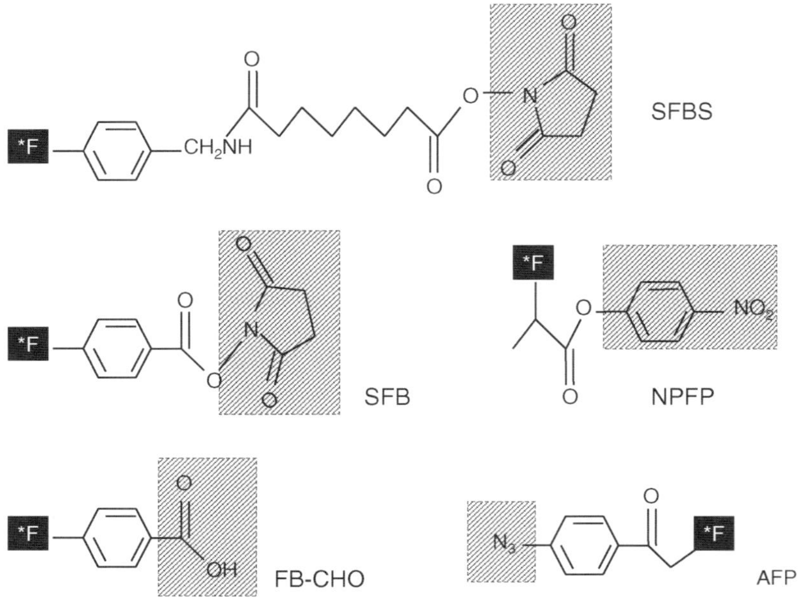

FIGURE 2.5 Prosthetic precursors used for radiofluorination of monoclonal antibody fragments and peptides. Chemically reactive groups on the precursors are shaded, and the site for radiofluorination of the precursor is indicated in black and with an asterisk. SFBS: 4-fluorobenzylamine succinimidyl ester; SFB: N-succinimidyl-4-(fluoromethyl)benzoate; NPFP: 4-nitrophenyl-2-fluoropropionate; FB-CHO: 4-fluorobenzaldehyde; AFP: 4-azido-fluorophenacyl.

substitution reactions are required; direct electrophilic radiofluorination requires carrier fluoride, which yields low specific activity that is unsuitable for imaging tumor-associated epitopes/receptors (63). To solve this problem, antibody fragments or peptides are labeled with ^{18}F by first substituting ^{18}F onto a prosthetic group and then chemically linking this group through a reactive side chain to the proteins (Fig. 2.5) (63). One of the primary considerations in designing methods for labeling proteins and peptides with ^{18}F is the amount of time required to complete the labeling and conjugation procedures due to the short 110 min half-life of ^{18}F. It is desirable that the labeling procedure including quality control testing be completed in 1–1.5 h, to minimize losses in yield, simply due to the physical decay of ^{18}F.

Garg et al. (64) described an approach for labeling F(ab′)$_2$ and Fab fragments of the antimyosin antibody R11D10 using 4-[^{18}F]fluorobenzylamine succinimidyl ester ([^{18}F]SFBS) (Fig. 2.5). Although this method was rapid with a total labeling time of 1.5 h, synthesis of [^{18}F]SFBS required three separate reactions followed by HPLC purification of the reagent; the yield of [^{18}F]SFBS was 25–40%. Conjugation of [^{18}F]SFBS to antibody fragments through reaction of the succinimidyl ester with ε-amino groups on lysines was performed in a fourth step that required purification of the ^{18}F-labeled fragments from excess [^{18}F]SFBS by size-exclusion chromatography (SEC). The immunoreactivity of the ^{18}F-labeled F(ab')$_2$ and Fab fragments of R11D10 were relatively preserved (89% and 75%, respectively). Using [^{18}F]SFBS as a prosthetic group, Zalutsky et al. (65) labeled F(ab')$_2$ fragments of the antiglioma mAb Mel-14, and obtained high T/NT ratios (up to 14 : 1 at 4 h p.i.) in a s.c. glioblastoma mouse tumor xenograft model. They also labeled Fab fragments of the TP-3 antibody with [^{18}F]SFBS, which permitted PET of osteosarcoma tumors in dogs (66). An analogous approach for labeling biomolecules with ^{18}F uses the prosthetic group, [^{18}F]-N-succinimidyl 4-(fluoromethyl)benzoate ([^{18}F]SFB) (Fig. 2.5) (67). The synthesis of [^{18}F]SFB was optimized (68) and the reagent was used for labeling F(ab')$_2$ fragments of the antiglioma mAb Mel-14. The immunore-activity of ^{18}F-labeled Mel-14 F(ab')$_2$ fragments using [^{18}F]SFBS or [^{18}F]SFB was indistinguishable (65% versus 64%, respectively) and the radiolabeled fragments demonstrated similar tissue distribution and pharmacokinetics in dogs (69). The 13-amino acid α-MSH peptide analogue, N-acetyl-Ser-Tyr-Ser-NorLeu-Glu-His-D-Phe-Arg-Trp-Gly-Lys-Pro-Val-NH$_2$, has been labeled with [^{18}F]SFB with good preserva-tion of receptor binding affinity (70).

Another reagent that has been used for labeling with ^{18}F is 4-nitrophenyl 2-[^{18}F] fluoropropionate ([^{18}F]NPFP) (Fig. 2.5) (71). [^{18}F]NPFP is a smaller prosthetic group that may have less impact on the receptor binding properties and/or physicochemical properties of peptides than [^{18}F]SFBS or [^{18}F]SFB. Accordingly, [^{18}F]NPFP was used for labeling the octreotide peptide analogue, SDZ 223-228 with ^{18}F resulting in full retention of biological activity (72). It was necessary to Boc-protect the ε-amino group on lysine-5 in SDZ 223-228 during labeling with [^{18}F]NPFP and then deprotect afterward, since this residue is critical for somatostatin receptor (SMSR) binding. Nevertheless, a radiofluorination technique using 4-[^{18}F]fluorobenzaldehyde ([^{18}F] FB-CHO) (Fig. 2.5) that allows site-specific labeling of aminoxyacetic acid functio-nalized octreotide analogues without the need to protect/deprotect the ε-amino group on ysine-5 has been reported (73). This method allowed PET imaging of s.c. transplantable AR42J rat pancreatic tumors in athymic mice at 1 h p.i. of the ^{18}F-labeled octreotide derivatives. In another study, D-Phe^1-octreotide labeled with ^{18}F using [^{18}F]NPFP exhibited higher liver accumulation and hepatobiliary clearance of radioactivity than the corresponding radioiodinated derivative in Lewis rats bearing pancreatic islet cell tumors, suggesting that [^{18}F]NPFP may increase the hydropho-bicity of peptides (74). Nevertheless, T/B ratios were 5.2 : 1 at 1 h p.i. and 4.2 : 1 at 2 h p.i., due to the rapid blood clearance of the ^{18}F-D-Phe^1-octreotide. [^{18}F]NPFP and [^{18}F]SFB have also been employed for ^{18}F labeling of peptides that contain the arginine-glycine-aspartic acid (RGD) sequence that binds to the αvβ3 integrin implicated in tumor angiogenesis (75, 76). In a study comparing labeling of proteins with [^{18}F]NPFP, [^{18}F]SFB, and another fluorinating reagent, 4-azidophenacyl-[^{18}F] AFP (Fig. 2.5), it was found that [^{18}F]NPFP-conjugated human serum albumin was partially unstable under slightly basic conditions (77). It was concluded that [^{18}F]SFB may be the most suitable radiofluorinating agent. Finally, [^{18}F]FB-CHO (Fig. 2.5) has been used to site-specifically label human serum albumin with ^{18}F at hydrazine nicotinamide (HYNIC) functional groups introduced into the protein using HYNIC N-hydroxysuccinimide ester (78). This method has similarities with the use of HYNIC for labeling biomolecules with ^{99m}Tc.

2.5.4 Astatine Radionuclides

Astatine-211 (^{211}At) (Table 2.3) is an α-particle emitter with a half-life of 7.2 h that is useful for labeling mAbs and peptides for targeted radiotherapy of malignancies (2, 79). ^{211}At is produced in a cyclotron using the ^{209}Bi$(\alpha,2n)^{211}$At reaction. However, the astatine-carbon bond is unstable (80) and direct electrophilic astatination of proteins at tyrosine residues using oxidizing agents such as chloramine-T or hydrogen peroxide results in low yield (81) and susceptibility *in vivo* to deastatination (82). Zalutsky et al. (83) addressed this problem by labeling *N*-succinimidyl 3-(tri-*n*-butylstannyl) benzoate (ATE) (Fig. 2.6) with ^{211}At. The *N*-succinimidyl 3-[^{211}At] astatobenzoate (SAB) was conjugated to ε-amino groups on lysine residues in antibodies, similar to the strategy used for radioiodination. Using SAB, intact mAb 81C6 IgG and F(ab')$_2$ fragments of mAb Mel-14 directed against glioblastoma were labeled with ^{211}At with preservation of immunoreactivity *in vitro* and excellent targeting *in vivo* to s.c. D-54 MG human glioblastoma xenografts in mice (84). Due to its short half-life (7.2 h) and the short range of the α-particles (50–100 μm), F(ab')$_2$ fragments are more suitable for labeling with ^{211}At than intact mAb IgGs since they penetrate deeper into tumors and are cleared more quickly from the blood (see Chapter 1) (7). Despite their more rapid blood clearance than F(ab')$_2$, Fab fragments are not appropriate for labeling with ^{211}At because they accumulate in high levels in the kidneys, thus posing a potential radiotoxicity hazard. Paired-label experiments in normal mice of the anti-CEA mAb C110 or its F(ab')$_2$ fragment labeled with ^{211}At or ^{131}I using the ATE method, revealed that there was greater tissue retention of ^{211}At than for ^{131}I (85). In another study, mice implanted s.c. with TK-82 human renal cell carcinoma xenografts showed similar tumor and normal tissue uptake following administration of ^{211}At or ^{125}I-labeled A6H F(ab')$_2$ (86). Tumor uptake in this mouse xenograft model was 30 % i.d./g for ^{211}At-A6H F(ab')$_2$ versus 19% i.d./g for ^{125}I-A6H F(ab')$_2$ at 19 h p.i.. T/B ratios were 3 : 1 for both ^{211}At and ^{125}I. ^{211}At-labeled antitenascin mAb 81C6 has been produced using the SAB reagent in sufficient quantities (2–10 mCi) under GMP conditions for use in a Phase I clinical trial in glioblastoma patients (87). Radiolytic decomposition of SAB due to the α-particles

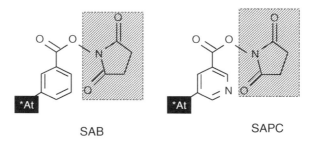

SAB   SAPC

FIGURE 2.6 Precursors used for residualizing radioastatination of monoclonal antibodies and peptides. Chemically reactive groups on the precursors are shaded, and the site for radioiodination of the precursor is indicated in black and with an asterisk. SAB: *N*-succinimidyl-3-astatobenzoate; SAPC: *N*-succinimidyl-5-astato-3-pyridinecarboxylate.

emitted by ^{211}At and subsequent low radiolabeling yields are nevertheless major challenges in producing clinical quality ^{211}At-labeled mAbs (88, 89). It appears that radiolytic decomposition is especially problematic when chloroform is used as a solvent for synthesizing SAB; benzene and methanol are alternatives that are less susceptible to the radiolytic effects of ^{211}At.

Zalutsky et al. (90) extended their previous residualizing radioiodination approach using *N*-succinimidyl 5-[^{131}I]iodo-3-pyridinecarboxylate (SIPC) to ^{211}At. Anti-EGFRvIII mAb L8A4 was labeled by conjugation with *N*-succinimidyl 5-[^{211}At] astato-3-pyridinecarboxylate (SAPC) (Fig. 2.6). Again, the premise is that proteolysis of the ^{211}At-labeled antibodies would yield a positively charged ^{211}At catabolite that would be unable to traverse the lysosomal membrane or be exocytosed. Immunoreactivity was maintained and ^{211}At- and ^{131}I-labeled L18A4 showed similar tumor and normal tissue accumulation in mice implanted s.c. with EGFRvIII-positive U87MG glioblastoma xenografts. Other biomolecules that have been labeled with ^{211}At include octreotide (91) and the anti-CD20 mAb rituximab (Rituxan; Roche Pharmaceuticals) directed against non-Hodgkin's B-cell lymphomas (92).

2.6 LABELING ANTIBODIES AND PEPTIDES WITH RADIOMETALS

2.6.1 Technetium Radionuclides

Technetium-99m (^{99m}Tc) (Table 2.1) is the most widely available, least expensive and most commonly used radionuclide in nuclear medicine and is thus an attractive candidate for labeling mAbs, their fragments as well as peptides for tumor imaging. ^{99m}Tc is a metastable form of ^{99}Tc, which decays by internal conversion to its ground state (^{99}Tc) with a half-life of 6 h, emitting a γ-photon of 140 keV that is easily imaged by gamma cameras available in all nuclear medicine facilities. ^{99m}Tc is produced from the decay of molybdenum-99 (^{99}Mo) using a commercially available ^{99}Mo/^{99m}Tc generator at very low cost. Its relatively short half-life minimizes the radiation exposure to patients undergoing imaging procedures, but necessitates rapid labeling procedures for antibodies or peptides. There are several approaches to labeling these molecules with ^{99m}Tc (93, 94). These include (i) direct methods that rely on the binding of ^{99m}Tc to endogenous thiols generated by reduction of disulfide linkages in mAbs or introduced chemically by reaction with thiolating agents such as 2-iminothiolane and (ii) indirect methods that involve binding of ^{99m}Tc to a chelating agent that is then conjugated to the mAb or peptide ("preformed chelator" approach) or to a chelator already incorporated into the biomolecule. There are advantages and disadvantages of each of these strategies with respect to ease of use, *in vivo* stability and amenability to kit formulation.

Intact mAb IgG can be labeled simply and directly with ^{99m}Tc by reduction of a small proportion ($<5\%$) of the inter- or intrachain disulfide bonds to free thiols by treatment with a 2000-fold molar excess of 2-mercaptoethanol (2-ME) for 30 min (Fig. 2.7) (94). The reduced IgGs are purified from excess 2-ME by SEC and labeled to high efficiency ($>90\%$) by transchelation of ^{99m}Tc from ^{99m}Tc-glucoheptonate

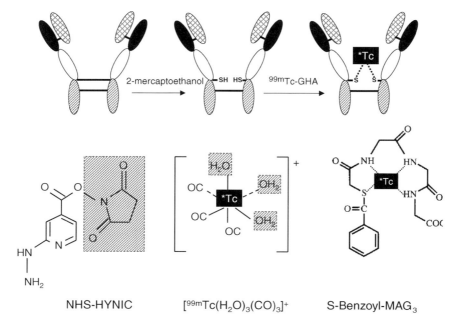

NHS-HYNIC   [⁹⁹ᵐTc(H₂O)₃(CO)₃]⁺   S-Benzoyl-MAG₃

FIGURE 2.7   Top panel: Method for direct labeling of intact monoclonal antibodies with
^{99m}Tc. Some disulfide linkages between the antibody chains are chemically reduced with
2-mercaptoethanol. The free thiols produced are sites for binding ^{99m}Tc-glucoheptonate
(^{99m}Tc-GHA). Bottom panel: Metal chelators that can be conjugated to monoclonal antibodies
or peptides for labeling with ^{99m}Tc. Chemically reactive groups on the precursors are shaded,
and complexation of ^{99m}Tc is indicated in black and with an asterisk. NHS-HYNIC: N-
hydroxysuccinimide ester of hydrazinenicotinamide; $[^{99m}$Tc(H₂O)₃(CO)₃]$^+$: ^{99m}Tc(I)-carbon-
yl complex; S-benzoyl-MAG3: S-benzoyl-mercaptoacetyl-glycyl-glycyl-glycine.

or ^{99m}Tc-methylene diphosphonate (^{99m}Tc-MDP). ^{99m}Tc-glucoheptonate or ^{99m}Tc-
MDP are prepared from commercial kits and ^{99m}Tc pertechnetate (^{99m}TcO$_4^-$) eluted
from a ^{99}Mo/^{99m}Tc generator. This method of labeling mAbs with ^{99m}Tc was
introduced by Schwartz and Steinstrasser in 1987 (95) and later optimized by Mather
and Ellison in 1990 (96). A modification of the Schwartz method employed exposure
of the anti-CD20 mAb rituximab (MabThera; Roche) to UV-light of wavelength
320 nm for 20 min to photoreduce some of the disulfide linkages to thiols that were
then labeled wth ^{99m}Tc by transchelation from ^{99m}Tc-MDP (97). This photoreduction
method was originally described by Stalteri et al. in 1996 (98); it does not require
removal of any chemical reducing agents and is somewhat simpler and potentially
more controllable than chemical reduction with 2-ME. The advantage of the Schwartz
technique is that purified reduced IgG can be dispensed into vials and stably
maintained either frozen or lyophilized until required for labeling with ^{99m}Tc, thus
generating a kit formulation (99). The major disadvantage of the method is that if too
great an amount of reducing agent is used, too many disulfides are reduced increasing
the risk for disrupting key linkages needed to maintain protein folding, integrity, and

immunoreactivity. This is especially a problem for mAb fragments such as $F(ab')_2$, Fab, and scFv. Likewise, the method cannot be used for labeling small peptides with ^{99m}Tc that harbor disulfide bonds required for maintaining biological activity (e.g., somatostatin analogues). Careful control and optimization of reduction conditions through measurement of the number of free thiols generated using Ellman's reagent (100, 101) are needed to successfully apply this method for labeling proteins or peptides with ^{99m}Tc. Another disadvantage is that IgG labeled with ^{99m}Tc through direct binding to thiols is subject to loss of the radiolabel *in vivo* over time by exchange with endogenous thiol containing molecules such as cysteine and glutathione (94). Nevertheless, the Schwartz method has been used for labeling several mAbs with ^{99m}Tc providing preserved immunoreactivity and ability to target and image human cancers in mouse xenograft models (102, 103) and in patients (104, 105).

Due to the limitations of the Schwartz method for labeling mAb fragments (e.g., $F(ab')_2$ or Fab) and peptides with ^{99m}Tc, alternative strategies have been investigated. One method utilizes HYNIC and coligands such as ethylenediaminodiacetic acid (EDDA), ethlendiaminetetraacetic acid (EDTA), tricine, or glucoheptonate to form a stable ^{99m}Tc complex with the biomolecule (Fig. 2.7) (106). This method was described by Abrams et al. in 1990 (107). HYNIC is conjugated to the mAb (or its fragments) or peptides using an excess of HYNIC *N*-hydroxysuccinimide ester (or other chemically reactive form), which forms an amide linkage with the *N*-terminus or ε-amino groups in lysines. After chromatographic purification, the HYNIC-derivatized biomolecule is labeled with ^{99m}Tc by incubation with $^{99m}TcO_4^-$ in the presence of $SnCl_2$ and coligand. Coligands are needed because HYNIC occupies only one or two of the six coordination sites of ^{99m}Tc (106). $SnCl_2$ is included to reduce ^{99m}Tc from its 7^+ valence state in $^{99m}TcO_4^-$ to a valence of 4^+ or 5^+ for complexation with HYNIC-coligands. Our group has labeled Fab fragments of the HER2/neu mAb trastuzumab (Herceptin) with ^{99m}Tc using HYNIC (108). ^{99m}Tc-HYNIC-trastuzumab Fab showed preserved binding affinity for HER2/neu receptors on SKBR-3 human breast cancer cells *in vitro* ($K_d = 1.6 \times 10^{-8} M^{-1}$), and demonstrated good tumor targeting *in vivo* in mice implanted s.c. with HER2/neu-positive BT-474 breast cancer xenografts (T/B ratio $= 3:1$ at 24 h p.i.). BT-474 tumors were visualized by imaging. Others have labeled mAb Fab fragments with ^{99m}Tc using HYNIC achieving preserved immunoreactivity and good tumor targeting (109, 110). As mentioned, the Schwartz method is not amenable to labeling peptides that harbor key intramolecular disulfide linkages. However, the somatostatin analogue, D-Phe^1-Tyr^3-octreotide (TOC) that contains a disulfide bond essential for SMSR binding, has been successfully labeled with ^{99m}Tc by conjugation to HYNIC using a variety of coligands (111–113). Similarly, the 53-amino acid peptide, EGF which harbors three disufide bonds required for maintenance of its tertiary structure and receptor binding properties has been labeled with ^{99m}Tc using HYNIC and tricine as coligand (114). In each case, receptor binding was preserved *in vitro* allowing tumor imaging in mouse xenograft models (113, 114) or in patients (111, 112).

An interesting finding is that the choice of coligand for HYNIC drastically affects the plasma protein binding, elimination from the blood and tissue distribution of the ^{99m}Tc-labeled biomolecules. This effect was first reported by Babich and Fischman in

1995 (115) who noted differences in radioactivity accumulation in the lungs, kidneys, liver, spleen, and gastrointestinal tract of rats for the chemotactic peptide (N-For-Met-Leu-Phe-Lys) conjugated to HYNIC and labeled with ^{99m}Tc using glucarate, glucoheptonate, mannitol, or glucamine as coligands. Use of EDDA as coligand instead of tricine for ^{99m}Tc labeling of HYNIC-TOC or the somatostatin analogue RC160 produced a more hydrophilic complex that cleared more quickly from the blood in rats and favored renal over hepatobiliary clearance (116). However, substitution of the backbone of EDDA with dimethyl, diethyl, or dibenzyl moieties increased the hydrophobicity of ^{99m}Tc-HYNIC-RC160, resulting in a five- to sixfold increased plasma protein binding *in vitro* and substantially increased liver uptake and hepatobiliary elimination in rats (117). Tricine as coligand for ^{99m}Tc-HYNIC-RC160 produced liver and intestinal radioactivity uptake intermediate between that of the EDDA and the substituted EDDA derivatives. HYNIC-TOC labeled with ^{99m}Tc using EDDA as coligand showed excellent tumor localization in a study of 10 patients with SMSR-positive tumors allowing imaging of lesions at 4 h p.i. (112). Similarly, in a study of 13 patients, 12 of whom had SMSR-positive malignancies, administered ^{99m}Tc-HYNIC-TOC labeled using tricine as coligand, tumors were imaged as early as 10 min p.i. (111). A comparison between these two studies revealed that the circulating blood background radioactivity was higher and urinary excretion of radioactivity was lower in patients receiving ^{99m}Tc-tricine-HYNIC-TOC than those administered ^{99m}Tc-EDDA-HYNIC-TOC, likely due to the increased plasma protein binding of ^{99m}Tc-tricine-HYNIC-TOC (118). Almost in all cases, the ^{99m}Tc-HYNIC-labeled analogues detected the same lesions as ^{111}In-DTPA-D-Phe^1-octreotide, the "gold standard" for SMSR tumor imaging, but the images were clearer and lesions were detected earlier using the ^{99m}Tc analogues. ^{99m}Tc-tricine-HYNIC-TOC missed one liver lesion in a patient, which was detected with ^{111}In-DTPA-D-Phe^1-octreotide, probably due to its higher liver uptake (111). The effect of coligand on plasma protein binding and blood clearance does not appear to be restricted to small peptides such as ^{99m}Tc-HYNIC-TOC. Ono et al. (119) found that 3-benzoylpyridine (BP) ternary complexes of ^{99m}Tc-HYNIC-conjugated Fab fragments [^{99m}Tc-HYNIC-Fab(tricine) (BP)] exhibited lower plasma protein binding *in vitro* than the corresponding binary complexes [^{99m}Tc-HYNIC-Fab(tricine)$_2$] and were cleared more quickly from the blood in mice (0.35 versus 0.98% i.d./g at 24 h p.i., respectively). ^{99m}Tc-HYNIC-Fab (tricine)(BP) also exhibited lower liver uptake than ^{99m}Tc-HYNIC-Fab(tricine)$_2$. They concluded that the ^{99m}Tc-tricine coligand exchanges *in vivo* with plasma proteins as well as lysosomal proteins in tissues causing slow blood clearance and high liver retention of radioactivity. This loss of radioactivity was decreased using the more stable ternary ^{99m}HYNIC-Fab(tricine)(BP) complexes. Our group similarly found a slower blood clearance in mice for trastuzumab Fab labeled with ^{99m}Tc through HYNIC using glucoheptonate as coligand than for ^{111}In-labeled trastuzumab Fab (3.2 versus 1.4% i.d./g at 24 h p.i., respectively) (108, 120). There was also higher liver uptake for ^{99m}Tc-HYNIC-trastuzumab Fab than for ^{111}In-trastuzumab Fab at 24 h p.i. (3.8 versus 2.4% i.d./g, respectively). Since ^{99m}Tc labeling efficiencies for antibodies and peptides are high (>90%) using HYNIC, kit formulation is possible and indeed kits have been created for labeling HYNIC-TOC with ^{99m}Tc (111).

A particularly useful method for labeling biomolecules with ^{99m}Tc, especially those produced by protein engineering techniques and that contain polyhistidine affinity tags, employs a ^{99m}Tc(I)-carbonyl complex (Fig. 2.7) (121, 122). This method, described by Waibel et al. in 1999 (122) involves a simple one-vial synthesis of the organometallic aqua-ion complex $[^{99m}$Tc(H$_2$O)$_3$(CO)$_3]^+$ that then efficiently complexes through release of its water molecules with imidazole nitrogens in histidine-containing biomolecules (Fig. 2.7). The $[^{99m}$Tc(H$_2$O)$_3$(CO)$_3]^+$ complex is formed by heating ^{99m}TcO$_4^-$, Na$_2$CO$_3$, and NaBH$_4$ flushed with carbon monoxide (CO) at 75°C for 30 min. The yield of $[^{99m}$Tc(H$_2$O)$_3$(CO)$_3]^+$ is >95%. A commercial kit for producing the $[^{99m}$Tc(H$_2$O)$_3$(CO)$_3]^+$ complex (Iso-Link; Mallinckrodt) is available. Labeling with ^{99m}Tc is achieved simply by mixing the $[^{99m}$Tc(H$_2$O)$_3$(CO)$_3]^+$ complex with the histidine-containing protein or peptide and heating for 20–30 min at 37°C. A scFv containing a polyhistidine tag was labeled with the ^{99m}Tc(I)-carbonyl complex resulting in a stable complex *in vitro* in human serum and retaining 87% of its radioactivity over 24 h at 37°C (122). It was also stable to challenge *in vitro* with a 5000-fold molar excess of histidine, retaining 94% of its radioactivity (122). This technique is attractive because recombinant antibodies and peptides often incorporate polyhistidine affinity tags for their purification, thus allowing them to be directly and easily labeled using the ^{99m}Tc(I)-carbonyl complex. Moreover, Re(I)-carbonyl complexes have also been prepared allowing this method, allowing extension to radionuclides of rhenium (e.g., ^{186}Re or ^{188}Re) (Table 2.3) for targeted radiotherapy of malignancies (123). More recently, it has been shown that these ^{99m}Tc(I)-carbonyl complexes can also be used to label peptides incorporating a novel single amino acid chelator (SAAC) (124). SAAC are synthetic nonnatural bifunctional amino acid derivatives that incorporate a tridentate chelate for binding $[^{99m}$Tc(H$_2$O)$_3$(CO)$_3]^+$, linked through a selectable sequence to an amino group for incorporation into the peptide during solid phase synthesis. The SAAC approach has proven extremely versatile for labeling many different types of biomolecules with good preservation of their receptor binding affinity, and allows site-specific introduction of the chelator for ^{99m}Tc (124).

Finally, tetradentate chelators such as N$_3$S (triamidothiols), N$_2$S$_2$ (diamido-dithiols), or N$_2$S$_4$ (diaminotetrathiols) have been used for labeling mAbs and peptides with ^{99m}Tc (Fig. 2.7) (106). The advantage of these chelators is that they form stable, well-defined ^{99m}Tc complexes, but they require introduction into antibodies or peptides through reaction of a chemically reactive ester in a side chain with ε-amine groups on lysines. This conjugation step is nonspecific and has the potential to target a key lysine required for maintenance of antigen or receptor binding. This problem can be solved in the case of small peptides by incorporating the tetradentate chelator at a specific position into the biomolecule during its solid-phase synthesis using sequences of amino acids such as cysteine-glycine-glycine-glycine (N$_3$S) (125). Moreover, there is the possibility that ^{99m}Tc will bind to low affinity endogenous metal binding sites in antibodies in addition to the high affinity sites introduced with the tetradentate chelator. Therefore, to avoid this possibility, a "preformed chelator" approach was reported by Fritzberg et al. (126). In this preformed chelator approach, a stable ^{99m}Tc-N$_2$S$_2$ complex is synthesized first and then conjugated to the mAbs through a reactive tetrafluorophenyl ester side chain. However, this strategy is complex and time consuming; the conjugation step is inefficient; and it does not lend itself easily to kit formulation. Despite these limitations, tetradentate chelators have been employed for labeling various mAbs and peptides with ^{99m}Tc (94, 106). An interesting application of a tetradentate chelator was described for ^{94m}Tc labeling of the somatostatin analogue, demotate (127). ^{94m}Tc is a positron emitter (β^+ 72%) ($E\beta^+_{max} = 2.5$ MeV) (Table 2.2) with a half-life of 52 min. Demotate incorporating the 1,4,8,11-tetraazaundecane (N$_4$) tetradentate chelator was labeled with ^{94m}Tc in the presence of SnCl$_2$. ^{94m}Tc-demotate exhibited preserved receptor binding to A-427 nonsmall cell lung cancer cells infected with the AdHASSTR2 adenovirus encoding the SMSR subtype 2. A-427 xenografts infected with AdHASSTR2 implanted s.c. into athymic mice were visualized by PET at 1 h p.i. of ^{94m}Tc-demotate.

2.6.2   Rhenium Radionuclides

Methods for labeling antibodies and peptides with radionuclides of rhenium have been developed from techniques used for labeling with ^{99m}Tc, due to the similarity in the chemistry of technetium and rhenium (128). There are two radionuclides of rhenium (^{186}Re and ^{188}Re) that are useful for targeted radiotherapy of cancer (Table 2.3). ^{186}Re decays with a half-life of 3.7 days emitting a β-particle with maximum energy (E_{max}) of 1.07 MeV as well as a low abundance (9%) γ-photon of 137 keV that is useful for imaging. ^{188}Re decays with a 17 h half-life emitting a β-particle with E_{max} of 2.12 MeV and a low abundance (15%) γ-photon of 155 keV that can be imaged. The twofold higher β-particle energy of ^{188}Re compared to ^{186}Re provides a threefold longer range in tissues (26 versus 9 mm, respectively), making it more useful for treating larger tumors (i.e., >1 cm in diameter) or tumors in which there is incomplete targeting of radiolabeled mAbs or peptides to cancer cells. On the other hand, the relatively short half-life of ^{188}Re (17 h) may limit its feasibility for labeling intact IgG mAbs for radiotherapeutic applications due to their slow kinetics of tumor uptake and clearance from the blood and normal tissues. Antibody fragments and peptides may be more suitable for labeling with ^{188}Re. A major advantage of ^{188}Re is that it can be produced carrier-free and in high purity using a ^{188}W/^{188}Re generator system (129), which would make the radionuclide available at low cost in any nuclear medicine facility.

Both direct and indirect methods have been used for labeling antibodies with ^{186}Re/^{188}Re (128). Direct methods rely on the binding of reduced ^{186}Re/^{188}Re to free thiols on the antibodies generated by reduction with 2-ME as described for ^{99m}Tc. However, the chemistries of technetium and rhenium are not identical. In particular, rhenium is more difficult to reduce from its 7^+ valence state to its lower valence states of 4^+ or 5^+ for labeling biomolecules, and it is more easily reoxidized (130, 131). Thus, a greater amount of SnCl$_2$ reducing agent is required. The optimal pH for labeling with ^{186}Re/^{188}Re is also lower than that for ^{99m}Tc-labeling (pH 4.5–5.0 versus pH 7.0–7.5, respectively) and pH values >5.0 can result in reoxidation of rhenium (130). Moreover, radiolysis of antibodies and peptides caused by the high-energy β-particles emitted by ^{186}Re/^{188}Re can be a problem, and thus radioprotectants such as ascorbic acid, gentisic acid, or human serum albumin are often incorporated into the formulations. In one study, ascorbic acid was employed both as a radioprotectant and a reducing agent for the disulfide bonds for labeling IgG with [186]Re (131). Much longer incubation times (17–24 h) have been used for direct labeling of mAbs with [186]Re/[188]Re (131, 132) than for ^{99m}Tc labeling (30 min) (133). However, in one study (134), a kit formulation was developed for labeling the anti-CD20 mAb ritiximab (Rituxan; Roche Pharmaceuticals) with [188]Re. A labeling efficiency >97% was obtained in only 1–1.5 h following addition of [188]Re perrhenate eluted from a [188]W/[188]Re generator to the kit.

Tetradentate chelators (e.g., N_3S) (Fig. 2.7) have also been used for labeling mAbs and their F(ab')$_2$ fragments (135, 136) as well as the somatostatin peptide RC-160 (137) with [186]Re/[188]Re. In the case of antibodies, the "preformed chelate" approach was used, whereas for RC-160 the N_3S chelator was introduced during solid phase peptide synthesis. One method that has been used for direct labeling of antibodies using [188]Re employs the hydroxamic acid, trisuccin (Fig. 2.7). Trisuccin differs from most tetradentate ligands (e.g., N_3S structures) in that it does not contain a free thiol. Therefore, it can be conjugated directly to a mAb and used for labeling with [186]Re/[188]Re without the risk of the free thiol reacting with the disulfides on the antibodies. Safavy et al. (138) labeled two different humanized forms of the tumor-associated glycoprotein-72 (TAG-72) mAb CC49 with [188]Re using trisuccin conjugated through a 6-oxoheptanoic acid linker. The labeling efficiency ranged from 80% to 98% and the immunoreactivity ranged from 69% to 77%. High tumor uptake of radioactivity (18–23% i.d./g) was found in mice implanted s.c. with TAG-72-positive LS174T human colon cancer xenografts. Finally, bombesin, a 14-amino acid analogue of gastrin-releasing peptide (GRP) that binds to GRP receptors on prostate, breast, lung and pancreatic cancers, has been labeled using a [188]Re(H_2O)(CO)$_3$ carbonyl complex (Fig. 2.7) similar to that described for labeling biomolecules with [99m]Tc (139). Targeting of PC-3 prostate cancer xenografts in mice was achieved with [188]Re-labeled bombesin.

2.6.3 Indium Radionuclides

Indium-111 (^{111}In) (Table 2.1) is a γ-emitting radionuclide ($E\gamma = 172$ and 245 keV) with a half-life of 67 h that is routinely used for labeling mAbs and peptides for SPECT imaging of tumors. The positron emitter, [110m]In has a half-life of 69 min and is useful for labeling peptides for PET. The β^+-energy of [110m]In (2.26 MeV) provides a spatial resolution of 3.0 mm compared to 0.7 mm for [18]F (13). Octreotide was labeled with [110m]In and used for PET of a patient with an SMSR-positive intestinal carcinoma metastasis (140). [114m]In and its daughter product, [114]In (Table 2.3) are long-lived radionuclide impurities in ^{111}In, but have potential for targeted radiotherapy of malignancies. [114m]In decays to [114]In with a half-life of 49.5 days emitting an imageable γ-photon of 190 keV, as well as Auger and conversion electrons that can kill cancer cells. [114]In decays to [114]Cd (0.5%) or [114]Sn (99.5%) with a half-life of 72 s emitting β^--particles ($E_{max} = 1.98$ MeV) that are also useful for treatment of tumors. Octreotide has been labeled with [114m]In (141). Antibodies and peptides are labeled

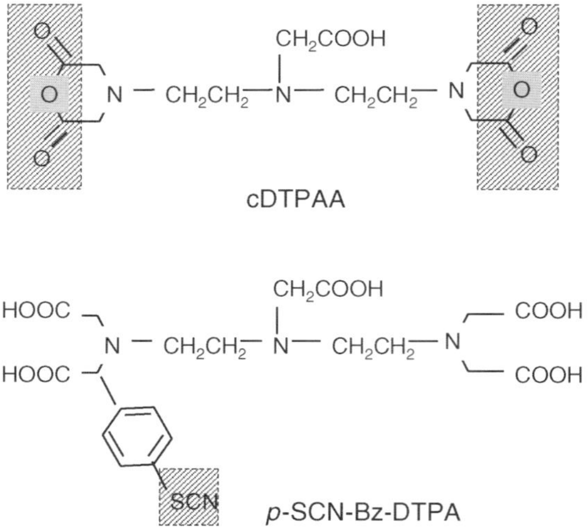

cDTPAA

p-SCN-Bz-DTPA

FIGURE 2.8   Metal chelators that can be conjugated to monoclonal antibodies or peptides for labeling with ^{111}In. cDTPAA: Bicyclic anhydride of diethylenetriaminepentaacetic acid; p-SCN-Bz-DTPA: p-isothiocyanatobenzyl-diethylenetriaminepentaacetic acid. Chemically reactive groups for conjugation to the antibodies or peptides are shaded.

with ^{111}In by introducing the chelator, DTPA through reaction of ε-amino groups on lysine residues or the N-terminal amine of the biomolecules with reactive forms of DTPA (Fig. 2.8) such as DTPA dianhydride (cDTPAA) (142), DTPA mixed anhydride (143) or DTPA p-benzylisothiocyanate (SCN-Bz-DTPA) (144, 145). Conjugation with cDTPAA involves suspending cDTPAA in anhydrous chloroform, dispensing an aliquot of the suspension containing a known amount of cDTPAA into a clean, dry glass tube and evaporating the chloroform to dryness using a gentle stream of nitrogen to leave a film of cDTPAA on the inside surface of the tube. The antibody or peptide dissolved in 50 mM sodium bicarbonate buffer, pH 7.5 is then added to the tube. Reaction with cDTPAA occurs rapidly within 1–2 min, although 15–30 min are often allowed. DTPA conjugation efficiency is dependent on protein concentration, pH and the molar ratio of cDTPAA: antibody/peptide. The DTPA-derivatized biomolecule is separated from excess DTPA by SEC and/or by ultrafiltration. Pure DTPA-conjugated antibodies or peptides can be dispensed into unit-dose vials to produce kits that can be labeled to high radiochemical purity (RCP >90%) simply by adding ^{111}In to the vial and incubating at room temperature for 15–30 min (146). Radiolabeling is achieved by transchelation of ^{111}In from acetate or citrate complexes to DTPA. The acetate or citrate counterions are used to maintain the solubility of ^{111}In at pH 5–7.5 required for labeling. ^{111}In-acetate or -citrate complexes are formed by mixing ^{111}InCl$_3$ with 0.5–1 M sodium acetate or citrate buffer, pH 5.0. One critical parameter that must be considered for labeling biomolecules with ^{111}In is the presence of trace amounts of divalent or trivalent metal ions (e.g., Fe, Al, Cd, and Zn) in the labeling reaction. These trace metals may exist at even higher levels than ^{111}In itself in commercial ^{111}InCl$_3$ solutions (147) and can interfere with labeling by occupying the small number of DTPA metal binding sites conjugated to the antibodies or peptides. Trace metal contamination is minimized by acid-washing of glassware, by using trace-metal free plasticware, through storing DTPA-conjugated proteins frozen (to avoid leaching of metals into the solution from the container) and by purification of conjugation buffers on a cation-exchange column (e.g., Chelex-100) (148). A method has been reported for ultrapurification of ^{111}In from trace metals in ^{111}InCl$_3$ solution by selective extraction of ^{111}In as an iodide complex into anhydrous diethyl ether (147).

Hnatowich et al. (142) found that the reaction between cDTPAA and IgG is dependent on pH (optimum between pH 7.5 and 8.5) and protein concentration (optimum >15 mg/mL). There was an inverse relationship between the molar ratio of cDTPAA:IgG and the conjugation efficiency with the greatest efficiency (>70%) found at a 1:1 ratio using a protein concentration of 15 mg/mL. However, it may not always be possible to achieve these high protein concentrations and thus, greater molar ratios of cDTPAA: antibody/peptide (e.g., 5 : 1 to 20 : 1) are often used to compensate for the lower conjugation efficiency at lower concentrations (e.g., 2–5 mg/mL). One of the limitations of the cDTPAA method is that the reagent contains two anhydride moieties that can react with ε-amino groups on lysines or the N-terminal amine of biomolecules. Reaction of cDTPAA with lysines on two biomolecules causes inter-molecular cross-linking leading to the fomation of dimers and higher molecular weight polymers. Reaction of cDTPAA with two lysines on the same molecule generates intramolecularly cross-linked species. This latter possibility is especially troublesome because in contrast to intermolecularly linked species, intramolecularly cross-linked biomolecules are not easily detected or measured by chromatographic techniques but may significantly diminish immunoreactivity or receptor binding properties due to protein misfolding (149). The proportion of intermolecularly linked molecules is directly proportional to the molar ratio of cDTPAA:IgG and protein concentration in the conjugation reaction, and the resulting substitution level (moles DTPA/mole biomolecule) of the bioconjugate. Hnatowich et al. (142) reported that the proportion of polymers increased from 0.3% when IgG was conjugated with cDTPAA at a 1 : 1 molar ratio (cDTPAA : IgG) to as much as 40% at a molar ratio of 10 : 1 when the IgG concentration was 15 mg/mL. Our group found less than 11% IgG dimers when the HER-2/neu mAb trastuzumab (Herceptin) was modified with a fourfold molar excess of cDTPAA using a protein concentration of 5 mg/mL. In this instance, there were two DTPA molecules/trastuzumab IgG (150). Another limitation of the cDTPAA method is that one of the five carboxylic acid groups of DTPA is used to form an amide linkage with the antibody or peptide. This converts DTPA to diethylentriaminetetraacetic acid (DTTA) that forms a much less stable heptadentate complex with ^{111}In. This instability is manifested by a moderate rate of transchelation *in vivo* (7–10% per day) of ^{111}In from DTPA-conjugated antibodies to transferrin, which causes deposition of radioactivity in the liver and bone marrow (151, 152). In contrast, the transchelation rate from the octadentate ^{111}In-DTPA complex to transferrin is 1–2% per day (152, 153). Despite these limitations, cDTPAA has been used for ^{111}In labeling of intact IgG mAbs (18, 120, 148, 152), single-chain Fv fragments (154) and peptides (18, 155, 156). In the case of peptides that harbor critical lysine residues necessary for receptor binding, these amino acids must be Boc-protected during conjugation of the N-terminal amine with DTPA, then deprotected afterward (156).

To address the limitations of cDTPAA, other forms of DTPA have been synthesized that position a reactive group on the methylene carbon of one of the carboxymethyl arms instead of using one of the carboxylic acid groups to link to an antibody or peptide (144, 145). One of these chelators is the p-benzylisothiocyanate derivative of DTPA (p-SCN-Bz-DTPA) (Fig. 2.8). The p-SCN-Bz-DTPA chelator reacts with ε-amino groups or the N-terminus of biomolecules to form a thiourea linkage. Since there is only one reactive moiety on the p-SCN-Bz-DTPA chelator, there is no possibility of inter- or intramolecular cross-linking, which helps to preserve immunoreactivity or receptor binding affinity (144, 145). Our group found that the immunoreactive fraction of mAb 2G3 reactive with a 330 kDa glycoprotein found on breast and ovarian cancer was 0.52 when conjugated with 0.5–1.5 mol of DTPA using cDTPAA, but was 0.77 when conjugated with p-SCN-Bz-DTPA (152). Importantly, the retention of all five carboxylic acid groups in the DTPA molecule for binding ^{111}In preserves the stable octadentate ^{111}In-DTPA complex, which diminishes the loss of ^{111}In to transferrin in plasma. The B-cell lymphoma antibody Lym-1 conjugated with p-SCN-Bz-EDTA (a chelator similar to p-SCN-Bz-DTPA) and labeled with ^{111}In showed a lower rate of loss of ^{111}In in serum than Lym-1 conjugated to DTPA using cDTPAA (<1% versus 14% over 5 days, respectively) (157). Similar results were obtained *in vitro* in serum and *in vivo* in plasma in mice for other ^{111}In-SCN-Bz-EDTA or SCN-Bz-DTPA immunoconjugates (158). However, our group did not observe increased stability of ^{111}In-labeled mAb 2G3 conjugated to SCN-Bz-DTPA *in vitro* in serum compared to ^{111}In-DTPA-mAb 2G3 (both had transchelation rates of 7% per day) but we did find a lower rate of loss of ^{111}In from ^{111}In-SCN-Bz-DTPA-2G3 to ascites fluid from ovarian cancer patients (5% versus 11% per day) (152). Avoidance of cross-linking and the increased serum stability of p-SCN-Bz-DTPA modified biomolecules decreases liver uptake of radioactivity, a common problem with ^{111}In-labeled mAbs and peptides (159).

Methods have been developed to site-specifically conjugate DTPA to the Fc domain of antibodies for labeling with ^{111}In in order to better preserve their immunoreactivity (160). Site-specific conjugation of DTPA was achieved by reaction of the N-terminal amine of the tripeptide, glycine-tyrosine-lysine-DTPA (GTK-DTPA) with aldehydes generated in the Fc-domain by sodium periodate oxidation (161). Stabilization of the resulting Schiff base linkage between GTK-DTPA and the antibodies was achieved by sodium borohydride reduction. This approach positions the DTPA chelator remote from the Fab antigen binding region. Nevertheless, the sodium periodate oxidation step itself can diminish antibody immunoreactivity if not controlled (162). The number of aldehydes generated by sodium periodate can be measured by a spectrophotometric assay which relies on the reaction of the aldehyde groups with dinitrophenylhydrazine (DNPH) generating a derivative that absorbs at 360 nm (163). In an analogous approach, our group reported a novel strategy for site-specific labeling of recombinant biomolecules with ^{111}In that takes advantage of the high affinity of the radiometal for transferrin ($K_a = 10^{28}$ L/mole). We fused the gene for the *n*-lobe of human transferrin (hn-Tf) through a DNA sequence that encoded a flexible polypeptide linker [(GGGGS)$_3$] to the gene for vascular endothelial growth factor (VEGF$_{165}$)

and expressed the recombinant fusion protein in *Pichia pastoris* (164). The hnTf-VEGF protein bound ^{111}In directly through its hnTf moiety and retained its binding affinity for VEGF receptors on human umbilical vascular endothelial cells (HUVECs). The protein did not bind to transferrin receptors on cells, since such binding requires both the *n*- and the *c*-lobes of transferrin. ^{111}In-hnTf-VEGF localized specifically in angiogenic U87MG glioblastoma xenografts implanted s.c. in athymic mice permitting imaging at 72 h p.i.. However, ^{111}In-hnTf-VEGF exhibited relatively high liver uptake and there was a moderately rapid loss of [111]In *in vitro* from the protein to transferrin in plasma (21% per day). This transchelation was not likely due to a lower affinity of the hnTf moiety for ^{111}In, but rather caused by competition with the relatively much higher concentrations of transferrin present in plasma.

Limiting the metal chelator substitution of a mAb or peptide to 1-2 DTPA or EDTA groups per molecule in order to preserve its immunoreactivity or receptor binding affinity restricts the specific activity that can be achieved for labeling with [111]In. For example, monosubstitution of EGF with DTPA restricts the maximum specific activity that can be practically achieved with ^{111}In to 40 MBq/µg (2.4×10^5 MBq/µmol). At this low specific activity, only 1 in 8 EGF molecules carry an ^{111}In atom and almost 90% of EGFR in a tumor would be targeted by nonradiolabeled EGF. In order to address this issue, Remy et al. (165) conjugated maleimide-derivatized EGF with a thiol-containing multibranched peptide containing 4 EDTA-like metal chelators for [111]In. However, the MCP-4-EDTA-*S*-MB-EGF conjugate showed a 40-fold decrease in receptor binding affinity in a competition assay with MDA-MB-468 human breast cancer cells compared to EGF. In contrast, our group found that derivatization of EGF directly with 1–2 DTPA metal chelators and labeling with ^{111}In yielded a radiopharmaceutical with receptor binding affinity identical to that of ^{125}I-labeled EGF ($K_a = 7.3 \times 10^8$ L/mol) (18). Greater success was achieved by our group in maximizing the specific activity of ^{111}In-EGF by conjugating maleimide-derivatized EGF to thiolated human serum albumin (HSA) that presents 60 lysine residues for DTPA conjugation (166). Conjugation of EGF to HSA diminished its receptor binding affinity 15-fold ($K_a = 5.1 \times 10^7$ L/mol) but there were no further decreases in affinity when up to 23 DTPA chelators were preferentially substituted into the HSA moiety. The specific activity of ^{111}In-DTPA-HSA-EGF was increased 10-fold compared to ^{111}In-DTPA-hEGF. ^{111}In-DTPA-HSA-EGF retained its receptor-mediated internalization and nuclear translocation properties in MDA-MB-468 cells and was fourfold more growth inhibitory toward the cells. An analogous strategy was reported by Manabe et al. (167) who used a polylysine peptide carrier to conjugate as many as 42 DTPA molecules to the anti-HLA mAb H-1 with >90% retention of immunoreactivity. Similarly, starburst dendrimers that display 64 amine groups on their surface were derivatized with up to 43 1B4M DTPA-like chelators, then conjugated through a maleimido bond to mAb OST7 (168). The maximum specific activity achieved for labeling these dendrimer immunoconjugates with [111]In (8.4 MBq/µg; 1.3×10^6 MBq/µmol) was 48-fold higher than directly conjugated 1B4M-OST7. In another study, polyethylene glycol (PEG) containing an amine functional group was used to conjugate DTPA chelators to the anti-EGFR mAb C225 with good preservation of immunoreactivity (169). This approach provides a method of labeling with ^{111}In while minimizing retention of radioactivity in the liver due to conjugation with PEG (170).

High retention of radioactivity in the liver with ^{111}In-DTPA-conjugated mAbs is thought to be due to intracellular trapping of ^{111}In-catabolites. ^{111}In-DTPA-labeled intact IgG antibodies interact with glycoprotein receptors on hepatocytes through their Fc-domain and are internalized into endosomes. The antibodies are routed to lysosomes for proteolytic degradation. The ultimate catabolite of proteolysis has been identified as ^{111}In-DTPA-ε-Lys (30, 171). This catabolite cannot easily cross the lysosomal membrane due to its positive charges or be exocytosed due to its poor recognition by cell membrane amino acid transporters and thus, becomes trapped. An analogous mechanism has been proposed for retention of ^{111}In-labeled mAb fragments and peptides by renal tubular cells. ^{111}In-labeled mAb fragments or peptides are filtered by the glomerulus and reabsorbed by renal tubular cells. Proteolytic catabolism within renal tubular cells of ^{111}In-DTPA-F(ab')$_2$ fragments yields ^{111}In-DTPA-ε-Lys, which is trapped within the cells (172). Proteolysis of ^{111}In-DTPA-D-Phe^1-octreotide or ^{111}In-DTPA-L-Phe^1-octreotide similarly results in retention of ^{111}In-DTPA-D-Phe^1-OH or ^{111}In-DTPA-L-Phe^1-OH catabolites, respectively, in renal tubular cells (173, 174).

The macrocyclic chelator, 1,4,7,10-tetraazacyclododecane N,N',N'',N'''-tetraacetic acid (DOTA) (Fig. 2.9) has been less commonly used for labeling mAbs and peptides with ^{111}In because the kinetics of binding ^{111}In by DOTA are slower than those of DTPA. Elevated temperatures of 37–43°C for mAbs (175) and heating at 100°C for peptides (176) combined with longer incubation times (30–45 min) are required to obtain complete incorporation of ^{111}In into DOTA-conjugated biomolecules. Nevertheless, intact IgG mAbs (175, 177), Fab fragments (178) and diabodies (179) as well as octreotide (176) have been successfully labeled with ^{111}In using DOTA. The stability of ^{111}In-DOTA-conjugated mAbs in serum is greater than that of

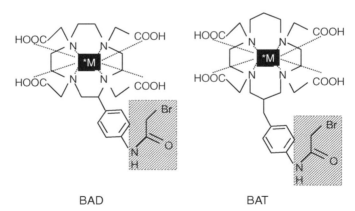

BAD   BAT

FIGURE 2.9 Chemically reactive forms of two macrocyclic chelators that can be conjugated to monoclonal antibodies or peptides for labeling with radiometals. Groups for conjugation to the antibodies or peptides are shaded. BAD: p-bromoacetamidobenzyl-DOTA complexes ^{90}Y, ^{67}Ga, ^{177}Lu, [213]Pb, [212]Bi or [225]Ac. BAT: bromoacetamidobenzyl-TETA complexes ^{64}Cu or [67]Cu.

^{111}In-DTPA-conjugated antibodies. Lewis et al. (175) found that there was less than 0.8% loss of ^{111}In in serum from ^{111}In-DOTA-cT84.66 mAb over 10 days compared to 13.0% for ^{111}In-DTPA-cT84.66. The advantage of conjugating mAbs or peptides with DOTA for labeling with ^{111}In is that it provides a close analogue of the corresponding yttrium-90 (^{90}Y)-DOTA-biomolecule used for targeted radiotherapy. Imaging with the ^{111}In-DOTA-biomolecule can thus be used to reliably predict radiation dosimetry estimates to tissues for the ^{90}Y-DOTA-biomolecule (180, 181). Use of an ^{111}In-DTPA-conjugated mAb or peptide for this purpose may yield inaccurate dosimetry estimates because of major differences in the stability *in vivo* between ^{111}In- and ^{90}Y-DTPA complexes.

2.6.4 Yttrium Radionuclides

Two radionuclides of yttrium: ^{86}Y and ^{90}Y are available for labeling mAbs or peptides for imaging or radiotherapeutic purposes. ^{86}Y (Table 2.2) is a positron emitter with a half-life of 14.7 h that is produced in a cyclotron using the ^{86}Sr$(p,n)^{86}$Y reaction (182). The spatial resolution for detection of the positron annihilation from ^{86}Y is 1.8 mm compared to 0.7 mm for ^{18}F (13). ^{90}Y (Table 2.3) is a pure β-emitter with a half-life of 2.7 days that is produced by a strontium-90 (^{90}Sr)/^{90}Y generator (183). ^{90}Y is an attractive radionuclide for targeted radiotherapy of cancer because it does not emit γ-radiation that minimizes the radiation exposure to healthcare personnel and family members from a patient administered high doses of ^{90}Y-labeled mAbs or peptides. For example, patients with non-Hodgkin's B-cell lymphoma treated with ^{90}Y-ibritumomab tiuxetan (Zevalin; IDEC Pharmaceuticals) do not need to be isolated for radiation safety reasons and can be treated as outpatients, whereas patients receiving ^{131}I-labeled tositumomab (Bexxar; GlaxoSmithKline) require special radiation safety precautions to be taken (184). However, since ^{90}Y does not emit γ-radiation, the tissue distribution of ^{90}Y-labeled mAbs and peptides cannot be easily imaged (only poor spatial resolution Bremstrahlung images can be acquired). Therefore, the radiation-absorbed doses to tissues in patients from ^{90}Y-labeled mAbs or peptides are often estimated by imaging with the ^{111}In-labeled analogues (180, 181). However, as previously mentioned, differences in the *in vivo* stability of ^{111}In- and ^{90}Y-labeled biomolecules can produce discrepancies in their pharmacokinetics and normal organ distribution yielding inaccuracies in the dosimetry estimates. PET using ^{86}Y-labeled analogues has been suggested as a means of more accurately estimating the radiation-absorbed doses from ^{90}Y-labeled biomolecules (185).

The macrocyclic chelator, DOTA (Fig. 2.9) is the chelating agent of choice for labeling mAbs and peptides with yttrium-90 (^{90}Y) due to the high stability of ^{90}Y-DOTA complexes ($K_a = 10^{24}$ L/mol) (186). Initially, DTPA was used as a chelator for ^{90}Y (187), but it was discovered that ^{90}Y-DTPA complexes were unstable *in vivo* that caused bone accumulation of free ^{90}Y released from the immunoconjugates. The stability of ^{90}Y-labeled biomolecules is of paramount importance because even a small amount of free ^{90}Y sequestered in the bone can contribute significantly to bone marrow toxicity due to the long range (10–12 mm) of the emitted β-particles ("cross-fire" effect). This is one of the reasons that the maximum tolerated dose of

^{90}Y-labeled mAbs (e.g., Zevalin) is 32 mCi, whereas up to 150 mCi of ^{131}I-labeled mAbs can be safely administered (2). Bz-SCN-DTPA (Fig. 2.8) yields more stable ^{90}Y complexes than those formed with DTPA, but the complexes are not as stable as the ^{90}Y-DOTA complexes. In one study (188), the bone uptake of radioactivity in mice implanted s.c. with SK-RC-52 renal cell carcinoma xenografts at 7 days p.i. of ^{90}Y-DOTA-conjugated chimeric G250 mAbs was only 0.4% i.d./g, compared to 1.2% i.d./g for Bz-SCN-DTPA- and 10.7% i.d./g for DTPA-conjugated mAbs. Similarly, a threefold lower bone uptake of radioactivity at 10 days p.i. in mice was observed for ^{88}Y-DOTA-hLL2 anti-CD22 mAbs compared to ^{88}Y-Mx-DTPA-hLL2 mAbs (189).

DOTA, which was first described by Meares et al. in 1990 (190) can be conjugated to mAbs by conversion to its reactive form, *p*-bromoacetamidobenzyl-DOTA (BAD) (Fig. 2.9) followed by reaction of BAD with free thiols introduced into the mAbs by reaction with 2-iminothiolane (2-IT) (191). 2-IT also creates a spacer between the mAbs and the DOTA chelator that allows more efficient labeling with ^{90}Y. An alternative approach involves reaction of an *N*-hydroxysuccinimidyl ester of DOTA with ε-amino groups on lysine residues or the N-terminus of biomolecules (192). Labeling of DOTA-conjugated mAbs is achieved by incubation with ^{90}Y chloride (^{90}YCl$_3$) mixed with 0.5 M ammonium acetate buffer, pH 7.0–7.5 for 30 min at 37°C (193). Similar conditions have been employed for labeling DOTA-conjugated peptides with ^{90}Y, except that a temperature of 80–100°C was used to accelerate the incorporation of ^{90}Y (194). Labeling of biomolecules with ^{90}Y using DOTA is highly susceptible to the effects of divalent trace metal ion contamination, especially Ca^{2+}, Fe^{2+} and Zn^{2+}. In fact, the affinity constant (K_a) for binding of Fe^{2+} by DOTA is 100,000 times higher than that for binding ^{90}Y (186). It is critical to exclude trace-metals as much as possible from the ^{90}Y-labeling step. One recent letter to the editor even suggests that minor trace-metal contamination of pipette tips can seriously diminish the labeling efficiency of DOTA-conjugated peptides with ^{90}Y (195). Another important issue in labeling biomolecules with ^{90}Y is the potential for radiolysis. Interaction of the moderate energy ($E_{max} = 2.2$ MeV) β-particles emitted by ^{90}Y with water molecules in the buffers used to formulate the radiopharmaceuticals generates highly reactive free radicals that degrade the metal chelator as well as the biomolecules. The radiolysis effect is dependent on specific activity and on radioactivity concentration. For example, ^{90}Y-BAD-2-IT-Lym-1 antibodies formulated at a specific activity of 1–2 mCi/mg remained pure and relatively immunoreactive (>75%) in storage over a 3-day period (196). However, at a specific activity of 4 mCi/mg, the radiochemical purity (RCP) of ^{90}Y-BAD-2-IT-Lym-1 dropped to 65% and the immunoreactivity decreased to 28%. At a specific activity of 9.4 mCi/mg, the RCP of ^{90}Y-BAD-2-IT-Lym-1 decreased to 21% and the immunoreactivity was virtually abolished (3%). Radiolysis can be minimized by inclusion of radioprotectants such as ascorbic acid, gentisic acid, or human serum albumin in the formulation and by freezing the radiolabeled biomolecules to minimize diffusion of free radicals in the aqueous solutions (196, 197). In one study, it was shown that ascorbic acid may also act as a suitable buffering agent, thereby removing the need for the ammonium acetate buffer (198).

A controversial issue with respect to the use of DOTA as a metal chelator for ^{90}Y and other radiometals is its potential immunogenicity in humans. In a Phase I/II clinical trial in which six ovarian cancer patients received ^{90}Y-DOTA-HMFG1 murine mAbs intraperitoneally, all patients developed anti-DOTA antibodies and three patients developed serum sickness (199). Four of eight patients who received ^{111}In-DOTA-conjugated mAbs intravenously developed anti-DOTA antibodies. The immune response appears to be directed against the DOTA ring structure and not the benzyl-containing side chain. The immune response to DOTA is also dependent on the antigenicity of the biomolecule to which it is conjugated, since rabbit IgG conjugated to DOTA did not induce anti-DOTA antibodies in rabbits, whereas mouse IgG-DOTA immunoconjugates administered to rabbits stimulated an immune response toward the chelators (200). Moreover, an immune response to DOTA has not been found with all mAbs. Anti-DOTA antibodies were not detected in the serum of 18 lymphoma patients administered multiple doses of ^{111}In-BAD-2-IT-Lym-1 mAbs, although these patients typically do not mount a strong immune response to radio-immunoconjugates (201). Similarly, ^{90}Y-DOTATOC, an octapeptide somatostatin analogue does not appear to be immunogenic, again suggesting that the biomolecule must be considered in assessing the immunogenicity of DOTA (202).

2.6.5   Gallium Radionuclides

Gallium-67 (^{67}Ga) (Table 2.1) is a cyclotron-produced radionuclide with a half-life of 78.2 h that has been used for many years in nuclear medicine as ^{67}Ga citrate for tumor imaging (203). Despite its history of use, the properties of ^{67}Ga are not ideal for imaging due to the high energy of two of its γ-photons ($E\gamma = 300$ and 393 keV) that make it difficult to collimate with the γ-camera. Nevertheless, ^{67}Ga has recently received attention as a radiolabel for mAbs for targeted radiotherapy of cancer, exploiting its abundant Auger and conversion electron emissions (see Chapter 9) (204). Most of the other recent interest in gallium radionuclides has been focused on labeling peptides with ^{68}Ga for PET (Table 2.2) (205). Antibodies have not been labeled with ^{68}Ga for PET, because their kinetics of tumor uptake and elimination from the blood and normal tissues is not compatible with the 68 min half-life of the radionuclide. ^{68}Ga is conveniently produced using a germanium-68 (^{68}Ge)/^{68}Ga generator system that allows production of the radionuclide for up to 1 year in a nuclear medicine facility from a single generator, due to the long half-life of ^{68}Ge (270 days) (206). ^{68}Ga decays to ^{68}Zn. Its positron energy is 1.92 MeV, which provides a spatial resolution of 2.4 mm versus 0.7 mm for ^{18}F (13). ^{66}Ga (Table 2.2) is a longer lived positron-emitting radionuclide of gallium (half-life 9.5 h) that has also been studied for labeling peptides for PET (207).

   Initially, desferrioxamine (DFO) was used as a chelator for labeling peptides with ^{68}Ga (74, 208), but more recently DOTA has been used (Fig. 2.9) (205, 209). Most PET studies with ^{68}Ga-labeled peptides have focused on DOTATOC, a synthetic octapeptide analogue of somatostatin. Labeling DOTATOC involves adjusting the pH of ^{68}GaCl$_3$ that is eluted in 0.5 M HCl from the ^{68}Ge/^{68}Ga generator to pH 4.8 with 50 mM sodium acetate buffer. The required amount of ^{68}Ga acetate complex is then mixed with DOTATOC (10–20 nmol) and heated at 95°C for 15 min in a heating block. Early studies in which DOTATOC was labeled with ^{68}Ga using this technique yielded labeling efficiencies that were only about 50%, thus requiring postlabeling purification of the radiopharmaceutical on a C-18 Sep-Pak cartridge (209, 210). The final radiochemical purity was >95%. The large volumes of ^{68}Ga eluates eluted from the ^{68}Ge/^{68}Ga generator as well as contamination with trace metals were believed to be responsible for the low labeling efficiencies for DOTATOC with ^{68}Ga. Improvements in the concentration of ^{68}Ga eluates using an ion-exchange cartridge combined with more homogeneous microwave heating of the labeling reaction for 10–20 min yielded almost complete incorporation of ^{68}Ga into extremely small quantities (<1 nmol) of DOTATOC, thus increasing the specific activity by almost 100-fold (211).

Interestingly, DOTATOC labeled with ^{68}Ga has a fivefold higher binding affinity for SMSR subtype 2 ($IC_{50} = 2.5$ nmol/L) compared to ^{90}Y-DOTATOC ($IC_{50} = 11$ nmol/L) and a ninefold higher affinity than that of ^{111}In-DTPA-octreotide ($IC_{50} = 22$ nmol/L) (212). Combined with the high spatial resolution and sensitivity of PET, this provided greater sensitivity for imaging small meningioma lesions (100% versus 85%) in patients compared to SPECT imaging with ^{111}In-DTPA-octreotide (210, 213). The somatostatin analogue DOTANOC that binds to SMSR subtypes 2 and 5 has been labeled with ^{68}Ga and used for PET of a patient with metastases from a neuroendocrine tumor (214). Preclinical studies have been performed in mouse tumor xenograft models using ^{68}Ga-DOTA-α-melanocyte-stimulating hormone (α-MSH) for imaging melanoma (215) or ^{68}Ga-DOTA-EGF to image gliomas or epidermoid carcinoma xenografts (216). Due to the availability of a generator system for ^{68}Ga, it is not necessary for a nuclear medicine facility to have access to a cyclotron, and therefore, it is likely that more studies leading to clinical application of many different ^{68}Ga-labeled peptides for PET imaging of tumors will be performed in the near future.

2.6.6 Copper Radionuclides

There are several radionuclides of copper that are suitable for labeling mAbs, their fragments or peptides for PET (e.g., ^{60}Cu, ^{61}Cu, or ^{64}Cu) (Table 2.2) (217). ^{64}Cu decays with a half-life of 12.7 h by three different pathways: (i) electron capture (41%) emitting Auger and conversion electrons, (ii) positron (β^+) emission (19%), and (iii) β^--emission (40%). The β^--emissions make the radionuclide attractive for radiotherapy of tumors in addition to its use in PET. Indeed, in one study, ^{64}Cu-labeled octreotide inhibited the growth of CA20948 rat pancreatic tumors in Lewis rats (218). In another study, ^{64}Cu-labeled 1A3 mAbs were used to treat hamsters implanted s.c. with GW39 human colon carcinoma xenografts (219). ^{64}Cu is most commonly produced in a biomedical cyclotron by the ^{64}Ni(p,n)^{64}Cu reaction using a ^{64}Ni-enriched target (220, 221). The positron energy of ^{64}Cu is almost identical to that of ^{18}F (0.657 MeV versus 0.635 MeV, respectively), and the intrinsic spatial resolution is indistinguishable (0.73 mm versus 0.70 mm, respectively) (13, 222). However, due to the lower abundance of positron emission with ^{64}Cu compared to ^{18}F (19% versus 97%), the sensitivity for PET with ^{64}Cu is about fivefold lower than that with ^{18}F (222). The short-lived positron emitter, ^{61}Cu (half-life 3.32 h) (Table 2.2) has a greater abundance of positron emission than ^{64}Cu (60% versus 19%) and has been used for labeling octreotide for PET of CA20948 tumors in rats (223). ^{67}Cu (Table 2.3) can be used for labeling mAbs and peptides for targeted radiotherapy. ^{67}Cu decays with a half-life of 2.6 days emitting β^- particles with energies of 0.395 (51%), 0.484 (28%), and 0.577 MeV (20%) as well as several γ-photons with energies ranging from 91 to 300 keV that are imageable (224). ^{67}Cu is produced in a biomedical cyclotron using the ^{nat}Zn$(p,2p)^{67}$Cu or ^{68}Zn$(p,2p)^{67}$Cu reactions or in a nuclear reactor using the ^{nat}Zn$(n,p)^{67}$Cu reaction (224).

DTPA and benzyl-EDTA (Fig. 2.8) were initially examined for chelating ^{67}Cu, but it was soon found that the complexes formed were unstable, releasing 70–95% of their radiolabel *in vitro* in serum over 5 days (225). In contrast, the macrocyclic chelator, 1,4,8,11-tetraazacyclotetradecane-N',N'',N'''-tetraacetic acid (TETA) (Fig. 2.9) provided a more stable ^{67}Cu complex that lost only 2–6% of the radiolabel in serum when conjugated to Lym-1 antibodies (225). This greater stability of ^{67}Cu-TETA versus ^{67}Cu-DTPA or Bz-EDTA complexes is not due to a higher thermodynamic stability, but rather is believed to be due to the structural rigidity of the ^{67}Cu-TETA complex that shields the radionuclide from attack by endogenous copper binding proteins. Copper is bound in plasma by albumin and transcuprein, which transport the metal to hepatocytes (217). Copper internalized by hepatocytes is used to synthesize metalloenzymes such as superoxide dismutase (SOD), is stored bound to metallothionein (MT) or is secreted back into the plasma complexed to ceruloplasmin. DOTA (Fig. 2.9) and its analogues form stable complexes with copper radionuclides, but TETA is the most widely used chelator (217).

BAT (Fig. 2.9) is a modified form of TETA that contains a bromoacetamidobenzyl side chain that is reactive with antibodies modified with 2-minothiolane (2-IT) to present a free thiol. The inclusion of the 2-IT spacer improves the labeling efficiency of the antibodies with ^{67}Cu and ^{64}Cu (226). DOTA and TETA chelators can be introduced into peptides during solid-phase synthesis for labeling with ^{64}Cu/^{67}Cu (227, 228). Labeling of TETA-conjugated mAbs or peptides is achieved by incubation with ^{64}Cu/^{67}Cu in 100 mM ammonium citrate buffer, pH 5.5 at room temperature for 30–60 min (227, 229). Disodium EDTA is added to chelate any unbound radiometal and the ^{64}Cu/^{67}Cu-labeled biomolecules are purified by SEC. ^{64}Cu/^{67}CuCu labeled peptides are usually purified on a C-18 Sep-Pak (Waters Associates, Inc., Milford, MA). Using TETA, mAb1A3 IgG and its F(ab')$_2$ fragments were labeled with ^{64}Cu for PET in a Phase I/II trial of 36 patients with colorectal carcinoma (230) and Lym-1 antibodies were labeled with ^{67}Cu for RIT of 12 patients with non-Hodgkin's lymphoma (229). Several different peptides have been labeled with ^{64}Cu using DOTA or TETA for PET of tumors including octreotide analogues (227, 228, 231, 232), RGD peptides that recognize $\alpha_v\beta_3$ integrins (233), vasoactive intestinal polypeptide (VIP) analogues (234) and bombesin derivatives (235).

^{64}Cu/^{67}Cu labeled intact IgG antibodies exhibit high accumulation in the liver, whereas ^{64}Cu/^{67}Cu-labeled F(ab')$_2$ fragments and peptides are retained by the kidneys. In rats administered 1A3 mAbs labeled with ^{67}Cu using four different chelators, including 1-[(1,4,8,11-tetraazacyclotetradec-1-yl)methyl]benzoic acid (CPTA), a macrocycle related to TETA, it was found that there was a slow rate of transchelation of ^{67}Cu to SOD in the liver (236). In contrast, ^{67}Cu-labeled F(ab')$_2$ fragments of mAb 1A3 were rapidly catabolized to ^{67}Cu-CPTA-ϵ-lysine. Similarly, transchelation of radiometal to SOD in the liver was observed for ^{64}Cu-TETA-octreotide in rats (237). However, in a clinical study, in which 10 patients received ^{67}Cu-2IT-BAT-Lym-1 antibodies (238), only a small fraction (0.8–7.8%) of the injected dose of ^{67}Cu was recycled by the liver, and in this case, mostly to ceruloplasmin. There was no transchelation to SOD detectable. The higher concentration of SOD in the liver of rats compared to humans may account for this species-dependent catabolic route for ^{64}Cu/^{67}Cu-labeled biomolecules (237).

2.6.7 Lutetium Radionuclides

Lutetium-177 (^{177}Lu) (Table 2.3) is a β-emitting radionuclide ($E\beta = 0.495$ MeV) with a half-life of 6.7 days that is useful for labeling mAbs and peptides for targeted radiotherapy. In addition, ^{177}Lu emits γ-photons of energies 113 and 208 keV that can be imaged. ^{177}Lu is produced in a nuclear reactor by neutron irradiation of ^{176}Lu$_2$O$_3$. Several chelators have been employed for labeling mAbs and peptides with ^{177}Lu including derivatives of DOTA (e.g., PA-DOTA and CA-DOTA) (Fig. 2.9) and analogues of DTPA (e.g., CHX-A-DTPA and SCN-Bz-DTPA) (Fig. 2.8) (239). The difference between the two DOTA chelators is that in PA-DOTA, the benzylisothiocyanate side chain used for conjugation is attached to one of the nitrogens in the macrocycle, whereas in CA-DOTA, it is attached to a methylene carbon. PA-DOTA forms a less stable complex with ^{177}Lu than CA-DOTA or CHX-DTPA (239). The instability of the ^{177}Lu-PA-DOTA complex may have accounted for the prolonged retention of radioactivity in the reticuloendothelial system (RES) including the bone marrow of patients administered ^{177}Lu-PA-DOTA-CC49 mAbs for RIT of TAG-72-positive malignancies and consequently, the observed dose-limiting hematopoietic toxicity (240). ^{177}Lu released from mAbs or peptides is expected to be sequestered in the skeleton since lutetium competes with calcium for bone deposition (241). It has been shown that by adding DTPA to ^{177}Lu-DOTA^0-Tyr^3]octreotate used for PDRT of SMSR-positive tumors, the small amount ($<$2–5%) of free ^{177}Lu impurity in the radiopharmaceutical can be rapidly eliminated from the body by renal excretion as ^{177}Lu-DTPA, thus minimizing bone uptake (242). In addition to labeling intact IgG CC49 mAbs with ^{177}Lu (240, 243), dimeric scFv fragments of this antibody have been labeled with ^{177}Lu for radiotherapeutic purposes (244).

2.6.8 Lead, Bismuth, and Actinium Radionuclides

DOTA (Fig. 2.9) is a also useful bifunctional chelator for labeling mAbs with radionuclides of lead, bismuth, and actinium for targeted radiotherapy (2). Lead-212 (^{212}Pb) decays to the α-emitter, bismuth-212 (^{212}Bi). The α-emitter, actinium-225 (^{225}Ac) decays to a series of daughter radionuclides that are α-emitters (^{221}Fr, ^{217}At, ^{213}Bi) or β-emitters (^{213}Po, ^{209}Tl, ^{209}Pb, ^{209}Bi). The use of ^{225}Ac as a radiolabel for mAbs has been termed an "atomic nanogenerator" since the decay of ^{225}Ac generates the daughter radionuclides locally in tumors that then emit several different forms of radiation that kill cancer cells (245). The main concern for employing radionuclides such as ^{212}Pb or ^{225}Ac that do not decay directly to stable elements is that the decay process may decrease the stability or even disrupt the metal chelate complex due to the transformation of one element into another (i.e., ^{212}Pb is converted into ^{212}Bi) with subsequent release of free daughter radionuclides. In particular, bismuth has an affinity for the kidneys, and it was shown that even low doses of ^{225}Ac-labeled HuM195 antibodies caused severe renal toxicity and anemia in monkeys, likely due to the release of ^{213}Bi and redistribution of the radionuclide to the kidneys (246). Bismuth radionuclides are also stably bound by CHX-A-DTPA and CHX-B-DTPA, two analogues of DTPA (Fig. 2.8) (247, 248). HER2/neu mAbs AE1 and trastuzumab (Herceptin, Roche Pharmaceuticals) have been labeled with ^{212}Pb or ^{225}Ac, respectively, using DOTA as the metal chelator (249, 250).

2.7   CHARACTERIZATION OF RADIOLABELED mAbs AND PEPTIDES

It is important to fully characterize the properties of radiolabeled mAbs and peptides intended for tumor imaging or targeted radiotherapy of malignancies. Characterization includes (i) analytical tests that evaluate homogeneity and radiochemical purity, (ii) radioligand binding assays that assess the ability to specifically bind target antigens/receptors *in vitro*, (iii) stability studies that assess the loss of radiolabel *in vitro* in biologically relevant media, and (iv) biodistribution, pharmacokinetic, dosimetry, and imaging studies that reveal tumor and normal tissue uptake *in vivo* in an animal model and predict radiation-absorbed doses in humans.

2.7.1   Evaluation of the Homogeneity of Radiolabeled mAbs and Peptides

Evaluation of homogeneity is especially important for mAbs or peptides conjugated to chelators for labeling with radiometals. Substitution of these biomolecules with too many chelators may significantly decrease their immunoreactivity or receptor binding affinity (160). The substitution level (moles of chelators/mole of biomolecule) may be measured by spectrophotometric, fluorescence or radiochemical assays. For example, HYNIC substitution in biomolecules intended to be labeled with ^{99m}Tc or ^{186}Re/^{188}Re can be measured by a colorimetric assay that relies on reaction of the HYNIC groups with *p*-nitrobenzaldehyde to form a complex that absorbs at 385 nm (163). DTPA substitution for labeling biomolecules with ^{111}In can be measured by several different techniques (i) radiochemical assays that rely on the binding of ^{111}In or ^{57}Co by DTPA (142, 251), (ii) spectrophotometric assays that generate a colored arsenazo III DTPA complex (252), (iii) fluorescence assays that measure the binding of europium (III) by DTPA (253), or (iv) immunoelectrofocusing (IEF) that reveals changes in pI associated with DTPA substitution (254). The most common method is simply to radiolabel an aliquot of the impure DTPA conjugation mixture with a trace amount of ^{111}In and separate the resulting ^{111}In-DTPA-biomolecule from free ^{111}In-DTPA by silica gel instant thin layer chromatography (ITLC-SG) developed in 100 mM sodium citrate, pH 5.0. The conjugation efficiency is then calculated from the ITLC-SG results and multiplied by the molar ratio of cDTPAA: antibody/peptide used in the reaction to estimate the average number of moles of DTPA per mole of biomolecule. Typically, a substitution level of 1–2 moles DTPA per mole of mAb or peptide is desirable to minimize polymerization and/or interference in immunoreactivity or receptor binding affinity (160). The number of DOTA chelators introduced into antibodies or peptides for labeling with ^{90}Y, ^{68}Ga, ^{177}Lu, ^{212}Bi, or ^{225}Ac can be measured by a spectrophotometric assay which measures the decrease in absorbance at 656 nm for a Pb(II)-arsenazo III complex upon transchelation of Pb^{2+} to DOTA (255).

The polymerization of biomolecules as a consequence of cross-linking due to bifunctional metal chelator substitution may be assessed by size-exclusion HPLC with UV-detection (Fig. 2.10) or by sodium dodecylsulfonate polyacrylimide gel electrophoresis (SDS-PAGE) (Fig. 2.10). Polymerization is most commonly due to cross-linking of mAbs or peptides by bifunctional chelators such as cDTPAA that contain two chemically reactive groups that interact with ε-amino groups on lysines on two separate biomolecules (intermolecular linkage) or within one biomolecule (intramolecular linkage) (142, 152). Such polymerization can be eliminated by using a bifunctional chelator that contains only a single reactive group (e.g., SCN-Bz-DTPA).

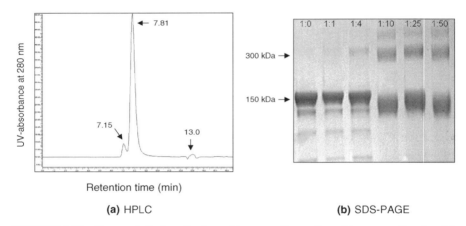

(a) HPLC   (b) SDS-PAGE

FIGURE 2.10 Two analytical methods used to assess the purity and homogeneity of metal-chelated monoclonal antibodies or peptides. (a) Size-exclusion HPLC analysis of trastuzumab (Herceptin) IgG conjugated with the bicyclic anhydride of DTPA (cDTPAA) on a Bio-Sep-SEC-S2000 column eluted with 150-mM sodium chloride/10-mM sodium phosphate buffer, pH 6.8, at a flow rate of 0.6 mL/min with UV-detection at 280 nm. The molar ratio of cDTPAA: IgG was 4 : 1. Peak with a retention time (t_R) of 7.81 min represents a DTPA-conjugated trastuzumab monomer. Peak with t_R of 7.15 min represents a DTPAA-trastuzumab dimer caused by cross-linking of the antibody molecules through the two chemically reactive groups on cDTPAA. Peak with t_R of 13.0 min represents a small amount of unconjugated DTPA (detected through a change in refractive index). (b) SDS-PAGE analysis of trastuzumab (Herceptin) IgG conjugated with cDTPAA on a 4–10% Tris HCl gradient minigel stained with Coomassie Brilliant Blue. There is an increase in the proportion of dimerized (300 kDa) and polymerized (>300 kDa) IgG species as the molar ratio of cDTPAA:IgG is increased from 1 : 1 to 50 : 1.

The level of polymerization that is tolerable depends on the biomolecule, but less than 10% higher molecular weight species is desirable to avoid diminishing immunoreactivity or receptor binding affinity, as well as to minimize sequestration by the reticuloendothelial system (159).

2.7.2 Measurement of Radiochemical Purity

In our laboratory, we determine the radiochemical purity of radiolabeled mAbs or peptides by size-exclusion HPLC on a BioSep SEC-S 2000 column (Phenomenex, Canada) eluted with 100 mM sodium phosphate buffer, pH 7.4 at a flow rate of 1.0 mL/ min and interfaced with a PerkinElmer diode array detector set at 280 nm and a flow scintillation analyzer (FSA; PerkinElmer) radioactivity detector (Fig. 2.11). The RCP of radiolabeled peptides could also be measured by reversed-phase HPLC on a C-18

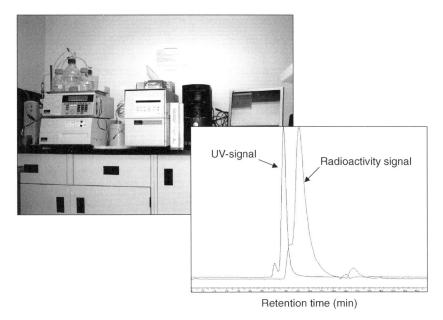

Retention time (min)

FIGURE 2.11   Left panel: A size-exclusion HPLC system with a diode array UV-detector and FSA flow-through radioactivity detector for evaluating the radiochemical purity and homogeneity of radiolabeled monoclonal antibodies and peptides. Right panel: Typical HPLC chromatograms obtained for analysis of ^{111}In-DTPA-conjugated trastuzumab (Herceptin). The radioactivity signal is offset from the UV-signal because of the distance of tubing between the two detectors, which are in sequence. The radioactivity signal has less resolution than the UV-signal due to its larger flow cell. ^{111}In-DTPA-trastuzumab IgG in this example has less than 5% free ^{111}In-DTPA impurity (indicated by the small peak on the radioactivity trace at $t_R = 13.8$ min). The major peak with a t_R of 7.7 min in the UV-trace and 9.5 min in the radioactivity trace represents a ^{111}In-DTPA-trastuzumab IgG monomer. The smaller peak with a t_R of 7.1 min in the UV-trace and 8.25 min in the radioactivity trace represents a ^{111}In-DTPA-trastuzumab IgG dimer.

column eluted with trifluoroacetic acid/methanol/water combinations. Through over-laying the UV and radioactivity traces, these HPLC chromatograms qualitatively confirm that the biomolecule is indeed radiolabeled. Furthermore, they identify and quantify any radiochemical impurities, such as free radiolabeled chelators or radio-nuclides. However, for rapid measurement of RCP, paper chromatography or ITLC systems are most commonly used. The level of RCP purity required for radiolabeled mAbs and peptides depends on their intended clinical application and the toxicity associated with any impurities. A RCP $>90\%$ would be acceptable for mAbs or peptides labeled with a single γ-photon emitter (e.g., ^{99m}Tc or ^{111}In) or a short-lived positron emitter (e.g., ^{18}F or ^{68}Ga) intended for tumor imaging. On the other hand, biomolecules labeled with α-emitters (e.g., ^{211}At) or β-emitters (e.g., ^{131}I or ^{90}Y) for targeted radiotherapy require a higher RCP ($>95\%$) because of the larger amounts of radioactivity administered and the inherent toxicity of the radiochemical impurities, particularly if they concentrate in a radiation sensitive organ (e.g., bone or thyroid). For example, free ^{90}Y is sequestered *in vivo* by bone, increasing the risk for bone marrow toxicity (256).

2.7.3 Measurement of Immunoreactivity/Receptor Binding Properties

The immunoreactivity or receptor binding characteristics of mAbs or peptides conjugated to chelators for radiometal labeling must be assessed since these affect tumor uptake *in vivo* (160). Similarly, these properties should be reevaluated for the final radiolabeled biomolecules. There are two parameters that are measured: (i) the antigen/receptor binding affinity and (ii) the immunoreactive fraction (IRF) or receptor binding fraction (RBF). The affinity constant (K_a) or dissociation constant (K_d) are measured in direct or indirect competition antigen/receptor binding assays and compared with those of the unmodified mAbs or peptides. Direct binding assays (previously known as Scatchard assays), are performed by incubating increasing concentrations of radiolabeled mAbs or peptides with tumor cells that express the target epitopes/receptors. These assays may also be performed by incubating the radiolabeled biomolecules with purified antigens/receptors coated onto wells in a microELISA plate. The radiopharmaceutical binding to the cells is measured in the absence (total binding (TB)) or presence (nonspecific binding (NSB)) of an excess (e.g., 100 nM) of nonradiolabeled biomolecules to saturate the epitopes/receptors. Specific binding (SB) is obtained by subtracting NSB from TB and is plotted versus the concentration of radiolabeled mAbs or peptides (Fig. 2.12). The data is fitted to a direct antigen/receptor binding model using nonlinear curve fitting software (e.g., Prism®, Graphpad Software, San Diego, CA) and the K_d and maximum number of binding sites/cell (B_{max}) estimated. Competition radioligand binding assays can be performed by measuring the binding of radiolabeled mAbs or peptides to cells displaying the target epitopes/receptors in the presence of increasing concentrations of unmodified biomolecules. The K_d is then estimated from the concentration of unmodified antibodies or peptides required to displace 50% of the initial binding of the radiolabeled biomolecules to the cells. Finally, the IRF or RBF of radiolabeled mAbs or peptides may be measured by the Lindmo assay (257). In this assay, a small

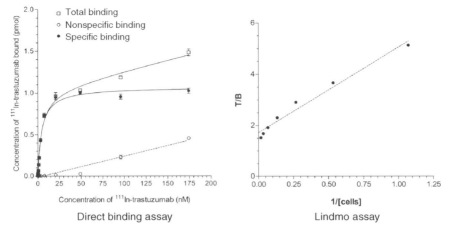

FIGURE 2.12   Left panel: Results for a direct binding assay for a ^{111}In-DTPA-trastuzumab (Herceptin) IgG incubated with SK-BR-3 human breast cancer cells overexpressing HER2/neu. The K_a and B_{max} in this example were 1.2×10^8 L/mol and 1.0×10^6 receptors/cell, respectively. Right panel: Results for measurement of the IRF of a ^{111}In-DTPA-trastuzumab IgG incubated with SK-BR-3 cells using the Lindmo method (257). In this example, the intercept on the ordinate (1/IRF) was 1.7 and the IRF was therefore 0.59.

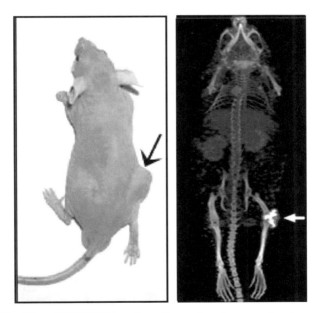

FIGURE 2.13   Micro-SPECT/CT imaging of an anthymic mouse bearing a subcutaneous MDA-MB-361 human breast cancer xenograft (arrow) at 72 h postintravenous (tail vein) injection of ^{111}In-DTPA-pertuzumab (Omnitarg), a second-generation antibody directed toward HER2/neu receptors. The tumor is clearly visualized using the radiopharmaceutical. The anatomy is shown by coregistration of the CT and nuclear medicine images.

amount of radiolabeled mAbs or peptides (e.g., 5–10 ng) is incubated with increasing concentrations of tumor cells that display the target epitopes/receptors. The total radioactivity counts added to the cells divided by the bound counts (T/B) is plotted versus the inverse of the cell concentration (1/[cells]). The intercept on the ordinate of this plot is 1/IRF or 1/RBF (Fig. 2.12). The IRF or RBF provides information that is different than that of K_a or K_d, in that these parameters describe the fraction of radiolabeled biomolecules that are able to bind their target epitopes or receptors, at "infinite antigen excess," irrespective of their binding affinity. Ideally, measurement of K_a or K_d as well as determination of IRF or RBF should be performed to assess immunoreactivity or receptor binding characteristics.

2.7.4   Evaluation of *In Vitro* and *In Vivo* Stability

Studies to evaluate the *in vitro* and *in vivo* stability of radiolabeled mAbs and peptides are an integral part of their development. Kits used to prepare radiolabeled biomolecules must meet specifications established for radiolabeling efficiency, immunoreactivity/receptor binding, homogeneity, sterility, and apyrogenicity and other quality control tests over the expected storage period (see Chapter 17) (146). Similarly, the radiolabeled biomolecule itself must maintain the specified level of RCP over the interval from the time of labeling until it is administered. The results of these stability studies will define the expiry times for the kits and/or radiolabeled mAbs or peptides (146). The stability of the radiolabeled biomolecules *in vitro* in biologically relevant media such as plasma or serum must also be determined. The anticipated mechanism of loss of radiolabel *in vivo* should be taken into account in designing these studies. For example, studies examining the transchelation of radiometal to transferrin in serum/plasma are necessary for [111]In- or ^{67}Ga-labeled biomolecules (152), whereas studies that measure transchelation to albumin or other copper binding proteins are required for ^{64}Cu/[67]Cu-labeled biomolecules (225). Cysteine-challenge studies are required for ^{99m}Tc-labeled biomolecules, especially those labeled using the Schwartz technique (94, 258). Frequently, stability *in vitro* does not predict stability *in vivo*, since the radiolabeled mAbs or peptides may be catabolized in tissues resulting in the release and redistribution of radionuclides. For example, radioiodinated mAbs and peptides are stable *in vitro* in serum/plasma but *in vivo* may be internalized by tumor and normal cells, in which they are rapidly catabolized by lysosomal proteases and deiodinases resulting in the release and redistribution of free radioiodine. This instability *in vivo* can be addressed by employing residualizing radioiodination techniques. Nevertheless, due to the limitations of *in vitro* stability studies, it is important to evaluate the stability of radiolabeled mAbs and peptides *in vivo* by chromatographic analysis of plasma/serum samples to identify any circulating radioactive catabolites, as well as by monitoring changes in tissue distribution of radioactivity over time by imaging or biodistribution studies. In these studies, it is critical to sample tissues that are known to sequester the radionuclide, should it be released from the biomolecule, that is, bone for ^{90}Y or ^{177}Lu and thyroid and stomach for iodine radionuclides. Noninvasive SPECT or PET imaging studies in humans allow these stability studies to be continued into clinical trials of the radiolabeled biomolecules.

2.7.5 Preclinical Biodistribution, Tumor Imaging, and Dosimetry Studies

It is important to evaluate the tumor and normal tissue distribution of radiolabeled biomolecules in an animal model (see Chapter 11). For radiolabeled mAbs and peptides intended to target solid tumors, these studies are performed in athymic mice implanted s.c. with human tumor xenografts (18, 108, 120, 133, 259). In the case of hematological malignancies (e.g., B-cell lymphoma or leukemias), athymic or non-obese diabetic severe combined immunodeficient (NOD-scid) mice may be engrafted s.c. or inoculated intravenously (i.v.) with malignant cells to establish a tumor model (260–262). Tumor-bearing mice are injected i.v. with 25–100 µCi (0.9–3.7 MBq) of radiolabeled mAbs or peptides. Control groups may consist of (i) mice bearing the same tumors but injected with radiolabeled nonspecific (irrelevant) mAbs or peptides that do not recognize the target epitopes/receptors, (ii) mice bearing these tumors but injected with the radiolabeled mAbs or peptides mixed with a large excess of the unlabeled biomolecules to saturate the epitopes/receptors, or (iii) mice implanted with a tumor that does not express the target epitopes/receptors but receiving the radiolabeled mAbs or peptides. Groups of 5–6 mice are sacrificed at selected times p. i. depending on the anticipated pharmacokinetics. The tumor and samples of blood and normal tissues are obtained, weighed and the radioactivity in each measured by γ-scintillation counting. For pure β-emitters (e.g., ^{90}Y), liquid scintillation counting may be used to measure the radioactivity in tissues or alternatively, the Bremstrahlung radiation caused by interaction of the β-particles with the tissues can be measured in a γ-counter. The tumor and normal tissue uptake is expressed as the percent injected dose per gram of tissue (% i.d./g). The T/B and T/NT ratios are calculated. Tumor uptake varies from 1–2% i.d./g for radiolabeled peptides to as high as 10–20% i.d./g for intact IgG mAbs (18). Tumor uptake should be significantly greater than that in the control groups of mice. The T/B ratio should be >2 : 1 and ideally, as high as 5 : 1 to 10 : 1. In addition to biodistribution studies, imaging studies of γ-emitting or positron-emitting biomolecules may be performed using dedicated high-resolution (1–2 mm) microSPECT (Fig. 2.13) or microPET small animal imaging devices (263, 264). These images can be quantified by region-of-interest (ROI) analysis and are useful to follow any changes in the biodistribution of radioactivity in an individual animal over time. In addition, preclinical small animal imaging studies provide "proof-of-principle" for radiolabeled biomolecules intended for tumor imaging applications in humans. The results of biodistribution and imaging studies can be used to predict the radiation-absorbed doses to organs in humans for subsequent clinical trials using computer software such as OLINDA (see Chapter 13) (265, 266).

2.7.6 Preclinical Studies to Evaluate Antitumor Effects and Normal Tissue Toxicity

Radiolabeled mAbs and peptides intended for RIT or PDRT of malignancies in humans need to be first evaluated preclinically in mouse tumor xenograft models. These studies involve administration of increasing doses of the radiolabeled mAbs or peptides to mice and monitoring the tumor growth or survival of the animals over time (259). Control groups of mice receive either the formulation vehicle (e.g., normal saline) or the nonspecific radiolabeled mAbs of the same class and isotype or peptides that do not recognize the target epitopes/receptors. Generalized and gastrointestinal toxicity is monitored by weighing the animals every few days to identify any significant weight loss (>10–20%), whereas bone marrow toxicity is assessed by determining leukocyte, red blood cell, and platelet counts in blood samples. Liver toxicity is evaluated by following serum transaminases, while kidney toxicity is determined by measuring serum creatinine. Hematopoietic toxicity, if present is usually observed within 2–3 weeks following injection of the radiopharmaceuticals and normalizes within 6–8 weeks, whereas, liver and kidney toxicity may require an extended observation period (4–8 weeks). An ideal radiotherapeutic agent will exhibit specific antitumor effects with minimal-moderate and manageable normal tissue toxicity (259).

2.7.7 Kit Formulation and Pharmaceutical Testing

Labeling methods for mAbs and peptides that involve the chelation of radiometals (e.g., ^{99m}Tc, ^{111}In, ^{68}Ga, ^{90}Y, ^{177}Lu, ^{64}Cu, and others) allow the formulation of radiopharmaceutical kits (99, 146). These kits consist of a unit dose of a solution of the metal chelator-biomolecule conjugates dispensed into a Type A pharmaceutical quality glass vial sealed with a rubber septum and aluminum crimp. The kits may be stored at 2–8°C, kept frozen at -10°C or lyophilized, depending on the particular mAb or peptide formulation. Radiolabeling is performed simply by adding a buffered solution of the radionuclide to the vial and incubating for a predetermined period of time and temperature to maximize incorporation. The kits are designed such that the labeling efficiency is >90–95%, thus requiring no postlabeling purification and only minimal final product quality control testing (e.g., assay for total radioactivity, pH measurement, and RCP testing). Other quality control testing procedures are performed on the kit formulation prior to its use. These tests include protein/peptide purity and homogeneity, evaluation of immunoreactivity or receptor binding properties, estimation of the level of chelator substitution, measurement of protein concentration and pH and stability studies. Key pharmaceutical tests such as the USP Sterility Test and USP Bacterial Endotoxins Test also need to be performed on the kits, since they produce injectable radiopharmaceuticals. The final radiolabeled biomolecules may be tested retrospectively for sterility and apyrogenicity providing that the kits have been previously validated using trial batches to without exception produce a sterile, apyrogenic product.

2.8 SUMMARY

Radiolabeled mAbs and peptides offer the opportunity for molecular imaging of tumors in order to probe their phenotypic properties. Various strategies have been developed for labeling these biomolecules with single γ-photon emitters or positron emitters for SPECT and PET imaging, respectively. These approaches can also be extended for labeling the biomolecules with α- or β-emitters or radionuclides that emit Auger and conversion electrons for targeted radiotherapy of malignancies. Establishment of specifications for radiolabeled mAbs and peptides supported by characterization assays assures the quality of these novel agents for human evaluation in clinical trials.

ACKNOWLEDGMENTS

The author would like to acknowledge the financial support for research originating in his laboratory at the University of Toronto and described in this chapter from the U.S. Department of Defense Breast Cancer Research Program, the Canadian Institutes of Health Research, the Canadian Breast Cancer Research Alliance, the Canadian Breast Cancer Foundation (Ontario Branch), the Cancer Research Society, Inc., the Susan G. Komen Breast Cancer Foundation, the Natural Sciences and Engineering Research Council of Canada, the James Birrell Neuroblastoma Research Fund, and the Ontario Institute of Cancer Research through funding provided by the Province of Ontario. The author would also like to acknowledge the outstanding contributions by the graduate students, postdoctoral fellows and research technicians who have been members of his research group over the years.

REFERENCES

1. Ferlay J, Bray F, Pisani P, Parkin DM. Cancer incidence, prevalence and mortality worldwide. *IARC CancerBase*, 5th edition. IARC Press, Lyon, France, 2000.
2. Reilly RM. Biomolecules as targeting vehicles for in situ radiotherapy of malignancies. In: Knaeblein J, Mueller R, editors. *Modern Biopharmaceuticals: Design, Development and Optimization*, Wiley-VCH, Weinheim, Germany, 2005, pp. 497–526.
3. Goldenberg DM. Future role of radiolabeled monoclonal antibodies in oncological diagnosis and therapy. *Semin Nucl Med* 1989;19:332–339.
4. Baselga J. Clinical trials of Herceptin (trastuzumab). *Eur J Cancer* 2001;37:S18–S24.
5. Behr TM, Béhé M, Wörmann B. Trastuzumab and breast cancer. *N Engl J Med* 2001; 345:995–996.
6. Krenning EP, Kwekkeboom DJ, Bakker WH, et al. Somatostatin receptor scintigraphy with [^{111}In-DTPA-D-Phe^1]- and [^{123}I-Tyr^3]-octreotide: the Rotterdam experience with more than 1000 patients. *Eur J Nucl Med* 1993;20:716–731.
7. Reilly RM, Sandhu J, varez-Diez TM, Gallinger S, Kirsh J, Stern H. Problems of delivery of monoclonal antibodies. Pharmaceutical and pharmacokinetic solutions. *Clin Pharmacokinet* 1995;28:126–142.
8. Yokota T, Milenic DE, Whitlow M, Schlom J. Rapid tumor penetration of a single-chain Fv and comparison with other immunoglobulin forms. *Cancer Res* 1992;52:3402–3408.
9. Saga T, Neumann RD, Heya T, Sato J, Kinuya S, Le N, et al. Targeting cancer micrometastases with monoclonal antibodies: a binding-site barrier. *Proc Natl Acad Sci USA* 1995;92:8999–9003.

10. van Osdol W, Fujimori K, Weinstein JN. An analysis of monoclonal antibody distribution in microscopic tumor nodules:consequences of a "binding-site barrier". *Cancer Res* 1991;51:4776–4784.

11. Behr TM, Goldenberg DM, Becker W. Reducing the renal uptake of radiolabeled antibody fragments and peptides for diagnosis and therapy: present status, future prospects and limitations. *Eur J Nucl Med* 1998;25:201–212.

12. Jamar F, Barone R, Mathieu I, Walrand S, Labar D, Carlier P, et al. ^{86}Y-DOTA^{0}-D-Phe^{1}-Tyr3-octreotide (SMT487)- a phase 1 clinical study: pharmacokinetics, biodistribution and renal protective effect of different regimens of amino acid co-infusion. *Eur J Nucl Med* 2003;30:510–518.

13. Pagani M, Stone-Elander S, Larsson SA. Alternative positron emission tomography with non-conventional positron emitters: effects of their physical properties on image quality and potential clinical applications. *Eur J Nucl Med* 1997;24:1301–1327.

14. Xu LY, Butler NJ, Makrigiorgos GM, Adelstein SJ, Kassis AI. Bystander effect produced by radiolabeled tumor cells in vivo. *J Nucl Med* 2002;43 (Suppl): 276P–277P.

15. Finn R, Cheung N-KV, Divgi C, St Germain J, Graham M, Pentlow K, et al. Technical challenges associated with the radiolabeling of monoclonal antibodies utilizing short-lived, positron emitting radionuclides. *Nucl Med Biol* 1991;18:9–13.

16. Hunter WM, Greenwood FC. Preparation of ^{131}I labeled human growth hormone of high specific activity. *Nature* 1962;194:495–496.

17. Fraker PJ, Speck JC, Jr. Protein and cell membrane iodinations with a sparingly soluble chloroamide, 1,3,4,6-tetrachloro-3α,6α-diphenylglycouril. *Biochem Biophys Res Commun* 1978;80:849–857.

18. Reilly RM, Kiarash R, Sandhu J, Lee YW, Cameron RG, Hendler A, et al. A comparison of EGF and MAb 528 labeled with ^{111}In for imaging human breast cancer. *J Nucl Med* 2000;41:903–911.

19. Zalutsky MR, Narula AS. A method for the radiohalogenation of proteins resulting in decreased thyroid uptake of radioiodine. *Appl Radiat Isot* 1987;38:1051–1055.

20. Zalutsky MR, Narula AS. Radiohalogenation of a monoclonal antibody using an N-succinimidyl 3-(tri-n-butylstannyl)benzoate intermediate. *Cancer Res* 1988;48:1446–1450.

21. Zalutsky MR, Noska MA, Colapinto EV, Garg PK, Bigner DD. Enhanced tumor localization and in vivo stability of a monoclonal antibody radioiodinated using N-succinimidyl 3-(tri-n-butylstannyl)benzoate. *Cancer Res* 1989;49:5543–5549.

22. Garg S, Garg PK, Zalutsky MR. N-Succinimidyl 5-(trialkylstannyl)-3-pyridinecarboxylates: a new class of reagents for protein radioiodination. *Bioconjug Chem* 1991; 2:50–58.

23. Reist CJ, Garg PK, Alston KL, Bigner DD, Zalutsky MR. Radioiodination of internalizing monoclonal antibodies using N-succinimidyl 5-iodo-3-pyridinecarboxylate. *Cancer Res* 1996;56:4970–4977.

24. Reist CJ, Archer GE, Wikstrand CJ, Bigner DD, Zalutsky MR. Improved targeting of an anti-epidermal growth factor receptor variant III monoclonal antibody in tumor xenografts after labeling using N-succinimidyl 5-iodo-3-pyridinecarboxylate. *Cancer Res* 1997;57:1510–1515.

25. Garg PK, Alston KL, Welsh PC, Zalutsky MR. Enhanced binding and inertness to dehalogenation of α-melanotropic peptides labeled using N-succinimidyl 3-iodobenzoate. *Bioconjug Chem* 1996;7:233–239.

26. Vaidyanathan G, Affleck DJ, Li J, Welsh P, Zalutsky MR. A polar substituent-containing acylation agent for the radioiodination of internalizing monoclonal antibodies: *N*-succinimidyl 4-guanidinomethyl-3-[^{131}I]iodobenzoate ([^{131}I]SGMIB). *Bioconjug Chem* 2001;12:428–438.

27. Vaidyanathan G, Affleck DJ, Bigner DD, Zalutsky MR. Improved xenograft targeting of tumor-specific anti-epidermal growth factor receptor variant III antibody labeled using *N*-succinimidyl 4-guanidinomethyl-3-iodobenzoate. *Nucl Med Biol* 2002; 29:1–11.

28. Shankar S, Vaidyanathan G, Affleck D, Welsh PC, Zalutsky MR. *N*-Succinimidyl 3-[^{131}I] iodo-4-phosphonomethylbenzoate ([^{131}I]SIPMB), a negatively charged substituent-bearing acylation agent for the radioiodination of peptides and mAbs. *Bioconjug Chem* 2003; 14:331–341.

29. Govindan SV, Mattes MJ, Stein R, McBride BJ, Karacay H, Goldenberg DM, et al. Labeling of monoclonal antibodies with diethylenetriaminepentaacetic acid-appended radioiodinated peptides containing D-amino acids. *Bioconjug Chem* 1999;10: 231–240.

30. Franano FN, Edwards WB, Welch MJ, Duncan JR. Metabolism of receptor targeted ^{111}In-DTPA-glycoproteins: Identification of ^{111}In-DTPA-ε-lysine as the primary metabolic and excretory product. *Nucl Med Biol* 1994;21:1023–1034.

31. Dumas P, Maziere B, Autissier N, Michel R. Specificite de L'iodotyrosine desiodase des microsomes thyroidiens et hepatiques. *Biochim Biophys Acta* 1973;293:36–47.

32. Stein R, Govindan SV, Mattes MJ, Chen S, Reed L, Newsome G, et al. Improved iodine radiolabels for monoclonal antibody therapy. *Cancer Res* 2003;63:111–118.

33. Govindan SV, Griffiths GL, Stein R, Andrews P, Sharkey RM, Hansen HJ, et al. Clinical-scale radiolabeling of a humanized anticarcinoembryonic antigen monoclonal antibody, hMN-14, with residualizing ^{131}I for use in radioimmunotherapy. *J Nucl Med* 2005; 46: 153–159.

34. Foulon CF, Reist CJ, Bigner DD, Zalutsky MR. Radioiodinatin via D-amino acid peptide enhances cellular retention and tumor xenograft targeting of an internalizing anti-epidermal growth factor receptor variant III monoclonal antibody. *Cancer Res* 2000; 60:4453–4460.

35. Thorpe SE, Baynes JW, Chroneos ZC. The design and application of residualizing labels for studies of protein catabolism. *FASEB* 1993;7:399–405.

36. Stein R, Goldenberg DM, Thorpe SR, Basu A, Mattes MJ. Effects of radiolabeling monoclonal antibodies with a residualizing iodine radiolabel in the accretion of radio-isotope in tumors. *Cancer Res* 1995;55:3132–3139.

37. Reist CJ, Archer GE, Kurpad SN, Wikstrand CJ, Vaidyanathan G, Willingham MC,et al. Tumor-specific anti-epidermal growth factor receptor variant III monoclonal antibodies: use of the tyramine-cellobiose radioiodination method enhances cellular retention and uptake in tumor xenografts. *Cancer Res* 1995;55:4375–4382.

38. Zalutsky MR, Xu FJ, Yu Y, Foulon CF, Zhao X-G, Slade SK, et al. Radioiodinated antibody targeting of the HER-2/neu oncoprotein: effects of labeling method on cellular Processing and tissue distribution. *Nucl Med Biol* 1999;26:781–790.

39. Sundberg ÄL, Blomquist E, Carlsson J, Steffen A-C, Gedda L. Cellular retention of radioactivity and increased radiation dose. Model experiments with EGF-dextran. *Nucl Med Biol* 2003;30:303–315.

40. Townsend DW, Carney JPJ, Yap JT, Hall NC. PET/CT today and tomorrow. *J Nucl Med* 2004;45 (Suppl): 4S–14S.

41. Chatziionnou AF. Molecular imaging of small animals with dedicated PET tomographs. *Eur J Nucl Med* 2002;29:98–114.

42. Gambhir SS, Czernin J, Schwimmer J, Silverman DH, Coleman RE, Phelps ME. A tabulated summary of the FDG PET literature. *J Nucl Med* 2001;42 (Suppl): 1S–93S.

43. Bakir MA, Eccles SA, Babich JW, Aftab N, Styles JM, Dean CJ, et al. c-erbB2 Protein overexpression in breast cancer as a target for PET using iodine-124-labeled monoclonal antibodies. *J Nucl Med* 1992;33:2154–2160.

44. Larson SM, Pentlow KS, Volkow ND, Wolf AP, Finn RD, Lambrecht RM, et al. PET scanning of iodine-124-3F9 as an approach to tumor dosimetry during treatment planning for radioimmunotherapy in a child with neuroblastoma. *J Nucl Med* 1992; 33:2020–2023.

45. Lee FT, Hall C, Rigopoulos A, Zweit J, Pathmaraj K, O'Keefe GJ, et al. Immuno-PET of human colon xenograft-bearing BALB/c nude mice using [124]I-CDR-grafted humanized A33 monoclonal antibody. *J Nucl Med* 2001;42:764–769.

46. Verel I, Visser GWM, Vosjan MJWD, Finn RD, Boellaard R, van Dongen GAMS. High-quality [124]I-labelled monoclonal antibodies for use as PET scouting agents prior to ^{131}I-radioimmunotherapy. *Eur J Nucl Med Mol Imaging* 2004;31:1645–1652.

47. Robinson MK, Doss M, Shaller C, Narayanan D, Marks JD, Adler LP, et al. Quantitative immuno-positron emission tomography imaging of HER2-positive tumor xenografts with an iodine-124 labeled anti-HER2 diabody. *Cancer Res* 2005; 65:1471–1478.

48. Sundaresan G, Yazaki PJ, Shively JE, Finn RD, Larson SM, Raubitschek AA, et al. [124]I-labeled engineered anti-CEA minibodies and diabodies allow high-contrast, antigen-specific small-animal PET imaging of xenografts in athymic mice. *J Nucl Med* 2003; 44:1962–1969.

49. Iozzo P, Osman S, Glaser M, Knickmeier M, Ferrannini E, Pike VW, et al. *In vivo* imaging of insulin receptors by PET: preclinical evaluation of iodine-125 and iodine-124 labelled human insulin. *Nucl Med Biol* 2002;29:73–82.

50. Eary JF. PET imaging for planning cancer therapy. *J Nucl Med* 2001;42:770–771.

51. Tolmachev V, Lovqvist A, Einarsson L, Schultz J, Lundqvist H. Production of [76]Br by a low-energy cyclotron. *Appl Radiat Isot* 1998;49:1537–1540.

52. Lovqvist A, Sundin A, Ahlstrom H, Carlsson J, Lundqvist H. [76]Br-labeled monoclonal anti-CEA antibodies for radioimmuno positron emission tomography. *Nucl Med Biol* 1995;22:125–131.

53. Sundin J, Tolmachev V, Koziorowski J, Carlsson J, Lundqvist H, Welt S, et al. High yield direct [76]Br-bromination of monoclonal antibodies using chloramine-T. *Nucl Med Biol* 1999;26:923–929.

54. Lovqvist A, Sundin A, Roberto A, Ahlstrom H, Carlsson J, Lundqvist H. Comparative PET imaging of experimental tumors with bromine-76-labeled antibodies, fluorine-18-fluorodeoxyglucose and carbon-11-methionine. *J Nucl Med* 1997;38:1029–1035.

55. Lovqvist A, Sundin A, Ahlstrom H, Carlsson J, Lundqvist H. Pharmacokinetics and experimental PET imaging of a bromine-76-labeled monoclonal anti-CEA antibody. *J Nucl Med* 1997;38:395–401.

56. Hoglund J, Tolmachev V, Orlova A, Lundqvist H, Sundin A. Optimized indirect [76]Br-bromination of antibodies using *N*-succinimidyl *para*-[[76]Br]bromobenzoate for radioimmuno PET. *Nucl Med Biol* 2000;27:837–843.

57. Wilbur DS, Hylarides MD. Radiolabeling of a monoclonal antibody with *N*-succinimidyl *para*-[[76]Br]bromobenzoate. *Nucl Med Biol* 1991;18:363–325.

58. Mume E, Orlova A, Malmström P-U, Lundqvist H, Sjoberg S, Tolmachev V. Radiobromination of humanized anti-HER2 monoclonal antibody trastuzumab using *N*-succinimidyl 5-bromo-3-pyridinecarboxylate, a potential label for immunoPET. *Nucl Med Biol* 2005;32:613–622.

59. Mume E, Orlova A, Larsson B, Nilsson A-S, Nilsson FY, Sjoberg S, et al. Evaluation of ((4-hydroxyphenyl)ethyl)maleimide for site-specific radiobromination of anti-HER2 affibody. *Bioconjug Chem* 2005;16:1547–1555.

60. Steffen A-C, Wikman M, Tolmachev V, Adams GP, Nilsson F, Stahl S, et al. *In vitro* characterization of a bivalent anti-HER-2 affibody with potential for radionuclide based diagnostics. *Cancer Biother Radiopharm* 2005;20:239–248.

61. Bruskin A, Sivaev I, Persson M, Lundqvist H, Carlsson J, Sjoberg S, et al. Radiobromination of monoclonal antibody using potassium [[76]Br] (4 isothiocyanatobenzyl-ammonio)-bromo-decahydro-closo-dodecaborate (bromo-DABI). *Nucl Med Biol* 2004; 31:205–211.

62. Winberg KJ, Persson M, Malmström P-U, Sjoberg S, Tolmachev V. Radiobromination of anti-HER2/neu/ErbB-2 monoclonal antibody using the p-isothiocyanatobenzene derivative of the [[76]Br]undecahydro-bromo-7,8-dicarba-nido-undecaborate(1-) ion. *Nucl Med Biol* 2004;31:425–433.

63. Okarvi SM. Recent progress in fluorine-18 labelled peptide radiopharmaceuticals. *Eur J Nucl Med* 2001;28:929–938.

64. Garg PK, Garg S, Zalutsky MR. Fluorine-18 labeling of monoclonal antibodies and fragments with preservation of immunoreactivity. *Bioconjug Chem* 1991;2:44–49.

65. Garg PK, Garg S, Bigner DD, Zalutsky MR. Localization of fluorine-18-labeled Mel-14 monoclonal antibody F(ab')$_2$ fragment in a subcutaneous xenograft model. *Cancer Res* 1992;52:5054–5060.

66. Page RL, Garg PK, Garg S, Archer GE, Bruland OS, Zalutsky MR. PET imaging of osteosarcoma in dogs using a fluorine-18-labeled monoclonal antibody Fab fragment. *J Nucl Med* 1994;35:1506–1513.

67. Lang L, Eckelman WC. One-step synthesis of ^{18}F labeled [^{18}F]-*N*-succinimidyl 4-(fluoromethyl)benzoate for protein labeling. *Appl Radiat Isot* 1994;45:1155–1163.

68. Vaidyanathan G, Zalutsky MR. Improved synthesis of *N*-succinimidyl 4-[^{18}F]fluorobenzoate and its application to the labeling of a monoclonal antibody fragment. *Bioconjug Chem* 1994;5:352–356.

69. Page RL, Garg PK, Vaidyanathan G, Zalutsky MR. Preclinical evaluation and PET imaging of ^{18}F-labeled Mel-14 F(ab')$_2$ fragment in normal dogs. *Nucl Med Biol* 1994; 21:911–919.

70. Vaidyanathan G, Zalutsky MR. Fluorine-18-labeled [Nle[4],D-Phe[7]]-α-MSH, an a-melanocyte stimulating hormone analogue. *Nucl Med Biol* 1997;24:171–178.

71. Guhlke S, Coenen HH, Stocklin G. Fluoroacylation agents based on small n.c.a. [^{18}F] fluorocarboxylic acids. *Appl Radiat Isot* 1994;45:715–727.

72. Guhlke S, Wester H-J, Bruns C, Stocklin G. (2-[^{18}F]fluoropropionyl-(D)phe^1)-octreotide, a potential radiopharmaceutical for quantitative somatostatin receptor imaging with PET: synthesis, radiolabeling, *in vitro* validation and biodistribution in mice. *Nucl Med Biol* 1994;21:819–825.

73. Poethko T, Schottelius M, Thumshirn G, Hersel U, Herz M, Henriksen G, et al. Two-step methodology for high-yield routine radiohalogenation of peptides: ^{18}F-labeled RGD and octreotide analogs. *J Nucl Med* 2004;45:892–902.

74. Wester H-J, Brockman J, Rosch F, Wutz W, Herzog H, Smith-Jones P, et al. PET-pharmacokinetics of ^{18}F-octreotide: a comparison with ^{67}Ga-DFO- and ^{86}Y-DTPA-octreotide. *Nucl Med Biol* 1997;24:275–286.

75. Chen X, Park R, Shahinian AH, Tohme M, Khankaldyyan V, Bozorgzadeh MH, et al. ^{18}F-labeled RGD peptide: initial evaluation for imaging brain tumor angiogenesis. *Nucl Med Biol* 2004;31:179–789.

76. Haubner R, Kuhnast B, Mang C, Weber WA, Kessler H, Wester H-J, et al. [^{18}F]galacto-RGD: synthesis, radiolabeling, metabolic stability, and radiation dose estimates. *Bioconjug Chem* 2004;15:61–69.

77. Wester H-J, Hamacher K, Stocklin G. A comparative study of n.c.a. fluorine-18 labeling of proteins via acylation and photochemical conjugation. *Nucl Med Biol* 1996;23:365–372.

78. Chang YS, Jeong JM, Lee Y-S, Kim HY, Rai GB, Lee SJ, et al. Preparation of ^{18}F-human serum albumin: a simple and efficient protein labeling method with ^{18}F using a hydrazone-formation method. *Bioconjug Chem* 2005;16:1329–1333.

79. Couturier O, Supiot S, Degraef-Mougin M, Faivre-Chauvet A, Cartier T, Chatal J-F,et al. Cancer radioimmunotherapy with alpha-emitting nuclides. *Eur J Nucl Med Mol Imaging* 2005;32:601–614.

80. Visser GWM, Diemer EL, Kaspersen FM. The nature of the astatine-protein bond. *Int J Appl Radiat Isot* 1981;32:905–912.

81. Aaij C, Tschroots WRJM, Lindner L, Feltkamp TEW. The preparation of astatine labelled proteins. *Int J Appl Radiat Isot* 1975;26:25–30.

82. Vaughan ATM, Fremlin JH. The preparation of astatine labelled proteins using an electrophilic reaction. *Int J Nucl Med Biol* 1978;5:229–230.

83. Zalutsky MR, Narula AS. Astatination of proteins using an *N*-succinimidyl tri-*n*-butylstannyl benzoate intermediate. *Appl Radiat Isot* 1988;39:227–232.

84. Zalutsky MR, Garg PK, Friedman HS, Bigner DD. Labeling monoclonal antibodies and F(ab')$_2$ fragments with the α-particle-emitting nuclide astatine-211: preservation of immunoreactivity and in vivo localizing capacity. *Proc Natl Acad Sci USA* 1989;86:7149–7153.

85. Garg PK, Harrison CL, Zalutsky MR. Comparative tissue distribution in mice of the alpha-emitter ^{211}At and ^{131}I as labels of a monoclonal antibody and F(ab')$_2$ fragment. *Cancer Res* 1990;50:3514–3520.

86. Wilbur DS, Vessella RL, Stray JE, Goffe DK, Blouke KA, Atcher RW. Preparation and evaluation of *para*-[^{211}At]astatobenzoyl labeled anti-renal cell carcinoma antibody A6H F(ab')$_2$. In vivo distribution comparison with para-[^{125}I]iodobenzoyl labeled A6H F(ab')$_2$. *Nucl Med Biol* 1993;20:917–927.

87. Zalutsky MR, Zhai X-C, Alston KL, Bigner D. High-level production of α-particle-emitting ^{211}At and preparation of ^{211}At-labeled antibodies for clinical use. *J Nucl Med* 2001;42:1508–1515.

88. Pozzi OR, Zalutsky MR. Radiopharmaceutical chemistry of targeted radiotherapeutics. Part I. Effects of solvent on the degradation of radiohalogenation precursors by [211]At α-particles. *J Nucl Med* 2005;46:700–706.

89. Pozzi OR, Zalutsky MR. Radiopharmaceutical chemistry of targeted radiotherapeutics. Part 2. Radiolytic effects of [211]At α-particles influence N-succinimidyl 3-[211]At-astato-benzoate synthesis. *J Nucl Med* 2005;46:1393–1400.

90. Reist CJ, Foulon CF, Alston K, Bigner DD, Zalutsky MR. Astatine-211 labeling of internalizing anti-EGFRvIII monoclonal antibody using N-succinimidyl 5-[[211]At]astato-3pyridinecarboxylate. *Nucl Med Biol* 1999;26:405–411.

91. Vaidyanathan G, Affleck D, Welsh P, Srinivasan A, Schmidt M, Zalutsky MR. Radio-iodination and astatination of octreotide by conjugation labeling. *Nucl Med Biol* 2000; 27:329–337.

92. Aurlien E, Larsen RH, Bruland OS. Demonstration of highly specific toxicity of the α-emitting radioimmunoconjugate [211]At-rituximab against non-Hodgkin's lymphoma cells. *Br J Cancer* 2000;83:1375–1379.

93. Liu S, Edwards DS, Barrett JA. [99m]Tc labeling of highly potent small peptides. *Bioconjug Chem* 1997;8:621–636.

94. Reilly RM. Immunoscintigraphy of tumours using [99]Tc[m]-labelled monoclonal antibodies: a review. *Nucl Med Commun* 1993;14:347–359.

95. Schwartz A, Steinstrasser A. A novel approach to Tc-99m labeled monoclonal antibodies. *J Nucl Med* 1987;28:721.

96. Mather SJ, Ellison D. Reduction-mediated technetium-99m labeling of monoclonal antibodies. *J Nucl Med* 1990;31:692–697.

97. Stopar TG, Milinaric-Rascan I, Fettich J, Hojker S, Mather SJ. ^{99m}Tc-rituximab radiolabelled by photoactivation: a new non-Hodgkin's lymphoma imaging agent. *Eur J Nucl Med Mol Imaging* 2006;33:53–59.

98. Stalteri MA, Mather SJ. Technetium-99m labelling of the anti-tumour antibody PR1A3 by photoactivation. *Eur J Nucl Med* 1996;23:178–187.

99. Morales AA, Nunez-Gandolff G, Perez NP, Veliz BC, Caballero-Torres I, Duconge J, et al. Freeze-dried formulation for direct ^{99m}Tc-labeling ior-egr/r3 mab: additives, biodistribution, and stability. *Nucl Med Biol* 1999;26:717–723.

100. Ellman GL. A colorimetric method for determining low concentrations of mercaptans. *Arch Biochem Biophys* 1958;74:443–450.

101. Iznaga-Escobar N, Morales A, Nunez G. Micromethod for quantification of SH groups generated after reduction of monoclonal antibodies. *Nucl Med Biol* 1996;23: 641–644.

102. Marks A, Ballinger JR, Reilly RM, Law J, Baumal R. A novel anti-seminoma monoclonal antibody (M2A) labelled with technetium-99m: potential application for radioimmunoscintigraphy. *Br J Urol* 1995;75:225–229.

103. Steffens MG, Oosterwijk E, Kranenborg MHGC, Manders JMB, Debruyne FMJ, Corstens FHM, et al. *In vivo* and *in vitro* characterizations of three ^{99m}Tc-labeled monoclonal antibody G250 preparations. *J Nucl Med* 1999;40:829–836.

104. Vallis KA, Reilly RM, Chen P, Oza A, Hendler A, Cameron R, et al. A phase I study of ^{99m}Tc-hR3 (DiaCIM), a humanized immunoconjugate directed towards the epidermal growth factor receptor. *Nucl Med Commun* 2002;23:1155–1164.

105. Torres LA, Perera A, Batista JF, Hernandez A, Crombet T, Ramos M, et al. Phase I/II clinical trial of the humanized anti-EGF-r monoclonal antibody h-R3 labelled with [99mTc] in patients with tumour of epithelial origin. *Nucl Med Commun* 2005; 26:1049–1057.

106. Fichna J, Janecka A. Synthesis of target-specific radiolabeled peptides for diagnostic imaging. *Bioconjug Chem* 2003;14:3–17.

107. Abrams MJ, Juweid M, tenKate CI, Schwartz DA, Hauser MM, Gaul FE, et al. Technetium-99m-human polyclonal IgG radiolabeled via the hydrazino nicotinamide derivative for imaging focal sites of infection in rats. *J Nucl Med* 1990;31:2022–2028.

108. Tang Y, Scollard D, Chen P, Wang J, Holloway C, Reilly RM. Imaging of HER2/neu expression in BT-474 human breast cancer xenografts in athymic mice using [99mTc]-HYNIC-trastuzumab (Herceptin) Fab fragments. *Nucl Med Commun* 2005; 26:427–432.

109. Ono M, Arano Y, Mukai T, Uehara T, Fujioka Y, Ogawa K, et al. Plasma protein binding of [99mTc]-labeled hydrazino nicotinamide derivatized polypeptides and peptides. *Nucl Med Biol* 2001;28:155–564.

110. Ultee ME, Bridger GJ, Abrams MJ, Longley CB, Burton CA, Larsen SK, et al. Tumor imaging with technetium-99m-labeled hydrazinenicotinamide-Fab' conjugates. *J Nucl Med* 1997;38:133–138.

111. Bangard M, Behe M, Guhike S, Otte R, Bender H, Maecke HR, et al. Detection of somatostatin receptor-positive tumours using the new [99mTc]-tricine-HYNIC-D-Phe[1]-Tyr[3]-octrotide: first results in patients and comparison with [111In]-DTPA-D-Phe[1]-octreotide. *Eur J Nucl Med* 2000;27:628–637.

112. Decristoforo C, Mather SJ, Cholewinski W, Donnemiller E, Riccabona G, Moncayo R. [99mTc]-EDDA/HYNIC-TOC: a new [99mTc]-labelled radiopharmaceutical for imaging somatostatin receptor-positive tumours: first clinical results and intra-patient comparison with [111In]-labelled octreotide derivatives. *Eur J Nucl Med* 2000;27:1318–1325.

113. Decristoforo C, Melendez-Alafort L, Sosabowski JK, Mather SJ. [99mTc]-HYNIC-[Tyr[3]]-octrotide for imaging somatostatin-receptor-positive tumors: preclinical evaluation and comparison with [111In]-octreotide. *J Nucl Med* 2000;41:1114–1119.

114. Cornelissen B, Kersemans V, Burvenich I, Oltenfreiter R, Vanderheyden JL, Boerman O, et al. Synthesis, biodistribution and effects of farnesyltransferase inhibitor therapy on tumour uptake in mice of [99mTc] labelled epidermal growth factor. *Nucl Med Commun* 2005;26:147–153.

115. Babich JW, Fischman AJ. Effect of "co-ligand" on the biodistribution of [99mTc]-labeled hydrazino nicotinic acid derivatized chemotactic peptides. *Nucl Med Biol* 1995;22:25–30.

116. Decristoforo C, Mather SJ. Technetium-99m somatostatin analogues: effect of labelling methods and peptide sequence. *Eur J Nucl Med* 1999;26:869–876.

117. Decristoforo C, Mather SJ. 99m-Technetium-labelled peptide-HYNIC conjugates: effects of lipophilicity and stability on biodistribution. *Nucl Med Biol* 1999; 26:389–396.

118. Decristoforo C, Cholewinski W, Donnemiller E, Riccabona G, Moncayo R, Mather SJ. Detection of somatostatin receptor-positive tumours using the new [99mTc]-tricine-HYNIC-D-Phe[1]-Tyr[3]-octreotide: first results in patients and comparison with [111In]-DTPA-D-Phe1-octreotide. *Eur J Nucl Med* 2000;27:1580.

119. Ono M, Arano Y, Mukai T, Fujioka Y, Ogawa K, Uehara T, et al. ^{99m}Tc-HYNIC-derivatized ternary ligand complexes for ^{99m}Tc-labeled polypeptides with low *in vivo* protein binding. *Nucl Med Biol* 2001;28(3): 215–224.

120. Tang Y, Wang J, Scollard DA, Mondal H, Holloway C, Kahn HJ, et al. Imaging of HER2/neu-positive BT-474 human breast cancer xenografts in athymic mice using ^{111}In-trastuzumab (Herceptin) Fab fragments. *Nucl Med Biol* 2005;32:51–58.

121. Sattelberger AP, Atcher RW. Nuclear medicine finds the right chemistry. *Nat Biotechnol* 1999;17:849–850.

122. Waibel R, Alberto R, Willuda J, Finnern R, Schibi R, et al. Stable one-step technetium-99m labeling of His-tagged recombinant proteins with a novel Tc(I)-carbonyl complex. *Nat Biotechnol* 1999;17:897–901.

123. Schibi R, Schwarzbach R, Alberto R, Ortner K, Schmalle H, Dumas C, et al. Steps toward high specific activity labeling of biomolecules for therapeutic application: preparation of precursor $[Re-188(H_2O)_3(CO)_3]^+$ and synthesis of tailor-made bifunctional ligand systems. *Bioconjug Chem* 2002;13:750–756.

124. Bartholoma M, Valliant J, Maresca KP, Babich J, Zubieta J. Single amino acid chelates (SAAC): a strategy for the design of technetium and rhenium radiopharmaceuticals. *Chem Commun* 2009;5:493–512.

125. Blok D, Feitsman HIJ, Kooy YMC, Welling MM, Ossendorp F, Vermeij P, et al. New chelation strategy allows for quick and clean ^{99m}Tc-labeling of synthetic peptides. *Nucl Med Biol* 2004;31:815–820.

126. Fritzberg AR, Abrams PG, Beamier PL, et al. Specific and stable labeling of antibodies with technetium-99m with a diamide diothiolate chelating agent. *Proc Natl Acad Sci USA* 1988;85:4025–4029.

127. Rogers BE, Parry JJ, Andrews R, Cordopatis P, Nock BA, Maina T. MicroPET imaging of gene transfer with a somatostatin receptor-based reporter gene and [94m]Tc-demotate. *J Nucl Med* 2005;46:1889–1897.

128. Griffiths GL, Goldenberg DM, Jones AL, Hansen HJ. Radiolabeling of monoclonal antibodies and fragments with technetium and rhenium. *Bioconjug Chem* 1992; 3:99.

129. Griffiths GL, Goldenberg DM, Knapp FF, Jr., Callahan AP, Chang CH, Hansen HJ. Direct radiolabeling of monoclonal antibodies with generator-produced rhenium-188 for radioimmunotherapy: labeling and animal biodistribution studies. *Cancer Res* 1991; 51:4594–4602.

130. Iznaga-Escobar N. Direct radiolabeling of monoclonal antibodies with rhenium-188 for radioimmunotherapy of solid tumors: a review of radiolabeling characteristics, quality control and *in vitro* stability studies. *Appl Radiat Isot* 2001;54:399–406.

131. John E, Thakur ML, DeFulvio J, McDevitt MR, Damjanov I. Rhenium-186-labeled monoclonal antibodies for radioimmunotherapy: preparation and evaluation. *J Nucl Med* 1993;34:260–267.

132. Rhodes BA, Lambert CR, Marek MJ, Knapp FF, Jr., Harvey EB. Re-188 labelled antibodies. *Appl Radiat Isot* 1996;47:7–14.

133. Reilly RM, Ng K, Polihronis J, Shpitz B, Ngai WM, Kirsh JC, et al. Immunoscintigraphy of human colon cancer xenografts in nude mice using a second-generation TAG-72 monoclonal antibody labelled with [99]Tc[m]. *Nucl Med Commun* 1994; 15:379–387.

134. Ferro Flores G, Torres-Garcia E, Garcia-Pedroza L, Arteage de Murphy C, Pedraza-Lopez M, Garnica-Garza H. An efficient, reproducible and fast preparation of ^{188}Re-anti-CD20 for the treatment of non-Hodgkin's lymphoma. *Nucl Med Commun* 2005;26:793–799.

135. Goldrosen MH, Biddle WC, Pancock J, Bakshi S, Vanderheyden JL, Fritzberg AR, et al. Biodistribution, pharmacokinetic, and imaging studies with ^{186}Re-labeled NR-LU-10 whole antibody in LS174T colonic tumor-bearing mice. *Cancer Res* 1990;24:7973–7978.

136. Visser GW, Gerretsen M, Herscheid JD, Snow GB, van Dongen G. Labeling of monoclonal antibodies with rhenium-186 using the MAG3 chelate for radioimmunotherapy of cancer: a technical protocol. *J Nucl Med* 1993;34:1953–1963.

137. Zamora PO, Gulhke S, Bender H, Diekmann D, Rhodes BA, Biersack HJ, et al. Experimental radiotherapy of receptor-positive human prostate adenocarcinoma with ^{188}Re-RC-160, a directly-radiolabeled somatostatin analogue. *Int J Cancer* 1996; 65:214–220.

138. Safavy A, Khazaeli MB, Safavy K, Mayo MS, Buchsbaum DJ. Biodistribution study of ^{188}Re-labeled trisuccin-HuCC49 and trisuccin-HuCC49ΔCH$_2$ conjugates in athymic nude mice bearing intraperitoneal colon cancer xenografts. *Clin Cancer Res* 1999;5 (Suppl): 2994s–3000s.

139. Smith CJ, Sieckman GL, Owen NK, Hayes DL, Mazuru DG, Volkert WA, et al. Radiochemical investigations of [^{188}Re(H$_2$O)(CO)$_3$-diaminopropionic acid-SSS-bombesin(7-14)NH$_2$]: synthesis, radiolabeling and *in vitro/in vivo* GRP receptor targeting studies. *Anticancer Res* 2003;23:63–70.

140. Lubberink M, Tolmachev V, Widström C, Bruskin A, Lundqvist H, Westlin J-E. ^{110m}In-DTPA-D-Phe^1-octreotide for imaging of neuoendocrine tumors using PET. *J Nucl Med* 2002;43:1391–1397.

141. Tolmachev V, Bernhardt P, Forssell-Aronsson E, Lundqvist H. ^{114m}In, a candidate for radionuclide therapy: low-energy cyclotron production and labeling of DTPA-D-Phe^1-octreotide. *Nucl Med Biol* 2000;27:183–188.

142. Hnatowich DJ, Childs RL, Lanteigne D, Najafi A. The preparation of DTPA-coupled antibodies radiolabeled with metallic radionuclides: an improved method. *J Immunol Meth* 1983;65:147–157.

143. Krejcarek GE, Tucker KL. Covalent attachment of chelating groups to macromolecules. *Biochem Biophys Res Commun* 1977;77:581–585.

144. Brechbiel MW, Gansow OA, Atcher RW, Schlom J, Esteban J, Simpson DE, et al. Synthesis of 1-(p-isothiocyanatobenzyl) derivatives of DTPA and EDTA. Antibody labeling and tumor-imaging studies. *Inorg Chem* 1986;25:2772–2781.

145. Westerberg DA, Carney PL, Rogers PE, Kline SJ, Johnson DK. Synthesis of novel bifunctional chelators and their use in preparing monoclonal antibody conjugates for tumor targeting. *J Med Chem* 1988;32:236–243.

146. Reilly RM, Scollard DA, Wang J, Mondal H, Chen P, Henderson LA, et al. A kit formulated under good manufacturing practices for labeling human epidermal growth factor with ^{111}In for radiotherapeutic applications. *J Nucl Med* 2004;45:701–708.

147. Zoghbi SS, Neumann RD, Gottschalk A. The ultrapurification of indium-111 for radiotracer studies. *Invest Radiol* 1986;21:710–713.

148. Reilly R, Sheldon K, Marks A, Houle S. Labelling of monoclonal antibodies 10B, 8C, and M2A with indium-111. *Int J Appl Radiat Isot* 1989;40:279–283.

149. Carney PL, Rogers PE, Johnson DK. Dual isotope study of iodine-125 and indium-111-labeled antibody in athymic mice. *J Nucl Med* 1989;30:374–384.

150. McLarty K, Cornelissen B, Scollard DA, Done SJ, Chun K, Reilly RM. Associations between the uptake of ^{111}In-DTPA-trastuzumab, HER2 density and response to trastuzumab (Herceptin) in athymic mice bearing subcutaneous human tumour xenografts. *Eur J Nucl Med Mol Imaging* 2009;36:81–93.

151. Hnatowich DJ, Griffin TW, Kosciuczyk C, Rusckowski M, Childs RL, Mattis JA, et al. Pharmacokinetics of an indium-111-labeled monoclonal antibody in cancer patients. *J Nucl Med* 1985;26:849–858.

152. Reilly RM, Marks A, Law J, Lee NS, Houle S. *In-vitro* stability of EDTA and DTPA immunoconjugates of monoclonal antibody 2G3 labelled with In-111. *Appl Radiat Isot* 1992;43:961–967.

153. Goodwin DA, Meares CF, McTigue M, McCall MJ, Chaovapong W. Metal decomposition rates of ^{111}In-DTPA and EDTA conjugates of monoclonal antibodies *in vivo*. *Nucl Med Commun* 1986;7:831–838.

154. Reilly RM, Maiti PK, Kiarash R, Prashar AK, Fast DG, Entwistle J, et al. Rapid imaging of human melanoma xenografts using an scFv fragment of the human monoclonal antibody H11 labelled with [111]In. *Nucl Med Commun* 2001;22(5): 587–595.

155. Orlova A, Briskin A, Sjöström A, Lundqvist H, Gedda L, Tolmachev V. Cellular Processing of [125]I- and ^{111}In-labeled epidermal growth factor (EGF) bound to cultured A431 tumor cells. *Nucl Med Biol* 2000;27:827–835.

156. Bakker WH, Albert R, Bruns C, et al. [^{111}In-DTPA-D-Phe[1]]-octreotide, a potential radiopharmaceutical for imaging of somatostatin receptor-positive tumors: synthesis, radiolabeling and *in-vitro* validation. *Life Sci* 1991;49:1583–1591.

157. Cole WC, DeNardo SJ, Meares CF, McCall MJ, DeNardo GL, Epstein AL, et al. Comparative serum stability of radiochelates for antibody radiopharmaceuticals. *J Nucl Med* 1987;28:83–90.

158. Deshpande SV, Subramanian R, McCall MJ, DeNardo SJ, DeNardo GL, Meares CF. Metabolism of indium chelates attached to monoclonal antibody: minimal transchelation of indium from benzyl-EDTA chelate *in vivo*. *J Nucl Med* 1990;31:218–224.

159. Blend MJ, Greager JA, Atcher RW, Brown JM, Brechbiel MW, Gansow OA, et al. Improved sarcoma imaging and reduced hepatic activity with indium-111-SCN-Bz-DTPA linked to MoAb 19-24. *J Nucl Med* 1988;29:1810–1816.

160. Reilly R. The immunoreactivity of radiolabeled antibodies: its impact on tumor targeting and strategies for preservation. *Cancer Biother Radiopharm* 2005;19:669–672.

161. Rodwell JD, Alvarez VL, Lee C, Lopes AD, Goers JWF, King HD, et al. Site-specific covalent modification of monoclonal antibodies: in vitro and in vivo evaluation. *Proc Natl Acad Sci USA* 1986;83:2632–2636.

162. Abraham R, Moller D, Gabel D, Senter P, Hellström I, Hellström KE. The influence of periodate oxidation on monoclonal antibody avidity and immunoreactivity. *J Immunol Meth* 1991;144:77–86.

163. King TP, Zhao SW, Lam T. Preparation of protein conjugates via intermolecular hydrazone linkage. *Biochemistry* 1986;25(19); 5774–5779.

164. Chan C, Sandhu J, Guha A, Scollard DA, Wang J, Chen P, et al. A human transferrin-vascular endothelial growth factor (hnTf-VEGF) fusion protein containing an intergrated binding site for ^{111}In for imaging tumor angiogenesis. *J Nucl Med* 2005;46:1745–1752.

165. Remy S, Reilly RM, Sheldon K, Gariépy J. A new radioligand for the epidermal growth-factor receptor - In-111 labeled human epidermal growth-factor derivatized with a bifunctional metal-chelating peptide. *Bioconj Chem* 1995;6(6): 683–690.

166. Wang J, Chen P, Su ZF, Vallis K, Sandhu J, Cameron R, et al. Amplified delivery of indium-111 to EGFR-positive human breast cancer cells. *Nucl Med Biol* 2001;28(8): 895–902.

167. Manabe Y, Longley C, Furnanski P. High-level conjugation of chelating agents onto immunoglobulins: use of an intermediary poly(L-lysine)-diethylenetriaminepentaacetic acid carrier. *Biochim Biophys Acta* 1986;883:460.

168. Kobayashi H, Sato N, Saga T, Nakamoto Y, Ishimori T, Toyama S, et al. Monoclonal antibody-dendrimer conjugates enable radiolabeling of antibody with markedly high specific activity with minimal loss of immunoreactivity. *Eur J Nucl Med* 2000; 27:1334–1339.

169. Wen X, Wu Q-P, Lu Y, Fan Z, Charnsangavej C, Wallace S, et al. Poly(ethylene glycol)-conjugated anti-EGF receptor antibody C225 with radiometal chelator attached to the termini of polymer chains. *Bioconjug Chem* 2001;12:545–553.

170. Wen X, Wu Q-P, Ke S, Ellis L, Charnsangavej C, Delpassand AS, et al. Conjugation with ^{111}In-DTPA-poly(ethylene glycol) improves imaging of anti-EGF receptor antibody C225. *J Nucl Med* 2001;42:1530–1537.

171. Duncan JR, Welch MJ. Intracellular metabolism of indium-111-DTPA labeled receptor targeted proteins. *J Nucl Med* 1993;34:1728–1738.

172. Rogers BE, Franano FN, Duncan JR, Edwards WB, Anderson CJ, Connett JM, et al. Identification of metabolites of ^{111}In-diethylenetriaminepentaacetic acid-monoclonal antibodies and antibody fragments *in vivo*. *Cancer Res* 1995;55:5714s–5720s.

173. Akizawa H, Arano Y, Uezono T, Ono M, Fujioka Y, Uehara T, et al. Renal metabolism of ^{111}In-DTPA-D-Phe[1]-octreotide *in vivo*. *Bioconjug Chem* 1998;9:662–670.

174. Bass LA, Lanahan MV, Duncan JR, Erion JL, Srinivasan A, Schmidt MA, et al. Identification of the soluble in vivo metabolites of indium-111-diethylenetriaminepen-taacetic acid-D-Phe[1]-octreotide. *Bioconjug Chem* 1998;9:192–200.

175. Lewis MR, Raubitschek A, Shively JE. A facile, water-soluble method for modification of proteins with DOTA. Use of elevated temperature and optimized pH to achieve high specific activity and high chelate stability in radiolabeled immunoconjugates. *Bioconjug Chem* 1994;5:565–576.

176. de Jong M, Bakker WH, Krenning EP, Breeman WAP, van der Pluijm ME, Bernard BF, et al. Yttrium-90 and indium-111 labelling, receptor binding and biodistribution of [DOTA[0],D-Phe[1],Tyr[3]]octreotide, a promising somatostatin analogue for radionuclide therapy. *Eur J Nucl Med* 1997;24:368–371.

177. Williams LE, Lewis MR, Bebb GC, Clarke KG, Odom-Maryon TL, Shively JE, et al. Biodistribution of [111]In- and ^{90}Y-labeled DOTA and maleimidocysteinamido-DOTA ocnjugated to chimeric anticarcinoembryonic antigen antibody in xenograft-bearing nude mice: comparison of stable and chemically labile linker systems. *Bioconjug Chem* 1998;9:87–93.

178. Tsai SW, Li L, Williams LE, Anderson AL, Raubitschek AA, Shively JE. Metabolism and renal clearance of ^{111}In-labeled DOTA-conjugated antibody fragments. *Bioconjug Chem* 2001;12:264–270.

179. Li L, Olafsen T, Anderson AL, Wu A, Raubitschek AA, Shively JE. Reduction of kidney uptake in radiometal labeled peptide linkers conjugated to recombinant antibody

fragments. Site-specific conjugation of DOTA-peptides to a Cys-diabody. *Bioconjug Chem* 2002;13:985–995.

180. Cremonesi M, Ferrari M, Zoboli S, Chinol M, Stabin MG, Orsi F, et al. Biokinetics and dosimetry in patients administered with ^{111}In-DOTA-Tyr^3-octreotide: implications for internal radiotherapy with ^{90}Y-DOTATOC. *Eur J Nucl Med* 1999; 26:877–886.

181. Onthank DC, Liu S, Silva PJ, Barrett JA, Harris TD, Robinson SP, et al. ^{90}Y and ^{111}In complexes of a DOTA-conjugated integrin $\alpha_v\beta_3$ receptor antagonist: different but biologically equivalent. *Bioconjug Chem* 2004;15:235–241.

182. Yoo J, Tang L, Perkins TA, Rowland DJ, LaForest R, Lewis JS, et al. Preparation of high specific activity ^{86}Y using a small biomedical cyclotron. *Nucl Med Biol* 2005; 32:891–897.

183. Chinol M, Hnatowich DJ. Generator-produced yttrium-90 for radioimmunotherapy. *J Nucl Med* 1987;28:1465–1470.

184. Harwood SJ, Gibbons LK, Goldner PJ, Webster WB, Carroll RG. Outpatient radioimmunotherapy with Bexxar. *Cancer* 2002;94:1358–1362.

185. Garmestani K, Milenic DE, Plascjak PS, Brechbiel MW. A new and convenient method for purification of ^{86}Y using a Sr(II) selective resin and comparison of biodistribution of ^{86}Y and ^{111}In labeled Herceptin. *Nucl Med Biol* 2002;29:599–606.

186. Liu S, Edwards DS. Bifunctional chelators for therapeutic lanthanide radiopharmaceuticals. *Bioconjug Chem* 2001;12:7–34.

187. Hnatowich DJ, Virzi F, Doherty PW. DTPA-coupled antibodies labeled with yttrium-90. *J Nucl Med* 1985;26:503–509.

188. Brouwers AH, van Eerd JEM, Frielink C, Oosterwijk E, Oyen WJG, Corstens FHM, et al. Optimization of radioimmunotherapy of renal cell carcinoma: labeling of monoclonal antibody cG250 with ^{131}I, ^{90}Y, ^{177}Lu or ^{186}Re. *J Nucl Med* 2004;45:327–337.

189. Griffiths GL, Govindan SV, Sharkey RM, Fisher DR, Goldenberg DM. ^{90}Y-DOTA-hLL2: an agent for radioimmunotherapy of non-Hodgkin's lymphoma. *J Nucl Med* 2003;44:77–84.

190. Deshpande SV, DeNardo SJ, Kukis DL, Moi MK, McCall MJ, DeNardo GL, et al. Yttrium-90-labeled monoclonal antibody for therapy: labeling by a new macrocyclic bifunctional chelating agent. *J Nucl Med* 1990;31:473–479.

191. McCall MJ, Diril H, Meares CF. Simplified method for conjugating macrocyclic bifunctional chelating agents to antibodies via 2-iminothiolane. *Bioconjug Chem* 1990;1:222–226.

192. Lewis MR, Kao JY, Anderson A-LJ Shively JE, Raubitschek A. An improved method for conjugating monoclonal antibodies with *N*-hydroxysulfosuccinimidyl DOTA. *Bioconjug Chem* 2001;12:320–324.

193. Kukis DL, DeNardo SJ, DeNardo GL, O'Donnell RT, Meares CF. Optimized conditions for chelation of yttrium-90-DOTA immunoconjugates. *J Nucl Med* 1998;39:2105–2110.

194. Breeman WAP, de Jong M, Visser TJ, Erion JL, Krenning EP. Optimising conditions for radiolabelling of DOTA-peptides with ^{90}Y, ^{111}In and ^{177}Lu at high specific activities. *Eur J Nucl Med Mol Imaging* 2003;30:917–920.

195. Hainsworth JES, Mather SJ. Regressive DOTA labelling performance with indium-111 and yttrium-90 over a week of use. *Eur J Nucl Med Mol Imaging* 2005;32:1348.

196. Salako QA, O'Donnell RT, DeNardo SJ. Effects of radiolysis on yttrium-90-labeled Lym-1 antibody preparations. *J Nucl Med* 1998;39:667–670.

197. Liu S, Edwards DS. Stabilization of ^{90}Y-labeled DOTA-biomolecule conjugates using gentisic acid and ascorbic acid. *Bioconjug Chem* 2001;12:554–558.

198. Liu S, Ellars CE, Edwards DS. Ascorbic acid: useful as a buffer agent and radiolytic stabilizer for metalloradiopharmaceuticals. *Bioconjug Chem* 2003;14:1052–1056.

199. Kosmas C, Snook D, Gooden CS, Courtenay-Luck NS, McCall MJ, Meares CF, et al. Development of humoral immune responses against a macrocyclic chelating agent (DOTA) in cancer patients receiving radioimmunoconjugates for imaging and therapy. *Cancer Res* 1992;52:904–911.

200. Watanabe N, Goodwin DA, Meares CF, McTigue M, Chaovapong W, Ransone CM, et al. Immmunogenicity in rabbits and mice of an antibody–chelate conjugate: comparison of (*S*) and (*R*) macrocyclic enantiomers and an acyclic chelating agent. *Cancer Res* 1994;54:1049–1054.

201. DeNardo GL, Mirick GR, Kroger LA, O'Donnell RT, Meares CF, DeNardo SJ. Antibody responses to macrocycles in lymphoma. *J Nucl Med* 1996;37:451–456.

202. Perico ME, Chinol M, Nacca A, Luison E, Paganelli G, Canevari S. The humoral immune response to macrocyclic chelating agent DOTA depends on the carrier molecule. *J Nucl Med* 2001;42:1697–1703.

203. Saha GB. Diagnostic uses of radiopharmaceuticals in nuclear medicine. *Fundamentals of Nuclear Pharmacy*, 5th edition, Springer, New York, 2004, pp. 247–329.

204. Michel RB, Brechbiel MW, Mattes MJ. A comparison of 4 radionuclides conjugated to antibodies for single-cell kill. *J Nucl Med* 2003;44:632–640.

205. Maecke HR, Hofmann M, Haberkorn U. ^{68}Ga-labeled peptides in tumor imaging. *J Nucl Med* 2005;46:172S–178S.

206. Schumacher J, Maier-Borst W. A new ^{68}Ge/^{68}Ga radioisotope generator system for production of ^{68}Ga in dilute HCl. *Int J Appl Radiat Isot* 1981;32:31–36.

207. Ugur O, Kothari PJ, Finn RD, Zanzonico P, Ruan S, Guenther H, et al. Ga-66 labeled somatostatin analogue DOTA-D-Phe1-Tyr3-octreotide as a potential agent for positron emission tomography imaging and receptor mediated internal radiotherapy of somatostatin receptor positive tumors. *Nucl Med Biol* 2002;29:147–157.

208. Smith-Jones PM, Stolz B, Bruns C, Albert R, Reist HW, Fridrich R, et al. Gallium-67/gallium-68-[DFO]-octreotide-a potential radiopharmaceutical for PET imaging of somatostatin receptor-positive tumors: synthesis and radiolabeling *in vitro* and preliminary *in vivo* studies. *J Nucl Med* 1994;35:317–325.

209. Breeman WAP, de Jong M, de Blois E, Bernhard BF, Konijnenberg M, Krenning EP. Radiolabelling DOTA-peptides with ^{68}Ga. *Eur J Nucl Med Mol Imaging* 2005; 32:478–485.

210. Henze M, Schuhmacher J, Hipp P, Kowalski J, Becker DW, Doll J, et al. PET imaging of somatostatin receptors using [^{68}Ga]DOTA-D-Phe^1-Tyr^3-octreotide: first results in patients with meningiomas. *J Nucl Med* 2001;42:1053–1056.

211. Velikyan I, Beyer GL, Langström B. Microwave-supported preparation of ^{68}Ga bioconjugates with high specific activity. *Bioconjug Chem* 2004;15:554–560.

212. Reubi JC, Schar J-C, Waser B, Wenger S, Heppeler A, Schmitt JS, et al. Affinity profiles for human somatostatin receptor subtypes SST1-SST5 of somatostatin radiotracers selected for scintigraphic and radiotherapeutic use. *Eur J Nucl Med* 2000;27:273–282.

213. Hofmann M, Maecke H, Börner AR, Weckesser E, Schöffski P, Oei ML, et al. Biokinetics and imaging with the somatostatin receptor PET radioligand ^{68}Ga-DOTATOC: preliminary data. *Eur J Nucl Med* 2001;28:1751–1757.

214. Wild D, Macke HR, Waser B, Reubi JC, Ginj M, Rasch H, et al. ^{68}Ga-DOTANOC: a first compound for PET imaging with high affinity for somatostatin receptor subtypes 2 and 5. *Eur J Nucl Med Mol Imaging* 2005;32:724.

215. Froideveaux S, Calame-Christe M, Schumacher J, Tanner H, Saffrich R, Henze M, et al. A gallium-labeled DOTA-α-melanocyte-stimulating hormone analog for PET imaging of melanoma metastases. *J Nucl Med* 2004;45:116–123.

216. Velikyan I, Sundberg ÄL Lindhe Ö, Höglund AU, Eriksson O, Werner E, et al. Preparation and evaluation of ^{68}Ga-DOTA-hEGF for visualization of EGFR expression in malignant tumors. *J Nucl Med* 2005;46:1881–1888.

217. Blower PJ, Lewis JS, Zweit J. Copper radionuclides and radiopharmaceuticals in nuclear medicine. *Nucl Med Biol* 1996;23:957–980.

218. Anderson CJ, Jones AL, Bass LA, Sherman EL, McCarthy DW, Cutler PD, et al. Radiotherapy, toxicity and dosimetry of copper-64-TETA-octreotide in tumor-bearing rats. *J Nucl Med* 1998;39:1944–1951.

219. Connett JM, Anderson CJ, Guo LY, Schwarz SW, Zinn KR, Rogers BE, et al. Radioimmunotherapy with a ^{64}Cu-labeled monoclonal antibody: a comparison with [67]Cu. *Proc Natl Acad Sci USA* 1996;93:6814–6818.

220. McCarthy DW, Shefer RE, Klinkowstein RE, Bass LA, Margeneau WH, Cutler CS, et al. Efficient production of high specific activity ^{64}Cu using a biomedical cyclotron. *Nucl Med Biol* 1997;24:35–43.

221. Obata A, Kasamatsu S, McCarthy DW, Welch MJ, Saji H, Yonekura Y, et al. Production of therapeutic quantities of ^{64}Cu using a 12 MeV cyclotron. *Nucl Med Biol* 2003;30:535–539.

222. Williams HA, Robinson S, Julyan P, Zweit J, Hastings D. A comparison of PET imaging characteristics of various copper radioisotopes. *Eur J Nucl Med Mol Imaging* 2005; 32:1473–1480.

223. McCarthy DW, Bass LA, Cutler PD, Shefer RE, Klinkowstein RE, Herrero P, et al. High purity production and potential applications of copper-60 and copper-61. *Nucl Med Biol* 1999;26:351–358.

224. Novak-Hofer I, Schubiger PA. Copper-67 as a therapeutic nuclide for radioimmunotherapy. *Eur J Nucl Med* 2002;29:821–830.

225. Cole WC, DeNardo SJ, Meares CF, McCall MJ, DeNardo GL, Epstein AL, et al. Serum stability of [67]Cu chelates: comparison with ^{111}In and [57]Co. *Nucl Med Biol* 1986; 13:363–368.

226. Anderson CJ, Schwarz SW, Connett JM, Cutler PD, Guo LY, Germain CJ, et al. Preparation, biodistribution and dosimetry of copper-64-labeled anti-colorectal carcinoma monoclonal antibody fragments 1A3-F(ab')₂. *J Nucl Med* 1995;36:850–858.

227. Lewis JS, Srinivasan A, Schmidt MA, Anderson CJ. *In vitro* and *in vivo* evaluation of ^{64}Cu-TETA-Tyr³-octreotate. A new somatostatin analog with improved target tissue uptake. *Nucl Med Biol* 1999;26:267–273.

228. Li WP, Lewis JS, Kim J, Bugaj JE, Johnson MA, Erion JL, et al. DOTA-D-Tyr¹-octreotate: a somatostatin analogue for labeling with metal and halogen radionuclides for cancer imaging and therapy. *Bioconjug Chem* 2002;13:721–728.

229. O'Donnell RT, DeNardo GL, Kukis DL, Lamborn KR, Shen S, Yuan A, et al. A clinical trial of radioimmunotherapy with [67]Cu-2-IT-BAT-Lym-1 for non-Hodgkin's lymphoma. *J Nucl Med* 1999;40:2014–2020.

230. Philpott GW, Schwarz SW, Anderson CJ, Dehdashti F, Connett JM, Zinn KR, et al. RadioimmunoPET: detection of colorectal carcinoma with positron-emitting copper-64-labeled monoclonal antibody. *J Nucl Med* 1995;36:1818–1824.

231. Anderson CJ, Dehdashti F, Cutler PD, Schwarz SW, LaForest R, Bass LA, et al. ^{64}Cu-TETA-octreotide as a PET imaging agent for patients with neuroendocrine tumors. *J Nucl Med* 2001;42:213–221.

232. Anderson CJ, Pajeau TS, Edwards WB, Sherman EL, Rogers BE, Welch MJ. *In vitro* and *in vivo* evaluation of copper-64-octreotide conjugates. *J Nucl Med* 1995;36:2315–2325.

233. Chen X, Hou Y, Tohme M, Park R, Khankaldyyan V, Gonzales-Gomez I, et al. Pegylated Arg-Gly-Asp peptide: [64]Cu labeling and PET imaging of brain tumor $\alpha_v\beta_3$-integrin expression. *J Nucl Med* 2004;45:1776–1783.

234. Thakur ML, Aruva MR, Gariépy J, Acton P, Rattan S, Prasad S, et al. PET imaging of oncogene overexpression using ^{64}Cu-vasoactive intestinal peptide (VIP) analog: comparison with ^{99m}Tc-VIP analog. *J Nucl Med* 2004;45:1381–1389.

235. Rogers BE, Bigott HM, McCarthy DW, Manna DD, Kim J, Sharp TL, et al. MicroPET imaging of a gastrin-releasing peptide receptor-positive tumor in a mouse model of human prostate cancer using a ^{64}Cu-labeled bombesin analogue. *Bioconjug Chem* 2003;14:756–763.

236. Rogers BE, Anderson CJ, Connett JM, Guo LW, Edwards WB, Sherman ELC, et al. Comparison of four bifunctional chelates for radiolabeling monoclonal antibodies with copper radioisotopes: biodistribution and metabolism. *Bioconjug Chem* 1996; 7:511–522.

237. Bass LA, Wang M, Welch MJ, Anderson CJ. *In vivo* transchelation of copper-64 from TETA-octreotide to superoxide dismutase in rat liver. *Bioconjug Chem* 2000;11:527–532.

238. Mirick GR, O'Donnell RT, DeNardo SJ, Shen S, Meares CF, DeNardo GL. Transfer of copper from a chelated [67]Cu-antibody conjugate to ceruloplasmin in lymphoma patients. *Nucl Med Biol* 1999;26:841–845.

239. Milenic DE, Garmestani K, Chappell LL, Dadachova E, Yordanov A, Ma D, et al. *In vivo* comparison of macrocyclic and acyclic ligands for radiolabeling of monoclonal antibodies with ^{177}Lu for radioimmunotherapeutic applications. *Nucl Med Biol* 2002; 29:431–442.

240. Mulligan T, Carrasquillo JA, Chung Y, Milenic DE, Schlom J, Feuerstein I, et al. Phase I study of intravenous ^{177}Lu-labeled CC49 murine monoclonal antibody in patients with advanced adenocarcinoma. *Clin Cancer Res* 1995;1:1447–1454.

241. Müller WA, Linzner U, Schäffer EH. Organ distribution studies of lutetium-177 in mouse. *Int J Nucl Med Biol* 1978;5:29–31.

242. Breeman WAP, van der Wansem K, Bernard BF, van Gameren A, Erion JL, Visser TJ, et al. The addition of DTPA to [^{177}Lu-DOTA^0, Tyr^3]octreotate prior to administration reduces rat skeleton uptake of radioactivity. *Eur J Nucl Med* 2003;30:315.

243. Alvarez RD, Partridge EE, Khazaeli MB, Plott G, Austin M, et al. Intraperitoneal radioimmunotherapy of ovarian cancer with ^{177}Lu-CC49: a phase I/II study. *Gynecol Oncol* 1997;65:94–101.

244. Chauhan SC, Jain M, Moore ED, Wittel UA, Li J, Gwilt PR, et al. Pharmacokinetics and biodistributon of ^{177}Lu-labeled multivalent single-chain Fv construct of the pancarcinoma monoclonal antibody CC49. *Eur J Nucl Med Mol Imaging* 2005;32: 264–273.

245. McDevitt MR, Dangshe M, Lai L, Simon J, Borchardt P, Frank KR, et al. Tumor therapy with targeted atomic nanogenerators. *Science* 2001;294:1537–1540.

246. Miederer M, McDevitt MR, Sgouros G, Kramer K, Cheung N-KV Scheinberg DA. Pharmacokinetics, dosimetry, and toxicity of the targetable atomic generator, [225]Ac-HuM195, in nonhuman primates. *J Nucl Med* 2004;45:129–137.

247. Milenic DE, Roselli M, Mirzadeh S, Pippin CG, Gansow OA, Colcher D, et al. *In vivo* evaluation of bismuth-labeled monoclonal antibody comparing DTPA-derived bifunctional chelates. *Cancer Biother Radiopharm* 2001;16:133–146.

248. Yao Z, Garmestani K, Wong KJ, Park LS, Dadachova E, Yordanov A, et al. Comparative cellular catabolism and retention of astatine-, bismuth-, and lead-radiolabeled internalizing monoclonal antibody. *J Nucl Med* 2001;42:1538–1544.

249. Ballangrud AM, Yang W-H, Palm S, Enmon R, Borchardt PE, Pellegrini V, et al. Alpha-particle emitting atomic generator (actinium-225)-labeled trastuzumab (Herceptin) targeting of breast cancer spheroids: efficacy versus HER2/neu expression. *Clin Cancer Res* 2004;10:4489–4497.

250. Horak E, Hartmann F, Garmestani K, Wu C, Brechbiel M, Gansow OA, et al. Radioimmunotherapy targeting of HER2/neu oncoprotein on ovarian tumor using lead-212-DOTA-AE1. *J Nucl Med* 1997;38:1944–1950.

251. Meares CF, McCall MJ, Rearden DT, Goodwin DA, Diamanti CI, McTigue M. Conjugation of antibodies with bifunctional chelating agents: isothiocyanate and bromoacetamide reagents: methods of analysis and subsequent addition of metal ions. *Anal Biochem* 1984;142:68.

252. Pippin CJ, Parker TA, McMurry TJ, Brechbiel MW. Spectrophotometric method for the determination of a bifunctional DTPA ligand in DTPA-monoclonal antibody conjugates. *Bioconjug Chem* 1992;3:342–5.

253. Hartikka M, Vihko P, Södervall M, Hakalahti L, Torniainen P, Vihko R. Radiolabelling of monoclonal antibodies: optimization of conjugation of DTPA to F(ab')$_2$-fragments and a novel measurement of the degree of conjugation using Eu(III) labelling. *Eur J Nucl Med* 1989;15:157–161.

254. Pham DT, Kaspersen FM, Bos ES. Electrophoretic method for the quantitative determination of a benzyl-DTPA ligand in DTPA monoclonal antibody conjugates. *Bioconjug Chem* 1995;6:313–315.

255. Dadachova E, Chappell LL, Brechbiel MW. Spectrophotometric method for determination of bifunctional macrocyclic ligands in macrocyclic ligand-protein conjugates. *Nucl Med Biol* 1999;26:977–982.

256. Breeman WAP, de jong MThM de Jong M, de Blois E, Bernard BF, de Jong M, et al. Reduction of skeletal accumulation of radioactivity by co-injection of DTPA in [^{90}Y-DOTA[0], Tyr[3]]octreotide solutions containing free [90]Y[3+]. *Nucl Med Biol* 2004; 31:821–824.

257. Lindmo T, Boven E, Cuttitta F, Fedorko J, Bunn PA Jr. Determination of the immunoreactive fraction of radiolabelled monoclonal antibodies by linear extrapolation of binding to infinite antigen excess. *J Immunol Meth* 1984;72:77–89.

258. Mardirossian G, Wu C, Rusckowski M, Hnatowich DJ. The stability of ^{99}Tcm directly labelled to an Fab' antibody via stannous ion and mercaptoethanol reduction. *Nucl Med Commun* 1992;13:503–512.

259. Chen P, Cameron R, Wang J, Vallis KA, Reilly RM. Antitumor effects and normal tissue toxicity of ^{111}In-labeled epidermal growth factor administered to athymic mice bearing epidermal growth factor receptor-positive human breast cancer xenografts. *J Nucl Med* 2003;44:1469–1478.

260. Sharkey RM, Karacay H, Chang CH, McBride WJ, Horak ID, Goldenberg DM. Improved therapy of non-Hodgkin's lymphoma xenografts using radionuclides pretargeted with a new anti-CD20 bispecific antibody. *Leukemia* 2005;19:1064–1069.

261. Michel RB, Rosario AV, Brechbiel MW, Jackson TJ, Goldenberg DM, Mattes MJ. Experimental therapy of disseminated B-cell lymphoma xenografts with ^{213}Bi-labeled anti-CD74. *Nucl Med Biol* 2003;30:715–723.

262. Bonnet D, Bhatia M, Wang JYC, Kapp U, Dick JE. Cytokine treatment or accessory cells are required to initiate engraftment of purified primitive human hematopoietic cells transplanted at limiting doses into NOD/SCID mice. *Bone Marrow Transplantation* 1999;23:203–209.

263. Iwata K, MacDonald LR, Hwang AB, et al. CT-SPECT for small animal imaging with A-SPECT. *J Nucl Med* 2002;43 (Suppl): 10P.

264. Lewis JS, Achilefu S, Garbow JR, LaForest R, Welch MJ. Small animal imaging: current technology and perspectives for oncological imaging. *Eur J Cancer* 2002;38:2173–2188.

265. Stabin MG, Sparks RB, Crowe E. OLINDA/EXM: the second-generation personal computer software for internal dose assessment in nuclear medicine. *J Nucl Med* 2005;46:1023–1027.

266. Reilly RM, Chen P, Wang J, Scollard D, Cameron R, Vallis KA. Preclinical pharmacokinetic, biodistribution, toxicology and dosimetry studies of ^{111}In-DTPA-hEGF: an Auger electron-emitting radiotherapeutic agent for EGFR-positive breast cancer. *J Nucl Med* 2006;47:1023–1031.

CHAPTER 3

The Design of Radiolabeled Peptides for Targeting Malignancies

LEONARD G. LUYT

3.1 INTRODUCTION

The use of peptide-based agents for selective delivery of radionuclides, cytotoxic entities, and other agents to malignant tumors has proven to be a highly effective targeting approach. This chapter will explore the relationship of peptide structure and the ability of such agents to target tumors with the necessary specificity. The practical aspects of peptide radiotherapeutic design, synthesis, pharmacokinetics, and other critical considerations will be discussed.

3.2 PEPTIDE TARGETS

The classical receptor targets for peptide-based radiopharmaceuticals are the 7-transmembrane (7TM) proteins. This family of receptors, often referred to as G-protein-coupled receptors (GPCRs), is the largest class of human receptors of pharmacological importance with an estimated 50% of clinically relevant drugs acting on this receptor class (1). Through recent advances in proteomics and genomics, several hundred GPCRs have been identified and a number of these have peptides as their endogenous ligands, providing the medicinal chemist with peptides as lead compounds from which to design targeting entities (2). The most prominent examples include peptide-based drugs and imaging agents for the somatostatin receptor and the bombesin receptor family. However, many GPCRs have known peptide or protein ligands and there is the suggestion that peptide-based agents could be designed for many receptors in this superfamily. A review by Tyndall et al. suggests that the commonality in this receptor class is the requirement of a turn conformation to be

Monoclonal Antibody and Peptide-Targeted Radiotherapy of Cancer, Edited by Raymond M. Reilly
Copyright © 2010 John Wiley & Sons, Inc.

TABLE 3.1   **Examples of G-Protein-Coupled Receptors Expressed in Human Cancer and of Possible Relevance for Peptide Radiotherapeutic Design**

Peptide	GPCR (Subtype)	Possible Cancer Target(s)	References
Bombesin/gastrin- releasing peptide	GRP-R (BB2)	Prostate	(4)
Cholecystokinin	CCK2-R	Medullary thyroid, small-cell lung	(5)
Glucagon-like peptide 1	GLP-1R	Insulinomas	(6)
α-Melanocyte-stimulating hormone	MC1R	Melanoma	(7)
Neurotensin	NTR1	Pancreatic, small-cell lung	(8, 9)
Neuropeptide Y	NPY1	Breast, Ewing sarcoma	(10, 11)
Somatostatin	SST2 and others	Neuroendocrine	(12)
Substance P	NK1R	Glioblastoma	(13)
Vasoactive intestinal peptide	VPAC1	Colorectal, breast, prostate	(14)

present in a ligand for affinity, proposing that ligand design using a turn scaffold could result in many additional peptide-based targeting entities (3).

Although GPCRs remain an important receptor family for cancer targeting, many other receptor families may also be of value. Considering the recent advances in peptide library screening, discussed later in this chapter, screening for receptor affinity of any protein target using large and entirely random libraries is now possible. While it is not feasible to discuss all potential cancer targets within the framework of this review, one could consider the most promising targets to be divided into two general classifications: tumor cell-surface receptors, which includes the GPCR family, examples of which are indicated in Table 3.1, and tumor vasculature targets.

The growth of solid tumors beyond a few cubic millimeters in size is dependent on adequate blood supply and thus angiogenesis is necessary for progression of the disease (15). The expression of many biochemical markers is indicative of angiogenesis including integrins, growth factors, proteases, and others (16). Cyclic-RGD (Arg-Gly-Asp) peptides have been shown to be effective vehicles for targeting $\alpha_V\beta_3$ integrins, which are associated with tumor angiogenesis, and many imaging probe candidates have been reported in the literature based on this RGD sequence. In addition, a peptidomimetic radiotherapeutic based on targeting $\alpha_V\beta_3$ integrin has been reported with a DOTA (1,4,7,10-tetraazacyclododecane-1,4,7,10-tetraacetic acid) chelator for complexing ^{86}Y (17, 18). Other integrins have also been proposed for tumor targeting, such as the epithelial-specific integrin $\alpha_V\beta_6$ (19).

3.3 PEPTIDES AS CANCER TARGETING AGENTS

Peptides as targeting vectors for radiotherapeutics provide a number of important advantages over other molecules, especially with respect to synthetic considerations and diversity. The synthesis of peptides is very well developed using automated solid-phase methodology allowing rapid preparation of peptide analogues in good yield. Modifying residues within a peptide sequence is also readily facilitated by this synthetic method providing libraries of potential cancer targeting derivatives. A vast collection of commercially available unnatural amino acids exists that can be readily incorporated into a sequence in order to generate chemical diversity. In addition, peptide-based compounds have low antigenicity, rapid access to tumors, good tumor penetration, and often have fast clearance from the body.

A serious limitation that prevents the widespread use of peptides as drug molecules is their poor oral bioavailability. Since peptide-based radiopharmaceuticals are typically administered intravenously, this is not a concern for the field of radiotherapeutics. A second significant problem with the use of peptide molecules *in vivo* is the poor metabolic stability of most natural (L-isomer amino acid) peptides. Fortunately, this can be overcome by using a variety of structural modifications known to be successful in preventing enzymatic degradation. The action of exopeptidases, enzymes that hydrolyze from the N- or C-termini of the peptide, may be avoided by structural modification of the termini. For example, the C-terminus may be synthesized as an amide and the N-terminus may be acetylated, thereby minimizing the action of the proteases. The action of endopeptidases, enzymes that hydrolyze amides within the peptide, may be prevented by modification of the specific amino acids recognized by the protease. For example, the amino acid preceding or following the point of hydrolysis can be replaced with a D-amino acid or unnatural amino acid, thereby rendering the site of hydrolysis unrecognizable to the protease. In this manner, a peptide with a very limited biological half-life can be converted into a modified peptide with minimal amino acid substitutions and retention of function, while significantly extending its stability in a biological environment. While conceptually simple, the art of converting natural peptide sequences to more stable entities is complex, as one must ensure that modifications to the structure do not render the molecule biologically inactive.

Another concern raised when discussing the administration of a peptide agent to a human, is the physiological response generated from the peptide if it is an agonist for the biological target. While this is a significant concern for a pharmaceutical peptide agent being administered in milligram quantities, it is typically less of a concern for a radiotherapeutic. For a radiotherapeutic that has been exhaustively purified (e.g., by HPLC), the final high specific activity product is obtained in picogram quantities. In almost all instances this incredibly low mass is unable to generate a noticeable physiological response. One reported exception is that of vasoactive intestinal peptide (VIP), where the administration of less than 1 µg quantity of HPLC purified ^{123}I labeled VIP was reported to have a biological effect with a transient drop in blood pressure (20). In contrast, a later report indicated no significant physiological effect for administration of the same compound, also HPLC purified (21). In the instances where a rigorous purification is not possible, as is often the case for larger peptide molecules where the chromatographic retention of the radiolabeled versus unlabeled material is nearly identical, then one is only able to separate peptide (labeled and unlabeled) from unbound radionuclides. The patient would then receive both the radiotherapeutic and the unlabeled starting peptide material, in most instances both of which have the potential to elicit a physiological response. Peptide radiolabeling is routinely carried out using between 0.05 and 0.5 mg of peptide starting material. Although still a very low mass of peptide, a physiological response is a possibility and needs to be carefully evaluated prior to clinical use. In the case of radiolabeled somatostatin analogues, the administration of 50 µg depreotide (Neospect) as a mixture of labeled and unlabeled peptide resulted in no adverse events (22). In contrast, the administration of 250 µg of ^{111}In-labeled human epidermal growth factor resulted in flushing and nausea in all patients (23).

In order to design effective targeting peptides, a number of features need to be considered: biological stability, strong target (receptor) affinity, target (receptor) specificity, and pharmacokinetic characteristics. These requirements lead to an iterative process being the ideal method for progressing from a "hit" peptide candidate to that of an effective targeting entity. The iterative steps typically consist of peptide design, synthesis, and biological evaluation. Initially, the biological evaluation should be a screening mechanism, whereby the rapid evaluation of many potential peptides is undertaken and provides data indicating in what direction the design process should proceed. This is often a competitive binding assay or similar, but could also be a cell-based assay. The emphasis at this stage is to create a molecule that has high affinity and specificity for the desired protein target, while having a construct that is likely to be stable *in vivo*. Common approaches to improving peptide stability include adding cyclic constraints and the addition of D-isomer or unnatural amino acids.

Once candidate molecules are discovered with high target affinity, *in vivo* evaluation can be initiated. It is becoming increasingly evident that results from screening do not necessarily translate into comparable *in vivo* results, although targeting ability in a molecule is a minimum starting point to ensure success. Evaluation of the radiolabeled peptide using an *in vivo* model then provides critical data regarding tumor uptake and pharmacokinetics. Clearance properties of the agent often need to be further refined by subsequent iterations of design, resynthesis, and further *in vivo* evaluation. Choosing appropriate *in vivo* models for evaluating radiolabeled peptides targeting tumors is of utmost importance for further translation to the clinical realm and is discussed in detail in Chapter 11.

Following this iterative design, synthesis, and evaluation protocol is absolutely necessary in order to develop a highly effective targeting agent for cancer. The literature has many examples of noniterative design, where an endogenous peptide is radiolabeled with little thought given to the effect of modifying the peptide structure or the anticipated biological stability. As is well documented for the design of therapeutic drugs, a linear, natural peptide does not typically yield an effective *in vivo* agent, although there are certainly many peptides currently in clinical use (24).

3.3.1 Discovery of Novel Cancer Binding Peptides

The discovery of cancer targeting peptides can be accomplished by either a rational, designed approach or by random, combinatorial methods (Fig. 3.1). The rational method is often initiated by starting with an endogenous peptide known to interact with

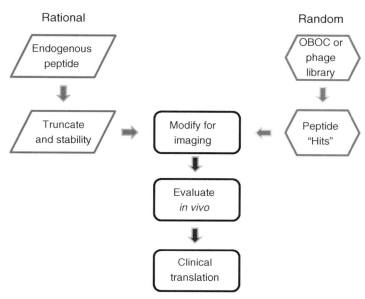

FIGURE 3.1   Flowchart showing rational and random approaches to peptide-based imaging probe discovery.

a receptor target. This starting point is typically a large peptide, which is not particularly suitable for *in vivo* use, as previously discussed. Standard peptide chemistry approaches are then used to determine the region of the peptide that is critical for receptor binding affinity. The first step is to carry out truncation of the peptide from both the N- and C-termini. This will minimize the length of amino acid sequence being used for designing the radiotherapeutic molecule and will determine if either terminal region of the peptide is involved in receptor binding. Once the minimal sequence length for biological activity is determined, then an alanine scan may be performed. This involves the systematic replacement of a single amino acid residue with L-alanine. Evaluation of the resultant alanine containing peptide indicates whether the amino acid in that particular position is important for receptor binding. A D-amino acid scan may also be undertaken, whereby a single amino acid is replaced by its complementary D-amino acid. This indicates the importance of conformation for that specific residue site. This methodology has been well described in the literature and will not be discussed in this chapter in detail.

There are many examples where the rational approach has been used successfully to develop an imaging probe for a peptide receptor target. The peptide hormone somatostatin, found in the body as both a cyclic 14-mer and 28-mer, was successfully truncated to an 8-mer (Octreotide) and even a 6-mer (MK-678) form (25). Systematic alanine and D-amino acid scans led to the determination that having a D-Trp in the peptide sequence actually increased its potency (26). From this fundamental knowledge of peptide structure–activity relationship, the design of radiolabeled somatostatin analogues became a reality leading to both diagnostic and therapeutic derivatives (see Chapter 4).

The advantages of solid-phase peptide synthesis are of particular benefit in proceeding through this type of rational design and synthesis. Due to the iterative nature of this process, the synthesis of many peptides is required and this is readily accomplished by the creation of focused libraries, where each well of an automated synthesizer prepares a single entity of known composition. With current technology, one can readily prepare 96 peptides or more at one time using this method.

Combinatorial peptide libraries enable the discovery of high-affinity binding ligands due to the vast number of entities that can be prepared. These libraries can be divided into three general classifications: focused libraries of discrete compounds, as described in the rational approach; one-bead one-compound (OBOC) libraries as a chemical approach; and phage-display libraries as a biological approach. The latter two methods generate large random libraries of compounds for evaluation against a tumor cell target (see Chapter 1). This can include receptor targets where the endogenous peptide ligand is not known.

The synthesis and screening of OBOC libraries has emerged as one of the most powerful techniques for exploring chemical diversity. First reported by the group of Kit Lam in 1991 (27), OBOC libraries have been utilized for the discovery of many cancer targeting peptides (28). Most well developed are random peptide libraries, although small-molecule libraries have also been explored. In brief, the technique involves the creation of an entirely random library on a polymer support using split-mix synthesis. This method starts by dividing the resin beads into a number of equal portions, with each portion having a single amino acid coupled to it. After reaction completion, all of the resin is recombined, thoroughly mixed and once again divided equally followed by coupling of the second amino acid. If one uses 19 amino acids (Cys is typically avoided due to the potential for disulfide formation) then after creating a 5-residue peptide library there is the potential for up to 2,476,099 individual sequences. The preferred polymer for this method is TentaGel, since this support allows for the screening of the library in an aqueous medium, a necessity for evaluation of protein–ligand interactions. The resin beads are typically 90 μm in diameter, which permits visual detection and manual or automated separation if necessary (Fig. 3.2). Each individual resin bead is covered with a single peptide sequence, while every other bead contains a different peptide sequence.

One of the particular advantages of OBOC methods is the ability to include nonnatural amino acids or even to create libraries based on small-molecule scaffolds. A significant limitation that exists for OBOC libraries is the ability to accurately and cost-effectively deconvolute the "hit" beads that are isolated. The initial and most commonly reported approach to deconvolute OBOC libraries utilizes automated Edman degradation as the deconvolution method. Although effective, there is limited throughput possible using this technique and if the deconvolution of hundreds of hit beads is required, this can become an expensive venture. Mass spectrometry (MS) has recently emerged as a cost and time effective means for determining OBOC hit identity (29). One method of doing this is to encode the beads with a mass tag on the interior of the resin bead, which is used to keep track of the structure being synthesized

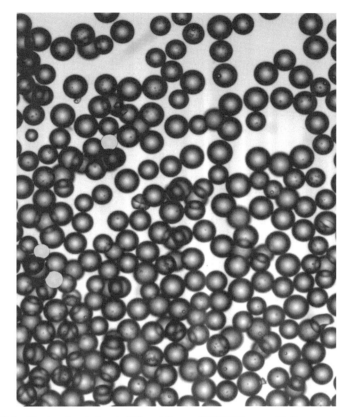

FIGURE 3.2   Tentagel resin beads (90 μm) from a OBOC library with fluorescence detection. (See insert for color representation of the figure.)

for subsequent deconvolution. This mass tag encodes the bead based on the sequence that it contains and is less likely to interact during screening due to being on the interior of the bead, while the exterior contains the actual entity being evaluated for target interaction (30, 31).

OBOC libraries have had success in generating lead peptides for the development of tumor targeting radiopeptides. Of course, since the library is providing only the targeting component, it is still necessary to further modify the structure determined from the OBOC screen in order to incorporate a radionuclide and in the case of a radiometal, a suitable chelator. A more efficient procedure would be to create an OBOC library that includes the radionuclide or more likely, a nonradioactive surrogate atom, that is, stable isotope of the corresponding radionuclide, within the library sequence. In this manner, one would be effectively screening a library of radiolabeled peptides instead of only screening the peptide receptor targeting moiety. This would be a significant step forward as hit sequences would represent the final radiolabeled peptide, likely resulting in a reduction in the degree of further modification required in order to arrive at a suitable agent for tumor targeting for radiotherapeutic or imaging applications.

Another way to screen OBOC libraries is to evaluate the ability of the bead to bind to cell-surface receptors of living cells (32, 33). Tumor cells are incubated with the OBOC beads with gentle agitation and beads that are coated with a monolayer of cells are considered hits. These positive beads are then removed and deconvoluted. One limitation of this approach is that the actual receptor target with which the peptide is interacting will need to be determined and this can be accomplished by using nontarget control cells or by blocking the suspected target with natural ligands.

A biological approach to creating random combinatorial peptide libraries is based on phage display and was first reported by George Smith in 1985 (34). This method uses a biochemical approach whereby phage, a virus that infects bacterial cells and most commonly being the filamentous bacteriophage M13, is used to present proteins or peptides on its surface after inserting an expression vector. A phage-display library is created by preparing a random mixture of phage clones with each displaying a single peptide on the surface. This library is then exposed to the protein target of interest, typically presented on a solid support. A washing step then removes any phage that did not bind to the target, followed by recovery of the bound phage by elution under nonspecific (all phage is removed) or specific (removes only phages bound to the target protein) methods. Amplification of the hit peptides allows for an adequate quantity of phage material to be produced and the encoding region of the phage is then sequenced.

One limiting aspect of the phage-display method, which is particularly relevant to the discovery of suitable *in vivo* tumor targeting entities, is that only natural amino acids will be present in the hit peptides. It is generally not possible to utilize D-isomer and unnatural amino acids, however, it is readily possible to create cysteine-cyclized peptide libraries using phage display, thereby adding a conformational constraint to the library (35, 36). There are currently no effective means of creating head-to-tail or other homodectic cyclic structures. There is a report of creating an entirely D-isomer peptide through a mirror image method (37–40), however, this does require the chemical synthesis of the D-enantiomer target protein, which limits the applicability of the concept. In this instance the phage library, expressing the random L-isomer peptides, will generate L-isomer peptides that bind to the D-isomer protein target. Thus, the enantiomer (D-isomer) peptide will interact with the natural L-isomer protein target.

3.3.2 The Addition of a Radionuclide to a Peptide

Historically, the radiolabeling of peptides and proteins for targeted radiotherapy has focused on radioiodination of available tyrosine residues. The therapeutic radionuclide ^{131}I has been most widely utilized. Although proven effective for therapy, radioiodinated peptides have limitations for both their synthesis and *in vivo* stability. In terms of synthesis, more than one tyrosine residue being present in the peptide results in a possible mixture of products, while if there are no tyrosine residues present, then a tyrosine residue needs to be added to the peptide structure. In addition iodoaromatics are typically not adequately stable *in vivo*, often resulting in deiodination (see Chapter 2) (41, 42). Methods have been developed to minimize the loss of the radioiodine, for example if an iodotyrosine is located at the C-terminus of the peptide, then N-terminal amide formation has been demonstrated to decrease the rate of deiodination (43). A method to avoid the synthetic issues of radioiodinated peptides is to use a prosthetic group approach, similar in concept to methods used for fluorine-18 peptide labeling (see Chapter 2). For example, radioiodinated *N*-succinimidyl iodobenzoate has been reported for conjugation to a peptide with an available amine, either on N-terminal or on a lysine residue (44, 45). Other indirect labeling methods have also been reported (46).

Many of the most promising therapeutic radionuclides, as discussed in Chapter 2, are radiometals (such as ^{67}Cu, ^{90}Y, ^{111}In, ^{177}Lu, or ^{188}Re) and require a chelator to be incorporated into the structure of the targeting peptide. The addition of the chelator may be accomplished in two ways, either in a conjugate (also referred to as pendant) design or through an integrated design. The vast majority of radiometal-labeled peptides utilize the conjugate design, as this is far simpler in terms of design and synthesis, but more importantly, tends to be more straightforward in creating peptide radiometal complexes that retain target binding affinity.

In the conjugate design, a bifunctional metal chelator (capable of coupling to a peptide and able to coordinate a radiometal) is directly incorporated into the peptide structure in a pendant fashion and subsequently the isotope is coordinated to the peptide–chelator entity. The intention in designing such molecules is to ensure that the metal complex is sufficiently removed from the receptor binding portion of the peptide, so that there is no negative effect on receptor–ligand interaction. This may even necessitate the addition of a spacer between the peptide-backbone and the isotope complex, in order to remove the metal from the pharmacophore region of the peptide. This can be especially prudent if the radiometal complex is charged. The use of a flexible spacer permits the radiometal complex to orient away from the receptor binding region. Many examples of adding a flexible linker exist in the literature and only a few will be mentioned here. For gastrin-releasing peptide (GRP) receptor peptides, Volkert and coworkers added an aliphatic chain between the bombesin peptide and a DOTA chelator, varying the length through the use of either β-Ala (β-alanine), 5-Ava (5-aminovaleric acid), 8-Aoc (8-aminooctanoic acid), or 11-Aun (11-aminoundecanoic acid) and determined that the optimal spacer length is 5–8 carbons in length (47). Rogers and coworkers also explored the use of linkers for radiolabeled bombesin, comparing various tripeptide sequences made up of Gly, Ser, and Glu for the spacer, concluding that Gly-Ser-Ser gave a compound with the best affinity for GRP receptors (48). Maecke and coworkers reported on a variety of spacers for DOTA-somatostatin analogues, including short polyethylene glycol (PEG) spacers, amino acid-based spacers, and even a sugar-modified amino acid (49).

In synthesizing a conjugated peptide, the chelator is most readily added to the N-terminus of the peptide. This step may even be included during the automated synthesis of the peptide, by simply having the chelator available with a free carboxylic acid and following standard peptide coupling methods. To place the chelator at other locations of the peptide, an orthogonal protecting group strategy must be employed. For example, a lysine residue may be inserted with a very acid labile ε-amine

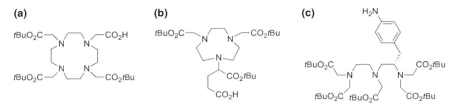

FIGURE 3.3 Commercially available bifunctional chelators: (a) DOTA-tris(*t*-Bu ester);(b) NODA-tris(*t*-Bu ester); (c) *p*-NH$_2$-Bn-DTPA-penta(*t*-Bu ester).

protecting group such as 4-methyltrityl (Mtt) or 1-(4,4-dimethyl-2,6-dioxo-cyclo-hexylidene)-3-methyl-butyl (ivDde) that can be deprotected selectively with hydrazine. Upon completion of the peptide sequence, this labile group is removed (leaving all other side-chain protecting groups intact) and the metal chelator is then coupled at that location. Similar strategies may be employed for unnatural amino acids with side-chain amines including diaminopropanoic acid and ornithine. One may also envision an orthogonally protected carboxylic acid as a side chain from aspartic or glutamic acid, in order to add a chelator containing a free amine. Nonnatural amino acids such as α-aminoadipic acid could also be used. A wide range of suitable chelators are now available commercially for coupling to a peptide sequence and are provided in a protected form to prevent any unwanted side reactions during the coupling procedure. For example, the commonly used DOTA chelator may be purchased with three of the carboxylic acids groups protected as *t*-butyl esters, while one carboxylic acid remains free for coupling to the peptide. Examples of commercially available, appropriately protected bifunctional chelators for peptide synthesis are indicated in Fig. 3.3.

In some instances a peptide sequence itself may be used as a chelation unit. This is especially the case for Re(V)–oxo complexes, where a carefully devised combination of amino, amido, and thiol moieties results in a tetradentate coordination sphere for the metal. This method was extensively developed in the instance of somatostatin analogues where both natural and nonnatural (e.g., diaminopropanoic acid) amino acids were inserted to create the peptide-backbone chelation unit, which upon coordination provided either negatively charged or neutral complexes (50).

A more convenient approach to incorporating a pendant chelator into a peptide is to add a modified amino acid that contains the chelator as part of the side chain, during the routine automated peptide synthesis. We first reported this concept as it relates to medically important radionuclides with the preparation of a modified lysine amino acid, where the ε-amine was functionalized with a N$_2$S$_2$ chelator suitable for Tc(V) or Re(V)–oxo complexes (51). This chelator was fully protected so that the amino acid may be incorporated at any residue position of a peptide (Fig. 3.4a). Since then, others have further elaborated on this concept and demonstrated its applicability for a number of radionuclide-containing peptide derivatives, including those containing radionuclides of Tc or Re, as well as a DOTA chelator for Y, In, Lu, and so on. For Tc and Re, a Fmoc-Lys(HYNIC-Boc)-OH amino acid was reported by Blower and coworkers for the use of the well-established hydrazinonicotinamide ligand (Fig. 3.4b) (52). Valliant and coworkers described a modified lysine with a tridentate

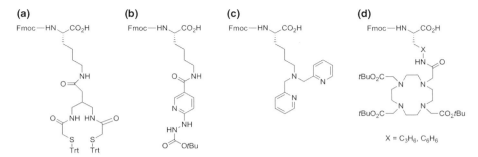

FIGURE 3.4 Modified amino acids incorporating a metal chelator and suitably protected for solid-phase peptide synthesis: (a) N_2S_2 chelator for Tc(V)/Re(V); (b) HYNIC; (c) bis-pyridyl chelator for Tc(I)/Re(I); (d) DOTA.

chelator appropriate for use with Tc and Re tricarbonyl species (Fig. 3.4c) (53). The commonly utilized chelator DOTA has been modified as a lysine analogue with the chelator attached to the ε-amine (54) as well as in the form of an unnatural aromatic amino acid (Fig. 3.4d) (55). The acronym SAAC (single amino acid chelator) has been used to describe these unnatural amino acid chelator constructs (53).

One of the significant disadvantages of adding a metal complex to a peptide sequence is that there is a large increase in the molecular weight of the final product. This will likely not be a concern for proteins, but for a small peptide of 5–10 residues, the metal complex can make up a quarter or more of the total mass. For improved behavior *in vivo* it is often assumed that maintaining a low molecular weight is advantageous. Evidence to support this includes a study comparing intact IgG versus Fab fragments of ^{99m}Tc-labeled anti-CEA monoclonal antibodies, where the antibody fragments demonstrated superior tumor imaging due to faster background clearance (56) (see Chapter 1). Small peptides, on the order of 5–12 residues in length, have repeatedly been shown to have better elimination properties than antibodies, even though the antibodies may have superior target binding affinity (57). Thus, peptides provide better tumor to background tissue ratios, which could potentially be of value in limiting the toxicity of a radiotherapeutic. Smaller engineered antibody fragments, such as diabodies and minibodies, may also achieve similar results by rapid elimination through the kidneys (58) (see Chapter 1).

An integrated design for radionuclide incorporation is one method that can circumvent this issue of increased molecular weight of the radiopharmaceutical due to the addition of a radionuclide–chelator complex. In this design, the metal complex becomes an integral part of the peptide and is able to be situated within the critical space of ligand–receptor interaction. Although an idealized concept, the integrated design has proven very difficult to exploit for peptide-based targeting entities. The best example for incorporating a potential therapeutic atom in an integrated fashion is in the realm of estrogen receptor binding compounds. Katzenellenbogen and coworkers reported a cyclopentadienyl-rhenium tricarbonyl complex where the metal is situated within the ligand binding domain of the estrogen receptor, while maintaining high estrogen receptor affinity (Fig. 3.5a) (59). We have recently demonstrated the ability to

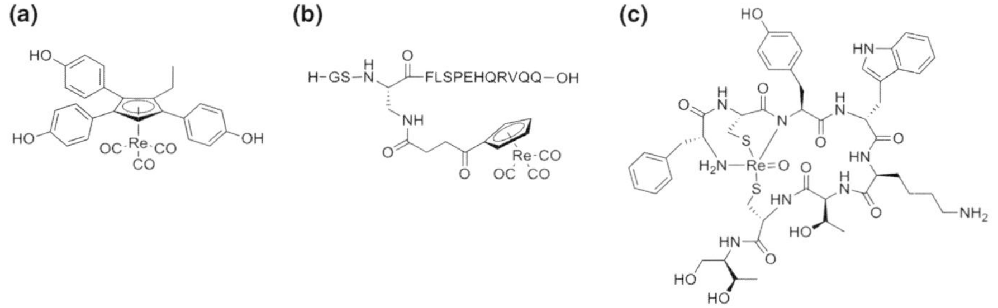

FIGURE 3.5 Rhenium complexes as integrated designs: (a) cyclopentadienyl-rhenium(I) tricarbonyl scaffold for the estrogen receptor; (b) 14-mer ghrelin analogue for the growth hormone secretagogue receptor; (c) rhenium-cyclized peptide for the somatostatin receptor.

include the same organometallic species as part of a required lipophilic section of the peptide ghrelin, targeting the growth hormone secretagogue receptor (60). In this instance, the lipophilic cyclpentadienyl-rhenium tricarbonyl core is able to function similarly to an aliphatic chain that is critical for receptor binding affinity (Fig. 3.5b). Rhenium has also been used to form cyclic peptide entities through the use of cysteine residues, where a thiolate-metal-thiolate bond creates the cyclic entity. Examples include an α-MSH (α-melanocyte-stimulating hormone) analogue (61), a GnRH (gonadotropin-releasing hormone) analogue (62), and a somatostatin derivative depicted in Fig. 3.5c (63). Although in these instances the radiometal is not necessarily present in the receptor binding pocket, it does provide a clever method of incorporating a metal complex into a required element of the peptide, in this instance through coordination assisted peptide cyclization.

One additional concern when the radiopharmaceutical design requires the addition of a metal radioisotope to a peptide is the possibility of creating isomers due to the newly formed metal complex. This is often the case for Re(V)–oxo complexes where an *anti* or *syn* configuration is possible with respect to the metal–oxo bond and the pendant bioconjugate. A number of methods to limit isomer formation have been reported. For example, we reported on a symmetrical chelator design, which limited the number of isomers possible and further discovered that under basic conditions a single isomer product was preferentially formed (64, 65). Identification of *syn/anti* isomers for amino acid-based chelators relevant for clinically used imaging agents have also been reported (66). Diastereomer formation is also a concern for some Re(I) tricarbonyl complexes where tridentate chelators upon complexation result in a mixture of isomers (67). Most cyclen type chelators result in single isomer metal complexes and therefore avoid this issue. However, the method used for conjugation to a peptide and the point of attachment to the cyclen unit, must be such that a single isomer product is maintained. It is certainly preferred if the formation of multiple isomers can be avoided, as this will simplify product characterization and regulatory approval of a radiotherapeutic (see Chapter 17).

3.3.3 Improving the *In Vivo* Behavior of Peptides

Structure–activity studies of a peptide sequence permit the development of peptide analogues with improved affinity and specificity for a receptor target. However, it is well established that the ability of a peptide to target a receptor alone is not sufficient for adequate uptake in a tumor *in vivo*. Modification of the peptide to improve pharmacokinetic behavior is almost always necessary and minor changes to the structure can have significant effects on biodistribution. A method that has proven useful for improving pharmacokinetic behavior of a radiolabeled small peptide is to add one or more saccharide units to the molecule. This can result in improved water solubility and a change in how the compound is cleared from the body. The cyclic RGD pentapeptide targeting various integrins is a good example of this approach, where the addition of a modified galactose to the peptide resulted in predominately renal excretion and low uptake in nontarget tissues (Fig. 3.6a) (68, 69). A glucosamine-modified somatostatin analogue has also been reported to improve the tumor to kidney ratio (Fig. 3.6b) (49). Pharmacokinetic properties can also be modified by replacing amino acid residues within a peptide sequence, when those residues are not required for receptor binding. The radiotherapeutic somatostatin analogue P2045, containing [188]Re, was designed with the addition of serine and threonine residues at the C-terminus (Fig. 3.6c), one factor that contributed to its optimized clearance properties (70).

One of the most significant toxicity issues of radiotherapeutics is elevated renal uptake and the retention of the isotope resulting in kidney damage (see Chapter 4) (71). The presence of lysine or other charged amines is believed to be associated with

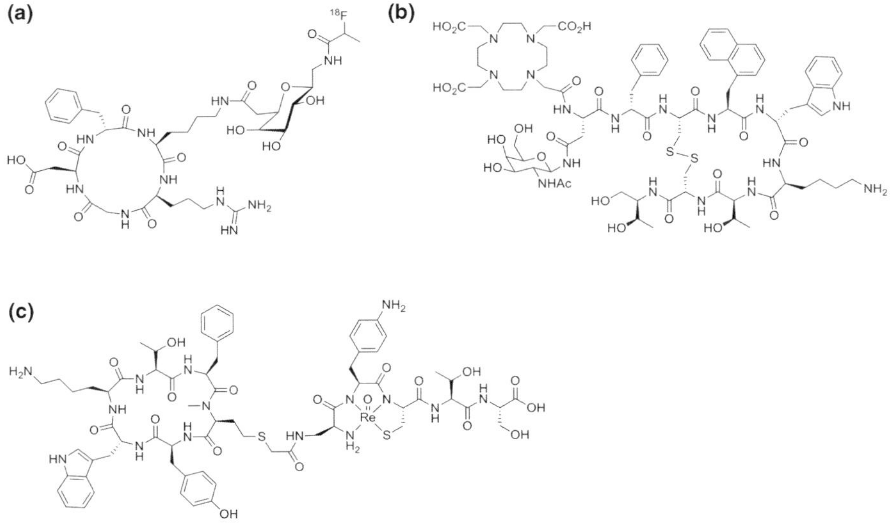

FIGURE 3.6 Modified peptides for improved *in vivo* behavior: (a) cyclic RGD peptide modified with galactose; (b) DOTA-somatostatin derivative containing *N*-acetyl glucosamine; (c) Re containing P2045 with a Thr-Ser C-terminus.

retention of the peptide in the proximal tubules of the kidneys and these residues should be limited or nonexistent in radiotherapeutic peptides except where necessary for receptor interactions (70, 72). Alternatively, one can acetylate or otherwise block the free ε-amine of lysine in order to reduce the effect of the lysine on kidney retention, as was demonstrated for DOTA-α-MSH (73). The substitution of a lysine with arginine proved to be effective in the instance of a ^{188}Re-labeled α-MSH derivative (74).

If efforts to reduce kidney retention through peptide structural modifications are not adequate, the coadministration of an agent to inhibit tubular reabsorption can be considered. In 1993, it was reported that lysine infusion in conjunction with an ^{111}In-somatostatin analogue was able to reduce the uptake of radioactivity in the kidneys (75). Since then, the coadministration of D- or L-isomers of lysine, arginine, amino-sugars, or the chemotherapy cytoprotective adjuvant amifostin have all proven beneficial for limiting the retention of radiometal-labeled peptides in the kidneys, resulting in a decrease in renal toxicity (74, 76).

In designing a tumor targeting peptide, the pharmacological requirements for the peptide–receptor interaction must be considered. While there is certainly no need for a biological response from the peptide–receptor interaction, it has been demonstrated that an agonist is often preferred in order to achieve the desired accumulation in tumors. Quantitative analysis of internalization using green fluorescent protein (GFP)-labeled receptors permits correlation of the EC_{50} of a ligand to its ability to internalize the receptor (77). The internalization of the radiotherapeutic into a tumor cell, by means of the peptide–receptor complex endocytosis and eventual receptor recycling to the cell surface, permits prolonged cellular retention of the radionuclide at the tumor with nonbound peptide clearing from the body (78). This was demonstrated with a somatostatin antagonist LS172 labeled with ^{64}Cu and with ^{177}Lu, both of which did not satisfactorily accumulate in tumors due to a lack of internalization (79). Radiolabeled somatostatin derivatives that are agonists are well established as being able to internalize and accumulate (80). A few exceptions have recently been reported, the first of which was an ^{111}In-labeled somatostatin antagonist that provided higher tumor uptake than an agonist, even though it did not elicit significant receptor internalization (81). Regardless, caution must be used in assuming the generality of an agonist requirement as it is likely to be receptor dependent. For example, a possible tumor target, the cholecystokinin receptor, was demonstrated to internalize from the binding of an antagonist (82).

3.4   MULTIMODALITY AGENTS

Considering the amount of effort required in order to develop a peptide targeting entity, it is becoming increasingly apparent that understanding the detailed biological mechanism of peptide ligand uptake in cells is of great value early in the development stage. For this, optical agents are very useful as they enable detailed evaluation at a much improved resolution compared to radionuclide-based imaging. One approach to accomplish this would be to create a dual-modality probe, where the peptide is conjugated both to a fluorescent dye and to a radionuclide or nonradioactive surrogate.

The potential radiotherapeutic drug may be thoroughly investigated at the microscopic level for its interactions with the cancer target using the fluorescence properties of the probe. However, the addition of both a dye and a radionuclide adds undesirable molecular weight to the entity. Valliant and coworkers proposed an elegant solution to this using a rhenium chelation system based on a diquinoline framework, whereby upon rhenium coordination the agent absorbs maximally at 301 nm and emits with peak fluorescence at 425 and 580 nm (83). They further demonstrated the ability to incorporate this fluorescent metal complex into a peptide targeting the formyl peptide receptor and showed uptake in human leukocytes using fluorescence microscopy.

3.5 FUTURE OUTLOOK

Peptide-based targeting has an excellent track record for its ability to develop entities that specifically interact with cell-surface receptors relevant to cancer, such as somatostatin, bombesin, and other receptor families. The tumor-specific targeting of peptides and peptidomimetics, combined with the relatively low molecular weight and therefore good pharmacokinetic properties of tissue penetration and whole body clearance, provide a class of molecules that are very appropriate for use as radiotherapeutics. As the interest in molecular imaging probes continues to increase, there is the belief that what can be used to image can certainly be expected to have value as a radiotherapeutic. This is especially the case in instances where similar radiometal–chelator complexes are available for either imaging or targeted radiotherapy. For example, DOTA chelators are appropriate for ^{111}In as well as for ^{177}Lu, and ^{99m}Tc-labeled peptides suitable for imaging may also be applicable using ^{186}R or ^{188}Re for radiotherapy. While this "piggy-back" notion of diagnostic-therapeutic pairs is valid, it is even more important in designing radiotherapeutics to have adequate tissue clearance to prevent toxicity. The application of well-established medicinal chemistry techniques to the field of radiotherapeutic development is an important tool to provide products of potential value in the clinic. Even more exciting is the ability to use peptides, as well as engineered antibody fragments, as a means of accessing new tumor targets. The high specificity for a receptor target that is possible through evaluating peptide libraries and by carrying out structure–activity studies gives a significant advantage for peptide-based radiotherapeutics.

REFERENCES

1. Klabunde T, Hessler G. Drug design strategies for targeting G-protein-coupled receptors. *ChemBioChem* 2002;3:929–944.

2. Reubi JC. Peptide receptors as molecular targets for cancer diagnosis and therapy. *Endocr Rev* 2003;24:389–427.

3. Tyndall JDA, Pfeiffer B, Abbenante G, Fairlie DP. Over one hundred peptide-activated G protein-coupled receptors recognize ligands with turn structure. *Chem Rev Mar* 2005;105:793–826.

4. Reile H, Armatis PE, Schally AV. Characterization of high-affinity receptors for bombesin/gastrin releasing peptide on the human prostate-cancer cell-lines PC-3 and DU-145-internalization of receptor-bound ^{125}I-(Tyr^4) bombesin by tumor-cells. *Prostate* 1994;25:29–38.

5. Reubi JC, Schaer JC, Waser B. Cholecystokinin (CCK)-A and CCK-B gastrin receptors in human tumors. *Cancer Res* 1997;57:1377–1386.

6. Wicki A, Wild D, Storch D, et al. [Lys^{40}(Ahx-DTPA-In-111)NH$_2$]-Exendin-4 is a highly efficient radiotherapeutic for glucagon-like peptide-1 receptor-targeted therapy for insulinoma. *Clin Cancer Res* 2007;13:3696–3705.

7. Salazar-Onfray F, Lopez M, Lundqvist A, et al. Tissue distribution and differential expression of melanocortin I receptor, a malignant melanoma marker. *Br J Cancer* 2002;87:414–422.

8. Reubi JC, Waser B, Friess H, Buchler M, Laissue J. Neurotensin receptors: a new marker for human ductal pancreatic adenocarcinoma. *Gut* 1998;42:546–550.

9. Zhang KJ, An R, Gao ZR, Zhang YX, Aruva MR. Radionuclide imaging of small-cell lung cancer (SCLC) using Tc-99m-labeled neurotensin peptide 8-13. *Nucl Med Biol* 2006;33:505–512.

10. Reubi JC, Gugger M, Waser B, Schaer JC. Y-1-mediated effect of neuropeptide Y in cancer: breast carcinomas as targets. *Cancer Res* 2001;61:4636–4641.

11. Korner M, Waser B, Reubi JC. High expression of neuropeptide Y1 receptors in Ewing sarcoma tumors. *Clin Cancer Res* 2008;14:5043–5049.

12. Froidevaux S, Eberle AN. Somatostatin analogs and radiopeptides in cancer therapy. *Biopolymers* 2002;66:161–183.

13. Hennig IM, Laissue JA, Horisberger U, Reubi JC. Substance-P receptors in human primary neoplasms—tumoral and vascular localization. *Int J Cancer* 1995;61:786–792.

14. Reubi JC, Laderach U, Waser B, Gebbers JO, Robberecht P, Laissue JA. Vasoactive intestinal peptide/pituitary adenylate cyclase-activating peptide receptor subtypes in human tumors and their tissues of origin. *Cancer Res* 2000;60:3105–3112.

15. Folkman J. Angiogenesis in cancer, vascular, rheumatoid and other disease. *Nat Med* 1995;1:27–31.

16. Ahlskog J, Paganelli G, Neri D. Vascular tumor targeting. *Q J Nucl Med Mol Imaging* 2006;50:296–309.

17. Harris TD, Kalogeropoulos S, Nguyen T, et al. Design, synthesis, and evaluation of radiolabeled integrin alpha(v)beta(3) receptor antagonists for tumor imaging and radiotherapy. *Cancer Biother Radiopharm* 2003;18:627–641.

18. Harris TD, Cheesman E, Harris AR, et al. Radiolabeled divalent peptidomimetic vitronectin receptor antagonists as potential tumor radiotherapeutic and imaging agents. *Bioconjug Chem* 2007;18:1266–1279.

19. Hausner SH, DiCara D, Marik J, Marshall JF, Sutcliffe JL. Use of a peptide derived from foot-and-mouth disease virus for the Noninvasive Imaging of human cancer: generation and evaluation of 4-[F-18]fluorobenzoyl A20FMDV2 for *in vivo* imaging of integrin alpha(v)beta(6) expression with positron emission tomography. *Cancer Res* 2007; 67:7833–7840.

20. Virgolini I, Kurtaran A, Raderer M, et al. Vasoactive-intestinal-peptide receptor scintigraphy. *J Nucl Med* 1995;36:1732–1739.

21. Hessenius C, Bader M, Meinhold H, et al. Vasoactive intestinal peptide receptor scintigraphy in patients with pancreatic adenocarcinomas or neuroendocrine tumours. *Eur J Nucl Med* 2000;27:1684–1693.

22. Blum JE, Handmaker H, Rinne NA. The utility of a somatostatin-type receptor binding peptide radiopharmaceutical (P829) in the evaluation of solitary pulmonary nodules. *Chest* 1999;115:224–232.

23. Vallis KA, Reilly RM, Scollard DA, et al. A Phase I Trial of ^{111}In-human epidermal growth factor in patients with metastatic EGFR-positive breast cancer. *Int J Radiat Oncol Biol Phys* 2008;72 (Suppl. 1): S178–S179.

24. Adessi C, Soto C. Converting a peptide into a drug: strategies to improve stability and bioavailability. *Curr Med Chem* 2002;9:963–978.

25. Janecka A, Zubrzycka M, Janecki T. Somatostatin analogs. *J Pept Res* 2001;58:91–107.

26. Rivier J, Brown M, Vale W. D-Trp^8-somatostatin—analog of somatostatin more potent than native molecule. *Biochem Biophys Res Commun* 1975;65:746–751.

27. Lam KS, Salmon SE, Hersh EM, Hruby VJ, Kazmierski WM, Knapp RJ. A new type of synthetic peptide library for identifying ligand-binding activity. *Nature* 1991;354:82–84.

28. Aina OH, Liu RW, Sutcliffe JL, Marik J, Pan CX, Lam KS. From combinatorial chemistry to cancer-targeting peptides. *Mol Pharm* 2007;4:631–651.

29. Franz AH, Liu R, Song A, Lam KS, Lebrilla CB. High-throughput one-bead-one-compound approach to peptide-encoded combinatorial libraries: MALDI-MS analysis of single TentaGel beads. *J Comb Chem* 2003;5:125–137.

30. Liu R, Marik J, Lam KS. A novel peptide-based encoding system for "one-bead one-compound" peptidomimetic and small molecule combinatorial libraries. *J Am Chem Soc* 2002;124:7678–7680.

31. Liu RW, Wang XB, Song AM, Bao T, Lam KS. Development and applications of topologically segregated bilayer beads in one-bead one-compound combinatorial libraries. *QSAR Comb Sci* 2005;24:1127–1140.

32. Park SI, Renil M, Vikstrom B, et al. The use of one-bead one-compound combinatorial library method to identify peptide ligands for alpha 4 beta 1 integrin receptor in non-Hodgkin's lymphoma. *Lett Pept Sci* 2001;8:171–178.

33. Peng L, Liu R, Marik J, Wang X, Takada Y, Lam KS. Combinatorial chemistry identifies high-affinity peptidomimetics against alpha4beta1 integrin for *in vivo* tumor imaging. *Nat Chem Biol* 2006;2:381–389.

34. Smith GP. Filamentous fusion phage—novel expression vectors that display cloned antigens on the virion surface. *Science* 1985;228:1315–1317.

35. Oneil KT, Hoess RH, Jackson SA, Ramachandran NS, Mousa SA, Degrado WF. Identification of novel peptide antagonists for GPIIb/IIIa from a conformationally constrained phage peptide library. *Proteins: Struct Funct Genet* 1992;14:509–515.

36. Hoffman JA, Giraudo E, Singh M, et al. Progressive vascular changes in a transgenic mouse model of squamous cell carcinoma. *Cancer Cell* 2003;4:383–391.

37. Schumacher TNM, Mayr LM, Minor DL, Milhollen MA, Burgess MW, Kim PS. Identification of D-peptide ligands through mirror-image phage display. *Science* 1996;271:1854–1857.

38. Wiesehan K, Buder K, Linke RP, et al. Selection of D-Amino-Acid peptides that bind to Alzheimer's disease amyloid peptide A beta(1-42) by mirror image phage display. *ChemBioChem* 2003;4:748–753.

39. Wiesehan K, Willbold D. Mirror-image phage display: aiming at the mirror. *Chem-BioChem* 2003;4:811–815.

40. Welch BD, VanDemark AP, Heroux A, Hill CP, Kay MS. Potent D-peptide inhibitors of HIV-1 entry. *Proc Nat Acad Sci USA* 2007;104:16828–16833.

41. Bakker WH, Krenning EP, Breeman WA, et al. Receptor scintigraphy with a radio-iodinated somatostatin analog—radiolabeling, purification, biologic activity, and *in vivo* application in animals. *J Nucl Med* 1990;31:1501–1509.

42. Bakker WH, Krenning EP, Breeman WA, et al. *In vivo* use of a radioiodinated somatostatin analog—dynamics, metabolism, and binding to somatostatin receptor-positive tumors in man. *J Nucl Med* 1991;32:1184–1189.

43. Sun L, Chu TW, Wang Y, Wang XG. Radiolabeling and biodistribution of a nasopharyngeal carcinoma-targeting peptide identified by *in vivo* phage display. *Acta Biochim Biophys Sin* 2007;39:624–632.

44. Zalutsky MR, Narula AS. A method for the radiohalogenation of proteins resulting in decreased thyroid uptake of radioiodine. *Appl Radiat Isot* 1987;38:1051–1055.

45. Santos JS, Muramoto E, Colturato MT, Silva CPG, Araujo EB. Radioiodination of proteins using prosthetic group: a convenient way to produce labelled proteins with *in vivo* stability. *Cell Mol Biol* 2002;48:735–739.

46. Wilbur DS. Radiohalogenation of proteins—an overview of radionuclides, labeling methods, and reagents for conjugate labeling. *Bioconjug Chem* 1992;3:433–470.

47. Hoffman TJ, Gali H, Smith CJ, et al. Novel series of In-111-labeled bombesin analogs as potential radiopharmaceuticals for specific targeting of gastrin-releasing peptide receptors expressed on human prostate cancer cells. *J Nucl Med* 2003;44:823–831.

48. Parry JJ, Kelly TS, Andrews R, Rogers BE. *In vitro* and *in vivo* evaluation of Cu-64-Labeled DOTA-Linker-Bombesin(7-14) analogues containing different amino acid linker moieties. *Bioconjug Chem* 2007;18:1110–1117.

49. Antunes P, Ginj M, Walter MA, Chen JH, Reubi JC, Maecke HR. Influence of different spacers on the biological profile of a DOTA-somatostatin analogue. *Bioconjug Chem* 2007;18:84–92.

50. Pearson DA, ListerJames J, McBride WJ, et al. Somatostatin receptor-binding peptides labeled with technetium-99m: chemistry and initial biological studies. *J Med Chem* 1996;39:1361–1371.

51. Hunter DH, Luyt LG. Lysine conjugates for the labelling of peptides with technetium-99m and rhenium. *J Label Compd Radiopharm* 2000;43:403–412.

52. Greenland WEP, Howland K, Hardy J, Fogelman I, Blower PJ. Solid-phase synthesis of peptide radiopharmaceuticals using Fmoc-N-epsilon-(Hynic-Boc)-lysine, a technetium-binding amino acid: application to Tc-99m-labeled salmon calcitonin. *J Med Chem* 2003;46:1751–1757.

53. Stephenson KA, Zubieta J, Banerjee SR, et al. A new strategy, for the preparation of peptide-targeted radiopharmaceuticals based on ion of peptide-targeted an Fmoc-lysine-derived single amino acid chelate (SAAC). Automated solid-phase synthesis, NMR characterization, and *in vitro* screening of fMLF(SAAC)G and fMLF[(SAAC-Re (CO)₃)⁺]G. *Bioconjug Chem* 2004;15:128–136.

54. Lewis MR, Jia F, Gallazzi F, et al. Radiometal-labeled peptide-PNA conjugates for targeting bcl-2 expression: preparation, characterization, and *in vitro* mRNA binding. *Bioconjug Chem* 2002;13:1176–1180.

55. De Leon-Rodriguez LM, Kovacs Z, Dieckmann GR, Sherry AD. Solid-phase synthesis of DOTA-peptides. *Chem Eur J* 2004;10:1149–1155.

56. Behr TM, Becker WS, Bair HJ, et al. Comparison of complete versus fragmented Tc-99m-labeled anti-CEA monoclonal antibodies for immunoscintigraphy in colorectal cancer. *J Nucl Med* 1995;36(3): 430–441.

57. Buchsbaum DJ, Rogers BE, Khazaeli MB, et al. Targeting strategies for cancer radiotherapy. *Clin Cancer Res* 1999;5:3048S–3055S.

58. Holliger P, Hudson PJ. Engineered antibody fragments and the rise of single domains. *Nat Biotechnol* 2005;23:1126–1136.

59. Mull ES, Sattigeri VJ, Rodriguez AL, Katzenellenbogen JA. Aryl cyclopentadienyl tricarbonyl rhenium complexes: novel ligands for the estrogen receptor with potential use as estrogen radiopharmaceuticals. *Bioorg Med Chem* 2002;10:1381–1398.

60. Rosita D, Dewit MA, Luyt LG. Fluorine and rhenium substituted ghrelin analogues as potential imaging probes for the growth hormone secretagogue receptor. *J Med Chem* 2009;52:2196–2203.

61. Giblin MF, Wang N, Hoffman TJ, Jurisson SS, Quinn TP. Design and characterization of alpha-melanotropin peptide analogs cyclized through rhenium and technetium metal coordination. *Proc Nat Acad Sci USA* 1998;95:12814–12818.

62. Barda Y, Cohen N, Lev V, et al. Backbone metal cyclization: novel Tc-99m labeled GnRH analog as potential SPECT molecular imaging agent in cancer. *Nucl Med Biol* 2004;31:921–933.

63. Bigott-Hennkens HM, Junnotula S, Ma L, Gallazzi F, Lewis MR, Jurisson SS. Synthesis and *in vitro* evaluation of a rhenium-cyclized somatostatin derivative series. *J Med Chem* 2008;51:1223–1230.

64. Luyt LG, Jenkins HA, Hunter DH. An N2S2 bifunctional chelator for technetium-99m and rhenium: complexation, conjugation, and epimerization to a single isomer. *Bioconjug Chem* 1999;10:470–479.

65. Hunter DH, Luyt LG. Single isomer technetium-99m tamoxifen conjugates. *Bioconjug Chem* 2000;11:175–181.

66. Cantorias MV, Howell RC, Todaro L, et al. MO tripeptide diastereomers (M = Tc-99/99m, Re): models to identify the structure of Tc-99m peptide targeted radiopharmaceuticals. *Inorg Chem* 2007;46:7326–7340.

67. Mundwiler S, Waibel R, Spingler B, Kunze S, Alberto R. Picolylamine-methylphosphonic acid esters as tridentate ligands for the labeling of alcohols with the fac-[M$(CO)_3$]$^+$ core (M = Tc-99m, Re): synthesis and biodistribution of model compounds and of a Tc-99m-labeled cobinamide. *Nucl Med Biol* 2005;32:473–484.

68. Haubner R, Wester HJ, Weber WA, et al. Noninvasive imaging of alpha(v)beta(3) integrin expression using F-18-labeled RGD-containing glycopeptide and positron emission tomography. *Cancer Res* 2001;61:1781–1785.

69. Haubner R, Kuhnast B, Mang C, et al. [F-18]Galacto-RGD: Synthesis, radiolabeling, metabolic stability, and radiation dose estimates. *Bioconjug Chem* 2004;15:61–69.

70. Cyr JE, Pearson DA, Wilson DM, et al. Somatostatin receptor-binding peptides suitable for tumor radiotherapy with Re-188 or Re-186. Chemistry and initial biological studies. *J Med Chem* 2007;50:1354–1364.

71. Behr TM, Goldenberg DM, Becker W. Reducing the renal uptake of radiolabeled antibody fragments and peptides for diagnosis and therapy: present status, future prospects and limitations. *Eur J Nucl Med* 1998;25:201–212.

72. Melis M, Krenning EP, Bernard BG, Barone R, Visser TJ, de Jong M. Localisation and mechanism of renal retention of radiolabelled somatostatin analogues. *Eur J Nucl Med Mol Imaging* 2005;32:1136–1143.

73. Froidevaux S, Calame-Christe M, Tanner H, Eberle AN. Melanoma targeting with DOTA-alpha-melanocyte-stimulating hormone analogs: structural parameters affecting tumor uptake and kidney uptake. *J Nucl Med* 2005;46:887–895.

74. Miao YB, Owen NK, Whitener D, Gallazzi F, Hoffman TJ, Quinn TP. *In vivo* evaluation of [188]Re-labeled alpha-melanocyte stimulating hormone peptide analogs for melanoma therapy. *Int J Cancer* 2002;101:480–487.

75. Hammond PJ, Wade AF, Gwilliam ME, et al. Amino-acid infusion blocks renal tubular uptake of an indium-labeled somatostatin analog. *Br J Cancer* 1993;67:1437–1439.

76. Rolleman EJ, Forrer F, Bernard B, et al. Amifostine protects rat kidneys during peptide receptor radionuclide therapy with [Lu-177-DOTA[0],Tyr[3]]octreotate. *Eur J Nucl Med Mol Imaging* 2007;34:763–771.

77. Fukunaga S, Setoguchi S, Hirasawa A, Tsujimoto G. Monitoring ligand-mediated internalization of G protein-coupled receptor as a novel pharmacological approach. *Life Sci* 2006;80:17–23.

78. Hofland LJ, Lamberts SWJ. The pathophysiological consequences of somatostatin receptor internalization and resistance. *Endocr Rev* 2003;24:28–47.

79. Edwards WB, Xu B, Akers W, et al. Agonist-antagonist dilemma in molecular imaging: evaluation of a monomolecular multimodal imaging agent for the somatostatin receptor. *Bioconjug Chem* 2008;19:192–200.

80. Hofland LJ, Lamberts SWJ, van Hagen PM, et al. Crucial role for somatostatin receptor subtype 2 in determining the uptake of [In-111-DTPA-D-Phe[1]]octreotide in somatostatin receptor-positive organs. *J Nucl Med* 2003;44:1315–1321.

81. Ginj M, Zhang HW, Waser B, et al. Radiolabeled somatostatin receptor antagonists are preferable to agonists for *in vivo* peptide receptor targeting of tumors. *Proc Nat Acad Sci USA* 2006;103:16436–16441.

82. Roettger BF, Ghanekar D, Rao R, et al. Antagonist-stimulated internalization of the G protein-coupled cholecystokinin receptor. *Mol Pharmacol* 1997;51:357–362.

83. Stephenson KA, Banerjee SR, Besanger T, et al. Bridging the gap between *in vitro* and *in vivo* imaging: Isostructural Re and Tc-99m complexes for correlating fluorescence and radioimaging studies. *J Am Chem Soc* 2004;126:8598–8599.

CHAPTER 4

Peptide Receptor Radionuclide Therapy in Patients with Somatostatin Receptor-Positive Neuroendocrine Tumors

MARTIJN van ESSEN, DIK J. KWEKKEBOOM, WOUTER W. de HERDER, LISA BODEI, BOEN L. R. KAM, MARION de JONG, ROELF VALKEMA, AND ERIC P. KRENNING

4.1 INTRODUCTION

The group of neuroendocrine gastroenteropancreatic (GEP) tumors consists of clinically functioning or nonfunctioning endocrine pancreatic tumors and gastrointestinal and bronchial neuroendocrine tumors, also known as carcinoids. GEP tumors usually are growing slowly and often are metastasized at diagnosis. These tumors can produce a variety of bioactive substances (e.g., serotonin, gastrin, insulin). Biotherapy with somatostatin analogues like octreotide and lanreotide can reduce hormonal overproduction and often results in symptomatic relief in patients with metastasized disease. However, objective regressions have seldom been achieved (1–3).

Peptide receptor radionuclide therapy (PRRT) with radiolabeled somatostatin analogues is a promising treatment modality for patients with inoperable or metastasized endocrine GEP tumors. The majority of endocrine GEP tumors abundantly express somatostatin receptor (sst) and these tumors can be visualized in patients using the radiolabeled somatostatin analogue [^{111}In-DTPA^0]octreotide (OctreoScan®). With OctreoScan, also other sstr positive tumors, for example, paragangliomas and differentiated thyroid carcinomas can be visualized. After successful studies to visualize somatostatin receptor-positive tumors, studies to determine the effects of therapies with radiolabeled somatostatin analogues were performed. Results are discussed in this chapter, as well as future options to improve these results.

Monoclonal Antibody and Peptide-Targeted Radiotherapy of Cancer, Edited by Raymond M. Reilly
Copyright © 2010 John Wiley & Sons, Inc.

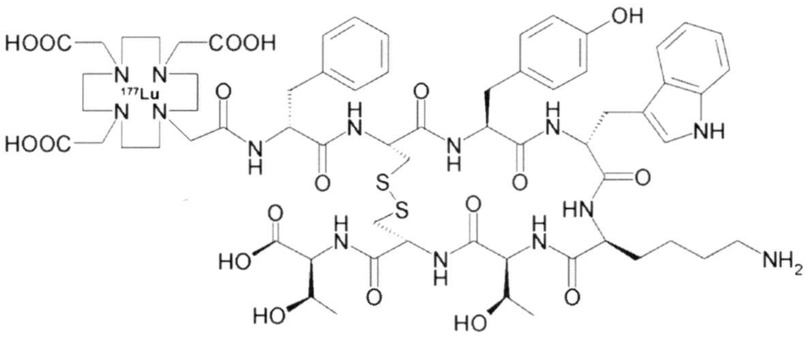

DOTA-DPhe-Cys-Tyr-DTrp-Lys-Thr-Cys-Thr
DOTA-DF-C-Y-DW-K-T-C-T

FIGURE 4.1   Structure of ^{177}Lu-octreotate.

Radiolabeled somatostatin analogues in general consist of three major components: a cyclic octapeptide (e.g., Tyr^3-octreotide or Tyr^3-octreotate), a chelator (e.g.,
DTPA or DOTA), and a radioactive element (e.g., ^{111}In, ^{90}Y, or ^{177}Lu). Figure 4.1
shows the structure of [^{177}Lu-DOTA^0,Tyr^3]octreotate (^{177}Lu-octreotate). Information
on the characteristics of these radionuclides that have been used for PRRT is provided
in Chapter 2 (Table 2.3).

4.2   RADIOTHERAPY WITH ^{111} IN-OCTREOTIDE

Initial studies used the Auger electron emitting [^{111}In-DTPA^0]octreotide (^{111}In-oc-
treotide) for PRRT, because somatostatin analogues labeled with beta-emitting
radionuclides were not available for clinical use (see Chapter 9). Treatment with
high doses of radioactivity of ^{111}In-octreotide in patients with metastasized neuroendocrine tumors often resulted in symptomatic improvement, but tumor size reduction
was rare (4–7). In the Rotterdam study, 5 of 26 patients with GEP tumors had a decrease
in tumor size of between 25% and 50% (minor response, MR), as measured on CT
scans, but none had partial remission (PR, tumor reduction of more than 50%, but not
complete) (4). These patients received a total cumulative dose of at least 20 GBq
(550 mCi). In another study in 27 patients with GEP tumors, PR was reported in 2 of 26
patients (8%) with measurable disease (5). In both studies, many patients were in a
rather poor clinical condition and many had progressive disease (PD) at baseline. A
third study in 12 patients with carcinoid tumors, gastrinoma or glucagonoma, reported
PR in 2 patients (6). Delpassand et al. recently published their results with administration of high doses of ^{111}In-octreotide (7). Thirty-two patients were treated. Eighteen
patients received 2 cycles of 500 mCi (18.5 GBq). Fourteen patients had only 1 cycle:
four had died before a 2nd cycle, seven withdrew their consent and three were not yet
eligible for the 2nd cycle at time of analysis. One patient who was treated with 1 cycle
had a grade 3 thrombocytopenia. No other significant hematological toxicity occurred,

nor did renal toxicity occur with a follow-up up to 24 months (average 13 months). In the 18 patients treated with 2 cycles, 16 had stable disease (SD) and 2 had PR. Median time to progression was not reported.

Long-term side effects of ^{111}In-octreotide therapies were most commonly due to bone marrow toxicity. Serious side effects consisted of leukemia and myelodysplastic syndrome (MDS) in three patients who had been treated in Rotterdam with total cumulative doses of >100 GBq (2.7 Ci) and bone marrow radiation absorbed doses were estimated to be more than 3 Gy (4). One of these patients had also received chemotherapy previously and this may have contributed to or caused this complication. One patient in the study by Anthony et al. (5) developed renal insufficiency. This was probably due to preexistent retroperitoneal fibrosis and not treatment related. Transient hepatic toxicity was observed in three patients with widespread liver metastases. Other studies did not mention MDS or renal failure. The maximum cumulative dose in these studies was 33.6 GBq (908 mCi) (6) or 37 GBq (1 Ci) (7).

^{111}In-labeled somatostatin analogues are not ideal for PRRT because of the short particle range of Auger-electrons and therefore shorter tissue penetration compared to more energetic beta-particle emitters. The decay of ^{111}In therefore has to take place much closer to the cell nucleus than that of ^{90}Y or ^{177}Lu to be tumoricidal. So it is not unexpected that CT-assessed tumor regression was observed only in rare cases after therapy with ^{111}In-octreotide. Strategies to promote the nuclear uptake of ^{111}In-labeled peptides may circumvent this limitation and allow effective radiotherapy with this radionuclide (see Chapter 9).

4.3 RADIOTHERAPY WITH ^{90}Y-DOTATOC

The modified somatostatin analogue [DOTA^0,Tyr^3]octreotide (DOTATOC) labeled with the beta-emitting radionuclide ^{90}Y was the next generation of sstr targeted radionuclide therapy. This analogue has a higher affinity for sstr subtype-2, which is predominantly overexpressed on GEP tumors (8) and it has DOTA instead of DTPA as the chelator, which allows more stable complexing of ^{90}Y. ^{90}Y-DOTATOC (OctreoTher®) was used in several Phase 1 and Phase 2 PRRT trials in various countries.

In Switzerland, investigators performed different Phase 1 and Phase 2 studies in patients with neuroendocrine GEP tumors (9–12). Initially, they used a dose-escalating scheme of up to a cumulative radioactivity of 160 mCi (6 GBq)/m^2 divided over 4 cycles without amino acid infusion as renal protection in half of the patients. Four of 29 patients developed renal insufficiency and none of these had received renal protection. The overall response rate (CR or PR) was 24% in patients with GEP tumors who were either treated with 6 GBq (160 mCi)/m^2 (10), or, in a later study, with 7.4 GBq (200 mCi)/m^2 in 4 cycles (11). In a study, also with a radioactivity dose of 7.4 GBq (200 mCi)/m^2, but administered in two cycles, one-third of 36 patients had complete or partial remissions (12). Although the latter treatment protocol seemed more effective, it is important to realize that this was not a randomized trial comparing these two dosing schemes. Patient and tumor characteristics therefore were probably different.

In Milan, Italy, the research group there performed dosimetric and dose-escalating studies with ^{90}Y-DOTATOC with and without the administration of renal protecting agents (13). They observed no major acute reactions when administering doses of up to 5.6 GBq (150 mCi) per cycle, but reversible grade 3 hematological toxicity was found in 43% of patients treated with 5.2 GBq (140 mCi). This was defined as the maximum tolerated dose (MTD) per cycle. Acute or delayed kidney failure developed in none of the patients, although the time of follow-up was rather short. This included 30 patients in the first phase of the study who received 3 cycles of up to 2.59 GBq (70 mCi) per cycle without renal protection. Partial and complete remissions (CRs) were present in 28% of 87 patients with neuroendocrine tumors (14).

The same group (15) performed a subsequent Phase 1 study in 40 patients with sstr positive tumors, including 21 with GEP tumors. Two treatment cycles with cumulative total radioactivities ranging from 5.9 to 11.1 GBq (160–300 mCi) were administered. Tumor regression was achieved in 29% of patients and the median duration of response was 9 months.

^{90}Y-DOTATOC was also administered in a multicenter Phase 1 trial. Sixty patients received escalating doses up to 14.8 GBq (400 mCi)/m^2 in 4 cycles or up to 9.3 GBq (250 mCi)/m^2 as a single dose, without reaching the MTD (16). In all patients, amino acids were administered concomitantly with ^{90}Y-DOTATOC for renal protection. The cumulative radiation absorbed dose to the kidneys was limited to 27 Gy. This was based on positron emission tomography (PET) data using ^{86}Y-DOTATOC, also with concomitant amino acid infusion. Dose-limiting toxicity was observed in three patients: one patient had transient hepatic toxicity, one had thrombocytopenia grade 4 ($<25 \times 10^9$/L), and one had MDS. Fifty-eight patients had carcinoid or other GEP tumors: seven patients had a MR (12%) and five had PR (9%). Disease was stable in 29 patients (50%) and progressive in 14 (24%). Treatment outcome could not be determined in three patients. Median time to progression was 29.3 months in the 41 patients with at least stable disease as treatment outcome. Considering all patients, median overall survival since the start of therapy was 36.7 months.

4.4 TARGETED RADIOTHERAPY STUDIES WITH ^{177}Lu-OCTREOTATE

[^{177}Lu-DOTA^0,Tyr^3]octreotate (^{177}Lu-octreotate, ^{177}Lu-DOTATATE) now is the third generation in sstr targeted radionuclide therapy and has been used in our hospital since 2000. [DOTA^0,Tyr^3]octreotate shows a considerable improvement in binding to sstr positive tissues compared with [DOTA^0,Tyr^3]octreotide. A nine-fold increase in affinity for the sstr subtype 2 for [DOTA^0,Tyr^3]octreotate was found in vitro compared with [DOTA^0,Tyr^3]octreotide (17) and six- to sevenfold for the ^{90}Y-labeled counterparts (17, 18). [DOTA^0,Tyr^3]octreotate, labeled with the beta- and gamma-emitting radionuclide ^{177}Lu, was very successful in terms of providing tumor regression and prolonging survival in a rat carcinoma model expressing sstr (19).

In a study in patients, uptake of radioactivity, expressed as percentage of the injected dose of ^{177}Lu-octreotate, was comparable to that of ^{111}In-octreotide for kidneys, spleen, and liver, but was three- to fourfold higher for four of five tumors (20). Figure 4.2 provides a clear example of this difference in tumor to liver ratio in one patient. Esser et al. (21) published the results of a comparison of the dosimetry in a therapeutic setting using 3.7 GBq (100 mCi) ^{177}Lu-octreotate and 3.7 GBq ^{177}Lu-DOTATOC in the same patients separated by an interval of 6–10 weeks. The mean radiation absorbed dose delivered to tumors with ^{177}Lu-octreotate was higher by a factor of 2.1 compared to ^{177}Lu-DOTATOC. This meant that higher absorbed doses could be achieved for most tumors with about equal doses to potentially dose-limiting normal organs. Also the physical properties of ^{177}Lu are potentially favorable because of the lower tissue penetration range of beta-particle emissions of ^{177}Lu compared with those of ^{90}Y, which results in lower radiation

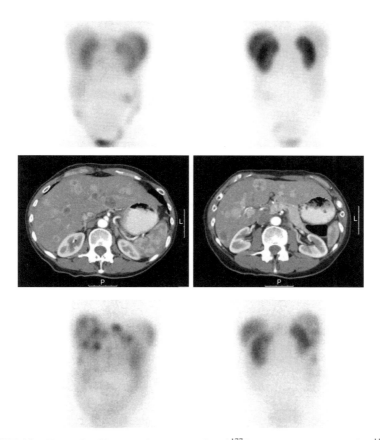

FIGURE 4.2   Example of improved tumor uptake of ^{177}Lu-octreotate compared to ^{111}In-octreotide. Upper row: Scintigraphy 24 h (anterior and posterior views) after injection of ^{111}In-octreotide did not clearly demonstrate enhanced uptake in liver metastases relative to normal liver parenchyma, but were clearly visible on CT (middle row). Lower row: On scintigraphy 24 h (anterior and posterior views) after therapy with ^{177}Lu-octreotate, liver metastases were easily recognized.

absorbed doses due to beta-irradiation of normal structures and organs adjacent to tissues with high uptake (i.e., "cross-fire effect").

^{177}Lu-labeled analogues have an important practical advantage over their ^{90}Y-labeled counterparts: ^{177}Lu is not a pure beta-emitter but also emits low-energy gamma-rays (113 and 208 keV, 10% abundance) and this allows post-therapy imaging and estimation of the dosimetry. ^{90}Y on the other hand is a pure beta-emitter and does not emit gamma rays or positrons for imaging and thus dosimetry is not easily measured (imaging of Bremstrahlung radiation resulting from the beta emissions is possible but the resolution is poor).

The investigators at Bad Berka, Germany (22), the European Institute of Oncology (IEO, Milano, Italy) (23) and Erasmus MC (Rotterdam, the Netherlands) (24) performed dosimetry studies on ^{177}Lu-octreotate. The data from these studies are summarized in Fig. 4.3. On average, the kidney dose was approximately 0.6–0.9 Gy/GBq, spleen dose 0.6–2.1 Gy/GBq, and bone marrow dose 0.04–0.07 Gy/GBq. However, interpatient variability was large. The Bad Berka group reported a mean absorbed dose in tumors of 9.7 Gy/GBq, which would result in 290 Gy with a cumulative administered activity of 30 GBq. The IEO group stated that the absorbed dose in tumors in their patients ranged from 0.6 to 56 Gy/GBq. Results from the Rotterdam group are pending.

The treatment effects of ^{177}Lu-octreotate therapy were recently published for a large group of patients (25), and earlier for a smaller number of patients (26, 27). The intended cumulative radioactivity dose was 22.2–29.6 GBq (600–800 mCi); however, if dosimetric calculations indicated that the radiation absorbed dose to the kidneys

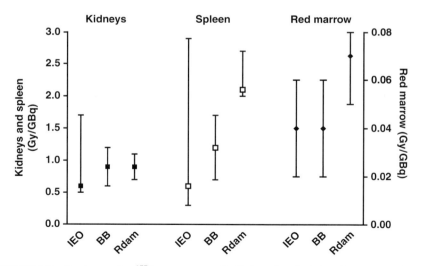

FIGURE 4.3   Dosimetry for ^{177}Lu-octreotate. Radiation absorbed dose is expressed in Gy/GBq, presented as median and range (error bars), for the Bad Berka study mean and standard deviation, respectively. BB: Bad Berka, Germany, Wehrmann et al. (22); IEO: European Institute of Oncology, Milano, Italy (23); Rdam: Rotterdam, Kwekkeboom et al. (24); RM: red marrow.

would exceed 23 Gy with a dose of 29.6 GBq, the cumulative radioactivity was reduced to 22.2–27.8 GBq. Treatment intervals in general were 6–10 weeks. Between January 2000 and August 2006, 504 patients were treated (a total of 1772 administrations of ^{177}Lu-octreotate). Side effects within the first 24 hours were nausea in 25% of administrations, vomiting in 10% and pain in 10%. Six patients with highly hormonally active neuroendocrine tumors developed a clinical crisis after administration due to massive increase in the release of bioactive substances. All patients recovered after adequate medical treatment (see Ref. 28 for details). Hematological side effects (WHO grade 3 or 4) occurred in 3.6% of administrations, or in 9.5% of patients. Mild and reversible alopecia was reported by 62% of patients. Serious delayed side effects occurred in nine patients: two had a decline in kidney function. Serum creatinine concentrations in the first patient were rising already in the year prior to the therapy. Creatinine clearance was 41 mL/min when starting therapy with ^{177}Lu-octreotate. This patient developed renal insufficiency 1.5 years after receiving her last treatment. Probably kidney failure was not related to the treatment with ^{177}Lu-octreotate because the deterioration had started before therapy. In the second patient, serum creatinine levels rose considerably 3 years after therapy with ^{177}Lu-octreotate. In this patient, a severely increasing tricuspid valve insufficiency with right-heart failure requiring administration of diuretics and other medication was more likely the cause of the renal insufficiency. Three patients with very extensive hepatic metastases had serious liver toxicity: one patient had diffuse, rapidly growing liver metastases from an aggressive endocrine pancreatic tumor. The patient developed hepatorenal syndrome after the first cycle and died 5 weeks thereafter. This was more likely to be attributable to the aggressive tumor growth. Two other patients had a temporary worsening of liver function who required hospitalization. Both patients recovered and therapy was resumed with a lower dose of ^{177}Lu-octreotate per cycle without serious side effects. Finally, four patients developed MDS. In one patient, MDS was more likely caused by prior chemotherapy with alkylating agents given the very short interval (several months) between the treatment with ^{177}Lu-octreotate and the diagnosis of MDS. In three other patients, MDS seemed related to the therapy with ^{177}Lu-octreotate. So, in summary, serious delayed toxicity probably attributable to the therapy with ^{177}Lu-octreotate was present in five patients (approximately 1%).

Therapy outcome was evaluated in 310 patients with GEP neuroendocrine tumors. Progressive disease was found in 61 (20%) patients, SD in 107 (35%), MR in 51 (16%), PR in 86 (28%), and CR was found in 5 (2%) patients. Higher remission rates were positively correlated with high tumor uptake during pretherapy ^{111}In-octreotide scintigraphy and a limited number of liver metastases, whereas a low Karnofsky Performance Score (KPS), extensive disease and weight loss were predictive factors for PD. Median time to progression was 40 months from the start of treatment in the 249 patients who had either SD or tumor regression (MR, PR, and CR).

This recent study reports data about survival after therapy with ^{177}Lu-octreotate as well. Median overall survival after beginning therapy was 46 months and median disease related survival was >48 months. Median progression free survival was 33 months. The most important predictive factor for survival was treatment outcome (PD versus non-PD). Other factors were extent of liver metastases, low KPS, baseline

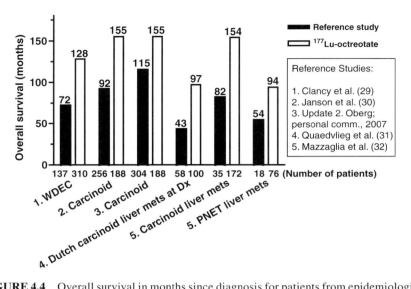

FIGURE 4.4   Overall survival in months since diagnosis for patients from epidemiological and interventional studies (black bars) and in similar patients with regard to tumor type and disease stage treated with ^{177}Lu-octreotate (white bars). Although these data are not based on randomized trials, there seems a survival benefit of 40–72 months in patients treated with ^{177}Lu-octreotate. (Adapted from Ref. 24.) Dx: diagnosis, WDEC: well-differentiated endocrine carcinoma, PNET: pancreatic neuroendocrine tumor.

weight loss and to a lesser degree, presence of bone metastases and diagnosis of gastrinoma/insulinoma/VIPoma. Compared to other treatments and epidemiological data (29–32), there seems to be a survival benefit of at least 3.5–6 years. The difference in survival since diagnosis between the patients treated with ^{177}Lu-octreotate and those from other studies is illustrated in Fig. 4.4. It is of note that these data are not based on randomized clinical trials.

Not only survival is important, but also the quality of life (QoL; see Chapter 16). QoL was analyzed in 50 patients with metastatic GEP tumors treated with ^{177}Lu-octreotate (33). The patients filled out the EORTC Quality of Life Questionnaire C30 before therapy and at follow-up visit 6 weeks after the last cycle. Global health status/ QoL scale significantly improved after therapy with ^{177}Lu-octreotate. The patients reported a significant improvement in symptom scores for fatigue, insomnia, and pain. Improvement of QoL domains was most frequently observed in patients with proven tumor regression.

4.5   PRRT WITH OTHER SOMATOSTATIN ANALOGUES

Lanreotide is another somatostatin analogue. It can be labeled with ^{111}In for diagnostic purposes and with ^{90}Y for therapeutic use. Its use has been advocated because of its increased affinity for sstr subtypes 3 and 4 compared to ^{111}In-octreotide (34), but this claim is questionable (17). Although radiolabeled lanreotide has been used to treat patients with GEP tumors, its affinity is poorer than that of radiolabeled $[DOTA^0,Tyr^3]$ octreotide/octreotate for the sstr subtype-2 (sst_2). In a study of 154 patients of whom 39 had carcinoid or other GEP tumors and were treated with $[^{90}Y$-$DOTA^0]$lanreotide, 8 of 39 patients exhibited regressive disease (defined as >25% reduction of tumor size) (21%), 17 had stable disease (44%), and 14 had progressive disease (36%) (34).

Recently, preliminary data have been presented using ^{90}Y-labeled $[DOTA^0,Tyr^3]$-octreotate (^{90}Y-octreotate) (35, 36). However, it is hard to draw conclusions as yet: the treatment protocols vary and the way of response evaluation was not clearly defined. An objective response rate (PR) of 37% (28/75) was noted and stabilization of disease occurred in 39/75 patients (52%). In the same study intra-arterial administration of ^{90}Y-octreotate in five patients was described. However, no detailed results for this application were provided. An important issue is that to date, reliable dosimetry of ^{90}Y-octreotate is lacking.

4.6 COMPARISON OF DIFFERENT PRRT STUDIES

Somatostatin receptor targeted radionuclide therapy, especially with peptide analogues labeled with a beta-emitter, is a promising new modality in the management of patients with inoperable or metastasized neuroendocrine tumors. Comparing the effects of treatment with ^{90}Y-DOTATOC and ^{177}Lu-octreotate is very difficult, because a randomized trial comparing these two treatments has never been done. The results that were obtained with both analogues nonetheless are very encouraging.

There is a clear difference in the reported tumor remission percentages between therapy studies with ^{90}Y-DOTATOC. The following factors may account for this. First of all, the administered radioactivities and dosing schemes were different: some patients received relatively low doses, because some studies used dose-escalating schemes, whereas others used fixed doses. Also several patient and tumor characteristics that determine treatment outcome may play a role, such as the amount of tumor uptake on ^{111}In-octreotide scintigraphy, the total tumor burden, the extent of liver involvement, weight loss, KPS, and the type of neuroendocrine tumor (25, 27). Therefore, differences in patient selection probably play an important role in determining treatment outcome. Also methodological factors may contribute to the different results in the different centers performing trials with the same agents, for example, differences in tumor response criteria (e.g., including MR or not) and the duration of follow-up. Therefore, it is essential to develop randomized trials using detailed protocols to establish which treatment scheme and which radiolabeled somatostatin analogues or combination of analogues are optimal. Also long-term follow-up is mandatory.

4.7 COMPARISON WITH CHEMOTHERAPY

In the early years of PRRT, it would have been preferable to have a randomized trial comparing PRRT to no further treatment at all. This would have allowed us to evaluate the effects of PRRT precisely, especially with regard to quality of life and survival. However, patients can now be treated with PRRT in several medical centers in the European Union and the results of such treatments are encouraging. Therefore, withholding such treatment to half of patients with symptomatic and/or progressive disease in an experimental setting cannot be ethically justified and so this is presently no longer possible, at least in the European Union. Treatment with nonradiolabeled somatostatin analogues and/or interferon alpha in patients with GEP tumors with PD at study entry resulted in tumor remission in 4 of 80 patients (5%) (37). This is less than the 46% with tumor remissions (including MR) in our 310 patients treated with ^{177}Lu-octreotate, whether they had PD at study entry or not (24). It seems highly improbable that such a clear difference is only caused by patient selection.

Response rates of 40–60% for single agent and combination chemotherapy were observed in well-differentiated pancreatic tumors and poorly differentiated tumors from any origin, whereas success rates for midgut tumors rarely exceed 20% in recent studies (see Refs 38–55 for review). Older series with "classical" chemotherapy reported high response rates (49–51), but these studies used not only imaging studies as a response criterion but also biochemical tests (measuring serum tumor marker levels) and physical examination for the evaluation of hepatomegaly. Differences in response criteria probably attribute much to the discrepancy between older and more recent studies. Multiple novel drugs, such as bevacizumab (Avastin), sunitinib (Sutent), sorafenib (Nexavar), vatalanib, thalidomide, and mTOR-inhibitors (ever-olimus (Rad001), temsirolimus) are under investigation as well, but these are not reviewed here.

Not only the proportion of patients with tumor remission and stabilization is important but also the duration of such a response and survival. The median time to progression for chemotherapy in most of the studies is less than 18 months, regardless of the varying percentages of objective responses. In this respect, treatment with ^{90}Y-DOTATOC or ^{177}Lu-octreotate compares favorably with a median time to progression of 30 and 40 months, respectively (16, 25).

4.8 OPTIONS FOR IMPROVING PRRT AND FUTURE DIRECTIONS

One way to improve the results of treatment with radiolabeled somatostatin analogues could be to reduce the amount of radiation deposited in normal tissues like the kidneys and bone marrow and reducing its unwanted side effects. This would allow an increase of the cumulative administered radioactivity. Renal protective agents (either lysine and arginine or a commercially available mixture of amino acids) should always be administered in clinical practice when using somatostatin analogues labeled with beta-emitters (56). These amino acids cause a reduction in renal uptake of radioactivity in the proximal tubules. Animal studies indicate that the addition of gelofusin to lysine and arginine can further decrease the renal uptake (57). This will be studied in future patients as well.

Another possibility to reduce the toxic effects of radiation on normal tissues may be administering amifostine. Amifostine is used in patients treated with external beam radiation therapy and reduces side effects, probably without affecting therapeutic results (58). In animal studies examining high doses of ^{177}Lu-octreotate, coadministration of amifostine clearly reduced functional renal damage. From that study it was not clear however if there were protective effects on bone marrow (59). At present, no studies combining ^{177}Lu-octreotate and amifostine in patients have been performed.

[90]Y and ^{177}Lu have different physical properties. The use of several different radiolabeled somatostatin analogues in the same patient can therefore be considered as well. From animal experiments, it became clear that ^{90}Y-labeled somatostatin analogues may be more effective for larger tumors, whereas ^{177}Lu-labeled somatostatin analogues may be more effective for smaller tumors. Their combination may be the most effective: in animals with various tumor sizes, therapy with both [90]Y- and ^{177}Lu-labeled octreotate had better remission rates than either [90]Y- or ^{177}Lu-labeled octreotate alone (60). Therefore, not only different radiolabeled peptides like octreotate and octreotide, and different radionuclides like ^{90}Y and ^{177}Lu, should be evaluated, but also PRRT with several combinations, preferably in a randomized clinical trial. However, reliable dosimetry with ^{90}Y-octreotate in humans is not available yet.

Investigating possible methods to increase the sstr density on tumors can be another interesting strategy, for example, by medication or irradiation. There are several studies *in vitro* and *in vivo* that indicate that irradiation from external beam radiotherapy and from PRRT might be used to upregulate receptor expression (61–64). At the moment however, human data are not available.

So far, all studies involving PRRT in patients have been done with somatostatin receptor agonists. Agonists are internalized by tumor cells and are therefore retained longer. Somatostatin receptor antagonists however are not internalized in cells and were therefore thought to be inappropriate for targeted radiotherapy. Recently however Ginj et al. (65) demonstrated almost twice as high tumor retention of the radiolabeled sst_2 antagonist ^{111}In-DOTA-sst_2-ANT during the first 24 h after administration in a tumor xenograft mouse model compared to the agonist ^{111}In-DTPA-octreotate, and this despite a lower affinity for the sst_2 of the antagonist. This possibly is explained by binding of the antagonist to a larger variety of receptor conformations. These results were unexpected and seem promising for trying to increase the radiation-absorbed dose to tumors. Unfortunately, no data were shown for the amount of radioactivity remaining in tumors later than 24 h after injection, which is a very important factor as well for estimating the total amount of radiation dose to the tumor. Also no data are available yet about toxicity and biodistribution in humans.

Adding radiosensitizing chemotherapeutic agents (e.g., 5-fluorouracil [5-FU] or capecitabine) may also be one future direction to improve PRRT. ^{90}Y-labeled antibody radioimmunotherapy in combination with 5-FU as radiosensitizer was found feasible and safe (66). PRRT using ^{111}In-octreotide combined with 5-FU resulted in symptomatic improvement in 71% of patients with neuroendocrine tumors (67), which is more frequent than in other studies using only ^{111}In-octreotide as treatment (4, 5). 5-FU was used in many of the numerous trials to investigate the effects of (fractionated) external beam radiotherapy with chemotherapy. More recent trials used capecitabine, a prodrug of 5-FU, which has the advantage of oral administration. Thymidine phosphorylase (TP) is one of the enzymes needed to convert the inactive form (capecitabine) into its active form (5-FU). Many tumors have a higher amount of TP than normal tissues and this results in a higher concentration of the active form in cancerous cells. In addition, irradiation can induce an upregulation of TP (68). This makes capecitabine a very interesting drug to combine with PRRT. Capecitabine in relatively low doses (1600–2000 mg/m^2/day), rarely provokes grade 3 hematologic or other toxicity such as severe hand-foot syndrome or stomatitis (68, 69). Recently, a pilot trial using capecitabine (1650 mg/m^2/day) and ^{177}Lu-octreotate was performed to evaluate if this new combination is safe and feasible. Seven patients were treated and 26 cycles of the combination were administered. Hematological toxicity was rare (one patient with grade 3 anemia and one patient with grade 3 thrombocytopenia, no grade 4 toxicity) and capecitabine-related side effects were rare as well (one patient with grade 1 stomatitis, no hand-foot syndrome) (70). Although it was not the aim of the study, it is reassuring that this new combination therapy was very effective. Figure 4.5 demonstrates a PR in one patient. With this knowledge, a randomized, clinical, multicenter trial comparing treatment with ^{177}Lu-octreotate with and without capecitabine in patients with GEP tumors was started in March 2007.

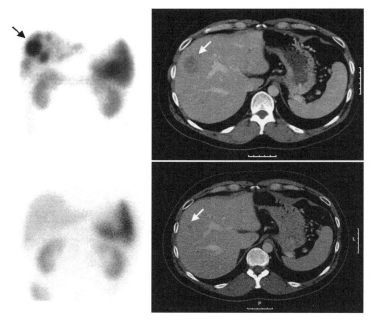

FIGURE 4.5   Imaging studies in a patient with a pancreatic tail neuroendocrine tumor with liver metastases, treated with 4 cycles of ^{177}Lu-octreotate of 7.4 GBq and capecitabine (1650 mg/m^2 per day for 2 weeks). Upper row: ^{111}In-octreotide scintigraphy (anterior view) and CT scan before starting therapy. Arrows indicate one of the several liver metastases. Lower row: ^{111}In-octreotide scintigraphy (anterior view) and CT scan 1 year after the final therapy cycle. There is a clear decrease in pathological uptake in liver lesions and pancreatic tail tumor on ^{111}In-octreotide scintigraphy and a partial remission is seen on the CT scan.

4.9 CONCLUSIONS

There are few effective therapies for patients with inoperable or metastasized neuroendocrine tumors. PRRT with radiolabeled somatostatin analogues is a promising treatment option for these patients, provided that pretherapy ssrt scintigraphy is positive. Treatment with any of the various ^{111}In, ^{90}Y, or ^{177}Lu-labeled somatostatin analogues that have been used can result in symptomatic improvement, but therapy with ^{111}In-labeled somatostatin analogues rarely results in tumor size reduction. Therefore, radiolabeled somatostatin analogues with beta-emitting isotopes like ^{90}Y and ^{177}Lu were developed. There is a large variation in the reported antitumor effects of ^{90}Y-DOTATOC between various studies: objective response was achieved in 9–33%. With ^{177}Lu-octreotate treatments, overall response (OR) was achieved in 29% of patients and MR in 16%, SD was present in 35% and progressive disease in 20%. High tumor uptake on somatostatin receptor scintigraphy and a limited amount of liver metastases were predictive factors for tumor remission. Side effects of PRRT are few and mostly mild, but renal protective agents are required; serious delayed side effects like MDS or renal failure (with appropriate renal protection) are rare ($<$1% for ^{177}Lu-octreotate). The median duration of therapy response for ^{90}Y-DOTATOC and ^{177}Lu-octreotate is 30 and 40 months respectively. Treatment with ^{177}Lu-octreotate seems to result in a survival benefit of several years. Patients with PD despite therapy with ^{177}Lu-octreotate clearly had a shorter survival than those who had tumor remission or stable disease. Quality of life improves significantly after treatment with ^{177}Lu-octreotate. These data about PRRT compare favorably with the limited number of alternative treatment approaches, like chemotherapy. Therefore, PRRT might become the therapy of first choice in patients with metastasized or inoperable GEP neuroendocrine tumors if more widespread use of PRRT is possible. Although not discussed in this chapter, the role of PRRT in somatostatin receptor expressing non-GEP tumors, like metastasized paraganglioma/pheochromocytoma (71, 72) and nonradioiodine-avid differentiated thyroid carcinoma (73) might also become more important. Further studies are needed to increase the antitumor effects of PRRT and to further reduce side effects. Also randomized clinical trials are essential to better determine the role of PRRT in the management of patients with sstr-positive tumors.

REFERENCES

1. Arnold R, Benning R, Neuhaus C, Rolwage M, Trautmann ME. Gastroenteropancreatic endocrine tumours: effect of Sandostatin on tumour growth. The German Sandostatin Study Group. *Digestion* 1993;54 (Suppl. 1):72–75.
2. Janson ET, Oberg K. Long-term management of the carcinoid syndrome. Treatment with octreotide alone and in combination with alpha-interferon. *Acta Oncol* 1993;32:225–229.
3. Ducreux M, Ruszniewski P, Chayvialle JA, et al. The antitumoral effect of the long-acting somatostatin analog lanreotide in neuroendocrine tumors. *Am J Gastroenterol* 2000;95:3276–3281.

4. Valkema R, de Jong M, Bakker WH, et al. Phase I study of peptide receptor radionuclide therapy with [^{111}In-DTPA^0]Octreotide: the Rotterdam experience. *Seminars Nucl Med* 2002;32:110–122.

5. Anthony LB, Woltering EA, Espanan GD, Cronin MD, Maloney TJ, McCarthy KE. Indium-111-pentetreotide prolongs survival in gastroenteropancreatic malignancies. *Semin Nucl Med* 2002;32:123–132.

6. Buscombe JR, Caplin ME, Hilson AJ. Long-term efficacy of high-activity ^{111}In-pentetreotide therapy in patients with disseminated neuroendocrine tumors. *J Nucl Med* 2003;44:1–6.

7. Delpassand ES, Sims-Mourtada J, Saso H, et al. Safety and efficacy of radionuclide therapy with high-activity In-111 pentetreotide in patients with progressive neuroendocrine tumors. *Cancer Biother Radiopharm* 2008;23:292–300.

8. Reubi JC, Waser B, Schaer JC, Laissue JA. Somatostatin receptor sst1-sst5 expression in normal and neoplastic human tissues using receptor autoradiography with subtype-selective ligands. *Eur J Nucl Med* 2001;28:836–846.

9. Otte A, Herrmann R, Heppeler A, et al. Yttrium-90 DOTATOC: first clinical results. *Eur J Nucl Med* 1999;26:1439–1447.

10. Waldherr C, Pless M, Maecke HR, Haldemann A, Mueller-Brand J. The clinical value of [^{90}Y-DOTA]-D-Phe^1-Tyr^3-octreotide (^{90}Y-DOTATOC) in the treatment of neuroendocrine tumours: a clinical phase II study. *Ann Oncol* 2001;12:941–945.

11. Waldherr C, Pless M, Maecke HR, et al. Tumor response and clinical benefit in neuroendocrine tumors after 7.4 GBq (90)Y-DOTATOC. *J Nucl Med* 2002;43:610–616.

12. Waldherr C, Schumacher T, Maecke HR, et al. Does tumor response depend on the number of treatment sessions at constant injected dose using ^{90}Yttrium-DO-TATOC in neuroendocrine tumors? [Abstract] *Eur J Nucl Med Mol Imaging* 2002;29: S100.

13. Chinol M, Bodei L, Cremonesi M, Paganelli G. Receptor-mediated radiotherapy with Y-DOTA-DPhe-Tyr-octreotide: the experience of the European Institute of Oncology group. *Semin Nucl Med* 2002;32:141–147.

14. Paganelli G, Bodei L, Handkiewicz Junak D, et al. ^{90}Y-DOTA-D-Phe1-Tyr3-octreotide in therapy of neuroendocrine malignancies. *Biopolymers* 2002;66:393–398.

15. Bodei L, Cremonesi M, Zoboli S, et al. Receptor-mediated radionuclide therapy with ^{90}Y-DOTATOC in association with amino acid infusion: a phase I study. *Eur J Nucl Med Mol Imaging* 2003;30:207–216.

16. Valkema R, Pauwels S, Kvols LK, et al. Survival and response after peptide receptor radionuclide therapy with [^{90}Y-DOTA0,Tyr^3]octreotide in patients with advanced gastro-enteropancreatic neuroendocrine tumors. *Semin Nucl Med* 2006;36:147–156.

17. Reubi JC, Schar JC, Waser B, et al. Affinity profiles for human somatostatin receptor subtypes SST1-SST5 of somatostatin radiotracers selected for scintigraphic and radio-therapeutic use. *Eur J Nucl Med* 2000;27:273–282.

18. Teunissen JJ, Kwekkeboom DJ, de Jong M, Esser JP, Valkema R, Krenning EP. Endocrine tumours of the gastrointestinal tract. Peptide receptor radionuclide therapy. *Best Pract Res Clin Gastroenterol* 2005;19:595–616.

19. de Jong M, Breeman WA, Bernard BF, et al. [^{177}Lu-DOTA^0,Tyr^3] octreotate for somatostatin receptor-targeted radionuclide therapy. *Int J Cancer* 2001;92:628–633.

20. Kwekkeboom DJ, Bakker WH, Kooij PP, et al. [^{177}Lu-DOTA^{0}Tyr^3]octreotate: comparison with [^{111}In-DTPA^0]octreotide in patients. *Eur J Nucl Med* 2001; 28:1319–1325.

21. Esser JP, Krenning EP, Teunissen JJ, et al. Comparison of ^{177}Lu-DOTA^0,Tyr^3]octreotate and ^{177}Lu-DOTA^0,Tyr^3]octreotide: which peptide is preferable for PRRT? *Eur J Nucl Med Mol Imaging* 2006;33:1346–1351.

22. Wehrmann C, Senftleben S, Zachert C, Müller D, Baum RP. Results of individual patient dosimetry in peptide receptor radionuclide therapy with ^{177}Lu DOTA-TATE and ^{177}Lu DOTA-NOC. *Cancer Biother Radiopharm* 2007;22:406–416.

23. Cremonesi M, Ferrari M, Bodei L, et al. Dosimetry in patients undergoing ^{177}Lu-DOTATATE therapy with indications for ^{90}Y-DOTATATE. *Eur J Nucl Med Mol Imaging* 2006;33:S102 (Abstract).

24. Kwekkeboom DJ, Bakker WH, Kooij PP, Konijnenberg MW, Srinivasan A, Erion JL, Schmidt MA, Bugaj JL, de Jong M, Krenning EP. [^{177}Lu-DOTA^{0}Tyr^3]octreotate: comparison with [^{111}In-DTPA^0]octreotide in patients. *Eur J Nucl Med* 2001;28:1319–25.

25. Kwekkeboom DJ, De Herder WW, Kam BL, et al. Treatment with the radiolabeled somatostatin analog [^{177}Lu-DOTA^0,Tyr^3]octreotate: toxicity, efficacy, and survival. *J Clin Oncol* 2008;26:2124–2130.

26. Kwekkeboom DJ, Bakker WH, Kam BL, et al. Treatment of patients with gastro-entero-pancreatic (GEP) tumours with the novel radiolabeled somatostatin analogue [^{177}Lu-DOTA^0,Tyr^3]octreotate. *Eur J Nucl Med Mol Imaging* 2003; 30:417–422.

27. Kwekkeboom DJ, Teunissen JJ, Bakker WH, et al. Treatment with the radiolabeled somatostatin analogue [^{177}Lu-DOTA^0,Tyr^3]octreotate in patients with gastro-entero-pancreatic (GEP) tumors. *J Clin Oncol* 2005;23:2754–2762.

28. de Keizer B, van Aken MO, Feelders RA, et al. Hormonal crises following receptor radionuclide therapy with the radiolabeled somatostatin analogue [^{177}Lu-DOTA^0, Tyr^3] octreotate. *Eur J Nucl Med Mol Imaging* 2008;35:749–755.

29. Clancy TE, Sengupta TP, Paulus J, Ahmed F, Duh MS, Kulke MH. Alkaline phosphatase predicts survival in patients with metastatic neuroendocrine tumors. *Dig Dis Sci* 2006;51:877–884.

30. Janson ET, Holmberg L, Stridsberg M, et al. Carcinoid tumors: analysis of prognostic factors and survival in 301 patients from a referral center. *Ann Oncol* 1997; 8:685–690.

31. Quaedvlieg PF, Visser O, Lamers CB, Janssen-Heijen ML, Taal BG. Epidemiology and survival in patients with carcinoid disease in The Netherlands. An epidemiological study with 2391 patients. *Ann Oncol* 2001;12:1295–1300.

32. Mazzaglia PJ, Berber E, Milas M, Siperstein AE. Laparoscopic radiofrequency ablation of neuroendocrine liver metastases: a 10-year experience evaluating predictors of survival. *Surgery* 2007;142:10–9.

33. Teunissen JJ, Kwekkeboom DJ, Krenning EP. Quality of life in patients with gastro-enteropancreatic tumors treated with [^{177}Lu-DOTA^0,Tyr^3]octreotate. *J Clin Oncol* 2004;22:2724–2729.

34. Virgolini I, Britton K, Buscombe J, Moncayo R, Paganelli G, Riva P. In- and Y-DOTA-lanreotide: results and implications of the MAURITIUS trial. *Semin Nucl Med* 2002;32:148–155.

35. Baum RP, Soldner J, Schmucking M, Niesen A. Peptidrezeptorvermittelte Radiotherapie (PRRT) neuroendokriner Tumoren Klinischen Indikationen und Erfahrung mit ^{90}Yttrium-markierten Somatostatinanaloga. *Der Onkologe* 2004;10:1098–1110.

36. Baum RP, Soldner J, Schmucking M, Niesen A. Intravenous and intra-arterial peptide receptor radionuclide therapy (PRRT) using Y-90-DOTA-Tyr3-octreotate (Y-90-DOTA-TATE) in patients with metastatic neuroendocrine tumors. *Eur J Nucl Med Mol Imaging* 2004;31:S238 [abstract].

37. Faiss S, Pape UF, Bohmig M, et al. Prospective, randomized, multicenter trial on the antiproliferative effect of lanreotide, interferon alfa, and their combination for therapy of metastatic neuroendocrine gastroenteropancreatic tumors—the International Lanreotide and Interferon Alfa Study Group. *J Clin Oncol* 2003;21:2689–2696.

38. Moertel CG. Treatment of the carcinoid tumor and the malignant carcinoid syndrome. *J Clin Oncol* 1983;1:727–740.

39. Moertel CG, Hanley JA, Johnson LA. Streptozocin alone compared with streptozocin plus fluorouracil in the treatment of advanced islet-cell carcinoma. *N Engl J Med* 1980;303:1189–1194.

40. Moertel CG, Lefkopoulo M, Lipsitz S, Hahn RG, Klaassen D. Streptozocin-doxorubicin, streptozocin-fluorouracil or chlorozotocin in the treatment of advanced islet-cell carcinoma. *N Engl J Med* 1992;326:519–523.

41. Cheng PN, Saltz LB. Failure to confirm major objective antitumor activity for streptozocin and doxorubicin in the treatment of patients with advanced islet cell carcinoma. *Cancer* 1999;86:944–948.

42. Van Hazel GA, Rubin J, Moertel CG. Treatment of metastatic carcinoid tumor with dactinomycin or dacarbazine. *Cancer Treat Rep* 1983;67:583–585.

43. Bukowski RM, Tangen CM, Peterson RF, et al. Phase II trial of dimethyltriazenoimidazole carboxamide in patients with metastatic carcinoid. *Cancer* 1994;73:1505–1508.

44. Ritzel U, Leonhardt U, Stockmann F, Ramadori G. Treatment of metastasized midgut carcinoids with dacarbazine. *Am J Gastroenterol* 1995;90:627–631.

45. Andreyev HJ, Scott-Mackie P, Cunningham D, et al. Phase II study of continuous infusion fluorouracil and interferon alfa-2b in the palliation of malignant neuroendocrine tumors. *J Clin Oncol* 1995;13:1486–1492.

46. Neijt JP, Lacave AJ, Splinter TA, et al. Mitoxantrone in metastatic apudomas: a phase II study of the EORTC Gastro-Intestinal Cancer Cooperative Group. *Br J Cancer* 1995;71:106–108.

47. Ansell SM, Pitot HC, Burch PA, Kvols LK, Mahoney MR, Rubin J. A Phase II study of high-dose paclitaxel in patients with advanced neuroendocrine tumors. *Cancer* 2001;91:1543–1548.

48. Kulke MH, Stuart K, Enzinger PC, et al. Phase II study of temozolomide and thalidomide in patients with metastatic neuroendocrine tumors. *J Clin Oncol* 2006;24:401–406.

49. Ekeblad S, Sundin A, Janson ET, et al. Temozolomide as monotherapy is effective in treatment of advanced malignant neuroendocrine tumors. *Clin Cancer Res* 2007;13:2986–2991.

50. Kulke MH, Kim H, Clark JW, et al. A Phase II trial of gemcitabine for metastatic neuroendocrine tumors. *Cancer* 2004;101:934–939.

51. Sun W, Lipsitz S, Catalano P, Mailliard JA, Haller DG, Eastern Cooperative Oncology Group. Phase II/III study of doxorubicin with fluorouracil compared with streptozocin

with fluorouracil or dacarbazine in the treatment of advanced carcinoid tumors: Eastern Cooperative Oncology Group Study E1281. *J Clin Oncol* 2005;23:4897–4904.

52. Kulke MH, Wu B, Ryan DP, et al. A phase II trial of irinotecan and cisplatin in patients with metastatic neuroendocrine tumors. *Dig Dis Sci* 2006;51:1033–1038.

53. Ansell SM, Mahoney MR, Green EM, Rubin J. Topotecan in patients with advanced neuroendocrine tumors: a phase II study with significant hematologic toxicity. *Am J Clin Oncol* 2004;27:232–235.

54. Kulke MH, Kim H, Stuart K, et al. A phase II study of docetaxel in patients with metastatic carcinoid tumors. *Cancer Invest* 2004;22:353–9.

55. O'Toole D, Hentic O, Corcos O, Ruszniewski P. Chemotherapy for gastro-enteropancreatic endocrine tumours. *Neuroendocrinology* 2004;80 (Suppl. 1): 79–84.

56. Rolleman EJ, Valkema R, de Jong M, Kooij PP, Krenning EP. Safe and effective inhibition of renal uptake of radiolabelled octreotide by a combination of lysine and arginine. *Eur J Nucl Med Mol Imaging* 2003;30:9–15.

57. Rolleman EJ, Bernard BF, Breeman WA, et al. Molecular imaging of reduced renal uptake of radiolabelled [$DOTA^0$,Tyr^3]octreotate by the combination of lysine and Gelofusine in rats. *Nuklearmedizin* 2008;47:110–115.

58. Sasse AD, Clark LG, Sasse EC, Clark OA. Amifostine reduces side effects and improves complete response rate during radiotherapy: results of a meta-analysis. *Int J Radiat Oncol Biol Phys* 2006;64:784–791.

59. Rolleman EJ, Forrer F, Bernard B, et al. Amifostine protects rat kidneys during peptide receptor radionuclide therapy with [^{177}Lu-$DOTA^0$,Tyr^3]octreotate. *Eur J Nucl Med Mol Imaging* 2007;34:763–771.

60. De Jong M, Valkema R, Jamar F, et al. Somatostatin receptor-targeted radionuclide therapy of tumors: preclinical and clinical findings. *Semin Nucl Med* 2002;32:133–140.

61. Behe M, Koller S, Püsken M, et al. Irradiation-induced upregulation of somatostatin and gastrin receptors *in vitro* and *in vivo*. *Eur J Nucl Med Mol Imaging* 2004;31:S237 [Abstract].

62. Oddstig J, Bernhardt P, Nilsson O, Ahlman H, Forssell-Aronsson E. Radiation-induced up-regulation of somatostatin receptor expression in small cell lung cancer *in vitro*. *Nucl Med Biol* 2006;33:841–846.

63. Capello A, Krenning E, Bernard B, Reubi JC, Breeman W, de Jong M, ^{111}In-labelled somatostatin analogues in a rat tumour model: somatostatin receptor status and effects of peptide receptor radionuclide therapy. *Eur J Nucl Med Mol Imaging* 2005;32:1288–1295.

64. Melis M, Forrer F, Capello A, et al. Up-regulation of somatostatin receptor density on rat CA20948 tumors escaped from low dose [^{177}Lu-$DOTA^0$,Tyr^3]octreotate therapy. *Q J Nucl Med Mol Imaging* 2007;51:324–333.

65. Ginj M, Zhang H, Waser B, et al. Radiolabeled somatostatin receptor antagonists are preferable to agonists for *in vivo* peptide receptor targeting of tumors. *Proc Natl Acad Sci USA* 2006;103:16436–16441.

66. Wong JY, Shibata S, Williams LE, et al. A Phase I trial of ^{90}Y-anti-carcinoembryonic antigen chimeric T84.66 radioimmunotherapy with 5-fluorouracil in patients with metastatic colorectal cancer. *Clin Cancer Res* 2003;9:5842–5852.

67. Kong G, Lau E, Ramdave S, Hicks RJ. High-dose In-111 octreotide therapy in combination with radiosensitizing 5-FU chemotherapy for treatment of SSR-expressing neuroendocrine tumors. *J Nucl Med* 2005;46 (Suppl. 2): 151P [Abstract].

68. Rich TA, Shepard RC, Mosley ST. Four decades of continuing innovation with fluorouracil: current and future approaches to fluorouracil chemoradiation therapy. *J Clin Oncol* 2004;22:2214–2232.

69. Dunst J, Reese T, Sutter T, et al. Phase I trial evaluating the concurrent combination of radiotherapy and capecitabine in rectal cancer. *J Clin Oncol* 2002;20:3983–3991.

70. van Essen M, Krenning EP, Kam BL, de Herder WW, van Aken MO, Kwekkeboom DJ. Report on short-term side effects of treatments with ^{177}Lu-octreotate in combination with capecitabine in seven patients with gastroenteropancreatic neuroendocrine tumours. *Eur J Nucl Med Mol Imaging* 2008;35:743–748.

71. van Essen M, Krenning EP, Kooij PP, et al. Effects of therapy with [^{177}Lu-DOTA[0], Tyr3] octreotate in patients with paraganglioma, meningioma, small cell lung carcinoma, and melanoma. *J Nucl Med* 2006;47:1599–160.

72. Forrer F, Riedweg I, Maecke HR, Mueller-Brand J. Radiolabeled DOTATOC in patients with advanced paraganglioma and pheochromocytoma. *Q J Nucl Med Mol Imaging* 2008;52:334–40.

73. Teunissen JJ, Kwekkeboom DJ, Krenning EP. Staging and treatment of differentiated thyroid carcinoma with radiolabeled somatostatin analogs. *Trends Endocrinol Metab* 2006;17:19–25.

CHAPTER 5

Targeted Radiotherapy of Central Nervous System Malignancies

MICHAEL R. ZALUTSKY, DAVID A. REARDON, AND DARELL D. BIGNER

5.1 MALIGNANT BRAIN TUMORS

Although significant advances have been made during the past decade in the treatment of many types of cancer, primary malignant brain tumors remain a major challenge to the oncologist. While the incidence of brain tumors is not as high as breast, prostate, lung, or colon cancer, they account for more than 1% of all cancers diagnosed in the United States (1). Each year, approximately 18,500 new cases of malignant glioma are diagnosed with about 13,100 patients dying of this disease. This annual mortality rate exceeds that of melanoma, Hodgkin's disease, multiple myeloma, and cancers of the uterus, esophagus, stomach, kidney, and bladder (1).

Despite aggressive multimodality treatment strategies, the outcome for most primary central nervous system (CNS) tumors, particularly, the most common and deadliest primary, adult malignant brain tumor, glioblastoma multiforme (GBM), remains unacceptably poor. For example, according to a recent population-based study, the overall survival of newly diagnosed GBM patients who received state-of-the-art surgery, imaging, radiotherapy, and chemotherapy was only 42.4% at 6 months, 17.7% at 1 year, and 3.3% at 2 years (2). Although chemotherapy has provided only marginal therapeutic benefit at best for patients with these tumors (3, 4), a recent, randomized, Phase III study demonstrated that temozolomide (a second-generation imidazotetrazine derivative that methylates specific DNA sites, including the O^6 position of guanine (5)) improved outcome when administered during and following radiotherapy compared with outcome with radiotherapy alone. However, this combination therapy, while more effective than external beam radiation alone, provided a median progression-free survival and overall survival of only 7.9 and 14.6 months, respectively (6).

Monoclonal Antibody and Peptide-Targeted Radiotherapy of Cancer, Edited by Raymond M. Reilly
Copyright © 2010 John Wiley & Sons, Inc.

The prognosis for patients with recurrent malignant brain tumors is even more dismal because salvage therapies following progression after conventional surgery, radiotherapy, and temozolomide are ineffective. Thus, there is no uniformly accepted standard of care for recurrent patients with the result that there is no well-defined benchmark for evaluating new treatments in this patient population. However, a useful comparator for this purpose are the cumulative results of eight consecutive clinical trials performed at a single institution that indicate a median progression-free survival of only 9 and 12 weeks for recurrent GBM and grade III malignant glioma patients, respectively (7). Somewhat more favorable outcomes have been reported in patients receiving temozolomide at first recurrence (median progression-free-survival for GBM of 12.4 weeks and 21.6 weeks for grade III patients (8, 9). Unfortunately, toxicity to normal CNS tissues from conventional radiotherapy and chemotherapy interventions contribute to an extremely limited quality of life among patients who survive beyond the median (10, 11).

Recursive partitioning analysis, a statistical methodology that creates a regression tree according to prognostic significance, has indicated some of the clinical variables associated with poor outcome including older age, poor performance status, and lack of resectability (12). For the purposes of developing new therapeutic strategies, it is important to focus on the biologic factors that interfere with the effective treatment of malignant brain tumors. Some of these are also operant in other types of cancer such as intratumor and intertumor heterogeneity, as well as *de novo* and acquired mechanisms of chemoresistance and DNA damage repair. Ineffective delivery of therapeutics into the CNS and tumor microenvironment is one of the most significant barriers to brain tumor treatment. The blood brain barrier drastically compromises delivery of drugs to brain tumors with the exception of small, lipid-soluble, or actively transported molecules (13). This magnifies the consequences of tumor hemodynamic parameters such as elevated interstitial pressure within the tumor and dysfunctional tumor vasculature (14), with the result that homogeneous delivery of a drug at therapeutically meaningful levels is very difficult to achieve. Finally, two characteristics of malignant gliomas—their high rates of genetic abnormalities and dysfunctional cell signaling pathways, result in very aggressive tumor cell proliferation, invasion, and angiogenesis (15).

5.2 RATIONALE FOR LOCOREGIONAL THERAPY

Local recurrence is the bane of brain tumor therapy and nearly all malignant gliomas recur. Despite the fact that surgery and radiotherapy are in essence localized approaches, the majority of malignant gliomas recur at or adjacent to the primary tumor site (16, 17), indicating that local control remains elusive. Because of the propensity of malignant gliomas to recur locally following conventional therapy, and the difficulties in achieving effective drug delivery following systemic administration as described above, several innovative treatment approaches have emerged that are focused on achieving better local control as a first step in improving overall outcome. Clearly, locoregional administration is a key aspect of these strategies because it provides the means to achieve a much higher concentration of the drug in the tumor by bypassing the blood–brain barrier while limiting systemic exposure and resultant toxicity. Moreover, local administration may also be of benefit in facilitating delivery of the therapeutic along the white matter tracts that are frequently followed by glioma cells as they infiltrate normal brain (18).

Locoregional approaches currently under active investigation vary with regard to both the nature of the local delivery method and the characteristics of the therapeutic itself. For example, carmustine-impregnated biodegradable wafers have been implanted into the wall of the resection cavity (19, 20). Other groups have utilized convection-enhanced delivery (microinfusion) to circumvent the diffusion limitation of macromolecules to distribute immunotoxins throughout the tumor bed (21–23). Another approach is to place an inflatable balloon device into the tumor resection cavity, which contains a solution of sodium 3-[^{125}I]-iodo-4-hydroxybenzenesulfonate to provide focal brachytherapy to the cavity margins (24–27).

5.3 TARGETED RADIOTHERAPY OF BRAIN TUMORS

A severe limitation of conventional radiotherapy for brain tumor treatment is its nonspecific nature, resulting in toxicity to normal brain within the radiation field. Because malignant glioma is a highly infiltrative disease, small clusters or single glioma cells can be present 4–7 cm from the primary tumor site and sometimes, even in the contralateral hemisphere (28, 29). Thus, it is exceedingly difficult to achieve tumor control without inducing excessive toxicity to normal brain, which is the predominant component in regions beyond the primary tumor margin. Targeted radiotherapy is an attractive approach to addressing this conundrum because it can increase the tumor selectivity of radiation dose deposition through the use of a molecular vehicle to selectively deliver a radionuclide to malignant glioma cells.

By targeting a receptor or other molecule that is overexpressed or uniquely expressed in brain tumors, it should be possible to achieve a significantly more favorable therapeutic index than possible with conventional external beam therapy. An important consideration is selecting a radionuclide with decay properties that are well matched to the characteristics of brain tumors (30). High energy β-emitters such as ^{90}Y (2.29 MeV $E_{\beta max}$) have relatively long ranges in tissue, which would maximize irradiation of receptor negative tumor cell populations through cross fire effects and also compensate for diffusion limitations of macromolecular carriers such as monoclonal antibodies (mAbs). However, there is a significant disadvantage to using long range β-emitters for treating brain tumors—the deposition of much of their decay energy in normal brain tissue. On the other hand, α-emitters such as ^{211}At have higher energies (5.87–7.45 MeV) yet have ranges in tissue of 55–80 μm, equivalent to only a few cell diameters, making them particularly well suited to the treatment of small foci of malignant glioma cells infiltrating normal brain. However, the disadvantage of α-particles is that they lack a significant cross fire effect, making homogeneous tumor delivery of the targeted radiotherapeutic a critical requirement. Most approaches for brain tumor targeted radiotherapy have utilized ^{131}I, presumably because this moderate-low energy β-emitter (0.606 MeV $E_{\beta max}$) offers a reasonable compromise between maximizing the homogeneity of tumor radiation dose deposition and minimizing radiation dose to normal brain.

Initial targeted radiotherapy trials on brain tumor patients involved intravenous administration of mAbs reactive with the epidermal growth factor (EGF) receptor (31, 32). A number of other molecules that are overexpressed on glioma cells compared with normal brain have been considered as targets for brain tumor radio-immunotherapy (RIT). These include the human transferrin receptor (33), the human neural cell adhesion molecule (34), the L1 cellular adhesion molecule (35), the lacto-series gangliosides 3′-isoLM1 and 3′,6′-isoLD1 (36), the chondroitin sulfate proteo-glycan GP240 (37), the transmembrane glycoprotein NMB (GPNMB) (38), and the multidrug resistance protein 3 (MRP3) (39). However, the preponderance of clinical RIT research in malignant glioma has involved radiolabeled mAbs reactive with the tenascin-C molecule administered, for reasons discussed in the previous section, directly into surgically created resection cavities. For this reason, this chapter predominantly focuses on the clinical development of radiolabeled anti-tenascin-C mAbs but also describes other targeted radiotherapeutics that have been investigated for the treatment brain tumor patients.

5.4 RATIONALE FOR TENASCIN-C AS A TARGET FOR RADIONUCLIDE THERAPY

Unlike most of the molecular targets exploited in radionuclide therapy, which are cell-surface markers, tenascin-C is found in the extracellular matrix. Although this large six-armed glycoprotein is expressed widely in the stroma and mesenchyme of tissues at various stages of differentiation, in normal adult tissues, it is primarily expressed in the liver, kidney, spleen, and papillary dermis (40, 41). Two problems for targeted radiotherapy are associated with tenascin-C expression in these normal organs—these organs represent highly accessible pools of target that can compete with tumor for uptake of the radiolabeled molecule and they could be subjected to dose-limiting radiotoxicity. For this reason, locoregional administration of tenascin-C targeted radiotherapeutics is recommended.

Tenascin-C expression is associated with a variety of tumors including breast, squamous cell, lung and prostate carcinomas, melanoma, and malignant glioma but also in other pathologic states such as inflammation and wound healing (42–44). The tenascin-C molecule has been well characterized by biochemical and ultrastructural analyses. The complete nucleotide and deduced amino acid sequence of full-length tenascin-C cDNA are known (45). Tenascin-C exists as a six-armed, disulfide-linked polymer with each arm ending in a characteristic knob, and consists of two T-shaped junctions, each joining three arms (46, 47). A central globular core joins the two trimers at their T-junction. Each arm contains multiple domains that can bind various cell-surface receptors such as integrins as well as extracellular matrix components, notably, fibronectin (FN) (48). As shown in Fig. 5.1, each 200–300 kDa arm contains 14 epidermal growth factor-like repeats, immediately adjacent to the amino terminus,

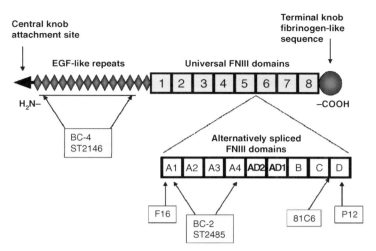

FIGURE 5.1 Schematic diagram of a single arm of the tenascin-C hexamer. The structural arrangement, beginning at the N-terminus, consists of the central knob attachment site (triangle), 14 epidermal growth factor-like domains, (diamonds), 8–15 fibronectin type III (FNIII) repeats, and a fibrinogen-like domain at the C-terminus. Multiple isoforms of tenascin-C exist due to variations in the alternatively spliced FNIII domains. The approximate epitope binding sites of the anti-tenascin-C mAbs discussed in the text are indicated by upward arrows with the mAb designations given in the boxes below the arrows.

followed by 8–15 FN type-III repeats, with a fibrinogen-like sequence located at the carboxy terminus.

Different forms of tenascin-C are generated by alternate splicing and variable transcription of up to 8 FNIII repeats that are inserted between the universally conserved repeat numbers 5 and 6. From a targeting perspective, it is important to note that there is variable expression of the different isoforms of tenascin-C in normal organs and tumors (49, 50). Tenascin-C is expressed in more than 90% of gliomas with the predominant isoforms in these tumors having molecular weights of 220, 230, and 320 kDa (50). The level of tenascin-C expression increases with degree of malignancy and is predominantly located around tumor blood vessels, with this feature becoming more striking with advancing tumor grade (51–55). Because of this perviascular pattern of expression, tenascin-C could serve as a target for achieving focal irradiation of tumor vasculature using short range radiation such as Auger electrons or α-particles (56), which could lead to the destruction of tumor cells not directly in the radiation path. Finally, numerous studies have documented that tenascin-C plays a key role in multiple physiological processes vital to tumor progression including angiogenesis, adhesion, migration, and proliferation (40, 41, 57–63).

5.4.1 Tenascin-C Targeting Vehicles

Tenascin-C is an attractive molecular target for the development of malignant glioma therapeutics because it is over expressed in the vast majority of brain tumors at levels that are considerably higher than most cell-surface targets. A number of murine mAbs that bind to different domains of the tenascin-C molecule have been developed with the intention of using them as targeting vehicles for brain tumor radionuclide therapy. The approximate binding epitopes for these mAbs are shown in Fig. 5.1. The BC-4 and next-generation ST2146 mAbs both bind to an epitope within the EGF-like repeat region that is present on all tenascin-C isoforms (63, 64). On the other hand, both the BC-2 and the ST2485 mAbs react with an epitope found on alternatively spliced FNIII A1 and A4 repeats, which share 83% homology (63, 65). In interpreting results obtained with these mAbs, it should be noted that following the completion of initial clinical trials with the BC-4 mAb, it was discovered that its hybridoma clone also generated an additional nonfunctional light chain. To circumvent this problem, ST2146 was developed as a BC-4 replacement (64). Subsequently, ST2485, which exhibits an elevated affinity to tenascin-C compared with BC-2, was developed to combine with ST2146 to enhance tumor targeting (65). Both ST2146 and ST2485 show promise as vehicles for targeted radiotherapy of brain tumors; however, clinical trials with these mAbs have yet to be reported. The anti-tenascin-C mAb developed by our group, 81C6, binds to an epitope within the alternatively spliced FNIII CD region (43).

Several promising alternative vehicles for targeting tenascin-C are in the preclinical development stage. Antibody phage display technology (see Chapter 1) has been utilized to develop two human recombinant mAbs, F16 and P12, which bind to epitopes within the A1 and D domains of the FNIII region, respectively (66). After affinity maturation, small $\sim$75 kDa immunoproteins (SIP), consisting of scFv- εCH4 dimers, were constructed, radioiodinated, and their biodistribution evaluated in a human glioma xenograft model. The SIP(F16) but not the SIP(P12) construct exhibited rapid preferential tumor targeting, albeit at levels about an order of magnitude lower than seen with the intact murine mAbs described above. Aptamers that bind with high affinity to tenascin-C have also been developed (67, 68). These oligonucleotide molecules, identified by the SELEX (systematic evolution of ligands by exponential enrichment) technology, have molecular weights of 8–15 kDa and affinities in the nanomolar range. In tumor bearing mice, ^{99m}Tc-labeled TTA1 anti-tenascin-C apatamer rapidly exhibited high tumor-to-normal tissue accumulation ratios (69). Again, the absolute magnitude tumor uptake was considerably lower than that achievable with intact mAbs. Thus, the pharmacokinetics observed to date with SIP and aptamers seem better suited to diagnostic rather than therapeutic applications.

In the sections that follow, the current status of clinical brain tumor targeted radiotherapy with these anti-tenascin-C mAbs, both directly labeled and as an essential component of a pretargeting strategy, will be reviewed. We shall focus primarily on the Phase I and II clinical studies at our institution performed with 81C6, which is now being investigated in a Phase III multicenter trial.

5.4.2   BC-2 and BC-4 mAbs: The Italian Experience

Riva and colleagues have conducted several clinical trials evaluating the locoregional administration of mAbs BC-2 and BC-4 labeled with either ^{131}I or ^{90}Y for the treatment of patients with malignant glioma (70–72). Unfortunately, no distinction was made between the two mAbs in these protocols, making it impossible to discern whether tenascin-C binding epitope played any role in therapeutic efficacy or normal tissue toxicity. Differentiation of response results for newly diagnosed and recurrent patients also was not done. On the other hand, in most of these trials, patients were evaluated based on tumor size at the time of treatment.

In a Phase II study with ^{131}I-labeled anti-tenascin-C BC-2 and BC-4 mAbs, a total of 91 patients were treated, including 74 with GBM, 9 with anaplastic astrocytoma (AA), 7 with anaplastic oligodendroglioma (AO), and 1 with oligodendroglioma (O) (73). Of these, 52 patients were classified as having small (defined as less than $2\,cm^3$) or undetectable residual tumor, with the remainder having larger tumors. The patient population was nearly equally divided between those with recurrent tumors ($n = 44$) and newly diagnosed malignancies ($n = 47$). Patients received between 3 and 10 locoregional injections of ^{131}I-labeled mAb with cumulative radioactivities of up to 2035 MBq. The median effective half-life for clearance of ^{131}I from the tumor cavity was 57.1 h and the mean radiation dose delivered to the surgically created resection cavity (SCRC) walls was 150 Gy. The median survival for patients with GBM, AA, and AO was 19, >46, and 23 months, respectively; no distinction was made between recurrent and newly diagnosed patient populations. In GBM patients, the response rate in patients with smaller volume disease, 56.7%, was more favorable than in those with larger tumors (17.8%).

The Italian group also performed a similar study with ^{90}Y-labeled BC-2 and BC-4 mAbs, investigating the potential effects of utilizing a radiometal emitting β-particles with a range greater than those emitted by [131]I (73). The patient cohort consisted of 43 evaluable patients (35 GBM, 6 AA, 2 O) and 16 had small or undetected residual tumor and 19 had larger lesions. Of these, 19 had recurrent tumors and 16 had newly diagnosed disease. The treatment protocol consisted of between 3 and 5 cycles of ^{90}Y-labeled anti-tenascin-C mAb up to a cumulative radioactivity of 3145 MBq. The median effective half-life of ^{90}Y in the tumor cavity was 43.2 h and the mean radiation dose delivered to the surgically created resection cavity interface was 280 Gy, a value nearly twice that observed for [131]I. The median survival from the time of initial diagnosis was 90 months for patients with AA and 20 months for patients with GBM. In patients with smaller volume disease, the response rate for ^{90}Y-labeled mAb treatment, 56.3%, was nearly identical to that observed with [131]I. However, in patients with larger tumors at the time of RIT, the response rate with ^{90}Y was somewhat higher that that observed with [131]I (26.3% versus 17.8%), which might reflect the longer β-particle range of [90]Y.

A more recent study was performed at hospitals in Cesna, Italy and Munich, Germany exclusively with anti-tenascin-C mAb BC-4 due to the lack of availability of the BC-2 mAb. A total of 37 patients with malignant brain tumors (13 AA, 24 GBM) received locoregional injections of either ^{131}I-labeled (Cesna and Munich) or ^{90}Y-labeled (Cesna) BC-4 (74). The treatment protocol in Cesna involved multiple cycles (mean 3, maximum 8) of radiolabeled mAbs at intervals of 6 to 8 weeks, while that in Munich utilized a single injection of 1100 MBq ^{131}I-labeled BC-4. The median survival for patients at the two institutions was not differentiated and was 17 months for those with GBM. Furthermore, survival results were not stratified according to radionuclide, and it was not mentioned whether the patient population consisted of those with recurrent, newly diagnosed brain tumors, or both. The 5-year survival probability reported for those with AA was about 85%. The low incidence of side effects, even in patients receiving multiple injections of radiolabeled mAbs, is encouraging.

5.4.3   Antitenascin-C mAb 81C6: The Duke Experience

5.4.3.1   *Rationale for Locoregionally Administered ^{131}I-Labeled 81C6 Targeted Radiotherapy*   With the goal of exploiting the overexpression of tenascin-C on brain tumors for molecularly targeted treatment of these malignancies, mAb 81C6 was developed (42, 43). This murine IgG$_{2b}$ binds to an epitope within the alternatively spliced FNIII CD region (Fig. 5.1) (75). A potential advantage of mAbs, like 81C6 that bind to alternatively spliced regions of the tenascin-C molecule versus those present on all isoforms is that this should increase the relative binding to tenascin-C in tumor compared with normal organs such as liver and spleen (76). Notably, 81C6 does not react with normal brain tissue (50).

A series of preclinical studies performed in athymic mice and rats with subcutaneous and intracranial human glioma xenografts provided evidence of the specificity of radioiodinated 81C6 for tenascin-C-expressing tumors *in vivo* (77–79). Many of these were done in paired label format, which provided critical documentation that radiolabeled mAb uptake in intracranial xenografts was specific and not just due to blood–brain barrier disruption. Experiments in athymic mice with subcutaneous D54 MG human glioma xenografts demonstrated the therapeutic efficacy of intravenously administered ^{131}I-labeled 81C6 (80) and in subsequent studies in athymic rats with intracranial tumors, significant prolongation of median survival, and even a few apparent cures, were observed (81).

Three diagnostic-level clinical studies performed in glioma patients with radioiodinated 81C6 provided critical information that influenced the design of subsequent targeted radiotherapy protocols. In the first study, nine patients with malignant brain tumors received paired-label intravenous injection of ^{131}I-labeled 81C6 and ^{125}I-labeled 45.6 isotype control mAb 29–77 h prior to scheduled tumor resection (82). Tumor tissue obtained at surgery exhibited five times higher levels of ^{131}I-labeled 81C6 compared with coadministered ^{125}I-labeled 45.6 and had an average of 25 times higher levels of ^{131}I than in normal brain. This confirmed that ^{131}I-labeled 81C6 accumulation in malignant brain tumors was both selective and specific. Next, a paired injection protocol demonstrated that intracarotid administration offered no tumor delivery advantage compared with intravenous injection (83). Finally, an 81C6 protein dose escalation protocol, monitored by single photon emission computed tomography (SPECT) imaging, indicated that therapeutically relevant tumor radiation doses could not be obtained after intravenous mAb injection without subjecting tenascin-C expressing liver and spleen to excessive radiation doses (84). Based on these observations, our clinical targeted radiotherapy protocols with radiolabeled 81C6 mAb utilized locoregional administration, most frequently into surgically created glioma resection cavities.

5.4.3.2 *Phase I Evaluation of Locoregionally Administered ^{131}I-Labeled*

81C6 An essential feature of all the targeted radiotherapy protocols described in the following sections is the use of compartmental approaches for the administration of radiolabeled 81C6 mAbs, with locoregional delivery into either the SCRC, spontaneous tumor cysts, or the intrathecal space for treating neoplastic meningitis. In addition to rapidly achieving high local concentrations of the labeled mAb, regional administration also offered the means to minimize systemic toxicity, circumvent the blood–brain barrier, overcome potentially high tumor interstitial pressure, and minimize the possible impact of systemic catabolism of the labeled mAb on its availability for tumor targeting. In interpreting the results of these trials, it should be noted that in most cases, the treatment protocols also included standard external beam radiotherapy and chemotherapy, which represents standard of care for these patient populations.

Our initial Phase I study included 31 patients (18 with GBM) with either leptomeningeal neoplasms or brain tumor resection cavities with subarachnoid communication who were treated with a single injection of ^{131}I-labeled 81C6 into the intrathecal space (85). The administered ^{131}I activity levels ranged from 1480 to 3700 MBq, with the maximum tolerated dose (MTD) determined to be 2960 MBq. The dose-limiting toxicity was hematologic and no nonhematologic grade 3 or 4 toxicity was observed. A partial radiographic response was seen in one patient, and disease stabilization occurred in 13 patients (42%). Five patients remained progression free for more than 409 days after receiving ^{131}I-labeled 81C6 treatment.

All subsequent trials involved compartmental delivery by direct injection of ^{131}I-labeled 81C6 into a SCRC via a Rickham catheter inserted at the time of resection (Fig. 5.2). Using this approach, two Phase I studies were conducted in parallel to

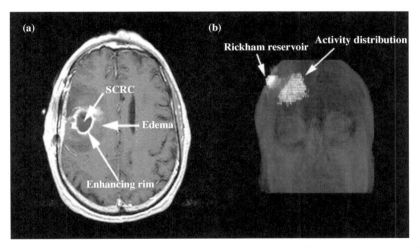

FIGURE 5.2 Locoregional monoclonal antibody radiotherapy following complete resection of brain tumor, involving surgically created resection cavity. (a) Gadolinium-enhanced T1-weighted MRI showing the SCRC following complete resection of a right parietal glioblastoma multiforme. (b) 3D view of registered MRI and single photon emission computed tomography images illustrating the typical distribution of ^{131}I-labeled 81C6 in the SCRC of a patient.

determine the MTD of ^{131}I-labeled 81C6 administered into the SCRC of patients with both recurrent and newly diagnosed malignant glioma. Entry criteria for these protocols included histopathologic confirmation of diagnosis, tumor localization within the supratentorial compartment, and a maximum of 1 cm residual enhancement on postoperative MRI. An important additional requirement was the demonstration of tumor reactivity with the 81C6 mAb by immunohistochemistry. Subsequent to the RIT procedure, most patients received systemic chemotherapy; those with newly diagnosed tumors also underwent conventional external beam radiotherapy.

The Phase I study in the recurrent malignant brain tumor population enrolled 34 patients including 26 (77%) with GBM (86). The administered dose of ^{131}I-labeled 81C6 ranged from 740 to 4440 MBq and the MTD was determined to be 3700 MBq. Dose-limiting toxicity in this patient population was neurologic. Even with the dose escalation design of this study, the median survival results were highly encouraging at 56 and 60 weeks for recurrent GBM patients and all treated patients, respectively.

Figure 5.3 presents serial MRI and [^{18}F]fluorodeoxyglucose PET images for a representative patient receiving ^{131}I-labeled 81C6 therapy. Following gross total resection, the rim minimally enhances on MRI while after radiolabeled mAb administration, the rim enhancement gradually becomes more prominent while the SCRC retracts. The corresponding PET images show an absence of metabolic activity in the region of the SCRC.

The parallel Phase I study in patients with newly diagnosed malignant glioma involved a total of 42 patients, 32 (76%) of which were diagnosed with GBM (87). In this trial, administered doses of ^{131}I-labeled 81C6 ranged from 740 to 6660 MBq. Dose-limiting toxicity again was neurologic and developed in one of seven patients at 4440 MBq , two of three patients at 5180 MBq, two of seven patients at 5920 MBq, and the one patient treated with 6660 MBq . The MTD of ^{131}I-labeled 81C6 administered into the SCRC for patients with newly diagnosed and previously untreated tumors was

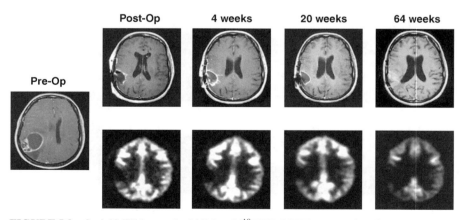

FIGURE 5.3   Serial MRI (top and middle) and [^{18}F]FDG PET scan results of a representative patient after [131]I 81C6 mAb therapy. Corresponding [^{18}F]FDG PET scan images (bottom) demonstrate a lack of increased metabolic activity in region of the surgically created resection cavity. Reproduced with permission from Ref. 100.

established to be 4440 MBq, a value 20% higher than that observed in patients with recurrent disease. Presumably, this difference reflects the fact that the recurrent population had received a standard course of external beam radiation prior to receiving ^{131}I-labeled 81C6. Noteworthy median survival results were observed in this trial as well, with values of 69 and 79 weeks measured for patients with GBM and for all patients, respectively. Patient-specific radiation dosimetry analysis was performed, which indicated that the 2 cm thick annular region surrounding the SCRC interface received an average radiation dose of 32 Gy, with the individual patient doses ranging between 2 and 59 Gy (88).

5.4.3.3 Phase II Evaluation of ^{131}I-Labeled 81C6 in Newly Diagnosed Malignant Glioma   A Phase II trial was then conducted with the same eligibility criteria as those in the Phase I studies. Patients with newly diagnosed malignant glioma received an administered dose of 4440 MBq of ^{131}I-labeled 81C6 into the SCRC and subsequently were treated with external beam radiation and systemic chemotherapy (89). Of the 33 total patients that were enrolled, 27 (82%) had GBM. Nine patients (27%) developed reversible hematologic toxicity, consistent with leakage of ^{131}I from the SCRC into the systemic circulation over time, and histologically confirmed, treatment-related neurotoxicity occurred in five (15%) patients. Irreversible neurologic toxicity was associated with the SCRC being immediately adjacent to or contiguous with the compromised CNS functional center: the SCRC abutted the motor strip in all five patients who developed irreversible neuromotor toxicity, while in the single patient who developed aphasia, the SCRC bordered Broca's area.

The radiation dose received by the SCRC margins was calculated based on the SCRC volume measured by MRI and the clearance half time of ^{131}I from the SCRC as determined by serial imaging (88). The average SCRC volume in these patients was 10.45 cm^3 (range, 0.5–30.5 cm^3), the average biologic half-life of ^{131}I-labeled 81C6 in the SCRC was 87 h (range, 26–282 h), and the average residence time was 78 h (range, 34–169 h) (89). Based on these data, the average absorbed dose to the 2 cm SCRC rim was calculated as 48 Gy with the dose in individual patients ranging from 24 to 116 Gy. The median survival achieved for all patients and those with GBM was 86.7 and 79.4 weeks, respectively.

To better understand the survival benefit obtained with ^{131}I-labeled 81C6 targeted radiotherapy, we compared our results with outcomes for conventional radiotherapy and chemotherapy predicted by a recursive partitioning model (12). This model is designed to assess the prognostic impact of several pretreatment characteristics and treatment-related variables among patients with newly diagnosed brain tumors. Even though only limited comparisons could be done because of the sample size of our Phase II study, the results suggest that ^{131}I-labeled 81C6 treatment compares favorably with the standard approach in similar subpopulations of patients. For example, in patients less than 50 years old with a newly diagnosed GBM, the recursive partitioning model predicted a median survival of 55 weeks for conventional treatment compared with 87 weeks for targeted radiotherapy. Likewise, patients with a newly diagnosed GBM who were over age 50 but with a Karnofsky performance status greater than 70%

were predicted to have a median survival of only 39 weeks by the recursive partitioning model compared with a median survival of 65 weeks observed in this subpopulation after radioimmunotherapy.

It is also important to evaluate ^{131}I-labeled 81C6 treatment in the context of other strategies for delivering a boost radiation dose to brain tumors such as interstitial brachytherapy and stereotactic radiosurgery. These comparisons should be done not solely based on survival prolongation but also need to consider minimizing side effects, particularly those that compromise quality of life. The prolonged survival achieved with radioimmunotherapy on our study compared favorably to those reported with ^{125}I-interstitial brachytherapy or stereotactic radiosurgery (90–92). Moreover, 33–64% of patients treated with either of these latter approaches required reoperation to debulk radiation necrosis and relieve symptomatic mass effect; in marked contrast, only 2 of the 109 patients (1.8%) combined from our Phase I and II ^{131}I-labeled 81C6 studies required reoperation for symptomatic radiation necrosis.

It is also important to compare the result we obtained with radiolabeled anti-tenascin-C 81C6 mAb with those associated with new therapeutic strategies that do not involve radiation. Notably, the median survival achieved on our Phase II study exceeded that reported following the placement of carmustine-loaded polymers (Gliadel®; MGI Pharmaceuticals, Bloomington, MN, USA) into SCRC of patients with newly diagnosed malignant glioma (20). Finally, the results of our Phase II trial compare favorably with those obtained in patients receiving temozolomide concurrent with their course of external beam radiation (93), which has emerged as the most recent standard of care therapy for patients with malignant glioma.

5.4.4 Strategies for Improving the Efficacy of 81C6-Based Targeted Radiotherapy

5.4.4.1 *Patient-Specific Dosing to Deliver 44 Gy to the SCRC*   Because of regulatory restraints, all of the protocols described above involved administration of ^{131}I-labeled 81C6 based on the dose of radioactivity of [131]I. However, dosimetric analyses performed on the 42 newly diagnosed malignant glioma patients treated on our Phase I study revealed that this resulted in a wide range of radiation absorbed doses delivered to the 2 cm cavity margin due to variations in cavity volume and residence time (88). In this group of patients, the SCRC ranged from 2 to 81 cm^3 while the SCRC residence time ranged from 10 to 113 h. The average absorbed dose delivered from a fixed 4440 MBq radioactivity dose of ^{131}I-labeled 81C6 to the SCRC interface and 2 cm SCRC perimeter was 1435 and 32 Gy, respectively; however, the range in values for the interface (46–9531 Gy) and 2 cm perimeter (3–59 Gy) doses were extensive.

This motivated us to examine the relationship between radiation absorbed dose to the 2 cm cavity margin and histopathology results among 16 patients in whom progressive changes on serial MRI scans were observed following ^{131}I-labeled 81C6 administration. Stereotactic biopsies were obtained from the SCRC interface in 15 patients and one at autopsy, and the results were classified as either (a) tumor, (b) radionecrosis, or (c) a mixture of tumor and radionecrosis. In the five patients with biopsies revealing only tumor, the average absorbed dose to the 2 cm cavity margin for these patients was 25 Gy (range, 12–44 Gy), but none of these patients developed any type of toxicity. In contrast, in the five patients whose biopsy results demonstrated solely radionecrosis, the average absorbed dose to the 2 cm shell was 47 Gy (range, 34–55 Gy); two of these patients developed delayed neurotoxicity, and one developed both acute and delayed neurotoxicity.

Graphical analysis of these results suggested a qualitative relationship between absorbed dose to the 2 cm cavity margin and outcome (88): patients who received less than 44 Gy were most likely to develop tumor recurrence with no radionecrosis or treatment-related toxicity while those who received more than 44 Gy were most likely to develop radionecrosis with possible signs of clinical neurotoxicity. On this basis, it was determined that the optimal radiation boost dose to the 2 cm shell from ^{131}I-labeled 81C6 for patients with newly diagnosed malignant glioma is 44 Gy. Moreover, results from our Phase II clinical study revealed a causal relationship between boost absorbed dose and median survival for those patients as well as for patients undergoing other radiation-based treatment modalities (95). This analysis confirms the importance of tailoring the boost radiation dose to the 2 cm cavity margin in order to achieve an effective balance between maximizing tumor control and minimizing normal brain toxicity.

A pilot study has recently been completed to evaluate a patient-specific dosing strategy for ^{131}I-labeled 81C6 that was designed to achieve a 44 Gy boost radiation absorbed dose to the 2 cm SCRC margin (96). The trial was conducted in adult patients who were newly diagnosed with malignant glioma with the objectives of determining the feasibility, efficacy, and toxicity of administering ^{131}I-labeled 81C6 into the SCRC at a radioactivity dose that would deliver 44 Gy boost to the 2 cm SCRC margin. In order to determine the dose of ^{131}I-labeled 81C6 that would be required in each patient, 37–111 MBq of ^{123}I-labeled 81C6 was administered 3–7 days postoperatively into the SCRC and gamma camera imaging was done immediately and 2, 24, and 48 h later. From these, the effective half-life and biological clearance half-life of radioiodinated 81C6 was determined, and after correcting for the differences in physical half-life of ^{123}I and ^{131}I, the SCRC residence time of ^{131}I-labeled 81C6 was calculated. A three-dimensional reconstruction of the head and SCRC was performed based on 2 mm thick postoperative MRI images. The calculated SCRC volume was used to estimate the initial SCRC activity, where a uniform radioactivity concentration was assumed. The radioactivity dose of ^{131}I-labeled 81C6 predicted to achieve a 44 Gy boost to the 2-cm SCRC margin was calculated for each patient based on the measured SCRC residence time and cavity volume using previously described methods (88).

Subsequent to receiving the individualized radioactivity dose of ^{131}I-labeled 81C6, patients received conventional external beam radiotherapy and chemotherapy. A total of 21 patients were enrolled, including 16 with GBM and 5 with anaplastic astrocytoma. With the exception of one patient, it was possible to achieve the targeted boost dose of 44 Gy $\pm$ 10% to the SCRC. It should be noted that in this patient, it was necessary to deliberately decrease the amount of ^{131}I-labeled 81C6 from the desired level because of technical factors affecting fluid aspiration at the time of mAb administration. Toxicities attributable to this targeted radiotherapy protocol were mild and limited to reversible grade 3 neutropenia or thrombocytopenia ($n = 3$), CNS wound infections

($n = 3$), and headache ($n = 2$). Patients on this study have been followed for a median of 151 weeks, and 87% of GBM patients were alive at one year. The median overall survival for all patients and those with GBM were 96.6 and 90.6 weeks, respectively.

An important outcome of this pilot study is the demonstration that consistently achieving a 44 Gy boost dose to the SCRC margin is feasible with patient-specific dosing of ^{131}I-labeled 81C6. This protocol, which includes both a diagnostic level and a therapeutic level dose of radiolabeled mAb, followed by conventional external beam radiation therapy and systemic chemotherapy, was well tolerated. Furthermore, patient-specific dosing of ^{131}I-labeled 81C6 resulted in a median survival that exceeds that of historical controls treated with surgery plus carmustine-impregnated biodegradable wafers (20).

Based on the encouraging results of RIT with ^{131}I-labeled 81C6, particularly with the latest modification to deliver a targeted 44 Gy boost to the 2 cm SCRC perimeter, a multi-institutional Phase III study has been initiated. This study, known as the GLASS-ART trial (http://www.glassarttrial.com/), is sponsored by Bradmer Pharmaceuticals (Toronto, ON, Canada). Patients will be randomized to a regimen of temozolomide plus standard external beam radiation (6) versus that regimen plus patient-specific ^{131}I-labeled 81C6, which is being commercialized as Neuradiab®.

5.4.5 Evaluation of More Stable Constructs: Human/Mouse Chimeric 81C6

The most common rationale for generating human/mouse chimeric mAbs, which consist of murine antigen-binding domains and human immunoglobulin constant domains, is that the presence of human constant region domains potentially decreases immunogenicity (see Chapter 1). This is probably not as important a consideration when mAbs are administered into SCRC where HAMA titers would be expected to be lower than in serum. However, until recently, it was difficult to produce the amount of murine 81C6 to support a multi-institutional randomized trial, and so a chimeric 81C6 (ch81C6) construct that was able to be produced in bulk was developed (97). Because the goal from the outset was to utilize this mAb for targeted radiotherapy rather than immunotherapy, it was possible to select a human IgG constant region that was well matched to this purpose. Because of the low affinity of human IgG_2 for Fc receptors, this IgG constant region was utilized, with the goal of minimizing the radiation dose to normal organs such as the liver that have high levels of Fc receptors.

Although the specificity and binding affinity of ch81C6 for tenascin-C were virtually identical to those of murine 81C6 (mu81C6), radioiodinated ch81C6 unexpectedly was shown to accumulate and be retained in human glioma xenografts at levels that were significantly higher than its murine parent (98). Subsequent *in vitro* and *in vivo* radiolabeled mAb catabolism investigations demonstrated that this unexpected behavior was due to enhanced stability of ch81C6 compared with mu81C6, which we hypothesized was due to the increased rigidity of the hinge region of this construct relative to that of its murine counterpart (99). Our results indicated that not only was ch81C6 less susceptible than mu81C6 to proteolysis but also to deiodination *in vivo*. We speculated that the increased stability of ^{131}I-labeled ch81C6 could lead to prolonged residence times in the SCRC, leading to higher radiation absorbed doses to the tumor cavity margins with lower radiation doses to systemic organs.

To investigate this hypothesis, we conducted a Phase I study to determine the maximum tolerated dose, pharmacokinetics, dosimetry, and evidence of therapeutic benefit of ^{131}I-labeled ch81C6 (100). Patient eligibility was the same as that described above for ^{131}I-labeled mu81C6 and the ^{131}I-labeled ch81C6 was administered through a Rickham catheter into the SCRC. However, in this trial, patients were classified into three strata: Stratum A: newly diagnosed and untreated; Stratum B: newly diagnosed but following external beam radiotherapy; or Stratum C: recurrent. Independent dose escalation in each stratum was performed. A total of 47 patients received radioactivity doses of up to 4440 MBq ^{131}I-labeled ch81C6 with 35 having newly diagnosed tumors (Strata A and B) and 12 having recurrent disease (Stratum C).

Regardless of the treatment stratum, dose-limiting toxicity was hematologic and defined the MTD as 2960 MBq. Although three patients developed neurologic dose-limiting toxicity, none required reoperation to debulk radiation necrosis. The median survival observed for newly diagnosed and recurrent patients was 88.6 and 65.0 weeks, respectively. The survival probabilities observed for newly diagnosed patients, stratified according to histology, are shown in Fig. 5.4. The residence time for ^{131}I-labeled ch81C6 in the SCRC was longer than that of its murine counterpart, consistent with the observation that therapeutic effectiveness was comparable at a lower administered radioactivity dose of ^{131}I-labeled mAb. However, the human IgG constant region and enhanced stability of the chimeric construct resulted in slower blood and whole-body elimination, which presumably accounted for its increased dose-limiting hematologic toxicity (due to increased irradiation of the bone marrow

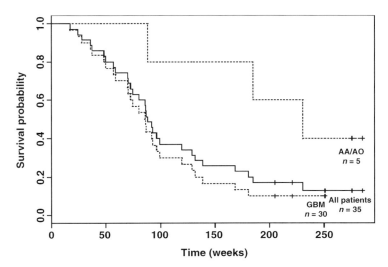

FIGURE 5.4   Kaplan–Meier overall survival estimates for newly diagnosed patients after stratification by histology. AA, anaplastic astrocytoma; AO, anaplastic oligodendroglioma; GBM, glioblastoma multiforme. Reproduced with permission from Ref. 100.

from circulating radiolabeled mAb). Because the mu81C6 hybridoma has now been stabilized and the amounts of mu81C6 needed for Phase III multi-institutional trials can now be produced, no further studies with ^{131}I-ch81C6 are planned.

5.4.6 Evaluation of More Potent Radionuclides: Astatine-211-Labeled CH81C6

One of the most important properties of a targeted radiotherapeutic is the nature of the radiation emitted by the radionuclide (see Chapter 2). [131]I and ^{90}Y, the radionuclides used in all of the clinical trials described above, decay by the emission of β-particles, a type of radiation with similar radiobiological characteristics (linear energy transfer (LET), oxygen and cell cycle dependence of cytotoxicity) as conventional external beam radiation. The cytotoxicity of these low linear energy transfer radiations is highly dependent on oxygen concentration, cell cycle position and dose rate. Alpha particles represent an intriguing alternative for targeted radiotherapy because they are high linear energy transfer radiation and their cytotoxicity is nearly independent of these clinically relevant variables. Furthermore, α-particles have a range in tissue equivalent to only a few cell diameters, a characteristic that is well matched to the treatment of small foci of tumor cells such as those found in the invasive front of gliomas. Studies performed in cell culture have demonstrated that human glioma cell lines could be killed as a result of only a few α-particle traversals per cell (101, 102).

These characteristics provided motivation for clinical evaluation of an 81C6 construct labeled with an α-particle-emitting radionuclide for the locoregional treatment of patients with malignant brain tumors. Astatine-211 ([211]At) was selected as the radionuclide because of its promising properties for targeted radiotherapy. These include a 7.2 h half-life, absence of long-lived α-particle-emitting daughter radionuclides, α-emission associated with each decay of [211]At, halogen chemistry, and an electron capture branch that provides polonium K X-rays (77–92 keV), which can be exploited for imaging the distribution of [211]At in patients. As a tenascin-C avid construct, ch81C6 was selected because of its enhanced stability, which was particularly important because of the total lack of clinical experience with [211]At-labeled targeted radiotherapeutics before initiation of this Phase I trial. Because the physical half-life of [211]At is more than 25 times shorter than that of ^{131}I, the hematological toxicity observed in patients receiving ^{131}I-labeled ch81C6 (100) would not be expected to be problematic because most of the [211]At decays should occur prior to significant egress of radioactivity from the SCRC.

We recently reported the results of this pilot study, which represents the first investigation of an [211]At-labeled radiotherapeutic in humans (103). It was designed to evaluate the feasibility, pharmacokinetics, and toxicity of [211]At-labeled ch81C6 administered into the SCRC as salvage therapy in patients with recurrent malignant brain tumors. The [211]At was produced on the Duke University Medical Center cyclotron by bombarding natural bismuth targets with 28 MeV α-particles and labeling was accomplished by reaction of ch81C6 with N-succinimidyl 3-[[211]At]astatobenzo-ate produced via an astatodestannylation reaction (see Chapter 2) (104). In the clinical pilot study, 18 patients received single doses of 10 mg of [211]At-labeled chimeric 81C6 mAb labeled with escalating doses of ^{211}At administered into the SCRC. Patients received 74 MBq ($n = 5$), 148 MBq ($n = 7$), 248 MBq ($n = 5$), and 370 MBq ($n = 1$) of ^{211}At-labeled ch81C6. The patient population consisted of 14 GBM, 3 anaplastic astrocytoma, and 1 patient with anaplastic oligodendroglioma.

Serial gamma camera imaging and blood sampling were performed over the first 24 h after injection to determine SCRC and whole-body pharmacokinetics during the interval over which more than 90% of administered ^{211}At atoms would decay. Based on gamma camera imaging, it was demonstrated that leakage of ^{211}At from the cavity was slow such that an average of $96.7 \pm 3.6\%$ of the ^{211}At decays occurred within the cavity. Furthermore, the maximum radioactivity found in the blood pool was $0.26 \pm 0.43\%$ of the injected dose. The results of these measurements suggested that ^{211}At-labeled ch81C6 was quite stable *in vivo*, at least administered into the SCRC and over this short time period.

Dose-limiting toxicities were not observed in any of these patients and thus, the MTD was not identified. Toxicities were limited to six patients experiencing grade 2 neurotoxicity within 6 weeks of receiving ^{211}At-labeled ch81C6, which resolved fully in all but one case; no attributable grade 3 or greater toxicities were observed. No patient required reoperation for radionecrosis. The average radiation absorbed dose delivered to the tumor SCRC margin was 2764 Gy and ranged between 155 and 35,000 Gy, reflecting the fact that cavity volumes for these patients varied from 0.2 to 37.2 cm^3. The median survival for GBM patients ($n = 14$) and all patients ($n = 18$) was 52 and 56.5 weeks, respectively. Particularly encouraging is the fact that two of these recurrent GBM patients survived for 150 and 151 weeks after ^{211}At-labeled ch81C6 treatment, which is markedly better than that obtained with conventional treatments and biodegradable polymer imbedded chemotherapeutics (Gliadel) (20). A generalized ramification of this study is that it provided proof-of-concept for the use of ^{211}At-labeled targeted radiotherapeutics in patients with cancer. More specifically, it demonstrated that regional administration of ^{211}At-ch81C6 is feasible, safe, and associated with encouraging antitumor benefit in patients with malignant CNS tumors.

5.4.7 Pretargeted Radioimmunotherapy

Pretargeted RIT will be discussed extensively in Chapter 8. This approach, also known as pretargeted antibody-guided radioimmunotherapy (PAGRIT), is a multistep procedure designed to compensate for the pharmacokinetic consequences (slow elimination from the blood pool and limited diffusion within the tumor) of binding the radiolabel to a macromolecular carrier such as an antibody (105). In this strategy, the mAb is not directly radiolabeled and is administered first in an unlabeled form, and after sufficient time has elapsed to allow it to bind to tumor and clear from normal tissues, a radiolabeled, low molecular weight hapten is injected. The approach that has been evaluated clinically in patients with malignant brain tumors exploits the exceptionally high binding affinity of avidin or streptavidin for the 244 Da vitamin, biotin. In particular, a three-step protocol is utilized: patients first receive biotinylated BC-4 mAb, followed 24 h later by avidin (to bind to the biotinylated mAb), and after an additional 18 h, a ^{90}Y-labeled biotin conjugate (to bind to the avidin).

In the first clinical trial evaluating this approach, the three reagents were adminis-tered into the SCRC via a catheter to 24 recurrent malignant glioma patients (GBM, $n = 16$; WHO grade III malignant glioma, $n = 8$) (106). The therapeutic regimen was repeated 8–10 weeks later. A MTD of 1110 MBq of ^{90}Y-labeled 1,4,7,10-tetraaza-cyclododecane-N,N',N'',N'''-tetraacetic acid (DOTA)-biotin was determined and the dose-limiting toxicity was neurologic. Median survival was 19 and 11.5 months for patients with recurrent grade III malignant glioma and GBM, respectively. In a subsequent report, 73 patients with recurrent GBM were analyzed to determine the effect of adding temozolomide to the treatment protocol (107). Overall and progres-sion-free survival were compared in patients treated with the ^{90}Y-DOTA-biotin PAGRIT approach ($n = 38$) and those treated with ^{90}Y-DOTA-biotin PAGRIT plus temozolomide ($n = 35$). Patients received an average of 3 and up to 7 cycles of treatment. The authors demonstrated that ^{90}Y-DOTA-biotin PAGRIT could be combined with temozolomide chemotherapy without significantly increasing toxicity. Overall survival (calculated from the time of surgical removal of tumor when newly diagnosed, not at the time of second surgery for recurrence) was 25 months in patients who received temozolomide compared with 17.5 months for those who did not.

The PAGRIT protocol has also been evaluated in newly diagnosed malignant brain tumor patients as an adjuvant treatment following surgery and conventional radio-therapy (108). An important difference from the procedures utilized in the studies in recurrent patients described above is that all reagents were administered intravenously instead of directly into the SCRC. A total of 37 patients were enrolled, including 17 with grade III malignant glioma and 20 with GBM. Nineteen patients received the PAGRIT regimen, with the ^{90}Y-labeled biotin now administered on the basis of body surface area (2200 MBq/m^2), while 18 patients were treated solely with surgery and radiotherapy, and served as controls. In GBM patients, treatment with PAGRIT resulted in a median overall survival of 33.5 months compared with 8 months for the control group.

5.4.8   Receptor-Targeted Peptides

As noted in Chapters 3 and 4, due to their smaller size and more rapid diffusion, peptides can offer significant advantages compared with mAbs for the treatment of solid tumors including CNS malignancies. The most clinically mature approach for peptide-based targeted radiotherapy is the treatment of tumors that over express somatostatin type 2 receptors (see Chapter 4). Unfortunately, this receptor is not expressed at high levels in the most malignant forms of brain tumors; however, it is present on lower grade gliomas. The Basel group has been evaluating the feasibility of treating patients with progressive glioma with ^{90}Y-DOTA^0-D-Phe^1-Tyr^3-octreotide (DOTATOC) (109–111).

In the most recent extended pilot study, five patients with progressive glioma (two with WHO grade II and three with WHO grade III) and five patients with surgically debulked WHO grade II gliomas were treated with between 1 and 5 cycles of ^{90}Y-DOTATOC at a cumulative radioactivity of 555–7030 MBq. Durations of response between 13 and 45 months were observed in the progressive glioma patients and disease stabilization was seen in the five newly diagnosed low-grade glioma patients that received ^{90}Y-DOTATOC following resection. However, it is not clear whether targeted radiotherapy will have a role in the treatment of lower grade gliomas because most clinicians favor a less aggressive approach to the treatment of patients with these malignancies.

Another molecular target that is being explored for targeted radiotherapy of brain tumors is the neurokinin type 1 receptor that is over expressed on both high- and low-grade gliomas (112). The major ligand for this receptor is substance-P, and this 11-mer peptide has been complexed with different radiometals using the DOTA macrocycle and these radiolabeled peptides have been evaluated in patients with malignant brain tumors (113). A total of 20 patients were evaluated with the treatment group consisting of 14 with GBM and 6 with WHO grade II or III malignancies. The radiolabeled peptide was administered through a catheter connected to the tumor or the SCRC. In the majority of cases, the peptide was labeled with ^{90}Y while ^{177}Lu and ^{213}Bi was utilized in three and two patients, respectively. Median survival following targeted radiotherapy was 11 months in patients with GBM and 16 months in patients with lower grade malignancies. The number of patients was too small to discern the extent to which there was a survival advantage related to the chemical stability or radiation range characteristics of the different radiometals used in this study.

5.4.9 Chlorotoxin

Chlorotoxin is a neurotoxin that was isolated from the giant yellow Israeli scorpion *Leiurus quinquestriatus* (114). This 36-amino acid peptide has been shown to bind to a variety of human tumors including glioma but not to normal brain. A synthetic form of chlorotoxin, TM-601, has been manufactured by TransMolecular, Inc. (Cambridge, MA) and is being evaluated as a targeted radiotherapeutic for malignant brain tumors. Serial SPECT imaging of nine recurrent glioma patients after receiving ^{131}I-labeled TM-601 (370 MBq, 0.25–1.0 mg) administered into the SCRC was performed to determine the clearance of the radiolabeled peptide from the cavity and its uptake into normal tissues (115, 116). Elimination of ^{131}I exhibited two-exponential behavior, with about 85% remaining after 1 h and 21% remaining after 24 h. The biological clearance thereafter was slower, with a biological half-life for SCRC clearance of 39 ± 6 h being reported. Loss of radioactivity from the cavity was attributed to either enzymatic deiodination of the peptide, release of the peptide from the tumor site, or peptide degradation.

A three-institution Phase I study was then carried out with the recurrent patients again treated with a single dose of ^{131}I-labeled TM-601 (370 MBq, 0.25–1.0 mg) administered into the SCRC (116). No dose-limiting toxicities were observed in the 17 GBM and 1 anaplastic astrocytoma patient receiving the labeled peptide although 4 patients were described as having adverse events. For unplanned reasons, three patients received a second dose of ^{131}I-labeled TM-601. There was no clear relationship between the mass of the peptide dose (mg) and adverse events or median survival. Patients were followed for 180 days with 7 deaths during this period and the median survival for all patients was 27.0 weeks. A Phase II trial of ^{131}I-labeled TM-601 involving higher activity levels of ^{131}I and multiple direct administrations of the labeled peptide into the SCRC is currently underway (114).

5.5   PERSPECTIVE FOR THE FUTURE

Malignant brain tumors present unique challenges for targeted radiotherapy. Unlike most other types of cancer, recurrence at or near the surgical resection site is the leading cause of fatality. Complete surgical removal of brain tumors is frequently exceedingly difficult due to their infiltrative nature and the need to avoid excessive insult to invaded regions of normal brain, which could markedly compromise quality of life. Thus, there are two challenges for brain tumor targeted radiotherapy that will undoubtedly require different approaches—achieving local control and destruction of tumor cells infiltrating the normal brain. Recent and ongoing clinical trials indicate that locoregionally delivered targeted radiotherapeutics can offer promising therapeutic benefits in a patient population in dire need of more meaningful treatments. Furthermore, this treatment strategy has been associated with a lower rate of radiation necrosis and the need for reoperation than brachytherapy or radiosurgery techniques.

It is important to note that cogent evaluation of the current state of the art as a springboard for future development is not an easy task. Although most clinical trials express therapeutic benefit in terms of overall survival, this end point has been defined in different ways: from initial diagnosis or surgery, from surgery to recurrence, and from the date of administration of the targeted radiotherapeutic. In some trials, the definition of overall survival is not presented. Given the rapid progression of GBM, one approach could appear more promising than another based on a different starting point for determining overall survival rather than improved therapeutic effectiveness. Another complicating factor is that in some studies, results for single and varying numbers of multiple administrations of the radiotherapeutic are pooled. Finally, largely due to regulatory constraints, in most clinical trials, the targeted radiotherapeutic was given at a predetermined level of radioactivity with the result that the average radiation dose delivered to the tumor cavity margins varied considerably due to differences in cavity size and radiopharmaceutical residence time.

It is envisioned that advances in targeted radiotherapy of brain tumors will most likely result from the development of patient-specific treatment strategies. High resolution anatomical imaging will be utilized to provide information about the size and location of the tumor to guide in radionuclide selection. Genetic profiling will indicate the molecular targets that are present on the tumor in high enough concentration to be exploited for targeted therapy. Finally, quantitative PET imaging will allow determination of patient-specific pharmacokinetic profile using a positron-emitting analogue of the radiotherapeutic. These data will then be utilized to determine the administered radioactivity dose to achieve a target radiation absorbed dose that can be correlated with response in light of individual differences in target molecule expression, catabolism, and delivery.

Implicit in this vision for the future is the recognition that in most cases, this will be best accomplished using a radiotherapeutic "cocktail" formulated from multiple radionuclides, molecular carriers with different biological properties (antibodies, peptides, organic molecules), and binding to multiple tumor-associated targets. Low energy β-emitters might be best for treating the region immediately adjacent to the cavity margin. One possibility under investigation is to replace ^{131}I with ^{177}Lu for this purpose because of the lower β-energy and less penetrating γ-ray component of the latter (117). For treating the highly aggressive tumor cells infiltrating normal brain, the use of short range radiation of high linear energy transfer, namely α-particles and Auger electrons (see Chapter 9) should be considered. An intriguing possibility is to utilize intracerebral infusion to deliver the targeted radiotherapeutic to distant tumor cells (118), an approach that is probably limited to use with radiation of cellular or subcellular range. We have begun to explore this possibility using multifunctional polypeptides, known as modular recombinant transporters, which can be utilized to deliver drugs to the nucleus of cancer cells expressing target receptors (119). An attractive feature of the modular recombinant transporter delivery system is its flexible nature, allowing modification of the ligand module to bind to different tumor associated targets as well as alteration of the radionuclide or radionuclides bound to the polypeptide.

ACKNOWLEDGMENTS

Work performed at Duke University Medical Center was supported in part by grants from the National Cancer Institute, the National Institute of Neurological Disease and Stroke, and the Department of Energy. Our clinical studies described herein reflect the efforts of more than 25 individuals working in the Preston Robert Tisch Brain Tumor Center at Duke.

REFERENCES

1. ACS. *Cancer Facts & Figures 2007*. American Cancer Society Inc., Atlanta, Georgia, 2007, pp. 1–52.
2. Ohgaki H, Dessen P, Jourde B, et al. Genetic pathways to glioblastoma: a population-based study. *Cancer Res* 2004;64:6892–6899.
3. Stewart LA. Chemotherapy in adult high-grade glioma: a systematic review and meta-analysis of individual patient data from 12 randomised trials. *Lancet* 2002;359:1011–1018.
4. Medical Research, Council. Randomized trial of procarbazine, lomustine, and vincristine in the adjuvant treatment of high-grade astrocytoma: a Medical Research Council trial. *J Clin Oncol* 2001;19:509–518.
5. Denny BJ, Wheelhouse RT, Stevens MFG, Tsang LLH, Slack JA. NMR and molecular modeling investigation of the mechanism of activation of the antitumor drug temozolomide and its interaction with DNA. *Biochemistry* 1994;33:9045–9051.
6. Stupp R, Mason WP, van den Bent MJ, et al. Radiotherapy plus concomitant and adjuvant temozolomide for glioblastoma. *New Engl J Med* 2005;352:987–996.

7. Wong ET, Hess KR, Gleason MJ, et al. Outcomes and prognostic factors in recurrent glioma patients enrolled onto phase II clinical trials. *J Clin Oncol* 1999;17:2572–2578.

8. Yung WKA, Prados MD, Yaya-Tur R, et al. Multicenter phase II trial of temozolomide in patients with anaplastic astrocytoma or anaplastic oligoastrocytoma at first relapse. *J Clin Oncol* 1999;17:2762–2771.

9. Yung WK, Albright RE, Olson J, et al. A phase II study of temozolomide vs. procarbazine in patients with glioblastoma multiforme at first relapse. *Br J Cancer* 2002;83:588–593.

10. Imperato JP, Paleologos NA, Vick NA. Effects of treatment on long-term survivors with malignant astrocytomas. *Ann Neurol* 1990;28:818–822.

11. Vick NA, Paleologos NA. External beam radiotherapy: hard facts and painful realities. *J Neuro-Oncol* 1995;24:93–95.

12. Curran WJ, Jr., Scott CB, Horton J, et al. Recursive partitioning analysis of prognostic factors in three Radiation Therapy Oncology Group malignant glioma trials. *J Natl Cancer Inst* 1993;85:704–710.

13. Neuwelt EA. Mechanisms of disease: the blood–brain barrier. *Neurosurgery* 2004; 54:131–142.

14. Jain RK. Normalization of tumor vasculature: an emerging concept in antiangiogenic therapy. *Science* 2005;307:58–62.

15. Rich JN, Bigner DD. Development of novel targeted therapies in the treatment of malignant glioma. *Nat Rev Drug Discov* 2004;3:430–446.

16. Hochberg FH, Pruitt A. Assumptions in the radiotherapy of glioblastoma. *Neurology* 1980;30:907–911.

17. Gaspar LE, Fisher BJ, Macdonald DR. Supratentorial malignant glioma: patterns of recurrence and implications for external beam local treatment. *Int J Radiat Oncol Biol Phys* 1992;24:55–57.

18. Enam SA, Rosenblum ML, Edvardsen K. Role of extracellular matrix in tumor invasion: migration of glioma cells along fibronectin-positive mesenchymal cell processes. *Neurosurgery* 1998;42:599–607.

19. Westphal M, Hilt DC, Bortey E, et al. A phase 3 trial of local chemotherapy with biodegradable carmustine (BCNU) wafers (Gliadel wafers) in patients with primary malignant glioma. *Neurooncology* 2003;5:79–88.

20. Brem H, Piantadosi S, Burger PC, et al. Placebo-controlled trial of safety and efficacy of intraoperative controlled delivery by biodegradable polymers of chemotherapy for recurrent gliomas. *Lancet* 1995;345:1008.

21. Sampson JH, Akabani G, Friedman AH, et al. Comparison of intratumoral bolus injection and convection-enhanced delivery of radiolabeled antitenascin monoclonal antibodies. *Neurosurg Focus* 2006;20:E14.

22. Sampson JH, Akabani G, Archer GE, et al. Progress report of a Phase I study of the intracerebral microinfusion of a recombinant chimeric protein composed of transforming growth factor (TGF)-alpha and a mutated form of the Pseudomonas exotoxin termed PE-38 (TP-38) for the treatment of malignant brain tumors. *J Neurooncology* 2003;65:27–35.

23. Sampson JH, Reardon DA, Friedman AH, et al. Sustained radiographic and clinical response in patient with bifrontal recurrent glioblastoma multiforme with intracerebral infusion of the recombinant targeted toxin TP-38: case study. *Neurooncology* 2005; 7:90–96.

24. Tatter SB, Shaw EG, Rosenblum ML, et al. An inflatable balloon catheter and liquid ^{125}I radiation source (GliaSite Radiation Therapy System) for treatment of recurrent malignant glioma: multicenter safety and feasibility trial. *J Neurosurgery* 2003;99:297–303.

25. Chan TA, Weingart JD, Parisi M, et al. Treatment of recurrent glioblastoma multiforme with GliaSite brachytherapy. *Int J Radiat Oncol Biol Phys* 2005;62:1133–1139.

26. Gabayan AJ, Green SB, Sanan A, et al. GliaSite brachytherapy for treatment of recurrent malignant gliomas: a retrospective multi-institutional analysis. *Neurosurgery* 2006; 58:701–709; discussion 701–709.

27. Rogers LR, Rock JP, Sills AK, et al. Brain Metastasis Study Group, Shaw, E.G. Results of a phase II trial of the GliaSite radiation therapy system for the treatment of newly diagnosed, resected single brain metastases. *J Neurosurg* 2006;105:375–384.

28. Burger PC, Heinz ER, Shibata T, Kleihues P. Topographic anatomy and CT correlations in the untreated glioblastoma multiforme. *J Neurosurg* 1988;68:698–704.

29. Goldbrunner RH, Bernstein JJ, Tonn JC. Cell-extracellular matrix interaction in glioma invasion. *Acta Neurochir* 1999;141:295–305.

30. Zalutsky MR, Radionuclide therapy. In: Roesch F, editor. *Handbook of Nuclear Chemistry Volume 4: Radiochemistry and Radiopharmaceutical Chemistry in Life Sciences*, Kluwer Academic Publishers, Dordrecht, The Netherlands, 2003, pp. 315–348.

31. Kalofonos HP, Pawlikowska TR, Hemingway A, et al. Antibody guided diagnosis and therapy of brain gliomas using radiolabeled monoclonal antibodies against epidermal growth factor receptor and placental alkaline phosphatase. *J Nucl Med* 1989; 30:1636–1645.

32. Quang TS, Brady LW. Radioimmunotherapy as a novel treatment regimen: ^{125}I-labeled monoclonal antibody 425 in the treatment of high-grade brain gliomas. *Int J Radiat Oncol Biol Phys* 2004;58:972–975.

33. Weaver M, Laske DW. Transferrin receptor ligand-targeted toxin conjugate (Tf-CRM107) for therapy of malignant gliomas. *J Neuro-Oncol* 2003;65:3–13.

34. Hopkins K, Chandler C, Eatough J, Moss T, Kemshead JT. Direct injection of ^{90}Y MoAbs into glioma tumor resection cavities leads to limited diffusion of the radioimmunoconjugates into normal brain parenchyma: a model to estimate absorbed radiation dose. *Int J Radiat Oncol Biol Phys* 1998;40:835–844.

35. Bourne S, Pemberton L, Moseley R, Lashford LS, Coakham HB, Kemshead JT. Monoclonal antibodies M340 and UJ181.4 recognize antigens associated with primitive neuroectodermal tumours/tissues. *Hybridoma* 1989;8:415–426.

36. Wikstrand CJ, Fredman P, Svennerholm L, Bigner DD. Detection of glioma-associated gangliosides GM2, GD2, GD3, 3'-isoLM1 3',6'-isoLD1 in central nervous system tumors *in vitro* and *in vivo* using epitope-defined monoclonal antibodies. *Prog Brain Res* 1994;101:213–223.

37. Krizan Z, Murray JL, Hersh EM. Increased labeling of human melanoma cells *in vitro* using combinations of monoclonal antibodies recognizing separate cell surface antigenic determinants. *Cancer Res* 1985;45:4904–4909.

38. Weterman MAJ, Ajubi N, van Dinter IMR. nmb, a novel gene, is expressed in low-metastatic human melanoma cell lines and xenografts. *Int J Cancer* 1995;60:73–81.

39. Kool M, van der Linden M, de Haas M. MRP3, an organic anion transporter able to transport anti-cancer drugs. *Proc Natl Acad Sci USA* 1999;96:6914–6919.

40. Erickson HP. Tenascin-C, tenascin-R and tenascin-X: a family of talented proteins in search of functions. *Curr Opin Cell Biol* 1993;5:869–876.

41. Jones FS, Jones PL. The tenascin family of ECM glycoproteins: structure, function, and regulation during embryonic development and tissue remodeling. *Develop Dynam* 2000;218:235–259.

42. Bourdon MA, Wikstrand CJ, Furthmayr H, Matthews TJ, Bigner DD. Human glioma-mesenchymal extracellular matrix antigen defined by monoclonal antibody. *Cancer Res* 1983;43:2796–2805.

43. Bourdon MA, Matthews TJ, Pizzo SV, Bigner DD. Immunochemical and biochemical characterization of a glioma-associated extracellular matrix glycoprotein. *J Cell Biochem* 1985;28:183–195.

44. Howeedy AA, Virtanen I, Laitinen L, Gould NS, Koukoulis GK, Gould VE. Differential distribution of tenascin in the normal, hyperplastic, and neoplastic breast. *Lab Invest* 1990;63:798–806.

45. Nies DE, Hemesath TJ, Kim JH, Gulcher JR, Stefansson K. The complete cDNA sequence of human hexabrachion (Tenascin). A multidomain protein containing unique epidermal growth factor repeats. *J Biol Chem* 1991;266:2818–2823.

46. Erickson HP, Inglesias JL. A six-armed oligomer isolated from cell surface fibronectin preparations. *Nature* 1984;311:267–269.

47. Pas J, Wyszko E, Rolle K, et al. Analysis of structure and function of tenascin-C. *Int J Biochem Cell Biol* 2006;38:1594–1602.

48. Trebaul A, Chan EK, Midwood KS. Regulation of fibroblast migration by tenascin-C. *Biochem Soc Trans* 2007;35:695–697.

49. Borsi L, Carnemolla B, Nicolo G, Spina B, Tanara G, Zardi L. Expression of different tenascin isoforms in normal, hyperplastic and neoplastic human breast tissues. *Int J Cancer* 1992;52:688–692.

50. Ventimiglia JB, Wikstrand CJ, Ostrowski LE, Bourdon MA, Lightner VA, Bigner DD. Tenascin expression in human glioma cell lines and normal tissues. *J Neuroimmunol* 1992;36:41–55.

51. Zagzag D, Friedlander DR, Miller DC, et al. Tenascin expression in astrocytomas correlates with angiogenesis. *Cancer Res* 1995;55:907–914.

52. Herold-Mende C, Mueller MM, Bonsanto MM, Schmitt HP, Kunze S, Steiner HH. Clinical impact and functional aspects of tenascin-C expression during glioma progression. *Int J Cancer.* 2002;98:362–369.

53. Behrem S, Zarkovic K, Eskinja N, Jonjic N. Distribution pattern of tenascin-C in glioblastoma: correlation with angiogenesis and tumor cell proliferation. *Pathol Oncol Res* 2005;11:229–235.

54. Hau P, Kunz-Schughart LA, Rummele P, et al. Tenascin-C protein is induced by transforming growth factor-beta1 but does not correlate with time to tumor progression in high-grade gliomas. *J Neuro-Oncol* 2006;77:1–7.

55. Zagzag D, Friedlander DR, Dosik J, et al. Tenascin-C expression by angiogenic vessels in human astrocytomas and by human brain endothelial cells *in vitro*. *Cancer Res* 1996;56:182–189.

56. Akabani G, McLendon RE, Bigner DD, Zalutsky MR. Vascular targeted endoradiotherapy using alpha-particle emitting compounds: theoretical analysis. *Int J Radiat Oncol Biol Phys* 2002;48:1259–1275.

57. Zagzag D, Shiff B, Jallo GI, et al. Tenascin-C promotes microvascular cell migration and phosphorylation of focal adhesion kinase. *Cancer Res* 2002;62:2660–2668.

58. Garcion E, Halilagic A, Faissner A, French-Constant C. Generation of an environmental niche for neural stem cell development by the extracellular matrix molecule tenascin C. *Development* 2004;131:3423–3432.

59. Garwood J, Garcion E, Dobbertin A, et al. The extracellular matrix glycoprotein Tenascin-C is expressed by oligodendrocyte precursor cells and required for the regulation of maturation rate, survival and responsiveness to platelet-derived growth factor. *Eur J Neurosci* 2004;20:2524–2540.

60. Ilunga K, Nishiura R, Inada H, et al. Co-stimulation of human breast cancer cells with transforming growth factor-beta and tenascin-C enhances matrix metalloproteinase-9 expression and cancer cell invasion. *Int J Exp Pathol* 2004;85:373–379.

61. Nishio T, Kawaguchi S, Yamamoto M, Iseda T, Kawasaki T, Hase T. Tenascin-C regulates proliferation and migration of cultured astrocytes in a scratch wound assay. *Neuroscience* 2005;132:87–102.

62. Ahmed I, Liu HY, Mamiya PC, et al. Three-dimensional nanofibrillar surfaces covalently modified with tenascin-C-derived peptides enhance neuronal growth *in vitro*. *J Biomed Mater Res A* 2006;76:851–860.

63. Swindle CS, Tran KT, Johnson TD. Epidermal growth factor (EGF)-like repeats of human tenascin-C as ligands for EGF receptor. *J Cell Biol* 2001;154:459–468.

64. Balza E, Siri A, Ponassi M, et al. Production and characterization of monoclonal antibodies specific for different epitopes of human tenascin. *FEBS Lett* 1993; 332:39–43.

65. De Santis R, Anastasi AM, D'Alessio V, et al. Novel antitenascin antibody with increased tumour localisation for Pretargeted Antibody-Guided RadioImmuno-Therapy (PAGRIT). *Br J Cancer* 2003;88:996–1003.

66. Petronzelli F, Pelliccia A, Anastasi AM, et al. Improved tumor targeting by combined use of two antitenascin antibodies. *Clin Cancer Res* 2005;11:7137s–7145s.

67. Brack SS, Silacci M, Birchler M, Neri D. Tumor-targeting properties of novel antibodies specific to the large isoform of tenascin-C. *Clin Cancer Res* 2006;12:3200–3208.

68. Hicke BJ, Marion C, Chang YF, et al. Tenascin-C aptamers are generated using tumor cells and purified protein. *J Biol Chem* 2001;276:48644–48654.

69. Daniels DA, Chen H, Hicke BJ, Swiderek KM, Gold L, A tenascin-C aptamer identified by tumor cell SELEX: systematic evolution of ligands by exponential enrichment. *Proc Natl Acad Sci USA* 2003;100:15416–15421.

70. Hicke B, Stephens AW, Gould T, et al. Tumor targeting by an aptamer. *J Nucl Med* 2006;47:668–678.

71. Riva P, Arista A, Franceschi G, et al. Local treatment of malignant gliomas by direct infusion of specific monoclonal antibodies labeled with [131]I: comparison of the results obtained in recurrent and newly diagnosed tumors. *Cancer Res* 1995;55:5952s–5956s.

72. Riva P, Franceschi G, Frattarelli M, et al. [131]I radioconjugated antibodies for the locoregional radioimmunotherapy of high-grade malignant glioma–phase I and II study. *Acta Oncol* 1999;38:351–359.

73. Riva P, Franceschi G, Frattarelli M, et al. Loco-regional radioimmunotherapy of high-grade malignant gliomas using specific monoclonal antibodies labeled with [90]Y: a phase I study. *Clin Cancer Res* 1999;5:3275s–3280s.

74. Riva P, Franceschi G, Riva N, Casi M, Santimaria M, Adamo M. Role of nuclear medicine in the treatment of malignant gliomas: the locoregional radioimmunotherapy approach. *Eur J Nucl Med* 2000;27:601–609.

75. Goetz C, Riva P, Poepperl G, et al. Locoregional radioimmunotherapy in selected patients with malignant glioma: experiences, side effects and survival times. *J Neuro-Oncol* 2003;62:321–328.

76. Murphy-Ullrich JE, Lightner VA, Aukhil I, Yan YZ, Erickson HP, Höök M. Focal adhesion integrity is downregulated by the alternatively spliced domain of human tenascin. *J Cell Biol* 1991;115:1127–1136.

77. Wikstrand CJ, Zalutsky MR, Bigner DD. Therapy of brain tumors with radiolabeled antibodies. In: Liau LM, Becker DP, Cloughsey TF, Bigner DD, editors. *Brain Tumor Immunotherapy*. Humana Press, Totowa, New Jersey, 2001, pp. 205–229.

78. Bourdon MA, Coleman RE, Blasberg RG, Groothuis DR, Bigner DD. Monoclonal antibody localization in subcutaneous and intracranial human glioma xenografts: paired-label and imaging analysis. *Anticancer Res* 1984;4:133–140.

79. Bullard DE, Wikstrand CJ, Humphrey PA, et al. Specific imaging of human brain tumor xenografts utilizing radiolabelled monoclonal antibodies (MAbs). *Nuklearmedizin* 1986;25:210–215.

80. Colapinto EV, Lee YS, Humphrey PA, et al. The localisation of radiolabelled murine monoclonal antibody 81C6 and its Fab fragment in human glioma xenografts in athymic mice. *Br J Neurosurg* 1988;2:179–191.

81. Lee YS, Bullard DE, Zalutsky MR, et al. Therapeutic efficacy of antiglioma mesenchymal extracellular matrix ^{131}I-radiolabeled murine monoclonal antibody in a human glioma xenograft model. *Cancer Res* 1988;48:559–566.

82. Lee YS, Bullard DE, Humphrey PA, et al. Treatment of intracranial human glioma xenografts with ^{131}I anti-tenascin monoclonal antibody (Mab) 81C6. *Cancer Res* 1988;48:2904–2910.

83. Zalutsky MR, Moseley RP, Coakham HB, Coleman RE, Bigner DD. Pharmacokinetics and tumor localization of ^{131}I-labeled anti-tenascin monoclonal antibody 81C6 in patients with gliomas and other intracranial malignancies. *Cancer Res* 1989;49:2807–2813.

84. Zalutsky MR, Moseley RP, Benjamin JC, et al. Monoclonal antibody and F(ab')$_2$ fragment delivery to tumor in patients with glioma: comparison of intracarotid and intravenous administration. *Cancer Res* 1990;50:4105–4110.

85. Schold SC, Jr., Zalutsky MR, Coleman RE, et al. Distribution and dosimetry of I-123-labeled monoclonal antibody 81C6 in patients with anaplastic glioma. *Invest Radiol* 1993;28:488–496.

86. Brown MT, Coleman RE, Friedman AH, et al. Intrathecal ^{131}I-labeled antitenascin monoclonal antibody 81C6 treatment of patients with leptomeningeal neoplasms or primary brain tumor resection cavities with subarachnoid communication: phase I trial results. *Clin Cancer Res* 1996;2:963–972.

87. Bigner DD, Brown MT, Friedman AH, et al. Iodine-131-labeled anti-tenascin monoclonal antibody 81C6 treatment of patients with recurrent malignant gliomas: Phase I trial results. *J Clin Oncol* 1998;16:2202–2212.

88. Cokgor I, Akabani G, Kuan CT, et al. Phase I trial results of iodine-131-labeled antitenascin monoclonal antibody 81C6 treatment of patients with newly diagnosed malignant gliomas. *J Clin Oncol* 2000;18:3862–3872.

89. Akabani G, Cokgor I, Coleman RE, et al. Dosimetry and dose-response relationships in newly diagnosed patients with malignant gliomas treated with iodine-131-labeled antitenascin monoclonal antibody 81C6 therapy. *Int J Radiat Oncol Biol Phys* 2000;46:947–958.

90. Reardon DA, Akabani G, Coleman RE, et al. Phase II trial of murine ^{131}I-labeled antitenascin monoclonal antibody 81C6 administered into surgically created resection cavities of patients with newly diagnosed malignant gliomas. *J Clin Oncol* 2002;20:1389–1397.

91. Scharfen CO, Sneed PK, Wara WM, et al. High activity iodine-125 interstitial implant for gliomas. *Int J Radiat Oncol Biol Phys* 1992;24:583–591.

92. Wen PY, Alexander E, III, Black PM, et al. Long term results of stereotactic brachytherapy used in the initial treatment of patients with glioblastomas. *Cancer* 1994;73:3029–3036.

93. Shrieve DC, Alexander E, III, Wen PY, et al. Comparison of stereotactic radiosurgery and brachytherapy in the treatment of recurrent glioblastoma multiforme. *Neurosurgery* 1995;36:275–282; discussion 282–274.

94. Stupp R, Dietrich PY, Ostermann Kraljevic S, et al. Promising survival for patients with newly diagnosed glioblastoma multiforme treated with concomitant radiation plus temozolomide followed by adjuvant temozolomide. *J Clin Oncol* 2002; 20:1375–1382.

95. Akabani G, Reardon DA, Coleman RE, et al. Dosimetry and radiographic analysis of iodine-131-labeled anti-tenascin 81C6 murine monoclonal antibody in newly diagnosed patients: a phase II study. *J Nucl Med* 2005;46:1042–1051.

96. Reardon DA, Zalutsky MR, Akabani G, et al. Patient-specific radioimmunotherapy with murine ^{131}I-labeled anti-tenascin monoclonal antibody 81C6 to achieve a 44 Gy boost to the resection cavity perimeter of patients with newly diagnosed primary malignant brain tumors: a pilot feasibility study. *Neuro-Oncol* 2008;10:182–189.

97. He X, Archer GE, Wikstrand CJ, et al. Generation and characterization of a mouse/human chimeric antibody directed against extracellular matrix protein tenascin. *J Neuroimmunol* 1994;52:127–137.

98. Zalutsky MR, Archer GE, Garg PK, Batra SK, Bigner DD. Chimeric anti-tenascin antibody 81C6: increased tumor localization compared with its murine parent. *Nucl Med Biol* 1996;23:449–458.

99. Reist CJ, Bigner DD, Zalutsky MR. Human IgG$_2$ constant region enhances *in vivo* stability of anti-tenascin antibody 81C6 compared with its murine parent. *Clin Cancer Res* 1998;4:2495–2502.

100. Reardon DA, Quinn JA, Akabani G, et al. Novel human IgG2b/murine chimeric antitenascin monoclonal antibody construct radiolabeled with ^{131}I and administered into the surgically created resection cavity of patients with malignant glioma: phase I trial results. *J Nucl Med* 2006;47:912–918.

101. Vaidyanathan G, Larsen RH, Zalutsky MR. 5-[[211]At]astato-2′-deoxyuridine, an α-particle emitting endoradiotherapeutic agent undergoing DNA incorporation. *Cancer Res* 1996;56:1204–1209.

102. Larsen RH, Akabani G, Welsh P, Zalutsky MR. The cytotoxicity and microdosimetry of [211]At-labeled chimeric monoclonal antibodies on human glioma and melanoma cells *in vitro*. *Radiat Res* 1998;149:155–162.

103. Zalutsky MR, Reardon DA, Akabani G, et al. Clinical experience with α-emitting astatine-211: treatment of recurrent brain tumor patients with [211]At-labeled chimeric 81C6 anti-tenascin monoclonal antibody. *J Nucl Med* 2008;49:30–38.

104. Zalutsky MR, Zhao X-G, Alston KL, Bigner DD. High-level production of α-particle-emitting [211]At and preparation of [211]At-labeled antibodies for clinical use. *J Nucl Med* 2001;42:1508–1515.

105. Reilly RM. Radioimmunotherapy of solid tumors: the promise of pretargeting strategies using bispecific antibodies and radiolabeled haptens. *J Nucl Med* 2006;47:196–199.

106. Paganelli G, Bartolomei M, Ferrari M, et al. Pre-targeted locoregional radioimmunotherapy with ^{90}Y-biotin in glioma patients: phase I study and preliminary therapeutic results. *Cancer Biother Radiopharm.* 2001;16:227–235.

107. Bartolomei M, Mazzetta C, Handkiewicz-Junak D, et al. Combined treatment of glioblastoma patients with locoregional pre-targeted ^{90}Y-biotin radioimmunotherapy and temozolomide. *Quat J Nucl Med Mol Imaging* 2004;48:220–228.

108. Grana C, Chinol M, Robertson C, et al. Pretargeted adjuvant radioimmunotherapy with Yttrium-90-biotin in malignant glioma patients: A pilot study. *Br J Cancer* 2002;86:207–212.

109. Merlo A, Hausmann O, Wasner M, et al. Locoregional regulatory peptide receptor targeting with the diffusible somatostatin analogue ^{90}Y-labeled DOTA[0]-D-Phe[1]-Tyr[3]-octreotide (DOTATOC): a pilot study in human gliomas. *Clin Cancer Res* 1999;5:1025–1033.

110. Hofer S, Eichhorn K, Freitag P, et al. Successful diffusible brachytherapy (dBT) of a progressive low-grade astrocytoma using the locally injected peptidic vector and somatostatin analogue [^{90}Y]-DOTA[0]-D-Phe[1]-Tyr[3]-octreotide (DOTATOC). *Swiss Med Wkly* 2001;131:640–644.

111. Schumacher T, Hofer S, Eichhorn K, et al. Local injection of the ^{90}Y-labelled peptidic vector DOTATOC to control gliomas of WHO grades II and III: an extended pilot study. *Eur J Nucl Med* 2002;29:486–493.

112. Hennig IM, Laissue JA, Horisberger U, Reubi JC. Substance-P receptors in primary human neoplasms: tumoral and vascular localization. *Int J Cancer* 1995;61:786–792.

113. Kneifel S, Cordier D, Good S, et al. Local targeting of malignant gliomas by the diffusible peptidic vector 1,4,7,10-tetraazacyclododecane-1-glutaric acid-4,7,10-triacetic acid-substance P. *Clin Cancer Res* 2006;12:3843–50.

114. Mamelak AN, Jacoby DB. Targeted delivery of antitumoral therapy to glioma and other malignancies with synthetic chlorotoxin (TM-601). *Expert Opin Drug Delivery* 2007;4:175–186.

115. Hockaday DC, Shen S, Fiveash J, et al. Imaging glioma extent with ^{131}I-TM-601. *J Nucl Med* 2005;46:580–586.

116. Mamelak AN, Rosenfeld S, Bucholz R, et al. Phase I single-dose study of intracavitary-administered iodine-131-TM-601 in adults with recurrent high-grade glioma. *J Clin Oncol* 2006;24:3644–3650.

117. Yordanov AT, Hens M, Pegram C, Bigner DD, Zalutsky MR. Antitenascin antibody 81C6 armed with [177]Lu: *in vivo* comparison of macrocyclic and acyclic ligands. *Nucl Med Biol* 2007;34:173–184.

118. Sampson JH, Akabani G, Berger MS, et al. Phase I study of intracerebral infusion of an epidermal growth factor receptor targeted toxin in recurrent malignant brain tumors. *Neurooncology* 2008;10:320–329.

119. Rosenkranz AA, Vaidyanathan G, Pozzi OR, Lunin VG, Zalutsky MR, Sobolev AS. Engineered modular recombinant transporters: application of a new platform for targeted radiotherapeutics to α-particle emitting [211]At. *Int J Radiat Oncol Biol Phys* 2008;72:193–200.

CHAPTER 6

Radioimmunotherapy for B-Cell Non-Hodgkin Lymphoma

THOMAS E. WITZIG

6.1 INTRODUCTION

The ultimate goal of effective cancer therapy is to deliver treatment to malignant cells while sparing normal cells. Although the development of therapeutic antibodies directed at tumor cells has truly accomplished the goal of targeting tumor cells, it has not consistently led to eradication of tumor cells. The efficacy of antibody therapy can be improved by adding a radionuclide to the antibody to form a radioimmunoconjugate (RIC). The use of RICs in cancer therapy is referred to as radioimmunotherapy (RIT), the subject of this chapter.

The first approved use of RIT was for lymphoma, the fifth most common tumor in men and women. In 2008, there were estimated to be 66,000 new cases of lymphoma and 19,000 deaths caused by the disease (1). Lymphoma was one of the early success stories of chemotherapy. By the 1990s, chemotherapy agents that had been developed, beginning in the late 1940s, were all in widespread use as single agents and in combination for treatment of lymphoma. In 1993, a large four-arm study comparing cyclophosphamide, doxorubicin, vincristine, and prednisone (CHOP) with three other more intense regimens concluded no advantage to the expanded regimens (2). This failure to demonstrate improvement of any of the new regimens over CHOP was discouraging to investigators in the field and occurred at a time when the incidence of lymphoma was steadily rising. This set the stage for the testing of unlabeled and radiolabeled monoclonal antibodies for lymphoma.

Monoclonal Antibody and Peptide-Targeted Radiotherapy of Cancer, Edited by Raymond M. Reilly
Copyright © 2010 John Wiley & Sons, Inc.

6.2 RADIOIMMUNOTHERAPY

6.2.1 Historical Background of RIT

Both unlabeled and radiolabeled antibodies have been important additions to the treatment of lymphomas and they were developed in parallel. The field of RIT began with the use of ^{131}Iodine (^{131}I) or ^{90}Yttrium (^{90}Y) conjugated to polyclonal antibodies (3–5). The antibodies were directed to carcinoembryonic antigen (CEA) or alphafetoprotein (AFP) for treatment of solid tumors and against ferritin for Hodgkin's disease (HD) and hepatoma. Although this initial approach did yield some antitumor activity, especially in relapsed HD (4, 5), the initial excitement of antibody treatment (both unlabeled and radiolabeled) waned because of the development of an immune response (antibodies) to the polyclonal agents. These "anti-antibodies" were a limitation because they hampered the ability to give repeated (fractionated) doses of the RIC (6).

The development of the technology to manufacture monoclonal antibodies (mAbs) revolutionized not only the field of laboratory diagnostics but also the clinical therapeutics (7). The first trials used anti-idiotype antibodies that were personalized to individual patients (8). The development of monoclonal antibodies to HLA-DR10 on malignant B-cells (Lym-1) (9) and later to the pan-B cell antigens such as B1 avoided the need to manufacture antibodies for each patient (10). In the 1990s, the development of anti-CD20 mAbs produced the first FDA-approved agents—rituximab (11, 12) and ^{131}I-tositumomab and ^{90}Y-ibritumomab tiuxetan (13, 14). ^{131}I-tositumomab was originally developed by Corixa Corporation (Seattle, WA) and is marketed under the trade name Bexxar® by GlaxoSmithKline. Bexxar is FDA-approved and is indicated for the treatment of patients with CD20 antigen-expressing relapsed or refractory, low-grade, follicular, or transformed non-Hodgkin's lymphoma (NHL), including patients with rituximab refractory NHL. ^{90}Y-ibritumomab tiuxetan was developed by IDEC Pharmaceuticals (San Diego, CA) and is marketed under the trade name Zevalin®. IDEC later merged with Biogen to form BiogenIdec (Cambridge, MA) and Zevalin is currently marketed by Spectrum Pharmaceuticals (Irvine, CA). Zevalin is FDA-approved for the treatment of patients with relapsed or refractory, low-grade or follicular B-cell NHL, including patients with rituximab refractory follicular NHL. There are no active trials with Lym-1 undergoing at this time.

Rituximab is an unlabeled mAb that targets the CD20 antigen on benign and malignant B-cells. After the initial phase I trials of rituximab (11, 15, 16), in a pivotal phase II clinical trial 166 patients with relapsed B-cell NHL were treated with rituximab 375 mg/m^2 weekly for 4 weeks and demonstrated an overall response rate (ORR) of 48% with 6% complete remission (CR) and a 13-month time to progression (TTP) (17). This led to the approval in 1997 of rituximab by the U.S. FDA for relapsed B-cell NHL. Rituximab is comarketed in the United States by BiogenIdec and Genentech (South San Francisco, CA). The approval of rituximab was a milestone for NHL since it was the first mAb to be approved for the treatment of this disease. The use of rituximab has increased rapidly since its approval and it is now used for B-cell malignancies and autoimmune conditions. Rituximab monotherapy has also been shown to be useful in previously untreated patients with follicular NHL (18, 19). Randomized trials have demonstrated the superiority of rituximab chemotherapy combinations (chemoimmunotherapy) in both indolent (20, 21) and aggressive NHL (22–25).

The application of RIT to treat B-cell NHL was a logical choice because NHL is sensitive to radiation delivered by conventional external sources (26, 27). Indeed, external beam radiation therapy (EBRT) has been a mainstay in the treatment of bulky masses that produce normal organ compromise and is effective at relieving pain from spinal cord compression. The difficulty in applying EBRT to most cases of indolent NHL is related primarily to the widespread nature of these tumors even at initial diagnosis. In addition, extensive EBRT damages normal marrow making it difficult to provide an effective chemotherapy or collect stem cells at a later date. For these reasons, conventional EBRT is typically limited to involved field applications in NHL.

The addition of rituximab and RIT to the treatment of NHL has led to an improvement in the overall survival (OS) of these patients (28–30). Despite this progress, patients cannot be promised cure neither with single-agent rituximab nor even with chemoimmunotherapy, and there remains room for improvement. This chapter will focus on studies that demonstrated the safety and efficacy of RIT and how RIT is integrated into the care of patients with lymphoma.

6.2.2 Radionuclides Used in RIT

In RIT, the goal is to utilize a radionuclide with high energy, but short path length, conjugated to an antibody to focus radiation on the target cell population while sparing the effects of radiation on the nearby normal tissues (Fig. 6.1). There are several possible radionuclides that emit alpha or beta particles, or gamma radiation, that can be attached to antibodies for therapeutic intent (Table 6.1 and reviewed in Ref. (31, 32)). Each of these alpha or beta particles has different energies and path lengths (33). Although many different radionuclides have undergone testing in clinical trials, the only ones in commercial use for FDA-approved indications are ^{90}Y or ^{131}I. There are several differences in the characteristics of the ^{131}I and ^{90}Y radionuclides that are important for the method of delivery to the patient; however, these differences have not translated into any apparent clinical advantage of one radionuclide over another. The ^{131}I-labeled antibodies can be used for imaging and dosimetry because they are both gamma and beta emitters. In contrast, since ^{90}Y is a pure beta emitter, imaging is performed with the corresponding gamma-emitting ^{111}Indium (^{111}In)-labeled antibodies.

These radionuclides can be potentially attached to any antibody. The choice of antibody depends on the antigenic profile of the tumor cell to be targeted. Ideal targets are those antigens that are preferentially expressed on tumor cells but not expressed or expressed at much lower density on normal cells so as to avoid toxicity to normal organs. Cell surface antigens that are not internalized or shed from the cell surface are often preferred. The microscopic intratumoral dosimetry of the radionuclide also appears to be important (34). The mechanism of antitumor activity of RIT is a

Ibritumomab tiuxetan

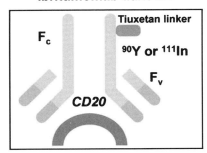

Tositumomab

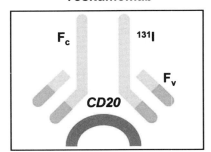

Radiolabeled rituximab

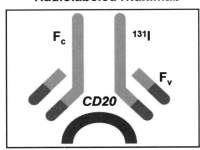

Epratuzumab

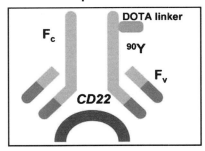

FIGURE 6.1   Radioimmunoconjugates in use for B-cell lymphomas. Ibritumomab tiuxetan is the murine parent antibody from which the human chimeric antibody rituximab was engineered. Tiuxetan is a linker/chelator that complexes ^{111}In (for imaging and dosimetry) or ^{90}Y (for therapy). Tositumomab is a murine anti-CD20 antibody to which the ^{131}I is directly attached through substitution into tyrosine amino acids to form Bexxar. The resulting RIC is used for both imaging and dosimetry. Rituximab is a human chimeric anti-CD20 antibody that is usually used in an unlabeled form. However, it can be directly labeled with ^{131}I for radioimmunotherapy. Epratuzumab is a humanized anti-CD22 antibody that is in clinical trials in both unlabeled and radiolabeled forms. The ^{90}Y is complexed by epratuzumab through a DOTA chelator.

combination of the effects of the antibody itself (35) and the targeted radiation delivered by the RIC. The antibody induces antibody-dependent cellular cytotoxicity (ADCC) or complement-dependent cytotoxicity (CDC). Anti-CD20 antibodies are classified as type 1 and type 2 on the basis of their ability to redistribute CD20 into lipid rafts and their potency *in vitro* (36–42). Rituximab is a type 1 anti-CD20 antibody that can produce clustering of lipid rafts when bound to CD20 but does not typically induce apoptosis. Tositumomab is a type 2 CD20 antibody that does not induce clustering but does induce apoptosis. *In vitro* studies have shown that radiation therapy increases tumor cell death when added to tositumomab but not when added to rituximab (43). The mechanism of this finding was dependent on mitogen-activated protein kinase (MAPK)/extracellular signal-regulated kinase (ERK). This may explain how RIT can kill cells that are rituximab-resistant.

TABLE 6.1   Characteristics of Radionuclides Used in Radioimmunotherapy for Non-Hodgkin Lymphoma

Parameter	^{131}Iodine	^{90}Yttrium
Gamma emission	Yes	No
Beta emission	Yes	Yes
Beta emission path length	0.8 mm	5 mm
Theoretical half-life	8 days	2.4 days
Localization of free radionuclide	Thyroid/stomach	Bone
Administration	Outpatient	Outpatient
Pretreatment unlabeled antibody	Yes	Yes
Imaging	Yes	No (^{111}In required as a surrogate)

RIC labeled with ^{131}I or ^{90}Y kills tumor cells by delivering high-energy radiation over a several millimeter path length enabling killing of adjacent tumor cells even if those cells are not actually bound to antibody ("cross-fire" effect). This is in contrast to unlabeled mAbs that theoretically must target each tumor cell in order to kill by ADCC or CDC or by inducing apoptosis. Recent studies by Jacobs et al. (44) using autoradiography demonstrate that Zevalin does target tumors preferentially over uninvolved tissue (such as negative bone marrow) and the RIC is localized to the tumor cell membrane rather than stroma (44). Although the RIC does deliver targeted radiation, the long path length can be toxic to nearby benign cells. As discussed in the following section, this has been an important issue only with bone marrow.

6.2.3   Administration of RIT: General Principles and Practice

The administration of RIT is a team effort (Fig. 6.2) (45, 46). There are some differences in administration of the RIC depending on whether Zevalin or Bexxar is used and these are discussed in more detail in the sections devoted to each agent. The hematologist/oncologist first identifies a potential patient and establishes eligibility. Patients undergo routine complete blood counts (CBC) to ensure that the absolute neutrophil count (ANC) is $\geq 1500 \times 10^6$/L and the platelet count is $\geq 100,000 \times 10^6$/L. Computerized tomography (CT) scans or a PET/CT are obtained to document the disease state at baseline so that response to RIT can be assessed (see Chapter 15). A bone marrow aspiration and biopsy with cytogenetic analysis is performed to make sure that marrow involvement with NHL occupies less than 25% of the cellularity and that there is no morphologic or cytogenetic evidence of myelodysplasia (MDS). The patient's insurance company is contacted to precertify the RIT. The RIC must be handled and injected by personnel certified by the U.S. Nuclear Regulatory Commission (NRC). The radiolabeled Zevalin and Bexxar are usually administered in the Nuclear Medicine or Radiation Oncology Department. However, the unlabeled antibody predose can be given in the oncology outpatient area and then the patient transferred to the RIT suite or alternatively both unlabeled and radiolabeled antibodies can be delivered in the nuclear medicine treatment area. In most academic centers,

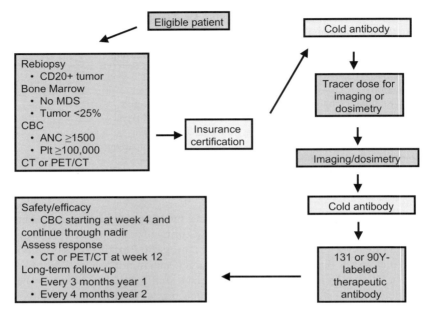

FIGURE 6.2   Administration of radioimmunotherapy is a team effort involving the hematologist/oncologist who identifies the eligible patient and performs the pretherapy evaluation. The radioimmunoconjugate is handled and administered by a nuclear medicine physician or radiation oncologist.

both unlabeled antibody and RIC are given on-site, whereas in the private practice sector the offices of the oncologist and the nuclear medicine site may be separate. This situation still can work for both offices because the RIC needs to be administered within 24 h of the unlabeled antibody. After the RIT has been administered, the patient is discharged and is followed by the oncologist. Myelosuppression begins to appear by week 4, nadirs by week 6–8, and then recovers. For clinical administration, there is no need to perform CBC prior to week 4 or after the counts have gone through the nadir and are in the safe range. Patients who are on anticoagulants such as heparin, coumadin, or aspirin should have these agents discontinued or monitored very closely during the period of maximal thrombocytopenia. The combination of thrombocytopenia and a prolonged prothrombin time places the patient at high-risk for bleeding. Prophylactic antibiotics are not needed for the typical patient. The patient is typically restaged at week 12 with a CBC and LDH (if initially elevated), CT scans for lesion assessment, and repeat marrow only if the pretreatment marrow was involved with lymphoma. In general, CTs prior to week 12 are not required unless the patient is under a clinical trial or progression is suspected. If performed prior to week 12, they will show response with the tumors becoming necrotic. It is possible that late responses not seen on week 12 scans will occur; however, this is uncommon. After week 12, the patient can return to routine tumor monitoring.

The rationale for the administration of unlabeled antibody prior to the RIC is to deplete normal blood B-cells and block nonspecific binding sites with the presumed net result of improved tumor targeting. This concept has been demonstrated in a mouse model to improve marrow biodistribution and to decrease nonspecific uptake of the RIC (47). The issue of whether administration of the unlabeled antibody interferes with RIC binding is an important one. Recent studies in a mouse xenograft model showed that rituximab pretreatment reduced targeting and tumor control by RIT (48). This has not been demonstrated in humans, and proving it is difficult because of differences between patients and the requirement for repeated imaging with and without a predose of unlabeled antibody. There have been no studies addressing this issue for Zevalin since the initial phase 1 studies (14, 49) tested different doses of the unlabeled antibody before the ^{90}Y-labeled Zevalin. In the study by Knox et al. (14), the addition of unlabeled murine ibritumomab improved biodistribution. In the second phase I study (49), two different levels, 125 mg/m^2 and 250 mg/m^2, of the unlabeled antibody rituximab were tested before Zevalin administration; smaller doses were not studied. In order to thoroughly test whether smaller doses of unlabeled antibody or different types of unlabeled antibodies are better than the ones used in the Bexxar and Zevalin treatment protocols would take an inordinate number of patients and is not practically possible. Because the response rates to both agents are so high (80%), a large number of patients would be needed.

6.2.4   Characteristics of Radiolabeled Monoclonal Antibodies to CD20

As noted above, initial studies of RIT in lymphoma used polyclonal antibodies (4); however, most RICs today are murine mAbs (50, 51). Recent studies of RIT using a variety of tumor antigen targets on NHL cells have indeed demonstrated tumor regressions with very few side effects in normal organs other than myelosuppression (13, 14, 29, 49, 52–74). This review will focus primarily on RICs that target CD20 that are now in clinical use for FDA-approved indications (50, 75–82) (Fig. 6.1). The CD20 antigen has proven to be an excellent target for RIT because CD20 expression is restricted to normal B cells, almost all B-cell NHL express CD20, the antigen is not internalized or expressed on other normal tissues (including stem cells), and depletion of normal B-cells by these antibodies has not led to significant short or long-term side effects.

6.2.4.1   Ibritumomab Tiuxetan (Zevalin)   Ibritumomab is a murine anti-CD20 antibody from which the human chimeric antibody rituximab was engineered. Ibritumomab is conjugated to tiuxetan, an MX-DTPA linker-chelator for ^{90}Y that forms Zevalin™ (CTI, Seattle, WA). Tiuxetan forms a covalent, urea-type bond with ibritumomab and chelates the radionuclide via 5-carboxyl groups. Zevalin is then reacted with either ^{111}In for tumor imaging and dosimetry or with ^{90}Y for RIT. ^{90}Y emits pure beta radiation with a path length of approximately 5 mm. The beta irradiation emitted by Zevalin is largely dissipated within 8 days of injection due to the short half-life of ^{90}Y (64 h). Because there is no gamma emission, useful tumor and normal organ images cannot be obtained with ^{90}Y-Zevalin. ^{111}In emits gamma rays; therefore, ^{111}In-Zevalin is used to produce high-quality images of the tumor and

normal organs for dosimetry and biodistribution studies (46, 83–86). Patients are imaged in anterior and posterior projections in a whole-body area mode between day 1 (the day of ^{111}In-Zevalin injection) and day 7. In clinical trials of Zevalin, if the ^{111}In-Zevalin scans predicted that the delivered dose of radiation to any nontumor organ was >2000 cGy, or if the dose to the bone marrow was >300 cGy, then no treatment with ^{90}Y-Zevalin was administered (87, 88). No patient in these trials failed dosimetry using these parameters. The ^{111}In-Zevalin images in the nontransplant setting are used only for safety. There is no correlation with tumor uptake of ^{111}In-Zevalin and ultimate tumor response to ^{90}Y-Zevalin or with toxicity (87, 89). The ^{111}In-Zevalin scans are also not clinically useful for lesion identification. For example, when PET scans were compared with ^{111}In-Zevalin images performed at the same time, the PET scan demonstrated more lesions with superior clarity than the ^{111}In-Zevalin images (44).

The first phase I trial of Zevalin used unlabeled ibritumomab before ^{90}Y-Zevalin and stem cells were cryopreserved as a precaution for prolonged myelosuppression. Patients were treated with single doses of 20–50 mCi of Zevalin and doses ≤40 mCi were not myeloablative (14). The second phase I trial used rituximab 125–250 mg/m^2 as the unlabeled antibody prior to Zevalin and patients were treated with one of the three dose levels – 0.2, 0.3, or 0.4 mCi/kg (49). If the patient weighed over 80 kg, the dose was capped at 32 mCi. Higher doses were not tested because stem cell cryopreservation was not required as part of the protocol eligibility. There was no provision for retreatment in this trial. In the third phase I trial, performed in Japan, rituximab was also used as the unlabeled antibody predose and two dose levels, 0.3 and 0.4 mCi/kg (90), were tested. Ten patients were enrolled and one had immediate progression leaving nine available for safety analysis. Three patients received 0.3 mCi/kg and six received 0.4 mCi/kg. This trial recommended 0.4 mCi/kg for routine use in Japan.

The U.S. FDA approved the Zevalin treatment program (Fig. 6.3) in February 2002, involving administration of rituximab and ^{111}In-Zevalin on day 1 followed by one imaging session to determine biodistribution 48–72 h later (images at other time points are optional). The criteria for an altered or unexpected biodistribution are diffuse increased uptake in normal lungs, kidneys with greater intensity than the liver, and intense areas of uptake throughout the normal bowel. If there is normal ^{111}In-Zevalin biodistribution, on day 8 the patient receives a second dose of rituximab followed by ^{90}Y-Zevalin infused over 10 min. The dose of ^{90}Y is weight-based for nonmyeloablative applications since only 7% of ^{90}Y is excreted by the kidneys over 7 days. Zevalin can be safely administered in the outpatient setting without radiation exposure hazard because it does not emit gamma radiation. Patients with platelet counts between 100–150,000 cells/mm^3 are dosed at 0.3 mCi/kg (capped at 32 mCi) and those with platelet counts >150,000 cells/mm^3 are administered 0.4 mCi/kg (capped at 32 mCi). In clinical studies of Zevalin, the dose for patients over 80 kg in weight was capped at 32 mCi because of concern for excessive myelosuppression if larger doses were administered. The issue of capping at 32 mCi for patients over 80 kg has made some investigators question whether this was underdosing these patients and resulting in a lower overall response rate (ORR). Wiseman et al. recently investigated this issue by reviewing the treatment results in 67 patients >80 kg (91). The ORR and CR rates were 79% and 28% versus 70% and 34% for the <80 kg and ≥80 kg groups,

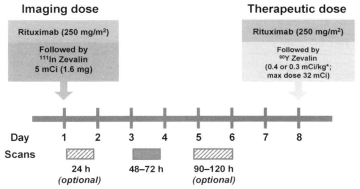

FIGURE 6.3   Treatment schedule for ^{90}Y-ibritumomab tiuxetan (Zevalin).

respectively. The median time to progression values were also similar at 8.9 and 9.5 months, respectively. There were no significant differences in the measures of efficacy or grade 3/4 nonhematologic or hematologic adverse events. Taken together, approximately 40% of patients will have their dose of Zevalin capped; however, this does not appear to have an impact on the ORR, duration of response (DR), or OS.

The tumor response to Zevalin typically becomes evident at week 4, and nearly all responders will demonstrate this by week 12. Although the radiation emitted by beta emitters such as ^{90}Y is high energy, patients should not expect immediate shrinkage as they often do in response to chemotherapy. This clinical observation is supported by a recent [^{18}F]-thymidine (FLT) PET study that demonstrated little change in tumor cell proliferation within 48 h of Zevalin injection (92) (see Chapter 15). In this study, mice bearing a follicular NHL xenotransplant were imaged with FLT PET pretreatment and 48 h after treatment with Zevalin. There was minimal decrease in lymphoma cell proliferation at 48 h in mice treated with Zevalin compared to a substantial reduction in those treated with cyclophosphamide.

6.2.4.2   *Tositumomab (Bexxar*^{TM}*)*   Tositumomab is an IgG$_{2a}$ murine mAb directed against CD20 (Fig. 6.1). It was previously referred to as anti-B1 (13, 93) before receiving the generic name tositumomab. For RIT, tositumomab has been radiolabeled with ^{131}I. Because ^{131}I emits both gamma and beta radiations, ^{131}I-tositumomab can be used for dosimetry that requires both imaging and treatment, that is, there is no need for ^{111}In-labeling (94). The Bexxar treatment regimen consists of unlabeled tositumomab and ^{131}I-tositumomab (GlaxoSmithKline, NC). Thus, both the unlabeled and the radiolabeled mAbs in the Bexxar regimen are the same and both are murine. Tositumomab is administered intravenously on 2 treatment days with each day consisting of two separate infusions (Fig. 6.4). Although both Zevalin and Bexxar require tumor imaging, for Bexxar the

Bexxar™ dosing and administration regimen

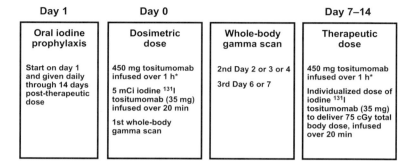

Day 1	Day 0		Day 7–14
Oral iodine prophylaxis	**Dosimetric dose**	**Whole-body gamma scan**	**Therapeutic dose**
Start on day 1 and given daily through 14 days post-therapeutic dose	450 mg tosumomab infused over 1 h* 5 mCi iodine 131I tositumomab (35 mg) infused over 20 min 1st whole-body gamma scan	2nd Day 2 or 3 or 4 3rd Day 6 or 7	450 mg tositumomab infused over 1 h* Individualized dose of iodine 131I tositumomab (35 mg) to deliver 75 cGy total body dose, infused over 20 min

* Patients are pretreated with acetaminophen 650 mg and
diphenhydramine 50 mg before unlabeled predose of tositumomab

FIGURE 6.4 Treatment schema for ^{131}I-tositumomab (Bexxar).

dosimetry estimates are used to calculate a patient-specific therapeutic dose (95). This is necessary because of the interpatient differences in body mass (weight), spleen size, tumor burden, and the metabolism and renal excretion of [131]I (13, 94–97). Because of the propensity of radioiodine to concentrate in the thyroid, SSKI solution (Lugol's) is started 1 day before the unlabeled tositumomab and is continued for 2 weeks after administration of the radiolabeled tositumomab. The first treatment is referred to as the dosimetric dose. It consists of tositumomab 450 mg infused over 1 h followed by 5 mCi of ^{131}I-tositumomab. Gamma camera scans to measure whole-body counts are performed on day 1 and repeated within 2–4 days and 6–7 days after the dose. The biodistribution of ^{131}I-tositumomab should be assessed by determination of total body residence time and visual examination of whole-body images from the first image taken at the time of count 1 (within an hour of the end of the infusion) and from the second image taken at the time of count 2 (at 2–4 days after administration). An evaluation of the third image at the time of count 3 (6–7 days after administration) is necessary and may help resolve ambiguities. The results of the dosimetry determine the amount of radioactivity (mCi of ^{131}I) needed for each patient to receive a specified total body radiation absorbed dose—75 cGy for patients who have platelet counts >150,000 cells/mm^3 or 65 cGy for patients with platelets between 100,000–150,000 cells/mm^3. The therapeutic dose is given within 7–14 days of the dosimetric dose and consists of 450 mg of tositumomab followed by a 20-min infusion of the patient-specific mCi amount of ^{131}I-tositumomab. Bexxar can be given as an outpatient procedure to a vast majority of patients utilizing the revised U.S. NRC regulations 10CFR 35.75, which allow outpatient release if the total effective dose equivalent to another person who is exposed to the treated patient is <500 mrem (0.5 cGy) (98, 99).

6.2.5 Clinical Results of Anti-CD20 RIT for Relapsed NHL

The clinical trials of the two anti-CD20 RIC that were performed to assess toxicity and efficacy have primarily been limited to patients with relapsed disease and excellent bone marrow and normal organ function. The eligibility requirements for RIT trials have been similar except for certain studies as mentioned in this section. Patients were to have measurable disease, bone marrow with less than 25% involvement with lymphoma, ANC $\geq$1500, platelet count $\geq$100,000 cells/mm^3, normal renal and liver function, and less than 25% of the marrow previously treated with EBRT. Patients were excluded from these trials if they had CNS lymphoma, HIV infection or HIV-related NHL, chronic lymphocytic leukemia, pleural or peritoneal fluid that was positive for lymphoma, known myelodysplasia, or a history of allogeneic or autologous stem cell transplant. Tables 6.2 and 6.3 describe the patient population included in each of these

TABLE 6.2   Summary of Clinical Trials of Ibritumomab Tiuxetan (Zevalin)

Trial	n	Goal	Reference
Phase I	14	• Used unlabeled ibritumomab prior to ^{90}Y-ibritumomab • Determine MTD • Indolent and aggressive NHL	(14)
Phase I/II	51	• Determine dose of rituximab prior to ^{111}In-ibritumomab • Determine MTD • Indolent and aggressive NHL including mantle cell	(49)
Phase I	9	• Determine safety of Zevalin in Japanese population • Select dose for future trials • Follicular and mantle cell NHL	(90)
Phase III	143	• Randomized trial of rituximab versus ^{90}Y-ibritumomab to determine if efficacy of ^{90}Y-ibritumomab is superior • Indolent and transformed NHL	(67)
Phase II	30	• Efficacy and toxicity of 0.3 mCi/kg ^{90}Y-ibritumomab for patients with platelet count of 100,000–149,000 $\times$ 10^6/L • Indolent and transformed NHL	(68)
Phase II	54	• Efficacy and toxicity of 0.4 mCi/kg ^{90}Y-ibritumomab for patients refractory to rituximab • Follicular NHL	(122)
Expanded safety analysis	349	• Evaluate the side effects experienced by patients treated with ^{90}Y-ibritumomab in clinical trials • Indolent or transformed NHL	(109)

^{90}Y: yttrium-90; ^{111}In: indium-111; MTD: maximum tolerated dose.

TABLE 6.3   Clinical Trials of ^{131}I-Tositumomab in Patients Without the Use of Stem Cell Support

Trial #	n	Goal	References
Phase I	59	• Used single or multiple doses trace-labeled tositumomab prior to therapeutic ^{131}I-tositumomab • Determine MTD of ^{131}I-tositumomab • Indolent, transformed, aggressive NHL	(13, 58, 103)
Phase II	86	• Validate dose determined in phase I • Indolent and transformed NHL	(70, 104)
Pivotal	60	• Compared response to last qualifying chemotherapy with that obtained with ^{131}I-tositumomab • Indolent and transformed NHL	(65)
Unlabeled versus radiolabeled	48	• Establish improved efficacy of radiolabeled tositumomab compared to unlabeled tositumomab • Indolent and transformed NHL	(121)
Rituximab pretreated	40	• Evaluated response to ^{131}I-tositumomab in patients failing rituximab	(123)
Previously untreated	76	• Evaluated overall and complete response rate in patients treated with ^{131}I-tositumomab as their first therapy • Previously untreated follicular NHL	(193)
Expanded access	368	• To treat patients with relapsed NHL with standard ^{131}I-tositumomab • Indolent and transformed	(105, 106)

clinical trials, Table 6.4 summarizes the efficacy results, and Table 6.5 the hematologic toxicity. In general, both Bexxar and Zevalin as single agents in relapsed indolent NHL produced an ORR of 80% with 30% CR, and 20% of patients achieved long-term remissions without relapse. These agents can produce impressive responses in both nodal and extranodal disease without damage to adjacent normal structures. The only side effect is reversible myelosuppression; there is no significant organ toxicity. In all trials of RIT to date, response assessment was performed with CT scans. The recent integration of PET scanning into response assessment guidelines (100) may change the response rates seen with RIT. Recent studies have been integrating PET into RIT trials (101, 102) (see Chapter 15). For example, in the study by Ulaner et al. (102), PET scans were performed on patients receiving RIT and demonstrated how patients classified as a PR by CT were PET negative and experienced long-term disease-free survival.

6.2.5.1   *Phase I ^{131}I-Tositumomab (Bexxar)*   The phase I trial of ^{131}I-tositumomab was designed to determine a dose that could be administered without stem

TABLE 6.4 Response Rates in Trials Without Stem Cell Support

Trial	Agent	n	ORR	CR	DR	DR/CR	References
Phase 1	^{131}I-tositumomab	59	71	34	8.9	18.3	(13, 58, 103)
Phase I	^{90}Y-ibritumomab	14	79	36	–	–	(14)
Phase I/II	^{90}Y-ibritumomab	51	67	26	11.7+		(49)
Phase I	^{90}Y-ibritumomab	10	70	50	–	–	(90)
Phase II	^{131}I-tositumomab	45	57	32	9.9	19.9	(70)
Phase II	^{131}I-tositumomab	41	76	49	1.3	>2.5 years	(104)
Pivotal	^{131}I-tositumomab	60	65	17	6.5		(65)
Randomized	^{90}Y-ibritumomab	73	80	30	14.2 (0.9–28.9)		(67)
Randomized	^{131}I-tositumomab	42	67	33	NR		(121)
Previously untreated	^{131}I-tositumomab	76	95	74	NR		(193)
Transformed	^{131}I-tositumomab	71	39	25	20	36.5	(127)
Rituximab refractory	^{90}Y-ibritumomab	54	74	15	6.4 (0.5–≥24.9)		(122)
Rituximab refractory	^{131}I-tositumomab	24	62	25	22.4	NR	(123)
Relapsed after rituximab but not refractory	^{131}I-tositumomab	16	69	56	24.9	NR	(123)
Expanded access	^{131}I-tositumomab	273	58	27	NR		(105)
Phase II for patients with thrombocytopenia	^{90}Y-ibritumomab	30	83	43	11.7 (3.6–≥23.4)		(68)

ORR: overall response rate; Cr: complete remission; DR: duration of response; DR/CR: duration of response in CR patients; NR: not reached.

TABLE 6.5 Hematologic Toxicity Experienced with Anti-CD20 Radioimmunoconjugates

Study	Neutrophils		Platelets		Hemoglobin		Reference
	Nadir $\times10^6$/L	% Grade 4 (<500 $\times10^6$/L)[a]	Nadir $\times10^6$/L	% Grade 4 (<10,000 $\times10^6$/L)[b]	Nadir (g/dL)	% Grade 4 (<6.5 g/dL)	
Zevalin studies							
Phase I	1100	27	49,500	10	–	–	(49)
Phase I	–	33	–	0	–	0	(90)
Randomized trial	900	32	42,000	6	10.8	1	(67)
Phase II for patients with thrombocytopenia	600	33	26,500	13	10.1	3	(68)
Rituximab refractory	700	35	33,000	9	9.9	4	(122)
Bexxar studies							
Pivotal trial	800	18	50000	2	10.2	0	(65)
Previous rituximab therapy	1200	18	85,000	10	11	0	(123)
Combined analysis	1060	16	70000	2	11.1	1	(107)
Previously untreated	1300	34 (grades 3/4)	83000	0	NA		(193)
Phase II in early relapse	1200	20	78000	5	11.3	0	(104)

[a] Grade 4 neutropenia is <500 $\times10^6$/L.
[b] Grade 4 thrombocytopenia is <10,000 $\times10^6$/L.

cell support (13, 58, 103). Patients were treated with 15–20 mg of intravenous anti-B1 (anti-CD20) mouse mAb trace-labeled with ^{131}I (5 mCi) over 30 min. The first 10 patients were administered ^{131}I-tositumomab without unlabeled tositumomab pretreatment. The next eight patients were given a small dose (135 mg) of unlabeled tositumomab pre-^{131}I-tositumomab administration and another two received 685 mg pretreatment (13). After the radiotracer dose, serial quantitative gamma-camera images and measurements of whole-body radioactivity were obtained, and with this approach tumors >2 cm were visualized on the scans. Thirty-four patients were included in the next phase I report (58). Seventeen (50%) had low-grade NHL, eight (24%) were transformed, and nine (26%) were intermediate grade. Bone marrow involvement was present in 26% and 62% were resistant to the last chemotherapy regimen. All patients received at least one radiotracer dose and 68% (23/34) had more than one radiotracer dose. It was concluded that indeed the unlabeled predose did improve tumor/normal organ biodistribution. Of the 34 patients entering the trial, 28 (82%) received a therapeutic dose; 3 patients developed human antimouse antibodies (HAMA) and 3 had rapid progression during the radiotracer studies that precluded treatment. Patients were retained in radiation isolation after the therapeutic dose for about 3 days. This trial treated patients with doses calculated to deliver 25–85 cGy to the whole body.

The dose-limiting toxicity (DLT) was myelosuppression and occurred at 85 cGy when two of the three patients treated at this dose experienced grades 3 and 4 neutropenia and thrombocytopenia. The maximum tolerated dose (MTD) was therefore determined to be 75 cGy. Three patients in the trial who had been previously treated with a stem cell transplant had more severe myelosuppression. No opportunistic infections were documented and no changes in serum Ig levels occurred. One patient developed myelodysplastic syndrome (MDS) resulting in acute leukemia and six patients developed HAMA. The ORR in 28 patients receiving a therapeutic dose was 79% (22/28) with 50% CR (14/28). Six patients who had achieved a PR or minor response were administered a second dose and three had a further response, but none of the three patients converted to a CR. When the ORR was examined by tumor type, 100% (13/13) of the low-grade NHL responded with 77% (10/13) CR; 75% (6/8) of the patients with transformed NHL with 38% (3/8) CR; and 43% (3/7) of the patients with intermediate NHL responded with 14% (1/7) CR. The DR for CR patients was 16.5+ months; neither the DR nor the TTP was reported for PR patients. When patients relapsed, it was in areas not previously involved with NHL in 75% (6/8) patients. Four patients were retreated at relapse and all responded again (2 CR; 2 PR) (58).

The next report on this phase I/II trial concluded that a predose of 475 mg of unlabeled tositumomab produced optimal tumor/normal organ biodistribution, and indeed that was the dose used in all other trials discussed in the following sections (103). The ORR was 71% (42/59) with 34% (20/59) CR. The patients with low-grade or transformed NHL had an ORR of 83% compared to 41% in those with an intermediate-grade NHL. The median DR for all responders was 8.9 months (18.3 months for CR patients). Additional long-term follow-up of these 59 patients was described in 2000 (73).

6.2.5.2 *Phase II Trials of Bexxar*

Several dedicated phase II trials were conducted to validate the high ORR to Bexxar determined in phase I/II trials (13, 58, 103). Vose et al. (70) studied 47 patients with relapsed low-grade or transformed NHL between December 1995 and November 1996; 45 received treatment. One patient progressed rapidly and was not given the therapeutic dose and the other did not receive the therapeutic dose at the protocol-specified time. Patients received either 75 cGy or 65 cGy depending on baseline platelets. The median age was 49 years (range, 23–74). The histology was low grade in 79% (37/47) and transformed in 21% (10/47); 92% (43/47) had stages III/IV disease. The patients had previously received a median of four prior chemotherapy regimens (range, 1–8). LDH was elevated in 38% (18/47) and bone marrow was involved in 51% (24/47). The calculated dose to the normal organs for patients receiving the 75 cGy total body dose was 499 cGy to the kidneys, 383 spleen, 225 liver, 214 bladder, and 183 to the lungs. The tumors were calculated to receive a mean dose of 795 cGy. The ORR was 57% (27/45) and 32% (15/45) had a CR. The median DR was 9.9 months for all responders and 19.9 months for responders that were CR. The median TTP for all patients was 5.3 months and 11.6 months for responders. The median OS from study entry was 36 months with no deaths due to the treatment. As of 2000, six patients remained in CR with durations of CR 26.9+ to 33.8+ months. The median ANC nadir was 800 cells/mm^3, platelets

43,000 cells/mm^3, and hemoglobin 10.2 g/dL. There was no mention of any patient developing MDS or acute leukemia. One patient (2%) developed HAMA. A more recent phase II trial of single-agent Bexxar restricted the patient population to only one or two prior therapies (104). The ORR was 76% (31/41) with 49% CR/Cru (CR unconfirmed); the median DR for all responders was 1.3 years and for CR patients more than 2.5 years. Seven of the patients had transformed histology and the ORR was 71% (5/7) with 29% (2/7) CR/CRu. Patients with bulky disease (>5 cm) had a lower chance of ORR but a similar DR.

6.2.5.3 *Pivotal Trial of Bexxar*   This trial enrolled 60 patients with relapsed low-grade ($n = 36$), transformed NHL ($n = 23$), or mantle cell ($n = 1$) lymphoma, and compared the ORR and DR to Bexxar with that achieved by the patient's last qualifying chemotherapy. The patients had received a median of four prior chemotherapies (range, 2–13) and were required to have failed to respond or progressed within 6 months of the last chemotherapy regimen. Thirteen (22%) patients had been previously treated with a fludarabine-containing regimen. The patients were a median of 54 months from initial diagnosis, 44% had an elevated LDH, and 55% had bulky disease (≥ 5 cm). The primary end point was DR. Nineteen patients in the trial had equivalent responses. Of the 41 cases where DR was not equivalent, 32 (78%) patients experienced a longer DR with Bexxar compared to 9 (22%) patients who experienced a longer DR with chemotherapy ($p < 0.001$). The median DR from Bexxar was 6.5 months compared to 3.4 months with chemotherapy. The ORR was a secondary end point in the trial; it was 65% (39/60) with Bexxar compared to 28% (17/60) with chemotherapy ($p < 0.001$). Seventeen percent (10/60) had a CR with Bexxar compared to three percent (2/60) with chemotherapy ($p = 0.01$). The ORR with Bexxar differed between the low-grade and transformed patient groups: 81% (29/36) of the low-grade patients had a tumor response with 19% CR (7/36) compared to 39% (9/29) and 13% CR (3/29) in the patients with transformed NHL (65).

After the initial trials of Bexxar demonstrated a high ORR with acceptable toxicity, an expanded access study treated patients with relapsed low-grade or transformed NHL with standard doses of Bexxar (total body dose of 75 cGy for patients with platelets >150,000 cells/mm^3 and 65 cGy for platelets 100,000–150,000 cells/mm^3) (105, 106). Three hundred sixty-eight patients were enrolled in fifty-three community and academic sites from July 1998 to March 2000. Bexxar was received by 98% (359/368) and 273 could be evaluated for efficacy. The median age was 58 years (range, 32–87). These patients had received a median of two prior therapies (range, 1–9) and 45% had failed prior rituximab. The histology at study entry was small lymphocytic lymphoma in 10%, follicular grade I in 37%, follicular grade II in 29%, and transformed in 21%. The patients were 90% stage III/IV, 60% had elevated LDH, 43% had bone marrow involvement, and 44% presented with bulky disease (>5 cm). The ORR was 58% and 27% had a CR. The ORR in rituximab failures was 47% (55/118) with 19% CR (23/118). In patients with bulky tumors, the ORR was 47% (57/121) with 17% (20/121) CR. The TTP for all patients was 7.1 months, and for responders both the median TTP and the median DR have not yet been reached.

6.2.6 Safety of Bexxar RIT

The potential adverse events with Bexxar treatment include a transient flu-like syndrome and bone marrow suppression. There is typically minimal infusion-related toxicity with unlabeled tositumomab, and only 7% and 2% of patients experienced an adverse event that required an adjustment to the rate of infusion during the dosimetric and therapeutic doses, respectively. The marrow toxicity has been summarized for 215 patients who received a 75 cGy total body dose (107). The median ANC nadir was 1060 cells/mm^3 occurring at day 46 and recovering by day 66. Sixteen percent of patients developed grade IV neutropenia with a median duration of 11 days. The median time to platelet nadir was 32 days, the platelet nadir was 70,000/mm^3, and only 2% of the patients developed grade IV thrombocytopenia. The duration of grade IV thrombocytopenia was 14 days and the platelets recovered in a median time of 50 days from the day of treatment. The median hemoglobin nadir was 11.1 g/dL on day 46. Only 1% of the patients developed grade IV hemoglobin and the hemoglobin recovered with a median of 61 days following the therapeutic dose. Seventeen percent of patients required hematologic supportive care with 8% receiving a platelet transfusion, 9% red blood cell transfusion, and 12% growth factors (G-CSF, GM-CSF, or erythropoietin). The hematologic nadirs varied by patient characteristics. For example, patients with a negative marrow pre-Bexxar had a 12% incidence of grade IV neutropenia compared to 21% if the marrow was positive for NHL. The presence of marrow lymphoma likely attracts higher amounts of the radioimmunoconjugate leading to increased normal marrow toxicity. Patients with no prior chemotherapy had a 5% incidence of grade IV neutropenia compared to 21% if they had 1–3 prior regimens and 23% if ≥ 4 prior regimens. The need for hematologic supportive care ranged from 0% in previously untreated patients to 32% in patients with ≥ 3 prior therapies. There was no difference in the incidence of grade IV toxicity by age.

The toxicity of Bexxar in the expanded access study showed similar results (106). The median ANC nadir was 1300 cells/mm^3, median platelets were 68,000 cells/mm^3, and median hemoglobin 11.2 g/dL. Grade IV neutropenia developed in 14% of patients and 2% experienced grade IV thrombocytopenia. Hospitalization for serious infection occurred in 4%. The HAMA rate was 8%. HAMA has not been reported to influence OS in the trials of Bexxar and Zevalin; however, in the trials of the Lym-1 RIC, patients who developed HAMA actually had a superior OS (108).

6.2.6.1 *Phase I Trials of ^{90}Y-Zevalin* Six separate clinical trials of ^{90}Y-Zevalin have been conducted over the past 10 years (Table 6.2). There were three phase I trials of Zevalin. The first study enrolled 14 patients with relapsed low or intermediate CD20-positive B-cell NHL (14). The patients were imaged twice with ^{111}In-Zevalin— the first imaging was performed without unlabeled ibritumomab; the second was performed following unlabeled ibritumomab. A comparison of the two sets of ^{111}In-Zevalin images demonstrated that predosing with unlabeled ibritumomab improved the biodistribution of the Zevalin. Patients were then treated with ^{90}Y-Zevalin with doses ranging from 13.5 to 50 mCi. All patients had stem cells harvested from peripheral blood or marrow prior to treatment with ^{90}Y-Zevalin; however, only

two patients (both had received 50 mCi of ^{90}Y-Zevalin) required reinfusion of stem cells. The ORR was 79% (11/14) with 36% CR, and 43% partial remission (PR).

A second phase I study was conducted to test the use of rituximab rather than unlabeled ibritumomab before Zevalin. It was felt that rituximab was less likely to cause a human antimouse or chimeric antibody (HAMA or HACA) response compared to the murine ibritumomab. Another goal of the second phase I trial was to determine the MTD of ^{90}Y-Zevalin that could be given to patients without using stem cells or prophylactic growth factors. Fifty-one patients were enrolled and the study concluded that 250 mg/m^2 was the optimal dose of rituximab to be used before ^{111}In-Zevalin imaging and ^{90}Y-Zevalin therapy (49). Dosimetry predicted that all patients were eligible for ^{90}Y-Zevalin; that is, all normal nontumor-bearing organs were predicted to receive <2000 cGy and the bone marrow <300 cGy (88). The median age was 60 years and 24% of patients were >70 years of age. Sixty-six per cent of patients had low-grade NHL, 28% intermediate grade, and 6% were mantle cell lymphoma (MCL). In the low-grade group, 6% were diffuse small lymphocytic lymphoma, 27% follicular small cleaved, and 33% had follicular mixed lymphoma. All patients had received prior chemotherapy (median of two prior regimens) and 92% had received an anthracycline. Thirty-seven percent had received prior external beam radiotherapy; twenty-seven percent had ≥2 extranodal sites of disease; fifty-nine percent had bulky disease (mass ≥5 cm); and forty-three percent were bone marrow positive for NHL.

The doses of ^{90}Y-Zevalin used in the phase I/II trial were 0.2 mCi–0.4 mCi/kg; 5 patients received 0.2 mCi/kg, 15 received 0.3 mCi/kg, and 30 patients received 0.4 mCi/kg. All patients who received 0.4 mCi/kg were able to recover bone marrow function without prophylactic growth factors or stem cells. The dose was not increased to more than 0.4 mCi/kg because substantial myelosuppression was already being obtained with 0.4 mCi/kg and stem cells had not been collected pre-Zevalin. The efficacy component of the phase I/II trial demonstrated a 67% ORR in all patients with 26% CR. In patients with low-grade NHL, the ORR was even higher at 82% with 26% CR (49). The median TTP for responders was 15.4 months; the DR was 11.7+ months.

The third phase 1 study was performed in Japan (90). Two doses of Zevalin were tested—0.3 mCi/kg (11.1 MBq/kg) or 0.4 mCi/kg (14.8 MBq/kg). Rituximab was given on day 1 and 8 at a dose of 250 mg/m^2. Ten patients were treated—four at 0.3 mCi/kg dose and six patients at the 0.4 mCi/kg dose level with an ORR of 70% (7/10). No patient developed HAMA/HACA. The main toxicity was myelosuppression and the recommended dose for future studies was 0.4 mCi/kg.

6.2.6.2 Phase II Trial of Zevalin in Patients with Mild Thrombocytopenia

Patients with mild thrombocytopenia from previous therapy are at an increased risk of myelosuppression from RIT. A separate phase II trial using a reduced dose of ^{90}Y-Zevalin (0.3 mCi/kg) for patients with relapsed NHL and a platelet count between 100,000–149,000 cells/mm^3 was designed. Thirty patients were treated and the ORR was 83% with 43% CR/CRu. The TTP was 9.4 months in all patients and 12.6 months in responders (68). The median DR was 11.7 months (3.6–≥23.4). Hematologic toxicity was the primary toxicity with a median nadir ANC of 600 × 10^6/L (grade IV in

33% of patients) and the median nadir platelet count 26,500 cells/mm^3 (grade IV in 13% of the patients).

6.2.6.3 Safety of Zevalin RIT   The safety of Zevalin RIT in 349 patients treated in all reported studies to date has been summarized (109). Patients were followed for up to 4 years after therapy or until progressive disease. Infusion-related toxicities were typically grade 1 or 2 and were associated with rituximab infusion; there were no further infusion-related reactions when ^{111}In- or ^{90}Y-Zevalin was administered at the conclusion of the rituximab. No significant normal organ toxicity was noted. The main toxicity noted was myelosuppression (Table 6.5) with the nadir hemoglobin, WBC, and platelet counts typically occurring at 7–9 weeks and lasting approximately 1–4 weeks depending on the method of calculation. Following the 0.4 mCi/kg dose, grade 4 neutropenia, thrombocytopenia, and anemia occurred in 30%, 10%, and 3% patients, respectively, and following the 0.3 mCi/kg dose in 35%, 14%, and 8% patients.

Bone marrow involvement with NHL at study entry was present in 146 patients (42%). Patients with any degree of marrow involvement had a significantly greater incidence of grade 4 neutropenia ($p = 0.001$), thrombocytopenia ($p = 0.013$), and anemia ($p = 0.040$) than patients with no marrow involvement. The incidence of grade 4 hematologic toxicity increased with increasing levels of marrow involvement at baseline. Since in all these trials prophylactic growth factors were not prescribed, it is not yet known whether grade 3/4 myelosuppression can be prevented if growth factors were integrated into the Zevalin program. Despite the substantial myelosuppression observed with RIT, only 7% of patients were hospitalized with infection (3% with neutropenia) and only 2% had grade 3 or 4 bleeding events. A recent case study reported skin necrosis after extravasation of Zevalin (110); therefore, RICs should be considered vesicants.

6.2.7 Radiolabeled Rituximab

Another approach to anti-CD20 RIT is to radiolabel rituximab with ^{131}I. Leahy et al. (74) conducted a trial that enrolled 91 patients with relapsed CD20-positive B-cell NHL. The entry criteria were similar to other RIT trials except there was no restriction on the level of marrow involvement. The ORR was 76% with 53% CR/CRu and the median DR was 10 months (20 months for CR patients). Nine patients were enrolled with >25% marrow involvement with NHL. These patients had a lower platelet nadir but the incidence of grade 4 myelosuppression was no different from those patients with less than 25% marrow involvement. This treatment can also be repeated due to the humanized nature of rituximab (111).

6.3 ANTIBODIES AGAINST CD22

6.3.1 Radiolabeled Epratuzumab

Epratuzumab (Immunomedics, Inc., Morris Plains, NJ, U.S.A.) is a humanized mAb to CD22, another cell surface antigen commonly found on B-cell lymphomas. The unlabeled antibody has been used in a number of clinical trials as a single-agent and shown to provide an ORR of 10% with an excellent toxicity profile in a phase I/II trial (112). A subsequent trial combined epratuzumab with rituximab (113) and with standard rituximab, cyclophosphamide, vincristine, and prednisone (RCHOP) (114) with acceptable safety.

This agent has also been developed as an RIC. Studies initiated in 1991 used ^{131}I (115). More recently, epratuzumab has been conjugated to the macrocyclic chelate 1,4,7,10-tetraazacyclododecane-N,N',N'',N'''-tetraacetic acid (DOTA) that then chelates ^{90}Y to form the RIC for RIT (116). The use of the more stable DOTA chelator results in less ^{90}Y being deposited in bone compared to DTPA chelators used to complex the radiometal potentially reducing myelosuppression (116). Clinical trials of ^{90}Y-epratuzumab have focused on fractionated therapy (117, 118). This approach differs from Bexxar and Zevalin where the entire amount of radioactivity is given as a single dose. Linden et al. (117) treated 16 patients with 2–4 weekly injections of ^{90}Y-epratzumab. The two patients who received four doses had dose-limiting myelosuppression. The ORR was 62% with 75% of indolent patients responding compared to 50% of those with aggressive disease; the CR rate was 25%. As expected, demonstration of CD22 staining on tumor cells was a strong predictor of response. Advantages of using a humanized antibody include the ability to give fractionated doses; only a small amount of unlabeled antibody is mixed with the RIC. A more recent report (118) focused on the use of PET to predict response. An update of this trial was recently reported (119). Sixty-four patients have now been enrolled. Eligibility included relapsed disease with no masses >10 cm, platelet count $>100,000/\text{mm}^3$, and $\leq 25\%$ marrow involvement. The therapy was well tolerated with myelosuppression the only significant toxicity. The ORR for 17 patients treated with $<10\,\text{mCi/m}^2$ total dose was 41% compared to 55% for those 29 patients treated with $>20\,\text{mCi/m}^2$. The ORR was 50% (7/14) in mantle cell lymphoma, 30% (3/10) in diffuse large B-cell lymphoma (DLBCL), and 71% (22/31) in follicular NHL patients. The current approach of $20\,\text{mCi/m}^2$ weekly $\times$ 2 doses is tolerable and effective and delivers more mCi of ^{90}Y than a single 32 mCi dose of Zevalin. Whether the fractionated dosing will improve ORR and DR and be financially feasible awaits further results. It is clear that ^{90}Y-epratuzumab represents a safe and novel approach to lymphoma RIT.

6.4   RIT VERSUS IMMUNOTHERAPY

At the conclusion of the initial phase I/II trials of RIT for NHL, it appeared that the ORR was higher with RIT than with the same corresponding unlabeled mAb. Two key trials were designed to further address the question as to whether RIT was superior to immunotherapy with a similarly targeted antibody. IDEC 106-04 was a prospective, randomized trial of RIT versus rituximab in patients with relapsed CD20-positive NHL who had never received rituximab. Patients were randomized to receive either 0.4 mCi/kg (maximum of 32 mCi) Zevalin or rituximab 375 mg/kg weekly for 4 weeks (67). Patients were eligible for this trial if they had a biopsy-proven low-grade, follicular, or transformed NHL, a performance status of 0–2, an absolute neutrophil count of $\geq 1,500 \times 10^6$/L, and a platelet count of $\geq 150,000 \times 10^6$/L. One-hundred forty-three patients were randomized in this trial—seventy-three received Zevalin and seventy rituximab. To achieve balance, the patients were stratified by disease type (small lymphocytic, follicular, and transformed). The analysis of all 143 patients found an ORR (International Workshop NHL criteria (120)) of 80% with ^{90}Y-Zevalin compared to 56% for rituximab ($p = 0.002$). The CR rate of 30% in the Zevalin arm was also higher than the 16% found with rituximab ($p = 0.04$). The median DR was 14.2 months (0.9–28.9). The trial was not statistically powered to detect a difference in TTP. The Kaplan–Meier estimated median TTP was 11.2+ months (range, 0.8–31.5+) for the Zevalin group compared to 10.1+ months (range, 0.7–26.1) for the rituximab group ($p = 0.173$). However, the estimated time to next therapy (TTNT) for patients with nontransformed histology indicated a significantly longer TTNT for Zevalin patients (17.8+ months; range, 2.1–21.7+) than for rituximab patients (11.2 months; range, 1.3–19.0+) ($p = 0.040$).

The second key trial utilized Bexxar. Seventy-eight patients with relapsed low-grade or transformed NHL were randomized to receive Bexxar as described above or two doses of unlabeled tositumomab (121). The median age of the patients was 55 years (range, 28–85), they had received a median of two prior regimens (range, 1–5); 88% were stages III/IV; 40% had elevated LDH; and 41% presented with bulky disease (size ≥ 5 cm). Seventeen percent had transformed NHL. The ORR was 67% (28/42) for Bexxar compared to 28% (10/36) for those who received unlabeled tositumomab ($p < 0.001$). The CR rate was 33% (14/42) versus 8% (3/36) ($p = 0.01$) in the Bexxar and tositumomab arms, respectively. The DR was 18 months for tositumomab and has not been reached for Bexxar. Nineteen patients crossed over to the Bexxar arm after progression and seventeen responded (89%) with 42% (8/19) CR. Thirteen patients did not cross over—eight because they developed HAMA, three received other therapy, one received no further treatment, and one died. The hematologic toxicity was higher in the RIT arm—17% (7/42) had grade IV ANC and 5% (2/42) grade IV thrombocytopenia. The HAMA rate was 17% (7/41) for the patients treated with Bexxar; 25% (9/36) in the unlabeled tositumomab group; and 5% (1/19) of the crossover patients.

6.5 RIT IN RITUXIMAB REFRACTORY PATIENTS

In 2009, the current treatment of new and relapsed NHL usually includes rituximab and this agent is often used as a maintenance therapy for up to 2 years. How will patients respond to CD20-directed RIT after extended treatment with rituximab to the point that the patient is rituximab refractory? It should be noted that rituximab refractory is defined as either no response to a standard course of rituximab or a response that lasts less than 6 months. Several studies have addressed this question and demonstrated that the ORR to RIT in this patient population is high, but the TTP is shorter than when the patient was not rituximab refractory. The trial of Zevalin in this patient population included 54 patients meeting this definition (122). All were treated with a single dose of 0.4 mCi/kg of ^{90}Y-Zevalin and were followed up without further therapy. The median age was 54 years (range, 34–73), 95% of patients had follicular NHL, 32% had bone marrow involvement, and 74% had bulky disease ($\geq$5 cm). This patient group was heavily pretreated with a median of four prior therapies. The dosimetry determined by ^{111}In-Zevalin was acceptable in all 27 cases in which it was performed. The median nadir ANC was 700×10^6/L, and in 35% of the patients it was grade IV. The median nadir platelet count was $33,000 \times 10^6$/L and was grade IV in 9%. The ORR using International Workshop criteria (120) was 74% with 15% CR. The estimated median TTP was 6.8 months (range, 1.1–25.9+) with 30% of data censored. Median TTP in the 40 responders was 8.7 months (range, 1.7–25.9+), with 28% of data censored. The estimated median DR was 6.4 months (range, 0.5–24.9+).

In their study, Horning et al. (123) administered a standard dose of Bexxar to 40 patients who had relapsed after rituximab. This trial was slightly different from the Zevalin trial discussed above in that although all patients previously had rituximab, only 24 were rituximab refractory. The median age of all patients was 57 years, the median number of prior regimens was 4, and 30% had marrow involvement. The ORR was 65% with 38% CR. In the 24 patients who were rituximab refractory (no prior response), the ORR was 62% (15/24) and 25% (6/24) obtained a CR. The median progression-free survival (PFS) was 10.4 months for all patients and 24.5 months for confirmed responders. Fifty per cent of patients had grade III/IV hematologic toxicity. The median ANC was 1200×10^6/L and the median platelet count nadir was $85,000 \times 10^6$/L. No patient developed HAMA.

It is apparent from these two important studies that the radiation component of RIT is important to its efficacy. Anti-CD20 RIT can overcome resistance in rituximab refractory tumors; however, the responses are shorter than in patients who are rituxmab naïve or relapsed after rituximab. It is to be noted that these studies for the most part enrolled patients who were rituximab refractory. This patient group is not the same as patients today who are treated with rituximab-based chemotherapy upfront and then relapse several years later. These patients are not considered to be rituxmab refractory and are very likely to respond to RIT.

6.6 RIT FOR PREVIOUSLY UNTREATED PATIENTS

The high ORR to RIT in relapsed NHL patients makes it an attractive option for new, untreated patients with advanced stage, asymptomatic indolent NHL. These patients often discover the NHL at a time when they have no symptoms, yet extensive disease. For example, many patients palpate a small node in the neck or groin, which upon biopsy is indolent NHL. Even though the node is discovered early, they are often stage III/IV after formal staging. The asymptomatic patient may want to avoid chemotherapy due to the potential side effects and the lack of curability. They are candidates for observation, rituximab immunotherapy, or a clinical trial. There has been sporadic use of RIT in this patient population (124) and one large, prospective, phase II trial (54). Kaminski et al. (54) treated 76 previously untreated patients with advanced stage (stages III or IV) follicular (70% grade 1; 29% grade 2) NHL (one patient had mantle cell NHL) with a single dose of Bexxar RIT. Sixty-four percent of the patients had bone marrow involvement, 30% had an elevated LDH, and 43% had bulky disease (defined as ≥5 cm). The median age was 49 years (range, 23–69). The ORR was 95% (72/76) with 75% (57/75) CR. Five-year PFS for all patients was 59% with a median PFS of 6.1 years. The five-year PFS for CR patients was 77% and 70% of the CR patients remained in CR for 4.3–7.7 years after treatment. Patients with only a PR had a median TTP of 7 months and none became long-term responders. Patients with bulky disease (>5 cm) and marrow involvement with NHL had a lower chance of obtaining a CR. There are limited data on the use of Zevalin for the untreated population. In the report on pulmonary lymphomas, there were two patients treated initially with Zevalin and they attained a CR (124). Other studies of Zevalin in the untreated population are being conducted at the universities of Miami (124) and Wisconsin (Brad Kahl, personal communication, September 2008).

It is important to follow this previously untreated group for the development of myelodysplasia and HAMA. To date, no patient in this previously untreated group has developed myelodysplastic syndrome; however, 63% developed HAMA. Whether the HAMA will wane with time and what impact this will have on the future ability to retreat patients is unknown. In the studies of Bexxar on previously treated patients, the HAMA rate has been about 10%. This is likely due to fact that when patients are previously treated with chemotherapy they are more immunosuppressed and less likely to develop HAMA. The development of HAMA was of initial concern because of the possibility that these antibodies might reduce the efficacy of other murine antibodies administered later, or cause anaphylactic reactions, but it has not been a major problem. The HAMA rate in the studies with Zevalin has been very low (<1%) (109).

6.7 RIT FOR RELAPSED LARGE-CELL LYMPHOMA

Patients with diffuse large-cell NHL who relapse after RCHOP and respond to salvage chemotherapy are typically treated with high-dose chemotherapy with stem cell support. Therefore, the available experience with RIT in aggressive NHL is limited to patients with relapsed or refractory disease that are not eligible for stem cell transplantation. In the initial phase I/II trial of Zevalin, the ORR was 43% for the 14 patients with intermediate-grade histology (49). A recent update on the long-term course of patients in that study demonstrated a 58% (7/12) ORR for the patients with DLBCL with a median DR of 49.8 months (1.3–67.6+ months) (125).

A more recent study evaluated Zevalin in the treatment of 104 elderly patients with relapsed and primary refractory DLBCL that were not candidates for SCT (126). The ORR in the entire group was 44%, but further analysis showed differences in ORR and TTP to Zevalin depending on whether the patient had previously received rituximab. Rituximab-naïve patients had a 52% ORR (24% CR/CRu) compared to a 19% ORR (12% CR/CRu) in the group that had previously been treated with rituximab. The latter group appears to have been particularly high-risk since 37% were reported to be refractory to RCHOP. The median TTP was 5.9 months for patients who were refractory to non-rituximab containing treatment regimens before Zevalin. The TTP was only 1.9 months for those patients who had received a rituximab-containing regimen prior to Zevalin. Although the treatment was generally well tolerated, two patients died of cerebral hemorrhage associated with grade 4 thrombocytopenia.

This is an important study because at least 40% of patients with DLBCL die of the disease, and many of these patients are not candidates for or relapse after SCT (23). There is an urgent need for new treatments for the relapsed DLBCL patient, and the excellent toxicity profile of RIT makes it an attractive option. This study indicates that Zevalin has reasonable activity in this population as a whole, but the fact that patients who had previously had a CD20-targeted treatment had a low ORR and a short TTP is important since all patients with DLBCL now receive rituximab with induction chemotherapy. Future studies need to address this by adding other agents to RIT or use RIT as adjuvant after chemotherapy for relapsed disease.

6.8 RIT FOR TRANSFORMED LYMPHOMA

Patients who transform from low-grade to large-cell NHL are a difficult group to treat. Eligible patients are typically recommended for SCT. However, these patients are often elderly, lack stem cell reserve, or have other comorbidities that make SCT impossible. The largest clinical experience of treating transformed NHL with RIT has been with Bexxar (127). There were 71 patients with transformed NHL who had been enrolled on the five studies since 1990. The median age was 59 years; the median time from diagnosis to study entry was 74 months (range, 8–334); and the median number of prior therapies was 4 (range, 1–11). In addition to the transformed histology, 28% of patients had marrow involvement; 70% bulky disease (>5 cm); 57% elevated LDH; and 52% with an International Prognostic Index (IPI) score of 3 or more. After Bexxar, the patients were followed for a median of 19.4 months (range, 0.5–101). The ORR was 39% with 25% CR. The median DR was 20 months (range, 10.8–not reached) for all responders and was 36.5 months (range, 14.7–not reached) for CR patients. The median TTP for all patients was 4.3 months (3.2–10.2) but for those who responded to treatment, the median TTP was 20.2 months (12.4–not reached). Five patients remain in remission beyond 40 months. Bexxar is the only RIT agent approved for transformed NHL.

6.9 RIT FOR MANTLE CELL LYMPHOMA

There have been limited trials of single-agent RIT for relapsed MCL. In the phase I/II trial of Zevalin, there were three patients with MCL and none responded (49). A dedicated trial of Zevalin on 15 patients with relapsed MCL showed a 33% ORR with a median DR of 5.7 months (128). There has been more success in MCL when the RIT has been delivered with SCT (129) or as adjuvant therapy (130). In their study, Smith et al. (130) administered a standard dose of Zevalin after four cycles of standard RCHOP to 56 patients with new, untreated MCL. Although the follow-up is short, the results appear superior to those seen with standard RCHOP without adjuvant therapy .

6.10 LONG-TERM RESULTS OF RIT

Prolonged follow-up of the initial studies provides useful data on the long-term prognosis of patients with relapsed indolent NHL treated with RIT (125, 131–133). Long-term responses (LTRs) have been defined as CR or PR for at least 12 months. Fisher et al. (131) reported results on 250 patients with relapsed or refractory low-grade or transformed NHL who had been treated with Bexxar. Long-term responses were achieved in 81 (32%) of 250 patients, 23% had a PFS >18 months, and 21% had a PFS >2 years. Patients who achieved LTR were more likely to have had a CR (77% versus 8%) and low-bulk disease (51% versus 31%) than those without an LTR. Other characteristics that were predictive of an LTR with Bexxar treatment were sensitivity to the last therapy, history of less than three previous therapies, follicular histology, normal LDH level, and modified IPI score ≤ 2. The LTR patient population had a median DR of 45.8 months with a median follow-up of 61 months. Those patients who attained a CR have done especially well; the median DR has not been reached and 47% remain in CR ranging from >2.7 years to >10.2 years. In the previously untreated study, the only patients with extended remissions were those who had a CR (54). A smaller series of 18 patients treated with Bexxar also demonstrated that 6 of the 8 patients with CR/CRu remained in unmaintained remission at 46–70 months (132). Long-term responses were seen in 37% (78/211) of the patients treated in Zevalin trials (133). Patients with bulky disease (masses >5 cm) were less likely to become a long-term responder; however, if they did respond, the DR was similar to those with low bulk disease. The achievement of a CR/CRu was a strong positive predictor of an LTR, with an odds ratio of 7.0 (95% CI, 3.4–14.5). At a median follow-up of 53.5 months (range, 12.7–88.9), the median DR was 28.1 months and the median TTP was 29.3 months. The findings in patients with follicular NHL ($n = 59$) were similar to those in the overall population of long-term responders. The estimated OS at 5 years was 53% for all patients treated with Zevalin and 81% for long-term responders. It is now apparent that even in the relapsed indolent patient population that LTR can be observed with single doses of RIT. Patients who are treated early in the relapse phase of disease have a higher rate of CR and a longer TTP than those who are treated after multiple relapses (134).

6.11 RISK OF MYELODYSPLASIA WITH RIT

The fact that myelosuppression is the primary toxicity of RIT raised concern about the risk of treatment-related myelodysplasia or acute leukemia (AL). In the Bexxar database of 995 patients who received a single dose of Bexxar for relapsed NHL, MDS or AL has been documented in 35 patients (3.5%). All 35 patients had also received chemotherapy for NHL. The annualized rate of MDS/AL was 1.6%, a rate similar to that observed in chemotherapy-alone treated patients (135). Patients with prior treatment with the purine nucleoside fludarabine had a relative risk of 3.1 for the development of MDS. Although follow-up is short, to date there has been no patient with MDS in the previously untreated group of patients treated with Bexxar (54, 135).

In the integrated safety analysis of Zevalin published in 2003 (109), MDS/AL was reported in five patients (1%) 8–34 months after treatment. All five of these patients had been previously treated with alkylating agents. A more recent analysis of 746 patients treated with Zevalin between 1996–2002 included patients in the trials discussed above as well as the compassionate use trials. A total of 17 cases of MDS/AL have now been identified for an incidence of 2.3% (17/746). The MDS/AL occurred at a median of 5.6 years (range, 1.2–13.9) after the diagnosis of NHL and 1.5 years (range, 0.1–5.8) after RIT. The annualized rates were 0.3% (95% CI, 0.2–0.4%) a year after the diagnosis of NHL and 0.7% (95% CI, 0.4–1.0%) a year after RIT (136). Those patients treated with prior fludarabine had a relative risk of MDS that was higher than with other chemotherapies (hazard ratio 3.5) as did patients with follicular disease type. One patient treated with ^{131}I-rituximab has now developed MDS followed by AL (111).

It is especially important to follow patients who were treated with RIT as adjuvant therapy. Four cases of MDS have now been reported in patients who were treated with only one prior therapy (137). Three of these patients were treated with fludarabine followed by Bexxar (53) and one patient treated in a clinical trial was administered cyclophosphamide, vincristine, and prednisone followed by Bexxar (138). The recent studies using Zevalin after fludarabine containing regimens (139, 140) will need to be observed closely over the next 10 years for the risk of MDS.

At this juncture, now over 10 years after the initiation of the RIT trials, there is no evidence yet that RIT increases the risk of MDS/AL over that of chemotherapy alone and no cases of MDS/AL in patients treated only with RIT. This issue will need ongoing investigation. Actually, the issue of bone marrow damage may become more relevant as RIT is used earlier in the disease course because patients will have longer OS post-RIT with more time for MDS/AL to become apparent.

6.12 FEASIBILITY OF TREATMENT AFTER RIT FAILURE

Since the CR rate with RIT when used for relapsed NHL is about 30%, most patients will at some point relapse again and require additional therapy. The myelosuppression universally observed after RIT has raised the question of feasibility of chemotherapy following RIT. Ansell et al. (141, 142) examined subsequent therapy administered to 58 patients who had relapsed after receiving 0.4 mCi/kg of Zevalin. The median age was 56 years (range, 26–77) and 48% (28/58) had marrow involvement. The median number of therapies pre-Zevalin was two (range, 1–7): 93% CHOP, 60% CVP, 41% rituximab, 34% chlorambucil, 19% fludarabine, 14% ProMACE/CytoBOM, and 22% (13/58) received miscellaneous regimens. Twenty-eight (48%) of the patients had transformed NHL—fourteen had the transformation before Zevalin and in fourteen patients the transformation developed after Zevalin (eight at first relapsed after Zevalin; six during chemotherapy that followed Zevalin). The median WBC at the time of the next therapy after Zevalin was 4.2×10^9/L (range, 0.9–8.7), the median hemoglobin was 11.6 g/dL (range, 7.2–14.7), and the median platelet count was 163,000/mm^3 (range, 14,000–292,000). The median number of subsequent treatments received was two (range, 1–7). Subsequent chemotherapy was feasible and tolerable. Twenty-eight percent of the patients received growth factors with their next chemotherapy and two patients required reduced doses due to persistent myelosuppression. Peripheral blood stem cell collection was also feasible, with one of the eight patients requiring marrow harvest after failing peripheral blood stem cell collection. All eight patients engrafted. One more patient received an allogeneic SCT with adequate engraftment. The issue of treatment after RIT has also recently been addressed using the Bexxar patient database. Sixty-eight patients who relapsed a median of 168 days after RIT were reviewed (143). At the time of relapse, the median WBC was 4900×10^6 cells/L (range, 1.1–21.4) and the median platelet count was $130,000 \times 10^6$/L (range, 9000–440,000); only the platelet count was significantly lower than the pre-RIT value. The 65% (44/68) of the patients who received further chemotherapy were able to receive a median of two (range 1–4) additional regimens using typical myelosuppressive agents. Thirteen patients went on to SCT, three had stem cells harvested prior to RIT; ten after RIT. In summary, in this selected group of patients who had met all the criteria for inclusion into an RIT trial, subsequent chemotherapy was feasible and tolerable. Stem cells were able to be collected and successful SCT performed. Although this data is encouraging, it should not be interpreted that stem cells will always be able to be collected on all patients after RIT. Patients that have received extensive chemotherapy or EBRT can be difficult to collect even in the absence of RIT (144). If the patient is considered to be a strong candidate for SCT, then stem cells should be collected before RIT.

6.13   COMBINATIONS OF RIT AND CHEMOTHERAPY

The effectiveness of RIT in relapsed NHL has led to studies that use RIT as adjuvant therapy after induction chemotherapy. Leonard et al. (53) treated 35 patients with untreated follicular NHL with three cycles of fludarabine (25 mg/m^2/d $\times$ 5 days every 5 weeks for 3 cycles) followed by a single dose of Bexxar. The ORR to fludarabine was 89% (31/35) with a CR of 9% (3/35). After Bexxar was administered, the ORR was 100% with 86% CR (30/35). Five of the six patients who prefludarabine had $>25\%$ marrow NHL were effectively cytoreduced to less than 25% following fludarabine and were thus eligible for RIT. Two patients developed human antimouse antibody. In a median follow-up of nearly 5 years, the median PFS had not been reached and will be >48 months. Patients with a high-risk Follicular International Prognostic Index (FLIPI) score had a median PFS of only 2 years and only 27% were estimated to be in remission at 5 years.

Press et al. (29) have recently reported long-term results of a phase II trial of CHOP chemotherapy for six cycles followed by a single dose of adjuvant Bexxar. Ninety patients were treated and 91% responded with 69% CR. After approximately 5 years of follow-up, the estimated 5-year PFS rate was 67% and the OS rate was 87%. These impressive results are being subjected to a randomized trial where patients with new, untreated follicular NHL receive either CHOP/Bexxar or RCHOP. A similar approach using rituximab-based chemoimmunotherapy followed by adjuvant RIT with Zevalin has been reported in an abstract form (145). Adjuvant Zevalin in patients older than 60 years with new, untreated DLBCL was recently reported (146). Twenty patients received six cycles of CHOP followed 6–10 weeks later by a single dose of Zevalin. The ORR was 100% (95% CR; 5% PR) with four of the five patients in PR after CHOP converting to CR after Zevalin. The 2-year PFS was estimated to be 75% and 2-year OS 95% with a median follow-up of 15 months. The Zevalin was well-tolerated and only one patient required transfusion of RBC and platelets. The authors indicate that they are now conducting a similar trial in the elderly but are using RCHOP induction rather than CHOP and have reduced the number of cycles from six to four before Zevalin.

Zinzani et al. (139) reported excellent results with a combined chemotherapy followed by RIT approach in 26 patients with new, untreated indolent nonfollicular NHL. The patients were treated with six cycles of fludarabine and mitoxantrone (FM) followed by adjuvant Zevalin 6–10 weeks later for patients who had achieved a response (CR or PR) to FM. The dose of fludarabine was 25 mg/m^2 on days 1–3 and mitoxantrone 10 mg/m^2 on day 1 every 28 days. It is important to note that the type of NHL treated were 10 lymphoma of mucosa-associated lymphoid tissue (MALT), 8 lymphoplasmacytic, and 8 small lymphocytic lymphoma (SLL); there were no follicular NHL in this study. The ORR to the FM was high (81%; 21/26) as expected. The CR rate was 50% (13/26) and the PR rate 31% (8/26). Twenty of the twenty-six patients (13 CR; 7 PR) went on to receive adjuvant Zevalin and at the end of the Zevalin 100% (20/20) were in CR. One of the patients with a PR did not receive Zevalin because the marrow had >25% involvement with SLL. The treatment was well tolerated with myelosuppression the only significant toxicity. One patient had febrile neutropenia and required hospitalization; there have been no second malignancies. Although the median follow-up is short at 20 months, 90% of the patients were estimated to be progression-free. There have been two patients who progressed and one has died. This study is important not only because of the demonstration of the effectiveness of adjuvant RIT in converting PR to CR but also because of the effectiveness in the more uncommon types of indolent NHL.

A similar trial was conducted in patients with new, untreated follicular NHL (140). Sixty-one patients with stages III/IV received fludarabine 40 mg/m^2/day on days 1–3 with mitoxantrone 10 mg/m^2 on day 1 every 28 days for six cycles. Patients then received a single dose of adjuvant Zevalin 6–10 weeks after the last dose of FM if they had responded to FM with a CR or PR and met the CBC requirements of platelets >100 × 10^9/L and ANC >1.5 × 10^9/L and marrow with less than 25% involvement with NHL. Of the original 61 patients, 60 responded to FM and 93% (57/61) proceeded to Zevalin. At the time of restaging after Zevalin, 86% (12/14) of the patients in PR after FM converted to CR after Zevalin for a total CR rate of 96% (55/57) or 90% (55/61) of enrolled patients. With a median follow-up of 30 months, the 3-year PFS is estimated to be 76% (95% CI, 72.3–82.4) and 3-year OS 100%. This protocol produced substantial myelosuppression with 63% (36/57) developing grade 3 or 4 hematologic toxicity and 37% (21/57) required blood product transfusions (140).

A large randomized study of first-line indolent therapy (FIT) with Zevalin RIT has recently been reported (147). Patients with stages III/IV untreated follicular grade I/II were treated with the induction regimen of the physician's choice. After therapy was concluded, all patients with a CR/PR were randomized to observation versus a single dose of 0.4 mCi/kg (maximum of 32 mCi) Zevalin RIT; there was no rituximab maintenance. Four hundred fourteen patients were enrolled with two hundred eight randomized to Zevalin and two hundred six to observation. The induction regimen was not prescribed, but 71% (294/414) received either CHOP or CVP; only 14% (59/414) received rituximab with their induction treatment. At a median follow-up of 3.5 years, the median PFS was 13.3 months for observation patients versus 36.5 months for those who received Zevalin ($p < 0.0001$; hazard ratio (HR) 0.46). In those patients who were in PR at the time of randomization to Zevalin or control, the median PFS was 29.3 months versus 6.2 months ($p = 0.0001$; HR, 0.3). In those patients who achieved CR at the time of randomization to Zevalin or control, the median PFS was 53.9 months versus 29.5 months ($p < 0.015$; HR, 0.6). In the Zevalin consolidation arm, 77% (78/101) of the patients with a PR after induction therapy converted to CR/Cru compared to 18% (17/97) of those in the control arm.

6.14 HIGH-DOSE RIT WITH STEM CELL SUPPORT

The use of high-dose RIT and SCT has been extensively studied but remains investigational (148–154). The advantage of using RIT in the context of SCT is that the risk of myelosuppression is reduced because stem cells are reinfused; rather, the dose-limiting toxicity becomes other organ toxicity. In a phase I study, Press et al. (155) entered 43 patients and 19 were able to receive therapeutic infusions of Bexxar. The ORR was 95% (18/19) and 84% (16/19) had a CR. Another 25 patients were enrolled into the phase II trial (60) and 84% (21/25) were treated. Tumor response was observed in 86% (18/21) with 76% CR. The stem cells were infused 12–18 days after Bexxar; the source of stem cells was marrow in 19 and blood in other 2. Neutrophil recovery to >500 and platelet count recovery to >20,000 occurred a median of 23 days and 22 days after stem cell infusion, respectively. HAMA was documented in 16% patients. Long-term follow-up (median 42 months) on these patients and those from the phase 1 trial demonstrated that the projected OS and PFS was 68% and 42%, respectively (57).

These studies have been expanded to combine Bexxar with cyclophosphamide and etoposide with autologous stem cell support (64). Fifty-two of the fifty-five enrolled patients received treatment. The median age was 47 years (range, 34–58). All patients were required to have a marrow that had a less than 25% involvement with NHL. The study determined that a dose of Bexxar that delivered 25 Gy to critical normal organs such as lung, liver, or kidneys could be safely combined with chemotherapy and stem cells. The estimated OS and PFS at 2 years was 83% and 68%, respectively. Gopal et al. (156) recently compared the long-term results of patients with relapsed follicular NHL treated with SCT with or without RIT. The 27 patients conditioned with high-dose Bexxar RIT had a superior OS (unadjusted HR for death = 0.4 (95% CI, 0.2–0.9), $P = 0.02$; adjusted HR, 0.3, $P = 0.004$) and PFS (unadjusted HR = 0.6 (95% CI, 0.3–1.0), $P = 0.06$; adjusted HR, 0.5, $P = 0.03$) compared to the 98 patients treated with chemotherapy and total body irradiation (TBI) or chemotherapy alone. The estimated 5-year OS and PFS were 67% and 48%, respectively, for high-dose Bexxar conditioning, and 53% and 29%, respectively, for chemotherapy alone or chemotherapy combined with TBI. The group at the University of Washington has evaluated SCT with conditioning with high-dose RIT and found it to be well tolerated even in older adults (129, 157). In a recent study, patients older than 60 years of age with relapsed B-cell NHL were treated with high-dose Bexxar at a dose designed to deliver 25–27 Gy to the critical normal organ receiving the highest radiation dose. Stem cells were infused approximately 2 weeks after the dose of RIT was administered. Twenty-four patients were treated and 54% went into the protocol with chemotherapy-resistant disease. Two patients experienced grade IV nonhematologic toxicity; otherwise, the only toxicity was myelosuppression. There were no treatment related deaths. Sixteen patients with mantle cell NHL have been treated with a similar protocol (129) and all patients had a tumor response with 91% CR. The OS and PFS at 3 years were 93 and 61%, respectively.

When Zevalin is used in the context of SCT, two approaches have been taken. Some studies have simply added a standard dose (0.4 mCi/kg) to a standard conditioning regimen. The other approach escalates the dose of Zevalin so that liver becomes the organ of interest for potential toxicity. Dosimetry (86, 153, 158–160) is required in this situation to calculate the doses of Zevalin that will deliver a specific dose of radiation to the liver. The first phase I/II trial of high-dose Zevalin was conducted by Nademanee et al. (161). Patients with relapsed NHL received a dose of Zevalin calculated to deliver 1000 cGy to the highest normal organ followed by etoposide 40–60 mg/kg on day 4 (prior to SCT) and cyclophosphamide 100 mg/kg day 2. Thirty-one patients were treated. Disease types were follicular NHL ($n = 12$), DLBCL ($n = 14$), and MCL ($n = 5$). The median dose of Zevalin calculated to deliver 1000 cGy to the highest normal organ was 71.6 mCi (2649.2 MBq; range, 36.6–105 mCi; range, 1354.2–3885 MBq). Engraftment was prompt with the median times to reach an ANC greater than 500×10^6/L and platelet counts more than $20,000 \times 10^6$/L were 10 days and 12 days, respectively. During follow-up, there have been two deaths and five patients have relapsed. The 2-year estimated OS is 92% and the PFS is 78%. It is to be noted that in this trial the dose of Zevalin was not dose escalated beyond a dose calculated to deliver 1000 cGy to the liver. There is another trial combining Zevalin with BCNU, etoposide, cytosine arabinoside, melphalan (BEAM) conditioning chemotherapy, and stem cells that escalates the dose of Zevalin in a typical phase I protocol. That trial has been published in abstract form (162). Chiesa et al. are conducting a trial of high-dose Zevalin with stem cell support (163). They have treated four patients with doses of 1.2 mCi/kg (45 MBq/kg) without extramedullary toxicity. In the trial reported by Ferrucci et al. (164), patients received single-agent high-dose Zevalin with stem cell support for relapsed NHL. Three dose levels were tested in a total of 13 patients—30 MBq/kg (0.8 mCi/kg), 45 MBq/kg (1.2 mCi/kg), and 56 MBq/kg (1.5 mCi/kg). ^{111}In-ibritumomab was delivered on day 20 and the ^{90}Y-ibritumomab on day 13; both doses were preceded by rituximab 250 mg/m². The disease types of the 13 patients were 8 DLBCL, 3 MCL, 1 transformed, and 1 follicular. The treatment was well tolerated with myelosuppression being the most common toxicity. The median time to engraftment was 12 days (range, 0–22) and 19 days (range, 0–48) for platelets and ANC, respectively. One patient reactivated herpes zoster and one patient had reversible grade 3 hepatotoxicity. One patient reactivated hepatitis C at day +50 and died in NHL CR due to hepatitis C. One patient with MCL had a CR and then developed MDS 2 years after SCT. No patient has yet developed HAMA. The ORR was 62% (8/13) with 6 CR and 2 PR. This study, although small, is important because it demonstrates the potential to use high-dose RIT as a single agent along with stem cells in patients who are not eligible for SCT using high-dose chemotherapy such as BEAM. It also provides guidance on the type of doses that can be tolerated in this SCT population. Virus reactivation is an issue with rituximab (165) and it appears that it will also be a potential toxicity for patients receiving high-dose RIT.

Krishnan et al. (151) used a standard dose of Zevalin (0.4 mCi/kg) 14 days before a standard autologous SCT. Forty-one patients were treated with a median age of 60 years (range, 19–78). The median number of previous therapies was two. Twenty patients had DLBCL, thirteen had MCL, four follicular, and four transformed. All patients engrafted and the estimated 2-year OS and PFS were 89% and 70%, respectively. When compared to historical controls, the adverse events were similar to patients receiving high-dose BEAM chemotherapy alone. Patients entering the study with a negative PET scan had a longer DFS.

In a recent study, Devizza et al. (153) took a slightly different approach. They harvested stem cells after nonmyeloablative chemotherapy that included rituximab and then administered Zevalin at a dose of 0.8 mCi/kg ($n = 13$) or 1.2 mCi/kg ($n = 17$). The stem cells were infused in two aliquots on day +7 and day +14 after Zevalin RIT. The hematologic toxicity was manageable, and although 27% of patients had an infection there were no grade IV infections. The patients have been followed for a median of 30 months and 87% are alive and 67% remain in remission. There has been no case of MDS.

It is clear that RIT can be used in the context of SCT. The use of standard, nonmyeloablative doses of RIT is simpler because the previously established procedures and protocols can be followed. High-dose RIT with SCT requires expert dosimetry and is costly but may provide more benefit for long-term survival and cure. The answers to these questions will be difficult to sort out without large randomized trials. A randomized phase 3 study of Bexxar-BEAM versus BEAM alone is ongoing and should answer some of these questions.

6.14.1 Use of RIT After Stem Cell Transplantation

During initial RIT trials, patients with prior SCT were excluded. There have now been two studies that have evaluated the use of Zevalin in this patient population. The first study was conducted by Vose et al. (166) that included 19 patients who were administered 0.1–0.2 mCi/kg as a single dose. The median time from SCT to RIT was 28 months. The main toxicity was hematologic with 53% of patients experiencing grade III/IV thrombocytopenia, 32% neutropenia, and 21% anemia. The ORR was 47% (9/19) and some of these have been durable. For example, the 1-year event-free survival (EFS) and OS rates were 26% and 57%, respectively. The recommended dose for future studies was 0.2 mCi/kg. Peyrade et al. (167) reported the results of Zevalin RIT in eight patients with follicular NHL who had relapsed after SCT. These investigators followed the non-SCT guidelines of Zevalin dosage: 0.4 mCi/kg for platelet counts $>150,000/mm^3$ and 0.3 mCi/kg for platelet counts 100,000–150,000/ mm^3. The median interval from SCT was 26 months (range, 14–72 months). Four patients had marrow involvement. The dose was 0.4 mCi/kg in five cases and 0.3 mCi/ kg in four cases. One patient developed grade IV thrombocytopenia and grade IV neutropenia and died of septic shock 6 months after RIT. All but two patients responded, four patients had a CR, and two patients had a PR; one progressed and one was stable. The median DFS was 12 months (range 0–25 months) with a median follow-up of 17 months. The 1-year OS was 83% (7/8). There have been no cases of MDS/AL. In general, these results are similar to those of Vose et al. (166). It is difficult to know from these small studies what the appropriate dose of RIT is for patients with relapsed NHL after SCT and who have marginal marrow reserve.

6.15 RIT FOR CENTRAL NERVOUS SYSTEM LYMPHOMA

Primary central nervous system lymphomas (PCNSLs) are difficult to treat if the patient relapses after initial therapy. Whole-brain radiation therapy (WBRT) is useful but can produce long-term side effects. There has been minimal use of RIT for PCNSL but several case studies show promise. One of the first studies used ^{131}I-labeled antibodies to B-cell antigens (CD19 in two patients, anti-CD10 in four, and both antibodies in one patient) injected intrathecally for 7 children 3–16 years of age who were in meningeal relapse of B-cell ALL (168). Six of the seven patients responded. The dose of radiation delivered to the subarachnoid CSF was between 12.2 and 25.3 Gy that was six times higher than that to the surface of the brain. Shah et al. (169) reported the first successful case of PCNSL treated with single-agent Zevalin. More recently, Pitini et al. (170) treated two patients with relapsed PCNSL with the standard Zevalin protocol and both entered CR. After recovering from myelosuppression, the patients were treated with maintenance temozolomide and remain in CR. Iwamoto et al. reported the results on six patients with relapsed PCNSL who received a single course of Zevalin (171). All patients had relapsed after definitive treatment and two patients had received WBRT. There were no unexpected toxicities and the ORR was 33% (2/6); one patient was stable and two progressed on therapy. Although the ORR was reasonable, the TTP was very short at 6.8 weeks (range, 4.7–14.3) and the median OS was also short at 14.3 weeks (range, 11.4–94). The ^{111}In images were evaluated and it was estimated that the median absorbed dose delivered to the CNS lymphoma was 701 cGy compared to 70 cGy to normal brain. This study is important in the sense that it demonstrates that RIT agents can cross the blood–brain barrier and deliver radiation to the tumor in higher amounts than the surrounding brain. Doolittle et al. suggest that the blood–brain barrier may need modification at the time of RIT for better tumor penetration (172).

6.16 RETREATMENT WITH RIT

The high ORR observed with single-agent RIT in relapsed NHL and the excellent toxicity profile has led to increased interest in retreatment. In the original phase I/II trials of Bexxar, 16 patients that responded to Bexxar later received a second dose of Bexxar at the time of relapse. The ORR with the second dose was 56% with 31% CR (73). These encouraging results led to a multicenter phase II trial that treated 32 patients (55). The ORR was again 56% (18/32) with 25% of patients attaining a CR. Interestingly, 10 of the 18 responders had a longer DR after the second course of Bexxar than they experienced with the first dose of Bexxar.

There has been a limited experience with treating patients with a second dose of Zevalin (14). Wiseman et al. (173) reported preliminary results on a phase I trial where all patients were treated with two sequential doses of Zevalin. The first dose was 0.4 mCi/kg and the phase I dose levels for the second dose delivered 3–6 months after the first dose were 0.2, 0.3, and 0.4 mCi/kg. The study demonstrated that it was safe to administer two doses of Zevalin at 0.4 mCi/kg/dose with the use of prophylactic growth factors when the doses are administered 6–9 months apart. It has yet to be determined whether this strategy produces a longer TTP or OS. In a study of ^{131}I-rituximab, six patients with response to the first dose were retreated at the time of relapse and four responded with a median DR of 11 months (74). Peyrade et al. (174) described a single case of a patient who received two doses of Zevalin. With the first dose a 12-month TTP was obtained. A second dose was administered and the patient achieved a CR that has been durable for 6+ months. Retreatment with ^{131}I-rituximab has also recently been shown to be feasible (111). In this series, 16 patients were retreated with a second dose of ^{131}I-rituximab and 88% (14/16) responded with 56% (9/16) achieving a CR. The length of time between the first and the second dose of RIT ranged from 12 to 54 months. Median DR for all responding patients was 10.5 months. Twenty-five percent (4/16) of the patients remained in an ongoing remission and two patients remain in an ongoing PR. Some of these responses have been durable up to 25+ months. The median TTP with the first dose was 14 months. Six of the thirteen responders had the same or greater DR with a second dose. The toxicity was similar and was primarily reversible myelosuppression. One patient has proceeded to receive a third dose of ^{131}I-rituximab and another patient has received a total of four doses. One patient with previous treatment with alkylating agent (chlorambucil) has developed MDS with subsequent evolution to AL.

6.17   RIT IN CHILDREN WITH RELAPSED NHL

There has been only one study of RIT in children (175). A phase 1 study was done in five patients and reported acceptable toxicity. Patients received rituximab 250 mg/m^2 followed by Zevalin at 0.4 mCi/kg ($n = 3$) or 0.1 mCi/kg ($n = 2$) in those patients who had previously underwent SCT. None of the five patients developed HAMA/HACA. This study has now proceeded to a phase II to evaluate ORR and DR.

6.18   RIT IN PATIENTS WITH LUNG INVOLVEMENT

The high-energy and short path length of RIT make it an ideal treatment for patients with extranodal disease in the lung. In all RIT trials, lung toxicity in the nontransplant setting has not been an issue. Stefanovic et al. (124), in their excellent review of pulmonary marginal zone NHL arising from bronchial-associated lymphoid tissue (BALT), mention that two of their patients were treated with Zevalin and achieved a CR. Most lung lesions were treated without any damage to surrounding lung tissue. There was a patient with a lung lesion that was successfully treated with Zevalin RIT (176). An area of PET positivity was later observed at the site and a biopsy showed this to be fibrosis.

6.19   RIT IN PATIENTS WITH SKIN LYMPHOMA

Maza et al. (177) treated 10 patients with primary cutaneous B-cell NHL—8 patients had follicular NHL and 2 patients had leg-type large-cell NHL. All patients except one had relapsed from prior treatment. The ORR was 100% and all patients attained CR within 8 weeks. The median time to response was 2 weeks. Four patients remain in remission; the other six patients have relapsed. Interestingly, in patients with follicular disease the relapses occurred in areas that were previously uninvolved. Both patients with leg-type large-cell NHL have relapsed after a DR of 7 and 12 months, respectively. Although these patients are a rare case, it is obvious that RIT offers a novel and very useful approach for these patients.

6.20   RIT IN PATIENTS WITH >25% MARROW INVOLVEMENT

All initial studies of RIT excluded patients with greater than 25% involvement of the marrow with NHL. This eligibility requirement was based on concern that there would be an excessive marrow toxicity in those patients. This requirement effectively limited the number of patients with small lymphocytic lymphoma because such patients often have heavy marrow involvement. All studies with Zevalin have kept that requirement. A recent phase 1 Bexxar study accepted patients with >25% marrow involvement if they had platelet counts greater than $150,000 \times 10^6/L$ (178). Two dose levels of Bexxar were tested, 45 cGy ($n = 8$) and 55 cGy ($n = 3$). All patients by definition had >25% marrow involvement and the median percent involvement was 40 (range, 30–65%). One of the six evaluable patients treated at 45 cGy had a DLT and one of the three patients treated at 55 cGy experienced DLT. The ORR was 18% (2/11) and one of these responses has been durable at 42+ months.

6.21   RIT IN OLDER PATIENTS

Age has never been a restriction in trials of RIT. Emmanouilides et al. (179) recently reviewed the Zevalin database ($n = 211$) to learn whether the ORR or toxicity was different in older patients divided into three groups: <60, 60–69, and 70 years. The ORR was no different between the three groups (range, 71–80%) and the DR was also no different (median of 9.9, 11.0, and 9.4 months). The incidence of myelosuppression was similar in the older group.

6.22   RIT IN HODGKIN'S DISEASE

There has been one report describing the successful treatment of a patient with relapsed Hodgkin's disease with Zevalin (180). The rationale for this was the high ORR seen with rituximab in patients with lymphocyte-predominant HD (181).

6.23   VIRAL INFECTIONS AFTER RIT

Most of the trials to date have shown a low rate of infection despite the neutropenia that accompanies RIT. This has been primarily attributed to the low rate of mucosal damage that is induced by RIT. There have been several instances of viral reactivation in patients who received Zevalin (182). This of course has been observed with rituximab and it is not surprising that it is now reported in patients receiving Zevalin. It is to be remembered that patients who receive Zevalin also receive two doses of rituximab; therefore, it is difficult to attribute the virus reactivation to RIT.

6.24   RADIATION THERAPY AFTER RIT

A recent study has demonstrated that patients who relapse or progress with RIT can respond to external beam radiation therapy. Justice et al. (183) reviewed 19 patients who received radiation therapy after Zevalin was administered and they found a 90% (26/29 sites radiated) ORR with 12 (41%) CR, 7 (24%) clinical complete remission (CCR), 7 (24%) PR, and 3 (10%) stable disease. Toxicities were generally transient, reversible, and corresponded to the anatomic regions irradiated.

6.25   SUMMARY

The results of the phase I, phase II, and randomized trials demonstrate that single doses of RIT are safe and efficacious in patients with relapsed B-cell NHL. In addition, some RICs such as ^{90}Y-epratuzumab can be safely administered in a fractionated dosing schema to increase the total amount of radioactivity delivered. At present, RIT is used worldwide primarily for patients with indolent NHL that has relapsed after conventional therapy (184). RIT produces a response rate of approximately 80%, 30% of the patients obtain a CR, and 20% have not yet relapsed after years of follow-up. The primary toxicity is myelosuppression and this is dose limiting if stem cell support is not used. The advantages to the patient are an excellent quality of life and all treatment being delivered in 1 week.

RIT produces a higher response rate than unlabeled rituximab in rituximab-naïve patients. There is also a very high (75–80%) response rate in rituximab-refractory patients but a shorter TTP in this patient group. RIT requires a longer time to next chemotherapy than rituximab. The rate of HAMA is variable between the two agents—less than 2% with Zevalin and approximately 10% with Bexxar. HAMA has not been a major issue for the RIC at present in use. Some patients who have received RIT have developed MDS/AL following therapy. All these patients were previously treated with chemotherapy that included alkylating agents and/or purine nucleoside analogues. Previously untreated patients who received RIT have not developed MDS/AL. Calculations suggest that the rate of MDS is no different in patients who received RIT from those who were previously treated with chemotherapy, although the time of follow-up is still relatively short.

There is a lower (approximately 40%) but significant response rate in patients with relapsed DLBCL. The response rate in bulky disease (>5 cm) is lower than those patients with nonbulky disease. Preliminary studies indicate that patients who relapse after RIT can undergo SCT once their counts have recovered from RIT. Patients who relapse after SCT can also receive RIT after counts have recovered.

6.26 FUTURE DIRECTIONS

The field of RIT has now emerged from the initial wave of clinical trials, and two agents have been approved by the U.S. FDA. It is clear that these agents have the highest ORR of any single agent for relapsed NHL and they are well tolerated. The next generation of clinical trials is focusing on moving RIT earlier in the disease course with the goal of increasing the rate of CR and to prolong OS (Table 6.6). For example, randomized trials are testing RIT as adjuvant therapy after induction of rituximab-based chemotherapy for indolent and aggressive NHL. The safety and efficacy of RIT with stem cell support have led to a randomized phase III trial testing Bexxar-BEAM versus rituximab-BEAM with stem cell support. This large trial will determine if RIT adds to TTP and OS in this patient population. Trials combining agents simultaneously with RIT are somewhat more difficult because of the issue of myelosuppression. The agents to combine must also be chosen carefully, and the results may not be predictable (185). These combinations are now being tested in phase I trials to maximize patient safety.

Despite the high ORR and excellent toxicity profile of the anti-CD20-targeted RIC, these agents have not been commercially successful (see Chapter 16) and this has dampened enthusiasm for development of other RICs that target other antigens. However, there are some RICs under development that target other antigens on the tumor cell such as CD22 mentioned above. Recent studies have also demonstrated the effectiveness of unlabeled anti-CD30 antibodies (186) and these are now being radiolabeled with ^{90}Y (187). Studies combining chemotherapy with RIC have demonstrated promise in mouse models of solid tumors and may have relevance to future studies on lymphoma (188). CD45 is an antigen expressed on lymphomas (T- and B-cells), leukemias, and some myelomas (189). This is an attractive target for

TABLE 6.6   Future Directions for Radioimmunotherapy

Issue	Trials
Adjuvant use of radioimmunotherapy after conventional chemoimmunotherapy for indolent or large cell non-Hodgkin lymphoma	• First-line Indolent Trial (FIT) demonstrated benefit of adjuvant Zevalin (147) • S9911- CHOP/Bexxar phase II better than historical CHOP in untreated follicular (29) • S0016 – CHOP/Bexxar versus RCHOP in untreated follicular, accrual completed; results pending • Adjuvant RIT after fludarabine-based induction regimens—promising phase II results (53, 139, 140) • Adjuvant Zevalin after CHOP for elderly DLBCL (146) • Adjuvant Zevalin after RCHOP for stages I/II DLBCL—ECOG 3402—accrual ongoing
Radioimmunotherapy with stem cell support	• Pilot phase I/II studies demonstrate safety of combining RIT with stem cell support (64, 129, 156, 157, 164, 194–198) • Randomized trial of Bexxar-BEAM versus BEAM conditioning ongoing
Addition of radiosensitizers or other immunostimulatory agents to the RIT program	• Studies ongoing

RIT in that CD45-targeted RICs would be potentially useful for all these diseases (190, 191). The *in vitro* studies are promising and trials have been initiated (192).

Although there are many studies evaluating the role of rituximab maintenance after chemotherapy or after response to rituximab, to date, there have been no reported studies providing data on rituximab maintenance following RIT. Although this is certainly feasible, it is not known whether response can be prolonged. The results of these exciting studies will establish the role of RIT in the overall treatment plan of patients with B-cell NHL.

REFERENCES

1. Jemal A, Siegel R, Ward E, et al. Cancer statistics, 2008. *CA Cancer J Clin* 2008;58:71–96.
2. Fisher RI, Gaynor ER, Dahlberg S, et al. Comparison of a standard regimen (CHOP) with three intensive chemotherapy regimens for advanced non-Hodgkin's lymphoma. *N Engl J Med* 1993;328:1002–1006.
3. Jones RJ, Piantadosi S, Mann RB, et al. High-dose cytotoxic therapy and bone marrow transplantation for relapsed Hodgkin's disease. *J Clin Oncol* 1990;8:527–537.

4. Vriesendorp HM, Herpst JM, Germack MA, et al. Phase I–II studies of yttrium-labeled antiferritin treatment for end-stage Hodgkin's disease, including Radiation Therapy Oncology Group 87-01. *J Clin Oncol J* 1991;9:918–928.

5. Vriesendorp HM, Herpst JM, Leichner PK, Klein JL, Order SE. Polyclonal [90]Yttrium labeled antiferritin for refractory Hodgkin's disease. *Int J Radiat Oncol Biol Phys* 1989;17:815–821.

6. Klein JL, Sandoz JW, Kopher KA, Leichner PK, Order SE. Detection of specific anti-antibodies in patients treated with radiolabeled antibody. *Int J Radiat Oncol Biol Phys* 1986;12:939–943.

7. Kohler G, Milstein C. Continuous cultures of fused cells secreting antibody of predefined specificity. *Nature* 1975;256:495–497.

8. Miller RA, Maloney DG, Warnke R, Levy R. Treatment of B-cell lymphoma with monoclonal anti-idiotype antibody. *N Engl J Med* 1982;306:517–522.

9. Lewis JP, Denardo GL, Denardo SJ. Radioimmunotherapy of lymphoma: a UC Davis experience. *Hybridoma* 1995;14:115–120.

10. Stashenko P, Nadler LM, Hardy R, Schlossman SF. Characterization of a human B lymphocyte-specific antigen. *J Immunol* 1980;125:1678–1685.

11. Maloney DG, Liles TM, Czerwinski DK, et al. Phase I clinical trial using escalating single-dose infusion of chimeric anti-CD20 monoclonal antibody (IDEC-C2B8) in patients with recurrent B-cell lymphoma. *Blood* 1994;84:2457–2466.

12. Reff ME, Carner K, Chambers KS, et al. Depletion of B cells *in vivo* by a chimeric mouse human monoclonal antibody to CD20. *Blood* 1994;83:435–445.

13. Kaminski MS, Zasadny KR, Francis IR, et al. Radioimmunotherapy of B-cell lymphoma with [^{131}I]anti-B1 (anti-CD20) antibody. *N Engl J Med* 1993;329:459–465.

14. Knox SJ, Goris ML, Trisler K, et al. Yttrium-90-labeled anti-CD20 monoclonal antibody therapy of recurrent B-cell lymphoma. *Clin Cancer Res* March 1996;2:457–470.

15. Maloney DG, Grillo-Lopez AJ, White CA, et al. IDEC-C2B8 (Rituximab) anti-CD20 monoclonal antibody therapy in patients with relapsed low-grade non-Hodgkin's lymphoma. *Blood* 1997;90:2188–2195.

16. Maloney DG, Grillo-Lopez AJ, Bodkin DJ, et al. IDEC-C2B8: results of a phase I multiple-dose trial in patients with relapsed non-Hodgkin's lymphoma. *J Clin Oncol* 1997;15:3266–3274.

17. McLaughlin P, Grillo-Lopez AJ, Link BK, et al. Rituximab chimeric anti-CD20 monoclonal antibody therapy for relapsed indolent lymphoma: half of patients respond to a four-dose treatment program. *J Clin Oncol* 1998;16:2825–2833.

18. Witzig TE, Vukov AM, Habermann TM, et al. Rituximab Therapy for Patients With Newly Diagnosed, Advanced-Stage, Follicular Grade I Non-Hodgkin's Lymphoma: A Phase II Trial in the North Central Cancer Treatment Group. *J Clin Oncol* 2005;23:1103–1108.

19. Colombat P, Salles G, Brousse N, et al. Rituximab (anti-CD20 monoclonal antibody) as single first-line therapy for patients with follicular lymphoma with a low tumor burden: clinical and molecular evaluation. *Blood* 2001;97:101–106.

20. Marcus R, Imrie K, Belch A, et al. CVP chemotherapy plus rituximab compared with CVP as first-line treatment for advanced follicular lymphoma. *Blood* 2005;105:1417–1423.

21. Hochster HS, Weller E, Gascoyne RD, et al. Maintenance Rituximab after CVP results in superior clinical outcome in advanced follicular lymphoma (FL): results of the E1496

phase III trial from the Eastern Cooperative Oncology Group and the Cancer and Leukemia Group B. *ASH Annual Meeting Abstracts* November 16, 2005;106(11): 349.

22. Coiffier B, Lepage E, Briere J, et al. CHOP chemotherapy plus rituximab compared with CHOP alone in elderly patients with diffuse large-B-cell lymphoma. *N Engl J Med* 2002;346:235–242.

23. Habermann TM, Weller EA, Morrison VA, et al. Rituximab-CHOP versus CHOP alone or with maintenance rituximab in older patients with diffuse large B-cell lymphoma. *J Clin Oncol* 2006;24:3121–3127.

24. Feugier P, Van Hoof A, Sebban C, et al. Long-term results of the R-CHOP study in the treatment of elderly patients with diffuse large B-cell lymphoma: a study by the Groupe d'Etude des Lymphomes de l'Adulte. *J Clin Oncol* 2005;23:4117–4126.

25. Pfreundschuh M, Trumper L, Osterborg A, et al. CHOP-like chemotherapy plus rituximab versus CHOP-like chemotherapy alone in young patients with good-prognosis diffuse large-B-cell lymphoma: a randomised controlled trial by the MabThera International Trial (MInT) Group. *Lancet Oncol* 2006;7:379–391.

26. Gospodarowicz MK, Bush RS, Brown TC, Chua T. Prognostic factors in nodular lymphomas: a multivariate analysis based on the Princess Margaret Hospital experience. *Int J Radiat Oncol Biol Phys* 1984;10:489–497.

27. Gustavsson A, Osterman B, Cavallin-Stahl E. A systematic overview of radiation therapy effects in non-Hodgkin's lymphoma. *Acta Oncologica* (Stockholm) 2003; 42:605–619.

28. Fisher RI, LeBlanc M, Press OW, Maloney DG, Unger JM, Miller TP. New treatment options have changed the survival of patients with follicular lymphoma. *J Clin Oncol* 2005;23:8447–8452.

29. Press OW, Unger JM, Braziel RM, et al. Phase II trial of CHOP chemotherapy followed by tositumomab/iodine I-131 tositumomab for previously untreated follicular non-Hodgkin's lymphoma: five-year follow-up of Southwest Oncology Group Protocol S9911. *J Clin Oncol* 2006;24:4143–4149.

30. Illidge T, Chan C. How have outcomes for patients with follicular lymphoma changed with the addition of monoclonal antibodies? *Leuk Lymphoma* 2008; 49:1263–1273.

31. Fritzberg A, Meares C, Metallic radionuclides for radioimmunotherapy. In: Abrams PaF AR, editor. *Radioimmunotherapy of Cancer*, Marcel Dekker, Inc., New York, 2000, pp. 57–79.

32. Milenic DE, Brady ED, Brechbiel MW. Antibody-targeted radiation cancer therapy. *Nat Rev Drug Discov* 2004;3:488–499.

33. Boswell CA, Brechbiel MW. Development of radioimmunotherapeutic and diagnostic antibodies: an inside-out view. *Nucl Med Biol* 2007;34:757–778.

34. Du Y, Honeychurch J, Glennie M, Johnson P, Illidge T. Microscopic intratumoral dosimetry of radiolabeled antibodies is a critical determinant of successful radioimmunotherapy in B-cell lymphoma. *Cancer Res* 2007;67:1335–1343.

35. Maloney DG. Concepts in radiotherapy and immunotherapy: anti-CD20 mechanisms of action and targets. *Semin Oncol* 2005;32 (1 Suppl. 1): S19–S26.

36. Beers SA, Chan CH, James S, et al. Type II (tositumomab) anti-CD20 monoclonal antibody outperforms Type I (rituximab-like) reagents in B-cell depletion regardless of complement activation. *Blood* 2008;112:4170–4177.

37. Chan HT, Hughes D, French RR, et al. CD20-induced lymphoma cell death is independent of both caspases and its redistribution into triton X-100 insoluble membrane rafts. *Cancer Res* 2003;63:5480–5489.

38. Cragg MS, Bayne MB, Tutt AL, et al. A new anti-idiotype antibody capable of binding rituximab on the surface of lymphoma cells. *Blood* 2004;104:2540–2542.

39. Cragg MS, Glennie MJ. Antibody specificity controls *in vivo* effector mechanisms of anti-CD20 reagents. *Blood* 2004;103:2738–2743.

40. Cragg MS, Walshe CA, Ivanov AO, Glennie MJ. The biology of CD20 and its potential as a target for mAb therapy. *Curr Dir Autoimmun* 2005;8:140–174.

41. Glennie MJ, French RR, Cragg MS, Taylor RP. Mechanisms of killing by anti-CD20 monoclonal antibodies. *Mol Immunol* 2007;44:3823–3837.

42. Walshe CA, Beers SA, French RR, et al. Induction of cytosolic calcium flux by CD20 is dependent upon B Cell antigen receptor signaling. *J Biol Chem* 2008; 283:16971–16984.

43. Ivanov A, Krysov S, Cragg MS, Illidge T. Radiation therapy with tositumomab (B1) anti-CD20 monoclonal antibody initiates extracellular signal-regulated kinase/mitogen-activated protein kinase-dependent cell death that overcomes resistance to apoptosis. *Clin Cancer Res* 2008;14:4925–4934.

44. Jacobs SA, Harrison AM, Swerdlow SH, et al. Radioisotopic localization of (90)yttrium-ibritumomab tiuxetan in patients with CD20+ non-Hodgkin's lymphoma. *Mol Imaging Biol* 2008;11:39–45.

45. Macklis RM. Radioimmunotherapy in a radiation oncology environment: building a multi-specialty team. *Int J Radiat Oncol Biol Phys* 2006;66 (2 Suppl.): S4–S6.

46. Macklis RM, Pohlman B. Radioimmunotherapy for non-Hodgkin's lymphoma: a review for radiation oncologists. *Int J Radiat Oncol Biol Phys* 2006;66:833–841.

47. Coulot J, Camara-Clayette V, Ricard M, et al. Imaging of the distribution of ^{90}Y-ibritumomab tiuxetan in bone marrow and comparison with pathology. *Cancer Biother Radiopharm* 2007;22:665–671.

48. Gopal AK, Press OW, Wilbur SM, Maloney DG, Pagel JM. Rituximab blocks binding of radiolabeled anti-CD20 antibodies (Ab) but not radiolabeled anti-CD45 Ab. *Blood* 2008;112:830–835.

49. Witzig TE, White CA, Wiseman GA, et al. Phase I/II trial of IDEC-Y2B8 radioimmunotherapy for treatment of relapsed or refractory CD20(+) B-cell non-Hodgkin's lymphoma. *J Clin Oncol* 1999;17:3793–3803.

50. Santos ES, Kharfan-Dabaja MA, Ayala E, Raez LE. Current results and future applications of radioimmunotherapy management of non-Hodgkin's lymphoma. *Leuk Lymphoma* 2006;47:2453–2476.

51. Kalofonos HP, Grivas PD. Monoclonal antibodies in the management of solid tumors. *Curr Top Med Chem* 2006;6:1687–1705.

52. Vose JM, Bierman PJ, Enke C, et al. Phase I trial of iodine-131 tositumomab with high-dose chemotherapy and autologous stem-cell transplantation for relapsed non-Hodgkin's lymphoma. *J Clin Oncol* 2005;23:461–467.

53. Leonard JP, Coleman M, Kostakoglu L, et al. Abbreviated chemotherapy with fludarabine followed by tositumomab and iodine I 131 tositumomab for untreated follicular lymphoma. *J Clin Oncol* 2005;23:5696–5704.

54. Kaminski MS, Tuck M, Estes J, et al. ^{131}I-tositumomab therapy as initial treatment for follicular lymphoma. *N Engl J Med* 2005;352:441–449.

55. Kaminski MS, Radford JA, Gregory SA, et al. Re-treatment with I-131 tositumomab in patients with non-Hodgkin's lymphoma who had previously responded to I-131 tositumomab. *J Clin Oncol* 2005;23:7985–7993.

56. Parker BA, Vassos AB, Halpern SE, et al. Radioimmunotherapy of human B-cell lymphoma with ^{90}Y-conjugated antiidiotype monoclonal antibody. *Cancer Research* 1990;50 (3 Suppl.): 1022s–1028s.

57. Liu SY, Eary JF, Petersdorf SH, et al. Follow-up of relapsed B-cell lymphoma patients treated with iodine-131-labeled anti-CD20 antibody and autologous stem-cell rescue. *J Clin Oncol* 1998;16:3270–3278.

58. Kaminski MS, Zasadny KR, Francis IR, et al. Iodine-131-anti-B1 radioimmunotherapy for B-cell lymphoma. *J Clin Oncol* 1996;14:1974–1981.

59. Kaminski M, Gribbin T, Estes J, et al. I-131 anti-B1 antibody for previously untreated follicular lymphoma (FL): clinical and molecular remissions. *Proc ASCO* 1998;17:2a.

60. Press OW, Eary JF, Appelbaum FR, et al. Phase II trial of ^{131}I-B1 (anti-CD20) antibody therapy with autologous stem cell transplantation for relapsed B cell lymphomas. *Lancet* 1995;346:336–340.

61. Press O, Eary J, Liu S, et al. A phase I/II trial of high dose iodine-131-anti-B1 (anti-CD20) monoclonal antibody, etoposide, cyclophosphamide, and autologous stem cell transplantation for patients with relapsed B cell lymphomas. *Proceedings of the American Society of Clinical Oncology* 1998;17:3a.

62. DeNardo GL, DeNardo SJ, Goldstein DS, et al. Maximum-tolerated dose, toxicity, and efficacy of ^{131}I-Lym-1 antibody for fractionated radioimmunotherapy of non-Hodgkin's lymphoma. *J Clin Oncol* 1998;16:3246–3256.

63. DeNardo SJ, DeNardo GL, Kukis DL, et al. [67]Cu-2IT-BAT-Lym-1 pharmacokinetics, radiation dosimetry, toxicity and tumor regression in patients with lymphoma. *J Nucl Med* 1999;40:302–310.

64. Press O, Eary J, Gooley T, et al. A phase I/II trial of iodine-131-tositumomab (anti-CD20), etoposide, cyclophosphamide, and autologous stem cell transplantation for relapsed B-cell lymphomas. *Blood* 2000;96:2934–2942.

65. Kaminski MS, Zelenetz AD, Press OW, et al. Pivotal study of iodine I 131 tositumomab for chemotherapy-refractory low-grade or transformed low-grade B-cell non-Hodgkin's lymphomas. *J Clin Oncol* 2001;19:3918–3928.

66. White CA, Halpern SE, Parker BA, et al. Radioimmunotherapy of relapsed B-cell lymphoma with yttrium 90 anti-idiotype monoclonal antibodies. *Blood* 1996; 87:3640–3649.

67. Witzig TE, Gordon LI, Cabanillas F, et al. Randomized controlled trial of ^{90}Y-labeled ibritumomab tiuxetan (Zevalin) radioimmunotherapy versus rituximab immunotherapy for patients with relapsed or refractory low-grade, follicular, or transformed B-cell non-Hodgkin's lymphoma. *J Clin Oncol* 2002;20:2453–2463.

68. Wiseman GA, Gordon LI, Multani PS, et al. Ibritumomab tiuxetan radioimmunotherapy for patients with relapsed or refractory non-Hodgkin lymphoma and mild thrombocytopenia: a phase II multicenter trial. *Blood* 2002;99:4336–4342.

69. Vose JM, Colcher D, Gobar L, et al. Phase I/II trial of multiple dose [131]Iodine-MAb LL2 (CD22) in patients with recurrent non-Hodgkin's lymphoma. *Leuk Lymphoma* 2000;38:91–101.

70. Vose JM, Wahl RL, Saleh M, et al. Multicenter phase II study of iodine-131 tositumomab for chemotherapy-relapsed/refractory low-grade and transformed low-grade B-cell non-Hodgkin's lymphomas. *J Clin Oncol* 2000;18:1316–1323.

71. Juweid ME, Stadtmauer E, Hajjar G, et al. Pharmacokinetics, dosimetry, and initial therapeutic results with [131]I- and [111]In-/^{90}Y-labeled humanized LL2 anti-CD22 monoclonal antibody in patients with relapsed, refractory non-Hodgkin's lymphoma. *Clin Cancer Res* 1999;5 (10 Suppl.): 3292s–3303s.

72. Waldmann TA, White JD, Carrasquillo JA, et al. Radioimmunotherapy of interleukin-2R alpha-expressing adult T-cell leukemia with yttrium-90-labeled anti-Tac. *Blood* 1995;86:4063–4075.

73. Kaminski MS, Estes J, Zasadny KR, et al. Radioimmunotherapy with iodine ^{131}I tositumomab for relapsed or refractory B-cell non-Hodgkin lymphoma: updated results and long-term follow-up of the University of Michigan experience. *Blood* 2000; 96:1259–1266.

74. Leahy MF, Seymour JF, Hicks RJ, Turner JH. Multicenter phase II clinical study of iodine-131-rituximab radioimmunotherapy in relapsed or refractory indolent non-Hodgkin's lymphoma. *J Clin Oncol* 2006;24:4418–4425.

75. Dillman RO. Radioimmunotherapy of B-cell lymphoma with radiolabelled anti-CD20 monoclonal antibodies. *Clin Exp Med* 2006;6:1–12.

76. Schaefer-Cutillo J, Friedberg JW, Fisher RI. Novel concepts in radioimmunotherapy for non-Hodgkin's lymphoma. *Oncology* (Williston Park) 2007;21:203–212.

77. Cheung MC, Haynes AE, Stevens A, Meyer RM, Imrie K. Yttrium 90 ibritumomab tiuxetan in lymphoma. *Leuk Lymphoma* 2006;47:967–977.

78. Witzig TE. Radioimmunotherapy for B-cell non-Hodgkin lymphoma. *Best Pract Res Clin Haematol* 2006;19:655–668.

79. Pohlman B, Sweetenham J, Macklis RM. Review of clinical radioimmunotherapy. *Expert Rev Anticancer Ther* 2006;6:445–461.

80. Nowakowski GS, Witzig TE. Radioimmunotherapy for B-cell non-Hodgkin lymphoma. *Clin Adv Hematol Oncol* 2006;4:225–231.

81. Visser OJ, Perk LR, Zijlstra JM, van Dongen GA, Huijgens PC, van de Loosdrecht AA. Radioimmunotherapy for indolent B-cell non-Hodgkin lymphoma in relapsed, refractory and transformed disease. *BioDrugs* 2006;20:201–207.

82. Weigert O, Illidge T, Hiddemann W, Dreyling M. Recommendations for the use of yttrium-90 ibritumomab tiuxetan in malignant lymphoma. *Cancer* 2006;107:686–695.

83. Wiseman GA, Leigh B, Erwin WD, et al. Radiation dosimetry results for Zevalin radioimmunotherapy of rituximab-refractory non-Hodgkin lymphoma. *Cancer* 2002;94 (4 Suppl.): 1349–1357.

84. Wiseman G, Leigh B, Erwin W, et al. Radiation dosimetry results from a Phase II trial of ibritumomab tiuxetan (Zevalin) radioimmunotherapy for patients with non-Hodgkin's lymphoma and mild thrombocytopenia. *Cancer Biotherapy and Radiopharmaceuticals* 2003;18:165–178.

85. Wiseman GA, Kornmehl E, Leigh B, et al. Radiation dosimetry results and safety correlations from ^{90}Y-ibritumomab tiuxetan radioimmunotherapy for relapsed or

refractory non-Hodgkin's lymphoma: combined data from 4 clinical trials. *J Nucl Med* 2003;44:465–474.

86. Tennvall J, Fischer M, Bischof Delaloye A, et al. EANM procedure guideline for radio-immunotherapy for B-cell lymphoma with ^{90}Y-radiolabelled ibritumomab tiuxetan (Zevalin). *Eur J Nucl Med Mol Imaging* 2007;34:616–622.

87. Wiseman GA, White CA, Sparks RB, et al. Biodistribution and dosimetry results from a phase III prospectively randomized controlled trial of Zevalin radioimmunotherapy for low-grade, follicular, or transformed B-cell non-Hodgkin's lymphoma. *Crit Rev Oncol Hematol* 2001;39:181–194.

88. Wiseman G, White C, Stabin M, et al. Phase I/II ^{90}Y-Zevalin (yttrium-90 ibritumomab tiuxetan, IDEC-Y2B8) radioimmunotherapy dosimetry results in relapsed or refractory non-Hodgkin's lymphoma. *Eur J Nucl Med* 2000;27:766–777.

89. Gokhale AS, Mayadev J, Pohlman B, Macklis RM. Gamma camera scans and pretreatment tumor volumes as predictors of response and progression after Y-90 anti-CD20 radio-immunotherapy. *Int J Radiat Oncol Biol Phys* 2005;63:194–201.

90. Watanabe T, Terui S, Itoh K, et al. Phase I study of radioimmunotherapy with an anti-CD20 murine radioimmunoconjugate (^{90}Y-ibritumomab tiuxetan) in relapsed or refractory indolent B-cell lymphoma. *Cancer Sci* 2005;96:903–910.

91. Wiseman GA, Conti PS, Vo K, et al. Weight-based dosing of yttrium 90 ibritumomab tiuxetan in patients with relapsed or refractory B-cell non-Hodgkin lymphoma. *Clin Lymphoma Myeloma* 2007;7:514–517.

92. Buck AK, Kratochwil C, Glatting G, et al. Early assessment of therapy response in malignant lymphoma with the thymidine analogue [^{18}F]FLT. *Eur J Nucl Med Mol Imaging* 2007;34:1775–1782.

93. Nadler LM, Takvorian T, Botnick L, et al. Anti-B1 monoclonal antibody and complement treatment in autologous bone-marrow transplantation for relapsed B-cell non-Hodgkin's lymphoma. *Lancet* 1984;2:427–431.

94. Wahl RL, Zasadny KR, MacFarlane D, et al. Iodine-131 anti-B1 antibody for B-cell lymphoma: an update on the Michigan Phase I experience. *J Nucl Med* 1998;39 (8 Suppl.): 21S–27S.

95. Wahl RL, Kroll S, Zasadny KR. Patient-specific whole-body dosimetry: principles and a simplified method for clinical implementation. *J. Nucl. Med* 1998;39 (Suppl.): 14S–20S.

96. Koral KF, Dewaraja Y, Li J, et al. Update on hybrid conjugate-view SPECT tumor dosimetry and response in ^{131}I-tositumomab therapy of previously untreated lymphoma patients. *J Nucl Med* 2003;44:457–464.

97. Wahl RL. Iodine-131 anti-B1 antibody therapy in non-Hodgkin's lymphoma: dosimetry and clinical implications. *J Nucl Med* 1998;39 (8 Suppl.): 1S.

98. Siegel JA. Revised Nuclear Regulatory Commission regulations for release of patients administered radioactive materials: outpatient iodine-131 anti-B1 therapy. *J Nucl Med* 1998;39 (8 Suppl.): 28S–33S.

99. Siegel JA, Kroll S, Regan D, Kaminski MS, Wahl RL. A practical methodology for patient release after tositumomab and ^{131}I-tositumomab therapy. *J Nucl Med* 2002; 43:354–363.

100. Cheson BD, Pfistner B, Juweid ME, et al. Revised response criteria for malignant lymphoma. *J Clin Oncol* 2007;25:579–586.

101. Papajik T, Prochazka V, Raida L, et al. Relapsed follicular lymphoma sequentially treated with rituximab and ^{90}Y-ibritumomab tiuxetan. Case Report. *Biomedical Papers* 2007;151:109–112.

102. Ulaner GA, Colletti PM, Conti PS. B-cell non-Hodgkin lymphoma: PET/CT evaluation after ^{90}Y-ibritumomab tiuxetan radioimmunotherapy: initial experience. *Radiology* 2008;246:895–902.

103. Stagg R, Wahl RL, Estes J, Kaminski M, Phase I/II study of iodine-131 anti-B1 antibody for non-Hodgkin's lymphoma (NHL): final results. *Proc ASCO* 1998;17:39a (abstract 150).

104. Davies AJ, Rohatiner AZ, Howell S, et al. Tositumomab and iodine I 131 tositumomab for recurrent indolent and transformed B-cell non-Hodgkin's lymphoma. *J Clin Oncol* 2004;22:1469–1479.

105. Gockerman J, Gregory S, Harwood S, et al. Interim efficacy results of Bexxarä in a large multicenter expanded access study. *Proc ASCO* 2001;20:285a (abstract 1137).

106. Schenkein D, Leonard J, Harwood S, et al. Interim safety results of Bexxarä in a large multicenter expanded access study. *Proc ASCO* 2001;20 (Part 1 of 2): 285a (abstract 1138).

107. Fehrenbacher L, Radford JA, Kaminski M, Kroll S, Vose J, Patient-specific dosing of Bexxar™ is associated with tolerable and predictable hematologic toxicity in patients with known hematologic risk factors. *Proc ASCO* 2001;20 (Part 1 of 2): 287a (abstract 1144).

108. Azinovic I, DeNardo GL, Lamborn KR, et al. Survival benefit associated with human anti-mouse antibody (HAMA) in patients with B-cell malignancies. *Cancer Immunol Immunother* 2006;55:1451–1458.

109. Witzig TE, White CA, Gordon LI, et al. Safety of Yttrium-90 ibritumomab tiuxetan radioimmunotherapy for relapsed low-grade, follicular, or transformed non-Hodgkin's lymphoma. *J Clin Oncol* 2003;21:1263–1270.

110. Williams G, Palmer MR, Parker JA, Joyce R. Extravazation of therapeutic yttrium-90-ibritumomab tiuxetan (zevalin): a case report. *Cancer Biother Radiopharm* 2006;21:101–105.

111. Bishton MJ, Leahy MF, Hicks RJ, Turner JH, McQuillan AD, Seymour JF. Repeat treatment with iodine-131-rituximab is safe and effective in patients with relapsed indolent B-cell non-Hodgkin's lymphoma who had previously responded to iodine-131-rituximab. *Ann Oncol* 2008;19:1629–1633.

112. Leonard JP, Coleman M, Ketas JC, et al. Epratuzumab, a humanized Anti-CD22 antibody, in aggressive non-Hodgkin's lymphoma: phase I/II clinical trial results. *Clin Cancer Res* 2004;10:5327–5334.

113. Strauss SJ, Morschhauser F, Rech J, et al. Multicenter phase II trial of immunotherapy with the humanized anti-CD22 antibody, epratuzumab, in combination with rituximab, in refractory or recurrent non-Hodgkin's lymphoma. *J Clin Oncol* 2006;24:3880–3886.

114. Micallef IN, Kahl BS, Gayko U, et al. Initial results of a pilot study of epratuzumab and rituximab in combination with CHOP chemotherapy (ER-CHOP) in previously untreated patients with diffuse large B-cell lymphoma (DLBCL). *J Clin Oncol* (Meeting Abstracts). July 15 2004;22 (14_Suppl.): 577-b.

115. Goldenberg DM, Horowitz JA, Sharkey RM, et al. Targeting, dosimetry, and radioimmunotherapy of B-cell lymphomas with iodine-131-labeled LL2 monoclonal antibody. *J Clin Oncol* 1991;9:548–564.

116. Griffiths GL, Govindan SV, Sharkey RM, Fisher DR, Goldenberg DM. ^{90}Y-DOTA-hLL2: an agent for radioimmunotherapy of non-Hodgkin's lymphoma. *J Nucl Med* 2003;44:77–84.

117. Linden O, Hindorf C, Cavallin-Stahl E, et al. Dose-Fractionated Radioimmunotherapy in non-Hodgkin's lymphoma using DOTA-conjugated, ^{90}Y-radiolabeled, humanized anti-CD22 monoclonal antibody, epratuzumab. *Clin Cancer Res* 2005;11:5215–5222.

118. Bodet-Milin C, Kraeber-Bodere F, Dupas B, et al. Evaluation of response to fractionated radioimmunotherapy with ^{90}Y-epratuzumab in non-Hodgkin's lymphoma by ^{18}F-fluorodeoxyglucose positron emission tomography. *Haematologica* 2008;93:390–397.

119. Kraeber-Bodere F, Morschhauser F, Huglo D, et al. Fractionated radioimmunotherapy in NHL with DOTA-conjugated, humanized anti-CD22 IgG, epratuzumab: results at high cumulative doses of ^{90}Y. *J Clin Oncol.* 2008;26 (May 20 Suppl.): abstract 8502.

120. Cheson B, Horning S, Coiffier B, et al. Report of an international workshop to standardize response criteria for non-Hodgkin's lymphoma. *J Clin Oncol* 1999;17:1244–1253.

121. Davis T, Kaminski M, Leonard J, et al. Results of a randomized study of Bexxar™ (tositumomab and iodine 131 tositumomab) versus unlabeled tositumomab in patients with relapsed or refractory low-grade or transformed non-Hodgkin's lymphoma. *Blood* 2001;98 (Suppl. 1 part 1): 843a.

122. Witzig TE, Flinn IW, Gordon LI, et al. Treatment with ibritumomab tiuxetan radio-immunotherapy in patients with rituximab-refractory follicular non-Hodgkin's lympho-ma. *J Clin Oncol* 2002;20:3262–3269.

123. Horning SJ, Younes A, Jain V, et al. Efficacy and safety of tositumomab and iodine-131 tositumomab (Bexxar) in B-cell lymphoma, progressive after rituximab. *J Clin Oncol* 2005;23:712–719.

124. Stefanovic A, Morgensztern D, Fong T, Lossos IS. Pulmonary marginal zone lymphoma: a single centre experience and review of the SEER database. *Leuk Lymphoma* 2008;49:1311–1320.

125. Gordon LI, Molina A, Witzig T, et al. Durable responses after ibritumomab tiuxetan radioimmunotherapy for CD20+ B-cell lymphoma: long-term follow-up of a phase 1/2 study. *Blood* 2004;103:4429–4431.

126. Morschhauser F, Illidge T, Huglo D, et al. Efficacy and safety of yttrium-90 ibritumomab tiuxetan in patients with relapsed or refractory diffuse large B-cell lymphoma not appropriate for autologous stem-cell transplantation. *Blood* 2007;110:54–58.

127. Zelenetz A, Saleh M, Vose J, Younes A, Kaminski M. Patients with transformed low grade lymphoma attain durable responses following outpatient radioimmunotherapy with tositumomab and iodine I 131 tositumomab [Bexxar®]. *Blood.* 2002;100 (11 Suppl. Part 1 of 2): abstract.

128. Oki Y, Pro B, Delpassand E, et al. A phase II study of yttrium 90 (^{90}Y) ibritumomab tiuxetan (Zevalin®) for treatment of patients with relapsed and refractory mantle cell lymphoma (MCL). *Blood* 2004;104: Abstract 2632.

129. Gopal AK, Rajendran JG, Petersdorf SH, et al. High-dose chemo-radioimmunotherapy with autologous stem cell support for relapsed mantle cell lymphoma. *Blood* 2002;99:3158–3162.

130. Smith MR, Zhang L, Gordon LI, et al. Phase II study of R-CHOP followed by ^{90}Y-ibritumomab tiuxetan in untreated mantle cell lymphoma: Eastern Cooperative Oncology Group study E1499. *ASH Annual Meeting Abstracts* November 16, 2007;110(11):389.

131. Fisher RI, Kaminski MS, Wahl RL, et al. Tositumomab and iodine-131 tositumomab produces durable complete remissions in a subset of heavily pretreated patients with low-grade and transformed non-Hodgkin's lymphomas. *J Clin Oncol* 2005; 23:7565–7573.

132. Buchegger F, Antonescu C, Delaloye AB, et al. Long-term complete responses after ^{131}I-tositumomab therapy for relapsed or refractory indolent non-Hodgkin's lymphoma. *Br J Cancer* 2006;94:1770–1776.

133. Witzig TE, Molina A, Gordon LI, et al. Long-term responses in patients with recurring or refractory B-cell non-Hodgkin lymphoma treated with yttrium 90 ibritumomab tiuxetan. *Cancer* 2007;109:1804–1810.

134. Emmanouilides C, Witzig TE, Gordon LI, et al. Treatment with yttrium 90 ibritumomab tiuxetan at early relapse is safe and effective in patients with previously treated B-cell non-Hodgkin's lymphoma. *Leuk Lymphoma* 2006;47:629–636.

135. Bennett JM, Kaminski MS, Leonard JP, et al. Assessment of treatment-related myelodysplastic syndromes and acute myeloid leukemia in patients with non-Hodgkin lymphoma treated with tositumomab and iodine I[131] tositumomab. *Blood* 2005; 105:4576–4582.

136. Czuczman MS, Emmanouilides C, Darif M, et al. Treatment-related myelodysplastic syndrome and acute myelogenous leukemia in patients treated with ibritumomab tiuxetan radioimmunotherapy. *J Clin Oncol* 2007;25:4285–4292.

137. Roboz GJ, Bennett JM, Coleman M, et al. Therapy-related myelodysplastic syndrome and acute myeloid leukemia following initial treatment with chemotherapy plus radioimmunotherapy for indolent non-Hodgkin lymphoma. *Leuk Res* 2007;31:1141–1144.

138. Link B, Kaminski MS, Coleman M, Leonard JP, Phase II study of CVP followed by tositumomab and iodine I[131] tositumomab (Bexxar therapeutic regimen) in patients with untreated follicular non-Hodgkin's lymphoma (NHL). *J Clin Oncol* (Meeting Abstracts). 2004;22 (14_Suppl.): 562-c.

139. Zinzani PL, Tani M, Fanti S, et al. A phase 2 trial of fludarabine and mitoxantrone chemotherapy followed by yttrium-90 ibritumomab tiuxetan for patients with previously untreated, indolent, nonfollicular, non-Hodgkin lymphoma. *Cancer* 2008;112:856–862.

140. Zinzani PL, Tani M, Pulsoni A, et al. Fludarabine and mitoxantrone followed by yttrium-90 ibritumomab tiuxetan in previously untreated patients with follicular non-Hodgkin lymphoma trial: a phase II non-randomised trial (FLUMIZ). *Lancet Oncol* 2008;9:352–358.

141. Ansell SM, Ristow KM, Habermann TM, Wiseman GA, Witzig TE.Subsequent chemotherapy regimens are well tolerated after radioimmunotherapy with yttrium-90 ibritumomab tiuxetan for non-Hodgkin's lymphoma. *J Clin Oncol* 2002;20:3885–3890.

142. Ansell SM, Schilder RJ, Pieslor PC, et al. Antilymphoma treatments given subsequent to yttrium 90 ibritumomab tiuxetan are feasible in patients with progressive non-Hodgkin's lymphoma: a review of the literature. *Clin Lymphoma* 2004;5:202–204.

143. Dosik AD, Coleman M, Kostakoglu L, et al. Subsequent therapy can be administered after tositumomab and iodine I-131 tositumomab for non-Hodgkin lymphoma. *Cancer* 2006;106:616–622.

144. Micallef IN, Apostolidis J, Rohatiner AZ, et al. Factors which predict unsuccessful mobilisation of peripheral blood progenitor cells following G-CSF alone in patients with non-Hodgkin's lymphoma. *Hematol J* 2000;1:367–373.

145. Shipley DL, Spigel DR, Carrell DL, Dannaher C, Greco FA, Hainsworth JD. Phase II trial of rituximab and short duration chemotherapy followed by ^{90}Y-ibritumomab tiuxetan as first-line treatment for patients with follicular lymphoma: a Minnie Pearl Cancer Research Network phase II trial. *J Clin Oncol* (Meeting Abstracts). 2004;22 (14_Suppl.): 562-b.

146. Zinzani PL, Tani M, Fanti S, et al. A phase II trial of CHOP chemotherapy followed by yttrium 90 ibritumomab tiuxetan (Zevalin) for previously untreated elderly diffuse large B-cell lymphoma patients. *Ann Oncol* 2008;19:769–773.

147. Morschhauser F, Radford J, Van Hoof A, et al. Phase III trial of consolidation therapy with yttrium-90-ibritumomab tiuxetan compared with no additional therapy after first remission in advanced follicular lymphoma. *J Clin Oncol* 2008;26:5156–5164.

148. Inwards DJ, Cilley JC, Winter JN. Radioimmunotherapeutic strategies in autologous hematopoietic stem-cell transplantation for malignant lymphoma. *Best Pract Res Clin Haematol* 2006;19:669–684.

149. Cilley J, Winter JN. Radioimmunotherapy and autologous stem cell transplantation for the treatment of B-cell lymphomas. *Haematologica* 2006;91:114–120.

150. Nademanee A, Forman SJ. Role of hematopoietic stem cell transplantation for advanced-stage diffuse large cell B-cell lymphoma-B. *Semin Hematol* 2006;43:240–250.

151. Krishnan A, Nademanee A, Fung HC, et al. Phase II trial of a transplantation regimen of yttrium-90 ibritumomab tiuxetan and high-dose chemotherapy in patients with non-Hodgkin's lymphoma. *J Clin Oncol* 2008;26:90–95.

152. Shimoni A, Nagler A. Radioimmunotherapy and stem-cell transplantation in the treatment of aggressive B-cell lymphoma. *Leuk Lymphoma* 2007;48:2110–2120.

153. Devizzi L, Guidetti A, Tarella C, et al. High-dose yttrium-90-ibritumomab tiuxetan with tandem stem-cell reinfusion: an outpatient preparative regimen for autologous hemato-poietic cell transplantation. *J Clin Oncol* 2008;26:5175–5182.

154. Molina A, Krishnan A, Fung H, et al. Use of radioimmunotherapy in stem cell transplantation and posttransplantation: focus on yttrium 90 ibritumomab tiuxetan. *Curr Stem Cell Res Ther* 2007;2:239–248.

155. Press OW, Eary JF, Appelbaum FR, et al. Radiolabeled-antibody therapy of B-cell lymphoma with autologous bone marrow support. *N Engl J Med* 1993;329:1219–1224.

156. Gopal AK, Gooley TA, Maloney DG, et al. High-dose radioimmunotherapy versus conventional high-dose therapy and autologous hematopoietic stem cell transplantation for relapsed follicular non-Hodgkin lymphoma: a multivariable cohort analysis. *Blood* 2003;102:2351–2357.

157. Gopal AK, Rajendran JG, Gooley TA, et al. High-dose [^{131}I]tositumomab (anti-CD20) radioimmunotherapy and autologous hematopoietic stem-cell transplantation for adults > or = 60 years old with relapsed or refractory B-cell lymphoma. *J Clin Oncol* 2007;25:1396–1402.

158. He B, Wahl RL, Du Y, et al. Comparison of residence time estimation methods for radioimmunotherapy dosimetry and treatment planning: Monte Carlo simulation studies. *IEEE Trans Med Imaging* 2008;27:521–530.

159. Urbano N, Modoni S. Evaluation of different beta-counting systems involved in ^{90}Y-Zevalin quality control. *Nucl Med Commun* 2007;28:943–950.

160. Cremonesi M, Ferrari M, Grana CM, et al. High-dose radioimmunotherapy with ^{90}Y-ibritumomab tiuxetan: comparative dosimetric study for tailored treatment. *J Nucl Med* 2007;48:1871–1879.

161. Nademanee A, Forman S, Molina A, et al. A phase 1/2 trial of high-dose yttrium-90-ibritumomab tiuxetan in combination with high-dose etoposide and cyclophosphamide followed by autologous stem cell transplantation in patients with poor-risk or relapsed non-Hodgkin lymphoma. *Blood* 2005;106:2896–2902.

162. Winter JN, Inwards DJ, Spies S, et al. Yttrium-90 ibritumomab tiuxetan doses calculated to deliver up to 15 Gy to critical organs may be safely combined with high-dose BEAM and autologous transplantation in relapsed or refractory B-cell non-Hodgkin's lymphoma. *J Clin Oncol* 2009;27:1653–1659.

163. Chiesa C, Botta F, Di Betta E, et al. Dosimetry in myeloablative ^{90}Y-labeled ibritumomab tiuxetan therapy: possibility of increasing administered activity on the base of biological effective dose evaluation. Preliminary results. *Cancer Biother Radiopharm* 2007;22:113–120.

164. Ferrucci PF, Vanazzi A, Grana CM, et al. High activity ^{90}Y-ibritumomab tiuxetan (Zevalin) with peripheral blood progenitor cells support in patients with refractory/resistant B-cell non-Hodgkin lymphomas. *Br J Haematol* 2007;139:590–599.

165. He YF, Li YH, Wang FH, et al. The effectiveness of lamivudine in preventing hepatitis B viral reactivation in rituximab-containing regimen for lymphoma. *Ann Hematol* 2008;87:481–485.

166. Vose JM, Bierman PJ, Loberiza FR, Jr., Bociek RG, Matso D, Armitage JO. Phase I trial of ^{90}Y-ibritumomab tiuxetan in patients with relapsed B-cell non-Hodgkin's lymphoma following high-dose chemotherapy and autologous stem cell transplantation. *Leuk Lymphoma* 2007;48:683–690.

167. Peyrade F, Triby C, Slama B, et al. Radioimmunotherapy in relapsed follicular lymphoma previously treated by autologous bone marrow transplant: a report of eight new cases and literature review. *Leuk Lymphoma* 2008;49:1762–1768.

168. Pizer BL, Kemshead JT. The potential of targeted radiotherapy in the treatment of central nervous system leukaemia. *Leuk Lymphoma* 1994;15:281–289.

169. Shah JJ, Meredith R, Shen S, et al. Case report of a patient with primary central nervous system lymphoma treated with radioimmunotherapy. *Clin Lymphoma Myeloma* 2006;7:236–238.

170. Pitini V, Baldari S, Altavilla G, Arrigo C, Naro C, Perniciaro F. Salvage therapy for primary central nervous system lymphoma with ^{90}Y-ibritumomab and temozolomide. *J Neurooncol* 2007;83:291–293.

171. Iwamoto FM, Schwartz J, Pandit-Taskar N, et al. Study of radiolabeled indium-111 and yttrium-90 ibritumomab tiuxetan in primary central nervous system lymphoma. *Cancer* 2007;110:2528–2534.

172. Doolittle ND, Jahnke K, Belanger R, et al. Potential of chemo-immunotherapy and radioimmunotherapy in relapsed primary central nervous system (CNS) lymphoma. *Leuk Lymphoma* 2007;48:1712–1720.

173. Wiseman G, Witzig T. Yttrium-90 Zevalin phase I sequential dose radioimmunotherapy trial of patients with relapsed low grade and follicular B-cell non-Hodgkins lymphoma (NHL): preliminary results. *Blood* 2002;100:358a (Abstract 1387).

174. Peyrade F, Italiano A, Fontana X, Peyrottes I, Thyss A. Retreatment with ^{90}Y-labelled ibritumomab tiuxetan in a patient with follicular lymphoma who had previously responded to treatment. *Lancet Oncol* 2007;8:849–850.

175. Cooney-Qualter E, Krailo M, Angiolillo A, et al. A phase I study of [90]Yttrium-ibritumomab-tiuxetan in children and adolescents with relapsed/refractory CD20-

positive non-Hodgkin's lymphoma: a Children's Oncology Group study. *Clin Cancer Res* 2007;13 (18 Pt 2): 5652s–5660s.

176. DeMonaco NA, McCarty KS, Jr., Joyce J, Jacobs SA. Focal radiation fibrosis after radioimmunotherapy for follicular non-Hodgkin lymphoma. *Clin Lymphoma Myeloma* 2007;7:369–372.

177. Maza S, Gellrich S, Assaf C, et al. Yttrium-90 ibritumomab tiuxetan radioimmunotherapy in primary cutaneous B-cell lymphomas: first results of a prospective, monocentre study. *Leuk Lymphoma* 2008;49:1702–1709.

178. Mones JV, Coleman M, Kostakoglu L, et al. Dose-attenuated radioimmunotherapy with tositumomab and iodine 131 tositumomab in patients with recurrent non-Hodgkin's lymphoma (NHL) and extensive bone marrow involvement. *Leuk Lymphoma* 2007; 48:342–348.

179. Emmanouilides C, Witzig TE, Wiseman GA, et al. Safety and efficacy of yttrium-90 ibritumomab tiuxetan in older patients with non-Hodgkin's lymphoma. *Cancer Biother Radiopharm* 2007;22:684–691.

180. Schnell R, Dietlein M, Schomacker K, et al. Yttrium-90 ibritumomab tiuxetan-induced complete remission in a patient with classical lymphocyte-rich Hodgkin's lymphoma. *Onkologie* 2008;31:49–51.

181. Ekstrand BC, Lucas JB, Horwitz SM, et al. Rituximab in lymphocyte-predominant Hodgkin disease: results of a phase 2 trial. *Blood* 2003;101:4285–4289.

182. Cil T, Altintas A, Tuzun Y, Pasa S, Isikdogan A. Hepatitis B virus reactivation induced by yttrium-90-ibritumomab-tiuxetan. *Leuk Lymphoma* 2007;48: 1866–1868.

183. Justice T, Martenson JA, Jr., Wiseman G, Witzig T. Safety and efficacy of external beam radiation therapy for non-Hodgkin lymphoma in patients with prior ^{90}Y-ibritumomab tiuxetan radioimmunotherapy. *Cancer* 2006;107:433–438.

184. Dreyling M, Trumper L, von Schilling C, et al. Results of a national consensus workshop: therapeutic algorithm in patients with follicular lymphoma-role of radioimmunotherapy. *Ann Hematol* 2007;86:81–87.

185. Dearling JL, Qureshi U, Begent RH, Pedley RB. Combining radioimmunotherapy with antihypoxia therapy 2-deoxy-d-glucose results in reduction of therapeutic efficacy. *Clin Cancer Res* 2007;13:1903–1910.

186. Bartlett NL, Younes A, Carabasi MH, et al. A phase 1 multidose study of SGN-30 immunotherapy in patients with refractory or recurrent CD30+ hematologic malignancies. *Blood* 2008;111:1848–1854.

187. Zhang M, Yao Z, Patel H, et al. Effective therapy of murine models of human leukemia and lymphoma with radiolabeled anti-CD30 antibody, HeFi-1. *Proc Natl Acad Sci USA* 2007;104:8444–8448.

188. Milenic DE, Garmestani K, Brady ED, et al. Multimodality therapy: potentiation of high linear energy transfer radiation with Paclitaxel for the treatment of disseminated peritoneal disease. *Clin Cancer Res* 2008;14:5108–5115.

189. Kumar S, Rajkumar SV, Kimlinger T, Greipp PR, Witzig TE. CD45 expression by bone marrow plasma cells in multiple myeloma: clinical and biological correlations. *Leukemia* 2005;19:1466–1470.

190. Zhang MM, Gopal AK. Radioimmunotherapy-based conditioning regimens for stem cell transplantation. *Semin Hematol* 2008;45:118–125.

191. Glatting G, Muller M, Koop B, et al. Anti-CD45 monoclonal antibody YAML568: a promising radioimmunoconjugate for targeted therapy of acute leukemia. *J Nucl Med* 2006;47:1335–1341.

192. Vallera DA, Sicheneder AR, Taras EP, et al. Radiotherapy of CD45-expressing Daudi tumors in nude mice with yttrium-90-labeled, PEGylated anti-CD45 antibody. *Cancer Biother Radiopharm* 2007;22:488–500.

193. Kaminski M, Tuck M, Regan D, Kison PV, Wahl RL. High response rates and durable remissions in patients with previously untreated, advanced-stage, Follicular lymphoma treated with tositumomab and iodine I-131 tositumomab (Bexxar®). *Blood* November 16, 2002;100 (11 Suppl. part 1 of 2): 356a abstract 1381.

194. Winter JN. Combining yttrium 90-labeled ibritumomab tiuxetan with high-dose chemotherapy and stem cell support in patients with relapsed non-Hodgkin's lymphoma. *Clin Lymphoma* 2004;5 (Suppl. 1): S22–S26.

195. Flinn I, Kahl BS, Frey E, et al. Dose finding trial of yttrium 90 (^{90}Y) ibritumomab tiuxetan with autologous stem cell transplantation (ASCT) in patients with relapsed or refractory B-cell non-Hodgkin's lymphoma (NHL). *Blood* 2004;104(11): Abstract 897.

196. Nademanee A, Krishnan A, Tsai N, et al. ^{90}Y-ibritumomab tiuxetan (Zevalin) in combination with high-dose therapy (HDT) followed by autologous stem cell transplant (ASCT) may improve survival in patients with poor-risk follicular lymphoma (FL) and diffuse large B-cell lymphoma (DLBCL): results of a retrospective comparative analysis. *Blood.* 2006;108: Abstract 327.

197. Nademanee AP, Molina A, Forman SJ, et al. A phase I/II trial of high-dose radio-immunotherapy (RIT) with Zevalin in combination with high-dose etoposide (VP-16) and cyclophosphamide (CY) followed by autologous stem cell transplant (ASCT) in patients with poor-risk or relapsed B-cell non-Hodgkin's lymphoma (NHL). *Blood* 2002;100(11): 182a (Abstract 679).

198. Shimoni A, Zwas ST, Oksman Y, et al. Yttrium-90-ibritumomab tiuxetan (Zevalin) combined with high-dose BEAM chemotherapy and autologous stem cell transplantation for chemo-refractory aggressive non-Hodgkin's lymphoma. *Exp Hematol* 2007;35:534–540.

CHAPTER 7

Radioimmunotherapy of Acute Myeloid Leukemia

TODD L. ROSENBLAT AND JOSEPH G. JURCIC

7.1 INTRODUCTION

Although standard therapy with cytarabine and an anthracycline for patients with acute myeloid leukemia (AML) is associated with complete response rates between 50% and 70%, long-term disease-free survival is seen in only 20–40% of patients. Following relapse, additional chemotherapy generally produces remissions in only 20–25% of patients. While allogeneic hematopoietic stem cell transplantation (HSCT) can produce long-term remissions in approximately 30% of patients with relapsed AML, most patients are not appropriate candidates due to age, comorbidities, or lack of a suitable donor. Therefore, new therapies are needed to improve overall survival and to reduce therapy-related toxicity.

The use of unlabeled monoclonal antibodies (mAbs) in the treatment of leukemias has met with mixed success. While the anti-CD20 antibody rituximab (1) and the anti-CD52 antibody alemtuzumab (2) have displayed significant activity in chronic lymphocytic leukemia, the lack of potency of many unconjugated mAbs has led many investigators to use antibodies as targeting vehicles for cytotoxic agents. Significant antileukemic effects have been observed with the anti-CD33-calicheamicin conjugate gemtuzumab ozogamicin in acute myeloid leukemia (3) and the anti-CD22-pseudomonas exotoxin construct BL22 in hairy cell leukemia (4). In an alternative strategy, antibodies can be used to target radionuclides directly to tumor cells. The leukemias are ideally suited for radioimmunotherapy (RIT) for several reasons. First, because of their location in the blood, bone marrow, spleen, and lymph nodes, leukemic blasts are easily accessible to circulating antibodies. Second, target antigens on blasts and other hematopoietic cells are well known and can be characterized for individual patients using flow cytometry. Finally, leukemias are radiosensitive tumors. This chapter focuses on issues of target antigen and radionuclide

Monoclonal Antibody and Peptide-Targeted Radiotherapy of Cancer, Edited by Raymond M. Reilly
Copyright © 2010 John Wiley & Sons, Inc.

selection, radiolabeling, antibody pharmacokinetics, and dosimetry. We will also review the results of recent preclinical and clinical trials in the RIT of leukemia.

7.2   ANTIGENIC TARGETS

Immunophenotypic characterization of the lineages and stages of hematopoietic differentiation provides the rationale for the selection of antigenic targets and associated carrier molecules for RIT. Most of these antigens, however, are neither lineage- nor tumor specific. For example, CD10, found on pre-B-cell acute lympho-blastic leukemia (ALL), is also found on follicular lymphomas and T-cell ALL. AML is characterized by the expression of the myeloid-associated antigens myeloperox-idase, CD13, CD15, CD33, and CD117 (5). HLA-DR is typically found on all subtypes of AML except acute promyelocytic leukemia (APL). Monocytic leukemias express antigens associated with more mature granulocytes and monocytes, including CD11a/18, CD11c, CD14, and CD15.

Target antigens investigated most extensively for RIT of myeloid leukemias include CD33, the pan-leukocyte antigen CD45, and CD66, which is found on granulocytes. CD33 is a 67 kDa cell-surface glycoprotein expressed on most myeloid and monocytic leukemia cells. It is also found on committed myelomonocytic and erythroid progeni-tor cells, but not on the earliest pluripotent stem cells, mature granulocytes, lymphoid cells, or nonhematopoietic cells (6, 7). Several anti-CD33 antibodies, including the murine antibody M195, its humanized counterpart lintuzumab (HuM195), and p67 have been used for RIT of AML. CD45 is a tyrosine phosphatase expressed by virtually all leukocytes, including myeloid and lymphoid precursors, mature lymphocytes, and myeloid and lymphoid blasts. It is not expressed on mature erythrocytes or platelets. Because of the broad expression of CD45, radioimmunoconjugates directed against this antigen should eliminate not only leukemic blasts but also normal leukocytes in the marrow, limiting its application to HSCT. RIT using the murine anti-CD45 antibody BC8 has been studied extensively at the Fred Hutchinson Cancer Research Center as part of transplant conditioning regimens. Finally, CD66c, also known as nonspecific cross-reacting antigen (NCA), is a 95 kDa glycoprotein expressed on myeloid cells but not on leukemia cells. Because of this expression pattern, any antileukemic effect of an anti-CD66 radioimmunoconjugate must rely on the "cross fire" of radionuclide emissions to reach untargeted blasts. As with BC8, use of BW 250/183, a murine antibody targeting CD66, has been limited to the transplant setting because of its expression on normal cells.

7.3   RADIONUCLIDE SELECTION

The choice of a radionuclide for RIT depends on various factors, including the emission characteristics of the radionuclide, its physical and biological half-life, the stability of the immunoconjugate *in vivo*, and the clinical setting for which the therapy is intended. Beta particles have a longer range in tissue (up to 12 mm) with a lower linear energy transfer (LET) ($\sim$0.2 keV/μm) than α particles, which have a short range (50–80 μm) and a high LET ($\sim$100 keV/μm) (8, 9). Because of the large size and high molecular weight of most antibodies, their diffusion into sites of bulky disease can be limited. The longer range of β-emissions, however, can overcome this limitation by creating a field effect to destroy tumor cells to which the radioimmunoconjugates are not directly bound. Therefore, therapy with β-emitters is likely to be most useful in the setting of extensive disease and HSCT, where the preparative regimen should eliminate both malignant and nonmalignant hematopoietic cells.

Most RIT trials for leukemia have used β-emitting isotopes such as iodine-131 (^{131}I), yttrium-90 (^{90}Y), and rhenium-188 (^{188}Re) (Table 7.1). ^{131}I has a relatively long half-life of 8.1 days and emits a low-energy β particle ($\sim$600 keV). The γ emissions from ^{131}I penetrate the body and can be detected by a gamma camera, thereby facilitating biodistribution and radiation dosimetry studies. Despite these benefits, there are a number of limitations associated with the use of ^{131}I. First, treatment at high doses requires patient isolation and can result in significant exposure to hospital staff (due to the penetrating nature of the gamma emissions). Second, because approximately one-third of the tyrosine residues, to which ^{131}I binds, are located in the hypervariable region of most mAbs, radioiodination at high specific activities may impair the ability of these antibodies to bind to their target antigen (10). Finally, when used as part of a preparative regimen for HSCT, sufficient time must separate ^{131}I treatment from stem cell infusion in order to prevent injury to the grafted cells from retained ^{131}I within the marrow. In transplantation trials with ^{131}I-labeled M195 and lintuzumab, these factors added up to 16 days to the preparative regimen (11).

Many of these limitations of ^{131}I can be overcome by the use of radiometals such as ^{90}Y and ^{188}Re. Unlike ^{131}I, antibody labeling with most radiometals employs bifunctional chelators, thereby permitting facile radiolabeling at higher specific activities and radiochemical purity using a "kit" formulation, thus simplifying preparation and minimizing quality control testing. Since ^{90}Y does not emit γ rays,

TABLE 7.1   Characteristics of Selected Radionuclides for Radioimmunotherapy

Radionuclide	Particle(s) Emitted	Half-Life	Particulate Energy (keV)	Mean Range of Emission (mm)
β-Emitters				
Iodine-131	β, γ	8.1 days	610	0.8
Yttrium-90	β	2.5 days	2280	2.7
Copper-67	β, γ	2.6 days	580	0.9
Rhenium-188	β, γ	17 h	2100	2.4
α-Emitters				
Bismuth-212	1 α, 1 β, 1 γ	1 h	7800	0.04–0.10
Bismuth-213	1 α, 2 β, 1 γ	46 min	8400	0.05–0.08
Astatine-211	1 α, 1 γ	7.2 days	6800	0.04–0.10
Actinium-225	4 α, 2 β, 2 γ	10 days	6000–8400	0.04–0.08
Radium-223	4 α, 2 β, 3 γ	11.4 days	6000–7000	0.04–0.08
Lead-212	1 α, 2 β, 1 γ	10.6 h	7800	0.04–0.10

hospitalization and patient isolation are not necessary. Imaging for biodistribution and dosimetry studies, however, requires administration of mAbs radiolabeled with trace amounts of a surrogate radionuclide, usually indium-111 (^{111}In), the biodistribution of which is not identical to that of ^{90}Y (12). Positron emission tomography (PET) of ^{86}Y-labeled constructs is one strategy that may improve radiation dosimetry estimates for RIT with ^{90}Y (13).

Since the range of α particles is only a few cell diameters, RIT with α-emitters may result in more specific tumor cell kill and less damage to surrounding normal tissues. Because of their high LET, only one to two α particles traversing through the nucleus are needed to kill a targeted cell (9), whereas up to 10,000 traversals by β particles may be required for cell death (14). In addition, injured cells have a limited capacity to repair damage induced by α particles. These properties make α particles ideal for the treatment of cytoreduced or minimal residual disease (MRD). The physical properties among various α-emitters differ widely (Table 7.1) (15). Bismuth-213 (^{213}Bi) has a half-life of only 46 min; therefore, this isotope is likely to be most useful in systems where carrier molecules can rapidly target disease sites. Preparation for clinical use requires a generator consisting of its parent isotope actinium-225 (^{225}Ac) loaded onto a cation exchange resin from which ^{213}Bi can be eluted (16–18). ^{225}Ac has a longer half-life (10 days) and in its decay to stable ^{209}Bi, ^{225}Ac generates several α-emitting daughters, including francium-221 (^{221}Fr), astatine-217 (^{217}At), and ^{213}Bi for a total yield of four α particles (Fig. 7.1). In this way, ^{225}Ac bound to antibodies can serve as an *in vivo* α-particle generator. ^{225}Ac-labeled tumor-specific antibodies can kill multiple cell lines *in vitro* with LD_{50} values that are 1000 times lower than those of analogous ^{213}Bi constructs. *In vivo* studies in nude mice bearing human prostate carcinoma and lymphoma xenografts showed that single nanocurie doses of ^{225}Ac-labeled tumor-specific antibodies significantly improved survival compared to controls and cured a significant fraction of animals (19). Although this increased potency could make ^{225}Ac more effective than other α-emitters, the possibility of free daughter radioisotopes in circulation after decay of ^{225}Ac, with uptake of free ^{213}Bi by the kidney, raises concerns about the potential toxicity of this radionuclide (20, 21).

7.4 RADIOLABELING

A variety of methods are used to conjugate radionuclides to antibodies, depending primarily on the nature of the element (see Chapter 2). Since ^{131}I binds to tyrosine residues, it can be conjugated directly to mAbs using the chloramine-T method. Tumor cell catabolism due to internalization of the antigen–antibody complex, followed by rapid degradation of the radioimmunoconjugate and expulsion of radiolabeled metabolites, represents a significant disadvantage to therapy with some ^{131}I-labeled constructs. This problem could potentially be overcome by the use of radiometals, which are better retained within cells after catabolism than ^{131}I (22) or by novel radioiodination methods, such as tyramine cellobiose, resulting in more stable radioimmunoconjugates (23) (see Chapter 2). ^{211}At is a halogen like ^{131}I and is usually linked to antibodies through incorporation of an aryl carbon-astatine bond.

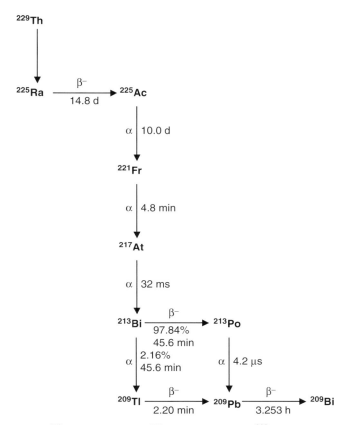

FIGURE 7.1   The ^{229}Th decay scheme. ^{225}Ac is isolated from ^{229}Th sources and decays by α emission through ^{221}Fr, ^{217}At, and ^{213}Bi, each of which also emits an α particle.

Methods used to create the aryl carbon-astatine bond usually involve an astatode-metallation reaction using a tin, silicon, or mercury precursor (24, 25). ^{188}Re has been directly labeled to the anti-CD66 antibody BW250/183 using tris-(2-carboxyethyl) phosphine as a reducing agent (26).

Other radionuclides require bifunctional chelators for complexation by antibodies (Fig. 7.2). The macrocyclic ligand 1,4,7,10-tetraazacyclododecane-1,4,7,10-tetraacetic acid (DOTA) and its derivatives have been used effectively for labeling antibodies with ^{90}Y (27), ^{212}Bi (28), and ^{225}Ac (29). However, in some experimental systems, DOTA can be immunogenic and is not suitable for conjugation of certain radiometals, such as ^{213}Bi. For these reasons, alternative chelators derived from diethylenetriamine pentaacetic acid (DTPA) have been developed. Tiuxetan, or 2-(p-isothiocyanatobenzyl)-5(6)-methyl-DTPA (Mx-DTPA), for example, is used to complex ^{90}Y and ^{111}In to the anti-CD20 antibody ibritumomab for treatment of B-cell lymphoma (30). Other derivatives of DTPA include the cyclic dianhydride derivative (31), 2-(p-isothiocyanatobenzyl)-DTPA (SCN-Bz-DTPA) (32), and the

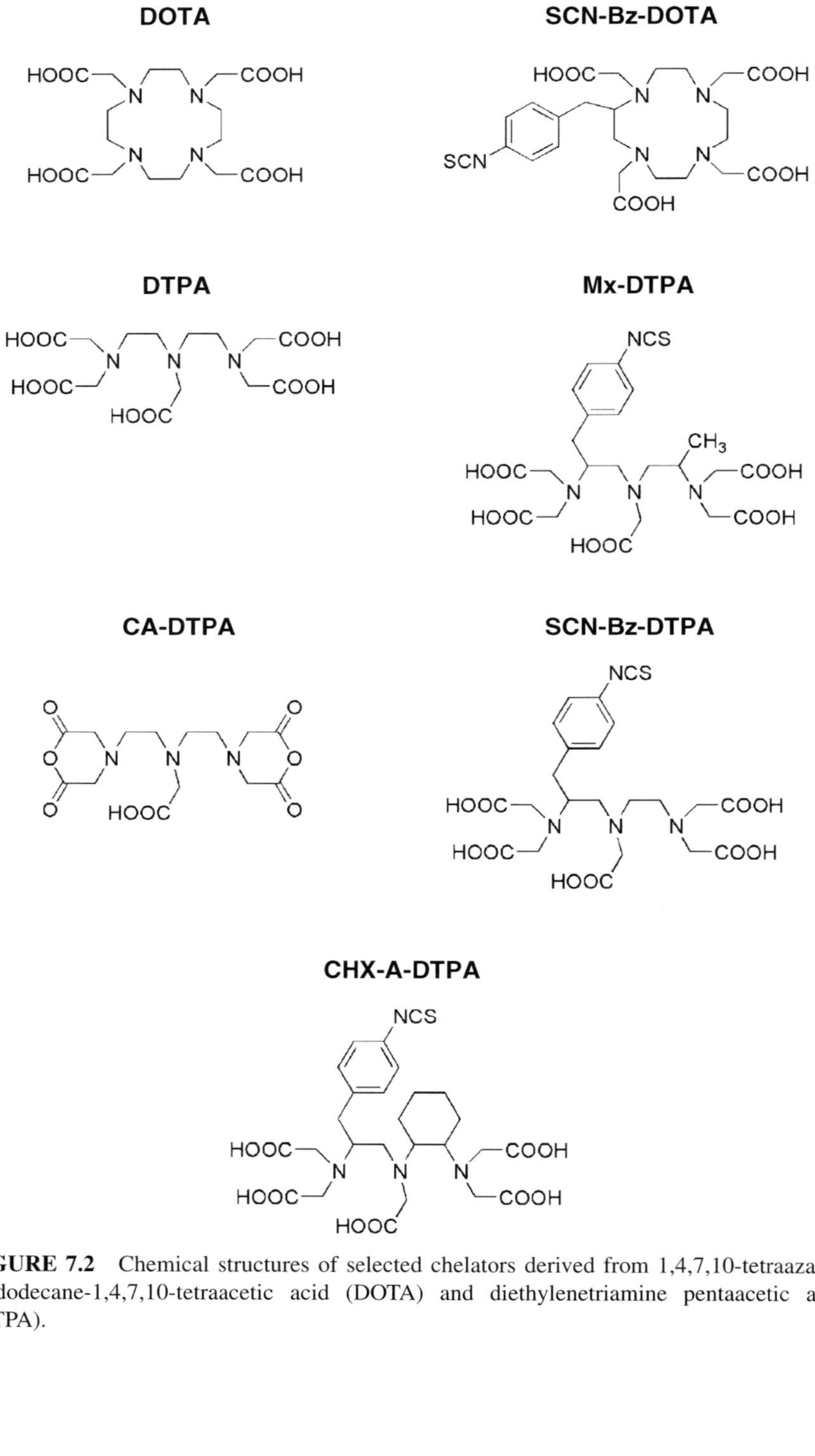

FIGURE 7.2 Chemical structures of selected chelators derived from 1,4,7,10-tetraazacy-clododecane-1,4,7,10-tetraacetic acid (DOTA) and diethylenetriamine pentaacetic acid (DTPA).

cyclohexylbenzyl derivative (CHX-A-DTPA) (16, 33), which is an effective chelator for both radionuclides of yttrium and bismuth (16, 34).

7.5 PHARMACOKINETICS AND DOSIMETRY

Factors such as variability in tumor burden and number of antibody binding sites per malignant cell in individual patients, antibody specificity and binding avidity, immunoreactivity, antigen–antibody internalization after binding, immunogenicity, and radionuclide half-life all contribute to the variable and sometimes suboptimal pharmacokinetics of radioimmunoconjugates. Careful biodistribution and dosimetry studies have lead to greater insights regarding the pharmacology of antibodies. For example, the influence of the number of available binding sites on antibody biodistribution was observed in a dose-escalation trial of trace-labeled ^{131}I-M195, where superior targeting to sites of disease as determined by gamma camera imaging was seen with a comparatively small dose of approximately 5 mg (35). At higher antibody doses, saturation of available antigen sites was demonstrated by flow cytometric analysis, and cardiac blood pooling was seen on gamma camera imaging, indicating excess circulating antibody. This may be explained in part by the relatively low number of binding sites (approximately 10,000–20,000) on each leukemia cell. Furthermore, in a Phase I study of ^{213}Bi-lintuzumab, decreasing levels of radioactivity in the liver and spleen were noted after multiple injections of subsaturating antibody doses ($\sim$1–2 mg), suggesting a "first-pass" binding effect with CD33 antigen saturation within the liver after administration of several milligrams of antibody (36).

Serial gamma camera imaging and measurements of plasma, urine, bone marrow, and tissue biopsy radioactivity are used to estimate absorbed radiation doses to different organs and tumor sites (see Chapter 13). The validity of these predictions, however, is limited by the accuracy in measuring radioactivity using gamma camera imaging and by the inability to visualize all sites of disease in patients. Single photon emission computed tomography (SPECT) may increase the accuracy of planar scintigraphy, especially when used in conjunction with computed tomography (CT) (37, 38). Models, based on this dosimetric data, may provide information about radiation doses delivered to tissues not directly sampled and may also be used to estimate total tumor burden and tumor burden in individual organs (39, 40).

RIT with short-lived α-particle emitters such as ^{213}Bi is associated with markedly different pharmacology than with longer lived β-emitters such as ^{131}I or ^{90}Y. With these longer lived radionuclides, pharmacokinetics are dominated by the biological clearance of the antibody. The distribution of the antibody within the first several minutes to hours after administration yields residence times in the blood that are negligible in proportion to the overall residence times achieved in target and normal organs. In contrast, for ^{213}Bi, with its 46 min half-life, 20% of the total α emissions occur within the first 15 min after injection, and after 3 h, only 6% of the total emissions remain. In addition, the higher linear energy transfer of α particles compared with β particles results in a relative biological effectiveness (RBE) for cell sterilization of 3–7, which must be considered in dosimetry estimates for α particle treatment (41). In contrast, beta particles have a RBE of 1. Given the high energy of α particles, which is deposited over a very short distance (50–100 μm), conventional methodologies that estimate mean absorbed dose over a specific organ volume may not always yield biologically meaningful information. While targeted cells may receive high absorbed radiation doses, adjacent cells may receive no radiation at all. Therefore, microdosimetric or stochastic analyses that account for the spatial distribution of various cell types and the distribution of α decays within the organ will be necessary to estimate the absorbed dose to tumor cells and normal tissues more accurately. Since the geometric relationship between the radionuclide and the target cell is not uniform, α particle hits cannot be assumed to be a Poisson distribution. Several distributions have been modeled, and microdosimetric spectra, expressed as specific energy probability densities, have been calculated. Based on this work, methods have been developed to perform basic microdosimetric assessments that account for the probability of the number of hits and the mean specific energy from a single hit (42).

7.5.1 Pretargeted Approaches

To improve the pharmacokinetics of RIT and tumor-to-normal tissue dose ratios, a novel "pretargeting" strategy has been developed that takes advantage of the rapid, high affinity, specific binding between streptavidin and biotin (43) (see Chapter 8). A mAb or fusion protein is first conjugated to the tetravalent streptavidin molecule and infused intravenously. Then, a biotinylated N-acetylgalactosamine-containing "clearing agent" is given to remove excess antibody–streptavidin conjugate circulating in the blood. In this step, the biotin component of the clearing agent binds to the streptavidin portion of the antibody construct and galactose receptors on hepatocytes remove the complexes from the circulation. Finally, therapeutically radiolabeled biotin is administered that binds to the "pretargeted" antibody–streptavidin conjugate on malignant cells. Unbound radiolabeled biotin is rapidly excreted in the urine.

Such a pretargeting approach has been applied to a mouse model of adult T-cell leukemia (ATL) (44). After administration of humanized anti-Tac (anti-CD25)-streptavidin (SA) and the biotinylated N-acetylgalactosamine-containing clearing agent, immunodeficient mice bearing human ATL xenografts received DOTA-biotin labeled with the α-emitter [213]Bi or the β-emitter [90]Y. Treatment with [213]Bi improved survival compared with controls, whereas ^{90}Y did not. Mice treated with [213]Bi by the pretargeting approach survived longer than those treated with [213]Bi conjugated directly to anti-Tac mAbs. This approach was also studied using an anti-Tac single chain Fv-SA fusion protein followed by [90]Y- or [213]Bi-labeled biotin to treat ATL in xenografted mice (45). Similar studies using an anti-CD45 antibody SA conjugate followed by [111]In- or ^{90}Y-labeled biotin demonstrated high target organ to nontarget organ radiation dose ratios and significant long-term survival in leukemic mice (46). The pretargeting approach is particularly useful for RIT with short-lived radionuclides such as [213]Bi, because the radiolabeled biotin molecule is quickly delivered to streptavidin-conjugated mAbs bound to leukemic cells maximizing the therapeutic effect, while excess [213]Bi-labeled biotin is rapidly eliminated, minimizing toxicity to the bone marrow or normal organs.

Pretargeted RIT has undergone significant improvement in the uniformity of the antibody–streptavidin targeting molecule and in simplification of the production process (47, 48). Using recombinant technology, an anti-CD45 tetravalent single-chain antibody-strepavidin fusion protein (scFv$_4$SA) has been developed. It retains full antigen- and biotin binding capabilities of its parent molecules and demonstrated promising results in preclinical testing for the treatment of leukemia (49). Clinical trials using this anti-CD45 scFv$_4$SA fusion protein for AML are planned.

7.6   RIT WITH β-PARTICLE EMITTERS

Most clinical RIT trials to date have used isotopes that emit β particles; however, α-particle RIT has been studied in patients with myeloid leukemias more recently. The results of selected RIT trials for leukemia are summarized in Table 7.2.

7.6.1 ^{131}I-M195 and ^{131}I-Lintuzumab

In a series of RIT studies, the murine anti-CD33 antibody, M195, and its humanized version, lintuzumab, were labeled with ^{131}I. In an early Phase I trial, 24 patients with relapsed or refractory myeloid leukemias were treated with escalating doses (50–210 mCi/m^2) of ^{131}I-M195 (50). Gamma camera images of the whole body demonstrated rapid uptake of the ^{131}I-M195 into areas of leukemic involvement, including the bone marrow, liver, and spleen. The radioisotope was retained at these sites for at least three days. The maximum tolerated dose (MTD) was not reached, but profound myelosuppression occurred at ^{131}I doses above 135 mCi/m^2, necessitating HSCT in eight patients. Twenty-two patients had reductions in the percentage of bone marrow blasts, and three achieved complete remissions (CR). Over one-third of patients developed human anti-mouse antibodies (HAMA). This study demonstrated that ^{131}I-M195 can deliver high radiation absorbed doses to the marrow with limited extramedullary toxicity and significant antileukemic effects.

To address the difficulties posed by the immunogenicity and lack of intrinsic antileukemic activity of M195 in a nonradiolabeled form, investigators developed a humanized version of this antibody, known as lintuzumab (HuM195). Lintuzumab maintained the binding specificity of M195 but, unlike M195, could induce antibody-dependent cell-mediated cytotoxicity (ADCC) against CD33-positive target cells using human peripheral blood mononuclear cells as effectors (51). Based on these results, ^{131}I-M195 and ^{131}I-lintuzumab were added to standard conditioning in order to intensify the preparative regimen for allogeneic HSCT (11). Thirty-one patients with overt relapsed or refractory AML, accelerated or myeloblastic CML, or advanced myelodysplastic syndrome (MDS) were treated with ^{131}I-M195 or ^{131}I-lintuzumab (120–230 mCi/m^2) followed by busulfan (16 mg/kg), cyclophosphamide (90 or 120 mg/kg), and infusion of related-donor bone marrow. Estimated absorbed radiation doses to the marrow ranged between 272 and 1470 cGy. Toxicities beyond those observed with the busulfan/cyclophosphamide conditioning regimen alone did not occur, and engraftment was not delayed. Twenty-eight of 30 patients achieved

TABLE 7.2   Selected Clinical Trials of Radiolabeled Antibodies for Treatment of AML

Radiolabeled antibody	Disease	Radioactivity Dose	No. of Patients	Results	Comments	References
[131]I-M195	Advanced AML, MDS, blastic CML	50–210 mCi/m^2	24	CR in 3 of 8 patients receiving BMT	5 patients received autologous HSCT; 3 received allogeneic HSCT	(50)
[131]I-M195, ^{131}I-lintuzumab	Advanced AML, MDS, blastic CML	120–230 mCi/m^2	30	24/25 evaluable patients had no evidence of leukemia; long-term DFS in 3 patients	Used with Bu/Cy before allogeneic HSCT	(11)
[90]Y-lintuzumab	Advanced AML	0.1–0.3 mCi/kg	19	13 patients had reductions in marrow blasts; 1 CR	Higher doses result in prolonged myelosuppression	(52)
[213]Bi-lintuzumab	Advanced AML, CMML	0.28–1 mCi/kg	18	14 patients had reductions in marrow blasts; no CRs	First demonstration of safety of α-particle therapy	(36)
[213]Bi-lintuzumab	AML	0.5–1.25 mCi/kg	31	2 CRs, 2 CRp, 2 PRs	Given after partial cytoreduction with Ara-C	(72)
[225]Ac-lintuzumab	Advanced AML	0.5–2 μCi/kg	9	4 patients had >33% reduction in marrow blasts	First study to demonstrate safety of an *in vivo* α-particle generator; accrual continues	(73)

Agent	Disease	Dose	No.	Results	Comments	Ref.
^{131}I-p67	AML	110–330 mCi	9	3 of 4 patients treated with therapeutic doses relapsed	Given with Cy/TBI before HSCT; many patients had unfavorable biodistribution	(54)
^{131}I-BC8	Advanced AML, ALL	76–612 mCi	44	7 of 25 patients with AML or MDS and 3 of 9 patients with ALL had long-term DFS	Given with Cy/TBI before HSCT	(55)
^{131}I-BC8	AML in first remission	101–298 mCi	46	61% 3-year DFS	Given with Bu/Cy prior to allogeneic HSCT	(56)
^{188}Re-BW 250/183	High-risk AML, MDS	11.1 GBq (mean)	36	45% DFS at median 18 months	Given as part of preparative regimen prior to HSCT	(58)
^{188}Re- or ^{90}Y -BW 250/183	AML, MDS over age 55 years	12.2 GBq for ^{188}Re; 3.8 GBq for ^{90}Y (mean)	20	52% 2-year survival	Given as part of a reduced-intensity conditioning regimen prior to HSCT	(59)

AML, acute myelogenous leukemia; MDS, myelodysplastic syndrome; CML, chronic myelogenous leukemia; CMML, chronic myelomonocytic leukemia; ALL, acute lymphoblastic leukemia; CR, complete remission; HSCT, hematopoietic stem cell transplantation; DFS, disease-free survival; PR, partial remission; Bu, busulfan; Cy, cyclophosphamide; Ara-C, cytarabine; TBI, total body irradiation.

remission, and 3 of 16 patients with refractory AML have remained in remission for 5+ to 8+ years following transplant.

7.6.2 ^{90}Y-Lintuzumab

^{90}Y offers several potential advantages to overcome the limitations associated with ^{131}I-labeled anti-CD33 antibodies. The higher energy and longer range β emissions of ^{90}Y permit a lower effective dose than ^{131}I, and the absence of γ emissions allows large doses of ^{90}Y to be given safely in the outpatient setting without posing a radiation hazard to family members or health care workers. Moreover, after internalization of antigen–antibody complexes into target cells, radiometals such as ^{90}Y are better retained by these cells than ^{131}I. In a Phase I trial, 19 patients with relapsed or refractory AML were treated with escalating doses of ^{90}Y-lintuzumab (0.1–0.3 mCi/kg), given as a single infusion without marrow support (52). Biodistribution and dosimetry studies were performed by coadministration of trace-labeled ^{111}In-lintuzmab. Up to 560, 880, and 750 cGy were delivered to the marrow, liver, and spleen, respectively. Myelosuppression was the dose-limiting toxicity, and the maximum tolerated dose of ^{90}Y-lintuzumab without stem cell rescue was 0.275 mCi/kg. Transient, low-grade liver function test abnormalities were also seen in 11 patients. Thirteen of 19 patients had reductions in bone marrow blasts. All patients treated at the highest dose level had markedly hypocellular bone marrow without evidence of leukemia up to 4 weeks after treatment. One of the seven patients treated at the maximum tolerated dose achieved CR lasting 5 months.

7.6.3 ^{131}I-Labeled p67

Using an individualized dosimetry approach, investigators at the Fred Hutchinson Caner Research Center studied another ^{131}I-labeled murine anti-CD33 antibody, p67, in patients with AML. In a Phase I trial, nine patients were initially treated with trace-radiolabeled doses of ^{131}I-p67 (53). While the radionuclide localized to the marrow in most patients, residence times were relatively short (9–41 h), likely due to rapid catabolism of the radioimmunoconjugate following internalization. Only four patients had "favorable biodistribution" with greater uptake of ^{131}I in the marrow and spleen than in nonhematopoietic organs. Those patients subsequently received therapeutic doses of ^{131}I-p67 (110–330 mCi), cyclophosphamide (120 mg/kg), and total body irradiation (TBI) (12 Gy), followed by allogeneic bone marrow transplantation. Although the therapy was well tolerated, three of the four patients eventually relapsed (54). Because of the unfavorable pharmacology and biodistribution of this radioimmunoconjugate, these investigators have since focused on the anti-CD45 antibody BC8.

7.6.4 ^{131}I-Labeled BC8

In a Phase I trial, 44 patients with advanced acute leukemia or MDS initially received BC8 labeled with radiotracer doses of ^{131}I (55). Favorable biodistribution occurred in

37 patients (84%). Thirty-four of these patients then received escalating therapeutic doses of ^{131}I-BC8 (76–612 mCi) followed by cyclophosphamide (120 mg/kg), total body irradiation (12 Gy), and allogeneic or autologous transplantation. Therapeutic doses of ^{131}I-BC8 were calculated to deliver 3.5–12.25 Gy to the liver, the normal organ receiving the highest dose in the dosimetry studies performed using the trace-labeled antibody. The MTD delivered an estimated radiation absorbed dose of 10.5 Gy to the liver when administered with cyclophosphamide and TBI. At this dose, an average of 24 and 50 Gy was delivered to the marrow and spleen, respectively (sites of leukemic involvement). Of the 25 patients with AML or MDS, 7 remained alive and disease-free at a median follow-up of 65 months. Of the nine patients with ALL, three remained alive and disease-free at 19, 54, and 66 months.

Based on these results, a Phase I/II trial using the preparative regimen of ^{131}I-BC8, busulfan, and cyclophosphamide in patients with AML in first remission was conducted (56). Fifty-nine patients received a trace-radiolabeled dose of ^{131}I-BC8 and 52 (88%) had favorable biodistribution. Forty-six patients then received a therapeutic dose of ^{131}I-BC8 (102–298 mCi) that was estimated to deliver 3.5 to 5.25 Gy to the liver. This resulted in a mean absorbed dose of 11.3 Gy to the marrow and 29.7 Gy to the spleen. The nonrelapse mortality and long-term disease-free survival were 21% and 61%, respectively. When compared to the outcome of 509 similar International Bone Marrow Registry patients, RIT-treated patients had a 0.65 hazard ratio for mortality. These results are sufficiently encouraging to warrant further study of this approach.

7.6.5 ^{188}Re-Anti-CD66

^{188}Re ($t_{1/2}$, 17 h) is a radiometal that emits both β particles as well as γ rays, which facilitate biodistribution and dosimetry imaging studies. In a pilot dosimetry trial, 12 patients with advanced leukemias received 6.5–12.4 GBq (175–335 mCi) of ^{188}Re-anti-CD66 followed by a standard preparative regimen and T-cell depleted allogeneic transplantation (26). Favorable biodistribution occurred in most patients, and a median of 14 Gy were delivered to the bone marrow (57). Subsequently, 36 patients with high-risk AML or MDS were treated with ^{188}Re-anti-CD66 followed by one of three preparative regimens: total body irradiation (12 Gy) plus cyclophosphamide (120 mg/kg), busulfan (12.8 mg/kg) plus cyclophosphamide (120 mg/kg), or TBI plus thiotepa (10 mg/kg) and cyclophosphamide (120 mg/kg) (58). Thirty-one patients received allogeneic grafts, one received a syngeneic graft, and four received autologous grafts. Favorable biodistribution occurred in all patients, and the median dose delivered to the bone marrow was 14.9 Gy (range, 8.1–28 Gy). In contrast to studies with ^{131}I-anti-CD45, in which the liver was the dose-limiting normal organ, the normal organ receiving the highest dose of radiation after ^{188}Re-anti-CD66 was the kidney (median dose, 7.2 Gy). Nephrotoxicity, likely due to the radiation, occurred in six patients (17%) between 6 and 12 months after transplantation. At a median follow-up of 18 months, disease-free survival was 45%. In addition, both ^{166}Re- and ^{90}Y-labeled anti-CD66 antibodies were shown to be safe as part of a reduced intensity preparative regimen in older patients with AML and MDS with a 2-year survival of 52% (59).

7.7 RIT WITH α-PARTICLE EMITTERS

The high energy and short range of α particles offer the possibility of more efficient and selective killing of tumor cells. Therefore, to increase the antitumor activity of native mAbs but avoid the nonspecific cytotoxicity of β-emitting radionuclides, α-particle RIT has been investigated.

7.7.1 Preclinical Studies

In a number of different rodent xenograft models, treatment with mAbs labeled with α-particle emitters has prolonged survival compared with relevant controls (19, 31, 60–62). In one of the first reports suggesting the feasibility of this approach, ^{212}Bi conjugated to the tumor-specific antibody 103A demonstrated activity against murine erythroleukemia (33). The results of many of these studies support the hypothesis that α-particle RIT may be more effective in the treatment of small-volume disease than in the treatment of more extensive tumors. For example, administration of ^{212}Bi-anti-Tac after inoculation of nude mice with a CD25-expressing lymphoma cell line led to prolonged tumor-free survival and prevented development of leukemia in some animals, whereas treatment of established tumors failed to produce responses (63). Similarly, in spheroid models, α-particle therapy has been more effective in reducing the volume of smaller spheroids compared with larger ones (64–66). In most of the animal models in which α-emitters and β-emitters have been directly compared, α-emitters have been more effective in preventing tumor growth and prolonging survival (67, 68).

^{213}Bi-labeled antibodies to CD45 (69) and the T-cell receptor (TCR)αβ (70) have been used for immunosuppression prior to nonmyeloablative bone marrow transplantation in a canine model. Both ^{213}Bi-labeled mAbs, when given prior to transplantation with mycophenoloate mofetil and cyclosporine, allowed for prompt engraftment of transplanted marrow and resulted in stable mixed chimerism. The high dose of radioactivity of ^{213}Bi (at least 2 mCi/kg) required for engraftment, however, may limit the utility of this approach in humans.

7.7.2 ^{213}Bi-Lintuzumab

In vitro, ^{213}Bi-lintuzumab killed cells expressing CD33 in a dose-dependent and specific activity-dependent fashion (16). Up to 10 mCi/kg of ^{213}Bi-lintuzumab could be injected intravenously into BALB/c mice without significant toxicity (71). Based on these preclinical studies, a Phase I clinical trial of ^{213}Bi-labeled lintuzumab was performed in patients with advanced myeloid leukemias (36). Eighteen patients with relapsed or refractory AML or chronic myelomonocytic leukemia were treated with 0.28–1.0 mCi/kg of ^{213}Bi-lintuzumab in 3–7 fractions over 2–4 days. Myelosuppression occurred in all patients, and transient minor liver function abnormalities occurred in six patients. Gamma camera images demonstrated uptake of ^{213}Bi in the bone marrow, liver, and spleen within 10 min of administration without significant uptake in any other organs, including the kidneys, which are known to be avid for free bismuth. Because of low whole-body radiation absorbed doses, ratios between the marrow, liver, and spleen and the whole body were 1000-fold greater than those seen with β-emitting lintuzumab constructs in similar patients. Fourteen patients (78%) had reductions in the percentage of bone marrow blasts, but no complete remissions occurred. Given that only 1 in 2700 HuM195 molecules carry the radiolabel at the specific activities injected and up to 10^{16} CD33 binding sites in total on malignant cells are available in patients with overt AML, extraordinarily high injected radioactivities of ^{213}Bi would be required to deliver one to two ^{213}Bi atoms per cell and produce complete remissions. Nevertheless, this trial provides the rationale for the continued investigation of this approach in a variety of cancers where minimal residual disease or micrometastatic disease may be present.

Since α-particle immunotherapy is likely to be most useful in the treatment of small-volume disease, a subsequent Phase I/II study was undertaken in which patients were first treated with chemotherapy to achieve partial cytoreduction of the leukemic burden followed by ^{213}Bi-lintuzumab (72). Thirty-one patients with newly diagnosed ($n = 13$) or relapsed/refractory ($n = 18$) AML (median age, 67 years) were treated with cytarabine (200 mg/m^2/day for 5 days) followed by ^{213}Bi-lintuzumab (0.5–1.25 mCi/kg). Prolonged myelosuppression was dose-limiting, and the MTD was 1 mCi/kg. Significant reductions in marrow blasts were seen at all dose levels, and clinical responses were observed in 6 of the 25 patients (24%) who received doses of at least 1 mCi/kg (2 CR, 2 CRp [CR with incomplete platelet recovery but transfusion-independence], 2 PR). The median response duration was 7.7 months (range, 2–12 months). In contrast to the result of the initial Phase I trial in which ^{213}Bi-lintuzumab was given without cytoreduction, pharmacokinetic and biodistribution studies suggested that saturation of all available CD33 sites by ^{213}Bi-lintuzumab was possible after treatment with cytarabine.

7.7.3 ^{225}Ac-Lintuzumab

The major obstacles to the widespread use of RIT with ^{213}Bi are its short half-life and the requirement of an on-site ^{225}Ac/^{213}Bi generator. One solution is to deliver the ^{225}Ac "generator" parent radionuclide to the target cell, allowing the production of atoms *in situ* that yield α emissions at or within the cancer cell. ^{225}Ac, which yields four α-emitting isotopes, can be complexed to a variety of antibodies using DOTA and can prolong the survival of animals in several xenograft models. In this ongoing Phase I trial, nine patients with relapsed/refractory AML were treated with a single infusion of ^{225}Ac-lintuzumab at doses of 0.5–2 μCi/kg. No dose-limiting toxicities have been seen, and there has been no evidence of radiation nephritis with follow-up to 10 months. Antileukemic effects have included elimination of peripheral blood blasts in three of six evaluable patients and dose-related reductions of at least 33% of bone marrow blasts in four patients. One patient had 3% bone marrow blasts after therapy (73). Dose escalation of ^{225}Ac, however, is likely to be limited by toxicities due to uncontrolled release of ^{225}Ac daughter isotopes, particularly accumulation of ^{213}Bi in the kidney. Renal irradiation from free, radioactive daughters of ^{225}Ac led to a time-dependent reduction in renal function in mice (74). Similarly, renal toxicity was seen 28 weeks after injection of ^{225}Ac-lintuzumab into cynomolgus monkeys after a cumulative dose of 4.5 µCi/kg (75). The longer serum half-life due to lack of target cell antigen in these animals, however, may increase toxicity compared to human application.

To allow further dose escalation of ^{225}Ac, several strategies were developed to limit the renal uptake of its daughters. Treatment with the dithiol chelators sodium 2,3-dimercapto-1-propane sulfonate (DMPS) or *meso*-2,3-dimercaptosuccinic acid (DMSA) before injection of ^{225}Ac-lintuzumab caused significant reductions in renal ^{213}Bi uptake in mice. Because francium, like potassium, is an alkali metal whose elimination is enhanced by high ceiling diuretics, pretreatment with furosemide and chlorothiazide also significantly reduced renal ^{221}Fr activity (a decay product of ^{225}Ac) and that of its daughter ^{213}Bi (76). The renin-angiotensin-aldosterone system (RAAS) has been implicated in the development of radiation nephropathy, and RAAS antagonism by spironolactone was also shown to prevent kidney damage compared with placebo in mice injected with ^{225}Ac-lintuzumab (77). Short-range Auger electron emitters (e.g., ^{111}In) conjugated to lintuzumab are also being studied for RIT of AML (see Chapter 9).

7.8 SUMMARY

RIT for leukemia is a promising strategy designed to increase the efficacy of native monoclonal antibodies, decrease the toxicity of therapy by targeting radiation to specific cell types or organ systems, and ultimately improve the long-term outcome for patients with leukemia. To date, most studies in leukemia have used the β-emitters ^{131}I, ^{90}Y, and ^{188}Re labeled to anti-CD33, anti-CD45, and anti-CD66 antibodies. These radioimmunoconjugates can eliminate large burdens of leukemia and can be given safely in conjunction with standard preparative regimens prior to HSCT. Whether preparative regimens that incorporate β-particle RIT improve outcomes compared with standard preparative regimens remains to be determined by randomized clinical trials. Alpha-emitters have promise in the treatment of small-volume disease as demonstrated by ^{213}Bi-labeled anti-CD33, which has been shown to have antileukemic activity and can produce complete remissions following cytoreduction with single-agent cytarabine in patients with advanced AML. Ongoing research could potentially result in therapies using more potent radionuclides, novel pretargeted methods of radiation delivery to improve tumor-to-normal tissue dose ratios, treatment of patients with less advanced disease, and randomized trials comparing RIT to more standard approaches alone and in combination.

REFERENCES

1. O'Brien SM, Kantarjian H, Thomas DA, et al. Rituximab dose-escalation trial in chronic lymphocytic leukemia. *J Clin Oncol* 2001;19:2165–2170.

2. Keating MJ, Flinn I, Jain V, Binet JL, et al. Therapeutic role of alemtuzumab (Campath-1H) in patients who have failed fludarabine: results of a large international study. *Blood* 2002;99:3554–3561.

3. Kreitman RJ, Wilson WH, Bergeron K, et al. Efficacy of the anti-CD22 recombinant immunotoxin BL22 in chemotherapy-resistant hairy-cell leukemia. *N Engl J Med* 2001;345:241–247.

4. Sievers EL, Larson RA, Stadtmauer EA, et al. Efficacy and safety of gemtuzumab ozogamicin in patients with CD33-positive acute myeloid leukemia in first relapse. *J Clin Oncol* 2001;19:3244–3254.

5. Todd WM. Acute myeloid leukemia and related conditions. *Hematol Oncol Clin North Am* 2002;16:301–319.

6. Andrews RG, Torok-Storb B, Bernstein ID. Myeloid-associated differentiation antigens on stem cells and their progeny identified by monoclonal antibodies. *Blood* 1983;62:124–132.

7. Griffin JD, Linch D, Sabbath K, et al. A monoclonal antibody reactive with normal and leukemic human myeloid progenitor cells. *Leuk Res* 1984;8:521–534.

8. Zalutsky MR, Bigner DD. Radioimmunotherapy with alpha-particle emitting radio-immunoconjugates. *Acta Oncol* 1996;35:373–379.

9. Macklis RM, Yin JY, Beresford B, Atcher RW, Hines JJ, Humm JL. Cellular kinetics, dosimetry, and radiobiology of α-particle radioimmunotherapy: induction of apoptosis. *Radiat Res* 1992;130:220–226.

10. Nikula TK, Bocchia M, Curcio MJ, Sgouros G, Ma Y, Finn RD, Scheinberg DA. Impact of the high tyrosine fraction in complementarity determining regions: measured and predicted effects of radioiodination on IgG immunoreactivity. *Mol Immunol* 1995;32:865–872.

11. Burke JM, Caron PC, Papadopoulos EB, et al. Cytoreduction with iodine-131-anti-CD33 antibodies before bone marrow transplantation for advanced myeloid leukemias. *Bone Marrow Transplant* 2003;32(6): 549–556.

12. Carrasquillo JA, White JD, Paik CH, et al. Similarities and differences in [111]In- and ^{90}Y-labeled 1B4M-DTPA antiTac monoclonal antibody distribution. *J Nucl Med* 1999;40:268–276.

13. Lovqvist A, Humm JL, Sheik A, et al. PET imaging of [86]Y-labeled anti-Lewis Y monoclonal antibodies in a nude mouse model: comparison between [86]Y and ^{111}In radiolabels. *J Nucl Med* 2001;42:1281–1287.

14. Humm JL. A microdosimetric model of astatine-211 labeled antibodies for radioimmu-notherapy. *Int J Radiat Oncol Biol Phys* 1987;13:1767–1773.

15. McDevitt MR, Sgouros G, Finn RD, et al. Radioimmunotherapy with alpha-emitting nuclides. *Eur J Nucl Med* 1998;25:1341–1351.

16. McDevitt MR, Finn RD, Ma D, Larson SM, Scheinberg DA. Preparation of alpha-emitting [213]Bi-labeled antibody constructs for clinical use. *J Nucl Med* 1999;40:1722–1727.

17. McDevitt MR, Finn RD, Sgouros G, Ma D, Scheinberg DA. An [225]Ac/[213]Bi generator system for therapeutic clinical applications: construction and operation. *Appl Radiat Isot* 1999;50:895–904.

18. Ma D, McDevitt MR, Finn RD, Scheinberg DA. Rapid preparation of short-lived alpha particle emitting radioimmunopharmaceuticals. *Appl Radiat Isot* 2001;55:463–470.

19. McDevitt MR, Ma D, Lai LT, et al. Tumor therapy with targeted atomic nanogenerators. *Science* 2001;294:1537–1540.

20. Jaggi JS, Seshan SV, McDevitt MR, et al. Renal tubulointerstitial changes after internal irradiation with α-particle-emitting actinium daughters. *J Am Soc Nephrol* 2005;16:2677–2689.

21. Miederer M, McDevitt MR, Sgouros G, et al. Pharmacokinetics, dosimetry, and toxicity of the targetable atomic generator, [225]Ac-HuM195, in nonhuman primates. *J Nucl Med* 2004;45:129–137.

22. Scheinberg DA, Strand M. Kinetic and catabolic considerations of monoclonal antibody targeting in erythroleukemic mice. *Cancer Res* 1983;43:265–272.

23. Ali SA, Warren SD, Richter KY, et al. Improving tumor retention of radioiodinated antibody: Aryl carbohydrate adducts. *Cancer Res* 1990;50 (Suppl.): 783s–788s.

24. Zalutsky MR, Vaidyanathan G. Astatine-211-labeled radiotherapeutics: an emerging approach to targeted alpha-particle radiotherapy. *Curr Pharm Des* 2000;6:1433–1455.

25. Zalutsky MR, Narula AS. Astatination of proteins using an N-succinimidyl tri-*n*-butylstannyl benzoate intermediate. *Int J Rad Appl Instrum A* 1988;39:227–232.

26. Seitz U, Neumaier B, Glatting G, Kotzerke J, Bunjes D, Reske SN. Preparation and evaluation of the rhenium-188-labelled anti-NCA antigen monoclonal antibody BW 250/183 for radioimmunotherapy of leukaemia. *Eur J Nucl Med* 1999;26:265–1273.

27. Deshpande SV, DeNardo SJ, Kukis DL, et al. Yttrium-90-labeled monoclonal antibody for therapy: labeling by a new macrocyclic bifunctional chelating agent. *J Nucl Med* 1990;31:473–479.

28. Junghans RP, Dobbs D, Brechbiel MW, et al. Pharmacokinetics and bioactivity of 1,4,7,10-tetra-azacylododecane off',N'',N'''-tetraacetic acid (DOTA)-bismuth-conjugated anti-Tac antibody for alpha-emitter ([212]Bi) therapy. *Cancer Res* 1993;53:5683–5689.

29. McDevitt MR, Ma D, Simon J, Frank RK, Scheinberg DA. Design and synthesis of [225]Ac radioimmunopharmaceuticals. *Appl Radiat Isot* 2002;57:841–847.

30. Roselli M, Schlom J, Gansow OA, et al. Comparative biodistribution studies of DTPA-derivative bifunctional chelates for radiometal labeled monoclonal antibodies. *Int J Radiat Appl Instrum B* 1991;18(4): 389–394.

31. Macklis RM, Kinsey BM, Kassis AI, et al. Radioimmunotherapy with alpha-particle-emitting immunoconjugates. *Science* 1988;240:1024–1026.

32. Ruegg CL, Anderson-Berg WT, Brechbiel MW, Mirzadeh S, Gansow OA, Strand M. Improved *in vivo* stability and tumor targeting of bismuth-labeled antibody. *Cancer Res* 1990;50:4221–4226.

33. Huneke RB, Pippin CG, Squire RA, Brechbiel MW, Gansow OA, Strand M. Effective alpha-particle-mediated radioimmunotherapy of murine leukemia. *Cancer Res* 1992;52:818–5820.

34. Camera L, Kinuya S, Garmestani K, et al. Evaluation of the serum stability and *in vivo* biodistribution of CHX-DTPA and other ligands for yttrium labeling of monoclonal antibodies. *J Nucl Med* 1994;35:882–889.

35. Scheinberg DA, Lovett D, Divgi CR, et al. A phase I trial of monoclonal antibody M195 in acute myelogenous leukemia: specific bone marrow targeting and internalization of radionuclide. *J Clin Oncol* 1991;9:478–490.

36. Jurcic JG, Larson SM, Sgouros G, et al. Targeted alpha particle immunotherapy for myeloid leukemia. *Blood* 2002;100:1233–1239.

37. Sgouros G, Chiu S, Pentlow KS, et al. Three-dimensional dosimetry for radioimmunotherapy treatment planning. *J Nucl Med* 1993;34:1595–1601.

38. Koral KF, Zasadny KR, Kessler ML, et al. CT-SPECT fusion plus conjugate views for determining dosimetry in iodine-131-monoclonal antibody therapy of lymphoma patients. *J Nucl Med* 1994;35:1714–1720.

39. Sgouros G, Graham MC, Divgi CR, Larson SM, Scheinberg DA. Modeling and dosimetry of monoclonal antibody M195 (anti-CD33) in acute myelogenous leukemia. *J Nucl Med* 1993;34:422–430.

40. Hamacher KA, Sgouros G. Theoretical estimation of absorbed dose to organs in radio-immunotherapy using radionuclides with multiple unstable daughters. *Med Phys* 2001;28:1857–1874.

41. Sgouros G, Ballangrud ÅM Jurcic JG, et al. Pharmacokinetics and dosimetry of an alpha-particle emitter labeled antibody: [213]Bi-HuM195 (anti-CD33) in patients with leukemia. *J Nucl Med* 1999;40:1935–1946.

42. Humm JL, Roeske JC, Fisher DR, Chen GT. Microdosimetric concepts in radioimmu-notherapy. *Med Phys* 1993;20:535–541.

43. Axworthy DB, Reno JM, Hylarides et al. Cure of human carcinoma xenografts by a single dose of pretargeted yttrium-90 with negligible toxicity. *Proc Natl Acad Sci USA* 2000;97:1802–1807.

44. Zhang M, Yao Z, Garmestani K, et al. Pretargeting radioimmunotherapy of a murine model of adult T-cell leukemia with the alpha-emitting radionuclide, bismuth 213. *Blood* 2002;100:208–216.

45. Zhang M, Zhang Z, Garmestan K, et al. Pretarget radiotherapy with an anti-CD25 antibody-stretavidin fusion protein was effective in therapy of leukemia/lymphoma xenografts. *Proc Natl Acad Sci USA* 2003;100:1891–1895.

46. Pagel JM, Hedin N, Drouet L, et al. Eradication of disseminated leukemia in a syngeneic murine leukemia model using pretargeted anti-CD45 radioimmunotherapy. *Blood* 2008;111:2261–2268.

47. Dubel S, Breitling F, Kontermann R, et al. Bifunctional and multimeric complexes of streptavidin fused to single chain antibodies (scFv). *J Immunol Methods* 1995;178:201–209.

48. Kipriyanov SM, Little M, Kropshofer H, et al. Affinity enhancement of a recombinant antibody: formation of complexes with multiple valency by a single-chain Fv fragment-core streptavidin fusion. *Protein Eng* 1996;9:203–211.

49. Yukang L, Pagel JM, Axworthy D, Pantelias A, Hedin N, Press OW. A genetically engineered anti-CD45 single-chain antibody-streptavidin fusion protein for pretargeted radioimmunotherapyof hematologic malignancies. *Cancer Res* 2006;66:3884–3892.

50. Schwartz MA, Lovett DR, Redner A, et al. Dose-escalation trial of M195 labeled with iodine 131 for cytoreduction and marrow ablation in relapsed or refractory myeloid leukemias. *J Clin Oncol* 1993;11:294–303.

51. Caron PC, Co MS, Bull MK, et al. Biological and immunological features of humanized M195 (anti-CD33) monoclonal antibodies. *Cancer Res* 1992;52:6761–6767.

52. Jurcic JG, Divgi CR, McDevitt MR, et al. Potential for myeloablation with yttrium-90-HuM195 (anti-CD33) in myeloid leukemia. *Proc Am Soc Clin Oncol* 2000;19:8a (abstract #24).

53. Appelbaum FR, Matthews DC, Eary JF. The use of radiolabeled anti-CD33 antibody to augment marrow irradiation prior to marrow transplantation for acute myelogenous leukemia. *Transplantation* 1992;54:829–833.

54. Ruffner KL, Matthews DC. Current uses of monoclonal antibodies in the treatment of acute leukemia. *Semin Oncol* 2000;27:531–539.

55. Matthews DC, Appelbaum FR, Eary J, et al. Phase I study of ^{131}I-anti-CD45 antibody plus cyclophosphamide and total body irradiation for advanced acute leukemia and myelodysplastic syndrome. *Blood* 1999;94:1237–1247.

56. Pagel JM, Appelbaum FR, Eary JF, et al. ^{131}I-anti-CD45 antibody plus busulfan and cyclophosphamide before allogeneic hematopoietic cell transplantation for treatment of acute myeloid leukemia in first remission. *Blood* 2006;107:2184–2191.

57. Kotzerke J, Glatting G, Seitz U, et al. Radioimmunotherapy for the intensification of conditioning before stem cell transplantation: differences in dosimetry and biokinetics of [188]Re- and ^{99m}Tc-labeled anti-NCA-95 MAbs. *J Nucl Med* 2000;41:531–537.

58. Bunjes D, Buchmann I, Duncker C, et al. Rhenium 188-labeled anti-CD66 monoclonal antibody to intensify the conditioning regimen prior to stem cell transplantation for patients with high-risk acute myeloid leukemia or myelodysplastic syndrome: results of a phase I–II study. *Blood* 2001;98:565–572.

59. Ringhoffer M, Blumstein N, Neumaier B, et al. [88]Re or ^{90}Y-labelled anti-CD66 antibody as part of a dose-reduced conditioning regimen for patients with acute leukaemia or myelodysplastic syndrome over the age of 55: results of a phase I–II study. *Br J Haematol* 2005;130:604–613.

60. Zalutsky MR, McLendon RE, Garg PK, Archer GE, Schuster JM, Bigner DD. Radioimmunotherapy of neoplastic meningitis in rats using an alpha-particle-emitting immunoconjugate. *Cancer Res* 1994;54:4719–4725.

61. Horak E, Hartmann F, Garmestani K, et al. Radioimmunotherapy targeting of HER2/neu oncoprotein on ovarian tumor using lead-212-DOTA-AE1. *J Nucl Med* 1997;38:1944–1950.

62. McDevitt MR, Barendswaard E, Ma D, et al. An alpha-particle emitting antibody ([213]Bi] J591) for radioimmunotherapy of prostate cancer. *Cancer Res* 2000;60:6095–6100.

63. Hartmann F, Horak EM, Garmestani K, et al. Radioimmunotherapy of nude mice bearing a human interleukin 2 receptor alpha-expressing lymphoma utilizing the alpha-emitting radionuclide-conjugated monoclonal antibody [212]Bi-anti-Tac. *Cancer Res* 1994;54:4362–4370.

64. Langmuir VK, Atcher RW, Hines JJ, Brechbiel MW. Iodine-125-NRLU-10 kinetic studies and bismuth-212-NRLU-10 toxicity in LS174T multicell spheroids. *J Nucl Med* 1990;31:1527–1533.

65. Kennel SJ, Stabin M, Roeske JC, et al. Radiotoxicity of bismuth-213 bound to membranes of monolayer and spheroid cultures of tumor cells. *Radiat Res* 1999;151:244–256.

66. Ballangrud ÅM Yang WH, Charlton DE, et al. Response of LNCaP spheroids after treatment with an alpha-particle emitter ([213]Bi)-labeled anti-prostate-specific membrane antigen antibody (J591). *Cancer Res* 2001;61:2008–2014.

67. Behr TM, Behe M, Stabin MG, et al. High-linear energy transfer (LET) alpha versus low-LET beta emitters in radioimmunotherapy of solid tumors: therapeutic efficacy and dose-limiting toxicity of [213]Bi- versus ^{90}Y-labeled CO17-1A Fab' fragments in a human colonic cancer model. *Cancer Res* 1999;59:2635–2643.

68. Andersson H, Palm S, Lindegren S, et al. Comparison of the therapeutic efficacy of [211]At- and ^{131}I-labelled monoclonal antibody MOv18 in nude mice with intraperitoneal growth of human ovarian cancer. *Anticancer Res* 2001;21:409–412.

69. Sandmaier BM, Bethge WA, Wilbur DS, et al. Bismuth 213-labeled anti-CD45 radio-immunoconjugate to condition dogs for nonmyeloablative allogeneic marrow grafts. *Blood* 2002;100:318–326.

70. Bethge WA, Wilbur DS, Storb R, et al. Selective T-cell ablation with bismuth-213 labeled anti-TCRαβ as nonmyeloablative conditioning for allogeneic canine marrow transplantation. *Blood* 2003;101:5068–5075.

71. Nikula TK, McDevitt MR, Finn RD, et al. Alpha-emitting bismuth cyclohexylbenzyl DTPA constructs of recombinant humanized anti-CD33 antibodies: pharmacokinetics, bioactivity, toxicity and chemistry. *J Nucl Med* 1999;40:166–176.

72. Rosenblat TL, McDevitt MR, Mulford DA, et al. Sequential cytarabine and alpha-particle immunotherapy with bismuth-213 (^{213}Bi)-labeled-Hum195 (lintuzumab) for acute myeloid leukemia (AML). *Blood* 2008;112:1025a [Abstract #2983].

73. Rosenblat TL, McDevitt MR, Pandit-Taskbar N, et al. Phase I Trial of the Targeted Alpha-Particle Nano-Generator Actinium-225 (^{225}Ac)-HuM195 (Anti-CD33) in Acute Myeloid Leukemia (AML). *Blood* 2007;110:277a [abstract #910].

74. Jaggi JS, Seshan SV, McDevitt MR, et al. Renal tubulointerstitial changes after internal irradiation with α-particle-emitting actinium daughters. *J Am Soc Nephrol* 2005;16:2677–2689.

75. Miederer M, McDevitt MR, Sgouros G, et al. Pharmacokinetics, dosimetry, and toxicity of the targetable atomic generator, ^{225}Ac-HuM195, in nonhuman primates. *J Nucl Med* 2004;45:129–137.

76. Jaggi JS, Kappel BJ, McDevitt MR, et al. Efforts to control the errant products of a targeted *in vivo* generator. *Cancer Res* 2005;65:4888–4895.

77. Jaggi JS, Seshan SV, McDevitt MR, et al. Mitigation of radiation nephropathy after internal α-particle irradiation of kidneys. *Int J Radiat Oncol Biol Phys* 2006;64:1503–1512.

CHAPTER 8

Pretargeted Radioimmunotherapy of Cancer

ROBERT M. SHARKEY AND DAVID M. GOLDENBERG

8.1 INTRODUCTION

For centuries, physicians have pursued effective and selective methods for treating diseases. Over just the past two centuries, it was discovered that there are substances in the blood that could fight infection, and by the turn of the last century, "magic bullets" were envisioned as a new frontier in medicine, particularly microbiology. However, it was not until the late 1940s that David Pressman, William Bale, Irving Spar, and other contemporaries provided the first evidence that antibodies developed against rodent tumors, and tagged with a radionuclide, could localize specifically to these targets (1–9). Today, we continue to face many of the same challenges that these early investigators encountered for radioconjugate targeting, such as antibody specificity, radiolabeling, and pharmacokinetics.

Specificity is the *Holy Grail* for all targeted compounds. While specificity conjures visions of uniqueness between the target and its environment, more often specificity is derived from quantitative differences between the target and other host tissues. Architectural separation that isolates presentation in the environment from the more accessible presentation in the target can also produce a level of specificity. The lack of suitably specific antibodies for targeting human tumors was the most prominent factor contributing to the waning interest in radioantibody targeting in the 1960s. However, a key realization that a human colonic tumor xenografted in a hamster cheek pouch continued to express a newly defined human oncofetal antigen, CEA (carcinoembryonic antigen), paved the way for resurgence of radioantibody targeting in the 1970s that led to the first successful demonstration of tumor localization in patients by external scintigraphy (10–12). Affinity-purified (monospecific) polyclonal antibodies used in these early clinical studies were quickly replaced with murine monoclonal antibodies, and more recently with less immunogenic humanized or fully human

Monoclonal Antibody and Peptide-Targeted Radiotherapy of Cancer, Edited by Raymond M. Reilly
Copyright © 2010 John Wiley & Sons, Inc.

antibodies and recombinant proteins discussed in Chapter 1. Monoclonal antibodies enhanced selectivity to a unique conformational determinant within a molecule, and recombinant antibodies have greatly amplified the repertoire of targeting structures with altered affinity/avidity and pharmacokinetic properties, but all base their binding on the guiding principles of antibodies and their specificity.

Of course, tumor detection and therapy with radiolabeled antibodies could never have advanced without the contributions of the radiochemists and chemists, who expanded our choices from the staple radionuclide used for many years, ^{131}I, which has poor imaging properties, to a host of new radionuclides with diverse imaging and therapeutic properties (13–22).

While specificity has a major role in defining the targeting utility of a given compound, its pharmacokinetic and biodistribution properties often have a more profound impact on tissue uptake than an agent's specificity. This is because targeting is a passive process, where a molecule injected into the bloodstream wanders in the vascular or extravascular fluid, being carried along by the natural flow of body fluids, until it encounters a cell bearing a target molecule it can bind to. Once bound, the antibody has the potential to be held there until the radionuclide decays, it may be released from the cell and return to extracellular fluid volume, it can be catabolized by local peptidases or internalized by the tumor cells and catabolized by lysosomal enzymes. From this process, the radionuclide is freed from the antibody, where it may be released back into the circulation or retained locally by the cells. Many free radionuclides have an affinity for certain body tissues, such as radiometals for bone or radioiodine for the thyroid, or they can bind to serum proteins and then can be redirected to other tissues. Only a small portion of the injected antibody is deposited within a tumor, while the rest of the antibody eventually extravasates from the blood, like all other proteins, and is catabolized by the body's tissues, mainly the liver, spleen, and other reticuloendothelial-rich organs. The delayed clearance from the tumor creates a differential uptake, where there is a higher concentration in the target than surrounding tissues, enabling visualization or potentially providing a therapeutic window of opportunity.

The body's blood vessels lead to well-defined regions of the body, but there are many tributaries that divert and dilute the radioimmunoconjugate within the total fluid volume of the body. This is where an agent's properties, such as its size and composition, determine its fate. A molecule's size/shape chiefly determines whether it will pass easily through the openings in the vascular channels and percolate into the extravascular space, where most diseases (targets) reside. Size also defines how quickly it will be sequestered from the blood by the reticuloendothelial system, primarily in the liver, spleen, and lymph nodes, where the larger sinusoidal openings between the endothelial cells enable more rapid extravasation, or if small enough (e.g., <60 kDa), to be removed via glomerular filtration (23). The composition of a molecule defines its charge and hydrophilicity, which can result in nonspecific binding to various tissues or even serum components. Composition naturally endows the conformational shape of the molecule that is key to binding to a target of interest, but other regions within the molecule's structure might bind unintentionally to other tissues that could diminish specificity. Glycosylation also impacts tissue uptake and blood clearance (24–27). The high degree of homology in the basic core structure of an IgG reduces the potential for variable tissue binding and clearance that otherwise might be encountered with *de novo* targeting agents, but as molecular engineering has revealed, changes in the core structure of an IgG primarily in the C_H2 and C_H3 regions can accelerate or delay the removal of IgG from the blood (see Chapter 1) (24, 28–32). Of course, agents can be administered in a defined compartment, such as the peritoneal cavity, by hepatic artery perfusion into the liver, or by the intrathecal route, or surgically resected cavity within the brain, to reduce the impact that a molecule's pharmacokinetic properties might have on targeting (33). Compartmental administration for tumors that are anatomically confined allows the antibody to be exposed first to the antigen, but often, a sizable portion of the agent escapes into the blood or lymphatic channels, free to travel throughout the body.

This chapter will discuss largely an innovative method for circumventing many of the pharmacokinetic challenges posed by a directly radiolabeled antibody, focusing on the technique called *pretargeting*.

8.2   THE CHALLENGE OF IMPROVING TUMOR/NONTUMOR RATIOS

IgG, the most commonly used form for antibody targeting, is designed by nature to remain in the plasma for several weeks, which gives it ample opportunity to circulate throughout the body before encountering an antigen. Early animal studies revealed that a radiolabeled antimouse kidney antibody could be selectively bound to the kidney as quickly as 15 min after an intravenous injection, but even in this system, kidney/blood ratios were <1 : 1 over the first 5 days (3). Enrichment of the specific IgG fraction by affinity purification enhances uptake and tumor/nontumor ratios to some degree, but still requires several days before the concentration of radioactivity in the blood is lower than that localized in a tumor (5, 34). In many respects, this protracted time for sufficient contrast to be developed influenced the choice of radionuclide, requiring one with a long enough half-life so that the signal in the tumor would be sufficiently strong by the time the ratios favored visualization. The first clinical studies shortened visualization time to 2–3 days after the injection of an ^{131}I-labeled antibody through technical "background subtraction" by administering ^{99m}Tc-albumin and ^{99m}Tc-pertechnetate prior to each imaging session to approximate nonspecific vascular and extravascular radioactivity concentrations, respectively (12). The ^{99m}Tc-image was subtracted from the ^{131}I-image to show areas of higher ^{131}I-antibody concentration in tumors. While this subtraction technique was successfully applied to detect a number of cancers, the method could be prone to technical artifacts generated by the subtraction method and by the operator's skill level.

The first clinical attempt to directly reduce excess radiolabeled antibody from the blood employed injection of an anti-antibody to form an immune complex that could be eliminated by the reticuloendothelial system. Preclinical and clinical testing showed this method cleared excess ^{131}I-IgG from the blood within 2 h, with deposition in the liver and spleen that took another 4–6 h to clear from the body (35, 36). This method was later re-evaluated and found to be a safe and effective means of reducing blood pool radioactivity, but the technique was restricted to radioiodinated antibodies, because when ^{131}I-IgG was catabolized in the liver, the radioiodine was eliminated from the body (37–41). However, with a radiometal-labeled antibody, the radiometal remains in the liver, resulting in highly unfavorable tumor/liver ratios. Although extracorporeal removal of radiolabeled antibodies provided an alternative way of promoting elimination of radioactivity without concern of forming immune complexes in the blood that could be deposited elsewhere in the body (42–44), other less intrusive methods have been sought. The simple answer was to use a fragment of an antibody. Portions of an IgG can be enzymatically removed to form F(ab$'$)$_2$ fragments that retain divalent binding to the antigen (see Chapter 1). Monovalent binding Fab$'$ fragments are formed by breaking the disulfide bonds holding the two heavy chains together. Today, molecular engineering is used to create molecules based largely on an antibody's heavy and light chain variable regions (Fv) that contain the framework to hold the complementarity-determining regions (CDRs) in the proper orientation for binding specificity. When these heavy and light chains of the Fv region are tethered together with different length amino-acid linkers, they can form monovalent single-chain Fv (scFv), divalent diabodies, trivalent tribodies, and so on. The scFv can be joined with other portions of an IgG to create a diverse repertoire of molecules with varying valency and pharmacokinetic and effector properties (45–48) (see Chapter 1).

Figure 8.1 schematically shows the structure of IgG along with its chemically and several recombinantly produced fragments, including a general overview of the *in vivo* targeting properties of some of these agents. IgG has the highest tumor uptake and retention of all antibody forms, but it takes the longest time to achieve maximum uptake, and it has the slowest elimination from the blood and tissues (49, 50). The slow elimination from the blood reduces visualization and increases radiation exposure to the highly sensitive bone marrow, resulting in dose-limiting myelosuppression for radioimmunotherapy (RIT). With few exceptions, faster blood and tissue clearance closely follows the molecular size of the antibody form, and as the molecule's time in the blood is reduced, a smaller fraction is available to localize in the target, and thus tumor uptake decreases. Furthermore, monovalent fragments have a shorter residence time in the tumor because of lower avidity. Complicating these relationships is how different types of radionuclides are handled by the tissues responsible for the antibody's removal from the body. Radioiodinated agents prepared by most standard methods are not retained by cells in the tissues (except the thyroid) or the tumor, and therefore when catabolized, the radioactivity is eliminated from the body. While there are forms of radioiodine suitable for imaging and therapy, certain aspects of their physical properties (half-life or decay emissions) are not ideally suited for these tasks. There are many other radiometals that have more ideal properties for imaging and/or therapy. When a radiometal-labeled antibody is taken up by a normal tissue, the chelated metal becomes entrapped within the cells, and thus only a small fraction of a radiometal-labeled antibody is removed from the body, with molecules >60 kDa being deposited in the liver, and smaller molecules in the kidneys. This normal tissue uptake is often much higher than that targeted at the tumor, yielding unfavorable conditions for imaging (e.g., tumors in the liver can be masked) or

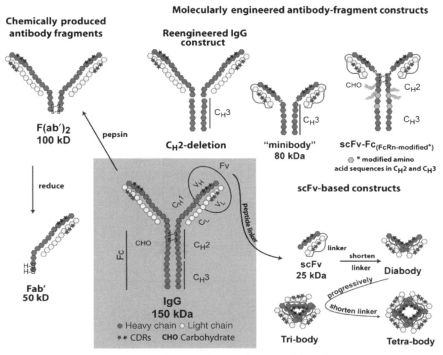

Chemically produced antibody fragments

Molecularly engineered antibody-fragment constructs

Reengineered IgG construct

F(ab')₂ 100 kD

pepsin

reduce

Fab' 50 kD

C_H2-deletion

"minibody" 80 kDa

scFv-Fc(FcRn-modified*)
○ * modified amino acid sequences in C_H2 and C_H3

scFv-based constructs

scFv 25 kDa

linker

shorten linker

Diabody

progressively

shorten linker

Tri-body

Tetra-body

Fv

peptide linker

Fc

CHO

C_H2

C_H3

IgG 150 kDa
● Heavy chain ○ Light chain
✳✳ CDRs   CHO Carbohydrate

In vivo targeting of antibodies and some of their fragments

	IgG	F(ab')₂	Fab'	scFv
Tumor uptake	Highest		→	Lowest
Tumor retention	Longest		→	Shortest
Time to maximum tumor uptake	Longest		→	Shortest
Tumor penetration	Least		→	Most
Blood clearance	Fastest		→	Slowest
Tissue uptake	Liver		→	Kidneys

FIGURE 8.1 Schematic representation of IgG, enzymatically prepared F(ab')₂ and Fab' fragments, and several engineered antibody fragments, with the inset table showing some of the more important *in vivo* targeting characteristics of these antibody forms. The C_H2-deletion was one of the earliest forms prepared, followed by the single-chain Fv-based constructs. The scFv is prepared by inserting the V_H and V_L portions of the IgG in a linear peptide structure with a 15–18 amino acid linker placed between them. Upon assembly, the protein folds to form a monovalent antigen-binding structure. By progressively shortening the linker, the molecules tend to form noncovalently linked constructs with 2, 3, or 4 binding sites. Single chains have also been fused to the C_H3 domain to form a minibody and placed onto the full IgG Fc portion. One of these latter constructs was prepared by modifying key amino acids involved with the recognition of FcRn receptor, which significantly enhances the clearance of this larger molecular weight protein compared to constructs of similar size. (See insert for the color representation of the figure.)

therapy (e.g., radiation dose to kidneys would damage this organ before achieving tumoricidal levels). Coupling the radionuclide to the antibody is nonetheless certainly the most efficient targeting mechanism. A number of direct radioimmunoconjugates have received the U.S. FDA approval (e.g., arcitumomab, capromab pendetide, ibritumomab tiuxetan, and tositumomab), yet efforts continue to focus on designing molecules that preserve the best properties of an intact IgG (high uptake and retention), but with the fast clearance kinetics of a small antibody fragment.

8.3   PRETARGETING: UNCOUPLING THE ANTIBODY–RADIONUCLIDE CONJUGATE

Perhaps, surprisingly, the biggest problem with most radionuclide–antibody conjugates is that they are highly stable in the blood, and thus, generally wherever the antibody is distributed, the radionuclide follows. A radionuclide that dissociates from the antibody would have very unfavorable properties. For example, cyclic DTPA (diethylene triamine pentaacetic acid) anhydride was one of the first chelate derivatives used to complex a radiometal (^{111}In/^{90}Y) to an antibody (13, 14), but the radiometals, particularly ^{90}Y, would slowly dissociate from the chelate, allowing the unbound metal to be taken up by other tissues, such as the cortical bone. When other chelates that bound the radiometal more tightly were developed (e.g., DOTA), they quickly replaced cyclic DTPA anhydride (see Chapter 2) (15, 16, 51–56). Other chemistries have been developed that allow selective cleavage of the chelate–radiometal complex from the antibody when it is catabolized in the liver (57, 58). This cleavable linkage improved tumor/liver ratios, but it did not address the high concentration of radioantibody remaining in the blood, and thus this method was unable to increase the radiation-absorbed dose delivered to the tumor. Molecular engineering (see Chapter 1) has made great strides in crafting constructs that try to strike a balance between optimal blood and tissue clearance with reasonable tumor uptake and retention (30, 59–64), but another approach had already achieved rapid blood clearance with high tumor uptake/good retention in the mid-1980s. This method is pretargeting, a technique that has evolved over the past 20 years to include three different approaches based on bispecific monoclonal antibodies (bsMAb) with radiolabeled haptens, avidin or streptavidin used for targeting radiolabeled biotin (two methods have been evaluated) (Fig. 8.2), and antibody–oligonucleotide conjugates for targeting radiolabeled complementary oligonucleotides (Fig. 8.3).

8.3.1   Bispecific Antibodies and Radiolabeled Haptens

If the primary difficulty with direct conjugates is their stability, might it be possible to separate the two components, yet still achieve selective targeting *in vivo*? There are a number of multistep methods for enhancing antigen detection commonly employed *in vitro*, such as radio- and enzyme-linked immunoassays, and immunohistochemistry, but could this process succeed *in vivo*? This is precisely the concept offered by Reardan et al., who reported on the binding properties of several monoclonal

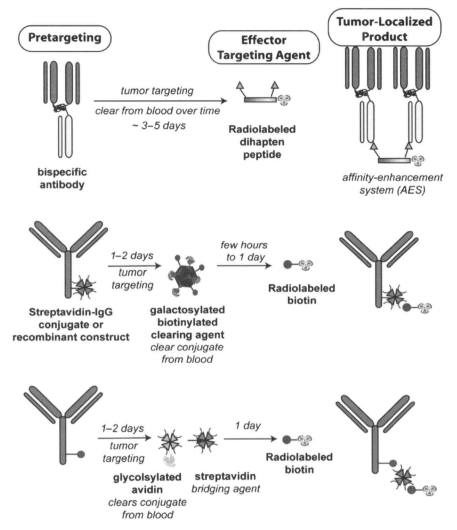

FIGURE 8.2   The basic steps used in three pretargeting approaches for localizing a radiolabeled effector. In the bsMAb approach, bsMAb is allowed several days to target the tumor and to clear from the blood. Once the level of bsMAb is low enough in the blood, a radiolabeled hapten-peptide is administered. Two hapten moieties on the peptide backbone allow the peptide to bind divalently to two closely associated bsMAb on the surface of the tumor, enhancing the binding affinity of the cross-linked pair. Avidin–biotin approaches are performed in two different ways. In the so-called "two-step" approach, an IgG-streptavidin conjugate is administered, and after 1–2 days to optimize tumor uptake, a clearing agent is administered to remove the immunoconjugate from the blood so that the radiolabeled biotin can be injected. The clearing agent has biotin to bind the streptavidin immunoconjugate and galactose residues that shuttle the complex rapidly to the liver. In the "three-step" approach, a biotin–IgG conjugate is given time to localize in the tumor and then is cleared from the blood using avidin. Shortly thereafter, streptavidin is administered, which will bind to the IgG–biotin in the tumor. After allowing time for the streptavidin to clear from the blood and normal tissues, radiolabeled biotin is finally administered. (See insert for the color representation of the figure.)

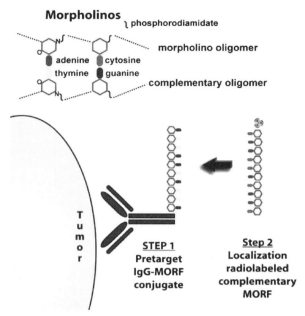

FIGURE 8.3 Pretargeting using an IgG–MORF–oligomer conjugate and a radiolabeled complementary oligomer. (See insert for the color representation of the figure.)

antibodies (MAbs) developed to metal-loaded EDTA (ethylenediaminetetraacetic acid) chelate derivatives coupled to a carrier protein, thereby making the metal chelate complex a *hapten* (defined as a substance that can bind to an antibody but not elicit the formation of an antibody unless attached to a carrier protein). They speculated that a new targeting method could be developed that employed a bsMAb made with one arm for binding a tumor antigen and the second arm that could be directed against a chelate loaded with a radiometal (65). Brennan et al. had published in the same year a chemical method for preparing bsMAb that could bind two different compounds (66). Shortly thereafter, bsMAbs were being prepared by hybridization methods (quadromas), and later by molecular engineering (67, 68).

The antichelate antibodies prepared by Reardan et al. showed remarkable preferential binding for what seemed to be rather subtle differences in the target, namely, simply by inserting different metals into the chelate (e.g., indium-EDTA versus gallium-, iron-, cobalt-, zinc-EDTA). Animal studies showed that premixing the anti-indium-EDTA antibody with ^{111}In-complexed EDTA altered the biodistribution of the ^{111}In-EDTA, and later studies showed that (^{111}In)EDTA could be dissociated from the antichelate antibody by adding nonradioactive (In)EDTA. Injecting (In)EDTA to animals given premixed ^{111}In-EDTA–anti-EDTA immune complexes could also dissociate the ^{111}In-EDTA from the complex, which resulted in a more rapid clearance of the ^{111}In-EDTA (69). These studies set the stage for the first pretargeting studies using bsMAb, where tumor-bearing animals were first given an antichelate antibody (70). Although the antichelate antibody was not specific for the implanted tumor, selective uptake occurred because tumors have "leaky" blood vessels that allowed higher amounts of IgG to localize in tumors than in normal tissues (71). One day after the antichelate antibody was injected, a transferrin–chelate conjugate was given to block and remove the antichelate antibody remaining in the blood before ^{111}In-chelate was given 1 h later. The antichelate antibody in the blood could have been blocked by simply administering unlabeled chelate, but these small complexes would remain in the circulation where they could continue to be accessible. By coupling the 4 mol of chelate per transferrin, small lattices of the antichelate antibody × transferrin-chelate could form, which would be filtered out by the liver. With most of the antichelate antibody removed from the blood, when ^{111}In-chelate was given 1 h later, it was able to circulate and bind to the antichelate antibody present primarily in the tumor, with the remaining free ^{111}In-chelate being eliminated very quickly from the blood and body by renal excretion. Administering various amounts of antibody or clearing agent (transferrin–chelate conjugate) resulted in expected differences in the biodistribution pattern of ^{111}In-chelate, depending on how well the antichelate antibody was blocked. However, under what was described as more optimal chase conditions, tumor/blood ratios were enhanced compared to when no chase was administered. There was also minimal uptake of radioactivity in the liver, suggesting that the antichelate antibody that was cleared from the blood to the liver was no longer accessible for binding to ^{111}In-chelate, either having been processed by the liver or blocked by the transferrin–chelate complex. The near absence of uptake in the liver was quite an accomplishment, since ^{111}In-labeled antibodies at that time had a very high hepatic uptake, and thus this procedure provided a mechanism for targeting radiometals at tumors without the high tissue uptake seen with directly conjugated radiometal-labeled antibodies.

A pretargeting approach based on another ^{111}In-EDTA derivative and an anti-CEA bsMAb (anti-CEA Fab′ coupled to an antichelate antibody) was clinically tested in patients with hepatic metastases from colorectal cancer (72). Patients were given 20–40 mg of the bsMAb and 4 days later were administered the ^{111}In-labeled chelate that was premixed with a small amount of bsMAb. Interestingly, this was done to increase the plasma half-life of ^{111}In-chelate, with the idea that the preformed ^{111}In-chelate–bsMAb complex would dissociate as it passed through the tumor, allowing ^{111}In-chelate to transfer to the bsMAb that was prelocalized to the tumor. With a dissociation half-life of 8.8 min (72), the antichelate–^{111}In-chelate complex in the blood and tissues would rapidly release ^{111}In-chelate, which would then be very quickly cleared by urinary excretion, reducing the blood pool radioactivity. In this clinical study, the bsMAb was given 4 days to clear from the blood before the radiolabeled chelate–bsMAb complex was given. Nearly all the cancer lesions in 14 patients were detected and, more impressively, this was the first ^{111}In-labeled antibody-based targeting system in which liver uptake was minimized to a level where hepatic metastases were seen as positive lesions, rather than negative lesions that do not take up radioactivity. Previous studies with an ^{111}In-anti-CEA IgG often failed to detect hepatic metastases or they were seen as negative defects on scans of the liver (73–78).

8.3.2 Pretargeting: Development of Avidin/Streptavidin and Radiolabeled Biotin

The next most significant advance in pretargeting occurred when Hnatowich et al. introduced a new method to bridge the antibody and radionuclide targeting steps based on the ultrahigh affinity of biotin for avidin/streptavidin (79). Such avidin/biotin binding methods were already in use for *in vitro* immunoassays and immunohistochemistry. These and other investigators explored several configurations, using an antibody conjugated with avidin paired with radiolabeled biotin, or biotin-conjugated antibodies used to capture radiolabeled avidin (79–85). This avidin–biotin affinity was nearly 6 logs higher than most antibody–antigen interactions ($K_d = 10^{-15}$ M), which essentially ensured an irreversible bond between biotin and avidin. Glycosylated avidin and its nonglycosylated counterpart, streptavidin, have four binding sites for biotin, which offers the potential that multiple radiolabeled biotins could be captured by a single pretargeted antibody–avidin conjugate. Importantly, the radiolabeled biotin had very rapid clearance from the body (biological half-life = 30 min), indicating that it had rapid extravasation and minimal binding to tissues as well as efficient renal excretion. The initial animal studies showed great potential of this method, and 2 years later, the first clinical studies were performed using the human milk fat globule-1 (HMFG1) IgG (anti-MUC1) coupled to streptavidin followed 2–3 days later with ^{111}In-biotin (86). In addition to establishing important safety data, this study also showed that endogenous biotin, which is present in the serum and tissues, did not block the streptavidin–IgG conjugate's ability to bind the subsequently administered ^{111}In-biotin. This was not the case in mice, where the concentration of endogenous biotin is much higher than in humans and requires animals to be fed a biotin-deficient diet for several days before initiating pretargeting studies (87, 88). Tumor targeting was observed in 8/10 patients but, disappointingly, similar uptake was seen in at least 5 of these patients given ^{111}In-biotin alone. This most likely represented blood pool radioactivity with slower washout from tumors due to their abnormal physiology. They also found that the streptavidin conjugate was immunogenic, with antibodies formed to both the murine IgG and the streptavidin. However, the more important finding from this study was the very low normal tissue and blood pool radioactivity (i.e., ^{111}In-biotin cleared with an alpha- and beta-phase half-life of 2.4 min and 4.2 h, respectively), suggesting that this method, when properly optimized, could greatly reduce blood and tissue background radioactivity.

This particular approach was perfected by the contributions of investigators at NeoRx Corp. (Seattle, WA). They focused their efforts on the development of a streptavidin conjugate prepared with the murine monoclonal antibody, NR-LU-10, as the pretargeting agent with ^{111}In/^{90}Y-DOTA (1,4,7,10-tetraazacyclododecane-*N,N′, N″,N‴*-tetraacetic acid)-biotin. They also introduced the use of a clearing agent, galactose-conjugated and biotinylated human serum albumin, to remove excess streptavidin–IgG from the blood before the radiolabeled biotin was injected. Hepatocytes have galactose receptors that effectively remove galactosylated antibody immune complexes from the blood (89, 90), and biotin not only serves as a specific binding ligand for the streptavidin conjugate but also, in sufficient excess, it would prevent the binding of the subsequently administered radiolabeled biotin, allowing it to flow relatively unabated through the body until it reaches the tumor. In their procedure, the streptavidin–IgG conjugate was allowed 2–3 days to achieve maximum tumor uptake, and then the clearing agent was given. One day later, the radiolabeled biotin was administered. Using this technique, Axworthy et al. (91) were the first to present provocative animal data indicating not only that a pretargeting approach could improve tumor/blood and tumor/tissue ratios, but also that the tumor uptake could be similar to that of a directly radiolabeled IgG. In some respects, this finding was puzzling since the pharmacokinetic behavior of directly labeled IgG and antibody fragments predicted that the more rapidly an agent cleared from the blood, the lower would be its tumor accumulation. The radiolabeled biotin cleared more rapidly from the blood than any other previously reported directly conjugated antibody fragment, but despite this, it was able to achieve IgG-like tumor uptake. So how did this pretargeting procedure accomplish this targeting panacea? A comparison of a typical directly radiolabeled antibody targeting procedure with pretargeting procedures reveals some very important differences (see Section 8.3.4).

8.3.3 Pretargeting with Oligonucleotide/Complementary Oligonucleotide Immunoconjugates

Another pretargeting approach being developed has used the complementary interaction of nucleic acid strands as a means of bridging the pretargeting agent with the effector. Initial studies employed antibodies conjugated to short (14–15-mer) DNA oligomers for pretargeting using a radiolabeled complementary oligomer effector molecule (92, 93). The main problem with these compounds is their nonspecific binding to serum proteins (especially for phosphorothioate oligonucleotides) and tissues and instability to phosphodiesterases (native phosphodiester oligomers), but a synthetic peptide nucleic acid (PNA) was found to have the stability and rapid clearance properties required for pretargeting (94, 95). Subsequently, investigators turned to more water-soluble morpholino oligomers (MORF) for pretargeting (96) (Fig. 8.3). A ^{99m}Tc-complementary MORF (cMORF) quickly cleared from the blood and most tissues, with only approximately 0.2%ID/g in the blood 1 h after injection (96). Renal uptake averaged approximately 6–7%ID/g at this time. Animals pretargeted 48 h earlier with an anti-CEA IgG–MORF conjugate had approximately 2% ID/g uptake of the ^{99m}Tc-cMORF 3 and 24 h after injection. However, tumor/blood ratios were no better than 3 : 1 by 24 h, most likely because a proportion of the ^{99m}Tc-cMORF effector bound to the IgG–MORF conjugate in the blood. Low uptake was also present in the liver, but renal retention reduced tumor/kidney ratios to <1 : 1. Later studies showed that a cMORF lacking cytosine had twofold lower renal uptake without affecting tumor uptake (97). Radiation-absorbed dose estimates in nude mice bearing a human colon cancer xenograft pretargeted using the anti-CEA IgG–MORF followed by administration of 0.38 mCi of ^{188}Re-cMORF revealed tumor/blood and tumor/kidney dose ratios of approximately 4 : 1. Tumor weights of treated animals necropsied 19 days after treatment were significantly lower than tumors removed from untreated animals. Thus, while this method is not as advanced in its development as the bsMAb and avidin–biotin approaches, it highlights another technique that could allow pretargeted delivery of radionuclides for radiotherapeutic applications.

8.3.4 Core Principles Associated with Pretargeting Procedures

With a directly radiolabeled antibody, the main goal should be to administer the smallest possible radioantibody protein dose prepared at the highest possible specific activity, which increases the probability that each molecule reaching the target would deliver a radioactive payload. Unfortunately, pharmacokinetic, biodistribution, and even specificity issues often dictate that additional unlabeled antibody be added (coadministered or preadministered) with the radiolabeled antibody to improve the antibody's biodistribution or targeting. Since a number of cells have Fc binding receptors, these cells can remove a portion of the IgG from the blood before it has an opportunity to reach the tumor, and the smaller the protein dose administered, the more substantial this portion will be. For example, in the case of the radiolabeled anti-CD20 antibodies used for lymphoma therapy, a predose of approximately 70–90 mg of the anti-CD20 MAbs can effectively reduce uptake in normal B-cells residing primarily in the spleen and bone marrow, but investigators ultimately elected to administer ≥400 mg of unlabeled anti-CD20, injected in advance of the radioimmunoconjugate, primarily because the unconjugated antibody was biologically active and also contributed to the antitumor effect (98–104). Any competition for radioantibody with an unlabeled antibody runs the risk of reducing the total amount of radioactivity delivered to the tumor, with a preinfusion having a greater risk of blocking some of the more accessible antigen-rich regions within the tumor before the radioantibody is given. Since uptake and tumor/blood ratios for an IgG are already low, it is not surprising that RIT with directly radiolabeled antibodies has had limited therapeutic impact, except for lymphomas, which have an inherently high radiosensitivity. As mentioned earlier, smaller antibody fragments bearing radiometals have exceptionally high renal uptake and retention that greatly exceeds that delivered to the tumor. There have been animal studies reporting improved therapeutic activity for antibody fragments versus IgG, and there even have been methods that reduce renal uptake of radiometals (105–112). While each of these has been clinically evaluated, neither provided a sufficient therapeutic boost to spur further clinical evaluation.

Pretargeting procedures often start with the injection of relatively high doses of the immunoconjugate (e.g., IgG-streptavidin or biotin conjugate) or bsMAb. This maximizes the conjugate loading in the tumor, which increases the number of binding sites available for the radiolabeled biotin or hapten (i.e., effector). Because the pretargeting immunoconjugate/bsMAb is not radiolabeled, there are no inherent radiotoxicity limitations to the amount administered. However, optimal pretargeting conditions do not demand administration of excessively high protein doses at levels that might saturate antigen binding sites in the tumor. Instead, optimal conditions will occur as long as the amount of conjugate or bsMAb prelocalized to the tumor is sufficient to capture the highest fraction of the effector that will ultimately reach the tumor. If an effector is radiolabeled at 2 mCi (74 MBq)/nmole, then often only 100 nmoles would be given. With such rapid elimination, only a small fraction of the effector will every pass through the tumor. Therefore, the administered amount of pretargeting agent needs only to be sufficient to optimize the capture of this small quantity of effector. In our experience, the optimal bsMAb dose can be reasonably approximated from the amount of hapten administered (which in our case is conjugated to a peptide) and is expressed as the mole ratio of the bsMAb to peptide-hapten.

Pagel et al. (113) provided additional insight into this point by showing in a human lymphoma xenograft model that the uptake of radiolabeled biotin (effector) could not be improved by pretargeting with immunoconjugates directed against multiple antigens, as compared to the best single antigen (e.g., CD20, CD22, or HLA-DR). By combining one immunoconjugate that had the highest uptake in the tumor with others that had lower uptake, they effectively reduced the number of moles of individual immunoconjugates in the tumor, thereby reducing the effector-capturing efficiency. To better understand this principle, envision the individual targeting of 90 mol of three antibodies to a tumor, where Ab1 delivered a maximum of 30 mol (30% uptake), Ab2 delivered only 20 mol (22.2% uptake), and Ab3 delivered 10 mol (11.1% uptake). Assuming the proportion of each antibody captured by the tumor would not be affected if its dose were changed, if we mixed these three antibodies in equal portions (30 mol each) and administered the same 90 mol, the amount of Ab1 delivered to the tumor would be only 10 mol (i.e., 30% of 30 nmol), of Ab2 there would be only approximately 7 mol (22% of 30 mol), and approximately 3 mol of Ab3 (11.1% of 30 mol) for a total of approximately 20 mol of Ab1A + b2A + b3. Thus, when given at the same dose, an antibody mixture would never be able to exceed the uptake achieved by the one antibody that has the highest uptake. Hence, under these conditions, mixtures of immunoconjugates or bsMAb would not load the tumor with a higher capacity for binding the radiolabeled effector. Had each antibody been given at its "saturating" dose, there would of course be more moles of antibody in the tumor; but, "overloading" a tumor with antibody will not necessarily lead to a higher uptake of the radiolabeled effector. Antibody mixtures could have a different benefit if they were directed against antigens on different cells within the tumor, providing a more uniform distribution to more cells within the population. However, it is important to remember that there will always be a finite capturing efficiency that cannot be significantly improved by simply loading more capturing agent in the tumor, unless there is a way to improve the delivery and percolation of the radiolabeled effector through the tumor.

In order to give the radiolabeled effector the best chance of binding to the prelocalized immunoconjugate in the tumor, the residual immunoconjugates in the body need to be blocked/cleared. The blood is the primary concern since any radiolabeled effector introduced by intravenous injection will encounter circulating immunoconjugates there first. However, other tissues also need to be considered, such as the liver that is the primary organ for removing IgG from the blood. Since most bsMAb studied to date have been F(ab')$_2$ fragments, investigators simply waited for the concentration of the bsMAb in the blood to be reduced to a level where interaction with the radiolabeled hapten-peptide was minimized. The molar ratios of immunoconjugate to hapten and degree of blood clearance required to minimize hapten–bsMAb interaction will vary depending on the binding affinity of the antihapten portion of bsMAb. Avidin–biotin approaches have relied on clearing agents that block and clear the immunoconjugates from the blood following a 1–3 day period during which the immunoconjugate has an opportunity to reach peak levels in the tumor. This is particularly important for these procedures, because the ultrahigh affinity of biotin for avidin will ensure stable binding, even if very small amounts of the immunoconjugates remain in the blood and tissues.

The timing requirements for a pretargeting procedure impose a restriction on the types of antigens that can be targeted; they must remain accessible (i.e., not internalized or catabolized) in sufficient quantity to bind the radiolabeled effector. For example, anti-CD20 and anti-CD45 antibodies are better pretargeting agents for lymphoma than anti-CD22, in part because CD22 will internalize when bound by an antibody, but CD20 and CD45 are also more plentiful on the cell surface, which will yield higher uptake than for a CD22-targeted agent (113, 114). A pretargeting agent that is internalized might require a clearing agent to shorten the interval in an effort to localize the effector to the pretargeting agent while it is still largely accessible. Once bound, the effector might then be internalized. Thus, an antibody that is internalized should not be dismissed, but it might not produce the best effector localization if other targets are available. Internalization of the pretargeted agent in normal tissues via nonspecific but receptor-targeted processes (e.g., binding to Fc receptors in the liver) is, however, desired since this would ensure it was not accessible when the radiolabeled effector was given, thus reducing unwanted radioactivity localization.

A key component in a pretargeting system is the specific activity of the radiolabeled effector. The effectors used for most pretargeting procedures are much less susceptible to damage to their epitope recognition properties during the radiolabeling procedure than an antibody, and therefore they can be prepared under conditions that favor higher incorporation of the radionuclide into the chelate bound to the hapten than when it is directly bound to an antibody. For example, antibodies are often radiolabeled at 5 mCi ^{90}Y/mg IgG. This represents a specific activity of 0.3 mCi/nmole. Very often, additional unlabeled antibody is given with radioimmunoconjugates, further reducing the effective specific activity. We typically prepare a ^{90}Y-hapten-peptide at specific activities >2 mCi/nmol or at least 6.5 times higher than a similarly radiolabeled IgG. Radiolabeling conditions are optimized to yield ≥97% binding to hapten-chelate, eliminating the need for postlabeling purification. Under these conditions, 1 in every 5 to 15 hapten-peptides harbors a radioactive element (depending on the radionuclide), with even higher yields limited only by residual competing trace metals in the radionuclide itself. For example, we have been able to achieve a specific activity of 48 mCi/nmol for a ^{68}Ga-hapten-peptide (unpublished data). At high specific activities, the amount of the pretargeting agent administered can be reduced. One alternatively could consider the opportunity of loading additional effector into the tumor by keeping the dose of the pretargeting agent high and giving multiple injections of a high-specific activity effector. We have shown that this is possible, but it is difficult to take advantage of this situation with radiolabeled effectors, since the total amount of radioactivity that can be administered is limited (115). However, one could envision the sequential use of a radiotracer and another nonisotopic labeled effector for dual-targeting purposes (e.g., combined SPECT/PET and optical imaging).

Another advantage of the pretargeting approach compared to the use of directly radiolabeled antibodies is how rapid the process of tumor targeting takes place, which is related to the small size of the radiolabeled effector molecule that has exceptionally fast extravasation, so that it can reach the pretargeted immunoconjugate/bsMAb in few minutes. Dynamic imaging of mice bearing a human colon cancer xenografts pretargeted with bsMAb followed by a ^{99m}Tc-labeled hapten-peptide showed ^{99m}Tc-uptake within 10 min, with imageable tumor/heart ratios in 20 min and even high tumor/kidney ratios within 40 min (116). Maximum uptake was achieved within 60 min, compared to 1–2 days for a radiolabeled IgG. Even the smallest antibody fragments can take several hours before they achieve maximum tumor uptake, and they clear much slower from the blood and normal tissues than the radiolabeled effector in a pretargeting approach. For example, the bladder begins to fill with radioactivity in just 5 min following injection of the radiolabeled hapten, with approximately 60% of the injected dose present in the bladder within 1 h. Radio-labeled biotin has minimal uptake and retention in the kidneys, and the peptide backbone of a peptide-conjugated hapten can be modified to favor renal elimination (as compared to hepatobiliary/gastrointestinal elimination) but still has minimal kidney retention (117). Some earlier studies attempted to localize radiolabeled avidin or streptavidin to pretargeted biotinylated antibodies (118), but streptavidin's physiochemical properties caused it to be trapped in the kidneys and avidin's glycosylation results in high hepatic uptake (119, 120). Thus, the ability of the radiolabeled hapten to be cleared quickly from the blood/body with minimal tissue retention is a key element for an effective pretargeting system.

However, the real strength of a pretargeting system is its ability to very efficiently trap the radiolabeled effector in the tumor. As with any injected agent, only a small fraction will percolate the vasculature of the tumor, so it is important to have the pretargeted immunoconjugates accumulate and be retained in the tumor in sufficient amounts to maximize the capture of the effector molecules. At least for the radiolabeled effectors commonly used in pretargeting, those leaving the vascular space will quickly encounter the pretargeted immunoconjugates bound to the surface of tumor cells. Binding to the effector-capturing agent, whether it is streptavidin or a bsMAb, will retain the effector in the tumor, while within minutes, the unbound effector molecules are rapidly eliminated from the body, creating almost immediate and exceptionally high tumor/blood ratios. The radiolabeled effector will persist in the tumor as long as it remains bound to the bsMAb's antieffector arm or to streptavidin, and as long as the immunoconjugate remains fixed to the tumor antigen. Unlike the initial bsMAb pretargeting configuration, where the effector was a single chelate–radiometal complex (i.e., a hapten), LeDoussal et al. found that tethering two haptens together with a small peptide linker enhanced uptake and retention of the effector—this was referred to as an *affinity enhancement system* (AES) (121). The principle was based on the fact that a divalent hapten would be retained longer by higher concentrations of a bsMAb bound to tumor cells than to lower concentrations of bsMAb in the circulation. With low bsMAb concentrations in the blood, less stable monovalent binding of the divalent hapten would be favored, and as the radiolabeled hapten-peptide is dissociated from bsMAb in the blood, it would allow the dissociated hapten to quickly clear and provide enhanced tumor/blood ratios. In the presence of locally higher concentrations of bsMAb in the tumor, more stable divalent binding would be favored. This initial finding was confirmed by this group in another model (122), and by two other groups using different antibodies and haptens (123, 124). Thus, today all bsMAb pretargeting procedures make use of a radiolabeled divalent hapten-peptide to enhance tumor localization. While the binding affinity of this enhancement cannot compete with the 10^{-15} M strength of a streptavidin–biotin bond, one has to keep in mind that the retention in the tumor for both of these cross-linking methods is limited by the weakest link (e.g., affinity/avidity of the pretargeted immunoconjugate for the target antigen on tumor cells).

Contrast this to an antibody or even a fragment, whose larger size impedes its transvascular movement into tumors, and rather than being easily eliminated, it continues to circulate in the body, awaiting catabolism for excretion of the radioactivity. While this might provide the potential for multiple tumor passes, the concentration of the antibody in the blood eventually decreases over time, being sequestered by other tissues in the body, thereby making it less available for tumor uptake. In addition, the longer the tumor uptake is delayed, the more of the radionuclide will be released by catabolism and be redistributed to other parts of the body, contributing to toxicity. Thus, pretargeting procedures are more highly adept at achieving exceptionally rapid tumor uptake and blood clearance than direct targeting of radioimmunoconjugates. The end result is that tumor radionuclide delivery is maximized, with possible higher radiation dose rates and often increased radiation-absorbed doses, while most normal tissues are spared prolonged radiation exposure (125, 126) (Fig. 8.4). As a consequence, significantly improved antitumor responses have been observed in a variety of animal models with pretargeting procedures compared to direct RIT (125, 127–140).

One of the more striking pretargeting results ever reported was achieved in a renal cell carcinoma model, where tumor uptake measured as high as $87.9 \pm 36\%$ injected dose per gram (%ID/g) 1 day after the injection of a divalent-^{111}In-DTPA(hapten)-peptide pretargeted with an anti-G250 (anticarbonic anhydrase IX) × anti-(In)DTPA bsMAb IgG, with tumor/blood ratios approximately 150 : 1 (123). Uptake of a monovalent ^{111}In-DTPA-peptide peaked 1 h after its injection at only $7.6 \pm 1.5\%$ ID/g, and continued to decrease over time to 1.3% at 24 h, with tumor/blood ratios of only 4 : 1. These results clearly illustrated the affinity enhancement advantage of a divalent hapten over a monovalent one. While tumor uptake of the pretargeted radiolabeled divalent hapten-peptide was higher than in other model systems, anti-G250 and other directly radiolabeled IgGs targeting renal cell carcinomas have also been reported to have much higher uptake than most other solid tumors (141–147). For example, in this tumor model, ^{111}In-G250 IgG uptake measured remarkably approximately 100% ID/g by 48 h, but tumor/blood ratios were only 8 : 1. Sands et al. (147) previously compared physiological properties of a renal cell carcinoma and a breast cancer xenograft in nude mice, showing that in the former unique

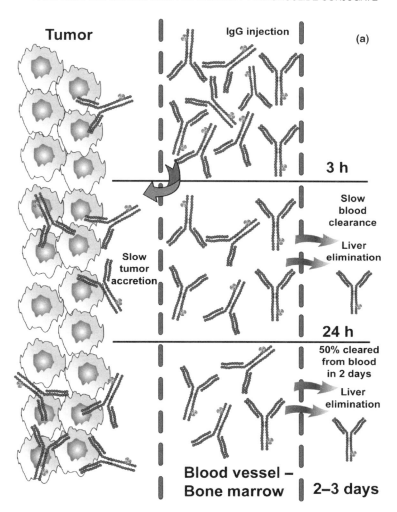

FIGURE 8.4 The pretargeting advantage over directly labeled immunoconjugates in localizing radionuclides to tumors. Radiolabeled IgG takes 1–2 days before achieving maximum tumor uptake, but it has a plasma clearance half-life of nearly 2 days (or longer). Thus, over the first 2–3 days, the bone marrow is exposed to more radiation than the tumor. The IgG is gradually removed from the blood, primarily by the liver, where some radionuclides, particularly radiometals, will accumulate. In pretargeting approaches, the radiolabeled effector is administered only after the antibody has been deposited in the tumor and cleared from the blood. The small-sized effector efficiently traverses the blood vessels, where it can bind to the pretargeted antibody. As quickly as it escapes the bloodstream, it is filtered from the body and removed by urinary excretion. The radiolabeled effectors are designed in a manner to reduce renal retention of the radionuclide. (See insert for the color representation of the figure.)

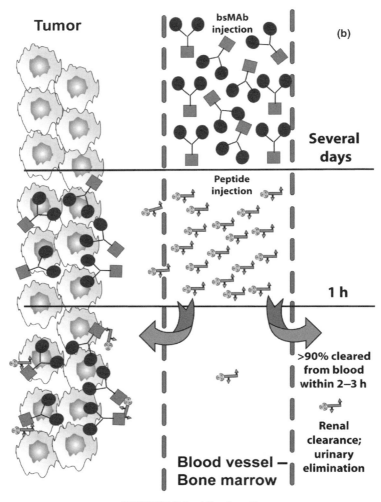

FIGURE 8.4   (*Continued*)

physiology (much higher blood flow and much greater vascular permeability) likely explained the enhanced uptake. The unique physiology of this model likely explains the exceptional uptake of radioimmunoconjugates despite a very low antigen density of just 4000 epitopes/cell. Using an anti-G250 × anti-DTPA bsMAb IgG (quadroma-based), Van Schaijk et al. (148) compared pretargeting of the divalent ^{111}In-DTPA-peptide hapten in three different renal cell cancer xenograft models that varied in antigen density from 4000 to 600,000 epitopes per cell. They found that the highest uptake occurred in the tumor xenograft with the lowest antigen density but with the highest vascular permeability, peaking at $278 \pm 130\%$ID/g at 1 h, while the xenograft with the highest antigen density but with the lowest permeability yielded the lowest uptake ($54 \pm 9\%$/g), peaking at 4 h. Since permeability in each tumor xenograft was based on measurements with an irrelevant IgG, permeability issues speak more directly of the ease of movement of the pretargeted bsMAb in and out of the tumor, and are not predictive of the radiolabeled hapten-peptide, which would be expected to extravasate and flow through the tumor interstitium easily. Enhanced radiolabeled hapten-peptide uptake was also observed in a renal xenograft model with an intermediate antigen density (80,000 epitopes/cell) and exhibiting similar permeability as the xenograft with the lowest uptake, but the kinetics of uptake and retention in this tumor were the slowest but highest of the three xenografts, increasing from 50% to 95%ID/g over 72 h. Because ^{125}I- and ^{111}In-anti-G250 IgG were retained similarly in this xenograft, they surmised that the bsMAb might have been retained better on the surface of tumor cells in this xenograft compared to the other two xenografts that had lower retention with the ^{125}I-IgG. These results clearly illustrate how tumor physiology is the main controlling factor for any form of immunotargeting, pretargeting included. Indeed, while the radiolabeled hapten molecule is very small and expected to penetrate more easily through tumors than an IgG, the molecular size of the pretargeted immunoconjugate and antigen binding-site barrier issues (149–152) will affect or restrict distribution of radioactivity within a tumor. However, since the pretargeted immunoconjugate is often given at higher protein doses than a directly radiolabeled IgG, this might foster a more uniform distribution within the tumor (153, 154). While one might think that this could potentially allow the radiolabeled effector to similarly distribute in a more uniform manner, the radiolabeled effector will then face its own binding-site barrier that would impede its migration through the tumor, particularly because, as noted earlier, the dose should be kept to the smallest possible amount to maximize specific activity and thus radioactivity delivered to each tumor cell, given a limited number of binding sites. Therefore, with such rapid distribution and elimination from the body, the radiolabeled effector will likely be deposited principally in a perivascular location, much like the microdistribution of a directly radiolabeled antibody. Saga et al. (118) showed that ^{125}I-streptavidin ($\sim$53 kD) could distribute uniformly in micrometastatic tumor nodules in the lungs of animals pretargeted 3 days earlier with a biotinylated antitumor IgG immunoconjugate. Small tumor nodules would not have the same physiological outward pressure gradients as larger masses that might otherwise act along with a binding-site barrier to impede protein penetration (155). Nevertheless, with preclinical and clinical studies increasingly moving toward RIT as a treatment for either small-volume disease or in a compartmental setting (156), the potential to distribute in a more uniform manner in these situations could only improve tumor responses.

Pretargeting methods have their greatest advantage when the radiolabeled effector is systemically administered, but they could offer certain advantages in a compartmental approach as well. For example, a streptavidin-conjugated antibody could be directly injected into a brain lesion or the surgical cavity left after the lesion has been excised, and then the radiolabeled biotin could be administered and held tightly, but an advantage over a direct radioimmunoconjugate in this situation would be difficult to envision. However, if the compartment were "leaky," perhaps in the case of peritoneal carcinomatosis, intraperitoneally injected radioactivity escaping from the peritoneal cavity into the blood would remain much longer in the blood using a directly labeled radioimmunoconjugate than if radiolabeled biotin were used and released.

8.4   CLINICAL STUDIES OF PRETARGETING

Three different pretargeting procedures have been evaluated in patients; two utilized different types of streptavidin/radiolabeled biotin approaches and the other employed a bsMAb/radiolabeled hapten-peptide (157).

8.4.1   "Two-Step" Pretargeting with Streptavidin Immunoconjugates and ^{90}Y-Biotin

This method was developed by the former NeoRx Corp. (Seattle, WA), and was based on preclinical testing showing improved therapeutic responses in comparison to directly radiolabeled IgG MAbs (125). The procedure involved the injection of an unlabeled streptavidin immunoconjugate of the murine NR-LU-10 antibody, which was later determined to bind to Ep-CAM (epithelial cell adhesion molecule; also known as EGP-2 and gp40), followed by a clearing step using galactosylated and biotinylated human serum albumin, and finally ^{111}In or ^{90}Y-labeled DOTA-biotin (158).

The optimization and radiation dose estimates for this procedure were described in two separate reports (158, 159). The initial studies showed the NR-LU-10 IgG-streptavidin immunoconjugates cleared at a slightly slower rate from the blood in humans than the unconjugated IgG, but it had similar tissue and tumor distribution (158). In contrast, ^{111}In-biotin alone cleared very quickly, with just 19% of the injected product remaining in the serum within 10 min of its injection, and 97% cleared from the body within 24 h, illustrating the lack of any specific tissue uptake for this agent (158). Optimization studies were conducted in 43 patients with diverse NR-LU-10-positive tumors given 168–600 mg of NR-LU-10-streptavidin immunoconjugate, and then between 24 and 72 h later, the galactosylated, biotinylated albumin-clearing agent was administered. The mass dose of the clearing agent was varied (110–600 mg), as well as the interval leading to the ^{111}In-biotin injection (4–24 h) and its mass dose (0.5–2 mg). Unfortunately, NR-LU-10 also bound to the gastrointestinal tract and kidneys (160), and investigators found increased renal uptake of ^{111}In-biotin when 600 mg of the NR-LU-10 immunoconjugate was given, which led to the selection of 400 mg as the optimal dose (the actual dosage was based on a target concentration of 125 µg/mL in the plasma). Attempts to block normal tissue uptake with excess unconjugated NR-LU-10 IgG in advance of the streptavidin immuno-conjugate injection reduced tumor uptake, and thus a blocking step was not used (158). With peak tumor uptake of the NR-LU-10 immunoconjugate occurring at 24–48 h, this group elected to use a 48-h interval before administering the clearing agent to promote its elimination from the blood. Administering the clearing agent by continuous infusion, bolus, or split into two fractions over 24 h did not have a significant impact on its effectiveness, and thus a bolus injection was ultimately used.

The clearing agent was best given at 10-fold mole excess compared to the streptavidin immunoconjugate in the blood ($\sim$350–400 mg). Higher doses of the clearing agent compromised the tumor uptake of ^{111}In-biotin. Under these conditions, approximately 93% of the immunoconjugate was cleared from the blood within 12 h, being limited apparently by the capacity of the liver to remove the formed complexes, as determined by monitoring the clearance of [186]Re-labeled immunoconjugate. When ^{111}In-biotin was injected 4 h after the clearing agent, about 12% was taken up by the liver, whereas if the interval was increased to 24 h, liver uptake was reduced to approximately 3%. Not surprisingly, the minimum mass of 0.5 mg of ^{111}In-biotin provided the best uptake of radioactivity in the tumor. Under optimal pretargeting conditions, the elimination rate of ^{111}In-biotin was slightly altered so that only about 75% of the injected dose was cleared from the blood within 1 day, whereas, as mentioned earlier, if given in the absence of the streptavidin immunoconjugate, more than 99% of ^{111}In-biotin was cleared by this time. Thus, it was not necessary to adjust the conditions such that ^{111}In-biotin following pretargeting clears at the same rate as ^{111}In-biotin alone, since the elimination rate was still significantly faster than that for a directly radiolabeled IgG.

A review of the radiation-absorbed doses predicted for ^{90}Y-biotin, based on imaging studies performed in patients given 0.5 mg of the ^{111}In-labeled biotin showed tumors, averaged 12.7 ± 9.5 cGy/mCi, with red marrow doses of only 0.37 ± 0.21 cGy/mCi (tumor/blood dose average of $\sim$35:1). These results were promising since radiation-absorbed doses to the red marrow of 200 cGy would likely be tolerated (depending on the method used for red marrow dose approximation), which would have allowed approximately 500 mCi of [90]Y-biotin to be injected with the corresponding delivery of approximately 6800 cGy to tumors. However, kidney doses were 13.3 cGy/mCi, resulting in tumor/kidney dose ratios of only about 1:1. If maximally tolerated renal doses were considered to be 2000 cGy, only 150 mCi would have been tolerated, which would have provided tumor doses of only approximately 2000 cGy. Clinical trials with an [131]I-labeled CC49 anti-TAG-72 IgG in colorectal cancer had achieved similar radiation-absorbed doses to colon cancer, but required administered activities as high as 300 mCi/m^2 with peripheral blood stem cell support (161). No significant antitumor responses were reported in that trial, and thus it was not surprising that no significant antitumor responses were reported in a phase II NR-LU-10-pretargeted [90]Y-biotin trial that was conducted in advanced colorectal cancer patients using the maximum tolerated dose of 100 mCi/m^2 of [90]Y-biotin (162). The phase I studies with the NR-LU-10-pretargeted [90]Y-biotin ultimately found doses were limited not by hematopoietic or renal toxicity but by severe (grade 3–4) gastrointestinal (GI) toxicity (diarrhea) due to cross-reactivity of the antibody with the GI tract, but there was late evidence of renal toxicity occurring at doses exceeding 100 mCi/m^2 (159). Radiation-absorbed doses to the GI tract ($\sim$10 cGy/mCi) were elevated in the NR-LU-10-pretargeted ^{90}Y-biotin protocol because of the selective uptake of NR-LU-10-streptavidin conjugate in this tissue, which subsequently led to an elevated uptake of ^{90}Y-biotin in the intestines.

Subsequently, clinical studies were initiated using a similar pretargeting procedure, but replacing the chemically linked streptavidin immunoconjugate with a

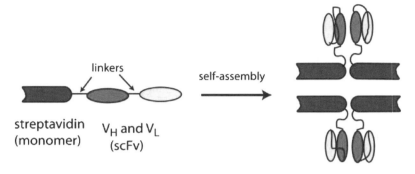

streptavidin - scFv (4)
fusion protein

FIGURE 8.5   Schematic of a streptavidin–scFv fusion protein. Streptavidin is composed of four subunits. Linking one of these subunits to an scFv of an antibody will result in a construct that when assembled will retain the four-subunit structures of streptavidin, but will now have four scFvs for binding to the target antigen.

streptavidin-anti-TAG-72 fusion protein ($\sim$170 kDa) (Fig. 8.5) and a newly designed galactosylated and biotinylated synthetic clearing agent in patients with colorectal cancer (163). Under the conditions tested in this pilot study, 90% of the ^{111}In-biotin was cleared from the blood within 8 h, somewhat better than that achieved with the previous pretargeting system, perhaps because of the faster clearance kinetics of the fusion protein or more efficient removal by the newly designed clearing agent, or both. The most important findings in this study were the tumor/normal tissue dosimetry estimates for ^{90}Y-biotin, which were based on imaging studies performed with the ^{111}In-labeled biotin. These calculations showed that in three of the seven patients, tumor/kidney dose ratios exceeded 3 : 1, with five patients having tumor/ kidney radiation-absorbed dose ratios ranging from 4.5 : 1 to 12 : 1 (e.g., average renal dose was 7 cGy/mCi, but tumor doses were highly variable, ranging from $\sim$4 to 120 cGy/mCi). Again, if the assumption is that the kidneys could tolerate a radiation-absorbed dose of 2000 cGy, tumors in these selected patients would have received well over 6000 cGy, which should be capable of eliciting a measurable response. Two other patients had a tumor/kidney dose ratio of approximately 2 : 1, which would deliver approximately 4000 cGy to the tumor and might also elicit a significant response. Toxicity and therapeutic benefit were not reported in this study, because patients received only 10 mCi/m^2 of ^{90}Y-biotin following pretargeting for dosimetry purposes.

This pretargeting procedure was also adapted for use in non-Hodgkin's lymphoma, first using a chemically synthesized rituximab anti-CD20 IgG-streptavidin immmu-noconjugate and later a similar type of fusion protein as mentioned above, based on the murine B9E9 anti-CD20 antibody (164, 165). Earlier animal studies showed these procedures easily outperformed treatment with a directly labeled ^{90}Y-anti-CD20 IgG, having less toxicity and far greater antitumor effects with cures of established tumors (133, 136). Although each of the previously mentioned clinical studies administered ^{90}Y-labeled biotin with therapeutic intent, both of these studies were performed primarily to evaluate optimal pretargeting conditions without fully exploring the maximum tolerated dose for the procedure. However, acceptable toxicity was observed and 3/14 patients experienced partial or complete antitumor responses. Interestingly, neither clinical study administered a predose of unlabeled anti-CD20 to reduce the possibility of unwanted immunoconjugate uptake in the normal spleen, as is routinely performed for RIT with radiolabeled anti-CD20 IgGs (e.g., ^{90}Y-ibritumomab tiuxetan or ^{131}I-tositumomab; see Chapter 6), yet the images provided did not show evidence of elevated uptake in this organ. However, since these trials were both limited in their scope, this issue would likely require closer inspection.

8.4.2 "Three-Step" Pretargeting with Streptavidin Immunoconjugates and ^{90}Y-Biotin

A different pretargeting procedure based on streptavidin-biotin was first proposed by Paganelli et al. (82), who compared a two-step and a three-step method for delivering either ^{131}I-streptavidin or ^{111}In-biotin, respectively, to athymic mice bearing peritoneal implants of a human colon cancer cell line (LoVo) pretargeted with a biotinylated antibody (AUA1). Radiolabeled streptavidin was localized to tumors pretargeted 1 day earlier with the biotinylated antibody. In the second case, instead of using a streptavidin–IgG conjugate to localize the radiolabeled biotin, the same antibody conjugated instead to biotin was administered. One day later, animals were injected with a 10-fold molar excess of unlabeled avidin, and then 2 h later with ^{111}In-biotin. ^{111}In-biotin was used to evaluate the procedure, but it is not a therapeutic analogue of biotin (^{90}Y-biotin would be required for treatment). All injections were given intraperitoneally. Tumors isolated from the peritoneal cavity 4 h after the radiopharmaceutical injection had an uptake of approximately 24%ID/g with the pretargeted approach and using ^{131}I-streptavidin; this was four times higher than with a directly radiolabeled IgG (MAb AUAI), and with improved tumor/blood ratios ($\sim 3.5:1$ versus $<1:1$). However, ^{131}I-streptavidin uptake in the liver was elevated, because streptavidin formed complexes with circulating biotinylated antibody in the blood, which were then deposited in the liver. They found that an intraperitoneal injection of unlabeled avidin could reduce the concentration of the biotinylated antibody in the blood, and so they decided in the second approach to administer an excess of unlabeled avidin 2 h before administering ^{111}In-biotin. In this procedure, the unlabeled avidin would bind first to the biotinylated antibody prelocalized to the peritoneal tumors, and as the excess avidin enters the bloodstream from the peritoneal cavity, it would bind to the circulating biotin IgG that would then be removed quickly by the liver (avidin is glycosylated, which accelerates its clearance by hepatic asialoglycoprotein receptors). When ^{111}In-biotin was subsequently injected i.p. 2 h later, tumor/blood and tumor/liver ratios were improved to approximately $50:1$ and $35:1$ just 2 h after radiopharmaceutical administration. However, tumor/kidney ratios were decreased to $<1:1$, reflecting the higher retention of ^{111}In-biotion in the kidneys compared to ^{125}I-streptavidin in the former approach.

This method was quickly examined in patients, but with all injections given intravenously (i.v.) (81). In this first clinical study, patients with a variety of CEA-producing tumors received 1 mg of a biotinylated anti-CEA IgG (murine MAb FO23C5), followed 3 days later by 4–6 mg of avidin, and finally 2 days later by ^{111}In-biotin. Tumor visualization was apparent on gamma camera scans with tumor/blood and tumor/liver ratios exceeding 5 : 1 within 20 min (tumor/kidney ratios were ~1 : 1). Two patients were imaged 1 month later without the benefit of pretargeting with the biotinylated antibody; one was administered ^{111}In-biotin alone, the other given the same dose of avidin followed by ^{111}In-biotin. In these follow-up studies, there was no obvious tumor targeting, whereas the pretargeting procedure localized their lesions. Imaging of one patient with ^{111}In-FO23C5 anti-CEA F(ab')$_2$ 2 weeks prior to the pretargeting study failed to disclose metastatic deposits in the liver because of elevated normal liver uptake, but with the pretargeting approach, these lesions were seen subsequently. These studies established the proof-of-principle for this new alternative cross-linking method based on streptavidin–biotin binding, and set the stage for future therapeutic use.

The first radiotherapeutic trial with the avidin–biotin pretargeting system was subsequently reported using this method but with a slight modification (166, 167). The trial, performed in patients with high-grade gliomas, involved the intravenous injection of an antitenascin IgG-biotin immunoconjugate (35 mg/m^2). The clearing step was modified to include first an injection with avidin (30 mg) that was followed 30 min later by streptavidin (50 mg). This modification first allowed avidin to clear the biotin conjugate from the blood by forming complexes that would be removed by the liver (avidin's glycosylation aids in hepatic removal of the complexes). By administering streptavidin later, when most of the circulating biotinylated antibody conjugate was already complexed with avidin and shuttled to the liver, it would circulate somewhat longer in the blood, allowing it enough time to bind to the pretargeted biotin immunoconjugate in the tumor. The next day, patients received ^{111}In or ^{90}Y-labeled biotin for binding to the streptavidin now prelocalized to the tumor and bound to the biotinylated antibody. Given the potential for limited transport across the blood–brain barrier, it was interesting to note that the investigators chose to administer all these agents intravenously. However, imaging studies clearly showed the tumors were localized. The maximal tolerated dose (MTD) was 80 mCi/m^2 (2.96 GBq/m^2), with hematologic toxicity being dose-limiting. Radiation-adsorbed doses to the kidneys and liver were estimated to be 2.7 and 1.5 cGy/mCi, respectively. Tumor doses averaged 15.2 ± 8.7 cGy/mCi, or a total of approximately 2100 cGy. Two months after treatment, 12 of the 48 patients showed objective responses (PR and CR), and another 52% had stable disease. Even 12 months after treatment, 17% of the patients still showed tumor reduction, and one patient was alive and without evidence of disease after a 10-year follow-up (Fig. 8.6).

These promising results led to a follow-up pretargeted RIT trial in grade III/IV glioma patients using the same protocol as above immediately after primary surgery and radiotherapy, to determine time to relapse and overall survival. Thirty-seven high-grade glioma patients (seventeen with three types of grade III glioma and twenty grade IV with glioblastoma multiforme) were enrolled in a nonrandomized,

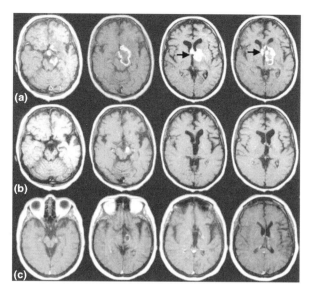

FIGURE 8.6 MRI study of a patient with anaplastic astrocytoma (arrow) before three-step radioimmunotherapy (a); same section of the brain (thalamic area) 1 year after therapy (b). Ten years after treatment, the patient is still in good condition and the MRI control clearly shows the excellent therapeutic effect of ^{90}Y-biotin treatment (c). (Reprinted with permission from Goldenberg DM et al. *J Clin Oncol.* 2006;24:823–834.)

two-arm study, with eight grade IV and eleven grade III glioma patients treated at a dose of approximately 60 mCi/m^2 (2.2 GBq/m^2) of ^{90}Y-biotin, while another twelve and six grade IV and III glioma patients, respectively, received no further treatment (168). The estimated median disease-free interval for grade III glioma patients was more than 56 months. More remarkable were the outcomes in the treated glioblastoma multiforme patients, in whom the disease-free interval and, consequently, life expectancy were much improved compared to the control group ($P < 0.01$). Survival time in treated glioblastoma multiforme patients was 33.5 months; in the corresponding control group, survival was not longer than 8 months.

This group has also reported a complete response for more than 1 year following a combination of external beam radiation therapy with 70 mCi (2.59 GBq) of pretargeted ^{90}Y-DOTA-biotin in a patient with an oropharyngeal carcinoma (169). A feasibility study was performed additionally in advanced ovarian cancer, where patients received an i.p. injection of a mixture of three biotinylated antibodies (anti-CEA, anti-TAG-72, and anti-folate receptor; 10–100 mg). One day later, they were administered avidin (50–350 mg) i.p. and then 12–18 h later received ^{90}Y-biotin either i.p. ($n = 16$) or i.v. ($n = 22$) (170). An imaging study with i.p. administered ^{99m}Tc-human albumin was performed in advance to decide which patients could receive the i.p. versus i.v. ^{90}Y-biotin injections, with patients having at least three-quarters of the abdomen filled with the ^{99m}Tc-albumin radiotracer determined to be most suitable for i.p. administration, while the others received the i.v. administered

^{90}Y-biotin. Most patients received 40–60 mCi, but an explanation of how doses were selected was not given, with the ^{90}Y-biotin dose varying from 10 to 100 mCi. Bremsstrahlung imaging (a type of low spatial resolution imaging of interaction of the β-particles from ^{90}Y with tissues in the body) showed most of the radioactivity was localized in the peritoneal disease. The route of ^{90}Y-biotin injection did not affect treatment outcome, with an objective response seen in 3/38 patients overall (8%), and stabilization in 12 patients (32%).

8.4.3 Bispecific Antibody-^{131}I-Hapten-Peptide

In addition to the one clinical study discussed previously with an anti-CEA × anti-chelate bsMAb followed by a monovalent radiolabeled chelate effector molecule (72), bsMAb-based pretargeting was advanced in human studies by the Immunotech, SA group (Marseille, France). The group primarily investigated an anti-CEA bsMAb-pretargeting system with the anti-hapten binding arm of the antibody specific for indium-complexed DTPA, but their system included the innovation of a divalent (In) DTPA hapten, with two (In)DTPAs bound to a tyrosine-lysine dipeptide, which enhanced tumor uptake and retention as compared to the monovalent hapten (122). Their initial preclinical studies had indicated that although tumor uptake with the pretargeting approach was lower than that achieved with a directly radiolabeled F(ab')$_2$, tumor/normal tissue ratios were more favorable for pretargeting, especially the tumor/blood ratios (122, 171). These improved tumor/blood ratios compared to direct targeting with ^{111}In-anti-CEA F(ab')$_2$ were confirmed in the first clinical testing in colorectal cancer patients (171, 172). Initial optimization studies in patients with primary and recurrent colorectal cancer suggested that a minimum dose of 0.06 mg/kg was required for the bsMAb, with the optimal interval between the bsMAb and the radiolabeled (^{111}In)-diDTPA-peptide being 4–5 days, and with a minimum dose of 0.1 nmol/kg of the hapten-peptide (172, 173). Although these initial studies noted appreciably reduced blood and liver radioactivity compared to a directly labeled anti-CEA F(ab')$_2$, they were unable to satisfactorily localize hepatic metastases. They suggested several possible explanations; for example, the bsMAb dose tested might have been too low (e.g., <10 mg) and its immunoreactivity was only 50%, or circulating CEA may have interfered, and the large tumors studied may have been necrotic (173). Indeed, this failure to image hepatic metastases, and the failure of other antibodies directly conjugated to radiometals at the time to achieve this, provided the opportunity for directly radiolabeled ^{99m}Tc-anti-CEA Fab' to become the preferred imaging agent for colorectal cancer since hepatic metastases that are common were visualized with this radiolabeled small antibody fragment, which does not accumulate extensively into a normal liver (174, 175). More recently, a pretargeting method utilizing recombinant anti-CEA bsMAb with a ^{99m}Tc- or an ^{124}I-labeled hapten-peptide showed significantly enhanced tumor uptake and tumor/normal tissue ratios in tumor-bearing animals, compared to the ^{99m}Tc- or ^{124}I-anti-CEA Fab' fragments, and less ambiguous and more sensitive visualization than PET with ^{18}F-FDG, suggesting a promising potential future role for pretargeted imaging (116, 176, 177).

While clinical studies were optimizing the pretargeting technique, additional preclinical studies were illustrating its therapeutic potential using the same diDTPA-peptide, except in this case the tyrosine in the peptide backbone was radioiodinated with [131]I (128). [90]Y was not used as a radiolabel for therapy studies, presumably because the anti-(In)DTPA antibody was highly specific for the (In) DTPA complex that included the metal, indium (122). Dosimetry estimates, based on animals treated with 3 mCi (111 MBq) of the ^{131}I-hapten-peptide following bsMAb pretargeting or receiving direct RIT with 0.325 mCi (12 MBq) of ^{131}I-anti-CEA IgG, predicted that the tumors would receive the same total radiation absorbed dose, but because pretargeted agents localize to their maximum level in 1 h or less, the average dose rate would be three times higher with the pretargeting approach. The dose to the blood (predictive of marrow toxicity) was about twofold lower for pretargeting, renal doses were approximately twofold higher, and liver doses were about the same. Measurement of blood counts and body weight indicated similar toxicity, but the pretargeted animals had nearly three times longer tumor-growth delay than the ^{131}I-IgG-treated animals. The pretargeting procedure with the ^{131}I-hapten-peptide was later shown to be significantly better than an ^{131}I-anti-CEA F(ab′)$_2$ in the same colon cancer xenograft model (178), and improved efficacy was also determined in animals bearing a CEA-expressing medullary thyroid cancer xenograft (179, 180).

Clinical trials using an anti-CEA bsMAb with an ^{131}I-labeled hapten-peptide have focused primarily on patients with small-cell lung cancer (SCLC) and medullary thyroid carcinoma (MTC) (181–183). SCLC is radiosensitive and at least one-third of the patients express CEA, making it a logical choice for therapy. While MTC is not radiosensitive, MTC is well vascularized and the majority of tumors express high levels of CEA. In addition, calcitonin is a highly sensitive and specific serum biomarker for MTC and could be used to assess response even of occult disease (183, 184). With no effective therapies for metastatic MTC, an examination of pretargeted therapy was pursued. Preliminary targeting studies indicated that radiation-absorbed doses to MTC tumors might range from 4.2 to as high as 174 cGy/mCi, and from 1.7 to 8 cGy/mCi in patients with SCLC, with high tumor uptake in small MTC tumors (0.1%ID/g at day 3 in a 0.8-g resected tumor), and moderate tumor uptake (0.009%ID/g) in larger SCLC tumors (11 ± 2 g) (185). In a phase I/II trial, 14 patients with SCLC received 1 mg anti-CEA bsMAb/4 nmol of the diDTPA-peptide followed 4 days later with the ^{131}I-diDTPA-peptide radiolabeled at a specific activity of 18.5 MBq/nmol (0.5 mCi/nmol). Doses started at 40 mCi and were escalated to 180 mCi (182). Hematologic toxicity was dose-limiting, with patients given ≥130 mCi experiencing severe thrombocytopenia and leukopenia 2–4 weeks after treatment. Peripheral blood stem cells were used in patients receiving ≥130 mCi to aid in recovery. Eight lesions in six patients received 79–3655 cGy (mean ± SD: 1289 ± 1318 cGy). Two patients had PR (one for 3 months and another for 17+ months), and one showed disease stabilization for 2 years, while the other patients progressed.

In a phase I/II trial, 26 MTC patients, mostly with large tumor burden, were sorted in terms of those showing bone marrow involvement by bone scan or MRI; these patients were started at an ^{131}I-diDTPA-peptide dose of 24 mCi/m^2 that was given

4 days after 12 mg of bsMAb/m^2. Patients with no evidence of bone marrow involvement were started at 60 mCi/m^2 of ^{131}I-diDTPA-peptide following 30 mg of bsMAb/m^2; ^{131}I-diDTPA-peptide escalation was in increments of 12 mCi/m^2 in both groups (179). Tumor uptake ranged from 0.003% to 0.26%ID/g and tumor radiation absorbed doses from 2.9 to 184 cGy/mCi. Severe hematological toxicity was observed, which was unexpected in a number of patients who were given relatively moderate amounts of the ^{131}I-diDTPA-peptide. Hematological toxicity was ultimately explained by previously undetected bone marrow involvement, which was seen on post-therapy images of the ^{131}I-diDTPA-peptide and confirmed by MRI. In subsequent clinical studies, such frequent and unexpected bone/bone marrow involvement has been confirmed in a higher number of patients (186). In terms of therapeutic outcome, among 17 assessable patients, 4 experienced pain relief, 5 had minor responses, and 4 demonstrated biological responses (increase in calcitonin doubling time by at least 100%) (181).

More recent studies have focused on the use of a new chimeric bsMAb immunoconjugate, composed of a humanized anti-CEA Fab' cross-linked to the murine anti-hapten Fab'. An evaluation of various pretargeting conditions in patients bearing diverse CEA-producing tumors concluded that 40 mg/m^2 of the chimeric bsMAb and a 5-day interval before administering the ^{131}I-diDTPA-peptide provided the best blood clearance and tumor uptake of radioactivity (187, 188). Hematologic toxicity again was found to be dose-limiting even under optimal pretargeting conditions, with tumors receiving an average of 5.2 cGy/GBq or 19.2 cGy/mCi. Nine cases of tumor stabilization for 3–12+ months were noted (45%), six in the MTC group and three in non-MTC patients (188). The rate of disease stabilization was significantly higher with 75 mg/m^2 of bsMAb (64%) than with 40 mg/m^2 (22%; $P = 0.04$).

A subsequent analysis of all MTC patients treated with this pretargeted RIT procedure indicated that overall survival (OS) was significantly longer in high-risk patients, who were considered to have a calcitonin doubling time less than 2 years than in high-risk, untreated patients (median overall survival 110 months versus 61 months; $P < 0.030$). Patients with bone/bone marrow disease had a longer survival than patients without such involvement (10-year OS, 83% versus 14%; $P < 0.023$). This latter finding suggested that pretargeted RIT might have been most effective against disease in the bone marrow, thereby extending survival. This analysis provided the basis on which a phase II trial in two groups of MTC patients, one group with tumor lesions documented by imaging (group I) and the other with no tumor lesions detected by imaging (occult disease), but with serum calcitonin levels >100 pg/ml and a calcitonin doubling time less than 2 years (group II). Recruitment to this trial was completed only recently, and interim results have not yet been reported.

8.5 PROSPECTS FOR COMBINATION THERAPIES

In non-Hodgkin's lymphoma, preclinical studies have clearly documented significant improvements in response with pretargeted approaches as compared to RIT with directly radiolabeled antibodies (133, 135, 136, 140, 189), but there is no doubt that successful treatment of solid tumors is more challenging. While studies of pretargeted RIT have been encouraging, in most clinical experiences to date, the radiation-absorbed dose delivered to tumors have not yet reached levels that are commonly achieved in the treatment of solid tumors with external beam radiation. Thus, if clinically significant objective responses are to occur, these treatments will need to be augmented. Preclinical studies combining directly radiolabeled antibody therapy with chemotherapy have shown that such augmentation is possible in solid tumors (190–194), yet clinical studies with some of these combinations in myeloablative (195, 196) and nonmyeloablative (197) settings are without evidence of response enhancement. These combinations will likely be compromised to some extent by the additive hematological toxicity associated with chemotherapy and RIT. In this regard, pretargeted RIT might have an advantage if hematological toxicity can be reduced at dose levels where radiation-absorbed doses to the tumor are maintained. In some cases, low doses of certain chemotherapeutic agents that are relatively nontoxic are radiosensitizing. The addition of a radiosensitizing amount of gemcitabine to a pretargeting protocol involving a streptavidin–anti-TAG-72 (CC49) immunoconjugate and ^{90}Y-biotin improved tumor response in animals bearing human LS174T colon cancer xenografts (198). Rather than using a suboptimal chemotherapy dose to enhance pretargeted RIT, Kraeber-Bodere et al. (199) combined a partial dose (65% of its MTD) of an ^{131}I-diDTPA-peptide pretargeted by an anti-CEA (MAb F6) × anti-DTPA bsMAb with a full MTD of either doxorubicin or paclitaxel in animals bearing TT MTC xenografts. The combination with paclitaxel, but not doxorubicin, resulted in a significant prolongation in tumor doubling time compared to any of the treatments alone, without increasing hematological toxicity or body weight loss.

Locoregional pretargeted RIT combined with temozolomide (TMZ) has been studied in patients with recurrent glioblastoma multiforme (200). Seventy-three patients treated with a second surgical debulking (plus catheter implantation) received at least two cycles of treatment separated by 2 months. Thirty-five of these patients were also treated with oral TMZ (two cycles of 200 mg/m^2/day for 5 days every 28 days) between each pretargeted RIT course. Doses of 2–5 mg of a biotinylated antitenascin IgG were injected into the surgical cavity, followed 18–24 h later by an injection of 5–15 mg of avidin, and then 14–16 h later by ^{90}Y-biotin (10–25 mCi, 370–925 MBq, depending on the size of the cavity), all through an in-dwelling catheter. Adding TMZ to the pretargeted therapy did not increase neurological toxicity, and no major hematological toxicity was observed. The majority of patients (75%) had stabilization of disease. In the group treated with pretargeted RIT alone, the OS and the progression-free survival (PFS) were 17.5 months (95% CI = 17–20 months) and 5 months (95% CI = 4–8 months), respectively. Patients treated with the combination had a statistically significant improved OS and PFS (median OS = 25 months; 95% CI = 23–30 months; and median PFS = 10 months; 95% CI = 9–18 months; log-rank test P-values <0.01 and <0.01). The previously described preclinical and clinical experiences indicate that chemotherapy can be added to pretargeted RIT without any significant impact on toxicity but with gains in therapeutic efficacy.

8.6 FUTURE INNOVATIONS

Despite the challenges that have faced RIT (156, 201), it remains a promising method for cancer treatment and diagnostic imaging. It is perhaps the only targeted cancer therapy that can kill cells without necessarily having to bind directly, be internalized, or otherwise interact with the tumor cell or the environment to exert its activity, especially if long-range β-particle-emitting radionuclides are used (e.g., ^{131}I or ^{90}Y). But this potential is limited by hematological toxicity associated with the administration of directly radiolabeled antibodies that are eliminated only slowly from the blood. This, combined with the poor penetration of these molecules into tumors, often delivers sublethal amounts of radiation. While adjustments in the antibody size and molecular composition can provide molecules with more favorable pharmacokinetic profiles, due to the stability of the bond between the radionuclide and the antibody, especially using radiometal chemistry, the majority of the radioactivity is deposited in the tissue responsible for elimination. The faster the clearance of the radiolabeled antibody from the blood, the lower the risk of hematological toxicity; but without equally efficient removal from the tissue responsible for elimination, the ability to achieve the necessary therapeutic window will continue to elude us.

In this sense, the separation of the radionuclide targeting from the antibody is highly logical, particularly when this separation allows innovation in designing effector molecules that can be more efficiently removed from the blood and tissues. Indeed, even though directly radiolabeled peptides can be used to target tumors with rapid elimination from the blood, often these peptides have undesirable uptake or retention in other tissues, and there may be little latitude in modifying them without reducing their receptor/antigen binding affinity. Biotin has already been shown to have excellent clearance properties with little tissue retention, and chemistries have been established to ensure a stable biotin conjugate. Although biotin is present in the body, clinical studies have shown that endogenous levels are not high enough to interfere with pretargeting using a biotin/avidin method. Radiolabeled hapten-peptides used for bsMAb pretargeting approaches require two hapten components situated on a peptide backbone that can be modified to encourage renal clearance. The positioning of the haptens on the backbone can affect binding to bsMAb, and D-amino acids increase the resistance to peptidase cleavage (121, 171).

Although the exceptionally fast and efficient clearance of these peptides from the blood and tissues reduces toxicity, it also leaves little time for the peptide to concentrate in the tumor, and thus substantially higher doses are required than with directly labeled radioimmunoconjugates. The effective capture and retention in the tumor of the radiolabeled hapten aids in achieving a higher concentration than in normal tissues, but even this advantage often falls short in its ability to combat large disease burdens that may be encountered, especially in clinical trial evaluation. To this end, investigators need to persevere and seek opportunities to evaluate these targeted therapies earlier in the treatment of the disease.

All pretargeting approaches have shown exceptional promise in preclinical testing, and the three methods examined clinically and discussed earlier have also fulfilled their expectation of improving tumor/blood ratios, but there is insufficient clinical evidence as yet to claim that these approaches can stand alone to bring about significant clinical responses. Perhaps the major issue that has impeded these techniques is the immunogenicity of the antibody immunoconjugates used in earlier studies. Streptavidin and avidin have proven immunogenic in most patients in all clinical settings, even in non-Hodgkin's lymphoma patients with a reduced capacity for eliciting antibody responses (164). Furthermore, the murine bsMAb used in the first clinical studies were similarly immunogenic in most patients, but the introduction of a humanized $\times$ murine Fab$'$ $\times$ Fab$'$ bsMab revealed only 1/12 (8%) developing an antimouse antibody response; however, 4/12 (33%) patients developed an antibody response to the humanized component (188). More recently, recombinant humanized bsMAb have been developed that should further reduce immunogenicity by eliminating more of the foreign protein scaffold sequences (202, 203). Approaches employing foreign proteins, such as streptavidin or avidin, will be more limited in their use than a bsMAb that can be rendered less immunogenic though protein engineering.

A potential limitation for the bsMAb pretargeting method is in the selection of the hapten and antihapten components. Most systems have used chelates that served as haptens that were also the carriers of the radionuclide (204). In accordance with the principles of the affinity enhancement system (121), two haptens are tethered together using a small (e.g., two–four amino acids) peptide backbone, such as the DTPA-tyrosine-lysine-DTPA peptide (diDTPA-peptide) used in a number of the previously mentioned studies. The anti-DTPA antibody was actually derived by immunization with indium-loaded DTPA, and subsequently a problem arose where the antihapten antibody bound most tightly to the specific metal–chelate complex (i.e., indium-DTPA), and where substituting another metallic radionuclide in the chelate (e.g., yttrium) decreased the stability of the chelate–metal complex or the binding to the antichelate antibody. However, the peptide backbone component allowed investigators to use creative chemistry to couple other radionuclide-binding sites to extend the number of radionuclides that could be used, such as incorporation of tyrosines for radioiodination or the coupling of ^{99m}Tc/^{188}Re-binding chelates to (In)DTPA hapten-peptides (205). Since the diDTPA-peptide was not optimized for carrying ^{90}Y, one would need to create a new antibody to a chelate, such as DOTA, that would have greater stability for ^{90}Y. Alternatively, one could try to insert two different chelating groups on one backbone, one serving primarily as the hapten to bind to the antihapten antibody, while the other to provide maximum stability for complexing the radiometal of interest. The latter approach would have considerable risk that the radionuclide would also be bound to the weaker of the two binding ligands. A better approach would have the hapten serve only for binding the antihapten antibody, allowing the peptide backbone to receive any number of binding ligands for radionuclides.

One such system was described by Janevik-Ivanovska et al. (206) for ^{131}I and was further developed by our group for ^{99m}Tc/^{188}Re, ^{111}In, ^{90}Y, ^{177}Lu, ^{68}Ga, and ^{124}I (115–117, 130, 135, 176, 177, 207, 208). The hapten, histamine-succinyl-glycine (HSG), was first developed as a derivative to be used in an immunoassay to measure

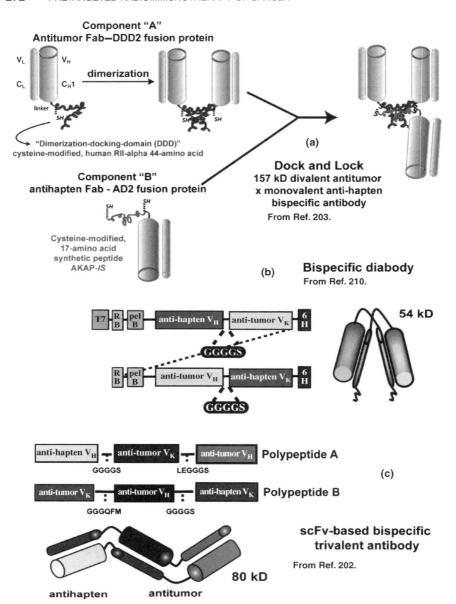

FIGURE 8.7 Examples of molecularly engineered bispecific antibodies. (a) A tri-Fab construct prepared by the dock-and-lock method. The antitumor Fab is linked to a short peptide linker that is then tethered to what is referred to as a dimerization docking domain. This amino acid sequence was modified to strategically insert a cysteine residue (DDD2). The antitumor Fab-DDD constructs naturally form noncovalent dimers, with the resulting complex forming a docking domain capable of binding the AD2 sequence that is linked to the antihapten Fab. The AD2 is a specially modified sequence with two cysteine residues inserted

histamine (209). The antibody m679 binds HSG with a nanomolar affinity, but approximately 10,000-fold lower affinity for histamine. Importantly, too, we have shown the HSG derivative does not have histamine-like pharmacologic activity *in vitro* or *in vivo*, nor is there any evidence for tissue binding of a DOTA-di-HSG-peptide that is planned for clinical studies (Sharkey, unpublished data). A variety of recombinant, humanized bsMAb based on the anti-HSG-HSG hapten-binding system have been constructed (202, 210) (Fig. 8.7). A novel "Dock and Lock" (DNL) procedure has been described recently that is capable, in a modular manner, of making humanized fusion protein constructs of anti-HSG with any tumor targeting agent (203, 211).

Pretargeting approaches are sometimes criticized for their inherent complexity, compared to a direct radioimmunoconjugate, since they require the optimization of two or more steps. If a direct radioimmunoconjugate were able to provide similar benefit as found with pretargeting approaches, the simpler technique would be used, but as already highlighted, pretargeting has proven to be more effective and less toxic for treatment of tumors, at least in preclinical models. The view that pretargeting is too complex is also overstated. While all these methods require somewhat more effort in their initial design, the preclinical and clinical experience gained to date has given investigators considerable insight into how these methods can be optimized with minimum of testing. As with the presently approved radioimmunoconjugates for the treatment of non-Hodgkin's lymphoma that require cooperation between the oncologist and the nuclear medicine physician to plan and administer a treatment that includes both unlabeled and radiolabeled anti-CD20 antibodies (see Chapter 6), in the case of pretargeting, the bsMAb can be given by the patient's oncologist, while the radiolabeled hapten could be administered by the nuclear physician. However, pretargeting procedures use considerably higher doses of radioactivity than treatment with direct radioimmunoconjugates, which would require the procedure to be given in radiation-qualified facilities and may require patient isolation for ^{131}I but not for the pure β-emitter, ^{90}Y. However, these procedures are analogous to Na^{131}I therapy, but since the radiolabeled haptens are rapidly cleared from the body within a few hours, the patient in most cases could be released from isolation the same day. Thus, there are already well-established clinical protocols in nuclear medicine clinics that could provide a working model for a pretargeted RIT program.

◄————————————————————————————————————

on either end of the sequence. When the antitumor-DDD2 is mixed with the antihapten-AD2, the molecules naturally form 157 kD complex with a specific orientation that places the cysteine residues in a position that allows the formation of covalent disulfide bonds. (b) Bispecific diabodies are formed by crafting a polypeptide that contains the V_H and V_L sequences for both the antitumor and the antihapten Fvs. (c) A trivalent scFv-based bispecific antibody has been formed by linking polypeptide chains that contain the two scFvs of the antitumor antibody and one antihapten. (See insert for the color representation of the figure.)

8.7 CONCLUSIONS

Pretargeting methods excel in their ability to rapidly deliver a high concentration of radioactivity specifically to tumors, while minimizing normal tissue uptake, and have the potential to improve cancer management in both radiotherapy and molecular imaging in the future. Recently developed two-step procedures that separate tumor targeting with a bsMAb from delivering the imaging or radio-therapeutic agent provide very high tumor/nontumor ratios within hours of injecting the radiolabeled hapten. Further humanization of the antibody molecule will potentially allow repeated treatment that could further enhance this approach for the treatment of cancer. We are thus optimistic for the use of this strategy for cancer therapy and diagnosis in the future.

REFERENCES

1. Korngold L, Pressman D. The localization of antilymphosarcoma antibodies in the Murphy lymphosarcoma of the rat. *Cancer Res.* 1954;14:96–99.
2. Pressman D. The zone of localization of antibodies: the specific localization of antibodies to rat kidney. *Cancer.* 1949;2:697–700.
3. Pressman D, Eisen HN, Fitzgerald PJ. The zone of localization of antibodies. VI. The rate of localization of anti-mouse-kidney serum. *J Immunol.* 1950;64:281–287.
4. Pressman D, Korngold L. The *in vivo* localization of anti-Wagner-osteogenic-sarcoma antibodies. *Cancer.* 1953;6:619–623.
5. Bale WF, Spar IL. Studies directed toward the use of antibodies as carriers of radioactivity for therapy. *Adv Biol Med Phys.* 1957;5:285–356.
6. Spar IL, Goodland RL, Bale WF. Localization of I131 labeled antibody of rat fibrin in transplantable rat lymphosarcoma. *Proc Soc Exp Biol Med.* 1959;100:259–262.
7. Bale WF, Spar IL, Goodland RL. Experimental radiation therapy of tumors with I-131-carrying antibodies to fibrin. *Cancer Res.* 1960;20:1488–1494.
8. Spar IL, Bale WF, Marrack D, Dewey WC, McCardle RJ, Harper PV. 131-I-labeled antibodies to human fibrinogen. Diagnostic studies and therapeutic trials. *Cancer.* 1967;20:865–870.
9. Bale WF, Contreras MA, Grady ED.Factors influencing localization of labeled antibodies in tumors. *Cancer Res.* 1980;40:2965–2972.
10. Goldenberg DM, Hansen HJ. Carcinoembryonic antigen present in human colonic neoplasms serially propagated in hamsters. *Science.* 1972;175:1117–1118.
11. Goldenberg DM, Preston DF, Primus FJ, Hansen HJ. Photoscan localization of GW-39 tumors in hamsters using radiolabeled anticarcinoembryonic antigen immunoglobulin G. *Cancer Res.* 1974;34:1–9.
12. Goldenberg DM, DeLand F, Kim E, et al. Use of radiolabeled antibodies to carcinoembryonic antigen for the detection and localization of diverse cancers by external photoscanning. *N Engl J Med.* 1978;298:1384–1386.
13. Scheinberg DA, Strand M, Gansow OA. Tumor imaging with radioactive metal chelates conjugated to monoclonal antibodies. *Science.* 1982;215:1511–1513.

14. Hnatowich DJ, Layne WW, Childs RL, et al. Radioactive labeling of antibody: a simple and efficient method. *Science*. 1983;220:613–615.

15. Meares CF, Moi MK, Diril H, et al. Macrocyclic chelates of radiometals for diagnosis and therapy. *Br J Cancer Suppl*. 1990;10:21–26.

16. Esteban JM, Schlom J, Gansow OA, et al. New method for the chelation of indium-111 to monoclonal antibodies: biodistribution and imaging of athymic mice bearing human colon carcinoma xenografts. *J Nucl Med*. 1987;28:861–870.

17. Yeh SM, Sherman DG, Meares CF. A new route to "bifunctional" chelating agents: conversion of amino acids to analogs of ethylenedinitrilotetraacetic acid. *Anal Biochem*. 1979;100:152–159.

18. Moi MK, Meares CF, McCall MJ, Cole WC, DeNardo SJ. Copper chelates as probes of biological systems: stable copper complexes with a macrocyclic bifunctional chelating agent. *Anal Biochem*. 1985;148:249–253.

19. Deshpande SV, DeNardo SJ, Meares CF, et al. Copper-67-labeled monoclonal antibody Lym-1, a potential radiopharmaceutical for cancer therapy: labeling and biodistribution in RAJI tumored mice. *J Nucl Med*. 1988;29:217–225.

20. Cole WC, DeNardo SJ, Meares CF, et al. Comparative serum stability of radiochelates for antibody radiopharmaceuticals. *J Nucl Med*. 1987;28:83–90.

21. Griffiths GL, Goldenberg DM, Jones AL, Hansen HJ. Radiolabeling of monoclonal antibodies and fragments with technetium and rhenium. *Bioconjug Chem*. 1992;3:91–99.

22. Pettit WA, DeLand FH, Bennett SJ, Goldenberg DM. Radiolabeling of affinity-purified goat anti-carcinoembryonic antigen immunoglobulin G with technetium-99m. *Cancer Res*. 1980;40:3043–3045.

23. Trejtnar F, Laznicek M. Analysis of renal handling of radiopharmaceuticals. *Q J Nucl Med*. 2002;46:181–194.

24. Leung SO, Qu Z, Hansen HJ, et al. The effects of domain deletion, glycosylation, and long IgG_3 hinge on the biodistribution and serum stability properties of a humanized IgG_1 immunoglobulin, hLL2, and its fragments. *Clin Cancer Res*. 1999;5:3106s–3117s.

25. Govindan SV, Griffiths GL, Michel RB, Andrews PM, Goldenberg DM, Mattes MJ. Use of galactosylated-streptavidin as a clearing agent with ^{111}In-labeled, biotinylated antibodies to enhance tumor/non-tumor localization ratios. *Cancer Biother Radiopharm*. 2002;17:307–316.

26. Sinclair AM, Elliott S. Glycoengineering: the effect of glycosylation on the properties of therapeutic proteins. *J Pharm Sci*. 2005;94:1626–1635.

27. Stork R, Zettlitz KA, Muller D, Rether M, Hanisch FG, Kontermann RE. N-glycosylation as novel strategy to improve pharmacokinetic properties of bispecific single-chain diabodies. *J Biol Chem*. 2008;283:7804–7812.

28. Olafsen T, Kenanova VE, Wu AM. Tunable pharmacokinetics: modifying the *in vivo* half-life of antibodies by directed mutagenesis of the Fc fragment. *Nat Protoc*. 2006;1:2048–2060.

29. Petkova SB, Akilesh S, Sproule TJ, et al. Enhanced half-life of genetically engineered human IgG1 antibodies in a humanized FcRn mouse model: potential application in humorally mediated autoimmune disease. *Int Immunol*. 2006;18:1759–1769.

30. Olafsen T, Kenanova VE, Sundaresan G, et al. Optimizing radiolabeled engineered anti-p185HER2 antibody fragments for *in vivo* imaging. *Cancer Res*. 2005;65:5907–5916.

31. Slavin-Chiorini DC, Kashmiri SV, Lee HS, et al. A CDR-grafted (humanized) domain-deleted antitumor antibody. *Cancer Biother Radiopharm.* 1997;12:305–316.

32. Mueller BM, Reisfeld RA, Gillies SD. Serum half-life and tumor localization of a chimeric antibody deleted of the CH2 domain and directed against the disialoganglioside GD2. *Proc Natl Acad Sci USA.* 1990;87:5702–5705.

33. Sharkey RM, Goldenberg DM. Use of antibodies and immunoconjugates for the therapy of more accessible cancers. *Adv Drug Deliv Rev.* 2008;60:1407–1420.

34. Primus FJ, Macdonald R, Goldenberg DM, Hansen HJ. Localization of GW-39 human tumors in hamsters by affinity-purified antibody to carcinoembryonic antigen. *Cancer Res.* 1977;37:1544–1547.

35. Spar IL, Goodland RL, Desiderio MA. Immunological removal of circulating I-131-labeled rabbit antibody to rat fibrinogen in normal and tumor-bearing rats. *J Nucl Med.* 1964;5:428–443.

36. McCardle RJ, Harper PV, Spar IL, Bale WF, Andros G, Jiminez F. Studies with iodine-131-labeled antibody to human fibrinogen for diagnosis and therapy of tumors. *J Nucl Med.* 1966;7:837–847.

37. Goldenberg DM, Sharkey RM, Ford E. Anti-antibody enhancement of iodine-131 anti-CEA radioimmunodetection in experimental and clinical studies. *J Nucl Med.* 1987;28:1604–1610.

38. Sharkey RM, Mabus J, Goldenberg DM. Factors influencing anti-antibody enhancement of tumor targeting with antibodies in hamsters with human colonic tumor xenografts. *Cancer Res.* 1988;48:2005–2009.

39. Begent RH, Ledermann JA, Green AJ, et al. Antibody distribution and dosimetry in patients receiving radiolabelled antibody therapy for colorectal cancer. *Br J Cancer.* 1989;60:406–412.

40. Pedley RB, Dale R, Boden JA, Begent RH, Keep PA, Green AJ. The effect of second antibody clearance on the distribution and dosimetry of radiolabelled anti-CEA antibody in a human colonic tumor xenograft model. *Int J Cancer.* 1989;43:713–718.

41. Goodwin D, Meares C, Diamanti C, et al. Use of specific antibody for rapid clearance of circulating blood background from radiolabeled tumor imaging proteins. *Eur J Nucl Med.* 1984;9:209–215.

42. Linden O, Kurkus J, Garkavij M, et al. A novel platform for radioimmunotherapy: extracorporeal depletion of biotinylated and ^{90}Y-labeled rituximab in patients with refractory B-cell lymphoma. *Cancer Biother Radiopharm.* 2005;20:457–466.

43. Martensson L, Nilsson R, Ohlsson T, Sjogren HO, Strand SE, Tennvall J. Reduced myelotoxicity with sustained tumor concentration of radioimmunoconjugates in rats after extracorporeal depletion. *J Nucl Med.* 2007;48:269–276.

44. Chen JQ, Strand SE, Tennvall J, Lindgren L, Hindorf C, Sjogren HO. Extracorporeal immunoadsorption compared to avidin chase: enhancement of tumor-to-normal tissue ratio for biotinylated rhenium-188-chimeric BR96. *J Nucl Med.* 1997;38:1934–1939.

45. Kim SJ, Park Y, Hong HJ. Antibody engineering for the development of therapeutic antibodies. *Mol Cells.* 2005;20:17–29.

46. Holliger P, Hudson PJ. Engineered antibody fragments and the rise of single domains. *Nat Biotechnol.* 2005;23:1126–1136.

47. Binz HK, Pluckthun A. Engineered proteins as specific binding reagents. *Curr Opin Biotechnol.* 2005;16:459–469.

48. Kipriyanov SM, Le Gall F. Recent advances in the generation of bispecific antibodies for tumor immunotherapy. *Curr Opin Drug Discov Devel.* 2004;7:233–242.

49. Covell DG, Barbet J, Holton OD, Black CD, Parker RJ, Weinstein JN. Pharmacokinetics of monoclonal immunoglobulin G1, F(ab')$_2$, and Fab' in mice. *Cancer Res.* 1986;46:3969–3978.

50. Sharkey RM, Motta-Hennessy C, Pawlyk D, Siegel JA, Goldenberg DM. Biodistribution and radiation dose estimates for yttrium- and iodine-labeled monoclonal antibody IgG and fragments in nude mice bearing human colonic tumor xenografts. *Cancer Res.* 1990;50:2330–2336.

51. Camera L, Kinuya S, Garmestani K, et al. Comparative biodistribution of indium- and yttrium-labeled B3 monoclonal antibody conjugated to either 2-(*p*-SCN-Bz)-6-methyl-DTPA (1B4M-DTPA) or 2-(*p*-SCN-Bz)-1,4,7,10-tetraazacyclododecane tetraacetic acid (2B-DOTA). *Eur J Nucl Med.* 1994;21:640–646.

52. Camera L, Kinuya S, Garmestani K, et al. Evaluation of the serum stability and *in vivo* biodistribution of CHX-DTPA and other ligands for yttrium labeling of monoclonal antibodies. *J Nucl Med.* 1994;35:882–889.

53. Deshpande SV, DeNardo SJ, Kukis DL, et al. Yttrium-90-labeled monoclonal antibody for therapy: labeling by a new macrocyclic bifunctional chelating agent. *J Nucl Med.* 1990;31:473–479.

54. Moi MK, DeNardo SJ, Meares CF. Stable bifunctional chelates of metals used in radiotherapy. *Cancer Res.* 1990;50:789s–793s.

55. Roselli M, Schlom J, Gansow OA, et al. Comparative biodistribution studies of DTPA-derivative bifunctional chelates for radiometal labeled monoclonal antibodies. *Int J Rad Appl Instrum B.* 1991;18:389–394.

56. Washburn LC, Sun TT, Lee YC, et al. Comparison of five bifunctional chelate techniques for ^{90}Y-labeled monoclonal antibody CO17–1A. *Int J Rad Appl Instrum B.* 1991;18:313–321.

57. Peterson JJ, Meares CF. Enzymatic cleavage of peptide-linked radiolabels from immunoconjugates. *Bioconjug Chem.* 1999;10:553–557.

58. Meares CF, McCall MJ, Deshpande SV, DeNardo SJ, Goodwin DA. Chelate radiochemistry: cleavable linkers lead to altered levels of radioactivity in the liver. *Int J Cancer Suppl.* 1988;2:99–102.

59. Seitz K, Zhou H. Pharmacokinetic drug–drug interaction potentials for therapeutic monoclonal antibodies: reality check. *J Clin Pharmacol.* 2007;47:1104–1118.

60. Batra SK, Jain M, Wittel UA, Chauhan SC, Colcher D. Pharmacokinetics and biodistribution of genetically engineered antibodies. *Curr Opin Biotechnol.* 2002;13:603–608.

61. Colcher D, Goel A, Pavlinkova G, Beresford G, Booth B, Batra SK. Effects of genetic engineering on the pharmacokinetics of antibodies. *Q J Nucl Med.* 1999;43:132–139.

62. Pluckthun A, Pack P. New protein engineering approaches to multivalent and bispecific antibody fragments. *Immunotechnology.* 1997;3:83–105.

63. Kenanova V, Wu AM. Tailoring antibodies for radionuclide delivery. *Expert Opin Drug Deliv.* 2006;3:53–70.

64. Kenanova V, Olafsen T, Crow DM, et al. Tailoring the pharmacokinetics and positron emission tomography imaging properties of anti-carcinoembryonic antigen single-chain Fv-Fc antibody fragments. *Cancer Res.* 2005;65:622–631.

65. Reardan DT, Meares CF, Goodwin DA, et al. Antibodies against metal chelates. *Nature*. 1985;316:265–268.

66. Brennan M, Davison PF, Paulus H. Preparation of bispecific antibodies by chemical recombination of monoclonal immunoglobulin G1 fragments. *Science*. 1985;229:81–83.

67. Karawajew L, Micheel B, Behrsing O, Gaestel M. Bispecific antibody-producing hybrid hybridomas selected by a fluorescence activated cell sorter. *J Immunol Methods*. 1987;96:265–270.

68. Songsivilai S, Clissold PM, Lachmann PJ. A novel strategy for producing chimeric bispecific antibodies by gene transfection. *Biochem Biophys Res Commun*. 1989;164:271–276.

69. Goodwin DA, Meares CF, David GF, et al. Monoclonal antibodies as reversible equilibrium carriers of radiopharmaceuticals. *Int J Rad Appl Instrum B*. 1986;13:383–391.

70. Goodwin DA, Meares CF, McCall MJ, McTigue M, Chaovapong W. Pre-targeted immunoscintigraphy of murine tumors with indium-111-labeled bifunctional haptens. *J Nucl Med*. 1988;29:226–234.

71. Primus FJ, Wang RH, Goldenberg DM, Hansen HJ. Localization of human GW-39 tumors in hamsters by radiolabeled heterospecific antibody to carcinoembryonic antigen. *Cancer Res*. 1973;33:2977–2982.

72. Stickney DR, Anderson LD, Slater JB, et al. Bifunctional antibody: a binary radiopharmaceutical delivery system for imaging colorectal carcinoma. *Cancer Res*. 1991;51:6650–6655.

73. Doerr RJ, Abdel-Nabi H, Merchant B. Indium 111 ZCE-025 immunoscintigraphy in occult recurrent colorectal cancer with elevated carcinoembryonic antigen level. *Arch Surg*. 1990;125:226–229.

74. Lamki LM, Patt YZ, Rosenblum MG, et al. Metastatic colorectal cancer: radioimmunoscintigraphy with a stabilized In-111-labeled F(ab')2 fragment of an anti-CEA monoclonal antibody. *Radiology*. 1990;174:147–151.

75. Abdel-Nabi HH, Schwartz AN, Higano CS, Wechter DG, Unger MW. Colorectal carcinoma: detection with indium-111 anticarcinoembryonic-antigen monoclonal antibody ZCE-025. *Radiology*. 1987;164:617–621.

76. Halpern SE, Haindl W, Beauregard J, et al. Scintigraphy with In-111-labeled monoclonal antitumor antibodies: kinetics, biodistribution, and tumor detection. *Radiology*. 1988;168:529–536.

77. Patt YZ, Lamki LM, Shanken J, et al. Imaging with indium111-labeled anticarcinoembryonic antigen monoclonal antibody ZCE-025 of recurrent colorectal or carcinoembryonic antigen-producing cancer in patients with rising serum carcinoembryonic antigen levels and occult metastases. *J Clin Oncol*. 1990;8:1246–1254.

78. Patt YZ, Lamki LM, Haynie TP, et al. Improved tumor localization with increasing dose of indium-111-labeled anti-carcinoembryonic antigen monoclonal antibody ZCE-025 in metastatic colorectal cancer. *J Clin Oncol*. 1988;6:1220–1230.

79. Hnatowich DJ, Virzi F, Rusckowski M. Investigations of avidin and biotin for imaging applications. *J Nucl Med*. 1987;28:1294–1302.

80. Paganelli G, Belloni C, Magnani P, et al. Two-step tumour targeting in ovarian cancer patients using biotinylated monoclonal antibodies and radioactive streptavidin. *Eur J Nucl Med*. 1992;19:322–329.

81. Paganelli G, Magnani P, Zito F, et al. Three-step monoclonal antibody tumor targeting in carcinoembryonic antigen-positive patients. *Cancer Res.* 1991;51:5960–5966.

82. Paganelli G, Pervez S, Siccardi AG, et al. Intraperitoneal radio-localization of tumors pretargeted by biotinylated monoclonal antibodies. *Int J Cancer.* 1990;45:1184–1189.

83. Paganelli G, Riva P, Deleide G, et al. *In vivo* labelling of biotinylated monoclonal antibodies by radioactive avidin: a strategy to increase tumor radiolocalization. *Int J Cancer Suppl.* 1988;2:121–125.

84. Paganelli G, Ferrari M, Ravasi L, et al. Intraoperative avidination for radionuclide therapy: a prospective new development to accelerate radiotherapy in breast cancer. *Clin Cancer Res.* 2007;13:5646s–5651s.

85. Pimm MV, Fells HF, Perkins AC, Baldwin RW. Iodine-131 and indium-111 labelled avidin and streptavidin for pre-targetted immunoscintigraphy with biotinylated antitumour monoclonal antibody. *Nucl Med Commun.* 1988;9:931–941.

86. Kalofonos HP, Rusckowski M, Siebecker DA, et al. Imaging of tumor in patients with indium-111-labeled biotin and streptavidin-conjugated antibodies: preliminary communication. *J Nucl Med.* 1990;31:1791–1796.

87. Hnatowich DJ. The *in vivo* uses of streptavidin and biotin: a short progress report. *Nucl Med Commun.* 1994;15:575–577.

88. Sharkey RM, Karacay H, Griffiths GL, et al. Development of a streptavidin-anticarcinoembryonic antigen antibody, radiolabeled biotin pretargeting method for radioimmunotherapy of colorectal cancer. Studies in a human colon cancer xenograft model. *Bioconjug Chem.* 1997;8:595–604.

89. Mattes MJ. Biodistribution of antibodies after intraperitoneal or intravenous injection and effect of carbohydrate modifications. *J Natl Cancer Inst.* 1987;79:855–863.

90. Ong GL, Ettenson D, Sharkey RM, et al. Galactose-conjugated antibodies in cancer therapy: properties and principles of action. *Cancer Res.* 1991;51:1619–1626.

91. Axworthy DB, Fritzberg AR, Hylarides MD, et al. Preclinical evaluation of an anti-tumor monoclonal antibody/streptavidin conjugate for pretargeted ^{90}Y radioimmunotherapy in a mouse xenograft model. *J Immunother.* 1994;16:158.

92. Kuijpers WH, Bos ES, Kaspersen FM, Veeneman GH, van Boeckel CA. Specific recognition of antibody-oligonucleotide conjugates by radiolabeled antisense nucleotides: a novel approach for two-step radioimmunotherapy of cancer. *Bioconjug Chem.* 1993;4:94–102.

93. Bos ES, Kuijpers WH, Meesters-Winters M, et al. *In vitro* evaluation of DNA–DNA hybridization as a two-step approach in radioimmunotherapy of cancer. *Cancer Res.* 1994;54:3479–3486.

94. Rusckowski M, Qu T, Chang F, Hnatowich DJ. Pretargeting using peptide nucleic acid. *Cancer.* 1997;80:2699–2705.

95. Wang Y, Chang F, Zhang Y, et al. Pretargeting with amplification using polymeric peptide nucleic acid. *Bioconjug Chem.* 2001;12:807–816.

96. Liu G, Mang'era K, Liu N, Gupta S, Rusckowski M, Hnatowich DJ. Tumor pretargeting in mice using ^{99m}Tc-labeled morpholino, a DNA analog. *J Nucl Med.* 2002;43:384–391.

97. Liu G, He J, Dou S, et al. Pretargeting in tumored mice with radiolabeled morpholino oligomer showing low kidney uptake. *Eur J Nucl Med Mol Imaging.* 2004;31:417–424.

98. Knox SJ, Goris ML, Trisler K, et al. Yttrium-90-labeled anti-CD20 monoclonal antibody therapy of recurrent B-cell lymphoma. *Clin Cancer Res.* 1996;2:457–470.

99. Buchsbaum DJ, Wahl RL, Glenn SD, Normolle DP, Kaminski MS. Improved delivery of radiolabeled anti-B1 monoclonal antibody to Raji lymphoma xenografts by predosing with unlabeled anti-B1 monoclonal antibody. *Cancer Res.* 1992;52:637–642.

100. Kaminski MS, Zasadny KR, Francis IR, et al. Radioimmunotherapy of B-cell lymphoma with ^{131}I-anti-B1 (anti-CD20) antibody. *N Engl J Med.* 1993;329:459–465.

101. Kaminski MS, Zasadny KR, Francis IR, et al. Iodine-131-anti-B1 radioimmunotherapy for B-cell lymphoma. *J Clin Oncol.* 1996;14:1974–1981.

102. Witzig TE, White CA, Wiseman GA, et al. Phase I/II trial of IDEC-Y2B8 radio-immunotherapy for treatment of relapsed or refractory CD20(+) B-cell non-Hodgkin's lymphoma. *J Clin Oncol.* 1999;17:3793–3803.

103. Wahl RL. Tositumomab and ^{131}I therapy in non-Hodgkin's lymphoma. *J Nucl Med. 46 Suppl.* 2005;1:128S–140S.

104. Wahl RL, Zasadny KR, MacFarlane D, et al. Iodine-131 anti-B1 antibody for B-cell lymphoma: an update on the Michigan Phase I experience. *J Nucl Med.* 1998;39:21S–27S.

105. Pedley RB, Boden JA, Boden R, Dale R, Begent RH. Comparative radioimmunotherapy using intact or F(ab′)₂ fragments of ^{131}I anti-CEA antibody in a colonic xenograft model. *Br J Cancer.* 1993;68:69–73.

106. Lane DM, Eagle KF, Begent RH, et al. Radioimmunotherapy of metastatic colorectal tumours with iodine-131-labelled antibody to carcinoembryonic antigen: phase I/II study with comparative biodistribution of intact and F(ab′)₂ antibodies. *Br J Cancer.* 1994;70:521–525.

107. Buchegger F, Pelegrin A, Delaloye B, Bischof-Delaloye A, Mach JP. Iodine-131-labeled MAb F(ab′)₂ fragments are more efficient and less toxic than intact anti-CEA antibodies in radioimmunotherapy of large human colon carcinoma grafted in nude mice. *J Nucl Med.* 1990;31:1035–1044.

108. Behr TM, Goldenberg DM. Improved prospects for cancer therapy with radiolabeled antibody fragments and peptides? *J Nucl Med.* 1996;37:834–836.

109. Behr TM, Sharkey RM, Sgouros G, et al. Overcoming the nephrotoxicity of radiometal-labeled immunoconjugates: improved cancer therapy administered to a nude mouse model in relation to the internal radiation dosimetry. *Cancer.* 1997;80:2591–2610.

110. Behr TM, Behe M, Stabin MG, et al. High-linear energy transfer (LET) alpha versus low-LET beta emitters in radioimmunotherapy of solid tumors: therapeutic efficacy and dose-limiting toxicity of ^{213}Bi- versus ^{90}Y-labeled CO17-1A Fab′ fragments in a human colonic cancer model. *Cancer Res.* 1999;59:2635–2643.

111. Behr TM, Blumenthal RD, Memtsoudis S, et al. Cure of metastatic human colonic cancer in mice with radiolabeled monoclonal antibody fragments. *Clin Cancer Res.* 2000;6:4900–4907.

112. Behr TM, Memtsoudis S, Sharkey RM, et al. Experimental studies on the role of antibody fragments in cancer radio-immunotherapy: influence of radiation dose and dose rate on toxicity and anti-tumor efficacy. *Int J Cancer.* 1998;77:787–795.

113. Pagel JM, Hedin N, Subbiah K, et al. Comparison of anti-CD20 and anti-CD45 antibodies for conventional and pretargeted radioimmunotherapy of B-cell lymphomas. *Blood.* 2003;101:2340–2348.

114. Pantelias A, Pagel JM, Hedin N, et al. Comparative biodistributions of pretargeted radioimmunoconjugates targeting CD20, CD22, and DR molecules on human B-cell lymphomas. *Blood*. 2007;109:4980–4987.

115. Sharkey RM, Karacay H, Richel H, et al. Optimizing bispecific antibody pretargeting for use in radioimmunotherapy. *Clin Cancer Res*. 2003;9:3897S–3913S.

116. Sharkey RM, Cardillo TM, Rossi EA, et al. Signal amplification in molecular imaging by pretargeting a multivalent, bispecific antibody. *Nat Med*. 2005;11:1250–1255.

117. Sharkey RM, McBride WJ, Karacay H, et al. A universal pretargeting system for cancer detection and therapy using bispecific antibody. *Cancer Res*. 2003;63:354–363.

118. Saga T, Weinstein JN, Jeong JM, et al. Two-step targeting of experimental lung metastases with biotinylated antibody and radiolabeled streptavidin. *Cancer Res*. 1994;54:2160–2165.

119. Schechter B, Silberman R, Arnon R, Wilchek M. Tissue distribution of avidin and streptavidin injected to mice. Effect of avidin carbohydrate, streptavidin truncation and exogenous biotin. *Eur J Biochem*. 1990;189:327–331.

120. Wilbur DS, Stayton PS, To R, et al. Streptavidin in antibody pretargeting. Comparison of a recombinant streptavidin with two streptavidin mutant proteins and two commercially available streptavidin proteins. *Bioconjug Chem*. 1998;9:100–107.

121. Le Doussal JM, Martin M, Gautherot E, Delaage M, Barbet J. *In vitro* and *in vivo* targeting of radiolabeled monovalent and divalent haptens with dual specificity monoclonal antibody conjugates: enhanced divalent hapten affinity for cell-bound antibody conjugate. *J Nucl Med*. 1989;30:1358–1366.

122. Le Doussal JM, Gruaz-Guyon A, Martin M, Gautherot E, Delaage M, Barbet J. Targeting of indium 111-labeled bivalent hapten to human melanoma mediated by bispecific monoclonal antibody conjugates: imaging of tumors hosted in nude mice. *Cancer Res*. 1990;50:3445–3452.

123. Boerman OC, Kranenborg MH, Oosterwijk E, et al. Pretargeting of renal cell carcinoma: improved tumor targeting with a bivalent chelate. *Cancer Res*. 1999;59:4400–4405.

124. Goodwin DA, Meares CF, McTigue M, et al. Pretargeted immunoscintigraphy: effect of hapten valency on murine tumor uptake. *J Nucl Med*. 1992;33:2006–2013.

125. Axworthy DB, Reno JM, Hylarides MD, et al. Cure of human carcinoma xenografts by a single dose of pretargeted yttrium-90 with negligible toxicity. *Proc Natl Acad Sci USA*. 2000;97:1802–1807.

126. Sharkey RM, Karacay H, Cardillo TM, et al. Improving the delivery of radionuclides for imaging and therapy of cancer using pretargeting methods. *Clin Cancer Res*. 2005;11:7109s–7121s.

127. Cheung NK, Modak S, Lin Y, et al. Single-chain Fv-streptavidin substantially improved therapeutic index in multistep targeting directed at disialoganglioside GD2. *J Nucl Med*. 2004;45:867–877.

128. Gautherot E, Bouhou J, Le Doussal JM, et al. Therapy for colon carcinoma xenografts with bispecific antibody-targeted, iodine-131-labeled bivalent hapten. *Cancer*. 1997;80:2618–2623.

129. Gautherot E, Le Doussal JM, Bouhou J, et al. Delivery of therapeutic doses of radioiodine using bispecific antibody-targeted bivalent haptens. *J Nucl Med*. 1998;39:1937–1943.

130. Karacay H, Brard PY, Sharkey RM, et al. Therapeutic advantage of pretargeted radio-immunotherapy using a recombinant bispecific antibody in a human colon cancer xenograft. *Clin Cancer Res*. 2005;11:7879–7885.

131. Pagel JM, Lin Y, Hedin N, et al. Comparison of a tetravalent single-chain antibody–streptavidin fusion protein and an antibody–streptavidin chemical conjugate for pretargeted anti-CD20 radioimmunotherapy of B-cell lymphomas. *Blood*. 2006;108:328–336.

132. Pagel JM, Pantelias A, Hedin N, et al. Evaluation of CD20, CD22, and HLA-DR targeting for radioimmunotherapy of B-cell lymphomas. *Cancer Res*. 2007;67:5921–5928.

133. Press OW, Corcoran M, Subbiah K, et al. A comparative evaluation of conventional and pretargeted radioimmunotherapy of CD20-expressing lymphoma xenografts. *Blood*. 2001;98:2535–2543.

134. Sato N, Hassan R, Axworthy DB, et al. Pretargeted radioimmunotherapy of mesothelin-expressing cancer using a tetravalent single-chain Fv-streptavidin fusion protein. *J Nucl Med*. 2005;46:1201–1209.

135. Sharkey RM, Karacay H, Chang CH, McBride WJ, Horak ID, Goldenberg DM. Improved therapy of non-Hodgkin's lymphoma xenografts using radionuclides pretargeted with a new anti-CD20 bispecific antibody. *Leukemia*. 2005;19:1064–1069.

136. Subbiah K, Hamlin DK, Pagel JM, et al. Comparison of immunoscintigraphy, efficacy, and toxicity of conventional and pretargeted radioimmunotherapy in CD20-expressing human lymphoma xenografts. *J Nucl Med*. 2003;44:437–445.

137. Yao Z, Zhang M, Axworthy DB, et al. Radioimmunotherapy of A431 xenografted mice with pretargeted B3 antibody-streptavidin and ^{90}Y-labeled 1,4,7,10-tetraazacyclododecane-N,N',N'',N'''-tetraacetic acid (DOTA)-biotin. *Cancer Res*. 2002;62:5755–5760.

138. Zhang M, Yao Z, Garmestani K, et al. Pretargeting radioimmunotherapy of a murine model of adult T-cell leukemia with the alpha-emitting radionuclide, bismuth 213. *Blood*. 2002;100:208–216.

139. Zhang M, Zhang Z, Garmestani K, et al. Pretarget radiotherapy with an anti-CD25 antibody–streptavidin fusion protein was effective in therapy of leukemia/lymphoma xenografts. *Proc Natl Acad Sci USA*. 2003;100:1891–1895.

140. Sharkey RM, Karacay H, Litwin S, et al. Improved therapeutic results by pretargeted radioimmunotherapy of non-Hodgkin lymphoma with a new recombinant, trivalent, anti-CD20, bispecific antibody. *Cancer Res*. 2008;68:5282–5290.

141. Stillebroer AB, Oosterwijk E, Oyen WJ, Mulders PF, Boerman OC. Radiolabeled antibodies in renal cell carcinoma. *Cancer Imaging*. 2007;7:179–188.

142. Steffens MG, Boerman OC, Oosterwijk-Wakka JC, et al. Targeting of renal cell carcinoma with iodine-131-labeled chimeric monoclonal antibody G250. *J Clin Oncol*. 1997;15:1529–1537.

143. Oosterwijk E, Bander NH, Divgi CR, et al. Antibody localization in human renal cell carcinoma: a phase I study of monoclonal antibody G250. *J Clin Oncol*. 1993;11:738–750.

144. Divgi CR, Bander NH, Scott AM, et al. Phase I/II radioimmunotherapy trial with iodine-131-labeled monoclonal antibody G250 in metastatic renal cell carcinoma. *Clin Cancer Res*. 1998;4:2729–2739.

145. Vessella RL, Lange PH, Palme DF, 2nd, Chiou RK, Elson MK, Wessels BW. Radio-iodinated monoclonal antibodies in the imaging and treatment of human renal cell carcinoma xenografts in nude mice. *Targeted Diagn Ther*. 1988;1:245–282.

146. Chiou RK, Vessella RL, Limas C, et al. Monoclonal antibody-targeted radiotherapy of renal cell carcinoma using a nude mouse model. *Cancer.* 1988;61:1766–1775.

147. Sands H, Jones PL, Shah SA, Palme D, Vessella RL, Gallagher BM. Correlation of vascular permeability and blood flow with monoclonal antibody uptake by human clouser and renal cell xenografts. *Cancer Res.* 1988;48:188–193.

148. van Schaijk FG, Oosterwijk E, Molkenboer-Kuenen JD, et al. Pretargeting with bispecific anti-renal cell carcinoma x anti-DTPA(In) antibody in 3 RCC models. *J Nucl Med.* 2005;46:495–501.

149. Saga T, Neumann RD, Heya T, et al. Targeting cancer micrometastases with monoclonal antibodies: a binding-site barrier. *Proc Natl Acad Sci USA.* 1995;92:8999–9003.

150. Fujimori K, Covell DG, Fletcher JE, Weinstein JN. Modeling analysis of the global and microscopic distribution of immunoglobulin G, $F(ab')_2$, and Fab in tumors. *Cancer Res.* 1989;49:5656–5663.

151. Weinstein H, Mazurek AP, Osman R, Topiol S. Theoretical studies on the activation mechanism of the histamine H2-receptor: the proton transfer between histamine and a receptor model. *Mol Pharmacol.* 1986;29:28–33.

152. Zhu H, Jain RK, Baxter LT. Tumor pretargeting for radioimmunodetection and radio-immunotherapy. *J Nucl Med.* 1998;39:65–76.

153. Boerman OC, Sharkey RM, Wong GY, Blumenthal RD, Aninipot RL, Goldenberg DM. Influence of antibody protein dose on therapeutic efficacy of radioiodinated antibodies in nude mice bearing GW-39 human tumor. *Cancer Immunol Immunother.* 1992;35:127–134.

154. Blumenthal RD, Fand I, Sharkey RM, Boerman OC, Kashi R, Goldenberg DM. The effect of antibody protein dose on the uniformity of tumor distribution of radioantibodies: an autoradiographic study. *Cancer Immunol Immunother.* 1991;33:351–358.

155. Jain RK. Physiological barriers to delivery of monoclonal antibodies and other macro-molecules in tumors. *Cancer Res.* 1990;50:814s–819s.

156. Sharkey RM, Goldenberg DM. Perspectives on cancer therapy with radiolabeled mono-clonal antibodies. *J Nucl Med. 46 Suppl.* 2005;1:115S–127S.

157. Goldenberg DM, Sharkey RM, Paganelli G, Barbet J, Chatal JF. Antibody pretargeting advances cancer radioimmunodetection and radioimmunotherapy. *J Clin Oncol.* 2006;24:823–834.

158. Breitz HB, Weiden PL, Beaumier PL, et al. Clinical optimization of pretargeted radio-immunotherapy with antibody–streptavidin conjugate and ^{90}Y-DOTA-biotin. *J Nucl Med.* 2000;41:131–140.

159. Breitz HB, Fisher DR, Goris ML, et al. Radiation absorbed dose estimation for [90]Y--DOTA-biotin with pretargeted NR-LU-10/streptavidin. *Cancer Biother Radiopharm.* 1999;14:381–395.

160. Breitz HB, Weiden PL, Vanderheyden JL, et al. Clinical experience with rhenium-186-labeled monoclonal antibodies for radioimmunotherapy: results of phase I trials. *J Nucl Med.* 1992;33:1099–1109.

161. Tempero M, Leichner P, Dalrymple G, et al. High-dose therapy with iodine-131-labeled monoclonal antibody CC49 in patients with gastrointestinal cancers: a phase I trial. *J Clin Oncol.* 1997;15:1518–1528.

162. Knox SJ, Goris ML, Tempero M, et al. Phase II trial of yttrium-90-DOTA-biotin pretargeted by NR-LU-10 antibody/streptavidin in patients with metastatic colon cancer. *Clin Cancer Res.* 2000;6:406–414.

163. Shen S, Forero A, LoBuglio AF, et al. Patient-specific dosimetry of pretargeted radio-immunotherapy using CC49 fusion protein in patients with gastrointestinal malignancies. *J Nucl Med.* 2005;46:642–651.

164. Forero A, Weiden PL, Vose JM, et al. Phase 1 trial of a novel anti-CD20 fusion protein in pretargeted radioimmunotherapy for B-cell non-Hodgkin lymphoma. *Blood.* 2004;104:227–236.

165. Weiden PL, Breitz HB, Press O, et al. Pretargeted radioimmunotherapy (PRIT) for treatment of non-Hodgkin's lymphoma (NHL): initial phase I/II study results. *Cancer Biother Radiopharm.* 2000;15:15–29.

166. Paganelli G, Grana C, Chinol M, et al. Antibody-guided three-step therapy for high grade glioma with yttrium-90 biotin. *Eur J Nucl Med.* 1999;26:348–357.

167. Cremonesi M, Ferrari M, Chinol M, et al. Three-step radioimmunotherapy with yttrium-90 biotin: dosimetry and pharmacokinetics in cancer patients. *Eur J Nucl Med.* 1999;26:110–120.

168. Grana C, Chinol M, Robertson C, et al. Pretargeted adjuvant radioimmunotherapy with yttrium-90-biotin in malignant glioma patients: a pilot study. *Br J Cancer.* 2002;86: 207–212.

169. Paganelli G, Orecchia R, Jereczek-Fossa B, et al. Combined treatment of advanced oropharyngeal cancer with external radiotherapy and three-step radioimmunotherapy. *Eur J Nucl Med.* 1998;25:1336–1339.

170. Grana C, Bartolomei M, Handkiewicz D, et al. Radioimmunotherapy in advanced ovarian cancer: is there a role for pre-targeting with (90)Y-biotin? *Gynecol Oncol.* 2004;93: 691–698.

171. Le Doussal JM, Barbet J, Delaage M. Bispecific-antibody-mediated targeting of radi-olabeled bivalent haptens: theoretical, experimental and clinical results. *Int J Cancer Suppl.* 1992;7:58–62.

172. Le Doussal JM, Chetanneau A, Gruaz-Guyon A, et al. Bispecific monoclonal antibody-mediated targeting of an indium-111-labeled DTPA dimer to primary colorectal tumors: pharmacokinetics, biodistribution, scintigraphy and immune response. *J Nucl Med.* 1993;34:1662–1671.

173. Chetanneau A, Barbet J, Peltier P, et al. Pretargetted imaging of colorectal cancer recurrences using an ^{111}In-labelled bivalent hapten and a bispecific antibody conjugate. *Nucl Med Commun.* 1994;15:972–980.

174. Moffat FL Jr, Pinsky CM, Hammershaimb L, et al. Clinical utility of external immu-noscintigraphy with the IMMU-4 technetium-99m Fab' antibody fragment in patients undergoing surgery for carcinoma of the colon and rectum: results of a pivotal, phase III trial. The Immunomedics Study Group. *J Clin Oncol.* 1996;14:2295–2305.

175. Hughes K, Pinsky CM, Petrelli NJ, et al. Use of carcinoembryonic antigen radio-immunodetection and computed tomography for predicting the resectability of recurrent colorectal cancer. *Ann Surg.* 1997;226:621–631.

176. McBride WJ, Zanzonico P, Sharkey RM, et al. Bispecific antibody pretargeting PET (immunoPET) with an [124]I-labeled hapten-peptide. *J Nucl Med.* 2006;47:1678–1688.

177. Sharkey RM, Karacay H, Vallabhajosula S, et al. Metastatic human colonic carcinoma: molecular imaging with pretargeted SPECT and PET in a mouse model. *Radiology.* 2008;246:497–507.

178. Gautherot E, Rouvier E, Daniel L, et al. Pretargeted radioimmunotherapy of human colorectal xenografts with bispecific antibody and ^{131}I-labeled bivalent hapten. *J Nucl Med*. 2000;41:480–487.

179. Kraeber-Bodere F, Faivre-Chauvet A, Sai-Maurel C, et al. Toxicity and efficacy of radioimmunotherapy in carcinoembryonic antigen-producing medullary thyroid cancer xenograft: comparison of iodine 131-labeled F(ab')$_2$ and pretargeted bivalent hapten and evaluation of repeated injections. *Clin Cancer Res*. 1999;5:3183s–3189s.

180. Kraeber-Bodere F, Faibre-Chauvet A, Sai-Maurel C, et al. Bispecific antibody and bivalent hapten radioimmunotherapy in CEA-producing medullary thyroid cancer xenograft. *J Nucl Med*. 1999;40:198–204.

181. Kraeber-Bodere F, Bardet S, Hoefnagel CA, et al. Radioimmunotherapy in medullary thyroid cancer using bispecific antibody and iodine 131-labeled bivalent hapten: preliminary results of a phase I/II clinical trial. *Clin Cancer Res*. 1999;5:3190s–3198s.

182. Vuillez JP, Kraeber-Bodere F, Moro D, et al. Radioimmunotherapy of small cell lung carcinoma with the two-step method using a bispecific anti-carcinoembryonic antigen/anti-diethylenetriaminepentaacetic acid (DTPA) antibody and iodine-131 Di-DTPA hapten: results of a phase I/II trial. *Clin Cancer Res*. 1999;5:3259s–3267s.

183. Chatal JF, Campion L, Kraeber-Bodere F, et al. Survival improvement in patients with medullary thyroid carcinoma who undergo pretargeted anti-carcinoembryonic-antigen radioimmunotherapy: a collaborative study with the French Endocrine Tumor Group. *J Clin Oncol*. 2006;24:1705–1711.

184. Barbet J, Campion L, Kraeber-Bodere F, Chatal JF. Prognostic impact of serum calcitonin and carcinoembryonic antigen doubling-times in patients with medullary thyroid carcinoma. *J Clin Endocrinol Metab*. 2005;90:6077–6084.

185. Bardies M, Bardet S, Faivre-Chauvet A, et al. Bispecific antibody and iodine-131-labeled bivalent hapten dosimetry in patients with medullary thyroid or small-cell lung cancer. *J Nucl Med*. 1996;37:1853–1859.

186. Mirallie E, Vuillez JP, Bardet S, et al. High frequency of bone/bone marrow involvement in advanced medullary thyroid cancer. *J Clin Endocrinol Metab*. 2005;90:779–788.

187. Kraeber-Bodere F, Faivre-Chauvet A, Ferrer L, et al. Pharmacokinetics and dosimetry studies for optimization of anti-carcinoembryonic antigen x anti-hapten bispecific antibody-mediated pretargeting of iodine-131-labeled hapten in a phase I radioimmunotherapy trial. *Clin Cancer Res*. 2003;9:3973s–3981s.

188. Kraeber-Bodere F, Rousseau C, Bodet-Milin C, et al. Targeting, toxicity, and efficacy of 2-step, pretargeted radioimmunotherapy using a chimeric bispecific antibody and ^{131}I-labeled bivalent hapten in a phase I optimization clinical trial. *J Nucl Med*. 2006;47:247–255.

189. Pagel JM, Hedin N, Drouet L, et al. Eradication of disseminated leukemia in a syngeneic murine leukemia model using pretargeted anti-CD45 radioimmunotherapy. *Blood*. 2008;111:2261–2268.

190. DeNardo SJ, Richman CM, Kukis DL, et al. Synergistic therapy of breast cancer with Y-90-chimeric L6 and paclitaxel in the xenografted mouse model: development of a clinical protocol. *Anticancer Res*. 1998;18:4011–4018.

191. DeNardo SJ, Kroger LA, Lamborn KR, et al. Importance of temporal relationships in combined modality radioimmunotherapy of breast carcinoma. *Cancer*. 1997;80:2583–2590.

192. Clarke K, Lee FT, Brechbiel MW, Smyth FE, Old LJ, Scott AM. Therapeutic efficacy of anti-Lewis(y) humanized 3S193 radioimmunotherapy in a breast cancer model: enhanced activity when combined with taxol chemotherapy. *Clin Cancer Res.* 2000;6:3621–3628.

193. Gold DV, Modrak DE, Schutsky K, Cardillo TM. Combined ^{90}Yttrium-DOTA-labeled PAM4 antibody radioimmunotherapy and gemcitabine radiosensitization for the treatment of a human pancreatic cancer xenograft. *Int J Cancer.* 2004;109:618–626.

194. Gold DV, Schutsky K, Modrak D, Cardillo TM. Low-dose radioimmunotherapy (^{90}Y-- PAM4) combined with gemcitabine for the treatment of experimental pancreatic cancer. *Clin Cancer Res.* 2003;9:3929s–3937s.

195. Richman CM, DeNardo SJ, O'Donnell RT, et al. High-dose radioimmunotherapy combined with fixed, low-dose paclitaxel in metastatic prostate and breast cancer by using a MUC-1 monoclonal antibody, m170, linked to indium-111/yttrium-90 via a cathepsin cleavable linker with cyclosporine to prevent human anti-mouse antibody. *Clin Cancer Res.* 2005;11:5920–5927.

196. Sharkey RM, Hajjar G, Yeldell D, et al. A phase I trial combining high-dose ^{90}Y-labeled humanized anti-CEA monoclonal antibody with doxorubicin and peripheral blood stem cell rescue in advanced medullary thyroid cancer. *J Nucl Med.* 2005;46:620–633.

197. Wong JY, Shibata S, Williams LE, et al. A phase I trial of ^{90}Y-anti-carcinoembryonic antigen chimeric T84.66 radioimmunotherapy with 5-fluorouracil in patients with metastatic colorectal cancer. *Clin Cancer Res.* 2003;9:5842–5852.

198. Graves SS, Dearstyne E, Lin Y, et al. Combination therapy with pretarget CC49 radioimmunotherapy and gemcitabine prolongs tumor doubling time in a murine xenograft model of colon cancer more effectively than either monotherapy. *Clin Cancer Res.* 2003;9:3712–3721.

199. Kraeber-Bodere F, Sai-Maurel C, Campion L, et al. Enhanced antitumor activity of combined pretargeted radioimmunotherapy and paclitaxel in medullary thyroid cancer xenograft. *Mol Cancer Ther.* 2002;1:267–274.

200. Bartolomei M, Mazzetta C, Handkiewicz-Junak D, et al. Combined treatment of glioblastoma patients with locoregional pre-targeted ^{90}Y-biotin radioimmunotherapy and temozolomide. *Q J Nucl Med Mol Imaging.* 2004;48:220–228.

201. Sharkey RM, Burton J, Goldenberg DM. Radioimmunotherapy of non-Hodgkin's lymphoma: a critical appraisal. *Expert Rev Clin Immunol.* 2005;1:47–62.

202. Rossi EA, Chang CH, Losman MJ, et al. Pretargeting of carcinoembryonic antigen-expressing cancers with a trivalent bispecific fusion protein produced in myeloma cells. *Clin Cancer Res.* 2005;11:7122s–7129s.

203. Rossi EA, Goldenberg DM, Cardillo TM, McBride WJ, Sharkey RM, Chang CH. Stably tethered multifunctional structures of defined composition made by the dock and lock method for use in cancer targeting. *Proc Natl Acad Sci USA.* 2006;103:6841–6846.

204. Chang CH, Sharkey RM, Rossi EA, et al. Molecular advances in pretargeting radioimmunotherapy with bispecific antibodies. *Mol Cancer Ther.* 2002;1:553–563.

205. Karacay H, McBride WJ, Griffiths GL, et al. Experimental pretargeting studies of cancer with a humanized anti-CEA x murine anti-[In-DTPA] bispecific antibody construct and a ^{99m}Tc-/^{188}Re-labeled peptide. *Bioconjug Chem.* 2000;11:842–854.

206. Janevik-Ivanovska E, Gautherot E, Hillairet de Boisferon M, et al. Bivalent hapten-bearing peptides designed for iodine-131 pretargeted radioimmunotherapy. *Bioconjug Chem.* 1997;8:526–533.

207. Griffiths GL, Chang CH, McBride WJ, et al. Reagents and methods for PET using bispecific antibody pretargeting and ^{68}Ga-radiolabeled bivalent hapten-peptide-chelate conjugates. *J Nucl Med*. 2004;45:30–39.

208. Sharkey RM, Karacay H, McBride WJ, Rossi EA, Chang CH, Goldenberg DM. Bispecific antibody pretargeting of radionuclides for immuno single-photon emission computed tomography and immuno positron emission tomography molecular imaging: an update. *Clin Cancer Res*. 2007;13:5577s–5585s.

209. Morel A, Darmon M, Delaage M. Recognition of imidazole and histamine derivatives by monoclonal antibodies. *Mol Immunol*. 1990;27:995–1000.

210. Rossi EA, Sharkey RM, McBride W, et al. Development of new multivalent-bispecific agents for pretargeting tumor localization and therapy. *Clin Cancer Res*. 2003;9:3886s–3896s.

211. Chang CH, Rossi EA, Goldenberg DM. The dock and lock method: a novel platform technology for building multivalent, multifunctional structures of defined composition with retained bioactivity. *Clin Cancer Res*. 2007;13:5586s–5591s.

CHAPTER 9

Targeted Auger Electron Radiotherapy of Malignancies

RAYMOND M. REILLY AND AMIN KASSIS

9.1 INTRODUCTION

In 1925, a young physicist named Pierre Victor Auger published a paper describing a new phenomenon that later became known as the Auger effect (1). He reported that when a cloud chamber was irradiated with low-energy X-ray photons, multiple electron tracts were observed and concluded that this event is a consequence of the ejection of inner-shell electrons from the irradiated atoms, the creation of primary electron vacancies within the shells of these atoms, a complex series of vacancy cascades, and the ejection of very low-energy electrons. It was later recognized that such low-energy electrons are also ejected by many radionuclides decaying by electron capture (EC) and/or internal conversion (IC), two processes that also introduce primary vacancies in the inner electronic shells of the daughter atoms. These vacancies, upon being filled by electrons from higher shells, move toward the outermost shell. These transitions are accompanied by the emission of either a characteristic atomic X-ray photon or as low-energy monoenergetic electrons (collectively known as Auger electrons). Typically, an atom undergoing EC and/or IC emits several (e.g., from 5 to $\sim$50) electrons (Table 9.1). As a consequence of their very low energies, these light, negatively charged particles travel in contorted paths and their range in water is very short (a few nanometers up to $\sim$0.5 μm). Furthermore, the ejection of these electrons leaves the decaying atoms transiently with a high positive charge and leads to the deposition of highly localized energy around the decay site. However, unlike energetic electron emitters (e.g., [131]I, [90]Y), whose linear energy transfer (LET) is low ($\sim$0.2 keV/μm) along most of their rather long linear path (millimeter to centimeter in tissue) with the ionizations occurring sparingly, the LET of Auger electrons is 10–100-fold higher (from $\sim$2 to $\sim$25 keV/μm) especially at very low energies (Fig. 9.1) with the ionizations clustered within several cubic nanometers around the point of

Monoclonal Antibody and Peptide-Targeted Radiotherapy of Cancer, Edited by Raymond M. Reilly
Copyright © 2010 John Wiley & Sons, Inc.

TABLE 9.1 Properties of Auger Electron-Emitting Radionuclides[a]

Radionuclide	Half-Life	Auger Electrons/ Decay	IC Electrons/ Decay	Total Electron Energy/Decay (keV)	Total γ- or X-Radiation/ Decay (keV)	Ratio of p/e Radiation/ Decay[b]
[125]I	57 days	24.9	0.9	19.4	42.0	2.2
[123]I	13 h	14.9	0.2	27.6	172.8	6.3
[67]Ga	78 h	4.7	0.3	34.4	167.2	4.9
[99m]Tc	6 h	4.0	1.1	16.3	126.3	7.7
[111]In	67 h	14.7	0.2	32.7	386.5	11.8
[201]Tl	73 h	36.9	1.1	45.5	93.0	2.0

[a] Adapted from Buchegger F et al. *Eur J Nucl Med Mol Imaging* 2006;33:1352–1363.
[b] Ratio of penetrating (γ- or X-) radiation to nonpenetrating (Auger and IC electron) energy emitted per decay.

decay (Table 9.2) (2, 3). The ejection of these electrons also leaves the decaying atoms transiently with a high positive charge and leads to the deposition of highly localized energy around the decay site (4–6). The dissipation of the potential energy associated with the high positive charge and its neutralization may, in principle, also act concomitantly and be responsible in part for the observed radiobiological effects.

9.2 RADIOBIOLOGICAL EFFECTS OF AUGER ELECTRONS

Throughout the first half of the twentieth century, the scientific community showed little interest in pursuing the radiobiological effects of low-energy electrons. In the

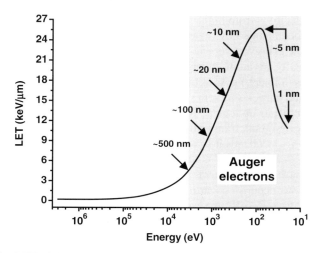

FIGURE 9.1 LET of electrons as a function of their energy. Arrows indicate the range of electrons at each specified energy.

TABLE 9.2   Electron Energies Required to Traverse Various Conformational States of DNA in Mammalian Cells and Their Respective LET

Conformational DNA States	Diameter (nm)	Electron Energy (eV)	LET[a] (keV/μm)
Double-stranded DNA	2	45	18
Two loops of double-stranded DNA on nucleosome	6	140	25
Nucleosome	11	245	21
Chromatin fiber	30	550	15

[a] Values derived from Fig. 9.1.

early 1960s, Carlson and White (7) demonstrated that the decay of the Auger electron emitter iodine-125 (^{125}I)—covalently bound to an ethyl or a methyl group—leads to extensive ionization and fragmentation of these molecules. By the late 1960s and thereafter, various groups began to report on the radiobiological effects of this and other Auger electron emitters in prokaryotic and eukaryotic organisms.

9.2.1 Relative Biological Effectiveness

The three-dimensional organization of chromatin within the mammalian cell nucleus involves many structural level compactions (e.g., chromatin fibers, nucleosomes, and double-stranded DNA). Since the dimensions of these DNA conformational states are all within the range of the high-LET, low-energy, short-range electrons (Fig. 9.1), the toxicity of Auger-emitting radionuclides is expected to be very high and will depend critically on the position of the decaying atom within the cell. This expectation has been substantiated by the results of *in vitro* studies showing that the decay of Auger electron emitters ^{77}Br, ^{123}I, and ^{125}I covalently bound to nuclear DNA leads to monoexponential decreases in mammalian cell survival (5, 8, 9). While each of these curves exhibited a unique slope, it soon became apparent (10) that a single slope is obtained when the dose to the cell nucleus was calculated (relative biological effectiveness (RBE) = ~7). In comparison, the decay of Auger electron emitters (e.g., ^{51}Cr, ^{67}Ga, ^{75}Se, ^{125}I, ^{201}Tl) within the cell cytoplasm (4, 6, 11–14), affixed to cell plasma membranes (14, 15), or located extracellularly (4–6, 11, 12, 14, 16) produces no extraordinary lethal effects (RBE < 2), and the survival curves (i) resemble those observed with X-ray (have a distinct shoulder) and (ii) exhibit shallow slopes.

Studies have also examined and compared the radiotoxicity of ^{125}I in mammalian cells when the radioelement is (i) incorporated into nuclear DNA consequent to *in vitro* incubations of mammalian cells, (ii) adjacent to DNA consequent to *in vitro* incubations of mammalian cells with ^{125}I-intercalators of DNA, ^{125}I-minor/major-groove binders of DNA, ^{125}I-oligonucleotide that forms a triplex with double-stranded DNA, and (iii) bound to transcriptional elements via ^{125}I-hormones. However, unlike the consistently high-LET-like survival curves obtained when the decaying atom is positioned within the double-stranded DNA, the prediction of radiotoxicity parameters (e.g., high/low-LET-like, D_0) is uncertain for situations in which the Auger electron-emitting atom is located in close proximity to nuclear DNA. For example, while exponential decreases in clonogenic survival are observed after incubation of mammalian cells with the DNA-intercalating ^{125}I-acridines (17, 18), the DNA-adduct-forming [195m]Pt-labeled *trans*-platinum (19), [123]I- and ^{125}I-labeled estrogens (20–29), and the DNA-minor-groove-binder ^{125}I-Hoechst ([125]IH) (30, 31), or with a triplex-forming ^{125}I-labeled synthetic oligonucleotide (32) results in a low-LET, linear-quadratic survival curve. In the latter case, the radiotoxicity is almost three orders of magnitude less than that of DNA-incorporated [125]I. This unpredictability in the radiobiologic response of mammalian cells has been underscored by recent studies showing that the incubation of Chinese hamster ovary cells (as opposed to Chinese hamster lung fibroblasts) with [125]IH produces a monoexponential decrease in survival (33).

9.2.2 DNA Damage

As mentioned above, internal emitters that undergo radioactive decay by EC and/or IC result in the emission of a surge of low-energy (<1 keV) Auger electrons. Since many of these electrons traverse a very short distance (few nanometers), the density of the hydroxyl radicals ($^{\bullet}OH$) generated will be very high around the decaying atom and will decrease drastically as a function of length of their tortuous path. Accordingly, when the radionuclide atoms are uniformly distributed in medium, their decay will result in the formation of randomly dispersed "hot spots" (volumes densely traversed by electrons and occupied by $^{\bullet}OH$ and other radicals) and "cold spots" (volumes sparsely traversed by electrons and deficient in $^{\bullet}OH$ and other radicals). Dosimetric calculations have also shown that, for example, when ^{125}I decays, (i) a very high dose ($\sim10^9$ cGy/decaying atom) is deposited in the immediate vicinity ($\sim2\,nm^3$) of the decay site (5), (ii) there is a sharp and significant drop in the energy deposited (from $\sim10^9$ to $\sim10^6$ cGy) as a function of increasing distance (few nanometers) from the decaying ^{125}I atom (3, 5, 6, 34, 35), and (iii) the DSB yield per decay drops precipitously as the ^{125}I atom is displaced a few angstroms from the central axis of DNA (36). Consistent with these dosimetric expectations, studies have demonstrated that when ^{125}I atoms are positioned in close proximity to the DNA molecules, either by the incorporation of an ^{125}I-labeled nucleoside analog into DNA or by the binding of an ^{125}I-labeled DNA groove binder or intercalator, its decay leads to the efficient induction of single-strand breaks (SSB) and double-strand breaks (DSB) in naked DNA (e.g., synthetic oligonucleotides, bacterial DNA, and plasmid DNA) and in eukaryotic chromatin (reviewed in Ref. 3). In contrast, ^{125}I is quite ineffective at inducing DSB when its atoms decay at a distance from the DNA molecule (e.g., ^{125}I-antipyrine) (37, 38). In the former case, the experimental results have also supported the mechanistic notions that (i) the bulk of SSB formed are principally consequent to indirect, $^{\bullet}OH$-mediated ionizations (38, 39); and (ii) DSB in naked

DNA molecules (~0.5–1.0 per decaying ^{125}I atom) are induced mainly by direct ionizations (38–41). However, this was found not to be true when the Auger electron emitter decays in eukaryotic chromatin whereby DSB has been shown to be caused by indirect, $^\bullet$OH-mediated ionizations (42–44).

Many questions concerning the biophysical mechanisms of DSB production by the decay of Auger electron-emitting ^{125}I remain, including the role of DNA supercoiling (bending/compaction). Previously, Kassis et al. had proposed that chromatin structure (highly compacted DNA) provides conditions conducive to the formation of >1 DSB per ^{125}I decay by indirect mechanism(s) since a cluster of $^\bullet$OH produced by decay of ^{125}I may attack a DNA site that is hundreds of nucleotides away from the decaying atom but placed in close proximity to it by the supercoiling of DNA (42). More recently, these authors used naked plasmid DNA to clarify further the role of DNA compaction in the production of radiation-induced DSB (45). Surprisingly, these studies have shown that the DSB yield consequent to the decay of ^{125}I is markedly *reduced* (approximately three times) by supercoiling of plasmid DNA, an observation that is contrary to the experimental findings in mammalian cells (42). Additionally, whereas the DSB in the supercoiled form are caused solely by direct ionizations, those induced in relaxed circular and linear plasmid DNA forms are consequent to direct ionizations and $^\bullet$OH-mediated indirect ionizations. These observations underscore the failure of current dosimetric methods to predict the magnitude of DSB and the need for developing more comprehensive models that include DNA topology for examining the biophysical mechanisms underlying DSB produced by Auger emitters.

9.2.3 Bystander Effects

A new finding that challenges the past half-century's central tenet (energy deposition within a cell is necessary for radiobiologic effects) is the existence of a bystander effect (46–51) (see Chapter 14). In this phenomenon, unirradiated cells react to signals that originate in nearby irradiated cells. For instance, investigators have demonstrated that mammalian cells display altered survival rates and increased genetic changes when α- and β-particles traverse a small fraction of the cell population. Increases in neoplastic transformation frequencies of bystander mammalian cells have also been reported.

The assessment of possible bystander effects consequent to the decay of low-energy electrons has also been pursued (49, 51–53). For example, because of their unique physical decay characteristics and the virtual absence of cross-fire irradiation of adjacent cells, the bystander effect consequent to the decay of the Auger electron emitters iodine-123 ($T_{1/2} = 13.3$ h) and iodine-125 ($T_{1/2} = 60.5$ days) have been assessed (49, 51). *In vivo* studies have shown that the growth of subcutaneously implanted human tumor cells was influenced by the presence of ^{125}I- or ^{123}I-labeled cells within the inoculated tumor cells. Despite the fact that the electron spectra of both radionuclides are identical, the injection of a mixture of unlabeled and ^{125}I-labeled cells in mice inhibited the growth of unlabeled, unirradiated cells (49), whereas the mixture with ^{123}I-labeled cells enhanced the growth of unlabeled cells (51).

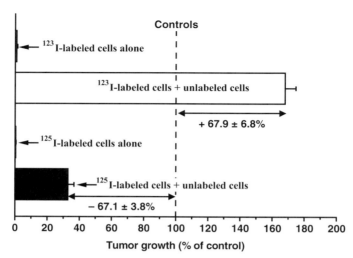

FIGURE 9.2 Inhibitory (^{125}IUdR) and stimulatory (^{123}IUdR) bystander effects consequent to decay of either Auger electron emitter within human tumors grown s.c. in mice. Adapted from Refs 13, 15.

Interestingly, the percentage decrease (^{125}I-induced) and increase (^{123}I-induced) in tumor growth were exactly the same for both isotopes (Fig. 9.2). Similar inhibitory (^{125}I) and stimulatory (^{123}I) bystander effects were also observed when the radio-labeled cells were incubated *in vitro* with unlabeled cells (51). It had been previously reported that cell survival in a three-dimensional tissue culture model was similarly compromised when cells were cocultured with ^{125}I-labeled cells (52).

The bystander effect induced by radioactive decay has impacted our views on risk assessment and therapeutic efficacy following the administration of radiopharmaceuticals to patients, particularly since many γ-emitting radionuclides used traditionally for imaging in nuclear medicine are also Auger electron emitters (e.g., ^{111}In, ^{123}I, ^{67}Ga, and ^{99m}Tc). Traditionally, dose estimations are carried out by averaging the radiation dose to cells within a tissue or tumor mass from radioactive atoms present on or within the cells (self-dose) and that from radionuclides present in/on other cells or in the extracellular fluids (cross-dose). Such absorbed dose estimates have played an important role in determining the amount of radioactivity to be administered to patients in diagnostic/therapeutic procedures (see Chapter 13). When a bystander effect is factored in, the actual radiobiological response will obviously be greater/less than that predicted by dosimetric estimates alone (see Chapter 14).

9.3 SELECTION OF AN AUGER ELECTRON-EMITTING RADIONUCLIDE

Many Auger electron-emitting radionuclides are available for radiolabeling monoclonal antibodies (mAbs), or peptides, and small organic molecules for targeted radiotherapy of cancer and are commercially produced (Table 9.1) (54). From a radiation physics point of view, their relative potency can be appreciated by examining the total number of electrons emitted per decay (both Auger and IC electrons) as well as the total energy released. ^{125}I has the greatest electron yield (25.8 electrons/decay) and total energy (19.4 keV). ^{111}In and ^{123}I have intermediate electron yield (14.9 and 15.1 electrons/decay, respectively) and total energy (32.7 and 27.6 keV, respectively). ^{67}Ga releases energy (34.4 keV) similar to ^{111}In or ^{123}I, but carried by fewer electrons (5.0 electrons/decay) thus providing a greater overall average electron energy. In particular, ^{67}Ga emits an abundant electron with energy of 8.4 keV that has a range in tissue of 2.0 μm, whereas the range of 99% of the electrons emitted by ^{111}In or ^{125}I is much less than 1 μm (55). It has been argued that ^{67}Ga is potentially a more useful Auger electron emitter than ^{111}In or ^{125}I for targeted radiotherapy, because these higher energy electrons could deposit DNA-damaging energy in the nucleus from the cytoplasm or cell surface, whereas ^{111}In and ^{125}I are most effective when they decay in the nucleus (56). ^{99m}Tc provides a similar electron yield (5.1 electrons/decay) as ^{67}Ga but twofold lower total energy (16.3 keV).

Besides the electron yield and energy, it is important to consider the ratio of penetrating (X- and γ-) to nonpenetrating (electron or β-particle) forms of radiation (p/e), since many Auger electron emitters also emit γ-radiation—this property has been exploited for decades in nuclear medicine for imaging. It has been proposed that an ideal therapeutic radionuclide should have a p/e ratio of ≤ 2 (57)—this is satisfied for ^{125}I (p/e = 2.2) but not for ^{111}In (p/e = 11.8) (Table 9.1). The p/e ratios are also reflected in the percentage of total energy that is emitted as Auger or IC electrons for the different radionuclides. Especially at high doses, the moderate-high energy but low LET γ-emissions from ^{111}In ($E\gamma = 171$ and 245 keV) can irradiate and potentially kill nontargeted normal cells including bone marrow stem cells—this would detract from the exquisite single cell killing that is theoretically possible for Auger electron emitting radiotherapeutics that are specifically inserted into tumor cells, due to the subcellular range of the electrons. Nonetheless, the γ-radiation from ^{111}In, ^{123}I, or ^{99m}Tc is helpful in that it can be exploited to visualize the tumor and normal tissue distribution of these radiotherapeutic agents in patients and in preclinical animal models by single photon emission computed tomography (SPECT).

Two other key factors to consider in selecting an Auger electron emitter for radiotherapy are (i) its physical half-life ($t_{1/2p}$) (Table 9.1) and (ii) the radiochemistry available for labeling the targeting vehicle (see Chapter 2). If the $t_{1/2p}$ is too long, the radiotherapeutic agent (or a radioactive catabolite) may redistribute away from the targeted cell before sufficient decays have occurred to cause lethal damage. One well-known example of this phenomenon is the use of ^{125}I-labeled estradiol (E$_2$) analogues for targeted radiotherapy of hormone-sensitive (i.e., ER-positive) tumors. The biological half-life ($t_{1/2b}$) of the E2/ER complex is only 4 h, whereas the $t_{1/2p}$ of ^{125}I is 57 days—less than 1% of the total radionuclide decays yielding Auger electrons would occur in this very short time frame. This realization led to the use of shorter $t_{1/2p}$ Auger electron emitters such as ^{80m}Br or ^{123}I for labeling the E2 analogues ($t_{1/2p}$ of 4.4 and 13.2 h, respectively). The maximum theoretical specific activity (SA) of a radionuclide is inversely proportional to its $t_{1/2p}$—this is an important consideration, especially for targeting low abundence receptors or epitopes on cancer cells. Additionally, when the radionuclide used to label these carrier molecules has low SA, the targeting vehicles carrying stable nuclides will not irradiate cells but will compete with radiolabeled molecules for binding to limited receptors or antigens on these cells. Finally, it is critical to consider the radiochemistry. Radioiodinated proteins and peptides are unstable *in vivo* unless "residualizing chemistry" is utilized (see Chapter 2). Deiodination *in vivo* will allow redistribution of released radionuclides to nontarget tissues, thereby diminishing their effectiveness for killing cancer cells while potentially increasing their toxicity to normal cells. In contrast, complexation of radiometals such as ^{111}In, ^{67}Ga, or ^{99m}Tc to targeting vehicles is much more stable *in vivo* and moreover, allows kit formulation for simple "instant" radiolabeling just prior to patient administration.

9.4 MICRODOSIMETRY

The radiation dosimetry of targeted radiotherapeutic agents is discussed more broadly and in detail in Chapter 13, but it is useful to highlight the microdosimetry of Auger electrons and compare this to the more well-known macrodosimetry of radiopharmaceuticals. The Medical Internal Radiation Dosimetry (MIRD) formalism states that the radiation absorbed dose deposited in a target compartment from radioactivity accumulated in a source compartment can be modeled and is described by the following equation:

$$\overline{D}_{T \leftarrow S} = \tilde{A}_S \times S$$

where, $\overline{D}_{T \leftarrow S}$ is the radiation absorbed dose deposited in the target compartment per unit of radioactivity in the source compartment (Gy/Bq), $\tilde{A}_S$ is the cumulated radioactivity in the source compartment (Bq·s), and S is a modeling factor based on an idealized phantom that takes into account the geometry of the source and target compartments as well as the physical decay properties of the radionuclide. While in macrodosimetry, the source and target compartments are whole organs or tissues (e.g., blood), those used to estimate the microdosimetry of Auger electrons are subcellular (i.e., the cell surface (CS), cytoplasm (C), or nucleus (N)). S values for phantom models of spherical idealized cells with different radii or having a different sized nuclei have been reported by Goddu et al.—these allow estimates of the microdosimetry for many radionuclides, including Auger electron emitters to be calculated (58). Consider the following simplified example of microdosimetry for an Auger electron-emitting radiotherapeutic agent. Assume that a cancer cell with radius of 5 μm and having a nucleus with radius of 3 μm expressing 1×10^6 epitopes was targeted to saturation by an ^{111}In-labeled mAb with a SA of 740 MBq/mg (1.1×10^8 GBq/mol). Using Avagadro's number, the maximum number of moles of ^{111}In-labeled mAbs that would bind to a single cell would be 1×10^6 receptors/6.02×10^{23} receptors/mol = 1.7×10^{-18} mol. At an SA of 1.1×10^8 GBq/mol, there would be a total of 0.2 Bq targeted to each cell. Assuming that this total radioactivity was bound and distributed almost instantaneously within the cell such that 20% localized to the cell surface, 70% was internalized into the cytoplasm and 10% was imported into the nucleus, there would be 0.04, 0.14, and 0.02 Bq in each of these source compartments, respectively. If there was no biological elimination of radioactivity (i.e., elimination was only by radioactive decay), the cumulative radioactivity in each of these source compartments can be calculated using the decay constant of ^{111}In ($\lambda = 2.87 \times 10^{-6}$ s) as follows:

$$\tilde{A}_{CS} = \frac{0.04\,\text{Bq}}{2.87 \times 10^{-6}\,\text{s}^{-1}} = 1.4 \times 10^4\,\text{Bq} \cdot \text{s}$$

$$\tilde{A}_{C} = \frac{0.14\,\text{Bq}}{2.87 \times 10^{-6}\,\text{s}^{-1}} = 4.9 \times 10^4\,\text{Bq} \cdot \text{s}$$

$$\tilde{A}_{N} = \frac{0.02\,\text{Bq}}{2.87 \times 10^{-6}\,\text{s}^{-1}} = 0.7 \times 10^4\,\text{Bq} \cdot \text{s}$$

Using the reported S values for ^{111}In for deposition of radiation absorbed dose in the nucleus (the critical radiosensitive target compartment) for a cell with the previously specified dimensions, the individual microdosimetry estimates can be obtained (58) as follows:

$$D_{N \leftarrow CS} = \tilde{A}_{CS} \times S_{N \leftarrow CS} = (1.4 \times 10^4\,\text{Bq} \cdot \text{s})(1.8 \times 10^{-4}\,\text{Gy/Bq} \cdot \text{s}) = 2.52\,\text{Gy}$$
$$D_{N \leftarrow C} = \tilde{A}_{CS} \times S_{N \leftarrow C} = (4.9 \times 10^4\,\text{Bq} \cdot \text{s})(3.18 \times 10^{-4}\,\text{Gy/Bq} \cdot \text{s}) = 15.58\,\text{Gy}$$
$$D_{N \leftarrow N} = \tilde{A}_{N} \times S_{N \leftarrow N} = (0.7 \times 10^4\,\text{Bq} \cdot \text{s})(6.0 \times 10^{-3}\,\text{Gy/Bq} \cdot \text{s}) = 42.21\,\text{Gy}$$

The estimated total radiation absorbed dose to the nucleus from these ^{111}In-labeled mAbs is 60.3 Gy. For comparison, fractionated doses up to a total of 50 Gy of external radiation are used for treatment of malignancies (59). It is important to recognize, however, that this is a simplified calculation. It is unlikely in reality that a cancer cell would be targeted to epitopic saturation by ^{111}In-labeled mAbs—this will substantially reduce the radiation absorbed dose deposited. Similarly, the radioactivity may take some time to accumulate in the cells or be biologically eliminated from them or redistribute to subcellular compartments, which would similarly modify the radiation absorbed dose. Geometric differences in the size of the cell or the intracellular compartments will also affect the dose deposited in the nucleus. Nonetheless, this simplified calculation provides an appreciation of the method of microdosimetry as well as a recognition of the effect of subcellular, especially nuclear localization on the radiation absorbed dose deposited in the nucleus (the "radiation sensitive target"). It can be seen from the above example, that the nuclear radioactivity which was only 10% of the total cellular radioactivity, was responsible for depositing 70% of the radiation absorbed dose in the nucleus. This clearly demonstrates that it is very important to achieve nuclear importation of Auger electron-emitting radiotherapeutics in order maximize their lethal effects on tumor cells. One of the major limitations of microdosimetry is that it is not possible to determine the subcellular distribution of radionuclides in a patient simply from an imaging study (e.g., SPECT or PET), whereas, such studies are commonly used to estimate macrodosimetry, since this requires only an estimate of the total cumulative radioactivity in a source organ (not its subcellular distribution) (60, 61). Thus, various assumptions based on preclinical biodistribution and subcellular fractionation data may ultimately need to be applied to estimate the radiation absorbed microdoses to cancer or normal cells in patients from Auger electron-emitting radiotherapeutic agents—these could yield significant errors, the magnitude of which would not be known with much certainty.

9.4.1   The "Cross-Dose" Contribution

The "cross-fire" effect describes the irradiation and killing of nontargeted cells by the moderate energy and long-range (2–10 mm) β-particles emitted by radionuclides such as ^{131}I or ^{90}Y that have been commonly conjugated to radiotherapeutic agents (62). This effect can be beneficial for eradicating nontargeted cancer cells in a large solid tumor mass in which there may be heterogeneous distribution of these agents. However, it is responsible for the nonspecific and dose-limiting myelotoxicity of radioimmunotherapy (RIT)—this is caused by the persistently high circulating levels of [131]I- or ^{90}Y-labeled mAbs perfusing the bone marrow. Due to their ultrashort nanometer–micrometer range, Auger electron-emitting radionuclides do not have such a cross-fire effect, which in theory should allow exquisite single-cell killing. Nonetheless, these radionuclides have an analogous more localized "cross-dose" (cd) effect, which describes the deposition of energy in the nucleus of proximal nontargeted cells by radioactivity on targeted cells within a small tumor cell cluster <400 μm (about 40 cells) in diameter (63). The radiation absorbed dose deposited in the nucleus of the targeted cell itself is known as the "self-dose" (sd). The sd/cd ratio reflects the proportion of radiation deposited in a cell that is contributed by the cross-dose—the smaller this ratio, the greater the proportion contributed by the cross-dose. Since the sd is highly restricted for Auger electron emitters, the proportion of cd increases dramatically as the tumor cluster diameter increases, and consequently, larger clusters of cells have the lowest sd/cd ratios (i.e., most of the radiation deposited in tumor cells is actually due to the cross-dose). The sd/cd ratios depend on the emission properties of the radionuclides—the highest ratios are found for ^{67}Ga; intermediate ratios for ^{111}In and ^{99m}Tc; and lowest ratios for ^{123}I and [125]I (63). The cross-dose effect may allow irradiation and killing of some very proximal nontargeted tumor cells by cells that have been targeted by Auger electron-emitting radiotherapeutic agents, thereby locally enhancing the antitumor effects, but is insufficient to kill more distant, nontargeted cells. In addition, it has been shown that Auger electron emitters have a "bystander" effect (see Chapter 14), which describes the killing of nontargeted tumor cells by biochemical mediators of cell death that are released by targeted and killed tumor cells (49, 51–53).

9.5   MOLECULAR TARGETS FOR AUGER ELECTRON RADIOTHERAPY OF CANCER

In the following sections, various molecular targets on cancer cells that have been exploited for Auger electron radiotherapy of maligancies using mAbs or peptides will be discussed.

9.5.1   DNA Synthesis Pathways as a Target

The very high *in vitro* radiotoxicity observed with DNA-incorporated Auger electron emitters has been exploited in experimental radionuclide therapy. In most of these *in vivo* studies, the thymidine analog 5-iodo-2′-deoxyuridine (^{123}IUdR/^{125}IUdR) has been used (64–67) and the results have been very promising. For example, the intraperitoneal (i.p.) injection of ^{125}IUdR (64) or ^{123}IUdR (65) into mice with i.p. ovarian cancer has led to a 4–5 log reduction in tumor cell survival. Similarly, the intrathecal (i.t.) administration of therapeutic doses of ^{125}IUdR into rats with i.t. tumors significantly delayed the onset of paralysis, especially when the radiopharmaceutical was coadministered with methotrexate (MTX), an antimetabolite that enhances IUdR uptake by DNA-synthesizing cells. In the latter cases, this was exemplified by a 5–6 log tumor cell kill and the curing of ∼30% of the tumor-bearing rats (66, 67). Consequent to these promising results, Kassis et al. administered MTX and ^{125}IUdR (1.85 GBq) to a patient with pancreatic cancer metastatic to the CNS who had failed to respond to conventional therapy (68). A dramatic drop in spinal fluid CA19.9 level was observed after a single treatment with the radiolabeled agent that was accompanied by clinical improvement (Fig. 9.3). These findings suggest that this may be an effective treatment for neoplastic meningitis.

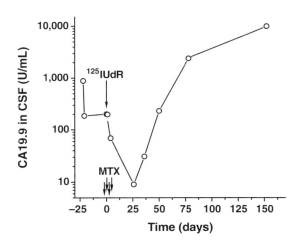

FIGURE 9.3   Therapeutic response of human antigen in patient with intrathecal tumor following intrathecal administration of MTX and ^{125}IUdR. Adapted from Ref. 68.

9.5.2 Somatostatin Receptors

Somatostatin receptors (SSTRs) are a family of 7-helix transmembrane G-protein-coupled receptors that specifically bind somatostatin (SMS), a naturally occurring cyclic 14- or 28-amino acid peptide (69, 70). There are five subclasses of SSTRs (sst_1, sst_2, sst_3, sst_4, and sst_5). SMS has many physiological actions including inhibition of exocrine and endocrine secretions, decreasing intestinal motility as well as antiproliferative effects (71). SMS_{14} and SMS_{28} have low nanomolar affinity for all five subclasses of SSTRs (72, 73). SSTRs are increased in most neuroendocrine gastroenteropancreatic (GEP) tumors as well as in some other malignancies including breast cancer, neuroblastoma, and lymphomas (74). Native SMS is rapidly degraded *in vivo* by proteases and has a very short circulation half-life of only a few minutes. Therefore, more stable SMS analogues have been constructed both for pharmacological treatment of SSTR-overexpressing malignancies and for radionuclide imaging and targeted radiotherapy. Octreotide (Sandostatin®, Sandoz, Fig. 9.4) is an octapeptide analogue of SMS used to treat SSTR-expressing tumors that maintains the pharmacophore required for receptor binding, but which incorporates two D-amino acids (D-Phe[1] and D-Trp[4]) that inhibit proteolysis (75). Pentetreotide (Octreoscan®, Mallinckrodt Medical) is diethylenetriaminepentaacetic acid (DTPA)-derivatized octreotide, an analogue that complexes ^{111}In, either for SPECT imaging (76) or for

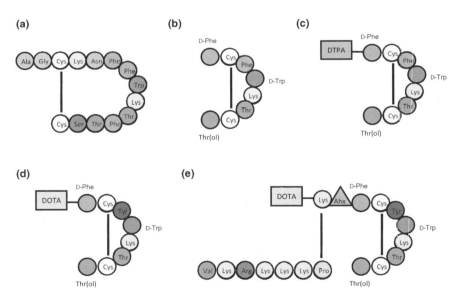

FIGURE 9.4 Amino acid sequences of somatostatin analogues. (a) Native somatostatin-14. (b) Octreotide. (c) Pentetreotide. (d) DOTATOC. (e) NLS-DOTATOC. Phe[1] and Trp[4] in octreotide and pentetreotide as well as DOTATOC are replaced with their corresponding D-isomers. The NLS peptide sequence in NLS-DOTATOC is derived from SV-40 large T-antigen and is linked along with DOTA to the somatostatin pharmacophore through a lysine-aminohexyl (Ahx) linker. (See insert for color representation of the figure.)

targeted Auger electron radiotherapy of SSTR2-expressing tumors. Other SMS analogues (Fig. 9.4) include [Tyr^3]-octreotide (TOC) in which Phe^3 is substituted by Tyr, and [Tyr^3]-octreotate (TATE) that is similarly substituted, but has a terminal threonine residue. These analogues have been derivatized with the chelator, 1,4,7,10-tetraazacyclododecane-1,4,7,10-tetraacetic acid (DOTA) to form DOTATOC and DOTATATE, which stably bind the β-emitters, ^{177}Lu or ^{90}Y for targeted radiotherapy (77), or the positron emitter, ^{68}Ga for positron emission tomography (PET) (78). Unlike SMS which binds with high affinity to all five SSTR subclasses, these analogues recognize mainly sst$_2$ and have five- to tenfold lower affinity than SMS (72). Lanreotide (Somatuline®, Beaufour Ipsen) is a long-acting form of SMS that additionally has moderate affinity for sst$_4$—this analogue has been conjugated to DOTA for labeling with ^{177}Lu or ^{90}Y (73, 79).

The majority of meningiomas, neuroblastomas, pituitary adenomas, small cell lung cancer, lymphomas, and breast cancer have sst$_2$ mRNA (80). SSTR protein is also strongest in these tumors for the sst$_2$ subclass, in agreement with the mRNA expression (81). Carcinoid malignancies, islet cell carcinomas, medullary thyroid cancer and ovarian tumors exhibit variable expression of sst$_1$, sst$_2$, or sst$_3$ mRNA (80) or protein (81). Prostate cancer and sarcomas display predominantly sst$_1$ receptors (81). In breast cancer, 21–46% of tumors are positive for sst$_2$ and there is an inverse correlation with expression of epidermal growth factor receptors (EGFRs) (82). Only about 25% of breast tumors coexpress sst$_2$ and EGFR, and interestingly, there appears to be histological separation with tumor cells expressing sst$_2$ but adjacent normal breast epithelial cells expressing EGFR. Coexpression of SSTR with receptors for vasoactive intestinal polypeptide (VIP), bombesin, cholecystokinin (CCK) or glucagon-like peptide (GLP) has been noted in many neuroendocrine tumors (83). Low levels of sst$_2$ are found on some normal tissues such as blood vessel endothelium, nerve cells, pancreatic islets, prostate epithelium, adrenal medulla, colonic mucosa, spleen, and lymphoid tissues (81). Of particular relevance is the almost exclusive expression of the sst$_2$ subclass on human bone marrow pluripotent stem cells as well as on committed progenitor cells (84)—this expression could cause myelosuppression and may increase the risk for myelodysplastic syndrome (MDS) in targeted Auger electron radiotherapy. SSTR have also been detected on normal lymphocytes (85). Nonetheless, the increased levels of sst$_2$ in tumors, especially those of neuroendocrine origin, combined with its restricted (or at least lower) expression in normal tissues makes it an attractive target for cancer radiotherapy and imaging. Indeed, octreotide and DOTATOC labeled with the β-emitters, ^{90}Y or ^{177}Lu have proven effective for treating SSTR-positive malignancies (62, 77) (see Chapter 4) and ^{111}In-pentetreotide is widely used for imaging SSTR-positive tumors, taking advantage of the two γ-photon emissions of ^{111}In [171 (90%) and 245 keV (94%)] (76). In this section, ^{111}In-pentetreotide will be discussed as a potential radiotherapeutic agent for tumors, exploiting its Auger electron emissions, rather than the γ-emissions used for tumor imaging.

9.5.2.1 *Internalization and Nuclear Importation of ^{111}In-Pentetreotide*

Internalization and nuclear importation of ^{111}In-pentetreotide are required to cause lethal DNA damage in SSTR-positive tumor cells, since 99% of the electrons emitted by ^{111}In have energies <3 keV and a range in tissues <1 μm (55). The higher energy IC electrons have a range up to 12 μm, and could cause DNA damage from the cell surface or cytoplasm, but these are much less abundant. The γ-emissions from ^{111}In are very low LET radiation and are not considered significantly damaging to DNA in comparison to the electron emissions. Internalization of native SMS and ^{111}In-pentetreotide likely occur through receptor-mediated endocytosis. Duncan et al. showed that ^{111}In-pentetreotide was rapidly internalized and delivered to lysosomes in pancreatic tumors and normal pancreatic tissues expressing SSTR in rats (86). ^{111}In-pentetreotide also distributes to renal tubular cell lysosomes following glomerular filtration with subsequent interaction of the filtered radiopharmaceutical with the renal brush border. Kidney uptake is dependent on the interaction of positive charges on ^{111}In-pentetreotide with the negatively charged membranes of renal tubular cells (87). Accumulation in hepatic lysosomes was similarly noted, but may not be SSTR-mediated. ^{111}In-pentetreotide is likely proteolytically degraded in lysosomes to ^{111}In-DTPA-(D)-Phe or ^{111}In-DTPA-Phe-Cys, two catabolites that remain trapped in the cells (86). This has been previously shown for other ^{111}In-labeled receptor binding proteins (88, 89) and is advantageous for targeted Auger electron radiotherapy because tumor cell retention of radioactivity will maximize the radiation absorbed dose and the resulting cytotoxicity.

There appear to be differences in the internalization and intracellular trafficking of native SMS following binding to different SSTR subclasses. Nouel et al. found that only 20–25% of ^{125}I-labeled SMS_{14} was internalized over 45 min at 37°C in COS-7 cells transfected with the sst_1 gene, whereas 75% was endocytosed by cells transfected with the sst_2 gene (90). Confocal microscopy with fluorescently labeled SMS_{14}, showed that most of the fluorescence remained just beneath the cell membrane in cells expressing sst_1, whereas sst_2-internalized ligands were endocytosed into vesicles that clustered around the nuclear membrane (an important finding for Auger electron radiotherapy). Cells expressing sst_3 or sst_5 also internalize SMS more efficiently than those expressing sst_1 or sst_4 (70). Cescato et al. showed that unlabeled pentetreotide as well as DOTATOC and DOTATATE ligands caused SSTR internalization in human embryonic kidney (HEK) and Chinese hamster ovary (CHO) cells expressing the sst_2 subclass (91). DOTA-conjugated analogues were much more potent inducers of endocytosis, possibly due to their higher receptor binding affinity (72). Agonist but not antagonist activity was required for internalization. ^{111}In-pentetreotide was reported to internalize into carcinoid and glucagonoma cells expressing SSTR with about a 1:10 ratio of radioactivity in the nucleus compared to the cytoplasm (92). However, there is variability in the extent of internalization of different radiolabeled SMS analogues in SSTR-expressing cells. De Jong et al. found that ^{111}In-pentetreotide was internalized 10-fold less efficiently than ^{125}I-Tyr^3-octreotide by sst_2-positive CA20948 and AR42J rat pancreatic tumor cells (93). [111]In- and ^{90}Y-DOTATOC were internalized two- and fourfold, respectively, more efficiently than ^{111}In-pentetreotide. It was speculated that this may be due to the higher receptor binding affinity of ^{125}I-Tyr^3-octreotide and radiolabeled DOTATOC analogues compared to ^{111}In-pentetreotide (72).

Wang et al. reported that both ^{111}In-pentetreotide and ^{64}Cu-labeled 1,4,8,11-tetraazacyclotetradecane-1,4,8,11-tetraacetic acid octreotide (TETA-OC) were internalized and translocated to the nucleus of AR42J tumor cells over a 24 h period at 37°C (94). However, maximum nuclear uptake was threefold higher for ^{64}Cu-TETA-OC than for ^{111}In-pentetreotide (19.5% versus 6.0%, respectively). It was proposed that nuclear importation may rely on the intracellular loss of radiometal from the chelator, and its redistribution to the nucleus (95). This may be greater for ^{64}Cu than ^{111}In due to the lower stability of the TETA versus DTPA radiometal complexes. Free ^{64}Cu cupric acetate similarly accumulated in the nucleus of AR42J cells. ^{64}Cu-TETA-OC was additionally taken up by mitochondria. Hornick et al. reported increasing nuclear uptake of ^{111}In-pentetreotide over 24 h in sst$_2$-positive IMR-32 human neuroblastoma cells, as well as binding of nuclear radioactivity to chromosomal DNA (96). The explanation for DNA binding was not clear, but we (93–99) and others (100) have similarly detected nuclear translocation and DNA binding of ^{111}In- and ^{125}I-labeled EGF in tumor cells overexpressing EGFR. Perinuclear and nuclear localization of radioactivity has been detected by autoradiography in tumor surgical specimens obtained from carcinoid patients receiving ^{111}In-pentetreotide (101, 102).

Based on the results described above, it is evident that ^{111}In-pentetreotide is internalized by receptor-mediated endocytosis in sst$_2$-positive tumor as well as normal cells and that a small proportion of the internalized radioactivity appears to be imported into the nucleus (and may even bind directly to DNA). The mechanism of nuclear importation however remains not well understood—factors such as the intracellular stability of the ^{111}In-DTPA complex may play a role (94, 95). Nonetheless, internalization and nuclear importation of at least a small proportion of radioactivity in tumor cells exposed to ^{111}In-pentetreotide have provided the rationale for exploring its use as an Auger electron-emitting radiotherapeutic agent for SSTR-overexpressing malignancies.

9.5.2.2 Preclinical Studies with ^{111}In-Pentetreotide

Interestingly, there have been relatively few preclinical studies that have examined the cytotoxicity *in vitro* and tumor growth-inhibitory effects *in vivo* of ^{111}In-pentetreotide in mouse tumor xenograft models compared to the many more clinical studies that have explored this strategy. Nonetheless, Capello et al. found that the clonogenic survival of CA20948 rat pancreatic cancer cells was virtually eliminated *in vitro* by exposing them for 5 h to 37 MBq (10^{-7} mol/L) of ^{111}In-pentetreotide (103). Higher surviving fractions (SFs) were found at lower amounts of radioactivity or by using shorter incubation times, while decreased survival was associated with increasing SA of ^{111}In-pentetreotide. Moreover, by designing the assay in such a way that the tumor cells were separated in culture by at least 1.2 mm, it was shown that the IC electrons from ^{111}In-DTPA (a control treatment) had no significant effect on survival—these electrons had a maximum range of 200–500 μm in the growth medium. Receptor-mediated binding and internalization of ^{111}In-pentetreotide were thus required for the cytotoxicity of the radionuclide.

SSTR-mediated tumor growth-inhibition by ^{111}In-pentetreotide was similarly shown *in vivo* in a rat hepatic metastasis model (104). Administration of 370 MBq

(a)   **(b)**

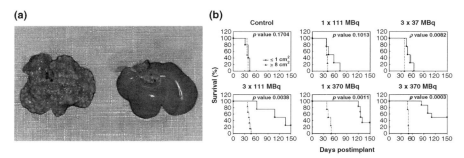

FIGURE 9.5   (a) Effect of treatment with two doses of 370 MBq (0.5 µg) of ^{111}In-pentetreo-
tide on day 1 and 8 on the formation of liver metastases from somatostatin-negative CC531
colon carcinoma cells (left) or SSTR-positive CA20948 pancreatic tumor cells (right) in rats.
Both cell lines were metastatic to the liver in untreated animals. Reprinted with permission from
Ref. 104. (b) Effect of treatment with different doses of ^{111}In-pentetreotide on the survival of
rats bearing small ($\leq$1 cm^2; solid line) or large ($\geq$8 cm^2; broken line) subcutaneous CA20948
tumors. Reprinted with permission from Ref. 105.

(0.5 µg) of ^{111}In-pentetreotide significantly decreased the number of hepatic metastases formed following inoculation of 5×10^5 SSTR-positive CA 20948 cells into the
portal vein of rats, with some animals developing no lesions and most demonstrating
<20 metastases. In contrast, all animals administered unlabeled DTPA-octreotide
(0.5 µg) developed more than 100 hepatic lesions, and there was no effect of [111]In--
pentetreotide on the formation of hepatic metastases in rats inoculated with SSTR-
negative CC-531 metastatic colon cancer cells (Fig. 9.5). Moreover, the effectiveness
of ^{111}In-pentetreotide was diminished by coadministration of 1 mg of octreotide to
block SSTR. Tumor size is an important predictor of the efficacy of [111]In--
pentetreotide (105). In rats bearing small ($\leq$1 cm^2) subcutaneous (s.c.) CA 20948
tumors, complete remissions (CR) were achieved in up to 50% of animals following
administration of three fractionated doses of 370 MBq of ^{111}In-pentetreotide (total
1110 MBq), whereas only partial remissions (PR) were obtained in rats with large
($\geq$8 cm^2) tumors. No responses were observed in rats with large tumors that received a
lower dose, whereas even a single dose of 111 MBq provided PR in 25% of those with
small tumors. In addition, almost a threefold increased survival was observed for rats
with small tumors compared to those with large tumors at the highest dose (Fig. 9.5).
Interestingly, ^{111}In-pentetreotide treatment led to an increase in SSTR density
measured in the explanted tumors by autoradiography—this may provide a rationale
for repeated as well as fractionated therapy. It is important to note that in this same CA
20948 tumor model, ^{90}Y-DOTATOC (370 MBq) was very effective for treatment of
large tumors (3–9 cm^2) providing CR in all animals but was less growth-inhibitory
toward small tumors ($\leq$1 cm^2), possibly due to deposition of the long range
(10–12 mm) β-particles outside the target volume (106). Thus, there is complementarity in the antitumor properties of Auger electron-emitting radiotherapeutics and
those labeled with much longer range β-emitting radionuclides, with the former more
effective for the treatment of small tumors, while the latter more useful for larger

tumors. Nonetheless, ^{90}Y-DOTATOC was not effective for treating very large tumors ($\geq 14\,cm^2$), likely due to the survival of nontargeted cells that were beyond the reach of "cross-fire" from the β-particles on cells that bound the radiotherapeutic agent (106).

9.5.2.3   Clinical Studies with ^{111}In-Pentetreotide

In 1994, Krenning et al. reported the first case of treatment of a patient with a metastatic neuroendocrine tumor with 7 doses of ^{111}In-pentetreotide ranging from 1590 to 4810 MBq (10–120 μg) over a period of 10 months (total 20,276 MBq) (107). A decrease in the size of liver metastases on CT was observed as well as transient decreases in tumor-associated glucagon and γ-glutamyltransferase levels. This report was followed by a case report in 1996, in which a 55-year old patient with a metastatic carcinoid tumor was treated with three doses of 3000, 3500, and 3100 MBq (40 μg each) of ^{111}In-pentetreotide over a period of 7 months (total 9600 MBq) (108). In addition, this patient received octreotide (300 μg) for 4 weeks prior to ^{111}In-pentetreotide therapy. Stable disease (SD) was achieved but no radiologically evident PR or CR were noted. However, octreotide caused a 14% decrease, and ^{111}In-pentetreotide, a further 31% decrease in the circulating levels of the tumor marker 5-hydroxyindoleacetic acid (HIAA). Plasma chromogranin A decreased more than threefold with combined octreotide and ^{111}In-pentetreotide therapy. Adverse events included a 10–45% decrease in leukocyte (WBC) counts and a minor reduction in platelet (Plt) counts after each ^{111}In-pentetreotide treatment, as well as a mild flushing sensation, and the development of severe pain over skeletal lesions. Subsequently, Tiensuu Janson et al. treated five patients with SSTR-positive carcinoid or pancreatic tumors with higher doses (6000 MBq; 40 μg) of ^{111}In-pentetreotide every 3 weeks for 3 cycles (109). Again, there was no change in tumor size, but two of three evaluable patients had a decrease in chromogranin A levels, while one patient showed an increase in this hormone. Decreased WBC and Plt counts were observed and one patient required a Plt transfusion. There was a slight decrease in hemoglobin (Hb) levels but no blood transfusion was needed. Decreased Plt counts as well as lymphocytes were similarly noted in another study by Caplin et al. in which eight patients with neuroendocrine tumors were treated with up to five doses of 1300–4600 MBq of ^{111}In-pentetreotide over 12 months (total 3100–15,200 MBq) (110). One patient exhibited a reduction in glomerular filtration rate (GFR) but had a low GFR prior to treatment (30 mL/min).

In a much larger clinical trial, Valkema et al. treated 50 patients with mainly carcinoid or medullary thyroid carcinomas with multiple doses of 6000–11,000 MBq (40–50 μg) of ^{111}In-pentetreotide separated by at least 2 weeks (total 20,000–160,000 MBq) (111). Therapeutic responses were seen in 21 of 40 evaluable patients and included one PR, 6 minor remissions (MR) and stabilization of previously progressive disease (SD) in 14 cases. Many of these patients were previously unresponsive to octreotide therapy. Decreased Plt and WBC counts were the main adverse effects observed, and lymphocyte counts often reached grade 3 (0.5–0.9×10^9/L) or 4 ($<0.5 \times 10^9$/L). Hb values decreased modestly, and in some patients reached grade 3. Particularly concerning was the development of MDS in three patients who received more than 100,000 MBq of ^{111}In-pentetreotide. MDS is a precursor to acute myelogenous leukemia (AML). However, no patients developed MDS at doses of ^{111}In-pentetreotide less than 100,000 MBq. MDS may be due to targeting of ^{111}In-pentetreotide to bone marrow progenitor cells that express sst_2 as mentioned earlier (84). No renal toxicity was noted as evidenced by no change in the serum creatinine (SCr) with follow-up evaluations for up to 3 years in some cases, despite calculated radiation absorbed doses to the kidneys as high as 30 Gy. A radiation absorbed dose of 23 Gy delivered by external radiation has been shown to cause renal failure in 5% of individuals after 5 years, and a 27 Gy dose results in renal failure in one of two patients (112). This contrasts with the serious renal toxicities reported for patients receiving high doses of ^{90}Y-DOTATOC without adequate renal protection (62). Decreased inhibin B levels that are associated with impaired spermatogenesis were noted in male patients receiving ^{111}In-pentetreotide.

Anthony et al. treated 27 patients with rapidly progressing GEP tumors that were unresponsive to conventional therapy with two doses of 6660 MBq each of ^{111}In-pentetreotide (113). Radiological PR were obtained in 2 cases (8%) and clinical improvement was noted in 16 (62%) patients, as well as decreased circulating levels of pancreastatin were measured in 22 (81%) patients. Decreased WBC, Plt and Hb reached NCI Common Toxicity Criteria grade 3 or 4 in only 1, 0, and 3 patients, respectively. One patient exhibited a grade 3 increased SCr. Grade 4 hepatotoxicity occurred in 3 patients who had extensive liver metastases. Increased survival relative to that expected for this group of treatment-refractory patients was noted (median survival 18 months versus 3–6 months). MDS was not found in 6 patients who were followed for up to 48 months. In a recently reported subsequent trial by this same group, 32 patients with progressive neuroendocrine malignancies were treated with one or two higher doses (18,500 MBq; 25 µg) of ^{111}In-pentetreotide separated by an interval of 10–12 weeks (114). SD was achieved in 16 (88%) patients and radiological PR in 2 (11%) cases (Fig. 9.6). Decreased levels of chromogranin A were measured in 14 of 18 patients (77%). Hematologic toxicities were reported in 5 of 13 patients (39%) receiving a single dose of ^{111}In-pentetreotide with 1 patient exhibiting grade 3 thrombocytopenia, while 7 of 18 (39%) receiving 2 doses exhibited these toxicities. There were no grade 3 hematological toxicities in patients receiving two doses of ^{111}In-pentetreotide. Hepatotoxicity was found in 3 of 14 patients (21%) treated with a single dose (two grade 1 and one grade 2) and in 8 of 18 patients (44%) receiving two doses (all Grade 1). There was no evidence of kidney toxicity despite administration of these very high doses of ^{111}In-pentetreotide without renal protection, and with follow-up as long as 2 years after initial treatment.

Nguyen et al. reported significantly prolonged survival in 20 patients with advanced metastatic neuroendocrine tumors treated with 3 monthly doses of ^{111}In-pentetreotide (7000 MBq) or ^{131}I-metaiodobenzylguanidine (^{131}I-mIBG; 3700–7400 MBq) compared to a similar group of 12 untreated patients (115). Radiologic SD was found in 13 of 15 (87%) patients receiving ^{111}In-pentetreotide and in 4 of 5 (80%) patients treated with ^{131}I-mIBG. One patient treated with ^{131}I-mIBG exhibited a PR and two patients receiving ^{111}In-pentetreotide progressed. The duration of tumor response was not significantly different for patients receiving ^{111}In-pentetreotide or those receiving ^{131}I-mIBG (13.2 versus 16.6 months, respectively). Stokkel et al. treated 11 patients

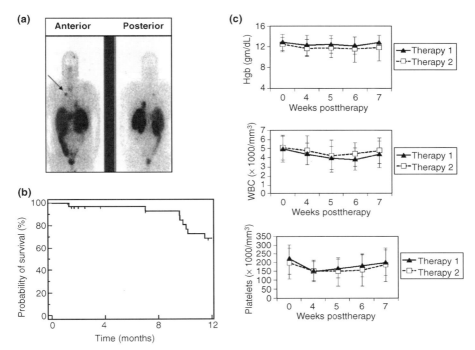

FIGURE 9.6   (a) Uptake of ^{111}In-pentetreotide in a metastatic lesion in the right middle lung (arrow) from a neuroendocrine tumor at 3 months after treatment with two doses of the radiopharmaceutical (total amount 36,300 MBq). Also seen on the images are a lesion in the left side of the brain and one in the lower left lung. (b) Kaplan-Meier plot of the overall survival for 32 patients receiving high dose ^{111}In-pentetreotide (18,090 or 26,300 MBq). (c) Effect of treatment with ^{111}In-pentetreotide on hemoglobin levels or WBC or platelet counts in this group of patients. Reprinted with permission from Ref. 114.

with progressive radioiodine nonresponsive nonmedullary thyroid cancer with 4 cycles of ^{111}In-pentetreotide (7400 MBq) separated by an interval of 2–4 weeks (total 14,300–33,100 MBq) (116). No patients demonstrated tumor regression by CT, but SD concurrent with stabilization of thyroglobulin (Tg) levels was achieved in four patients (36%). Only patients with low Tg levels (<1000 μg/L) prior to treatment (reflecting low tumor volume) responded to treatment with ^{111}In-pentetreotide. This, once again emphasizes that tumor size as an important predictor of the effectiveness of Auger electron-emitting radiotherapeutics (117). Limouris et al. administered up to 12 cycles of ^{111}In-pentetreotide (4070–7030 MBq; 40–50 μg) via hepatic catheter separated by 4–5 weeks to 13 patients with liver metastases from neuroendocrine malignancies (118). Ultrasound revealed decreases in the viability of the treated tumors.

Based on the clinical studies described above, it can be concluded that ^{111}In-pentetreotide provides stabilization of disease in previously progressive metastatic SSTR-positive tumors, but does not yield radiologically documented remissions (119).

Nonetheless, SD has been associated with decreased levels of circulating tumor-associated hormones and clinical improvement. Improvements in the outcome from ^{111}In-pentetreotide treatment could potentially be achieved by selecting patients with lower tumor volumes, or by increasing its uptake into the nucleus of SSTR-positive tumor cells, where the Auger electrons are most damaging to DNA and lethal.

9.5.2.4 Comparative Renal Toxicity of ^{111}In-Pentetreotide and ^{90}Y-DOTATOC
[111]In-pentetreotide and ^{90}Y-DOTATOC are both sequestered and retained by the kidneys but paradoxically, there is inconsequential renal toxicity resulting from administration of high doses of ^{111}In-pentetreotide in contrast to ^{90}Y-DOTATOC, which has been associated with severe renal impairment including failure in some patients who did not receive adequate kidney protection (62). Renal protection can be achieved by coinfusion of cationic D-amino acids (e.g. lysine or arginine) that inhibit the interaction of ^{90}Y-DOTATOC with proximal renal tubular cells (120). Megalin, a 600 kDa transmembrane glycoprotein on renal tubular cell membranes has been implicated in the reabsorption of ^{111}In-pentetreotide and many other radiopeptides (87, 121). One possible explanation for the lack of renal toxicity from ^{111}In-pentetreotide may be its microdistribution to regions of the kidneys that are relatively insensitive to radiation (e.g., renal tubules) combined with the nanometer–micrometer range of the emitted Auger and IC electrons—this prevents irradiation of distant radiosensitive areas such as the glomeruli through a "cross-fire" effect. In three patients receiving ^{111}In-pentetreotide and proceeding to nephrectomy, autoradiography of the excised kidneys showed that radioactivity was heterogeneously distributed mainly within the inner cortex of the kidneys, whereas most (85%) of the radiation sensitive glomeruli are located in the outer cortex (122). Microdosimetry modeling revealed that for an average cortical dose of 27 Gy, 60% of this region would receive a dose <17 Gy from ^{111}In-pentetreotide, whereas 60% would receive a dose >17.5 Gy from ^{90}Y-DOTATOC (123, 124). Moreover, there are steep gradients in the radiation absorbed doses delivered to regions within the kidneys that depend on the microdistribution of ^{111}In-pentetreotide, whereas ^{90}Y-DOTATOC provides a more homogeneous dose deposition (Fig. 9.7) (124). As previously mentioned, a radiation absorbed dose of 23–27 Gy delivered by external γ-radiation is predicted to cause renal failure in 5–50% of patients over 5 years (112). Heterogeneous dose deposition in the kidneys (124), resulting in lower renal toxicity is similarly seen with ^{177}Lu-DOTATATE than with ^{90}Y-DOTATOC, which emits β-radiation of lower energy and shorter range than [90]Y (0.5 MeV and 2 mm versus 2.3 MeV and 12 mm) (124, 125).

9.5.2.5 Novel Somatostatin Analogues for Auger Electron Radiotherapy
One approach to improving the effectiveness of Auger electron radiotherapy with SMS analogues would be to promote their importation into the nucleus of tumor cells where the Auger electrons are most damaging to DNA, and lethal. Ginj and Maecke reported several DOTATOC derivatives modified with heptapeptides [HPKKKRKV] harboring the nuclear localizing sequence (NLS; underlined) of SV40 large T antigen (126) (Fig. 9.4). NLS are recognized by importin-β in the cytoplasm

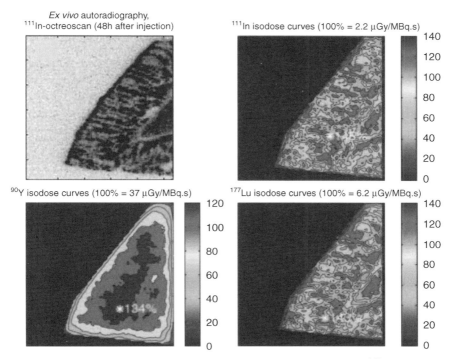

FIGURE 9.7   Radiation absorbed microdosimetry model predictions for ^{111}In (top right), ^{90}Y (bottom left), or ^{177}Lu (bottom right) in an explanted human kidney section from a patient receiving ^{111}In-pentetreotide. Also shown is the autoradiogram of the tissue section (top left). The homogeneity in radiation absorbed dose microdistribution was greatest for ^{90}Y and much less for ^{111}In or ^{177}Lu. Reprinted with permission from Ref. 124. (See insert for color representation of the figure.)

that interacts with importin-α; these complexes are actively shuttled across the nuclear membrane powered by Ran GTPase (127). DOTATOC modified at the C-terminus with NLS peptides exhibited a 100-fold decreased sst_2 binding affinity, but N-terminal modification preserved activity (128). ^{111}In-NLS-DOTATOC exhibited 6-fold higher internalization and 45-fold greater nuclear localization than ^{111}In-DOTATOC without NLS peptides in rat AR42J tumor cells but internalization remained receptor-mediated (128). Greater retention by the cells was found for ^{111}In-NLS-DOTATOC. The relative cytotoxicity of ^{111}In-NLS-DOTATOC and ^{111}In-DOTATOC unfortunately was not studied, but our group has shown that the mAb, trastuzumab (Herceptin; Roche Pharmaceuticals Ltd.) modified with this same NLS sequence and labeled with ^{111}In is at least sixfold more cytotoxic to SKBR-3 human breast cancer (BC) cells overexpressing HER2 than ^{111}In-trastuzumab without NLS modification (129). Similarly increased nuclear uptake and enhanced cytotoxicity toward myeloid leukemia cells were found for ^{111}In-anti-CD33 HuM195 mAbs modified with NLS peptides (130, 131). Graham et al., synthesized a DOTA conjugate of Tyr^3-octreotate

(TATE) modified with a DNA-intercalating dye (Hoechst 33258 or 33342) in order to similarly promote nuclear uptake but in addition provide DNA binding (132). Although not studied for Auger electron radiotherapy, these novel SMS analogues exhibited preserved binding to sst_2 on rat cortex membranes and demonstrated enhanced nuclear importation in AR42J cells visualized by confocal fluorescence microscopy.

Bernard et al., synthesized conjugates of ^{111}In-DTPA-Tyr^3-octreotate and the tripeptide Arg-Gly-Asp (RGD), which recognizes $\alpha_v\beta_3$ integrins (133). It was hypothesized that the ^{111}In-DTPA-Tyr^3-octreotate moiety would mediate internalization into sst_2-positive tumor cells while the RGD sequences would activate caspase-3 intracellularly, promoting apoptosis. Indeed, caspase-3 activity was increased 2-fold resulting in a 1.5-fold decreased survival of CA20948 rat pancreatic cancer cells *in vitro* compared to ^{111}In-DTPA-Tyr^3-octreotate without RGD peptide modification (134). High and sst_2-specific tumor uptake was observed *in vivo* for ^{111}In-RGD-DTPA-Ty^3-octreotate in rats bearing CA20948 tumors, but there was also high renal accumulation (135). The authors concluded that this would prevent its application for targeted radiotherapy due to the potential for kidney toxicity, however, as discussed earlier, this may not be manifested for the Auger electron emissions from ^{111}In, due to the microdosimetry of this radionuclide in the kidney.

9.5.3 Epidermal Growth Factor Receptors

The epidermal growth factor receptor is the first member of the Type 1 family of transmembrane peptide growth factor receptors; this family also includes the HER2, HER3, and HER4 receptors (136, 137). Binding of EGF, a 53-amino acid peptide ligand, by the EGFR causes receptor dimerization and autophosphorylation of its intracellular tyrosine kinase domain with subsequent activation of mitogenic signaling cascades. Overexpression of EGFR has been detected in many malignancies including cancers of the breast, ovary, head and neck, lung, bladder, and colon as well as in glioblastomas (138). In breast cancer, EGFR overexpression with densities up to 100-fold higher on tumor cells (10^5–10^6 receptors/cell) compared to normal epithelial tissues (10^3–10^4 receptors/cell) has been found in 30–50% of cases and is directly correlated with estrogen receptor (ER) negativity, insensitivity to hormonal therapy, as well as a poor long-term survival (139–141). EGF/EGFR complexes are rapidly internalized following binding of EGF to its cell-surface receptors, and these are thought to be routed to lysosomes for proteolytic degradation (142). However, a proportion of internalized EGF/EGFR complexes escape lysosomal destruction, and translocate to the nucleus, especially in cells highly overexpressing EGFR and with rapid proliferation rates (143). Nuclear localization of EGFR is believed to be mediated by a NLS [RRRHIVRKRTLRR] present in the transmembrane domain of the receptor (144).

A nonclassical role for the EGF/EGFR complex in the nucleus as a transcription factor for cyclin-D1 has recently been proposed to explain its nuclear importation (145). Lo et al. found nuclear EGFR in almost 40% of 130 breast cancer specimens by immunohistochemical staining, with 7% showing intense reactivity (146). Nuclear EGFR was inversely correlated with overall patient survival and was positively correlated with increased levels of cyclin D-1 and the proliferation marker, Ki-67. Others have similarly noted uptake of EGF in the nucleus of EGFR-positive tumor cells *in vitro* as well as some rapidly proliferating normal cells (100, 147, 148). Many other peptide growth factors and their receptors have similarly been found in the nucleus of cancer cells (127). Most normal epithelial cells exhibit very low levels of EGFR with the exception of hepatocytes and renal tubular cells that display moderate EGFR density (10^5 receptors/cell) (149–151). Importantly, <3% of the bone marrow stem cell population are EGFR-positive (152). This should obviate any lethal effects on these cells from Auger electrons emitted by targeted radiotherapeutic agents that bind EGFR on tumor cells, since receptor-mediated internalization and nuclear importation into bone marrow progenitor cells would be required for toxicity. Notwithstanding this restriction, it does not rule-out nonspecific irradiation and killing of bone marrow stem cells by the low LET γ-radiation emitted by Auger electron emitters circulating in the blood and perfusing the marrow, especially at high administered doses of radioactivity. The overexpression of EGFR in almost all ER-negative, hormone insensitive and poor prognosis breast cancers combined with the well-known internalization and recently reported nuclear translocation of EGF/EGFR complexes in some tumor cells makes the EGFR a very attractive target for Auger electron radiotherapy.

9.5.3.1 ^{111}In-*Labeled Human Epidermal Growth Factor (^{111}In-DTPA-hEGF)*

Our group has been exploring targeted Auger electron radiotherapy of EGFR-amplified breast cancer using human EGF complexed with ^{111}In by substitution with DTPA (^{111}In-DTPA-hEGF). ^{111}In-DTPA-hEGF was specifically bound with high affinity ($K_a = 7.5 \times 10^8$ L/mol) to EGFR overexpressed on MDA-MB-468 human breast cancer cells (1.3×10^6 receptors/cell) (153). The radiopharmaceutical was rapidly bound and internalized by MDA-MB-468 cells (>70% in 15 min) with about 15% of internalized ^{111}In deposited in the nucleus within 24 h. Two thirds of the nuclear radioactivity was directly associated with DNA (99). Recently, we confirmed that ^{111}In-DTPA-hEGF was imported into the nucleus bound to EGFR and importin-β_1, and the extent of nuclear localization was proportional to the EGFR density on breast cancer cells, with maximal nuclear uptake at 6×10^5 EGFR/cell (97). Exposure of MDA-MB-468 cells to increasing amounts of ^{111}In-DTPA-hEGF *in vitro* greatly reduced their clonogenic survival (Fig. 9.8), with <3% viable colonies present at 111–130 mBq/cell (2–3 pCi/cell) (99). The amount of ^{111}In-DTPA-hEGF required to reduce the SF to 37% (D_0 value) was 40 mBq/cell (1 pCi/cell), but when the total cell-bound radioactivity was corrected for the small proportion (10%) of radioactivity bound to DNA (the fraction thought to be most responsible for cell kill), the D_0 value was 4 mBq/cell (0.1 pCi/cell). A comparison of the growth-inhibitory properties *in vitro* of high SA ^{111}In-DTPA-hEGF (30 MBq/μg; 1.8×10^5 MBq/μmol) on MDA-MB-468 cells and commonly used chemotherapeutic agents for breast cancer revealed that the radiopharmaceutical was 85–300-fold more potent on a molar concentration basis than paclitaxel, methotrexate, or doxorubicin and several logarithms more effective than 5-fluorouracil (5-FU) (154) (Fig. 9.8). The IC$_{50}$ for

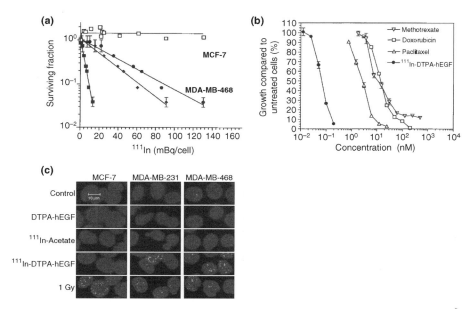

FIGURE 9.8 (a) Clonogenic survival of MDA-MB-468 overexpressing EGFR (1×10^6 receptors/cell) or MCF-7 human breast cancer cells with a 100-fold lower EGFR density, exposed to increasing amounts of ^{111}In-DTPA-hEGF. The clonogenic survival curve for MDA-MB-468 cells is plotted versus the amount of radioactivity per cell (circles), in the cytoplasm (diamonds), or in the nucleus (squares). Reprinted with permission from Ref. 99. (b) Cell growth inhibition of MDA-MB-468 cells treated with increasing concentrations of ^{111}In-DTPA-hEGF (specific activity 30 MBq/mg; 1.8×10^5 MBq/mmol), paclitaxel, methotrexate, or doxorubicin. Reprinted with permission from Ref. 154. (c) Nuclear γ-H2AX foci detected in MCF-7, MDA-MB-231, and MDA-MB-468 human breast cancer cells following exposure to normal saline, unlabeled DTPA-hEGF (21 nmol/L), ^{111}In-acetate (3.2 MBq/mL), ^{111}In-DTPA-hEGF (3.2 MBq/mL; 21 nmol/L), or γ-radiation (1 Gy). The nucleus was counterstained blue using DAPI. Reprinted with permission from Ref. 161. (See insert for color representation of the figure.)

^{111}In-DTPA-hEGF was <70 pmol/L and the IC$_{90}$ was 200 pmol/L. In contrast, the IC$_{50}$ values for paclitaxel, methotrexate, and doxorubicin were 6, 15, and 20 nmol/L, respectively, and were 4 μmol/L for 5-FU. The IC$_{90}$ values for paclitaxel, methotrexate, and doxorubicin were 20, 70, and 75 nmol/L, respectively, and for 5-FU were >10 μmol/L. Treatment of MDA-MB-468 cells with picomolar concentrations of ^{111}In-DTPA-hEGF provided the equivalent growth inhibition as several Gy of γ-radiation. Although MDA-MB-468 cells overexpress EGFR, they are paradoxically growth-inhibited by EGF at concentrations greater than 1 nmol/L (155), which is due to EGF-induced upregulated expression of the cyclin-dependent kinase inhibitor, p21$^{WAF-1/Cip-1}$. Increased p21$^{WAF-1/Cip-1}$ levels cause G$_1$/S-phase arrest and promote apoptosis (156, 157). Nevertheless, the IC$_{50}$ and IC$_{90}$ values for MDA-MB-468 cells exposed to unlabeled DTPA-hEGF were at least 7–10 times greater than for

^{111}In-DTPA-hEGF (500 pmol/L and 2 nmols/L, respectively). The antiproliferative effects of ^{111}In-DTPA-hEGF on breast cancer cells were dependent on EGFR density with lower potency observed for MDA-MB-231 cells (2×10^5 EGFR/cell) and no effect found for MCF-7 cells (1×10^4 EGFR/cell) (154). Greater antiproliferative potency toward MDA-MB-468 cells can be achieved by conjugating hEGF to human serum albumin which can then be modified with up to 23 DTPA chelators for binding ^{111}In in order to increase its SA (158). ^{111}In-DTPA-HSA-hEGF harboring nine DTPA chelators and labeled to a SA 10-times greater than ^{111}In-DTPA-hEGF, exhibited a fourfold greater potency toward MDA-MB-468 cells (IC$_{50}$ 15 versus 60 pmol/L). Conjugation of hEGF with HSA diminished its EGFR binding affinity 15-fold, but did not effect its internalization or nuclear translocation properties.

Exposure of MDA-MB-468 cells to 1 μmol/L of the EGFR-selective tyrosine kinase inhibitor, gefitinib (Iressa, Astra-Zeneca Pharmaceuticals) increased the nuclear uptake of ^{111}In-DTPA-hEGF twofold and diminished the survival of the cells twofold (98). In addition, a 1.5-fold increase in foci of immunofluorescence for phosphorylated histone-2AX (γ-H2AX) that accumulates at sites of double-strand DNA breaks was found. Compared to untreated MDA-MB-468 cells, the number of γ-H2AX foci was increased sixfold for ^{111}In-DTPA-hEGF. The increased nuclear localization of ^{111}In-DTPA-hEGF in the presence of gefitinib is speculated to be due to inhibition of phosphorylation of Tyr-1045 in the cytoplasmic domain of the EGFR, required for recognition by c-Cbl which ubiquitinates the receptor and directs it to lysosomes for degradation (159). Diminished lysosomal destruction of ^{111}In-DTPA-hEGF/EGFR complexes may allow more of these to translocate and accumulate in the nucleus. Increased DSBs and lethality of ^{111}In-DTPA-hEGF in the presence of gefitinib may thus be due to higher nuclear uptake of the radiopharmaceutical or to inhibition of EGFR-upregulated DNA repair pathways (160). Recently we found that the number of γ-H2AX foci caused by exposure of breast cancer cells to ^{111}In-DTPA-hEGF was directly proportional to their EGFR expression with the highest density of foci found in MDA-MB-468 cells, intermediate levels detected in MDA-MB-231 cells, and negligible foci noted in MCF-7 cells with 1.3×10^6, 2×10^5, and 1×10^4 EGFR/cell, respectively (Fig. 9.8) (161). In contrast, γ-radiation caused γ-H2AX foci in all three cell types, thus illustrating the principle of receptor-targeted Auger electron radiotherapy. Moreover, a strong inverse correlation was found between the SF of breast cancer cells and the extent of DSBs that were caused by ^{111}In-DTPA-hEGF (161).

9.5.3.2 Preclinical Studies with ^{111}In-DTPA-hEGF

The uptake of ^{111}In-DTPA-hEGF in small (6–30 mm^3) MDA-MB-468 xenografts at 72 h postin-travenous (tail vein) injection (p.i.) was relatively low (2–5 percent injected dose/g (% i.d./g) but tumor/blood ratios exceeded 12:1 (153). Tumor accumulation was specific as shown by coinjection of an excess of unlabeled EGF that diminished uptake twofold (153), as well as by a threefold lower accumulation found in MCF-7 tumors that express a 100-fold lower level of EGFR, at 24 h p.i. of ^{111}In-DTPA-hEGF (97). Tumor uptake was strongly dependent on lesion size with smaller tumors (<5 mm^3) exhibiting $>30\%$ i.d./g and very small tumors (1–2 mm^3) taking up as much as

80% i.d./g (162). ^{111}In-DTPA-hEGF demonstrated bi-exponential rapid elimination from the blood of nontumor bearing mice following i.v. (tail vein injection) with an α-phase half-life ($t_{1/2}$) of 3–6 min and a β-phase $t_{1/2}$ of 24–36 mins (161). The volume of distribution of the central compartment (V_1) and at steady-state (V_{ss}) were 340–370 mL/kg and 430–685 mL/kg, respectively, which were 5–10 times greater than the estimated plasma volume, probably reflecting extensive sequestration by some normal tissues such as the liver and kidneys. Hepatic uptake decreased fivefold from a maximum of 50% i.d./g at 1 h p.i. of ^{111}In-DTPA-hEGF to 12% i.d./g at 72 h (163). Kidney uptake similarly diminished sevenfold from 30% i.d./g at 1 h p.i. to 4% i.d./g at 72 h p.i. Macrodosimetry estimates predicted that the radiation absorbed doses from ^{111}In-DTPA-hEGF in humans would be highest to the kidneys (1.82 mSv/MBq), lower large intestine (1.12 mSv/MBq) and liver (0.76 mSv/MBq); the whole body dose would be 0.19 mSv/MBq (161).

^{111}In-DTPA-hEGF is absorbed rapidly from a s.c. injection in mice providing equivalent tumor and normal tissue distribution as from an i.v. (tail vein) injection (162). Administration of five weekly s.c. injections (on side contralateral to the tumor) of ^{111}In-DTPA-hEGF (18.5 MBq; 3.4 μg each) to athymic mice bearing established s.c. MDA-MB-468 xenografts provided a threefold decreased tumor growth rate compared to untreated mice (162) (Fig. 9.9). Nonestablished tumors in mice in which treatment was commenced 1 week after tumor cell inoculation, exhibited regression. MCF-7 tumors also appeared to be slightly growth-inhibited by ^{111}In-DTPA-hEGF, but the difference with untreated mice did not reach statistical

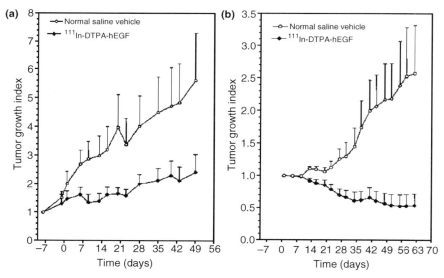

FIGURE 9.9 Inhibition of the growth of established 14–15 mm^3 subcutaneous MDA-MB-468 human breast cancer xenografts (a) or nonestablished 10 mm^3 tumor xenografts (b) following five weekly subcutaneous doses (at a different site than tumor implantation) of ^{111}In-DTPA-hEGF (cumulative dose 92.5 MBq; 17 μg).

significance. Unlabeled DTPA-hEGF had much weaker growth-inhibitory effects on MDA-MB-468 tumors than ^{111}In-DTPA-hEGF. The stronger antitumor effects of ^{111}In-DTPA-hEGF on nonestablished MDA-MB-468 xenografts may be due to its much higher accumulation in small tumors, perhaps combined with more complete targeting of the tumor cells.

Our group has conducted several studies examining the toxicity of ^{111}In-DTPA-hEGF, particularly toward tissues such as the liver and kidneys that accumulate large amounts of the radiopharmaceutical (99, 162, 163). These studies have consistently shown no significant increases in serum alanine aminotransferase (ALT) that would be associated with hepatotoxicity or elevated SCr associated with renal toxicity, despite administration of high doses of ^{111}In-DTPA-hEGF. Normal Balb/c mice were injected with 3.7 to 44 MBq of ^{111}In-DTPA-hEGF—only at the highest dose was there a slight increase in serum ALT over 72 h (99). There was also no significant increase in ALT in mice receiving two doses of ^{111}In-DTPA-hEGF (37 and 74 MBq) separated by a 4-week interval. In tumor-bearing athymic mice, there were no effects of ^{111}In-DTPA-hEGF (five doses of 18.5 MBq; total 92.5 MBq) on ALT, SCr, WBC, Plt, RBC, or Hb or any changes noted in body weight that would be associated with generalized toxicity from the radiopharmaceutical (162). Finally, in a more comprehensive study, there were no morphologic pathological changes found by light microscopy in 19 different tissues following administration of 44 MBq of ^{111}In-DTPA-hEGF to Balb/c mice or 85 MBq to white New Zealand rabbits (a nonrodent species) (163). There were minor clinically insignificant ultrastructural changes observed in the liver by electron microscopy. These amounts in mice and rabbits corresponded to 42 times and 1 times the planned maximum dose of ^{111}In-DTPA-hEGF for a Phase I clinical trial in humans. This investigation also confirmed earlier studies that the radiopharmaceutical exhibited no toxic effects on the liver, kidneys, or hematopoietic system. The absence of myelosuppression from ^{111}In-DTPA-hEGF could be explained by the absence of EGFR on hematopoietic stem cells that is required for its internalization and nuclear importation and consequent toxicity from the Auger electrons, combined with its rapid elimination from the blood which would minimize the DNA-damaging effects of the low LET γ-radiation from [111]In ($E\gamma = 171$ and 245 keV). However, the reason for its paradoxical growth-inhibitory effects on EGFR-positive tumors but apparent lack of toxicity toward the liver and kidneys, tissues that express moderate levels of EGFR and avidly sequester the agent is not known. One could speculate that there may be more efficient nuclear uptake of ^{111}In-DTPA-hEGF in tumors than in the liver or kidneys providing greater toxicity from the Auger electrons, but recent work by our group has revealed that in fact, there is greater nuclear localization *in vivo* in cells from the liver, kidneys, and spleen than in tumor cells (97). Another possible explanation may include differences in the cellular microdistribution of ^{111}In-DTPA-hEGF within these organs and the resulting microdosimetry—this was previously discussed as a possible explanation for the paradoxical absence of renal toxicity from ^{111}In-pentetreotide compared to that for ^{90}Y-DOTATOC. Finally, differences in cell turnover rates, radiosensitivity, DNA repair capacity as well as free radical scavenger concentrations may account for the lack of toxicity of ^{111}In-DTPA-hEGF toward these normal tissues that accumulate the radiopharmaceutical.

9.5.3.3 Translation of ^{111}In-DTPA-hEGF to Phase I Clinical Trial   Based
on the promising preclinical results for ^{111}In-DTPA-hEGF described above, our group
formulated a kit under good manufacturing practices (GMP) for preparation of the
radiopharmaceutical in a quality suitable for administration to patients (164). This kit
contains a unit dose of 250 µg of DTPA-hEGF in 1 mol/L sodium acetate buffer, pH 6.0
and can be labeled to high (>95%) efficiency with ^{111}InCl$_3$. The kit is pretested for
chemical and radiochemical purity, homogeneity, receptor binding affinity, sterility,
and apyrogenicity and is stable for at least 3 months. Following approval by Health
Canada, we initiated a Phase I clinical trial of ^{111}In-DTPA-hEGF in patients with
EGFR-positive and chemotherapy-resistant metastatic breast cancer (165). Sixteen
patients were treated with increasing single doses (355–2290 MBq; 250 µg) of
^{111}In-DTPA-hEGF. Pharmacokinetic studies revealed that the radiopharmaceutical
was rapidly eliminated from the blood with an α-phase $t_{1/2}$ of 10 mins and a β-phase
$t_{1/2}$ of 9.4 h. The volumes of distribution were large (V_1 of 460 mL/kg and V_{ss} of
2640 mL/kg), likely reflecting avid sequestration of ^{111}In-DTPA-hEGF by the liver
and kidneys. Confirming the results from the preclinical toxicity studies, there were no
significant treatment-related changes in ALT, SCr, RBC, WBC, Plt, or Hb at any dose
administered. Some minor adverse reactions were noted that were transient and
self-limited including facial flusing, nausea, vomiting, and hypotension (one patient).
These reactions are speculated to be due to the bioactivity of the EGF component of the
radiopharmaceutical and not related to the ^{111}In radiolabel. At the very modest single
doses administered (<2290 MBq), there were no tumor responses achieved in this
group of chemotherapy-refractory patients, but tumor uptake of ^{111}In-DTPA-hEGF
was observed by γ-camera imaging in several patients (Fig. 9.10). We are now
planning an extended Phase I trial of ^{111}In-DTPA-hEGF in which patients will receive
multiple treatments as were previously studied in athymic mice bearing EGFR-
positive breast cancer xenografts (162). This extended Phase I study will be supported
by Cancer Research U.K.

9.5.3.4 Auger Electron Radioimmunotherapy of EGFR-Positive Tumors
mAbs specific for EGFR have been studied preclinically *in vitro* using tumor cell
cultures, *in vivo* in xenograft mouse models (166–168) and clinically for Auger
electron RIT (169–171). Michel et al. showed that anti-EGFR mAb 528 labeled with
^{111}In or ^{125}I (1.5 MBq/mL) dramatically reduced the SF of A431 squamous carcinoma cells overexpressing EGFR (2–3 × 10^6 receptors/cell) *in vitro* by 97–100% (168).
^{111}In was more potent than ^{125}I, which was attributed to a greater SA for labeling mAb
528 and greater intracellular retention of ^{111}In compared to radioiodine (172). Mattes
and Goldenberg reported that in mice bearing A431 tumor xenografts, treatment
with ^{111}In-anti-EGFR mAb 525 significantly prolonged survival compared to ^{111}In--
labeled control mAbs or no treatment (167). A more rapid elimination of ^{111}In-mAb
225 than the control mAbs provided greater protection from systemic toxicity. The
maximum tolerated dose (MTD) was 68 MBq in nontumor bearing NCr nude mice
but was subsequently adjusted to 2.7 MBq/g of body weight, since some treated
mice died at this dose. ^{125}I-labeled F(ab')$_2$ fragments of the anti-EGFR mAb

(a)   (b)

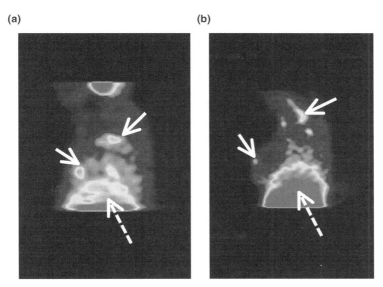

FIGURE 9.10   (a) Uptake of ^{111}In-DTPA-hEGF detected on a sagittal γ-camera image of a patient with metastases from EGFR-positive breast cancer to the lower lobe of the left lung as well as in the apex of the lung (solid white arrows). (b) Uptake of ^{111}In-DTPA-hEGF detected in a primary breast cancer on a sagittal image as well as in axillary lymph nodes (solid white arrows). In both patients, high uptake of ^{111}In-DTPA-hEGF into the liver (broken white arrow) that expresses moderate EGFR density is also visualized. (See insert for color representation of the figure.)

425 (1.5 MBq/mL) decreased the SF of U87MG glioblastoma cells overexpressing EGFR in clonogenic assays to <1% (166). Two doses of ^{125}I-mAb 425 F(ab')$_2$ (5.6 MBq each) administered intraperitoneally (i.p.) at 4 and 11 days after s.c. inoculation of EGFR-positive U87MG human glioblastoma cells in athymic mice were more effective than ^{131}I-mAb 425 F(ab')$_2$ fragments in controlling the growth of these tumors (166). There were no changes in WBC counts in mice receiving ^{125}I-mAb 425 F(ab')$_2$. It was speculated that the greater potency of ^{125}I-mAb 425 F(ab')$_2$ compared to the ^{131}I-labeled analogue may be due to the emission of the Auger electrons in close proximity to the nucleus. In a clinical trial of intact ^{125}I-mAb 425, Quang et al. administered multiple doses (1295–3330 MBq; total 1480–8288 MBq) to 180 patients with glioblastoma multiforme (GBM) or astrocytomas as an adjuvant treatment (171). A significant increase in the expected survival of these patients was noted with median survival of 4–150 months for those with GBM and 4–270 months for those with astrocytomas. Taken together, the studies described above show that anti-EGFR mAbs labeled with ^{111}In or ^{125}I exhibit specific and potent cytotoxicity toward tumor cells *in vitro* and prolong the survival of mice implanted with EGFR-overexpressing tumor xenografts *in vivo*, and may prolong survival in patients with EGFR-positive tumors.

9.5.4 Targeting HER2 Receptors

The human epidermal growth factor receptor-2 (HER2) is the second member of the EGFR family of transmembrane tyrosine kinases (137). HER2 is overexpressed up to 100-fold ($\sim 1 \times 10^6$ receptors/cell) compared to most normal epithelial tissues in 25% of breast cancers as a consequence of gene amplification, and is the target for immunotherapy with the humanized HER2 mAb trastuzumab (Herceptin; Roche Pharmaceuticals) (173). The HER2 receptor is believed to be internalization-impaired compared to other EGFR family members, but one of the proposed mechanisms of action of trastuzumab is promotion of HER2 downregulation (i.e., internalization) (174, 175). In addition, analogous to the EGFR, nuclear localization of HER2 has been reported (176). HER2 overexpression is assessed in tumor biopsies by immunohistochemistry (IHC; 0 to 3+ scale) that examines the receptor protein levels or by probing for increased copies of the HER2 gene by fluorescence *in situ* hybridization (FISH). Only patients with tumors exhibiting moderate-high (2+ to 3+) IHC scores or having more than three copies of the HER2 gene by FISH are predicted to respond to trastuzumab given in combination with anthracyclines or paclitaxel (177, 178). In fact, only about half of patients with HER2-overexpressing tumors respond to this treatment, demonstrating that some forms of breast cancer have an innate resistance to the drug (179). Moreover, in almost all initially responding patients, resistance to trastuzumab rapidly develops within a year (180). New strategies are needed to enhance the potency of trastuzumab toward HER2-amplified breast cancer in order to extend its range to tumors with low to moderate receptor density (i.e., 1+ to 2+) as well as to overcome these resistance mechanisms. Targeted Auger electron radiotherapy is one promising approach that could address these important challenges.

9.5.4.1 ^{111}In-Labeled Anti-HER2 Monoclonal Antibodies Our group has been studying Auger electron RIT of HER2-amplified breast cancer using ^{111}In-trastuzumab modified with 13-mer peptides [CGYGPKKKRKVGG] that harbor the NLS (underlined) of SV-40 large T antigen which promotes its nuclear importation following HER2-mediated internalization into breast cancer cells (129). Conjugation of 3 or 6 NLS peptides resulted in only a minor decrease in HER2 binding affinity compared to ^{111}In-trastuzumab without NLS modification ($K_a = 3.1$–5.9×10^9 mol/L versus 8.2×10^9 mol/L, respectively). However, NLS conjugation encouraged internalization and substantially promoted nuclear importation in MDA-MB-231, MDA-MB-361, and SK-BR-3 human breast cancer cells (Fig. 9.11). Internalization of ^{111}In-NLS-trastuzumab in SK-BR-3 cells could be blocked by coexposure to an excess of unlabeled trastuzumab, demonstrating that despite NLS conjugation, uptake remained HER2-mediated. The clonogenic survival of breast cancer cells exposed to ^{111}In-NLS-trastuzumab was significantly reduced compared to cells treated with ^{111}In-trastuzumab without NLS peptides or unlabeled trastuzumab, and was dependent on the level of HER2 expression (Fig. 9.12). ^{111}In-NLS-trastuzumab containing 6 NLS peptides was five- and sixfold more effective, respectively, than ^{111}In-trastuzumab or unlabeled trastuzumab at killing SK-BR-3 cells displaying the highest

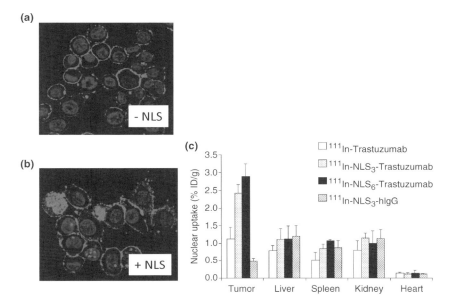

FIGURE 9.11   (a) Immunofluorescence confocal microscopy reveals that trastuzumab localizes mainly to the cell membrane of HER2-overexpressing SKBR-3 human breast cancer cells. (See insert for color representation of the figure.) (b) In contrast, trastuzumab modified with nuclear localization sequence (NLS)-containing peptides is internalized into the cytoplasm of SKBR-3 cells and is imported into the nucleus which is counterstained blue with DAPI. (See insert for color representation of the figure.) (c) NLS-conjugation significantly and preferentially increases the nuclear uptake of ^{111}In-labeled trastuzumab in subcutaneous HER2-overexpressing MDA-MB-361 human breast cancer xenografts *in vivo* in athymic mice compared to normal tissues. The number of NLS per trastuzumab molecule is shown. Reprinted with permission from Ref. 129.

HER2 density (2×10^6 receptors/cell). Importantly, ^{111}In-NLS-trastuzumab retained moderate toxicity toward MDA-MB-361 cells with intermediate HER2 expression (5×10^5 receptors/cell) and minor toxicity toward MDA-MB-231 cells with very low HER2 density (5×10^4 receptors/cell), whereas these cells were resistant to unlabeled trastuzumab (i.e., Herceptin). ^{111}In-NLS-trastuzumab exposure of breast cancer cells caused an increase in double-strand DNA breaks measured by the γ-H2AX assay (Fig. 9.12). Tumor uptake of ^{111}In-NLS-trastuzumab in s.c. MDA-MB-361 tumor xenografts *in vivo* in athymic mice was not significantly different than for ^{111}In-trastuzumab (12–13% i.d./g at 72 h p.i.) but was more than twofold higher than that of ^{111}In-NLS-conjugated human IgG (hIgG; 5% i.d./g), demonstrating specificity for HER2. Nuclear uptake in tumor cells *in vivo* was two- to threefold higher for ^{111}In-NLS-trastuzumab than ^{111}In-trastuzumab and five- to sixfold higher than for ^{111}In-NLS-hIgG.

More recently, we have found that ^{111}In-NLS-trastuzumab is able to kill breast cancer cells that are resistant to trastuzumab despite retaining their HER2 expression,

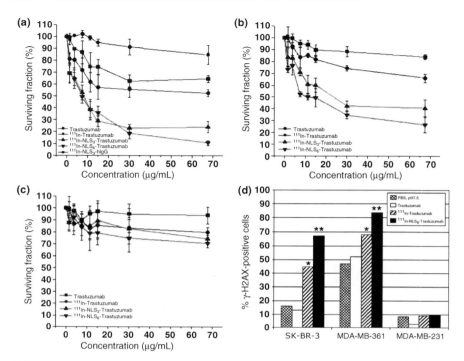

FIGURE 9.12 (a) Clonogenic survival of SKBR-3 human breast cancer cells with high HER2 density (2×10^6 receptors/cell) exposed to trastuzumab, ^{111}In-trastuzumab, and ^{111}In-NLS-trastuzumab with 3 or 6 NLS peptides or ^{111}In-NLS-hIgG nonspecific IgG. (b) Clonogenic survival of MDA-MB-361 breast cancer cells with intermediate HER2 expression (5×10^5 receptors/cell) exposed to these treatments. (c) Clonogenic survival of MDA-MB-231 breast cancer cells with very low HER2 expression (5×10^4 receptors/cell) exposed to these treatments. (d) γ-H2AX foci in the nucleus of SKBR-3, MDA-MB-361, or MDA-MB-231 cells exposed to these treatments. Reprinted with permission from Ref. 129.

and that the potency of the radiopharmaceutical can be enhanced by coexposing these cells to low, noncytotoxic concentrations of methotrexate (181). The effective concentration to reduce the SF of trastuzumab-resistant TrR1 breast cancer cells by 50% (EC_{50}) was 8 μmol/L for trastuzumab compared to 1 μmol/L for ^{111}In-trastuzumab, and 0.1 μmol/L for ^{111}In-NLS-trastuzumab. Methotrexate further reduced the EC_{50} for ^{111}In-NLS-trastuzumab against TrR1 cells to only 0.013 μmol/L. The radiosensitizing effect of methotrexate is believed to be due to its effect on depletion of reduced folates in tumor cells that are required for the synthesis of thymidylate and purine nucleotides needed for repair of DNA lesions inflicted by the Auger electrons from ^{111}In. Dose-escalation studies in nontumor bearing Balb/c mice showed that amounts less than 9.25 MBq caused no significant decrease in WBC, RBC, or Plt counts, or Hb levels, or increases in ALT or SCr or changes in body weight. Based on these findings, athymic mice bearing s.c. MDA-MB-361 tumors were injected with a single dose (9.25 MBq; 4 mg/kg) of ^{111}In-NLS-trastuzumab, ^{111}In-NLS-hIgG, or ^{111}In-trastuzumab, the equivalent amount of unlabeled trastuzumab or received no treatment. Strong antitumor effects were found for ^{111}In-NLS-trastuzumab that were significantly greater than those for control treatments (unpublished data). ^{111}In-NLS-hIgG had some minor tumor growth-inhibitory effects which may be due to the low LET γ-emissions of ^{111}In at these doses.

Michel at al. demonstrated 100% killing of HER2-overexpressing SKBR-3 cells *in vitro* by incubation with increasing concentrations (up to 7.4 MBq/mL) of a mixture of ^{111}In-labeled anti-HER2 mAbs 21.1 and 4D5 (the murine analogue of trastuzumab) and mAbs against epithelial glycoprotein-2 (EGP-2) (168). Use of a mixture of these mAbs increased the amount of ^{111}In bound to the cells compared to targeting with single mAbs. The cytotoxicity of these HER2 or EGP-2 mAbs was more than 4 logarithms greater than for ^{111}In-labeled nonreactive control mAbs, even at concentrations as high as 29.6 MBq/mL. Cytotoxicity of the ^{111}In-labeled HER2 mAbs was similarly observed for SK-OV-3 human ovarian carcinoma cells with high HER2 density, but their survival could not be eliminated completely; this was attributed to partial radioresistance of these cells since a similar level of ^{111}In targeting as on SK-BR-3 cells was found. In a follow-up *in vivo* study, Mattes and Goldenberg reported significantly prolonged survival in mice bearing s.c. SK-OV-3 tumor xenografts treated with 59 MBq of [111]In-4D5 compared to mice receiving the same dose of an irrelevant control mAb (167). At a slightly higher and more toxic dose (68 MBq) that caused death in two of nine mice, cures were obtained in five of seven surviving mice. No examination of the effects of treatment with these or lower amounts of ^{111}In-labeled HER2 mAbs on bone marrow, liver, or kidney function was performed. Based on the studies described above, HER2 also appears to be a very promising target for Auger electron radiotherapy of malignancies.

9.5.5 Other Antigens/Receptors in Solid Tumors

9.5.5.1 *Radiolabeled Monoclonal Antibodies*   MAbs labeled with ^{125}I and recognizing other epithelial-derived antigens, have been studied for Auger electron radiotherapy of solid tumors including colon cancer (182–189), lung cancer (190), and melanoma (191). Mab A33 recognizes a cell-surface epitope on human colon cancer cells and is internalized into cytoplasmic vesicles that distribute to perinuclear regions within 1–2 μm of the cell nucleus (182). ^{125}I-A33 was effective in controlling the growth of s.c. SW1222 colon cancer xenografts in athymic mice, but required a 4.5-fold higher dose than ^{131}I-A33 for equivalent antitumor effects. However, ^{125}I-A33 was 10 times less toxic than ^{131}I-A33 (MTD 185 MBq versus 18.5 MBq, respectively) providing an overall twofold improvement in the therapeutic index (183). In a Phase I/II clinical trial of 21 patients with chemotherapy-resistant colon cancer administered increasing doses of ^{125}I-A33 up to 12,950 MBq/m^2, decreased carcinoembryonic antigen (CEA) levels were noted in 3 patients, while one patient demonstrated a mixed radiological response on CT (184). There were no major toxicities, except in one

patient who had prior exposure to mitomycin and developed transient grade 3 thrombocytopenia.

In a separate study, ^{125}I-labeled CO17-1A mAbs recognizing a tumor-associated epitope on colorectal cancer, were internalized by SW1116 colon cancer cells, which caused chromosomal damage and decreased their clonogenic survival by 3 logarithms at concentrations up to 1.5 MBq/mL (186). Similar to ^{125}I-A33, the MTD for ^{125}I-CO17-1A (111 MBq) was 10-fold higher than for ^{131}I-CO17-1A (11 MBq) in athymic mice implanted s.c. with GW-39 human colon cancer xenografts (187). The dose-limiting toxicity was myelosuppression, but no significant changes in BUN or SCr or hepatic transaminases were found. At the MTD, ^{125}I-CO17-1A was much more effective at controlling tumor growth than ^{131}I-CO17-1A, with PR obtained in half of the mice; CR were achieved in 30% of the mice at higher doses (185 MBq) supported by bone marrow transplant. In this same study, the MTD of ^{111}In- and ^{90}Y-labeled CO17-1A were 4 and 85 MBq, respectively; myelosuppression was dose-limiting Myelosuppression was speculated to be due to the longer range IC electrons and γ-photons emitted by ^{125}I or ^{111}In, since the CO17-1A antigen is not expressed on bone marrow stem cells, which should obviate any toxicity from the nanometer-micrometer range Auger electrons. Again, CO17-1A labeled with the Auger electron-emitter, ^{111}In was more effective when administered at the MTD than CO17-1A labeled with the β-emitter, ^{90}Y, with CR achieved only with ^{111}In-CO17-1A (Fig. 9.13). The greater antitumor potency of ^{125}I compared to ^{131}I-labeled mAbs at equitoxic doses was confirmed in a separate study that examined both CO17-1A and the anti-CEA mAb F023C5 (188).

In a Phase I clinical trial of ^{125}I-CO17-1A administered in escalating single or multiple doses from 740 to 9250 MBq to 28 patients with metastatic colorectal cancer, there were no major hematological toxicities, although a few patients exhibited minor-moderate decreases in Plt and WBC counts, particularly at the highest dose (185). Unfortunately, at the doses examined, there were no objective responses observed; 10 patients had SD at 6 weeks follow-up. It was proposed that the lack of response may have been due to poor penetration of the radiolabeled mAbs into large tumor deposits combined with the treatment refractoriness of the patient population studied.

9.5.5.2 Radiolabeled Receptor Binding Peptides

The expression of glucagon-like peptide-1 receptors (GLP-1R) is increased on benign and malignant insulinomas often reaching greater density than that of SMSR (82). Lys^{40}(Ahx-DTPA-^{111}In)NH$_2$]-exendin-4 is a novel GLP-1 analogue with potential for Auger electron radiotherapy of insulinomas (192). This agent exhibited extraordinarily high tumor uptake (287% i.d./g) at 4 h p.i. in Rip1Tag2 transgenic mice that develop insulinomas (193). This very high tumor uptake was believed to be due to a dense expression of GLP-1R on the insulinoma cells (>10,000-fold greater compared to normal pancreatic β-cells), a high affinity of Lys^{40}(Ahx-DTPA-^{111}In)NH$_2$]-exendin-4 for GLP-1R ($K_d = 2$ nmol/L), slow elimination of radioactivity from tumors, as well as strong angiogenesis in this model providing high tumor blood flow (193). Strong antitumor effects were observed in Rip1Tag2 mice administered doses of Lys^{40}(Ahx-DTPA-^{111}In)NH$_2$]-exendin-4 ranging from 1.1 to 28 MBq of (192). At the highest

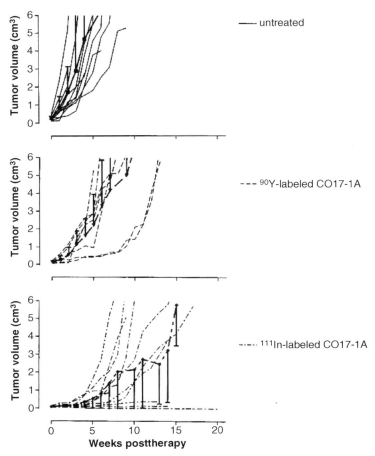

FIGURE 9.13   Tumor growth inhibition produced by equitoxic doses of ^{90}Y- and ^{111}In-labeled CO17-1A monoclonal antibodies (4 MBq and 85 MBq, respectively) in athymic mice implanted subcutaneously with GW-39 human colon cancer xenografts compared to control untreated mice. At the MTD, ^{111}In-CO17-1A was more effective than ^{90}Y-CO17-1A. Reprinted with permission from Ref. 187.

dose, <6% of the tumor volume remained at 8 days posttreatment. Response was associated with reduced tumor cell proliferation measured by decreased bromine deoxyuridine (BrdU) incorporation, as well as decreased cyclin D2 and increased apoptosis (TUNEL assay). Although these results were promising, severe radiation damage to the kidneys was found at 6 months following treatment with the highest dose of 28 MBq of [Lys^{40}(Ahx-DTPA-^{111}In)NH$_2$]-exendin-4. Kidney damage was attributed to the very high renal uptake (209% i.d./g at 4 h decreasing to 104% i.d./g at 48 h); kidney uptake was not GLP-1R-mediated (193). Reducing the dose to <11.2 MBq eliminated renal toxicity while still preserving therapeutic benefit, suggesting that this approach may yet be feasible.

Peptides and small molecules that recognize other receptors that are displayed on tumor cells and are internalized following ligand binding also have potential for Auger electron radiotherapy. These include [111]In- or ^{99m}Tc-labeled bombesin analogues that interact with gastrin-releasing peptide (GRP) receptors expressed in prostate and breast cancer (194, 195), as well as [111]In-, [99m]Tc-, or ^{67}Ga-labeled folates that bind folate receptors in ovarian cancer (196–198). The insulin-like growth factor-1 receptor (IGF-1R) may represent a particularly useful target for Auger electron radiotherapy of trastuzumab-resistant breast cancer using ^{111}In-labeled IGF-1(E3R) an IGF-1 analogue, but very high kidney uptake and low tumor accumulation may make this impractical (199). ^{111}In-labeled IGF-1R mAbs may be more feasible due to their higher tumor accumulation compared to that for small peptides and their much lower kidney uptake (200).

9.5.6   Cell-Surface Epitopes in Lymphomas

The better access of hematological malignancies compared to solid tumors to intravenously administered mAbs combined with their greater radiosensitivity has yielded very encouraging results in clinical trials of RIT of non-Hodgkin's lymphoma (NHL) as well as myeloid leukemias (AML), using β-emitting radionuclides such as ^{90}Y and [131]I (62) (see Chapter 6). This subsequently led to the regulatory approval and marketing of two radioimmunoconjugates for treatment of NHL: ^{90}Y-ibritumomab tiuxetan (Zevalin, IDEC Pharmaceuticals) and ^{131}I-tositumomab (Bexxar, Glaxo-SmithKline). MAbs labeled with subcellular range Auger electron emitters may be more suited to treating these amalignancies, since they achieve targeted single-cell killing, while minimizing or eliminating the "cross-fire" effect. While this effect is beneficial for enhancing the treatment of solid tumors, in which not all cells are targeted by the mAbs, it is responsible for dose-limiting toxicity to the bone marrow in RIT (62). In addition, due to the long range of the β-particles, most of the energy is deposited outside of the cell to which the mAbs are targeted (i.e., outside the target volume) (201); this is problematic for treating hematological malignances in which it is desirable to have all of the radiation deposited within a single targeted malignant cell. The use of nanometer–micrometer range Auger electron emitters conjugated to mAbs recognizing epitopes on NHL cells or leukemic cells would address these issues and could be very promising for treatment of these malignancies.

9.5.6.1   *Radiolabeled LL1 Monoclonal Antibodies*   MAb LL1 recognizes the major histocompatability (MHC) antigen class-II invariant chain (CD74) present on B-cells and macrophage-lineage cells. It is internalized by Raji B-cell lymphoma cells at an unusually high rate ($>10^7$ molecules per day). This is not due to a high expression level of CD74, which is only 5×10^4 molecules/cell, but rather to rapid repopulation of these epitopes to the cell-surface, allowing more radiolabeled mAb LL1 to bind and internalize (202). Griffiths et al. showed that concentrations up to 3.4 MBq/mL of LL1 labeled with ^{125}I, ^{111}In or ^{99m}Tc killed 99-100% of Raji cells *in vitro* (203). It was argued that the higher energy 22 keV Auger and 31 keV IC electrons of ^{125}I were most responsible for cell killing, since these mAbs were not

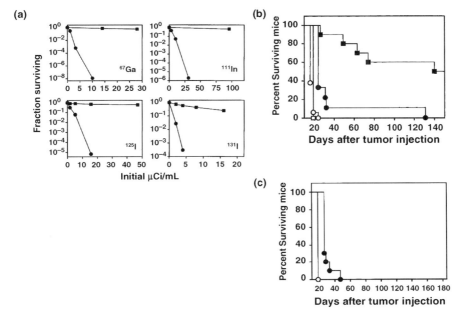

FIGURE 9.14 (a) Clonogenic survival of Raji B-lymphoma cells exposed to CD74-specific monoclonal antibody LL1 (closed circles) or a control antibody (closed squares) labeled with ^{67}Ga, ^{111}In, ^{125}I, or ^{131}I. (b) Survival of mice bearing disseminated Raji xenografts treated with 8.9 MBq (closed squares) or 2.8 MBq of ^{67}Ga-LL1 antibodies (closed circles) or with ^{67}Ga-labeled MN-14 control antibodies (open squares) or no treatment (open circles). (c) Survival of mice with Raji xenografts treated with the maximum tolerated dose of ^{90}Y-LL1 (0.9 MBq; closed circles) or receiving no treatment (open circles). Reprinted with permission from Ref. 205.

imported into the nucleus. Similarly, the 19–22 keV low abundance electrons of ^{111}In or the 119 keV electron emitted by ^{99m}Tc may be the most relevant. Govindan et al. compared the toxicity of mAb LL1 conjugated to ^{67}Ga, ^{125}I, or ^{111}In or the β-emitters, ^{131}I or ^{90}Y toward Raji cells (Fig. 9.14) (56). The concentration required for 99% cell kill (EC$_{99}$) ranged from 0.1 MBq/mL for ^{67}Ga to 0.3 MBq/mL for ^{125}I, and 0.5 MBq/mL for ^{111}In. The β-emitters, ^{131}I and ^{90}Y were more potent with EC$_{99}$ values of 0.08 and 0.16 MBq/mL, respectively. However, the specificity ratios, which described the killing of Raji cells by LL1 compared to that using a nonreactive control mAb were in most instances much higher for the Auger electron emitters (24:1 to 95:1) than for ^{131}I or ^{90}Y (31:1 and 8:1, respectively). ^{111}In had the lowest specificity ratio among the Auger electron emitters, possibly due to its energetic γ-emissions (172 and 245 keV). The greater specificity of ^{131}I compared to ^{90}Y-labeled LL1 may be due to its lower β-particle energy, allowing more radiation to be deposited within the target volume (i.e., a single cell). ^{67}Ga emerged as a potent and particularly promising Auger electron emitter conjugated to LL1; this was explained by its greater proportion of energetic electrons compared to ^{125}I or ^{111}In, which may deposit more radiation in the nucleus from the cell surface or cytoplasm (note that these antibodies were not accumulated in the nucleus). Nonetheless, it remains to be determined if this advantage would be maintained for mAbs conjugated to NLS-containing peptides that are able to translocate to the cell nucleus following receptor-mediated internalization. In addition, ^{111}In can be purchased in higher specific activity than ^{67}Ga and strong chelators for ^{67}Ga are not currently available (172). The effectiveness of ^{111}In and ^{125}I-LL1 for killing three other B-cell lymphoma cells (Ramos, RL and Daudi) *in vitro* has also been demonstrated (204).

The MTD of ^{111}In- and ^{67}Ga-labeled LL1 were 11–15 MBq and 8–11 MBq, respectively, in severe combined immunodeficiency (scid) mice whereas ^{90}Y-LL1 was 10-fold more toxic (MTD 0.9 MBq) (205). ^{111}In-LL1 (13 MBq) administered 5 days after i.v. inoculation of Raji cells produced compete cure in 40% of animals with no loss in body weight, that could indicate any generalized normal organ toxicity. This therapeutic effect was greatly diminished however, if treatment was given 9 days after Raji cell inoculation, illustrating once again, that Auger electron emitters are most effective for treating low-volume disease. ^{67}Ga-labeled LL1 (9 MBq) had similar tumor growth-inhibitory effects as ^{111}In-LL1, but ^{90}Y-LL1 administered at the MTD (0.9 MBq) provided no benefit (Fig. 9.14). It is important to appreciate that scid mice harbor a genetic mutation which radiosensitizes their tissues and that would have limited the dose that could be safely administered in this study (206). Michel et al. found that ^{111}In-LL1 was effective at controlling the growth of s.c. B-cell lymphoma xenografts if administered at a dose of 9 MBq within 3 days (but not 5 days) after Raji cell inoculation or within 5 days after Daudi cell inoculation (207). Daudi tumors were more responsive than Raji, and treatment started as late as 24 days after inoculation completely prevented their growth at the MTD (9 MBq). Even doses one-tenth of the MTD (1.2 MBq) were effective in preventing Daudi tumor growth. These results confirmed the strong antitumor effects of LL1 labeled with ^{111}In for treatment of B-cell lymphoma xenografts in scid mice but highlighted that some lymphoma phenotypes may be more responsive.

9.5.6.2 *Radiolabeled 1F5 and L243 Monoclonal Antibodies* Mab 1F5 recognizes CD20 while L243 reacts with more abundant MHC class II α/β chain epitopes on normal B-cells and B-cell lymphoma cells. Michel et al. compared mAbs 1F5 and L243 labeled with ^{67}Ga, ^{125}I, ^{111}In, or the β-emitter, ^{131}I for killing Raji cells *in vitro* (208). A fourfold higher concentration of ^{67}Ga-labeled 1F5 was required than L243 to reduce the SF of Raji cells to <0.0001% (4 MBq/mL versus <1 MBq/mL), possibly reflecting the greater epitopic density. ^{67}Ga-labeled nonreactive control mAbs were far less effective at reducing the SF of Raji cells (<1 logarithm decrease at 4 MBq/mL). In terms of cytotoxic potency, the order was ^{67}Ga > ^{125}I > ^{111}In, similar to that found for LL1. ^{131}I was more potent than the Auger electron emitters, but was associated with greater nonspecific toxicity than ^{125}I or ^{67}Ga, but not ^{111}In, which demonstrated some nonspecific cell killing at high concentrations. It was argued that since at least one of these antibodies (1F5) remains on the cell surface, internalization and nuclear importation were not absolute requirements for cytotoxicity from Auger electron emitters such as ^{67}Ga, ^{125}I, or ^{111}In. However, it is important to appreciate that cytotoxicity from surface-bound Auger electron emitters will depend on the far less

abundent higher energy (5–30 keV) Auger and IC electrons, limiting their potency. Furthermore, cellular dosimetry models predict that the radiation absorbed dose deposited in the nucleus from ^{111}In will be increased twofold from localization in the cytoplasm compared to the cell surface, but more than 35-fold when the radionuclide is located in the nucleus, itself (58). Also, it is not proven that 1F5 remains mostly on the cell surface, since confocal fluorescence microscopy of Raji cells showed internalization and accumulation to a juxtanuclear region (209). [111]In-1F5 (10.4 MBq) was potent at controlling the growth *in vivo* of s.c. RL B-cell lymphoma xenografts in scid mice but in contrast to ^{111}In-LL1, was unable to prevent the growth of Raji tumors in 80% of the animals (207). Rituxumab that binds CD20 and is used to treat NHL has also been labeled with ^{111}In and shown to kill Daudi cells *in vitro*, with induction of apoptosis being the primary mechanism of cell death (210).

9.5.7 Targeting CD33 Epitopes in Acute Myeloid Leukemia

CD33 is a 67 kDa cell-surface adhesion molecule that is expressed on acute myeloid leukemia (AML) cells as well as on committed myelomonocytic and erythroid progenitor cells, but is not present on normal lymphoid, nonhematopoietic or mature myeloid cells, or the earliest myeloid progenitor (stem) cells (211, 212). ^{131}I-labeled murine M195 or humanized HuM195 anti-CD33 mAbs have proven efficacy for eradicating AML cells in mouse leukemia models and killing large numbers of leukemic cells in patients (213, 214) (see Chapter 7). However, a major challenge has been their radiotoxicity to the bone marrow, likely due to the long-range (2 mm) β-particles emitted by ^{131}I that can nonspecifically irradiate and kill hematopoietic stem cells ("cross-fire" effect) (62). To improve the specificity of radiolabeled anti-CD33 mAbs and enhance their potency, McDevitt et al. labeled HuM195 with the short range (50–100 μm) α-emitter, [225]Ac, which decays to several daughter radionuclides that are themselves α-emitters, or β-emitters (i.e., "atomic nanogenerator") (215). However, one of these daughter decay products, [213]Bi, was found to redistribute to the kidneys in cynomolgus monkeys causing severe renal impairment (216). Other anti-CD33 mAbs have been conjugated to the chemotherapeutic drug, calicheamycin (gemtuzimab ozagamicin; Mylotarg, Wyeth Pharmaceuticals) for treatment of AML (217, 218). Unfortunately, these chemoimmunoconjugates are eliminated from leukemic cells by multidrug resistance (MDR) transporters, which diminishes their effectiveness (218, 219). Unlabeled HuM195 was not effective for immunotherapy of AML in Phase III clinical trials (220); this was suspected to be due to the low expression of CD33 epitopes on leukemic cells as well as rapid internalization that may have prevented activation of antibody-dependent cellular cytotoxicity (ADCC) or complement-mediated cellular toxicity (CMCC) (220, 221). One approach to improving the selectivity of radiolabeled anti-CD33 mAbs for killing leukemic cells, while minimizing their toxicity toward normal tissues such as the bone marrow, would be to conjugate them to subcellular range Auger electron-emitting radionuclides (e.g., ^{125}I or ^{111}In). The basis for this approach was reported more than 20 years ago for ^{125}I-T101 mAbs that bind T65, a 65 kDa epitope present on normal and malignant T cells; these Auger electron-emitting agents could kill human malignant T-cell leukemia cell lines in culture (222). Such radiolabeled anti-CD33 mAbs may also be able to overcome chemotherapy resistance in leukemia. Athough not a radiolabeled antibody, this has been shown in principle for 5-^{123}I-iodo-4'-thio-2'-deoxyuridine (^{123}I-ITdU), a thymidine analogue labeled with the Auger electron emitter, ^{123}I, that was able to kill leukemia cells that were resistant to doxorubicin (223).

9.5.7.1   *[111]In-Labeled M195 and HuM195 Monoclonal Antibodies*   Our group has been investigating targeted Auger electron radiotherapy of AML using ^{111}In-labeled M195 or HuM195 mAbs modified with 13-mer peptides [CGYGPKKKRKVGG] harboring the NLS of SV-40 large T-antigen (underlined) to direct the radiolabeled mAbs to the nucleus following CD33-mediated internalization (as previously described for ^{111}In-trastuzumab). There was a slight increase in the K_d value of HuM195 as the number of NLS peptides was increased from 4 ($K_d = 4.3 \times 10^{-9}$ mol/L) to 8 or 12 ($K_d = 6.3–6.9 \times 10^{-9}$ mol/L) (130). However, there was a strong direct correlation between the number of NLS peptides introduced into ^{111}In-HuM195 and the proportion of radioactivity that was imported into the nucleus of HL-60 human leukemic cells; a sixfold increase was found for ^{111}In-HuM195 modified with 8 NLS peptides compared to ^{111}In-HuM195 without NLS (65.9% versus 10.5% of the cell-bound radioactivity) (Fig. 9.15). In clonogenic assays, only 1.5–3.3 mBq/cell (<0.1 pCi/cell) of ^{111}In-NLS-HuM195 targeted to HL-60 cells was required to reduce their SF to <5–10%. Moreover, ^{111}In-NLS-HuM195 was more effective than ^{111}In-HuM195 without NLS in decreasing the survival of primary AML patient specimens *in vitro*, with a 2-fold decreased SF found in 7 of 9

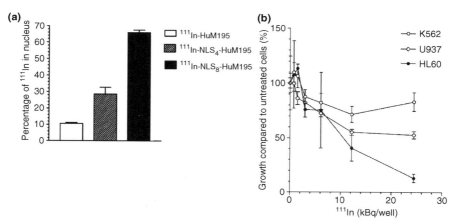

FIGURE 9.15   (a) Effect of conjugation of NLS peptides to ^{111}In-labeled anti-CD33 monoclonal antibody HuM195 on its nuclear uptake in HL-60 human myeloid leukemia cells. (b) Growth-inhibition by ^{111}In-NLS-HuM195 of HL-60 cells with high CD33 expression compared to that for U937 human histiocytic lymphoma cells with a threefold lower CD33 density and K562 human chronic myeloid leukemia cells with negligible CD33 expression. Reprinted with permission from Ref. 130.

FIGURE 3.2 (See page 107 for text discussion.)

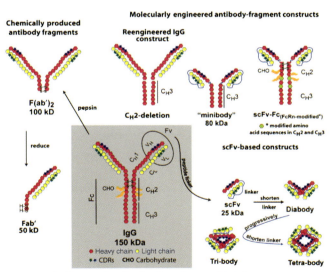

Chemically produced antibody fragments

Molecularly engineered antibody-fragment constructs

Reengineered IgG construct

F(ab')₂
100 kD

pepsin

C_H2-deletion

"minibody"
80 kDa

scFv-Fc_(FcRn-modified*)

CHO   C_H2

C_H3   C_H3

* modified amino acid sequences in C_H2 and C_H3

reduce

Fv

V_H

V_L

C_H1

C_L

Fc

CHO   C_H2

C_H3

IgG
150 kDa

● Heavy chain ○ Light chain
★✶ CDRs   CHO Carbohydrate

Fab'
50 kD

H-S
H-S

peptide linker

scFv-based constructs

scFv
25 kDa

linker

shorten
linker

Diabody

progressively
shorten linker

Tri-body   Tetra-body

In vivo targeting of antibodies and some of their fragments

	IgG	F(ab')₂	Fab'	scFv
Tumor uptake	Highest ——————→ Lowest			
Tumor retention	Longest ——————→ Shortest			
Time to maximum tumor uptake	Longest ——————→ Shortest			
Tumor penetration	Least ——————→ Most			
Blood clearance	Fastest ——————→ Slowest			
Tissue uptake	Liver ——————→ Kidneys			

FIGURE 8.1 (See page 245 for text discussion.)

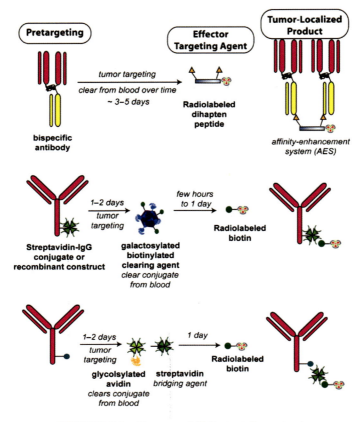

FIGURE 8.2 (See page 247 for text discussion.)

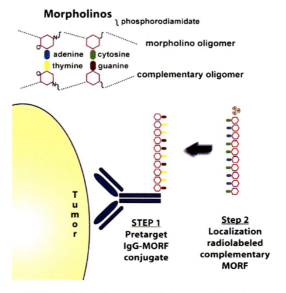

FIGURE 8.3 (See page 248 for text discussion.)

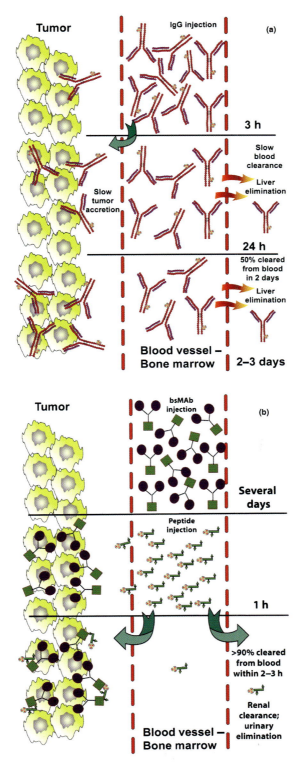

FIGURE 8.4 (See page 257 for text discussion.)

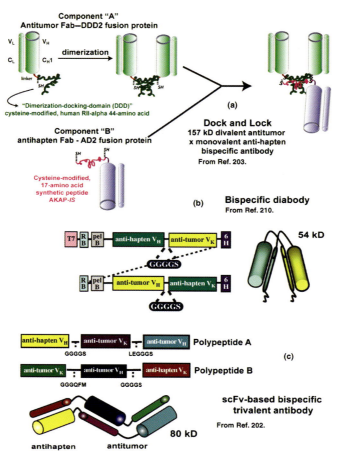

FIGURE 8.7 (See page 272 for text discussion.)

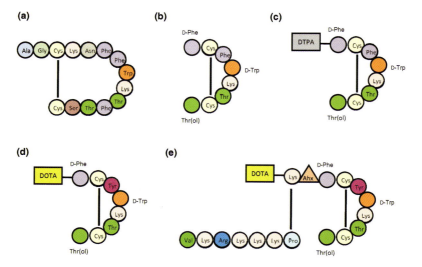

FIGURE 9.4 (See page 300 for text discussion.)

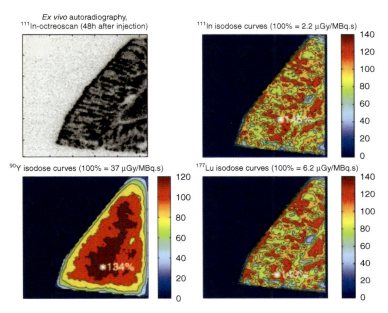

FIGURE 9.7 (See page 309 for text discussion.)

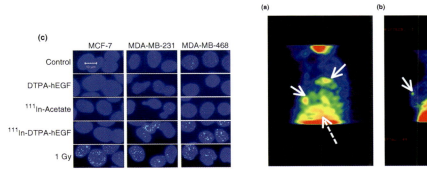

FIGURE 9.8c (See page 312 for text discussion.)

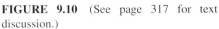

FIGURE 9.10 (See page 317 for text discussion.)

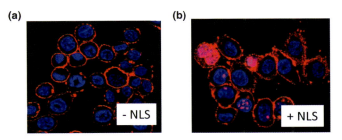

FIGURE 9.11 (See page 319 for text discussion.)

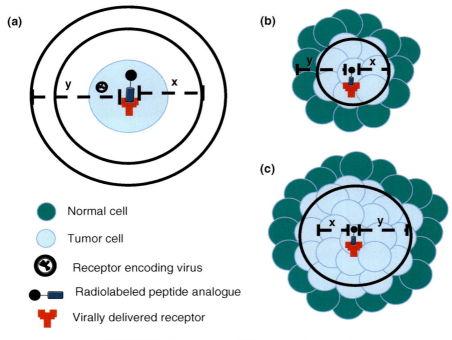

(a)

(b)

(c)

- Normal cell
- Tumor cell
- Receptor encoding virus
- Radiolabeled peptide analogue
- Virally delivered receptor

FIGURE 10.1 (See page 350 for text discussion.)

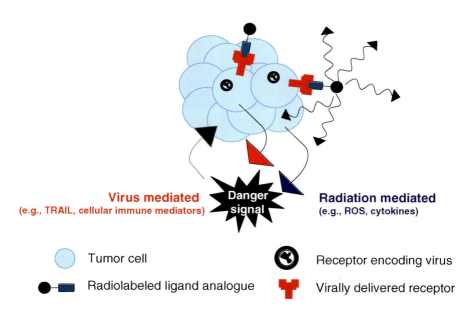

Virus mediated
(e.g., TRAIL, cellular immune mediators)

Danger signal

Radiation mediated
(e.g., ROS, cytokines)

- Tumor cell
- Radiolabeled ligand analogue
- Receptor encoding virus
- Virally delivered receptor

FIGURE 10.2 (See page 351 for text discussion.)

1.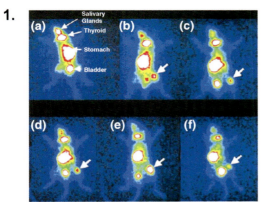

2.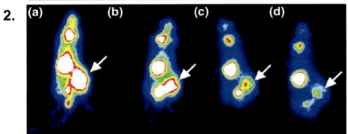

FIGURE 10.6 (See page 372 for text discussion.)

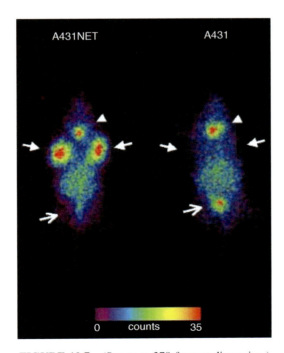

FIGURE 10.7 (See page 379 for text discussion.)

(a)

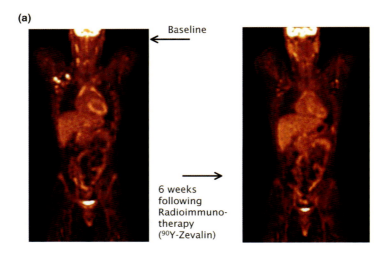

Baseline

6 weeks following Radioimmuno-therapy (^{90}Y-Zevalin)

(b) **FDG-PET/CT partial response**

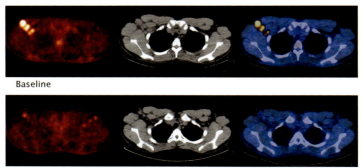

Baseline

6 weeks following radioimmunotherapy (^{90}Y-Zevalin)

FIGURE 15.4   (See page 537 for text discussion.)

FDG-PET/CT complete response

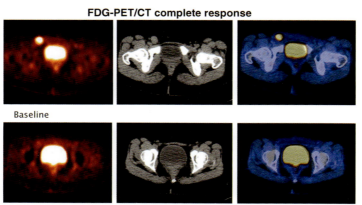

Baseline

8 weeks following radioimmunotherapy (^{90}Y-Zevalin)

FIGURE 15.5   (See page 538 for text discussion.)

specimens and >10-fold in 2 specimens. The toxicity of ^{111}In-NLS-HuM195 was dependent on the level of CD33 expression with U937 human histiocytic lymphoma cells having threefold lower CD33 expression compared to HL-60 cells that exhibited intermediate sensitivity while K562 human chronic myeloid leukemia cells with negligible CD33 were not sensitive (Fig. 9.15). In normal Balb/c mice administered 3.7 MBq (22 µg) of ^{111}In-NLS-HuM195, there were no changes in body weight, or decreased WBC, Plt, RBC counts, or decreased Hb or increases in SCr or ALT that would indicate toxicity. Although these doses were quite low, they were almost 500-fold higher on a kilobecquerel per kilogram basis than those of ^{225}Ac-HuM195 that were found to cause severe anemia and renal tubular damage in monkeys, illustrating the much greater safety margin provided by Auger electron-emitting radiotherapeutics (216). Importantly, we recently showed that ^{111}In-NLS-M195 or ^{111}In-NLS-HuM195 were able to kill HL-60/MX1 cells that were resistant to mitoxantrone, as well as primary AML specimens that displayed high levels of several MDR transporters including pgp-170, BCRP1 and MRP1, responsible for chemotherapy resistance in leukemia (131). A key finding was that the SA of ^{111}In-NLS-HuM195 or ^{111}In-NLS-M195 was critically important for cytotoxicity due to the low CD33 density (4×10^4 receptors/cell) on leukemic cells with a minimum SA of 2–3 MBq/µg required for effective cell killing. Multidrug resistance often leads to relapses following successful chemotherapy of leukemia; Auger electron radiotherapy may be one strategy to overcome this challenge (224).

9.6 SMALL-MOLECULE AUGER ELECTRON RADIOTHERAPY

This chapter focuses on the application of mAbs or peptides as targeting vehicles for Auger electron radiotherapy of malignancies, but it is useful to briefly review for comparison the use of small molecules as delivery systems for this class of radionuclides. These include estradiol analogues (20, 29, 225–228), metaiodobenzylguanidine (mIBG) (27, 229–234) as well as some novel agents that are able to intercalate directly into DNA (235–237).

9.6.1 Radiolabeled Estradiol Analogues

Estradiol diffuses into cells and specifically binds to estrogen receptors. ERs are transcription factors that translocate from the cytoplasm to the nucleus following binding of E2, where they interact with estrogen response elements (ERE) on DNA, increasing gene expression (238). This direct binding of E2 to ERE provides a route for insertion of Auger electron-emitting radionuclides into the nucleus and promotes their interaction with DNA, the most radiation sensitive target in the cell. Recognizing that ERs are present in more than 50% of breast cancers, this creates an opportunity for ER-targeted Auger electron radiotherapy. Beckmann et al. showed that DNA strand breaks were produced in ER-positive MCF-7 breast cancer cells incubated *in vitro* with 16-α-^{125}I-iodo-17-β-estradiol (^{125}I-E2) (225). The SF of the MCF-7 cells was decreased fivefold by exposure to <0.2 MBq/mL of ^{125}I-E2, whereas the SF of an ER-negative MCF-7 subclone was not effected. Yasui et al. showed differential toxicity of the estradiol analogue, E-17-α-^{125}I-iodovinyl-11-β-methoxyestradiol (^{125}I-VME2) toward MCF-7 cells compared to ER-negative MDA-MB-231 cells (29). Differential toxicity has similarly been noted for ^{125}I-labeled tamoxifen (a competitive ER antagonist) toward MCF-7 cells compared to V79 Chinese hamster cells that have a 15-fold lower ER expression (20).

The above studies are promising but a serious limitation with the use of ^{125}I-labeled E2 analogues for Auger electron radiotherapy is the long physical half-life ($t_{1/2}$) of ^{125}I (57 days) relative to the short biological $t_{1/2}$ (4 h) of the E2/ER complex in cells, combined with the low expression of ER in a tumor cell (239). The difference in the physical kinetics of the radionuclide and biological kinetics of the targeting vehicle would minimize the radiation absorbed dose deposited in DNA, and thus, the effectiveness for treatment of ER-positive tumors *in vivo*, despite the promising results *in vitro*. One strategy that has been explored to address this issue as well as the limited ER availability was to radiolabel the E2 analogues to maximum achievable SA with shorter-lived Auger electron emitters such as ^{80m}Br or ^{123}I ($t_{1/2}$ of 4.4 and 13.2 h, respectively). DeSombre et al. showed that ^{123}I-ME2 and the E2 analogue, 2-^{123}I-iodo-1,1-bis-(4-hydroxyphenyl)-2-phenylethylene (^{123}I-BHPE) (8.9 MBq/pmol) were toxic *in vitro* to CHO cells transfected with the ER gene after only a short incubation time of 30–60 min (228). The D_{37} (dose required to reduce the SF to 37%) was 300–600 decays per cell, which was well within the range of ER expression (1–3×10^4 molecules/cell). Similarly high potencies were found for these radiolabeled E2 analogues in ER-positive MCF-7 cells and multicellular spheroids (27). In a subsequent study, a 2-logarithm decrease in the tumorigenicity of MCF-7 cells *in vivo* in athymic mice was found if the cells were pretreated with ^{123}I-VME2 (227). The inverse correlations previously noted between EGFR and ER expressions in breast cancer would make Auger electron radiotherapy using radiolabeled E2 analogues a useful complementary strategy to an approach targeting EGFR (240). Moreover, combining radiotherapeutic agents targeting EGFR and ER would address issues of tumor cell heterogeneity as well as loss of ER expression by some malignant cells (241, 242).

9.6.2 Radiolabeled mIBG Analogues

Metaiodobenzylguanidine is a guanethidine analogue that is actively concentrated and stored in neuroblastoma cells via the norepinephrine transporter (NET); it also accumulates in these cells by diffusion across the cell membrane. ^{131}I-mIBG has been used for treatment of neuroblastoma, but a major limitation is its dose-limiting toxicity to the bone marrow, likely due to the long-range (2 mm) β-particles emitted by the radionuclide that nonspecifically irradiate and kill hematopoetic stem cells ("crossfire" effect) (243, 244). Radioiodinated mIBG is accumulated intracellularly mainly within mitochondria and there is some relocalization of vesicles incorporating the radiopharmaceutical to the nuclear membrane, but these are not imported into the nucleus itself (229). To diminish the bone marrow toxicity of ^{131}I-mIBG as well as improve its ability to eradicate residual micrometastases following chemotherapy of neuroblastoma that are not within the irradiation volume of the moderate energy (0.6 MeV) β-particles emitted by ^{131}I, the agent has been labeled with the Auger electron emitters, [125]I (230–234) or [123]I (245, 246). Sisson et al. reported almost 20 years ago ^{125}I-mIBG given in doses of 9657–15,059 MBq to three children for treatment of Stage III neuroblastoma (233). The main toxicities were decreased Plt and WBC counts as well as decreased Hb. Stabilization of disease was achieved in two patients, but one patient progressed and died. Hoefnagel et al. reported a mixed tumor response in one patient administered 7400 MBq of ^{125}I-mIBG and a temporary stabilization of disease progression in a second patient receiving a dose of 3700 MBq (230). Sisson et al. subsequently treated 10 children with persistent or recurrent Stage III or IV neuroblastoma with 8300–30,100 MBq of ^{125}I-mIBG (234). Bone marrow toxicity was dose-limiting and suspected to have contributed in part to the deaths of 2 of 5 patients who rapidly progressed. There was no radiologic tumor response but five patients exhibited an unusually prolonged survival considering their advanced stage of disease with 50% alive at 18 months.

Preclinically, ^{125}I-mIBG has been found effective for killing SK-N-MC neuroblastoma multicell spheroids *in vitro* (231). We and others have similarly determined that mIBG labeled with ^{125}I or ^{123}I can kill neuroblastoma cells and spheroids in culture (245, 246). The bone marrow toxicity observed clinically with ^{125}I-mIBG is somewhat surprising because our group found that bone marrow stem cells were spared from the toxicity *in vitro* of ^{123}I-mIBG *in vitro* due to the absence of NET, whereas SK-N-SH neuroblastoma cells were killed (246). However, Rutgers et al. reported that ^{131}I-mIBG was more effective than ^{125}I-mIBG administered at the same doses (60–100 MBq) for treatment of macroscopic s.c. SK-N-SH neuroblastoma xenografts in athymic mice, and the theoretical advantage of ^{125}I-mIBG over ^{131}I-mIBG for the treatment of microscopic hepatic metastases was not realized (232). However, it is likely that the doses of ^{125}I-mIBG administered in this study were too low for an appropriate comparison with ^{131}I-mIBG. As discussed previously for radiolabeled CO17-1A mAbs, Auger electron emitters have been found more effective than β-emitting radionuclides when administered at *equitoxic* doses (but not at *equal* doses)—the effective dose as well as the MTD of an Auger electron-emitting agent may be up to 10-fold higher than that of the corresponding β-emitting analogue (188).

9.6.3 DNA Intercalating Agents

Auger electron-emitting radiotherapeutic agents that can directly bind and intercalate into DNA would be the most effective for killing tumor cells. Ickenstein et al. designed several different ^{125}I-labeled duanorubicin analogues, one of which (^{125}I-Comp1) translocated to the nucleus (Fig. 9.11), intercalated directly into DNA and caused DNA DSBs (0.4 breaks per decay) (235). ^{125}I-Comp1 at a concentration of 0.5 ng/mL (50 kBq/mL) reduced the growth of SKBR-3 breast cancer cells *in vitro* more than 100-fold, whereas unlabeled daunorubicin or nonradioactive [127]I-Comp1 had no cytotoxic effect at these low mass concentrations. Anthracyclines including doxorubicin or daunorubicin are thought to translocate to the nucleus bound to cytoplasmic proteosomes which harbor an endogenous NLS on the α-subunit that interacts with the importin-α/β nuclear importation machinery (247). Häfliger et al. showed that DNA intercalators complexed to a ^{99m}Tc-carbonyl $[^{99m}Tc(OH_2)_3(CO)_3]^+$ complex caused DNA DSBs in φX174 double-stranded plasmid DNA *in vitro* due to the emission of Auger electrons (237). This group modified these ^{99m}Tc-intercalating complexes with octapeptides harboring the SV40 NLS [PKKKRKVGG] to encourage their nuclear importation (236). These ^{99m}Tc-NLS DNA-intercalating complexes accumulated in the nucleus of B16F1 melanoma cells, causing substantial DNA damage (micronuclei) and reducing the SF to virtually zero at a very low molar concentrations (17.5 nmol/L). However, these concentrations were associated with very high radioactivity concentrations (350 MBq/mL). ^{99m}Tc may not be an ideal Auger electron-emitter for this purpose, since as previously discussed, it has a low yield of electrons that have shorter range and deliver less total energy than either ^{111}In or ^{125}I.

9.7 SUMMARY AND CONCLUSIONS

Auger electron emissions from radionuclides commonly used in nuclear medicine for imaging studies have generally been ignored for decades, since their very low energy and ultrashort nanometer range suggested that they would deposit inconsequential doses of radiation in the nucleus of cells, and thus be relatively harmless. It is now being realized that these radionuclides actually have great potential for the treatment of cancer if they can be selectively delivered to tumor cells, internalized into the cytoplasm and transported in close proximity to the cell nucleus, where their high LET causes extensive and lethal DNA damage. The results of preclinical evaluations of Auger electron emitters such as ^{111}In, ^{67}Ga, or ^{125}I conjugated to mAbs and peptides that target and exploit the overexpression of tumor-associated antigens or receptors on malignant cells have been very encouraging, particularly for the eradication of small-volume tumors and micrometastases. Furthermore, clinical trials of these agents have demonstrated that they provide clinical improvement in cancer patients and significant decreases in tumor-associated biomarkers, although objective remissions have been rare. There is clearly a considerable amount of radiopharmaceutical design innovation that will be required to take full advantage of the exquisitely selective and highly potent single-cell killing that can be achieved with this intriguing class of targeted radiotherapeutics. These design aspects include selection of the most appropriate molecular targets and targeting vehicles, enhancing the nuclear uptake and DNA binding of the radiopharmaceuticals, as well as amplifying the amount of radioactivity delivered to tumor cells with each cell-surface antigen/receptor recognition event. Clinical trials in patients with low-volume disease who are most likely to respond to these agents, as well as in those with tumors resistant to other forms of therapy that may respond to targeted radiotherapy would provide great insight into the potential role of Auger electron emitters in the treatment of cancer in the future. The story is just unfolding now and much remains to be told.

ACKNOWLEDGMENTS

Financial support (R. Reilly) for the research described in this chapter was provided by the U.S. Department of Defense Breast Cancer Research Program, the Susan G. Komen Breast Cancer Foundation, the Canadian Breast Cancer Research Alliance, the Canadian Breast Cancer Foundation (Ontario Branch), the Canadian Institutes of Health Research and the Ontario Institute of Cancer Research. This work was also supported in part by NIH 5 R01 CA15523 (Amin I. Kassis).

REFERENCES

1. Auger P. Sur les rayons β secondaires produits dans un gaz par des rayons X. *Comptes Rendues Hebdomadaires des Seances de l'Académie des Sciences* 1925;180:65–68.
2. Cole A. Absorption of 20-eV to 50,000-eV electron beams in air and plastic. *Radiat Res* 1969;38:7–33.
3. Kassis AI. The amazing world of Auger electrons. *Int J Radiat Biol* 2004;80:789–803.
4. Kassis AI, Adelstein SJ, Haydock C, Sastry KSR. Radiotoxicity of ^{75}Se and ^{35}S: theory and application to a cellular model. *Radiat Res* 1980;84:407–425.
5. Kassis AI, Sastry KSR, Adelstein SJ. Kinetics of uptake, retention, and radiotoxicity of ^{125}IUdR in mammalian cells: implications of localized energy deposition by Auger processes. *Radiat Res* 1987;109:78–89.
6. Kassis AI, Fayad F, Kinsey BM, Sastry KSR, Taube RA, Adelstein SJ. Radiotoxicity of ^{125}I in mammalian cells. *Radiat Res* 1987;111:305–318.
7. Carlson TA, White RM. Formation of fragment ions from CH_3Te^{125} and $C_2H_5Te^{125}$ following the nuclear decays of CH_3I^{125} and $C_2H_5I^{125}$. *J Chem Phys* 1963;38:2930–2934.
8. Kassis AI, Adelstein SJ, Haydock C, Sastry KSR, McElvany KD, Welch MJ. Lethality of Auger electrons from the decay of bromine-77 in the DNA of mammalian cells. *Radiat Res* 1982;90:362–373.
9. Makrigiorgos GM, Kassis AI, Baranowska-Kortylewicz J, et al. Radiotoxicity of 5-[^{123}I]iodo-2′-deoxyuridine in V79 cells: a comparison with 5-[^{125}I]iodo-2′-deoxyuridine. *Radiat Res* 1989;118:532–544.
10. Kassis AI, Makrigiorgos GM, Adelstein SJ. Dosimetric considerations and therapeutic potential of Auger electron emitters. In: Adelstein SJ, Kassis AI, Burt RW, editors. *Frontiers in Nuclear Medicine: Dosimetry of Administered Radionuclides*, Proceedings of Symposium, 1989. American College of Nuclear Physicians, Washington, DC, 1990, pp. 257–274.
11. Hofer KG, Harris CR, Smith JM. Radiotoxicity of intracellular ^{67}Ga, ^{125}I and ^{3}H: nuclear versus cytoplasmic radiation effects in murine L1210 leukaemia. *Int J Radiat Biol* 1975;28:225–241.
12. Kassis AI, Adelstein SJ, Haydock C, Sastry KSR. Thallium-201: an experimental and a theoretical radiobiological approach to dosimetry. *J Nucl Med.* 1983;24:1164–1175.
13. Kassis AI, Sastry KSR, Adelstein SJ. Intracellular distribution and radiotoxicity of chromium-51 in mammalian cells: Auger-electron dosimetry. *J Nucl Med* 1985;26:59–67.

14. Miyazaki N, Shinohara K. Cell killing induced by decay of ^{125}I during the cell cycle: comparison of ^{125}I-antipyrine with ^{125}I-bovine serum albumin. *Radiat Res* 1993; 133: 182–186.

15. Warters RL, Hofer KG, Harris CR, Smith JM. Radionuclide toxicity in cultured mammalian cells: elucidation of the primary site of radiation damage. *Curr Top Radiat Res Q* 1977;12:389–407.

16. Kassis AI, Howell RW, Sastry KSR, Adelstein SJ. Positional effects of Auger decays in mammalian cells in culture. In: Baverstock KF, Charlton DE, editors. *DNA Damage by Auger Emitters*. Taylor and Francis, London, 1988, pp. 1–13.

17. Martin RF, Bradley TR, Hodgson GS. Cytotoxicity of an ^{125}I-labeled DNA-binding compound that induces double-stranded DNA breaks. *Cancer Res* 1979;39: 3244–3247.

18. Kassis AI, Fayad F, Kinsey BM, Sastry KSR, Adelstein SJ. Radiotoxicity of an [125]I-- labeled DNA intercalator in mammalian cells. *Radiat Res* 1989;118:283–294.

19. Howell RW, Kassis AI, Adelstein SJ, et al. Radiotoxicity of platinum-195m-labeled trans-platinum (II) in mammalian cells. *Radiat Res* 1994;140:55–62.

20. Bloomer WD, McLaughlin WH, Weichselbaum RR, et al. Iodine-125-labelled tamoxifen is differentially cytotoxic to cells containing oestrogen receptors. *Int J Radiat Biol* 1980;38:197–202.

21. Bloomer WD, McLaughlin WH, Milius RA, Weichselbaum RR, Adelstein SJ. Estrogen receptor-mediated cytotoxicity using iodine-125. *J Cell Biochem* 1983;21:39–45.

22. McLaughlin WH, Milius RA, Pillai KMR, Edasery JP, Blumenthal RD, Bloomer WD. Cytotoxicity of receptor-mediated 16α-^{125}I]iodoestradiol in cultured MCF-7 human breast cancer cells. *J Natl Cancer Inst* 1989;81:437–440.

23. Epperly MW, Damodaran KM, McLaughlin WH, Pillai KMR, Bloomer WD. Radio-toxicity of 17α-^{125}I]iodovinyl-11β-methoxyestradiol in MCF-7 human breast cancer cells. *J Steroid Biochem Mol Biol* 1991;39:729–734.

24. DeSombre ER, Hughes A, Shafii B, et al. Estrogen receptor-directed radiotoxicity with Auger electron-emitting nuclides: E-17α-^{123}I] iodovinyl-11β-methoxyestradiol and CHO-ER cells. In: Howell RW, Narra VR, Sastry KSR, Rao DV, editors. *Biophysical aspects of Auger processes, American Association of Physicists in Medicine*, Symposium Series No 8. American Institute of Physics, Woodbury, NY, 1992, pp. 352–371.

25. Yasui LS. Cytotoxicity of ^{125}I decay in the DNA double strand break repair deficient mutant cell line, xrs-5. *Int J Radiat Biol* 1992;62:613–618.

26. DeSombre ER, Hughes A, Landel CC, Greene G, Hanson R, Schwartz JL. Cellular and subcellular studies of the radiation effects of Auger electron-emitting estrogens. *Acta Oncol* 1996;35:833–840.

27. Kearney T, Hughes A, Hanson RN, DeSombre ER. Radiotoxicity of Auger electron-emitting estrogens in MCF-7 spheroids: a potential treatment for estrogen receptor-positive tumors. *Radiat Res* 1999;151:570–579.

28. Yasui LS, Hughes A, DeSombre ER. Relative biological effectiveness of accumulated [125]IdU and ^{125}I-estrogen decays in estrogen receptor-expressing MCF-7 human breast cancer cells. *Radiat Res* 2001;155:328–334.

29. Yasui LS, Hughes A, DeSombre ER. Cytotoxicity of ^{125}I-oestrogen decay in non-oestrogen receptor-expressing human breast cancer cells, MDA-231 and oestrogen receptor-expressing MCF-7 cells. *Int J Radiat Biol* 2001;77:955–962.

30. Walicka MA, Ding Y, Roy AM, Harapanhalli RS, Adelstein SJ, Kassis AI. Cytotoxicity of [^{125}I]iodoHoechst 33342: contribution of scavengeable effects. *Int J Radiat Biol* 1999;75:1579–1587.

31. Karagiannis TC, Lobachevsky PN, Martin RF. Cytotoxicity of an ^{125}I-labelled DNA ligand. *Acta Oncol* 2000;39:681–685.

32. Sedelnikova OA, Panyutin IG, Thierry AR, Neumann RD. Radiotoxicity of iodine-125-labeled oligodeoxyribonucleotides in mammalian cells. *J Nucl Med* 1998;39:1412–1418.

33. Yasui LS, Chen K, Wang K, et al. Using Hoechst 33342 to target radioactivity to the cell nucleus. *Radiat Res* 2007;167:167–175.

34. Sastry KSR, Rao DV. Dosimetry of low energy electrons. In: Rao DV, Chandra R, Graham MC, editors. *Physics of Nuclear Medicine: Recent Advances.* American Institute of Physics, Woodbury, NY, 1984, pp. 169–208.

35. Pomplun E, Booz J, Charlton DE. A Monte Carlo simulation of Auger cascades. *Radiat Res* 1987;111:533–552.

36. Charlton DE, Humm JL. A method of calculating initial DNA strand breakage following the decay of incorporated ^{125}I. *Int J Radiat Biol* 1988;53:353–365.

37. Kassis AI, Harapanhalli RS, Adelstein SJ. Comparison of strand breaks in plasmid DNA after positional changes of Auger electron-emitting iodine-125. *Radiat Res* 1999;151:167–176.

38. Kassis AI, Harapanhalli RS, Adelstein SJ. Strand breaks in plasmid DNA after positional changes of Auger electron-emitting iodine-125: direct compared to indirect effects. *Radiat Res* 1999;152:530–538.

39. Balagurumoorthy P, Chen K, Bash RC, Adelstein SJ, Kassis AI. Mechanisms underlying production of double-strand breaks in plasmid DNA after decay of ^{125}I-Hoechst. *Radiat Res* 2006;166:333–344.

40. Kandaiya S, Lobachevsky PN, D'Cunha G, Martin RF. DNA strand breakage by ^{125}I-- decay in a synthetic oligodeoxynucleotide: fragment distribution and evaluation of DMSO protection effect. *Acta Oncol* 1996;35:803–808.

41. Panyutin IG, Neumann RD. Radioprobing of DNA: distribution of DNA breaks produced by decay of ^{125}I incorporated into a triplex-forming oligonucleotide correlates with geometry of the triplex. *Nucleic Acids Res* 1997;25:883–887.

42. Walicka MA, Adelstein SJ, Kassis AI. Indirect mechanisms contribute to biological effects produced by decay of DNA-incorporated iodine-125 in mammalian cells *in vitro*: double-strand breaks. *Radiat Res* 1998;149:134–141.

43. Kassis AI, Walicka MA, Adelstein SJ. Double-strand break yield following ^{125}I decay: effects of DNA conformation. *Acta Oncol* 2000;39:721–726.

44. Elmroth K. Stenerlöw B. DNA-incorporated ^{125}I induces more than one double-strand break per decay in mammalian cells. *Radiat Res* 2005;163:369–373.

45. Balagurumoorthy P, Chen K, Adelstein SJ, Kassis AI. Auger electron-induced double-strand breaks depend on DNA topology. *Radiat Res* 2008;170(1): 70–82.

46. Nagasawa H, Little JB. Induction of sister chromatid exchanges by extremely low doses of α-particles. *Cancer Res.* 1992;52:6394–6396.

47. Hickman AW, Jaramillo RJ, Lechner JF, Johnson NF. α-Particle-induced p53 protein expression in a rat lung epithelial cell strain. *Cancer Res* 1994;54:5797–5800.

48. Azzam EI, de Toledo SM, Gooding T, Little JB. Intercellular communication is involved in the bystander regulation of gene expression in human cells exposed to very low fluences of alpha particles. *Radiat Res* 1998;150:497–504.

49. Xue LY, Butler NJ, Makrigiorgos GM, Adelstein SJ, Kassis AI. Bystander effect produced by radiolabeled tumor cells in vivo. *Proc Natl Acad Sci USA* 2002;99:13765–13770.

50. Barcellos-Hoff MH, Chaudhry MA, De Veaux L, et al. Meeting report: radiation-induced genomic instability and bystander effects; implications for evolutionary biology, report on an international workshop on radiation-induced low-dose effects held at McMaster University in Hamilton, Ontario, Canada, October 28–November 1, 2004. *Radiat Res* 2005;163:473–476.

51. Kishikawa H, Wang K, Adelstein SJ, Kassis AI. Inhibitory and stimulatory bystander effects are differentially induced by iodine-125 and iodine-123. *Radiat Res* 2006;165:688–694.

52. Howell RW, Bishayee A. Bystander effects caused by nonuniform distributions of DNA-incorporated ^{125}I. *Micron* 2002;33:127–132.

53. Boyd M, Ross SC, Dorrens J, et al. Radiation-induced biologic bystander effect elicited *in vitro* by targeted radiopharmaceuticals labeled with α-, β-, and Auger electron–emitting radionuclides. *J Nucl Med* 2006;47:1007–1015.

54. O'Donoghue JA, Wheldon TE. Targeted radiotherapy using Auger electron emitters. *Phys Med Biol* 1996;41:1973–1992.

55. Howell RW. Radiation spectra for Auger-electron emitting radionuclides: Report no. 2 of AAPM Nuclear Medicine Task group No. 6. *Med Phys* 1992;19:1371–1383.

56. Govindan SV, Goldenberg DM, Elsamra SE, Griffiths GL, Ong GL, Brechbiel MW, et al. Radionuclides linked to a CD74 antibody as therapeutic agents for B-cell lymphoma: comparison of Auger electron emitters with β-particle emitters. *J Nucl Med* 2000; 41:2089–2097.

57. Bernhardt P, Forssell-Aronsson E, Jacobsson L, Skamemark G. Low-energy electron-emitters for targeted radiotherapy of small tumours. *Acta Oncologica* 2001;40:602–608.

58. Goddu SM, Howell RW, Rao DV. Cellular dosimetry: absorbed fractions for mono-energetic electron and alpha particle sources and *S*-values for radionuclides uniformly distributed in different cell compartments. *J Nucl Med* 1994;35:303–316.

59. Richter MP, Horvick D. Principles of Radiation Oncology. In: Calabresi P, Schein PS, editors. *Medical Oncology: Basic Principles and Clinical Management of Cancer*, 2nd edition. McGraw-Hill, Inc., Toronto, 1993, pp. 253–262.

60. Stewart JSW, Hird V, Snook D, et al. Intraperitoneal ^{131}I- and ^{90}Y-labelled monoclonal antibodies for ovarian cancer: pharmacokinetics and normal tissue dosimetry. *Int J Cancer* 1988;3 (Suppl): 71–76.

61. Williams LE, Beatty BG, Beatty JD. Estimation of monoclonal antibody-associated ^{90}Y activity needed to achieve certain tumor radiation doses in colorectal cancer patients. *Cancer Res* 1990;50 (Suppl): 1029s–1030s.

62. Reilly RM. Biomolecules as targeting vehicles for in situ radiotherapy of malignancies. In: Knaeblein J, Mueller R, editors. *Modern Biopharmaceuticals: Design, Development and Optimization*. Wiley-VCH, Weinheim, Germany, 2005; pp. 497–526.

63. Goddu SM, Rao DV, Howell RW. Multicellular dosimetry for micrometastases: dependence of self-dose versus cross-dose to cell nuclei on type and energy of radiation and subcellular distribution of radionuclides. *J Nucl Med* 1994;35:521–530.

64. Bloomer WD, Adelstein SJ. 5-^{125}I-iododeoxyuridine as prototype for radionuclide therapy with Auger emitters. *Nature* 1977;265:620–662.

65. Baranowska-Kortylewicz J, Makrigiorgos GM. Van den Abbeele AD, Berman RM, Adelstein SJ, Kassis AI. 5-^{123}I]iodo-2'-deoxyuridine in the radiotherapy of an early ascites tumor model. *Int J Radiat Oncol Biol Phys* 1991;21:1541–1551.

66. Kassis AI, Dahman BA, Adelstein SJ. *In vivo* therapy of neoplastic meningitis with methotrexate and 5-^{125}I]iodo-2'-deoxyuridine. *Acta Oncol* 2000;39:731–737.

67. Kassis AI, Kirichian AM, Wang K, Safaie Semnani E, Adelstein SJ. Therapeutic potential of 5-^{125}I]iodo-2'-deoxyuridine and methotrexate in the treatment of advanced neoplastic meningitis. *Int J Radiat Biol* 2004;80:941–946.

68. Rebischung C, Hoffmann D, Stfani L, et al. First human treatment of resistant neoplastic meningitis by intrathecal administration of MTX Plus [125]IUdR. *Int J Radiat Biol* 2008;84 (12):1123–1129.

69. Guillermmet-Guibert J, Lahlou H, Cordelier P, Bousquet C, Pyronnet S, Susini C. Physiology of somatostatin receptors. *J Endocrinol Invest* 2005;28 (11 Suppl): 5–9.

70. Csaba Z, Dournand P. Cellular biology of somatostatin receptors. *Neuropeptides* 2001;35:1–23.

71. Florio T. Molecular mechanisms of the antiproliferative activity of somatostatin receptors (SSTRs) in neuroendocrine tumors. *Frontiers in Bioscience* 2008;13:822–840.

72. Reubi JC, Schar J-C, Waser B, Wenger S, Heppeler A, Schmitt JS, et al. Affinity profiles for human somatostatin receptor subtypes SST1-SST5 of somatostatin radiotracers selected for scintigraphic and radiotherapeutic use. *Eur J Nucl Med* 2000;27:273–282.

73. Virgolini I, Britton K, Buscombe J, Moncayo R, Paganelli G, Riva P. [111]In- and ^{90}Y-DOTA-lanreotide: results and implications of the MAURITIUS trial *Semin Nucl Med* 2002;32:148–155.

74. Kwekkeboom DJ, Krenning EP. Somatostatin receptor imaging. *Semin Nucl Med* 2002;32:84–91.

75. Lamberts SWJ, Reubi J-C, Krenning EP. The role of somatostatin analogs in the control of tumor growth. *Semin Oncol* 1994;21:61–64.

76. Krenning EP, Kwekkeboom DJ, Bakker WH, et al. Somatostatin receptor scintigraphy with [^{111}In-DTPA-D-Phe[1]]- and [^{123}I-Tyr[3]]-octreotide: the Rotterdam experience with more than 1000 patients. *Eur J Nucl Med* 1993;20:716–731.

77. Van Essen M, Krenning EP, de Jong M, Valkema R, Kwekkeboom D. Peptide receptor radionuclide therapy with radiolabelled somatostatin analogues in patients with somatostatin receptor positive tumours. *Acta Oncologica* 2007;46:723–34.

78. Maecke HR, Hofmann M, Haberkorn U. ^{68}Ga-labeled peptides in tumor imaging *J Nucl Med* 2005;46:172S–178S.

79. Smith-Jones PM, Bischof C, Leimer M, Gludovacz D, Angelberger P, et al. DOTA-lanreotide: a novel somatostatin analog for tumor diagnosis and therapy. *Endocrinol* 1999;140:5136–5148.

80. Reubi JC, Schaer JC, Waser B, Mengod G. Expression and localization of somatostatin receptor SSTR1, SSTR2, and SSTR3 messenger RNAs in primary human tumors using *in situ* hybridization. *Cancer Res* 1994;54:3455–3459.

81. Reubi JC, Waser B, Schaer JC, Laissue JA. Somatostatin receptor sst1-sst5 expression in normal and neoplastic human tissues using receptor autoradiography with subtype-selective ligands. *Eur J Nucl Med Mol Imaging* 2001;28:836–846.

82. Reubi JC, Waser B, Foekens JA, Klijn JGM, Lamberts SWJ, Laissue J. Somatostatin receptor incidence and distribution in breast cancer using receptor autoradiography: relationship to EGF receptors. *Int J Cancer* 1990;46:416–420.

83. Reubi JC, Waser B. Concomitant expression of several peptide receptors in neuroendcrine tumours: molecular basis for *in vivo* multireceptor tumour targeting. *Eur J Nucl Med Mol Imaging* 2003;30:781–793.

84. Oomen SPMA, Hofland LJ, van Hagen PM, Lamberts SWJ, Touw IP. Somatostatin receptors in the hematopoietic system. *Eur J Endocrinol* 2000;143:S9–S14.

85. Aguila MC, Rodriguez AM, Aguila-Mansilla HN, Lee WT. Somatostatin antisense oligodeoxynucleotide-mediated stimulation of lymphocyte proliferation in culture. *Endocrinol* 1996;137:1585–1590.

86. Duncan JR, Stephenson MT, Wu HP, Anderson CJ. Indium-111-diethylenetriaminepentaacetic acid-octreotide is delivered *in vivo* to pancreatic, tumor cell, renal, and hepatic lysosomes. *Cancer Res* 1997;57:659–671.

87. Gotthardt M, van Eerd-Vismale J, Oyen WJ, de Jong M, Zhang H, Rolleman E, et al. Indication for different mechanisms of kidney uptake of radiolabeled peptides. *J Nucl Med* 2007;48:596–601.

88. Duncan JR, Welch MJ. Intracellular metabolism of indium-111-DTPA labeled receptor targeted proteins. *J Nucl Med* 1993;34:1728–1738.

89. Franano FN, Edwards WB, Welch MJ, Duncan JR. Metabolism of receptor targeted ^{111}In-DTPA-glycoproteins: identification of ^{111}In-DTPA--lysine as the primary metabolic and excretory product. *Nucl Med Biol* 1994;21:1023–1034.

90. Nouel D, Gaudriault G, Houle M, Reisine T, Vincent J-P, Mazella J, et al. Differential internalization of somatostatin in COS-7 cells transfected with SST_1 and SST_2 receptor subtypes: a confocal microscopic study using novel fluorescent somatostatin derivatives. *Endocrinol* 1997;138:296–306.

91. Cescato R, Schutlz S, Waser B, Eltschinger V, Rivier JE, et al. Internalization of sst_2, sst_3, and sst_5 receptors: effects of somatostatin agonists and antagonists. *J Nucl Med* 2006;47:502–511.

92. Andersson P, Forssell-Aronsson E, Johanson V, Wangberg B, Nilsson O, Fjalling M, et al. Internalization of indium-111 into human neuroendocrine tumor cells after incubation with indium-111-DTPA-D-Phe[1]-octreotide. *J Nucl Med* 1996;37:2002–2006.

93. de Jong M, Bernard BF, De Bruin E, van Gameren A, Bakker WH, Visser TJ, et al. Internalization of radiolabelled [DTPA[0]]octreotide and [DOTA[0],Tyr[3]]octreotide: peptides for somatostatin receptor-targeted scintigraphy and radionuclide therapy. *Nucl Med Commun* 1998;19:283–288.

94. Wang M, Caruano AL, Lewis MR, Meyer LA, VanderWaal RP, Anderson CJ. Subcellular localization of radiolabeled somatostatin analogues: implications for targeted radiotherapy of cancer. *Cancer Res* 2003;63:6864–6869.

95. Eiblmaier M, Andrews R, LaForest R, Rogers BE, Anderson CJ. Nuclear uptake and dosimetry of ^{64}Cu-labeled chelator-somatostatin conjugates in an SSTr2-transfected human tumor cell line. *J Nucl Med* 2007;48:1390–1396.

96. Hornick CA, Anthony CT, Hughey S, Gebhardt BM, Espenan GD, Woltering EA. Progressive nuclear translocation of somatostatin analogs. *J Nucl Med* 2000; 41:1256–1263.

97. Hu M, Scollard D, Chan C, Chen P, Vallis K, Reilly RM. Effect of the EGFR density of breast cancer cells on nuclear importation, *in vitro* cytotoxicity, and tumor and normal-tissue uptake of [111In]DTPA-hEGF. *Nucl Med Biol* 2007;34: 887–896.

98. Bailey KE, Costantini DL, Cai Z, Scollard DA, Chen Z, Reilly RM, et al. Epidermal growth factor receptor inhibition modulates the nuclear localization and cytotoxicity of the Auger electron emitting radiopharmaceutical 111In-DTPA human epidermal growth factor. *J Nucl Med* 2007;48:1562–1570.

99. Reilly RM, Kiarash R, Cameron R, Porlier N, Sandhu J, Hill RP, et al. 111In- labeled EGF is selectively radiotoxic to human breast cancer cells overexpressing EGFR. *J Nucl Med* 2000;41:429–438.

100. Rakowicz-Szulczynska EM, Rodeck U, Herlyn M, Koprowski H. Chromatin binding of epidermal growth factor, nerve growth factor, and platelet-derived growth factor in cells bearing the appropriate surface receptors. *Proc Natl Acad Sci USA* 1986; 83:3728–3732.

101. Janson ET, Westlin J-E, Ohrvall U, Oberg K, Lukinius A. Nuclear localization of 111In after intravenous injection of [111In-DTPA-D-Phe1]-octreotide in patients with neuroendocrine tumours. *J Nucl Med* 2000;41:1514–1518.

102. Likinius A, Ohrvall U, Westlin J-E, Oberg K, Janson ET. *In vivo* cellular distribution and endocytosis of the somatostatin receptor–ligand complex. *Acta Oncologica* 1999; 38:383–387.

103. Capello A, Krenning EP, Breeman WAP. Bernard BF, de Jong M. Peptide receptor radionuclide therapy *in vitro* using [111In-DTPA0]octreotide. *J Nucl Med* 2003;44: 98–104.

104. Slooter GD, Breeman WAP, Marquet RL, Krenning EP, Van Eijck CHJ. Anti-proliferative effect of radiolabelled octreotide in a metastasis model in rat liver. *Int J Cancer* 1999; 81:767–771.

105. Capello A, Krenning E, Bernard B, Reubi J-C, Breeman W, de Jong M. 111In-labelled somatostatin analogues in a rat tumour model: somatostatin receptor status and effects of peptide receptor radionuclide therapy. *Eur J Nucl Med Mol Imaging* 2005;32: 1288–1295.

106. de Jong M, Breeman WAP, Bernard BF, Bakker WH, Visser TJ, Kooij PPM, et al. Tumor response after [90Y-DOTA0,Tyr3]-octreotide radionuclide therapy in a transplantable rat tumor model is dependent on tumor size. *J Nucl Med* 2001;42:1841–1846.

107. Krenning EP, Kooij PP, Bakker WH, Breeman WA, Postema PTE, Kwekkeboom DJ, et al. Radiotherapy with a radiolabeled somatostatin analogue, [111In-DTPA-D-Phe1]-octreotide. A case history. *Ann NY Acad Sci* 1994;733:496–506.

108. Fjalling M, Andersson P, Forssell-Aronsson E, Gretarsdottir J, Johansson V, Tissell LE, et al. Systemic radionuclide therapy using indium-111-DTPA-D-Phe1-octreotide in midgut carcinoid syndrome. *J Nucl Med* 1996;37:1519–1521.

109. Tiensuu Janson E, Eriksson B, Oberg K, Skogseid B, Ohrvall U, Nilsson S, et al. Treatment with high dose [111In-DTPA-D-Phe1]-octreotide in patients with neuroendocrine tumors. *Acta Oncologica* 1999;38:373–377.

110. Caplin ME, Mielcarek W, Buscombe JR, Jones AL, Croasdale PL, Cooper MS, et al. Toxicity of high-activity 111In-octreotide therapy in patients with disseminated neuroendocrine tumours. *Nucl Med Commun* 2000;21:97–102.

111. Valkema R, de Jong M, Bakker WH, Breeman WA, Kooij PPM, Lugtenburg PJ, et al. Phase I study of peptide receptor radionuclide therapy with [111In-DTPA 0]octreotide: the Rotterdam experience. *Semin Nucl Med* 2002;32(2): 110–122.

112. Emami B, Lyman J, Brown A, et al. Tolerance of normal tissue to therapeutic irradiation. *Int J Radiat Oncol Biol Phys* 1991;21:109–122.

113. Anthony LB, Woltering EA, Espenan GD, Cronin MD, Maloney TJ, McCarthy KE. Indium-111-pentetreotide prolongs survival in gastroenteropancreatic malignancies. *Semin Nucl Med* 2002;32:123–132.

114. Delpassand ES, Sims-Mourtada J, Saso H, Azhdarinia A, Ashoori F, Torabi F, et al. Safety and efficacy of radionuclide therapy with high-activity In-111 pentetreotide in patients with progressive neuroendocrine tumors. *Cancer Biother Radiopharm* 2008; 23:292–300.

115. Nguyen C, Faraggi M, Giraudet A-L, de Labriolle-Vaylet C, Aparicio T, Rouzet F, et al. Long-term efficacy of radionuclide therapy in patients with disseminated neuroendocrine tumors uncontrolled by conventional therapy. *J Nucl Med* 2004;45:1660–1668.

116. Stokkel MPM, Verkooijen RBT, Bouwsma H, Smit JWA. Six month follow-up after 111In-DTPA-octreotide therapy in patients with progressive radioiodine non-responsive thyroid cancer: a pilot study. *Nucl Med Commun* 2004;25:683–690.

117. O'Donoghue JA, Bardies M, Wheldon TE. Relationship between tumour size and curability for targeted radionuclide therapy with beta-emitting radionuclides. *J Nucl Med* 1995;36:1902–1909.

118. Limouris GS, Dimitropoulos N, Kontogeorgakos D, Papanikolos G, Koutoulidis V, et al. Evaluation of the therapeutic response to In-111-DTPA octreotide-based targeted therapy in liver metastatic neuroendocrine tumors according to CT/MRI/US findings. *Cancer Biother Radiopharm* 2005;20:215–217.

119. Kwekkeboom DJ, Mueller-Brand J, Paganelli G, Anthony LB, Pauwels S, et al. Overview of results of peptide receptor radionuclide therapy with 3 radiolabeled somatostatin analogs. *J Nucl Med* 2005;46:62S–66S.

120. Behr TM, Goldenberg DM, Becker W. Reducing the renal uptake of radiolabeled antibody fragments and peptides for diagnosis and therapy: present status, future prospects and limitations. *Eur J Nucl Med* 1998;25:201–12.

121. de Jong M, Barone R, Krenning E, Bernard B, Melis M, et al. Megalin is essential for renal proximal tubule reabsorption of 111In-DTPA-octreotide. *J Nucl Med* 2005; 46:1696–1700.

122. de Jong M, Valkema R, van Gameren A, van Boven H, Bex A, et al. Inhomogeneous localization of radioactivity in the human kidney after injection of [111In-DTPA]octreotide. *J Nucl Med* 2004;45:1168–1171.

123. Konijnenberg MW, Bijster M, Krenning EP, de Jong M. A stylized computational model of the rat for organ dosimetry in support of preclinical evaluations of peptide receptor radionuclide therapy with 90Y, 111In, or 177Lu. *J Nucl Med* 2004;45:1260–1269.

124. Konijnenberg M, Melis M, Valkema R, Krenning E, de Jong M. Radiation dose distribution in human kidneys by octreotides in peptide receptor radionuclide therapy. *J Nucl Med* 2007;48:134–142.

125. Valkema R, Pauwels S, Kvols LK, Kwekkeboom DJ, Jamar F, et al. Long-term follow-up of renal function after peptide receptor radiation therapy with 90Y-DOTA0,Tyr3-octreotide and 177Lu-DOTA0,Tyr3-octreotate. *J Nucl Med* 2005;48:83S–91S.

126. Ginj M, Maecke HR. Synthesis of trifunctional somatostatin based derivatives for improved cellular and subcellular uptake. *Tetrahedron Lett* 2005;46:2821–2824.

127. Costantini DL, Hu M, Reilly RM. Peptide motifs for insertion of radiolabeled biomolecules into cells and routing to the nucleus for cancer imaging or radiotherapeutic applications. *Cancer Biother Radiopharm* 2008;23:3–24.

128. Ginj M, Hinni K, Tschumi S, Schultz S, Maecke HR. Trifunctional somatostatin-based derivatives designed for targeted radiotherapy using Auger electron emitters. *J Nucl Med* 2005;46:2097–2103.

129. Costantini DL, Chan C, Cai Z, Vallis KA, Reilly RM. ^{111}In-labeled trastuzumab (Herceptin) modified with nuclear localization sequences (NLS): an Auger electron-emitting radiotherapeutic agent for HER2/neu-amplified breast cancer. *J Nucl Med* 2007; 48:1357–1368.

130. Chen P, Wang J, Hope K, Jin L, Dick J, Cameron R, Brandwein J, Minden M, Reilly RM. Nuclear localizing sequences (NLS) promote nuclear translocation and enhance the toxicity of the anti-CD33 monoclonal antibody HuM195 labeled with ^{111}In in human myeloid leukemia cells. *J Nucl Med* 2006;47:827–836.

131. Kersemans V, Cornelissen B, Minden M, Brandwein J, Reilly RM. Drug-resistant AML cells and primary AML specimens are killed by ^{111}In-anti-CD33 monoclonal antibodies modified with nuclear localizaing peptide sequences. *J Nucl Med* 2008;49:1546–1554.

132. Graham K, Wang Q, Boy RG, Eisenhut M, Haberkorn U, Mier W. Synthesis and evaluation of intercalating somatostatin receptor binding peptide conjugates for endoradiotherapy. *J Pharm Pharmaceut Sci* 2007;10:286s–297s.

133. Bernard B, Capello A, van Hagen M, Breeman W, Srinivasan A, et al. Radiolabeled RGD-DTPA-Tyr^3-octreotate for receptor-targeted radionuclide therapy. *Cancer Biother Radiopharm* 2004;19:173–180.

134. Capello A, Krenning EP, Bernard BF, Breeman WAP, van Hagen MP, de Jong M. Increased cell death after therapy with an Arg-Gly-Asp-linked somatostatin analog. *J Nucl Med* 2004;45:1716–1720.

135. Capello A, Krenning EP, Bernard BF, Breeman WAP, Erion JL, de Jong M. Anticancer activity of targeted proapoptotic peptides. *J Nucl Med* 2006;47:122–129.

136. Boonstra J, Rijken P, Humbel B, Cremers F, Verkleij A, van Bergen en Henegouwen P. The epidermal growth factor. *Cell Biol International* 1995;19:413–430.

137. Rubin I, Yarden Y. The basic biology of HER2. *Annals Oncol* 2001;12 (Suppl 1): S3–S8.

138. Ciardello F, Tortora G. A novel approach in the treatment of cancer: targeting the epidermal growth factor receptor. *Clin Cancer Res* 2001;7:2958–2970.

139. Battaglia F, Scambia G, Rossi S, et al. Epidermal growth factor receptor in human breast cancer: correlation with steroid hormone receptors and axillary lymph node involvement. *Eur J Cancer Clin Oncol* 1988;24:1685–1690.

140. Harris AL, Nicholson S, Sainsbury JRC, et al. Epidermal growth factor receptors in breast cancer: association with early relapse and death, poor response to hormones and interactions with neu. *J Steroid Biochem* 1989;34:123–131.

141. Klijn JGM, Berns PMJJ, Schmitz PIM, Foekens JA. The clinical significance of epidermal growth factor receptor (EGF-R) in human breast cancer: a review on 5232 patients. *Endocr Rev* 1992;13:3–17.

142. Carpenter G, Cohen S. ^{125}I-Labeled human epidermal growth factor. Binding, internalization and degradation in human fibroblasts. *J Cell Biol* 1976;71:159–171.

143. Lo H-W, Hsu S-C, Hung M-C. EGFR signaling pathway in breast cancers: form traditional signal transduction to direct nuclear translocation. *Breast Cancer Res Treat* 2005;95:211–218.

144. Laduron PM. Genomic pharmacology: more intracellular sites for drug action. *Biochem Pharmacol* 1992;44:1233–1242.

145. Lin S-Y, Makino K, Xia W, Matin A, Wen Y, Kwong KY, et al. Nuclear localization of EGF receptor and its potential new role as a transcription factor. *Nature Cell Biol* 2001; 3:802–808.

146. Lo H-W, Xia W, Wei Y, Ali-Seyed M, Huang S-F, Hung M-C. Novel prognostic value of nuclear epidermal growth factor receptor in breast cancer. *Cancer Res* 2005; 65:338–348.

147. Marti U, Ruchti C, Kampf J, Thomas GA, Williams ED, Peter HJ, et al. Nuclear localization of epidermal growth factor and epidermal growth factor receptors in human thyroid tissues. *Thyroid* 2001;11:137–145.

148. Schausberger E, Eferl R, Parzefall W, Chabikovsky M, Breit P, Wagner EF, et al. Induction of DNA synthesis in primary mouse hepatocytes is associated with nuclear pro-trans-forming growth factor α and erbb-1 and is independent of c-jun. *Carcinogenesis* 2003;24:835–841.

149. Gladhaug P, Christoffersen T. Kinetics of epidermal growth factor binding and processing in isolated intact rat hepatocytes: dynamic externalization of receptors during ligand internalization. *Eur J Biochem* 1987;164:267–275.

150. St. Hilaire RJ, Hradek GT, Jones AL. Hepatic sequestration and biliary secretion of epidermal growth factor: evidence for a high-capacity uptake system. *Proc Natl Acad Sci USA* 1983;80:3797–3801.

151. Fisher DA, Salido EC, Barajas L. Epidermal growth factor and the kidney. *Ann Rev Physiol* 1989;51:67–80.

152. Waltz TM, Malm C, Nishikawa BK, Wasteson A. Transforming growth factor-alpha (TGF-alpha) in human bone marrow: demonstration of TGF-alpha in erythroblasts and eosinophilic precursor cells and of epidermal growth factor receptors in blastlike cells of myelomonocytic origin. *Blood* 1995;85:2385–2392.

153. Reilly RM, Kiarash R, Sandhu J, Lee YW, Cameron RG, Hendler A, et al. A comparison of EGF and MAb 528 labeled with ^{111}In for imaging human breast cancer. *J Nucl Med* 2000;41:903–911.

154. Chen P, Mrkobrada M, Vallis KA, Cameron R, Sandhu J, Hendler A, et al. Comparative antiproliferative effects of ^{111}In-DTPA-hEGF, chemotherapeutic agents and gamma-radiation on EGFR-positive breast cancer cells. *Nucl Med Biol* 2002;29:693–699.

155. Filmus J, Trent JM, Pollak MN, Buick RN. Epidermal growth factor receptor gene-amplified MDA-468 breast cancer cell line and its non-amplified variants. *Molec Cellular Biol* 1987;7:251–257.

156. Armstrong DK, Kaufmann SH, Ottaviano YL, Furuya Y, Buckley JA, Isaacs JT, et al. Epidermal growth factor-mediated apoptosis of MDA-MB-468 human breast cancer cells. *Cancer Res* 1994;54:5280–523.

157. Thomas T, Balabhadrapathruni S, Gardner CR, Hong J, Faaland CA, Thomas TJ. Effects of epidermal growth factor on MDA-MB-468 breast cancer cells: alterations in polyamine biosynthesis and the expression of p21/CIP1/WAF1. *J Cell Physiol* 1999; 179:257–266.

158. Wang J, Chen P, Su ZF, Vallis K, Sandhu J, Cameron R, et al. Amplified delivery of indium-111 to EGFR-positive human breast cancer cells. *Nucl Med Biol* 2001; 28:895–902.

159. Grovdal LM, Stang E, Sorkin A, Madshus IH. Direct interaction of Cbl with pTyr 1045 of the EGF receptor (EGFR) is required to sort the EGFR to lysosomes for degradation. *Exp Cell Res* 2004;300:388–395.

160. Baumann M, Krause M. Targeting the epidermal growth factor receptor in radiotherapy: radiobiological mechanisms, preclinical and clinical results. *Radiother and Oncol* 2004;72:257–266.

161. Cai Z, Chen Z, Bailey KE, Scollard DA, Reilly RM, Vallis KA. Relationship between induction of phosphorylated H2AX and survival in breast cancer cells exposed to ^{111}In-DTPA-hEGF. *J Nucl Med* 2008;49:1353–1361.

162. Chen P, Cameron R, Wang J, Vallis KA, Reilly RM. Antitumor effects and normal tissue toxicity of ^{111}In-labeled epidermal growth factor administered to athymic mice bearing epidermal growth factor receptor-positive human breast cancer xenografts. *J Nucl Med* 2003;44:1469–1478.

163. Reilly RM, Chen P, Wang J, Scollard D, Cameron R, Vallis KA. Preclinical pharmacokinetic, biodistribution, toxicology, and dosimetry studies of ^{111}In-DTPA-human epidermal growth factor: an auger electron-emitting radiotherapeutic agent for epidermal growth factor receptor-positive breast cancer. *J Nucl Med* 2006;47:1023–1031.

164. Reilly RM, Scollard DA, Wang J, Mondal H, Chen P, Henderson LA, et al. A kit formulated under good manufacturing practices for labeling human epidermal growth factor with ^{111}In for radiotherapeutic applications. *J Nucl Med* 2004;45:701–708.

165. Vallis KA, Reilly RM, Scollard DA, Petronis J, Caldwell C, Hendler A, et al. A Phase I clinical trial of ^{111}In-human epidermal growth factor (^{111}In-hEGF) in patients with metastatic EGFR-positive breast cancer. *J Nucl Med* 2005;46:152P.

166. Bender H, Takahashi H, Adachi K, Belser P, Liang S, et al. Immunotherapy of human glioma xenografts with unlabeled, [131]I- or ^{125}I-labeled monoclonal antibody 425 to epidermal growth factor receptor. *Cancer Res* 1992;52:121–126.

167. Mattes MJ, Goldenberg DM. Therapy of human carcinoma xenografts with antibodies to EGFr and HER-2 conjugated to radionuclides emitting low-energy electrons. *Eur J Nucl Med Mol Imaging* 2008;35:1249–1258.

168. Michel RB, Castillo ME, Andrews PM, Mattes MJ. *In vitro* toxicity of A-431 carcinoma cells with antibodies to epidermal growth factor receptor and epithelial glycoprotein-1 conjugated to radionuclides emitting low-energy electrons. *Clin Cancer Res* 2004; 10:5957–5966.

169. Epenetos AA, Courtenay-Luck N, Pickering D, Hooker G, Durbin H, Lavender JP, et al. Antibody guided irradiation of brain glioma by arterial infusion of radioactive antibody against epidermal growth factor receptor and blood group A antigen. *Br Med J* 1985; 290:1463–1466.

170. Kalofonos HP, Pawlikowska TR, Hemingway A, et al. Antibody-guided diagnosis and therapy of brain gliomas using radiolabeled monoclonal antibodies against epidermal growth factor receptor and placental alkaline phosphatase. *J Nucl Med* 1989; 30:1636–1645.

171. Quang TS, Brady L. Radioimmunotherapy as a novel treatment regimen: ^{125}I-labeled monoclonal antibody 425 in the treatment of high-grade brain gliomas. *Int J Radiat Oncol Biol Phys* 2004;58:972–975.

172. Reilly RM. The radiopharmaceutical science of monoclonal antibodies and peptides for imaging and targeted in situ radiotherapy of malignancies. In: Gad SC, editor. *Handbook of Biopharmaceutical Technology.* John Wiley & Sons, Toronto, 2007, pp. 987–1053.

173. Baselga J, Albanell J. Mechanism of action of anti-HER2 monoclonal antibodies. *Annals Oncol* 2001;12 (Suppl 1.): S35–S41.

174. Cuello M, Ettenberg AA, Clark AS, Keane MM, Posner RH, Nau MM, et al. Down-regulation of the erbB-2 receptor by trastuzumab (Herceptin) enhances tumor necrosis factor-related apoptosis-inducing ligand-mediated apoptosis in breast and ovarian cancer cell lines that overexpress erbB-2. *Cancer Res* 2001;61:4892–4900.

175. Baselga J, Albanell J. Mechanism of action of anti-HER2 monoclonal antibodies. *Annals Oncol* 2001;12 (Suppl 1): S35–S41.

176. Xie Y, Hung M-C. Nuclear localization of p185^{NEU} tyrosine kinase and its association with transcriptional activation. *Biochem Biophys Res Commun* 1994;203:1589–1598.

177. van de Vijver MJ. Assessment of the need and appropriate method for testing for the human epidermal growth factor receptor-2 (HER2). *Eur J Cancer* 2001;27:S11–S17.

178. Seidman AD, Fornier MN, Esteva FJ, Tan L, Kaptain S, et al. Weekly trastuzumab and paclitaxel therapy for metastatic breast cancer with analysis of efficacy by HER2 immunophenotype and gene amplification. *J Clin Oncol* 2001;19:2587–2595.

179. Baselga J. Clinical trials of Herceptin (trastuzumab). *Eur J Cancer* 2001;37:S18–S24.

180. Picarte M. Circumventing de novo and acquired resistance to trastuzumab: new hope for the care of ErbB2-positive breast cancer. *Clin Breast Cancer* 2008;8 (Suppl 3): S100–S113.

181. Costantini DL, Bateman K, McLarty K, Vallis KA, Reilly RM. Trastuzumab-resistant breast cancer cells remain sensitive to the Auger electron-emitting radiotherapeutic agent ^{111}In-NLS-trastuzumab and are radiosensitized by methotrexate. *J Nucl Med* 2008; 49:1498–505.

182. Daghighian F, Barendswaard E, Welt S, Humm J, Scott A, Willingam MC, et al. Enhancement of radiation dose to the nucleus by vesicular internalization of iodine-125-labeled A33 monoclonal antibody. *J Nucl Med* 1996;37:1052–1057.

183. Barendswaard EC, Humm JL, O'Donoghue JA, Sgouros G, Finn RD, Scott AM, et al. Relative therapeutic efficacy of ^{125}I- and ^{131}I-labeled monoclonal antibody A33 in a human colon cancer xenograft model. *J Nucl Med* 2001;42:1251–1256.

184. Welt S, Scott AM, Chaitanya R, Divgi R, Kemeny NE, et al. Phase I/II study of iodine-125-labeled monoclonal antibody A33 in patients with advanced colon cancer. *J Clin Oncol* 1996;14:1787–1797.

185. Meredith RF, Khazaeli MB, Plott WE, Spencer SA, Wheeler RH, Brady LW, et al. Initial clinical evaluation of iodine-125-labeled chimeric 17-1A for metastatic colon cancer. *J Nucl Med* 1995;36:2229–2233.

186. Woo DV, Li D, Mattis JA, Steplewski Z. Selective chromosomal damage and cytotoxicity of a 125I-labeled monoclonal antibody 17-1A in human cancer cells. *Cancer Res* 1996;49:2952–2958.

187. Behr TM, Behe M, Lohr M, Sgouros G, Angerstein C, Wehrmann E, et al. Therapeutic advantages of Auger electron- over β-emitting radiometals or radioiodine when conjugated to internalizing antibodies. *Eur J Nucl Med* 2000;27:753–765.

188. Behr TM, Sgouros G, Vougioukas V, Memtsoudis S, Gratz S, Schmidberger H, et al. Therapeutic efficacy and dose-limiting toxicity of Auger-electron vs. beta emitters in

radioimmunotherapy with internalizing antibodies: evaluation of [125]I- vs. ^{131}I-labeled CO17-1A in a human colorectal cancer model. *Int J Cancer* 1998;76:738–748.

189. Forssell Aronsson E, Gretarsdottir J, Jacobsson L, Back T, Hertzman S, et al. Therapy with ^{125}I-labelled internalized and non-internalized monoclonal antibodies in nude mice with human colon carcinoma xenografts. *Nucl Med Biol* 1993;20:133–144.

190. Sugiyama Y, Chen F-A, Takita H, Bankert RB. Selective growth inhibition of human lung cancer cell lines bearing a surface glycoprotein gp160 by ^{125}I-labeled anti-gp160 monoclonal antibody. *Cancer Res* 1988;48:2768–2773.

191. Lindmo T, Boven E, Mitchell JB, Morstyn G, Bunn PA. Jr. Specific killing of human melanoma cells by ^{125}I-labeled 9.2.27 monoclonal antibody. *Cancer Res* 1985;45:5080–5087.

192. Wicki A, Wild D, Storch D, Seemayer C, Gotthardt M, et al. [Lys^{40}(Ahx-DTPA-^{111}In) NH$_2$]-exendin-4 is a highly efficient radiotherapeutic for glucagon-like peptide-1 receptor-targeted therapy for insulinoma. *Clin Cancer Res* 2007;13:3696–3705.

193. Wild D, Behe M, Wicki A, Storch D, Waser B, et al. [Lys^{40}(Ahx-DTPA-^{111}In)NH$_2$] exendin-4, a very promising ligand for glucagon-like peptide-1 (GLP-1) receptor targeting. *J Nucl Med* 2006;47:2025–2033.

194. Van de Wiele C, Dumont F, Vanden Broecke R, Oosterlinck W, Cocquyt V, et al. Technetium-99m RP527, a GRP analogue for visualisation of GRP receptor-expressing malignancies: a feasibility study. *Eur J Nucl Med* 2000;27:1694–1699.

195. Breeman WAP, de Jong M, Erion JL, Bugaj JE, Srinivasan A, et al. Preclinical comparison of ^{111}In-labeled DTPA- or DOTA-bombesin analogs for receptor-targeted scintigraphy and radionuclide therapy. *J Nucl Med* 2002;43:1650–1656.

196. Müller C, Schubiger PA, Schibli R. Synthesis and *in vitro/in vivo* evaluation of novel [99m]Tc(CO)$_3$-folates. *Bioconjug Chem* 2006;17:797–806.

197. Siegel BA, Dehdashti F, Mutch DG, Podoloff DA, Wendt R, et al. Evaluation of ^{111}In-DTPA-folate as a receptor-targeted diagnostic agent for ovarian cancer: initial clinical results. *J Nucl Med* 2003;44:700–707.

198. Wang S, Lee RJ, Mathias CJ, Green MA, Low PS. Synthesis, purification, and tumor cell uptake of ^{67}Ga-deferoxamine-folate, a potential radiopharmaceutical for tumor imaging. *Bioconj Chem* 1996;7:56–62.

199. Cornelissen B, McLarty K, Kersemans V, Reilly RM. The level of insulin growth factor-1 receptor (IGF-1R) expression is directly correlated with the tumor uptake of ^{111}In-IGF-1(E3R) *in vivo* and the clonogenic survival of breast cancer cells exposed *in vitro* to trastuzumab (Herceptin). *Nucl Med Biol* 2008;35:645–653.

200. Reilly RM, Sandhu J, varez-Diez TM, Gallinger S, Kirsh J, Stern H. Problems of delivery of monoclonal antibodies. Pharmaceutical and pharmacokinetic solutions. *Clin Pharmacokinet* 1995;8:126–142.

201. Sastry KSR, Haydock C, Basha AM, Rao DV. Electron dosimetry for radioimmunotherapy: optimal electron energy. *Radiat Prot Dosim* 1985;13:249–252.

202. Hansen HJ, Ong GL, Diril H, Roche PA, Griffiths GL, Goldenberg DM, et al. Internalization and catabolism of radiolabeled antibodies to the MHC class II invariant chain by B-cell lymphomas. *Biochemistry* 1996;320:293.

203. Griffiths GL, Govindan SV, Sgouros G, Ong GL, Goldenberg DM, Mattes MJ. Cytotoxicity with Auger electron-emitting radionuclides delivered by antibodies. *Int J Cancer* 1999;81:985–992.

204. Ong GL, Elsamra SE, Goldenberg DM, Mattes MJ. Single-cell cytotoxicity with radiolabeled antibodies. *Clin Cancer Res* 2001;7:192–201.

205. Ochakovskaya R, Osorio L, Goldenberg DM, Mattes MJ. Therapy of disseminated B-cell lymphoma xenografts in severe combined immunodeficient mice with an anti-CD74 antibody conjugated with ^{111}Indium, ^{67}Gallium, or ^{90}Yttrium. *Clin Cancer Res* 2001;7:1505–1510.

206. Kurimasa A, Nagata Y, Shimizu M, Emi M, Nakamura Y, Oshimura M. A human gene that restores the DNA-repair defect in SCID mice is located on 8p11.1 → q11.1. *Human Genetics* 1994;93:21–26.

207. Michel RB, Rosario AV, Andrews PM, Goldenberg DM, Mattes MJ. Therapy of small subcutaneous B-lymphoma xenografts with antibodies conjugated to radionuclides emitting low-energy electrons. *Clin Cancer Res* 2005;11:777–786.

208. Michel RB, Brechbiel MW, Mattes MJ. A comparison of 4 radionuclides conjugated to antibodies for single-cell kill. *J Nucl Med* 2003;44:632–640.

209. Michel RB, Mattes MJ. Intracellular accumulation of the anti-CD20 antibody 1F5 in B-lymphoma cells. *Clin Cancer Res* 2002;8:2701–2713.

210. Mattes MJ. The mechanism of killing of B-lymphoma cells by ^{111}In-conjugated antibodies. *Int J Radiat Biol* 2008;84:389–399.

211. Dinndorf PA, Andrews RG, Benjamin D, et al. Expression of normal myeloid-associated antigens by acute leukemia cells. *Blood* 1986;67:1048–1053.

212. Freeman SD, Kelm S, Barber EK, Crocker PR. Characterization of CD33 as a new member of the sialoadhesin family of cellular interaction molecules. *Blood* 1995;85:2005–2012.

213. Burke JM, Jurcic JG, Scheinberg DA. Radioimmunotherapy for acute leukemia. *Cancer Control* 2002;9:106–113.

214. Jurcic JG, Caron PC, Nikula TK, Papadopoulos EB, Finn RD, Gansow OA, et al. Radiolabeled anti-CD33 monoclonal antibody M195 for myeloid leukemias. *Cancer Res* 1995;55:5908s–5910s.

215. McDevitt MR, Dangshe M, Lai L, Simon J, Borchardt P, Frank KR, et al. Tumor therapy with targeted atomic nanogenerators. *Science* 2001;294:1537–1540.

216. Miederer M, McDevitt MR, Sgouros G, Kramer K, Cheung N-KV, Scheinberg DA. Pharmacokinetics, dosimetry, and toxicity of the targetable atomic generator, ^{225}Ac-HuM195, in nonhuman primates. *J Nucl Med* 2004;45:129–137.

217. Sievers EL, Larson RA, Stadtmauer EA, Estey E, Lowenberg B, Dombret H, et al. Efficacy and safety of gemtuzumab ozagamicin in patients with CD33-positive acute myeloid leukemia in first relapse. *J Clin Oncol* 2001;19:3244–3254.

218. Sievers EL, Appelbaum FR, Spielberger RT, Forman SJ, Flowers D, Smith FO, et al. Selective ablation of acute myeloid leukemia using antibody-targeted chemotherapy: a phase I study of an anti-CD33 calicheamicin immunoconjugate. *Blood* 1999;93:3678–3684.

219. Linenberger ML, Hong T, Flowers D, Sievers EL, Gooley TA, Bennett JM, et al. Multidrug-resistance phenotype and clinical responses to gemtuzumab ozogamicin. *Blood* 2001;98:988–994.

220. Feldman EJ, Brandwein J, Stone R, Kalaycio M, Moore J, et al. Phase III randomized multicenter study of a humanized anti-CD33 monoclonal antibody, lintuzumab, in

combination with chemotherapy, versus chemotherapy alone in patients with refractory or first-relapsed acute myeloid leukemia. *J Clin Oncol* 2005;23:4110–4116.

221. Caron PC, Co MS, Bull MK, Avdalovic NM, Queen C, Scheinberg DA. Biological and immunological features of humanized M195 (anti-CD33) for the therapy of acute myelogenous leukemia. *Cancer Res* 1992;52:6761–6762.

222. Boven E, Lindmo T, Mitchell JB, Bunn PA. Jr. Selective cytotoxicity of ^{125}I-labeled monoclonal antibody T101 in human malignant T cell lines. *Blood* 1986;67:429–435.

223. Reske SN, Deisenhofer S, Glatting G, Zlatopolskiy BD, Morgenroth A, Vogg TJ, et al. ^{123}I-ITdU-mediated nanoirradiation of DNA efficiently induces cell kill in HL-60 leukemia cells and in doxorubicin-, β-, or γ-radiation-resistant cell lines. *J Nucl Med* 2007;48:1000–1007.

224. Weisburg JH. Multidrug resistance in acute myeloid leukemia: potential new therapeutics. *J Nucl Med* 2008;49:1405–1406.

225. Beckman MW, Scharl A, Rosinsky BJ, Holt JA. Breaks in DNA accompany estrogen-receptor-mediated cytotoxicity from 16α[125I]iodo-17β-estradiol. *J Cancer Res Clin Oncol* 1993;119:207–214.

226. DeSombre ER, Mease RC, Hughes A, Harper PV, DeJesus OT, Friedman AM. Bromine-80m-labeled estrogens: Auger electron-emitting, estrogen receptor-directed ligands with potential for therapy of estrogen receptor-positive cancers. *Cancer Res* 1988;48:899–906.

227. DeSombre ER, Hughes A, Hanson RN, Kearney T. Therapy of estrogen receptor-positive micrometastases in the peritoneal cavity with Auger electron-emitting estrogens. *Acta Oncol* 2000;39:659–666.

228. DeSombre ER, Shaffii B, Hanson RN, Kuivanen PC, Hughes A. Estrogen receptor-directed radiotoxicity with Auger electrons: specificity and mean lethal dose. *Cancer Res* 1992;52:5752–5758.

229. Gaze MN, Huxham IM, Mairs RJ, Barrett A. Intracellular localization of metaiodobenzylguanidine in human neuroblastoma cells by electron spectroscopic imaging. *Int J Cancer* 1991;47:875–880.

230. Hoefnagel CA, Smets LA, Voute PA, de Kraku J. Iodine-125-MIBG therapy for neuroblastoma. *J Nucl Med* 1992;32:361–362.

231. Roa WHY, Miller GG, McEwan AJB, McQuarrie S, Tse J, Wu J, et al. Targeted radiotherapy of multicell neuroblastoma spheroids with high specific activity [125I]meta-iodobenzylguanidine. *Int J Radiat Oncol Biol Phys* 1998;41:425–432.

232. Rutgers M, Buitenhuis CKM, van der Valk MA, Hoefnagel CA, Voute PA, Smets LA. [131I]- and [125I]metaiodobenzylguanidine therapy in macroscopic and microscopic tumors: a comparative study in SK-N-SH human neuroblastoma and PC12 rat pheochromocytoma xenografts. *Int J Cancer* 2000;90:312–325.

233. Sisson JC, Hutchinson RJ, Shapiro B, Zasadny KR, Normolle DP, Wieland DM, et al. Iodine-125-MIBG to treat neuroblastoma: preliminary report. *J Nucl Med* 1990;31:1479–1485.

234. Sisson JC, Shapiro B, Hutchinson RJ, Shulkin BL, Zempel S. Survival of patients with neuroblastoma treated with ^{125}I-MIBG. *Am J Clin Oncol* 1996;19:144–148.

235. Ickenstein LM, Edwards K, Sjöberg S, Carlsson J, Gedda L. A novel ^{125}I-labeled daunorubicin derivative for radionuclide-based cancer therapy. *Nucl Med Biol* 2006;33:773–783.

236. Haefliger P, Agorastos N, Renard A, Giambonini-Brugnoli G, Marty C, Alberto R. Cell uptake and radiotoxicity studies of an nuclear localization signal peptide-intercalator conjugate labeled with $[^{99m}Tc(CO)_3]^{+\cdot}$ *Bioconjug Chem* 2005;16:582–587.

237. Häfliger P, Agorastos N, Spingler B, Georgiev O, Viola G, Alberto R. Induction of DNA-doule-strand breaks by Auger electrons from ^{99m}Tc complexes with DNA-binding ligands. *ChemBioChem* 2005;6:414–421.

238. Freiss G, Prebois C, Vignon F. Control of breast cancer growth by steroids and growth factors: interactions and mechanisms. *Breast Cancer Res Treat* 1993;27:57–68.

239. Nardulli AM, Katzenellenbogen BS. Dynamics of estrogen receptor turnover in uterine cells *in vitro* and in uteri *in vivo*. *Endocrinol* 1986;119:2038–2046.

240. Sainsbury JRC, Farndon JR, Sherbet GV, Harris AL. Epidermal growth factor receptors and oestrogen receptors in human breast cancer. *Lancet* 1985;1(8425): 364–366.

241. Chrysogelos SA, Yarden RI, Lauber AH, Murphy JM. Mechanisms of EGF receptor regulation in breast cancer cells. *Breast Cancer Res Treat* 1994;31:227–236.

242. King RJB. Progression from steroid sensitive to insensitive state in breast tumours. *Cancer Surv* 1992;14:131–146.

243. Matthay KK, DeSantes K, Hasegawa B, Huberty J, Hattner RS, Ablin A, et al. Phase I dose escalation of ^{131}I-metaiodobenzylguanidine with autologous bone marrow support in refractory neuroblastoma. *J Clin Oncol* 1998;16:229–236.

244. Tepmongkol S, Heyman S. ^{131}I MIBG therapy in neuroblastoma: mechanisms, rationale, and current status. *Med Pediatr Oncol* 1999;32:427–431.

245. Cunningham SH, Mairs RJ, Wheldon TE, Welsh PC, Vaidyanathan G, Zalutsky MR. Toxicity to neuroblastoma cells and spheroids of benzylguanidine conjugated to radio-nuclides with short-range emissions. *Br J Cancer* 1998;77:2061–2068.

246. He Y, Das B, Baruchel S, Kumar P, Wiebe L, Reilly RM. Meta-$[^{123}I]$iodobenzylguanidine is selectively radiotoxic to neuroblastoma cells at concentrations that spare cells of hematopoietic lineage. *Nucl Med Commun* 2004;25:1125–1130.

247. Kiyomiya K, Matsuo S, Kurebe M. Mechanism of specific nuclear transport of adria-mycin: the mode of nuclear translocation of adriamycin-proteosome complex. *Cancer Res* 2001;61:2467–2471.

CHAPTER 10

Viral Introduction of Receptors for Targeted Radiotherapy

KATHRYN OTTOLINO-PERRY AND JUDITH ANDREA McCART

10.1 INTRODUCTION

Using viruses to facilitate the introduction of receptors into tumor cells for use in targeted radiotherapy is one type of gene therapy. Gene therapy, broadly defined, is the introduction of foreign DNA into a cell for therapeutic gain. While generally thought of for gene replacement, as in the case of inherited disorders such as cystic fibrosis and replacement of the CFTR gene (1), in this chapter it will be discussed as a means of putatively permitting one to make any tumor "receptor-of-choice" positive. The overwhelming advantage is that this gives one the option to treat any tumor with targeted radiotherapy, regardless of its original receptor status. Receptor-negative tumors can be converted to a positive status. Even receptor-positive tumors can be enhanced by receptor overexpression and a more homogenous receptor density.

While easily achieved in cell culture, delivery remains one of the most difficult challenges to overcome *in vivo*. For applications that require all cells within a tumor or organ to be transduced, this remains generally insurmountable with present technologies. However, in the case of gene therapy strategies to deliver receptors to tumor cells for targeted radiotherapy, even a 10% transduction rate can lead to a therapeutic benefit. This is largely due to the radiation "cross-fire" (Fig. 10.1) and bystander (Fig. 10.2) effects. The radiation cross-fire effect occurs when emitted radiation, originating from a cell expressing the virally encoded receptor, is deposited in a neighboring cell. The path length and energy emitted are believed to determine the extent of the cross-fire effect mediated by any given radionuclide. Radionuclides that deposit the majority of their energy within the cell to which they are bound will obviously be less effective in killing uninfected cells than those with a longer path length. The radiation cross-fire effect is therefore particularly relevant when considering treatment with α- and β-emitting radionuclides, which have path lengths of

Monoclonal Antibody and Peptide-Targeted Radiotherapy of Cancer, Edited by Raymond M. Reilly
Copyright © 2010 John Wiley & Sons, Inc.

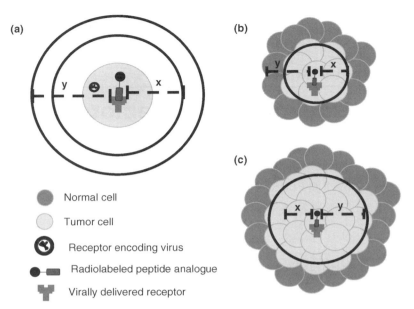

FIGURE 10.1 Maximizing the radiation cross-fire effect using rational radionuclide selection. The maximum distance radiation travels within tissues is defined by the radionuclide path length. Depending on the type of radiation emitted from a given radionuclide, the majority of energy will be deposited at the end of the path length or spread out along the entire distance of the path length. (a) The virus-encoded receptor (e.g., SSTR2, NIS, NET, D_2R, and GRPr), expressed on the surface of infected tumor cells, will be specifically bound by its radioligand analogue. Radiation emitted from the radionuclide will travel a given distance (x, y) and deposit its energy in cells along its path length, including uninfected and nonirradiated cells. While virus-mediated receptor expression is tumor-specific, radiation damage due to the cross-fire effect has the potential to equally effect both tumor and normal cells. Therefore, a radionuclide with a shorter path length would be more suitable for treatment of smaller tumors (b), whereas one with a longer path length would be more effective for larger tumors (c). Rational selection of radionuclides for radiopharmaceutical labeling based on the approximate size of the tumor and the radionuclide path length could minimize unwanted radiation damage to surrounding normal cells. (See insert for the color representation of the figure.)

$50–100\,\mu m$ and $2–12\,mm$, respectively. The biological bystander effect describes the indirect mechanism of cell death resulting from radiation- or virus-induced cell damage (see Chapter 14). These two effects are distinct and mediated by different factors. Distinct biological changes such as compromised genetic stability, radionuclide-dependent increases in cell death, or decreased clonogenic potential have been observed in nonirradiated cells cocultured with either irradiated cells (2) or the media from irradiated cells (3). The virus-mediated biological bystander effect is due to transmission of proapoptotic signals and stimulation of an antitumor immune response. It is expected that the radiation cross fire and biological bystander effects (radiation- and virus-mediated) together will significantly contribute to the death of noninfected receptor-negative cells.

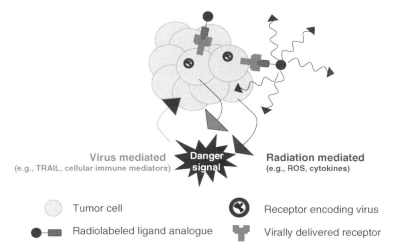

Virus mediated
(e.g., TRAIL, cellular immune mediators)

Danger signal

Radiation mediated
(e.g., ROS, cytokines)

Tumor cell

Radiolabeled ligand analogue

Receptor encoding virus

Virally delivered receptor

FIGURE 10.2   The biological bystander effect mediated by radiation damage or viral infection results in the death of uninfected and/or nonirradiated cells. The biological bystander effect, a process by which uninfected or nonirradiated cells are eliminated, can be mediated by irradiated or virally infected cells. The radiation-mediated biological bystander effect occurs when factors, such as cytokines or reactive oxygen species (ROS), released from cells directly damaged by radiation, act upon nonirradiated cells to induce cell death. The virus-mediated biological bystander effect leads to the death of uninfected cells due to (i) release of soluble factors, such as TRAIL, or (ii) induction of an immune response. TRAIL is a signalling molecule, which upon binding of the TRAIL receptor can lead to induction of apoptosis through activation of effector caspases. The cell-mediated immune responses, including infiltration of tumors by natural killer cells, neutrophils, and cytotoxic T-cells can also contribute to bystander killing of uninfected, nonirradiated tumor cells. (See insert for the color representation of the figure.)

A second challenge, particularly when delivering a toxin such as a therapeutic radionuclide, is adequate specificity of the gene delivery such that normal cells are minimally affected. In order to facilitate the delivery of foreign DNA, such as receptors, specifically into tumor cells, a variety of delivery vehicles or vectors have been used. These can be divided into viral or nonviral vectors. Viruses can be either replicating or nonreplicating. The most common virus used to date for gene therapy is a nonreplicating adenovirus. While nonreplicating viruses are safe, they can transduce only a select number of cells. Replicating viruses can spread to surrounding cells after replication. The obvious concern regarding replicating viruses is safety, and thus tissue targeting in this case is critical. As the majority of viral vectors have elicited an immune response leading to eradication of the virus and failed delivery of the genetic material, this led to the development of nonviral vectors such as liposomes that do not trigger an immune response. While the discussion of all gene therapy vectors is beyond the scope of this chapter, a brief overview of the viral vectors used to deliver receptors into cells for targeted radiotherapy will be given.

10.2 VIRAL VECTORS

10.2.1 Nonreplicating Viruses

10.2.1.1 Adenovirus   Oncolytic adenoviruses are large, double-stranded DNA viruses. Adenoviruses enter cells through the coxsackie and adenovirus receptor (CAR). Ubiquitous expression of CAR on cells leads to a widespread uptake of the virus. Efforts to achieve tumor targeting have utilized immunologic and genetic methods, both focused on altering the fiber knob (which binds the CAR receptor). For example, this can be done either by using immunoconjugates that bind both the knob and the specific receptor of choice (e.g., epidermal growth factor receptor (EGFR)) or genetically engineering ligands into the H-loop of the knob such as EGF (4). This permits entry only into EGFR-positive tumors. Nonreplicating adenoviruses have been extensively used as gene therapy vectors for many indications as they are easy to manipulate, can be produced to high titers, and do not cause major toxicity in humans (5).

10.2.1.2 Retrovirus   Retroviruses, including the HIV-derived lentivirus, are nonreplicating, single-stranded RNA viruses. They lack many genes encoding proteins required for assembly and thus must be produced from packaging cell lines that provide the necessary proteins for production of a mature virion. They are highly efficient at transducing dividing cells *in vitro*, where their RNA genomes are reverse transcribed and the DNA is then inserted into the host genome to provide long-term gene expression. Unfortunately, they are much less efficient *in vivo*. The use of lentiviruses, which can transduce nondividing cells, has led to improvements in gene transfer but *in vivo* tumor targeting remains difficult and thus they are generally used in conjunction with cell-based therapies that are transduced *in vitro* (5).

Retrovirus transduction, specificity, and efficiency can be improved through inclusion of tissue-specific promoters (6, 7). Immunoglobulin (Ig) enhancer and promoter elements inserted into a sodium iodide symporter (NIS) expressing, self-inactivating (SIN) lentivirus vector resulted in myeloma restricted infection *in vitro* with limited infection of other normal and cancerous hematological cell lines (6). Efficiency of transduction was further increased by the use of a vector containing a double Ig enhancer/promoter element (6). In another study, a lentivirus vector containing the ubiquitin C (UbC) promoter was used to increase the *in vivo* stability of human NIS (hNIS) expression in a murine colon cancer cell line (8). Previous studies indicated that cytomegalovirus (CMV) driven reporter gene expression was cell type restricted (9) and achieved only short-term *in vivo* expression in the liver following virus-mediated gene therapy due to promoter silencing (10).

10.2.2 Replicating Viruses: Mechanism of Tumor Cell Specificity

Replicating oncolytic viruses are innately tumor selective and thus an excellent platform for the delivery of receptors specifically to tumor cells. Oncolytic viruses selectively target, infect, and kill cancer cells leaving normal cells intact, thus toxicity to normal tissues is minimal. Because the viruses preferentially infect and replicate in tumor cells, this leads to strong and highly selective expression of any transgenes encoded within the virus. When these transgenes are sufficiently and selectively expressed in tumor cells this can yield very high tumor to normal tissue ratios and the protein products of expressed transgenes, such as the human somatostatin type 2 receptor (hSSTR2) gene or the NIS gene, can provide targets for imaging and radiotherapy. Several replicating viruses have been used for receptor delivery. These include replicating adenovirus, measles virus (MV), vesicular stomatitis virus (VSV), and vaccinia virus (VV). Each virus possesses several unique mechanisms leading to specific delivery of the receptor into tumors that will be briefly discussed.

10.2.2.1 *Replicating Adenovirus* Replicating adenoviruses have been engineered to be tumor-specific agents. Similar to nonreplicating viruses, the tumor targeting properties of adenoviruses have been engineered in three ways: (i) deletion of critical viral genes; (ii) insertion of tumor/tissue-specific promoters; and (iii) modification of the viral fiber knob used for cell entry as described above. The prototypical tumor-selective replicating adenovirus is ONYX 015, which was deleted from the E1B 55K gene (11). It was suggested that this virus could only replicate in tumors harboring p53 mutations (12). Other adenoviruses have been deleted from genes that similarly result in selective replication in tumors with retinoblastoma (Rb) family gene mutations (13) and tumors with defects in interferon (IFN) signaling (14). Many investigators have engineered essential viral genes under the control of tumor- or tissue-specific promoters such as tyrosinase for melanoma (15) or prostate-specific antigen (PSA) promoters for prostate cancer (16), and observed tumor/tissue-specific adenoviral replication.

10.2.2.2 *Measles Virus* Measles virus is a negative strand RNA virus (17). The vaccine strain Edmonston B (MV-Edm) is very attenuated in normal human cells, yet a potent oncolytic virus. The natural tumor selectivity of MV-Edm derives from its preferred use of the cellular receptor CD46 (rather than the SLAM receptor utilized by wild-type MV) (18). The high levels of CD46 expression (19) in many cancer cells compared to normal nontransformed cells, at least partially, explains the tumor specificity observed for MV-Edm. The development of a reverse genetics system (20) for engineering recombinant MV has enabled the creation of more potent MVs for radiovirotherapy (21).

10.2.2.3 *Vesicular Stomatitis Virus* VSV is a small, negative strand, RNA virus. While it naturally has a wide tissue tropism, it causes a very mild infection in humans perhaps due to its unique sensitivity to IFN (22). As many cancer cells have defects in their IFN pathways, they have been shown to be supportive of a productive VSV infection and hence are selectively killed (23). VSV has been previously shown to selectively replicate and kill tumors with aberrant p53, Ras, or Myc signaling (24) accounting for up to 90% of cancers. VSV can be grown to high titers (25) and easily engineered to express foreign genes; indeed, recently it has been engineered to express the hNIS gene for combined imaging and radiotherapy of myeloma (26).

10.2.2.4 Vaccinia Virus   Vaccinia virus is a large double-stranded DNA virus. It is a close cousin of smallpox (variola virus) (27). VV is possibly a mutated or laboratory version of the cowpox virus originally described by Jenner (28) to provide protection against smallpox infection. In order to enhance the tumor selectivity, the WR strain of VV was mutated by deletion of the thymidine kinase and vaccinia growth factor genes (29) to render it dependent upon dividing cells for its replication. It is easily grown to high titers and recombinant viruses are readily engineered (30). Large amounts of foreign DNA can be inserted into the VV genome by homologous recombination (31) without requiring viral deletions. A recombinant VV with a transgene encoding the hSSTR2 receptor to enable imaging of virus delivery and expression, and potentially tumor-targeted radiotherapy (32), has been engineered.

10.3   VIRALLY DELIVERED RECEPTORS

In the following section, the five receptors that have been investigated as potential imaging and/or radiotherapeutic targets in virus gene therapy studies will be discussed. The receptors are somatostatin receptor subtype 2 (SSTR2), sodium iodide symporter (NIS), gastrin-releasing peptide receptor (GRPr), norepinephrine transporter (NET), and dopamine receptor (DR-D2). Table 10.1 provides an overview of the receptors with particular emphasis on the details one may find useful when considering which receptor would be most suited to a given application.

10.3.1   Somatostatin Receptor

Somatostatin receptor scintigraphy using radiolabeled somatostatin analogues is a well-established and valuable tool for diagnosis, imaging, and monitoring of treatment response in SSTR-positive neuroendocrine tumors. Based on the successful track record of somatostatin receptor-based therapies and radiotherapies (see Chapter 4), novel viral vectors encoding the SSTR gene have been developed to allow imaging and treatment of a variety of SSTR-negative tumors in preclinical studies.

10.3.1.1 Receptor/Ligand Physiological Function in Normal Tissues
To date, five subtypes of the somatostatin receptor have been identified (SSTR1-5) (33–36). SSTR expression is rather ubiquitous throughout the body; however, each subtype has a different pattern of expression in normal and tumor tissues. One or more receptor subtype is expressed in the brain, pituitary, pancreas, stomach, salivary gland, liver, kidneys, lympocytes, lungs, testes, ovaries, thyroid, immune cells, intestines, and myocardium (37). Somatostatin (SS) is a small cyclic peptide that exists in two active forms, somatostatin-14 or somatostatin-28, each resulting from the proteolytic cleavage of the 116 amino acid propresomatostatin precursor (38, 39). Somatostatin signaling has a variety of biological activities including (i) inhibiting the secretion of numerous hormones, pancreatic enzymes, bile, and colonic fluid (40); (ii) inhibiting gastrointestinal (GI) motility (40); (iii) modulating CNS neurotransmission (41); and (iv) autocrine- and paracrine-mediated antiproliferative effects on normal (42) and

TABLE 10.1 Overview of Virally Delivered Receptors Used in Combination with PRRT

	SSTR2	NIS	NET	GRPr	D$_2$R	DAT
Distribution in normal tissue	Brain, pituitary, pancreas islets, stomach, kidney, spleen, adrenal glands	Thyroid, salivary gland, stomach, lactating mammary glands, nasal mucosa, placenta, thymus, hair follicles, prostate, ovaries	Brain, sympathetic nervous system, placenta, adrenal gland	Brain, brain stem, spinal cord, testis, stomach, smooth muscles of gastrointestinal and urogenital tracts	Brain, lymphocytes, lungs	Brain
Distribution in cancers	Neuroendocrine tumors (pituitary adenomas, endocrine pancreatic tumors, gastrointestinal and lung carcinoids, paragangliomas, pheochromocytomas, small-cell carcinomas, Merkel cell carcinomas, neuroblastomas, and medullary thyroid cancers), epithelial tumors (breast, lung, pancreatobiliary tract, liver, colorectal, follicular thyroid and prostate), meningioma, glioma, soft tissue sarcoma, malignant melanoma, lymphoma and thymoma medulloblastoma, renal cell carcinoma,	Differentiated thyroid cancer, breast cancer	Pheochromocytoma, gastrointestinal neuroendocrine tumors	Glioblastoma small-lung cell carcinoma, nonsmall lung cell carcinoma, gastrointestinal carcinoids, colon cancer, prostate cancer, cervical cancer, renal carcinoma	Esophageal squamous cell carcinoma, lung carcinoma	N/A

(continued)

TABLE 10.1 *(Continued)*

	SSTR2	NIS	NET	GRPr	D2R	DAT
	hepatoma, inactive pituitary adenoma, growth hormone-producing pituitary adenoma, gastric carcinoma					
cDNA size	3000 bp	3595 bp (hNIS)	1915 bp	1227 bp	1372 bp	2774 bp
Peptide ligands	Somatostatin-14, somatostatin-28, octreotide, lanreotide, SOM230, In-DTPA-octreotide (pentereotide), In-DOTA-[Tyr3]octreotide (DOTA-TOC), Y-DOTA-TOC, Ga-DOTA-TOC, DOTA-lanreotide (DOTA-LAN), DOTA-[Try3]octreotate (DOTA-TATE), In-DOTA-[1-Nal3] octreotide (DOTA-NOC), Y-DOTA-NOC, In-DOTA-NOC-ATE, In-DOTA-BOC-ATE, P2045, SMT 487	N/A	Metaiodobenyl-guanidine, m-hydroxy-ephedrine	Bombesin, Tyr$\wedge$4-bombesin, mIP-bombesin, gastrin releasing peptide	Raclopride, nemonapride, spiperone, N-methylspiperone, 3-(2-[^{18}F] fluoroethyl] spiperone (FESP)	TRODAT
Radionuclides	^{68}Ga, ^{90}Y, ^{94m}Tc, ^{99m}Tc, ^{99}Re, ^{111}In, ^{177}Lu, ^{188}Re	^{99m}Tc, ^{123}I, ^{124}I, ^{125}I, ^{131}I, ^{211}At	^{11}C, ^{123}I, ^{124}I, ^{131}I	^{18}F, ^{64}Cu, ^{68}Ga, ^{86}Y, ^{99m}Tc, ^{111}In, ^{125}I	^{3}H, ^{11}C, ^{18}F	^{99m}Tc
Nonreplicating Viral Vectors	Adenovirus	Adenovirus, retrovirus	Adenovirus, retrovirus	Adenovirus	Adenovirus	Adeno-associated viral vector
Replicating Viral Vectors	Vaccinia virus	Adenovirus, vesicular stomatitis virus, measles virus	N/A	N/A	N/A	N/A

tumor cells (43). Each SSTR subtype has been demonstrated to mediate unique downstream signaling cascades.

10.3.1.2 *Somatostatin and Somatostatin Receptors in Cancer*   SSTR expression has been detected in numerous primary tumors and cancer cell lines. High levels of SSTR expression are found in a majority of neuroendocrine and nervous system tumors (44) and variable rates of expression are found in a variety of epithelial tumors, soft tissue sarcomas, melanomas, lymphomas, and thymomas (45) (Table 10.1). Where the receptor is overexpressed, SSTRZ is the most commonly detected subtype (46).

Somatostatin (SS) signaling induces antiproliferative and apoptotic effects in tumor cells by direct and indirect mechanisms (43). Modulation of signal transduction pathways, including downregulation of the mitogen-activated protein kinase (MAPK) cascade, directly inhibits tumor cell proliferation (47). Alternatively, indirect mechanisms are mediated through inhibition of hormone and growth factor secretion from normal cells that in turn limits tumor growth (48). Interestingly, it has also been suggested that a loss of SSTR expression may be linked with a growth advantage in some tumors (49, 50), likely due to the loss of the inhibitory effects mediated by SS signaling.

The natural SS peptide has a short biological half-life (1 to 3 min), therefore to improve its pharmacological properties numerous SS analogues have been developed; several of which have been approved for use in humans (Table 10.1). In the early stages of SS analogue development, clinical use centered around the imaging of receptor-positive primary tumors and metastases using radiolabeled analogues and unlabeled SS analogue treatment of receptor-positive tumors. More recently, peptide receptor radiotherapy (PRRT) has been investigated using therapeutic doses of radiolabeled SS analogues (see Chapter 4).

Traditional PRRT is effective only in tumors with endogenous SSTR expression and is therefore significantly limited in its application as a broad-spectrum cancer therapy. Consequently, SSTR gene therapy strategies have been investigated to compliment PRRT and allow imaging and treatment of tumors without endogenous SSTR expression.

10.3.1.3 *SSTR Advantages and Disadvantages*   SSTR has many attributes that make it an ideal candidate for combined viral gene therapy and PRRT. First, SSTR delivery alone has been demonstrated to result in significant reductions in tumor growth *in vitro* (51) and *in vivo* (52, 53). Second, while natural SS binds each receptor subtype with approximately equal affinity, analogues have been designed to preferentially bind a specific receptor subtype (46). Given the ubiquitous SSTR tissue distribution, the ability to target only the subtype being overexpressed in the tumor (due to viral gene therapy) will significantly minimize uptake in nontarget tissues. Third, the impressive number of available SS analogues, each with differing pharmacokinetic properties, allows selection of ligands based on the properties best suited to each specific indication. Furthermore, the numerous radionclides used to label SS analogues, including ^{125}I, ^{99m}Tc, ^{111}In, ^{90}Y, ^{64}Cu, ^{188}Re, ^{66}Ga (54) and ^{225}Ac (55),

increase the number of choices regarding the desired imaging modality and the specific type of radiation being delivered to the tumor (Table 10.1). Finally, SSTR2 mRNA is approximately 3000 bp and can be easily incorporated into most viral backbones.

The disadvantages to selecting SSTR for gene therapy and PRRT include complex and expensive peptide labeling procedures, a lack of good specific anti-SSTR2 antibodies for evaluation of protein expression and localization in laboratory animal models and concerns regarding toxicity in other tissues normally expressing SSTR, such as the kidney, spleen, and bone marrow (56).

10.3.1.4   Viral Vectors for Delivery of SSTR2   Adenoviruses are the most widely studied vectors for delivery of the SSTR2 gene. AdCMV-SSTR2 is the most basic adenoviral vector to be investigated for both *in vitro* and *in vivo* gene delivery. This vector is replication incompetent due to variable deletions in E1 and/or E3 viral genes, and contains SSTR2 cDNA under the control of a CMV promoter. This strong viral promoter has been extensively used to achieve high levels of transgene expression in numerous gene therapy studies. Variations of the basic SSTR2 Ad vector encoding additional genes include AdCMV-SSTR2-TK (57-60), RGDTKSSTR (58, 60), and AdHAhSSTR2 (61). AdCMV-SSTR2-TK and RGDTKSSTR encode the herpes simplex virus (HSV) thymidine kinase (TK) protein that can be imaged using radioiodinated 2′-deoxy-2′-fluoro-β-D-arabinofuranosyl-5-iodouracil (FIAU). The TK gene also allows molecular chemotherapy using ganciclovir. While it has been demonstrated that the percentage of trapped ^{125}I-FIAU is significantly greater than that of the ^{99m}Tc-labeled somatostatin analogue P2045 in AdCMV-SSTR2-TK-infected A-427 lung cancer cells (57), *in vitro* gamma camera imaging of internalized ^{125}I-FIAU resulted in significantly increased background and decreased spatial resolution compared to ^{99m}Tc-labeled P2045 in infected A-427 (57, 62) and two ovarian cancer cell lines (SKOV3.ip1 and OVCAR3) (58, 62). This suggests that ^{125}I-FIAU/TK may be less suitable for *in vivo* imaging. Furthermore, treatment with ganciclovir was shown to decrease SSTR2 expression, indicating that it may not be a good marker for measuring delivery of the TK gene (60). RGDTKSSTR also encodes a modified adenoviral fiber knob (RGD) that enhances viral infectivity (63). AdHAhSSTR2 encodes an N-terminal hemaglutinin (HA) epitope-tagged somatostatin receptor. HA is an influenza surface protein that, unlike SSTR2, is not naturally found in humans. Investigators rationalized that including HA would provide a more accurate image of only those cells that have been infected; however, gamma camera imaging of tumor-bearing mice infected with AdHAhSSTR2 and injected with a ^{99m}Tc-labeled anti-HA antibody showed significant uptake in the liver, blood, lungs, and spleen (61), indicating that this vector may have limited *in vivo* potential.

SSTR2 gene delivery has also been investigated using an oncolytic vaccinia virus (32). This virus has deletions in two viral genes (TK and vaccinia growth factor) resulting in increased tumor specificity due to decreased viral replication in normal tissues (29). While all other vectors investigated are nonreplicating with limited tumor specificity, this recombinant VV expressing SSTR2 has all the advantages associated with replicating viruses in addition to high tumor specificity.

10.3.1.5 *In Vitro* Virus-Mediated SSTR2 Expression and Radionuclide

Uptake   *In vitro* studies have demonstrated adenovirus-delivery of SSTR2 with specific uptake of radiolabeled SS analogues in a number of tumor cell lines including nonsmall cell lung (57, 62, 64–67), pancreatic (51–53, 64), breast (64), ovarian (58, 64, 67, 68), and glioma (59). Specific analogue uptake has also been demonstrated following VV-mediated delivery of the STTR2 gene to monkey fibroblasts (32). Significant decreases in radiolabeled SS analogue uptake in the presence of competing unlabeled SS analogue or in cells infected with a receptor-negative control virus demonstrated specific binding between the radiolabeled SS analogue and somatostatin receptor, and confirmed internalization was due to the virally delivered receptor as opposed to an endogenous receptor (32, 57, 59, 62, 64–69).

Uptake of radiolabeled SS analogues in Ad-infected cells depends on viral dose and exhibits a maximum uptake at different multiplicities of infection (MOI) in different cell lines (59, 64, 68). Glioma cell lines infected with AdCMV-SSTR2-TK showed an increase in ^{111}In-DOTA-Tyr^3-octreotide uptake at increasing MOIs with a maximum uptake at MOI 100 (59). Zinn et al. demonstrated that cell lines less susceptible to adenovirus infection (SKOV3.ip1 and BxPC-3) require an MOI of 100 to achieve maximal uptake (64, 68), while those that are more susceptible (A-427 and MDA-MB-468) achieve maximal uptake at an MOI of 10. Interestingly, ^{99m}Tc-P2045 internalization in SKOV3.ip1 cells infected with the infectivity-enhanced RGDTKSSTR vector was significantly increased compared to those infected with the Ad-CMV-SSTR2-TK vector at the same MOI (58). This suggests that modification of virus cell surface proteins to increase infectivity in less susceptible cells may allow a reduction in the viral dose required to achieve maximal SSTR2 expression *in vivo*.

The percentage of internalization also depends on the length of incubation. Gamma camera imaging and gamma counter analysis of A-427 and SKOV3.ip1 cells infected with SSTR2 expressing Ad vectors and incubated with ^{99m}Tc-P2045, ^{94m}Tc-Demotate, or ^{188}Re-P829 for varying durations demonstrated internalization of the radioligands increased with increasing times of incubation (62, 64, 65, 68, 69). This implies that *in vivo* tumors expressing virally encoded SSTR2 will continue to accumulate radioactivity for a significant period of time given continued radiolabeled SS analogue bioavailability.

The majority of *in vitro* studies have used gamma-emitting ^{99m}Tc or ^{111}In, although other radionuclides including ^{90}Y (66, 70, 71), ^{125}I (72), ^{64}Cu (67), ^{94m}Tc (69), and ^{188}Re (65) have been investigated. Gamma emitting radionuclides have traditionally been used due to the availability of technologies that can detect and quantify their uptake *in vitro* and *in vivo*. Comparison of gamma camera and gamma counter analysis demonstrated no significant difference in the percent uptake of ^{99m}Tc-P2045 following viral delivery of SSTR2 (58, 62, 64). The concordance of these two measurements serves to validate the accuracy and rationale of gamma camera analyses for *in vivo* imaging.

10.3.1.6 *In Vivo* Virus-Mediated SSTR2 Expression and Radionuclide

Uptake   The critical *in vivo* experiments for evaluating the potential efficacy of combination receptor gene therapy and targeted PRRT are (i) assessment of radioligand and virus biodistribution, (ii) noninvasive imaging demonstrating tumor-specific localization of the radioligand, (iii) evaluation of tumor burden and response, and (iv) survival studies. Biodistribution studies are important for validating specific tumor uptake and assessing the potential for toxicity to normal tissues due to untoward accumulation of the virus and/or radionuclide. Tumor-specific targeting of radiolabeled analogues relies on virus delivery of the receptor specifically to cancerous cells. Since the adenoviral vectors used for SSTR2 receptor gene delivery are nonreplicating, it is not possible to assay for live viral particles as is typically done in studies using replicating viruses. Therefore, evaluation of radionuclide biodistribution by measuring radioactivity in a gamma counter is an indirect assessment of virus biodistribution. In an intraperitoneal (i.p.) tumor xenograft model, mice receiving i.p. Ad-CMV-hSSTR2 showed an increased concentration of the SS analogue, ^{99m}Tc-P2045, in the liver, spleen, large intestine, small intestine, and blood compared to animals receiving no virus injection (68). Similar increases in nontarget tissue uptake after Ad-CMV-hSSTr2 versus Ad-CMV-LacZ treatment were observed in mice bearing i.p. SKOV3.ip1 tumors receiving ^{64}Cu-TETA-octreotide (67). These data suggest that these vectors may be limited to intratumoral (i.t.) administration thereby significantly limiting their clinical applications. Gamma camera imaging with region-of-interest (ROI) analysis as well as gamma counter measurements consistently demonstrate significantly increased levels of radionuclide uptake in tumors injected with Ad vectors expressing SSTR2 relative to receptor-negative Ad control vectors (57, 60, 65–67, 72). However, there are several organs that often show significantly higher levels of radionuclide uptake relative to the tumor, including the kidneys, liver, spleen, lungs, pancreas, large intestines, and cecum (60, 65, 66, 68, 69, 72). ^{99m}Tc-P829 binding in the kidneys was found to depend on radiopharmaceutical dose whereas there was no significant difference in liver, spleen, or lung uptake at different doses (65). Interestingly, uptake in the liver, spleen, and lungs, but not in the tumor or kidneys, was decreased when ^{188}Re-P829 was used instead of ^{99m}Tc-P829, possibly indicating differences in the *in vivo* stability of these two radiolabeled analogues. Figure 10.3 shows the biodistribution of somatostatin radioligands in an i.p. xenograft model of human ovarian cancer. [^{111}In]-DTPA-D-Phe^1-octreotide resulted in higher tumor-specific localization compared to [^{125}I]-Tyr^1-somatostatin following i.p. injection with Ad-CMV-hSSTR2, again demonstrating improved *in vivo* stability of the former (72). Furthermore, ^{64}Cu-TETA-octreotide localization following i.p. Ad-CMV-hSSTR2 treatment was significantly lower in all nontarget tissues relative to tumor (67). These studies indicate that different radionuclides and analogues may exhibit different biodistribution patterns in the same animal models infected with virus. Tumor uptake has also been demonstrated to correlate positively with virus dose (57, 65). However, a higher virus dose was also associated with significantly increased uptake in the kidneys (65). In several studies, treatment with D- or L-lysine has been demonstrated to significantly decrease the levels of kidney uptake of radiolabeled somatostatin analogues without impacting tumor uptake (70, 72).

When replicating vectors are used for receptor delivery, it is possible to directly determine virus biodistribution by assaying for viable infectious particles. In contrast to adenovirus studies where kidney uptake appears to be mediated at least in part by

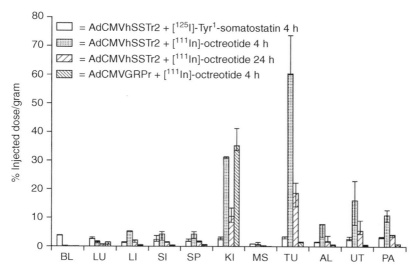

FIGURE 10.3   Radioligand biodistribution following treatment with an SSTR2-expressing adenovirus and different radiolabeled somatostatin analogues. Immune-compromised (athymic nude) mice bearing intraperitoneal (i.p.) ovarian cancer tumors were injected i.p. with 1×10^9 pfu of Ad-CMV-hSSTR2 or a receptor-negative control virus Ad-CMV-GRPr. Two days postinfection, mice were administered 2 μCi of either [^{125}I]-Tyr^1-somatostatin or [^{111}In]-DTPA-D-Phe^1-octreotide. Organs were harvested 4 h or 24 h later and the percentage of the injected dose absorbed in each organ was determined using a gamma counter. [^{111}In]-DTPA-D-Phe^1-octreotide resulted in significantly greater tumor uptake (60.4%ID/g) compared to [^{125}I]-Tyr^1-somatostatin (3.5%ID/g) at 4 h ($p = 0.008$) and [^{111}In]-DTPA-D-Phe^1-octreotide uptake remained high at 24 h (18.6%ID/g). Tumor uptake in mice administered Ad-CMV-GRPr (1.6%ID/g) was significantly decreased compared to those receiving Ad-CMV-hSSTR2 ($p = 0.016$) while uptake in the kidneys was similar in both groups (35.3%ID/g and 31.2% ID/g, respectively). BL: blood; LU: lung; LI: liver; SI: small intestine; SP: spleen; KI: kidney; MS: muscle; TU: tumor; AL: abdominal lining; UT: uterus; PA: pancreas. *Columns,* median tissue concentration from a group of five animals; *bars,* range from the 25th percentile to the 75th percentile. (Reprinted with permission from Rogers, BE et al. *Clin Cancer Res* 1999;5:383–393. Fig. 2.)

viral transduction, nontarget kidney uptake of ^{111}In-pentetreotide occurred following i.p. treatment with either the SSTR2-expressing or the receptor-negative vaccinia virus, indicating that this was not a result of virus-mediated gene expression (32). In addition, ROI analysis demonstrated tumor uptake was significantly increased in mice infected with the SSTR2 expressing virus compared to the control virus lacking the SSTR2 gene (Fig. 10.4). Similar to other studies, the uptake in the kidneys was found to be much higher than tumor uptake following treatment with the SSTR2 expressing virus and administration of a radiolabeled somatostatin analogue. *Ex vivo* virus biodistribution demonstrated significantly increased viral titers in the tumors relative to nontarget organs. While moderate viral titers were detected in the liver and kidneys, SSTR2 gene expression was below the limit of detection by RT-PCR. The intrinsic and

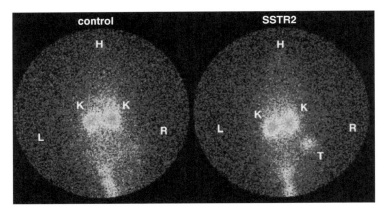

FIGURE 10.4   Comparison of vvDD-GFP and vvDD-SSTR2 imaging after ^{111}In-pentetreotide injection in tumor-bearing mice. Posterior whole-body images of tumor-bearing athymic mice 1 week after i.p. injection with vaccinia virus and 24 h after i.v. (tail vein) injection of ^{111}In-pentetreotide. Tumor (T) is visible on the right flank of vvDD-SSTR2-injected mouse (right) but not in the control vvDD-GFP-injected mouse (left). Visualization of both kidneys (K) is noted as well as the tail (site of injection). The right (R) and left (L) sides of the mouse as well as the head (H) are indicated. (Reprinted with permission from Ref. 32.)

genetically engineered tumor specificity of this VV makes it an ideal candidate for regional and systemic administration.

The timing of virus and radiopharmaceutical delivery may be critical to therapeutic success as it is expected that the levels of receptor expression will peak at a particular time point and eventually diminish below the limits of detection (and targeting). Radiopharmaceutical delivery should therefore coincide with maximal viral gene expression. The timing of maximal viral gene expression will depend both on the mode of delivery and on the kinetics of infection and gene expression for the given vector. No significant difference in ^{111}In]-DTPA-D-Phe[1]-octreotide uptake was observed in subcutaneous (s.c.) A427 or i.p. SKOV3.ip1 tumors when the timing of radiopharmaceutical delivery was varied between 1 and 4 days after AdCMVSSTR2 infection (66, 72). Similarly, no significant difference in tumor uptake was observed when ^{64}Cu-TETA-octreotide was delivered 2 or 4 days postinfection (p.i.) (67).

One of the disadvantages of using a nonreplicating virus is its poor ability to spread throughout the tumor, thus limiting subsequent targeting by the radiopharmaceutical. There is only one round of infection without the benefit of dose amplification that may be achieved when using replicating viruses. As a result, higher viral doses or multiple injections may be required to achieve maximal levels of infection. This limitation has been clearly demonstrated in several *in vivo* studies that, by using immunohistological staining of virus-infected tumors, have observed that SSTR2-positive cells are limited to the needle track of the virus injection (58, 66). Unfortunately, the use of multiple adenovirus doses has had little effect on improving radioligand uptake. Successful imaging of s.c. SKOV3.ip1 tumors administered two i.t. injections of the adenovirus RGDTKSSTR was achieved at 8 days p.i. and up to 15 days p.i. in some mice (60). In the s.c. A427 model, mice receiving two AdCMVSSTR2 doses, administered 1 week apart, showed no significant difference in tumor or nontumor tissue uptake of [^{111}In]-DTPA-D-Phe^1-octreotide by gamma counting compared to mice receiving one injection of virus (66). In the i.p. SKOV3ip.1 model, administration of a second virus dose 2 or 3 days after the first resulted in significantly increased tumor uptake compared to delivery of the two viral doses on consecutive days (72). However, even though the timing of the virus doses was important, the highest level of tumor uptake following two virus injections was not significantly different from that following a single injection.

To date, only one study has examined the use of virus-mediated SSTR2 gene therapy combined with PRRT (66). In this study, BALB/c nude mice bearing s.c. A-427 tumors were injected intratumorally with Ad-CMV-SSTR2 followed by a retro-ocular injection of 400 μCi or 500 μCi of [^{90}Y]-SMT-487 at 2 and 4 days p.i. This combined treatment was repeated at 7 days after the initial virus injection. Mice receiving the combined therapy showed a significant decrease in tumor volume and time to tumor quadrupling compared to controls (no treatment or targeted radiotherapy alone). This study serves as proof-of-principle that following successful imaging of tumor-specific virus-mediated SSTR2 delivery, administration of a therapeutic radiolabeled somatostatin analogue (^{90}Y is a β-emitter) at sufficient doses can inhibit tumor progression. Nonetheless, in addition to the effects on overall survival, details regarding the relationship between tumor size, location, type, and ideal radiopharmaceutical–virus combinations remain to be elucidated.

10.3.2 Sodium Iodide Symporter

Well before the sodium iodide symporter was identified, it was noted that the thyroid was capable of actively concentrating iodide. These observations led to the widespread use of radioiodine (^{123}I, ^{124}I, and ^{131}I) and other radioactive NIS substrates (e.g., ^{99m}TcO$_4^-$) for the diagnosis, treatment, and monitoring of residual or metastatic disease in a variety of thyroid malignancies. Since the cloning and characterization of the NIS gene in 1996 (73), it has been possible to target tumors lacking endogenous NIS expression for radioiodine imaging and therapy through viral delivery of the NIS gene. It is important to note that NIS is not a peptide-targeted receptor; however, it remains an important and relevant topic for discussion in the context of receptor gene therapy and targeted radiotherapy because it illustrates the principles involved. Extensive research has been undertaken to determine the feasibility and efficacy of using virally delivered NIS for radionuclide-mediated imaging and therapy in numerous tumor models. Preclinical studies using replication-deficient adenovirus or oncolytic adenovirus, measles virus, and vesicular stomatitis virus have demonstrated the great potential of this therapeutic approach and have led to clinical trials.

10.3.2.1 *Symporter/Substrate Physiological Function*   NIS is a transmembrane cell-surface glycoprotein responsible for active transport of iodide into thyroid follicular cells and several extrathyroidal tissues. In the thyroid, influx of iodide is a precursor to thyroid hormone biosynthesis and is therefore a critical

component of normal thyroid function. Through a process known as organification, iodide is oxidized and incorporated into the thyroglobulin protein that is the precursor to the thyroid hormones triiodothyronine (T_3) and thyroxine (T_4). This process is responsible for the thyroid's ability to store large quantities of iodide. Nonthyroidal tissues also known to accumulate radioiodine during whole-body scans include the salivary gland, stomach, lactating mammary glands, nasal mucosa, placenta, thymus, and hair follicles. Following the cloning of the human NIS gene (74), mRNA analysis confirmed previous tissue distribution findings in addition to identifying expression in tissues such as the prostate and ovaries that are thought not to concentrate radioiodine *in vivo* (75). NIS expression in the lactating mammary gland enriches the breast milk with iodide that is essential for proper development of the newborn. The role of NIS in other extrathyroidal tissues is not clear; however, studies of individuals lacking functional NIS expression indicate that it has little physiological significance in these tissues.

10.3.2.2 *NIS and Radioiodine Therapy in Cancer*   Radioiodine (^{131}I) therapy was first employed in 1942 for patients with Graves' disease (76). Successful imaging and elimination of the thyroid gland led to the use of radioiodine therapy for other thyroid disorders and malignancies. Today, radioiodine is used in the treatment of differentiated thyroid cancers to destroy residual thyroid tissue following thyroidectomy, to monitor for metastasis using full-body scans and to treat primary and metastatic tumors (77–79). ^{131}I is typically administered as an adjuvant therapy following surgical removal of thyroid tumors and complete thyroidectomy. In a retrospective study of 382 patients with differentiated thyroid cancers, postoperative radioiodine therapy was associated with a significantly improved 10-year local relapse-free rate (80). ^{131}I therapy was also found to significantly improve the 10-year survival rate of a select group of patients with highly differentiated metastatic tumors that accumulated radioiodine (79).

Apart from thyroid cancer, breast cancers are the only other malignancies known to express significant levels of NIS. Upon immunohistochemical analysis of human breast cancer specimens, more than 80% of patients were positive for NIS expression whereas only 0–13% of normal breast tissue stained positive (81, 82). However, less than 25% of NIS-positive tumors, as determined by reverse transcriptase (RT) PCR (83) or immunohistochemistry (84), demonstrated accumulation of iodide *in vivo*. While radioiodine is not used for imaging or treatment of human breast cancers, recent data from animal models and human breast cancer patients suggests that strategies designed to upregulate endogenous NIS expression and membrane trafficking could make radioiodine a useful diagnostic and therapeutic tool in the future.

Understanding the factors that regulate NIS expression has important clinical implications for radioiodine-based diagnostics and therapy. Pretreatment of patients with drugs that downregulate NIS expression in normal tissues minimizes their uptake, potentially improving diagnostic sensitivity by nuclear medicine imaging and minimizing radiation-associated toxicity to normal organs. Iodide uptake is decreased by iodine, T_3 and T_4, through feedback-decreased NIS expression, while the inhibitor perchlorate binds NIS to prevent iodide transport (85). In thyroid cancer patients with low levels of iodide uptake, drugs that increase tumor NIS expression could make radioiodine therapy significantly more successful. Retinoic acid (86) and histone deacetylase inhibitors (87–89) have been shown to induce endogenous NIS expression *in vitro* and *in vivo*. Many other hormones, cytokines, growth factors, and drugs may affect NIS expression (78) and therefore could be clinically relevant when considering NIS gene therapy and radioiodine therapy.

10.3.2.3 NIS Advantages and Disadvantages

In order to harness the significant benefits of radioiodine therapy, viral vectors encoding the hNIS gene have been used to drive protein expression in tumor cells lacking endogenous NIS. Although issues with NIS cellular localization and function are sometimes a barrier to treatment of endogenously positive NIS breast (90) and thyroid cancers (91), virally encoded NIS appears to be properly trafficked to the cell membrane in most studies (92–95). The complete hNIS mRNA is 3594 bp long encoding a 643 amino acid protein (91). Transduction of cells with a partial cDNA fragment (1929 bp) corresponding to amino acids 1-612 is sufficient to induce significant increases in $Na^{125}I$ accumulation (74). Given the moderate size of the NIS gene, it can be easily inserted into most viral vectors. Iodide is the main physiological ligand of NIS; however, NIS can also bind and accumulate, with decreasing affinity, ClO_4^-, ReO_4^-, SCN^-, ClO_3^-, and Br^- (96). ^{123}I, ^{124}I, ^{125}I, ^{131}I, and $^{99m}TcO_4^-$ (similar in charge to perchlorate and transported by NIS) are the most commonly used radionuclides, both in laboratory studies and in clinical practice. Recently, NIS has also been demonstrated to result in specific accumulation of ^{211}At, an α-emitter (97). The specific energies, type of radiation emitted, and path lengths of these radionuclides dictate their utility for different applications (Table 10.2) including imaging by different modalities (SPECT or PET) or treatment of different tumor sizes (Fig. 10.2). An advantage of using NIS over other receptors for targeted radiotherapy is that NIS binds directly to the radionuclide without the need to conjugate it to a targeting vehicle (i.e., peptide or monoclonal antibody), thereby significantly decreasing the cost and simplifying the approach. In addition, NIS is an active transport protein that should provide greater tumor accumulation of radioiodine than receptor proteins that bind and slowly internalize radiopharmaceutical ligands.

10.3.2.4 Tumor-Targeted Viral Delivery of NIS

Delivery of NIS to tumor cells has been investigated using a variety of recombinant viral vectors including replication-deficient and oncolytic adenoviruses, retrovirus, measles virus, and vesicular stomatitis virus. *In vitro* and *in vivo* investigations of these vectors in numerous cancer cell lines and animal models have provided significant evidence to support the use of virally delivered NIS for both imaging of gene delivery and targeted radiotherapy. As is true for all receptors discussed in this chapter, non-replicating adenoviruses are the most commonly investigated gene delivery vector. The main strategy for increasing the specificity of NIS encoding viral vectors has been to subclone the NIS gene under the control of a tumor-specific promoter, thereby limiting NIS expression to cells where the promoter is highly active. While the early vectors used the ubiquitous CMV promoter (92, 94, 98–101), newer vectors utilize

TABLE 10.2   Radionuclide Properties and Relevant Applications

Radionuclide	Physical half-life	Emitted radiation	Energy	Tissue distance (max.) of therapeutic radiation	Clinical use
^{3}H	12.3 years	β	18.6 keV (max.)		Autoradiography
^{11}C	20 min	Positron	960 keV (max.)	4.2 mm	PET imaging
^{18}F	1.82 h	Positron	630 keV	2.6 mm	PET imaging
^{64}Cu	12.7 h	Positron	3300 keV (max.)		
		β	511 keV, 8 keV		PET imaging
		γ	578 keV		Radiotherapy
^{68}Ga	67.6 min	Positron	2920 keV		
		β	1899 keV, 822 keV		PET imaging
		γ	511 keV, 1077 keV		Radiotherapy
^{90}Y	2.7 days	β	2300 keV (max.)	12 mm	Radiotherapy
^{94m}Tc	52 min	Positron	2470 keV (max.)		PET imaging
^{99m}Tc	6 h	γ	140 keV		SPECT imaging
		Auger			SPECT imaging
^{111}In	2.8 days	Auger		10 um	Radiotherapy
		γ	245 keV, 171 keV, 31 keV		SPECT imaging
^{123}I	13.13 h	Auger, γ	159 keV, 27 keV, 31 keV		SPECT imaging
^{124}I	4.2 days	Positron	1000 keV		PET imaging
		β	1532 keV, 2135 keV		
		γ	511 keV, 603 keV, 1691 keV		
^{125}I	60.2 days	Auger		10 nm	Radiotherapy
		γ	35 keV		
^{131}I	8.05 days	β	606 keV, 334 keV, 247 keV	2 mm	Radiotherapy
		γ	364 keV, 637 keV, 284 keV		SPECT imaging
^{177}Lu	6.7 days	β	490 keV (max.)	2 mm	Radiotherapy
		γ	133 keV, 208 keV		
^{188}Re	17 h	β			Radiotherapy
^{211}At	7.2 h	α		70 um	Radiotherapy

mucin 1 (MUC1) (92, 99, 100), ARP$_2$PB (95), carcinoembryonic antigen (CEA) (93), and UbC (8) promoters.

MUC1 is a glycoprotein that is overexpressed in a number of epithelial malignancies including breast (103), ovarian (104, 105), pancreatic (106, 107), and gastrointestinal (106, 108) cancers. Specific fragments at the 3′ end of the MUC1 promoter can drive transcription of exogenous genes specifically in cancer cells with high levels of endogenous MUC1 protein (100, 109–112). Radioiodide uptake in Ad-MUC1-hNIS infected cells was high in MUC1-positive pancreatic cell lines whereas uptake was significantly lower in a MUC1-negative cervical cancer cell line (92). While the specificity of the Ad-MUC1-hNIS vector was increased relative to Ad-CMV-hNIS, the latter vector consistently resulted in higher levels of radioiodide uptake *in vitro* (92, 99, 100). The decreased activity of the MUC1 vector may be due to improved membrane trafficking of NIS under control of the CMV promoter (100) or to the intrinsic strength of the CMV promoter. Similar results were observed *in vivo*. Gamma camera imaging of infected tumor-bearing mice following injection of ^{123}I$^-$ showed higher radioactivity uptake in Ad-CMV-hNIS relative to Ad-MUC1-hNIS infected tumors (92, 99, 100). Not surprisingly, in a study of an ovarian cancer model with i.t. virus injections, Ad-CMV-hNIS + ^{131}I was more effective than Ad-MUC1-hNIS + ^{131}I (100).

Probasin is a prostate-specific protein predominately expressed in the dorsolateral lobes of rodent prostates. ARP$_2$PB is a fragment of the prostate-specific probasin promoter that is activated in response to androgen. Ad-ARP$_2$PB-hNIS resulted in high radioiodide uptake in prostate cancer cells in the presence of androgen and low uptake in cancer cell lines of other origins (95). Furthermore, the prostate-specific, androgen-dependent uptake following infection with Ad-ARP$_2$PB-hNIS was significantly greater than that induced by Ad-CMV-hNIS. Interestingly, prostate cancer cells showed 90% membrane localization of NIS when infected with Ad-ARP$_2$PB-hNIS and less than 10% when infected with Ad-CMV-hNIS. In other cancer cell lines, Ad-ARP$_2$PB-hNIS showed less than 5% immunoreactivity reaffirming the specificity of this vector for androgen-dependent prostate cancer cells *in vitro*.

Carcinoembryonic antigen is a protein expressed in numerous human tumors as a result of upregulation of the CEA promoter (pCEA). Investigation of the Ad-pCEA-hNIS virus demonstrated decreased ^{123}I uptake and biological half-life in Ad-pCEA-hNIS-infected meduallary thyroid cancer xenografts relative to Ad-CMV-hNIS (93). When combined with ^{131}I therapy (3 mCi i.p.), mice receiving Ad-pCEA-hNIS showed a significant reduction in tumor growth and significantly improved survival compared to either therapy alone.

The CMV promoter is highly efficient *in vitro*; however, promoter silencing and loss of transgene expression in many tissues *in vivo* have prompted investigation of other ubiquitously expressed promoters, such as UbC, with activity in a wider range of tissues. A lentivirus vector with hNIS under control of the UbC promoter resulted in stable *in vivo* NIS expression in *ex vivo* transduced tumors for up to 27 days (8). ^{131}I (1mCi) therapy in mice bearing i.p. ovarian cancer tumors transduced with this vector resulted in a significant reduction in tumor growth compared to that in the absence of radiotherapy.

To improve tumor-targeted expression of NIS by oncolytic VSV, residue 51 of the matrix protein was deleted, thereby preventing inhibition of type 1 IFN production in normal cells and limiting viral replication to cancer cells with impaired IFN signaling pathways. *In vitro* infection with VSV(Δ51)-NIS resulted in a significantly decreased cell viability in myeloma cell lines and primary cells relative to normal cell lines and primary bone marrow progenitors (26). While virus-induced cell death was tumor specific, ^{125}I uptake relative to mock infected cells was higher in monkey kidney cells compared to myeloma cells indicating that VSV(Δ51)-NIS is still capable of infecting and expressing NIS in normal cells. The toxicity of VSV(Δ51)-NIS was significantly decreased compared to wild-type VSV in immunocompromised mice bearing s.c. myeloma tumors (26). VSV(Δ51)-NIS infection in this model resulted in significant tumor-specific ^{123}I accumulation determined by SPECT imaging 4 days p.i. Comparison of i.t. and i.v. injection of 0.5 mCi of ^{123}I demonstrated no significant difference in tumor uptake (8.1 ± 2.4 versus $5.5 \pm 0.9\%$ injected dose/g).

10.3.2.5 *Strategies to Improve Tumor Cell Killing*

Improving or overcoming low transduction efficiency is one of the major challenges facing gene therapy. This is a particular challenge for nonreplicating viruses as is illustrated in Fig. 10.5.

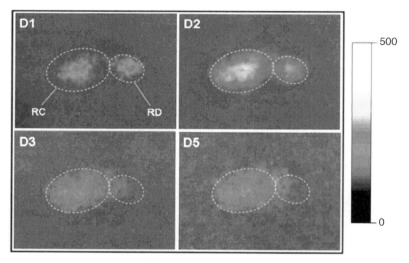

FIGURE 10.5 Comparison of $^{99m}TcO_4^-$ uptake following intraprostatic injection of a replication-defective or replication-competent adenovirus vector expressing hNIS (sodium iodide transporter). Nontumor-bearing dogs received intraprostatic injections of replication-competent Ad-yCD/mutTK$_{SR39}$$rep$-hNIS ($5 \times 10^{11}$ viral particles) and replication-defective (RD) Ad-CMV-FLhNIS (5×10^{11} viral particles) in opposite lobes. Postinfection (D1, D2, D3, and D5) dogs received 20–40 mCi Na^{99m}TcO$_4$ via a femoral catheter and were subjected to nuclear imaging 4 h later. The area and intensity of the signal are significantly greater in the lobe receiving Ad-yCD/mutTK$_{SR39}$$rep$-hNIS and uptake persists for up to 5 days postinfection. The gray scale indicates the level of gene expression in counts/pixel. (Reprinted by permission from Macmillan Publishers: Molecular Therapy, Ref. 114.)

[123]I scintigraphy following treatment with replication-deficient Ad-rNIS showed the highest tumor uptake along the needle track of the injection site (101). Infection of a small fraction of the total tumor limits the number of cells capable of accumulating radioiodine. However, through the radiation cross-fire effect, as previously discussed, uninfected cells may be killed by direct radiation damage from the β-particles emitted by ^{131}I in targeted cells (Fig. 10.1). Prostate cancer cells showed significantly increased cell death following ^{131}I exposure when retrovirus-transduced cells expressing rNIS were cultured as three-dimensional spheres where the cross-fire effect is more prominent, as opposed to monolayers (113). The efficacy of the radiation cross-fire effect was demonstrated *in vivo* using myeloma tumor cells transduced with an hNIS encoding lentivirus (6). Groups of mice received increasing percentages of transduced cells followed by treatment with [131]I (1 mCi). Tumors grew rapidly in mice with nontransduced cells, while those with 50 or 100% transduced cells achieved complete tumor regression, indicating that the 50% of cells lacking NIS expression were also killed, probably through a cross-fire effect from cells accumulating [131]I. This illustrates how the radiation cross-fire effect is a means by which a low transduction efficiency may be overcome to result in successful cancer therapy.

NIS-expressing adenovirus vectors have also been engineered to include other therapeutic genes such as yeast cytosine deaminase (yCD) and herpes simplex virus thymidine kinase (HSV-1 TK). Ad5-yCD/mutTK$_{SR39}$rep-hNIS is a replication-competent (RC) virus encoding a yCD and mutant HSV-1 TK fusion protein. The TK gene allows prodrug therapy with ganciclovir and PET imaging with radiopharmaceutical substrates of TK (^{18}F-FHBG: 9-(4-[^{18}F]fluoro-3-hydroxymethylbutyl)), while the yCD enzyme facilitates prodrug therapy with 5-fluorocytosine (5-FC). Ad5-yCD/mutTK$_{SR39}$rep-hNIS has recently been investigated in murine and canine models for treatment of prostate cancer (114–116) and is being developed for testing in a phase I clinical trial. While this vector has been demonstrated to facilitate tumor- or prostate-specific Na^{99m}TcO$_4$ accumulation *in vitro* and *in vivo*, no data exist demonstrating the feasibility or efficacy of combined targeted radiotherapy and prodrug therapy.

Several commonly used anticancer chemotherapeutics have demonstrated significant potential for (i) improving the efficacy of viral delivery of the NIS gene *in vitro* or *in vivo* or (ii) increasing radioiodine uptake. Retinoic acid (RA) has been found to increase endogenous NIS expression and radioiodine uptake in thyroid cancer patients (117), and it was later discovered that it could increase endogenous NIS expression in MCF-7 breast cancer cells *in vitro* (118). MCF-7 cells infected with Ad-CMV-NIS and treated with at least 10^{-7} M all-trans retinoic acid (ATRA) showed a significant increase in NIS mRNA compared to either mock-infected cells (118.5-fold) or Ad-CMV-NIS in the absence of ATRA (97.5-fold). Furthermore, ATRA was found to significantly increase radioiodine uptake in Ad-CMV-NIS-infected cells compared to those treated with ATRA alone (119).

Doxorubicin is an anthracycline antibiotic that both intercalates into DNA and activates the transcription factor NF-kB (120) for which there are binding sites in the CMV promoter (121). ARO anaplastic thyroid cancer cells infected with Ad-CMV-hNIS or Ad-CMV-Luc (incorporating the luciferase gene) and treated with doxorubicin showed significant increases in radioiodine uptake and bioluminescence, respectively, compared to untreated controls (122). Western blot analysis showed a dose-dependent decrease in IkB (an inhibitor of NF-kB), and electrophoretic mobility shift assays demonstrated an increase in specific NF-kB binding in ARO cells expressing hNIS. In s.c. ARO tumor xenograft models infected with Ad-CMV-hNIS, cotreatment with doxorubicin resulted in a significant increase in tumor uptake of $^{99m}TcO_4^-$ detected by gamma camera imaging and gamma counter analysis compared to virus alone.

Cyclophosphamide (CP) is an alkylating agent that acts to both inhibit cell growth and depress immune system responses. In the context of replicating viral therapy, CP may be used to slow the host's antiviral immune response, thereby allowing sufficient time for viral replication and gene expression to occur. CP pretreatment of mice infected with an NIS expressing measles virus (MV-NIS) was found to increase acute viremia and increase the rate of viral clearance from organs by an unknown mechanism (123). However, pretreatment also resulted in sporadic positive results in RT-PCR tests for virus mRNA in the brain. Interestingly, in immune-competent squirrel monkeys, CP pretreatment resulted in higher levels of viral mRNA and increased persistence compared to untreated controls.

Variations in the NIS gene inserted into the viral backbone have also been investigated as a means of improving radionuclide accumulation. A virus containing a C-terminal truncated human NIS gene (Ad-CMV-hNIS#9) resulted in significantly decreased *in vitro* radioiodine uptake and protein expression over time compared to a virus encoding the full-length NIS gene (Ad-CMV-FLhNIS) (124). Ad-CMV-hNIS#9 was able to induce hNIS expression and ^{125}I uptake in s.c. glioma tumors in mice; however, its *in vivo* efficacy was not compared to the Ad-CMV-FlhNIS virus. While the majority of studies have examined the use of hNIS, evidence exists to suggest that the rat NIS (rNIS) gene may be more effective in achieving higher transduction efficiency and radioiodine uptake. Retroviral transduction of either hNIS or rNIS into a panel of human or rat tumor cell lines resulted in a higher increase in ^{123}I uptake in rNIS-transduced cells (8–67-fold over nontransduced) compared to hNIS-transduced cells (3–22-fold over nontransduced) (125). Northern blot analysis demonstrated rNIS mRNA was significantly higher than hNIS mRNA in two of the six cell lines tested; however, results could not be confirmed by Western blot due to low hNIS antibody sensitivity. Despite these findings, no significant difference in clonogenic survival was observed between cells expressing rNIS or hNIS and treated with ^{131}I. Later studies examined the efficacy of Ad-CMV-rNIS for *in vivo* delivery of the NIS gene and radioiodine-mediated imaging and therapy. In a s.c. model of human prostate cancer, tumor volume was significantly decreased in mice receiving three i.t. injections of 10^8 pfu Ad-CMV-rNIS followed by 1 or 3 mCi ^{131}I compared to either therapy alone (101). However, no significant difference was observed between the mice receiving the lower and higher ^{131}I doses. In a study of rat hepatocellular carcinoma, a dose of 18 mCi of ^{131}I following i.t. Ad-CMV-rNIS (5×10^9 infectious particles) resulted in a 30% decrease in nodule size, significant reduction in new nodule formation, and a significant increase in long-term survival (up to 200 days) (94).

10.3.2.6   Effect of Targeted Radiation on Viral Replication   In principle, targeted radiation may either enhance or suppress viral replication depending on how the radionuclide affects both the cell and the virus. There are very few studies that have examined the effect of targeted radiation on viral replication. Understanding how different replicating viruses respond to radiation exposure is critical when selecting a suitable viral vector for gene delivery. External beam radiation has been shown to potentiate HSV-1 replication through induction of a cellular radiation-resistance protein with significant homology to a viral protein involved in virus replication (126, 127). Alternatively, viral replication can be inhibited by ultraviolet (UV) radiation through induction of changes in host cell conditions that do not favor viral replication (128). In a study using MV-NICE, a virus expressing NIS and CEA, investigators explored the possibility of combining viral therapy with Auger electron emitting ^{125}I as a safety measure to halt virus replication (129). CEA-production was analyzed as a marker of viral replication. Myeloma cells infected with MV-NICE and treated with ^{125}I showed a perchlorate-sensitive decrease in CEA production and a complete absence of viable virus particles *in vitro*. Despite these results, ^{125}I had no significant effect on CEA production in MV-CEA or MV-NICE infected s.c. models of myeloma *in vivo* (129).

10.3.2.7   Imaging NIS Expression   Gamma camera imaging is the most commonly utilized imaging modality for monitoring viral NIS expression; however, other important imaging techniques have been investigated, either for their ability to provide complementary data or as potentially superior alternatives. Positron emission tomography (PET) imaging has been used to quantify ^{76}Br$^-$ uptake *in vitro* (130) and ^{124}I uptake *in vivo* following Ad-CMV-hNIS (131, 132), MV-NIS infection (7), or lentiviral transduction (7). An *in vivo* ROI analysis of PET images showed a good linear fit with gamma counter analysis (131, 132). Tumor and stomach uptake were similar using PET (^{124}I) or gamma camera (^{123}I) imaging; however, PET/CT images provided improved anatomical resolution (7). In a canine model receiving intraprostatic injections of Ad5-yCD/mutTK$_{SR39}$rep-hNIS or rAd-CMV-FLhNIS, the resolution of tomographic SPECT imaging of resected prostates was not better than whole-body planar scintigraphy (114). SPECT/CT scans allow the radionuclide-based imaging of SPECT combined with the high spatial resolution of CT scans. This type of imaging has been successfully used to monitor NIS expression in mice bearing s.c. colon tumors infected with AdAM6 or AdIPI following ^{99m}TcO$_4^-$ administration (133). An interesting technique, which could have relevance in preclinical gene therapy studies, was used to estimate the volume and amount of viral gene expression (114, 115). Following ^{125}I administration, digitized autoradiographs of prostate sections were made into three-dimensional isodose diagrams demonstrating that the replication-competent Ad5-yCD/mutTK$_{SR39}$rep-hNIS had a twofold increase in the level of gene expression and a threefold increase in the volume of gene expression compared to the replication-deficient rAd-CMV-FLhNIS (114). These results indicate that Ad5-yCD/mutTK$_{SR39}$rep-hNIS is capable of disseminating over a larger volume and inducing a higher overall level of transgene expression. While this method of *ex vivo* tissue imaging provides more specific three-dimensional information, similar results were achieved by scintigraphy using ^{99m}Tc (Fig. 10.5) without requiring animal sacrifice and tissue removal (114). Figures 10.5 and 10.6 (1) also illustrate the long-term persistence of virus-mediated receptor expression that can be achieved by using a replication-competent virus.

As an alternative to imaging, viral delivery of the CEA protein has been investigated, as a secondary viral marker that can be detected through simple blood sampling, since CEA is secreted by tumors into the circulation. This approach is designed to decrease treatment costs and radiation exposure by potentially eliminating the

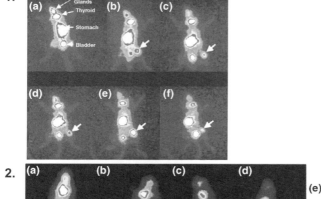

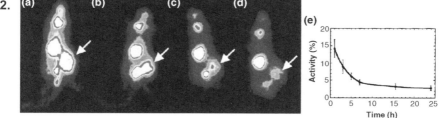

FIGURE 10.6 ^{123}I uptake in myeloma xenografts following intravenous administration of a measles virus expressing NIS (MV-NIS). (1) Immune-deficient CB17-SCID mice bearing different subcutaneous myeloma xenografts (ARH-77, a and c; MM1, b; KAS-6/1, d–f) were injected intravenously with 2×10^6 pfu of MV-NIS (b–f) or MV-Edm (a), a receptor-negative control virus. Mice received i.p. injections of ^{123}I (500 µCi) 3 days (d) postinfection (p.i.) (d), 9 days p.i. (a–c, e), and 17 days p.i. (f), and were imaged 1 h after radioligand injection using a gamma camera. No ^{123}I uptake was observed in the tumors of mice receiving the negative control virus (a) while tumor uptake was observed in all mice receiving MV-NIS. Maximum uptake in KAS-6/1 tumors was observed 9 days p.i. (e), likely corresponding to the timing of maximum virus-mediated NIS expression. (2) Immune-deficient CB17-SCID mice bearing subcutaneous MM1 xenografts were injected i.p. with ^{123}I (500 µCi) 9 days postintravenous MV-NIS infection (2×10^6 pfu). Gamma camera imaging was performed at 1 (a), 3 (b), 5 (c), 7 (d), 15, and 24 h after radioligand injection. Tumor uptake as a percentage of the injected dose was highest at 1 h (12–17%) and dissipated over a 24 h period (e). *Error bars*, SD; *magnification*, 0.55 mm per pixel. (This work was originally published in *Blood*. Dingli D et al. 2004; 103:1641–1646 © American Society of Hematology.) (See insert for the color representation of the figure.)

required imaging step. This has been achieved either through delivery of MV-NIS and MV-CEA together (134) or through construction of a novel MV-NICE virus expressing both CEA and NIS (129). Combined MV-NIS and MV-CEA therapy resulted in consistently lower levels of CEA production over time compared to MV-CEA alone in i.p. models of ovarian cancer (134). Despite lower CEA levels, significant tumor-specific ^{99m}Tc uptake was observed by gamma camera imaging in combined MV-NIS- and MV-CEA-infected mice. These mice also exhibited significantly increased survival rates compared to MV-CEA alone, although this may be due to the fact that they received double the overall virus dose of mice receiving MV-CEA alone. Plasma CEA levels were a relatively good marker of viral replication in mice receiving MV-CEA alone; however, the relationship was weaker in mice receiving both MV-CEA and MV-NIS. It is important to note, however, that mice are not susceptible to MV infection and therefore in this model CEA levels can be definitively attributed to virus replication in human tumors. Alternatively, humans are susceptible to MV infection; therefore, monitoring serum levels of a virally encoded marker protein will not necessarily represent tumor-specific replication and an imaging step may still be required.

Tumor uptake and retention of the radionuclide are required both for imaging and for treatment. Numerous studies have demonstrated that radionuclide uptake depends on virus and radionuclide dose. Viral dose dependency has been demonstrated using gamma counter analysis (124, 131, 132), gamma camera imaging (130), PET (130), and autoradiography (130) of transduced cells cultured *in vitro*. This relationship has been confirmed *in vivo* using gamma camera imaging (93, 114, 135), PET imaging (131, 132), SPECT/CT (133), and autoradiography (114). *Ex vivo* gamma counter analysis of organs following treatment with increasing doses (10^7–10^8 green fluorescent units, GFU) of i.v. Ad-CMV-hNIS showed dose-dependent uptake in the liver, adrenal glands, lungs, and pancreas while uptake in the spleen plateaued at doses higher then 5×10^7 GFU (132). It has also been demonstrated that uptake is not detectable below certain viral doses and the linearity of the relationship is lost at high viral doses (114, 130, 135). Viral dose also shows a positive correlation with hNIS expression by quantitative RT-PCR; that in turn shows a positive correlation with gamma counter analysis and gamma camera imaging (131, 135). One study demonstrated that in addition to increasing ^{123}I uptake, doubling the viral dose also resulted in a more than threefold increase in its biological half-life (93). Increasing the radioactivity dose also results in increased uptake *in vitro* using gamma camera imaging, PET, and autoradiography (130); however, there is no study that compares the *in vivo* tumor uptake in animals receiving different doses of radioactivity.

Retention times vary considerably between cell lines (98); however, a significant percentage (40–90%) of the radiopharmaceutical is typically lost during the first 5–10 min (8, 94, 99, 124, 136). *In vivo*, long-term retention in tumors may depend on the radiopharmaceutical. Figure 10.6 shows an example of the short-lived ^{123}I uptake in MV-NIS but not in MV-Edm-infected tumors. The majority of data suggest that ^{123}I tumor uptake is reduced to close to background levels within 48 h (6, 21, 99, 136); however, little imaging data exist past 48 h due to the short physical half-life of the radionuclide (13 h). However, there are also minimal data on the long-term persistence of ^{131}I that has a longer physical half-life of 8 days. One study has demonstrated that ^{131}I can persist for more than 11 days in tumors, and to a lesser extent in other tissues of rats infected with Ad-CMV-rNIS (94). Given that *in vitro* data suggest a rapid efflux of radioiodine from cells yet *in vivo* data show potential for long-term retention, it is possible that radioiodine is maintained within the cells through a dynamic reuptake process, possibly mediated by the NIS (94).

10.3.2.8 Optimizing Therapy   The optimal radionuclide dose and timing of radionuclide delivery was investigated in an orthotopic rat glioma model (137). The glioma cells stably expressed NIS due to retrovirus transduction prior to stereotactic implantation. The survival benefit of escalating doses of ^{131}I (4, 5.6, 8, and 16 mCi) was evaluated and there was a significant increase in the life span of rats receiving the highest (16 mCi) dose. Furthermore, rats treated early during tumor development had significantly improved survival compared to rats treated when tumor volume was approximately 2 mm^3 (137). Comparison of the effects of different radionuclides on survival showed that ^{125}I (16 mCi) had no therapeutic benefit while a combination of ^{125}I and ^{131}I (8 mCi each) was not significantly better than ^{131}I (16 mCi) alone (137).

10.3.2.9 Biodistribution and Dosimetry   The biodistribution of the radionuclide is an important consideration following gene delivery by either replicating or nonreplicating viruses given concerns about radiation-associated normal tissue toxicity. Nuclear imaging of mice and canines invariably demonstrates nontarget radionuclide uptake in the thyroid, stomach, urinary bladder, and occasionally in the salivary gland and intestines in the presence or absence of viral NIS delivery (6–8, 92–94, 99–102, 131–135, 138). Uptake is either due to endogenous NIS expression (thyroid, stomach and salivary glands) or elimination of the radionuclide from the body (bladder and intestines). Different approaches have been used to attempt to decrease nontarget organ uptake. Thyroxine (T$_4$) supplemented diets, administered for 11 days prior to imaging, resulted in decreased ^{99m}TcO$_4^-$ uptake in the thyroid on scintigraphic images (102). This was complemented by a Western blot analysis showing a decrease in NIS protein expression in animals receiving the supplemented diet. Tri-iodothyronine (T$_3$) supplemented or iodine-deficient diets are also commonly used to minimize thyroid uptake. Thyroid cancer-bearing mice pretreated with lithium chloride showed a significant decrease in ^{131}I uptake in the stomach relative to sodium chloride-pretreated mice, based on scintigraphy and ROI analysis (54).

Given that nontarget uptake of the radionuclide is so common, dosimetric calculations to estimate the radiation-absorbed dose to nontarget tissues is critical to minimize or at least predict radiation-associated toxicity. Before proceeding with therapy, it must be demonstrated that the desired therapeutic dose of radiation can be safely delivered to the target tissue without risk to other tissues. Generally, for radioiodine therapy of thyroid cancers, where dosimetric calculations are done, the maximum tolerated dose (MTD) is defined as less than 2 Gy to the blood and 24–27 Gy to the lungs (140). In a preclinical canine model, estimated radiation-absorbed doses to numerous tissues were calculated following intraprostatic injection of Ad5-vCD/mutTK$_{SR39}$rep-hNIS and i.v. Na^{99m}TcO$_4$ (32 mCi) (115). All tissues were found to receive less than 2% of the prostate dose of 0.23 Gy, with the urinary bladder wall being a dose-limiting tissue. In a similar model, ^{123}I (3 mCi) scintigraphy and ROI analysis following intraprostatic injections of Ad-CMV-hNIS were used to estimate the absorbed ^{131}I dose in various tissues per mCi ^{131}I (139). The thyroid was the dose-limiting tissue receiving a mean dose of 12.4 ± 2.8 Gy/mCi ^{131}I while the prostate received only 0.2 ± 0.02 Gy/mCi. In retrovirus-transduced thyroid cancer-bearing mice, dosimetric calculations following scintigraphy with diagnostic level ^{131}I (400 µCi, corresponding to 1650 MBq/m^2) estimated a tumor radiation-absorbed dose of 5.2–5.4 Gy (136). In a mouse myeloma tumor model, ^{123}I (0.5 mCi) for scintigraphy estimated a tumor radiation-absorbed dose of 11.6 and 18.4 Gy/mCi of ^{131}I following i.v. and i.t. injection of VSV(Δ51)-NIS, respectively (26). In a different mouse myeloma model, ^{123}I (0.5 mCi) for scintigraphy following MV-NIS infection estimated a tumor-absorbed dose of 0.4 Gy/mCi ^{131}I (21). For effective ablation of thyroid cancer metastases in patients, ^{131}I treatment resulting in radiation-absorbed doses of at least 80 Gy to the metastases is required (215). Based on the results presented here, relatively high doses of ^{131}I would have to be given to achieve a dose of 80 Gy in the tumor.

10.3.3 Other Receptors

Historically, development of radiolabeled peptide analogues for cancer imaging and therapy has been precipitated by observations of receptor overexpression in specific tumor types. The increasing availability of peptide analogues suitable for radionuclide labeling has driven the exploration of strategies that aim to expand the range of susceptible tumor types beyond those with endogenous receptor overexpression. Peptide targeting has been investigated in combination with other virally delivered receptors including gastrin-releasing peptide receptor (GRPr), norepinephrine transporter (NET; also known as noradrenalin transporter, NAT), dopamine receptor subtype 2 (D$_2$R), and dopamine transporter (DAT). While virally delivered GRPr and NET are used in tumor imaging or treatment studies, delivery of the D$_2$R receptor has been studied exclusively as a marker of gene therapy for neurodegenerative disorders.

10.3.3.1 *Gastrin-Releasing Peptide Receptor* Human GRPr (also known as bombesin receptor 2, BB$_2$) is a glycosylated 384 amino acid protein with seven transmembrane domains consistent with that of G-protein-coupled receptors (141). In humans, GRPr mRNA expression is highest in the pancreas, stomach, brain, and adrenal glands (142), while mouse mRNA is found mostly in areas of the digestive tract and regions of the brain (143). Protein expression is widespread in mouse brains and has been found in specific cell types within the gastrointestinal and urogenital tracts (144). The normal physiological function of GRPr is complex and not fully understood. GRPr activation is linked to many gastrointestinal (145) and CNS (146, 147) processes, and it seems to play an important role in immune responses (148, 149) and lung development (150). Receptor activation has been

demonstrated to induce tissue (151) and tumor growth (152, 153) as well as tumor cell invasion (154) and migration (155). GRPr expression has been reported in primary human ovarian (156), lung (157), prostate (158, 159), breast (160), cervical (161), pancreatic (162), colorectal (163), and gastrointestinal carcinoid (164) tumors. The role of GRPr in tumor development and progression is still under investigation; however, its overexpression has not been linked to changes in survival outcomes in colon cancer (165) and gastric adenocarcinoma patients (166). Encouraged by the success of somatostatin receptor-based imaging and therapy, GRPr analogues coupled to many different radionuclides have been developed and used to successfully image GRPr-positive tumors in numerous preclinical models (167–170) and in several preliminary human clinical trials (171–173). The desire to use these high-affinity radiopharmaceuticals for imaging and treating receptor-negative tumors (by introducing the target receptor using gene transduction) has led investigators to develop novel gene therapy vectors that result in tumor-specific expression of GRPr.

Delivery of GRPr by replication-deficient adenoviruses has been investigated in cervical carcinoma (174), ovarian adenocarcinoma (174–176), glioma (175), non-small cell lung cancer (174, 175), colon cancer (175), pancreatic cancer (177), and cholangiocarcinoma (177) cell lines, as well as in *in vivo* human ovarian carcinoma models (174–176). Ad-CMV-GRPr is the most commonly used vector (174–177), although vectors with different tumor-specific promoters have also been investigated (177). MUC1 and erb promoters were used to restrict GRPr expression to MUC1- and erbB2-positive tumors, respectively (177). Ad-erb-GRPr infection resulted in specific uptake of $[^{125}I]$-Tyr^4-bombesin in erbB2-positive, but not erbB2-negative, breast cancer cell lines. Ad-DF3-GRPr (DF3 = MUC1) infection resulted in specific uptake of $[^{125}I]$-Tyr^4-bombesin in a cholangiocarcinoma cell line expressing high levels of MUC1 protein, but not in those expressing low levels of the protein. However, Ad-DF3-GRPr also resulted in specific uptake in a pancreatic cancer cell line that expressed only low levels of MUC1, indicating that this type of tumor targeting may not be effective for all cell types. Uptake at any MOI for either Ad-MUC1-GRPr or Ad-erb-GRPr was significantly lower than that following infection with Ad-CMV-GRPr demonstrating that increased specificity typically comes with a decreased virus-mediated receptor gene expression.

The efficacy of the GRPr ligands $[^{125}I]$-Tyr^4-bombesin and $[^{125}I]$-mIP-bombesin has been compared *in vitro* and *in vivo* (174). In GRPr-positive cells, $[^{125}I]$-mIP-bombesin resulted in more surface-bound ligands after 1 h (48.8% versus 21.8%) and more retention of internalized radioligand at 4 h (32.0% versus 9%) relative to $[^{125}I]$-Tyr^4-bombesin. *In vivo* $[^{125}I]$-mIP-bombesin had a significantly increased tumor to blood ratio in mice with i.p. tumors infected with Ad-CMV-GRPr i.p. and administered the radioligand i.v. 2 days postinfection. Uptake in nontarget tissues of healthy mice was much higher in the small intestine and kidneys of mice injected with $[^{125}I]$-mIP-bombesin and significantly higher in the stomach and thyroids of mice receiving $[^{125}I]$-Tyr^4-bombesin. All later investigations used $[^{125}I]$-mIP-bombesin for *in vivo* studies due to its better *in vivo* tumor localization. $[^{125}I]$-mIP-bombesin has also been compared with a radiolabeled anti-erbB2 monoclonal antibody that was previously reported to induce tumor-specific cell death when its receptor gene was delivered by an adenovirus vector (178). When the anti-erbB2 antibody was labeled with ^{131}I and delivered i.v. in mice with erbB2-positive i.p. human ovarian cancer tumors, it had significantly higher blood and tumor half-lives but a lower tumor to blood ratio compared to [125I]-mIP-bombesin following Ad-CMV-GRPr treatment (175). In the absence of virus treatment, [125I]-mIP-bombesin (2 mCi, i.p.) showed significant uptake %ID/g in the small intestine (46.0%ID/g), kidney (20.8% ID/g), liver (8.3%ID/g), and spleen (4.9%ID/g). Interestingly, [125I]-mIP-bombesin biodistribution in some nontarget tissues was significantly increased after treatment with Ad-CMV-GRPr compared to infection with a control vector, Ad-CMV-LacZ. Uptake in the liver (174, 176), small intestine (174, 176), spleen (176), uterus (174, 176), kidney (177), and abdominal cavity lining (174, 176) was significantly increased due to virus-mediated GRPr expression. Given that tumor cells were administered i.p., likely metastasis to the abdominal organs may partially account for the increased uptake in these tissues. No significant difference in [125I]- and [131I]-mIP-bombesin biodistribution was observed in mice with i.p. human ovarian tumors treated with Ad-CMV-GRPr (176) indicating that imaging with the [125I]-labeled ligand could be used for dosimetric calculations before treatment with [131I]-mIP-bombesin. Historically, radiolabeled ligand development has focused on GRPr agonists (such as bombesin); however, recent evidence suggests that receptor antagonists may have increased tumor specificity and may therefore show superior biodistribution profiles (179).

10.3.3.2 Norepinephrine Transporter The norepinephrine transporter is an important protein in the CNS, facilitating neurotransmitter reuptake in noradrenergic neurons. NET has 12 transmembrane domains and belongs to a family of Na^+/Cl^--dependent monoamine transporters (180). The human NET (hNET) cDNA was first cloned in 1991 (181) and subsequent studies demonstrated hNET mRNA and protein expression in numerous regions of the CNS (182, 183), peripheral sympathetic nerve endings (of the heart and blood vessels) (184), lung endothelial cells (185), and the placenta (186). Norepinephrine (NE) plays a critical role in regulating mood, sleep, behavior, and sympathetic homeostasis (187). Abnormalities in hNET expression or function have been associated with dysautonomia (disorder of the autonomic nervous system), essential hypertension, variable risks and outcomes following myocardial ischemia, anorexia nervosa, attention-deficit hyperactivity disorder, depression, addiction, and pain (187). Considerable knowledge of the biochemistry of transporter–ligand interactions and functional transport has spurred the development of a wide range of NET synthetic ligands. NET binds its natural catecholamine ligands NE, dopamine (DA), and epinephrine (EPI) with differing affinities and efficacies of transport (DA > NE > EPI) (187, 188).

Similar to SSTR, NET expression is upregulated in neuroendocrine tumors (189); however, increased receptor expression appears to be variable and likely confined to a more specific subset of tumors relative to SSTR (190). Interestingly, *in vitro* treatment of neuroendocrine gastrointestinal tumor cell lines with the NET-specific ligand metaiodobenzylguanidine (MIBG) resulted in NET-dependent apoptosis and growth inhibition (191) and, when combined with interferon-γ, an additive antiproliferative effect was observed (192). Given the relatively localized distribution of NET expression to the brain and neuronal tissues and the availability of radioligands, NET may hold promise for viral-mediated gene delivery coupled with radiopharmaceutical targeting for imaging and treatment of NET-negative tumors.

Viral delivery of NET has been investigated as a reporter for gene therapy using retroviral and adenoviral vectors. Fibrosarcoma (193), epitheloid carcinoma (193), murine sarcoma (193), rat glioma (194), T-cell leukemia (194), and human neuro-blastoma (194) cells stably transduced with retroviral vectors expressing NET demonstrate significantly increased uptake of [^{123}I]- or [^{131}I]-MIBG compared to matched parental cells (193, 194). The level of NET expression and overall uptake varied significantly between the different transduced cells. MIBG retention over time was markedly high compared to that observed for the other receptors previously discussed. At 2 h post-treatment, 71–87% of baseline [^{131}I]-MIBG was retained with variability between cell lines (193). NET-transduced human epitheloid carcino-ma (193) and rat glioma (194) were used in *in vivo* mouse tumor xenograft models to evaluate the efficacy of ^{123}I-MIBG or ^{124}I-MIBG for imaging NET-positive tumors by SPECT or PET, respectively. Figure 10.7 shows a SPECT image demonstrating uptake of ^{123}I-MIBG in tumor cells transduced with a NET-expressing retrovirus but not in cells transduced with a control vector. Serial imaging of s.c. tumors demon-strated that the best tumor to background contrast was at 24 h post-^{123}I-MIBG injection by SPECT imaging, while the ideal window for ^{124}I-MIBG imaging by PET extended from 4–48 h postinjection (194). The maximum tumor to muscle (%ID/g) ratio was 25 : 1 for transduced epitheloid carcinoma xenografts as determined by gamma counting of harvested tissues (193) and 130 : 1 for transduced glioma xenografts as determined by PET imaging and ROI analysis (194). In the case of epitheloid carcinoma tumors, gamma camera imaging revealed only a threefold increase in tumor uptake relative to background radioactivity, indicating that imaging may underestimate the true increase in receptor gene expression due to circulating blood pool. In both models, the transduced tumor uptake was significantly higher than that in parental cell tumors.

AdTrack-hNET, a replication-deficient adenovirus expressing NET and enhanced green fluorescent protein (eGFP) under the control of CMV promoters, has been used in combination with ^{11}C-*m*-hydroxyephedrine (^{11}C-*m*HED), a radioligand with structural homology to norepinephrine. AdTrack-hNET resulted in viral dose-depen-dent increase in ^{11}C-*m*HED accumulation in green monkey kidney, human glioma, and human ovarian adenocarcinoma cell lines *in vitro* (195). Maximum uptake was highest in the glioma cells followed by the kidney and ovarian adenocarcinoma cells. *In vivo* PET imaging of rats bearing s.c. glioma tumors using ^{11}C-*m*HED (1 mCi) was unable to demonstrate any significant difference between tumors that had previously been injected with AdTrack-hNET and those injected with AdTrack-Luc. *Ex vivo* gamma counting of harvested tissues, however, revealed a 14–27% increase in radioactivity in AdTrack-hNET relative to AdTrack-Luc-infected tumors.

Imaging and *ex vivo* gamma counting of both retrovirus-transduced (194) and adenovirus-infected (195) athymic mice bearing s.c. glioma xenografts showed

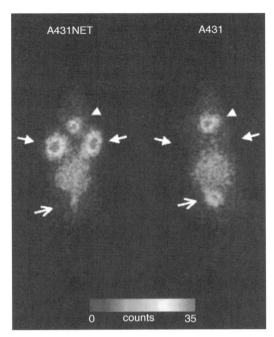

FIGURE 10.7   [^{123}I]MIBG uptake in epithelial carcinoma tumors retrovirally transduced to express NET. Immune-compromised mice (CD1 nu/nu) bearing subcutaneous A431 or A431NET tumors (transduced *in vitro* with a retrovirus containing the NET gene) were injected intravenously with 12 MBq of [^{123}I]MIBG. Gamma camera imaging performed 24 h after radioligand injection showed uptake in the mice bearing A431NET but not A431 tumors (closed arrows). Uptake was also observed in the thyroid (arrowheads) and bladder (open arrows) due to renal excretion. (Reprinted with permission from Ref. 193.) (See insert for the color representation of the figure.)

significant radioligand uptake in the bladder, kidney, intestine, liver, thyroid (194, 195), and heart (195). This nontarget uptake likely reflects endogenous NET expression or radioligand excretion given that retroviral transduction was performed *in vitro* and infection with the nonreplicating adenovirus was demonstrated to occur only in a small proportion of tumors. ^{123}I-MIBG was retained longer in retrovirus-transduced tumors relative to nontarget tissues indicating that it may be possible to achieve sufficient radiation doses to tumors without toxicity to other tissues (194).

10.3.3.3 Dopamine Receptor and Dopamine Transporter   The dopamine receptor (DR) family is comprised of five subtypes (D1–D5) belonging to the larger family of rhodopsin-like G-protein-coupled receptors (196). Particular attention will be paid to the D2 subtype given that the gene therapy strategies discussed below utilize only this receptor subtype. DR–D2 (D$_2$R) is expressed mainly in the CNS, and under normal physiological conditions DR signaling plays an important role in locomotion. Models of decreased receptor activation have implicated D_2R in motor disorders similar to Parkinson's disease (197). D_2R expression has also been reported in the pulmonary arteries (198), kidney nephrons, and renal cortex (199).

Delivery of DR has been investigated using several adenovirus and two vaccinia virus vectors. To date, these studies have mainly focused on D_2R gene therapy as a reporter gene for tumor imaging. In the context of malignant disease, viral delivery of the rat D_2R gene has been studied in combination with delivery of the mutant HSV-1 TK gene, sr39tk (200, 201). The mutant sr39tk protein serves as an improved PET reporter resulting from its ability to use positron-emitting acycloguanosine substrates more efficiently than the wild-type protein (202). Furthermore, TK can be used in therapeutic strategies using ganciclovir, as discussed in previous sections. D_2R and HSV1-sr39TK have been delivered in combination as two separate vectors (Ad-CMV-D_2R and Ad-CMV-HSV1-sr39TK) (201) or as one vector with the two genes separated by an internal ribosome entry site (IRES) (Ad.DTm) (200). *In vitro* infection of rat glioma and monkey kidney cells with Ad-CMV-D_2R or Ad-CMV-HSV1-sr39TK resulted in a viral dose-dependent increase in binding of ^{3}H-spiperone, a D_2R radioligand or ^{3}H-penciclovir (PCV) phosphorylation by the TK enzyme (201). In addition, cells coinfected with both viruses showed a good positive correlation between the levels of ^{3}H-spiperone D_2R binding and TK phosphorylation of ^{3}H-PCV (201). *In vivo*, systemic infection with Ad.DTm resulted in significantly decreased liver uptake of the D_2R ligand [^{18}F]-FESP and TK ligand [^{18}F]-FHBG relative to uptake of these respective ligands following infection with Ad-CMV-D_2R or Ad-CMV-HSV1-sr39TK alone (200). Nevertheless, uptake following Ad.DTm infection was sufficient for quantification using microPET and was found to depend on viral dose with HSV1-sr39TK detection only at doses higher than 5×10^6 pfu. MicroPET and *ex vivo* autoradiography of rat brains microperfused with Ad-CMV-D_2R showed increased uptake of radiolabeled D_2R ligands relative to those infected with a receptor-negative control virus (203–206). Examination of brain sections by *ex vivo* autoradiography suggests viral spread of only about 1 mm from the injection site reflecting a relatively low transduction efficiency (205). Histological analysis of brain sections demonstrated no significant pathological features in virus-injected brains (203), providing preliminary evidence of the safety of the virus for intrastriatum administration. These studies suggest D_2R gene therapy may have potential for imaging and treatment of brain tumors.

The dopamine transporter, a related receptor that also binds dopamine, has been investigated as a superior alternative to D_2R for dopamine radiopharmaceutical targeting (207). DAT, unlike D_2R, is not associated with any intracellular signaling pathways and is therefore less likely to have unwanted physiological responses to peptide binding. Delivery of DAT has been investigated using an adeno-associated virus (AAV2/5-CMV-DAT) injected intramuscularly into contralateral hind limbs of tumor-free immunocompromised mice. SPECT imaging performed at 4 weeks p.i. with [^{99m}Tc]TRODAT-1 (20 mCi), a radiolabeled DAT ligand, showed a 150–270% increased signal in AAV2/5-CMV-DAT compared to Ad-CMV-LacZ-treated limbs. A viral dose-dependent increase in the SPECT signal was observed for

AAV2/5-CMV-DAT with a minimum limit of detection greater than 10^8 pfu. Similar results were obtained in immune-competent rabbits.

These studies support the use of D_2R or DAT as a reporter for noninvasive imaging of gene therapy and highlight the potential for targeted radiotherapy if combined with therapeutic radioligands.

10.4   COMBINED ONCOLYTIC AND TARGETED RADIOTHERAPY

While the therapeutic potential of receptor delivery by nonreplicating viruses is solely due to the receptor-targeted radionuclide, employing oncolytic viruses as the gene therapy vector has the added benefit of virus-mediated cell death. Furthermore, it is postulated that the combined virotherapy and targeted radionuclide therapy will have a synergistic antitumor effect. The hypothesized mechanism of this synergistic antitumor effect is twofold mediated by both the radiation cross-fire effect (Fig. 10.1) and the biological bystander effect (Fig. 10.2), as discussed previously.

Nonreplicating viruses administered intratumorally can only achieve gene transfer in the primary tumor, while systemically delivered oncolytic viruses have the potential to infect both the primary tumor and the distant metastases for subsequent targeting by radioligands. Given that distant metastases are typically less likely to be treated by surgical resection compared to the primary tumor, targeting the metastases could have a significant impact on improving patient survival for a number of solid malignancies. In addition, systemically delivered oncolytic viruses could also prove effective in nonsolid tumors.

One of the interesting and not very well-studied aspects of combined oncolytic virotherapy and targeted radiotherapy is the potential interactions between the particulate radiation (i.e., α- or β-particles or Auger electrons) and the virus. Numerous studies have demonstrated the synergistic relationship between oncolytic viruses and external ionizing radiation (126, 127, 208–213). Ionizing radiation has been demonstrated to potentiate *in vitro* and *in vivo* herpes virus infection of lung cancer and mesothelioma cells through induction of the radiation resistance protein, growth arrest, and DNA damage inducible protein-34 (GADD34) (126, 213). GADD34 shows structural homology to ICP34.5, a herpes virus protein involved in translation of viral proteins and associated with neurotoxicity. Vectors with deletions of ICP34.5 have improved toxicity profiles but show attenuated replication in many tumor cell lines. In tumors infected with a ICP34.5-deleted virus, radiation-mediated induction of GADD34 functionally replaces ICP34.5 thereby enhancing viral replication. The synergism of this combination therapy allows dose reductions (2- to 6000-fold) of both agents without compromising the therapeutic effects. A similar synergistic antitumor effect was demonstrated using replication competent Ad-5Δ24RGD in irradiated glioma cells *in vitro* (208, 214); however, *in vivo* success has varied.

Despite the considerable knowledge gaps, the fact that oncolytic viruses combined with targeted radioligands have been successfully applied in both noninvasive tumor imaging (21, 26, 114, 116, 133, 134) and therapy (21, 26) suggests this appears to be a viable approach. As discussed above, radiovirotherapy has been investigated in different *in vivo* models of multiple myeloma using VSV(Δ51)-NIS or MV-NIS combined with [131]I. Immune-competent mice bearing s.c. or orthotopic myeloma tumors were treated i.v. or i.t. with VSV(Δ51)-NIS followed by i.p. [131]I (1 mCi) 1 day p.i. (26). Treatment with ^{131}I resulted in additional tumor growth suppression up to 15 days p.i. compared to virus alone. A significant increase in the survival benefit was observed for mice receiving both ^{131}I and oncolytic virus therapy. No significant difference was observed between animals receiving i.t. or i.v. virus injections indicating high tumor localization after systemic administration. Similar improved survival following radiovirotherapy was observed in the orthotopic model.

In the second study, immunodeficient mice bearing s.c. myeloma tumors were injected i.v. with 2×10^6 pfu MV-NIS or a receptor-negative control virus (21). [131]I (1 mCi) was administered i.p. 6 days postinfection. Tumor regression was observed in response to virus alone; however, the speed of regression was significantly increased in the combination therapy group. When MV-resistant myeloma cells were tested, neither ^{131}I nor MV-NIS treatment alone had any significant impact on tumor growth relative to untreated controls. However, the combination therapy resulted in complete tumor regression in four of the five mice. The promising results of this study has led to evaluation of the MV-NIS virus in a phase I clinical trial (Mayo Clinic, MC038C, http://clinicaltrials.mayo.edu) for treatment of multiple myeloma.

10.5 SUMMARY

The use of viral vectors to deliver receptors for targeted radiotherapy is very promising. While it is still not feasible to deliver a receptor to every cell within a tumor, the radiation cross-fire and bystander effects render this unnecessary, and effective therapy is still possible. A number of viruses have been used and viral selection clearly depends on several factors including the cell type to be targeted and the required duration of gene expression.

One of the most promising developments is the combination of oncolytic viral therapy and targeted radiotherapy for tumors, in which not only does the virus deliver the receptor of choice to the cancerous cell but also has antitumor effects of its own, leading to enhanced responses. While this topic is still in its infancy, over time the benefits of combination therapy will likely become apparent. Future studies addressing the biological interactions between replicating viruses and targeted radiotherapy are needed to further our understanding of these synergistic, tumor-targeting modalities.

REFERENCES

1. Rosenfeld MA, Yoshimura K, Trapnell BC et al. *In vivo* transfer of the human cystic fibrosis transmembrane conductance regulator gene to the airway epithelium. *Cell* 1992; 68:143–155.

2. Mitchell SA, Randers-Pehrson G, Brenner DJ, Hall EJ. The bystander response in C3H 10T1/2 cells: the influence of cell-to-cell contact. *Radiat Res* 2004;161:397–401.

3. Mairs RJ, Fullerton NE, Zalutsky MR, Boyd M. Targeted radiotherapy: microgray doses and the bystander effect. *Dose Response* 2007;5:204–213.

4. Curiel DT. Strategies to adapt adenoviral vectors for targeted delivery. *Ann NY Acad Sci* 1999;886:158–171.

5. Young LS, Searle PF, Onion D, Mautner V. Viral gene therapy strategies: from basic science to clinical application. *J Pathol* 2006;208:299–318.

6. Dingli D, Diaz RM, Bergert ER, O'Connor MK, Morris JC, Russell SJ. Genetically targeted radiotherapy for multiple myeloma. *Blood* 2003;102:489–496.

7. Dingli D, Kemp BJ, O'Connor MK, Morris JC, Russell SJ, Lowe VJ. Combined I-124 positron emission tomography/computed tomography imaging of NIS gene expression in animal models of stably transfected and intravenously transfected tumor. *Mol Imaging Biol* 2006;8:16–23.

8. Kim HJ, Jeon YH, Kang JH et al. *In vivo* long-term imaging and radioiodine therapy by sodium-iodide symporter gene expression using a lentiviral system containing ubiquitin C promoter. *Cancer Biol Ther* 2007;6:1130–1135.

9. Baskar JF, Smith PP, Nilaver G et al. The enhancer domain of the human cytomegalovirus major immediate-early promoter determines cell type-specific expression in transgenic mice. *J Virol* 1996;70:3207–3214.

10. Loser P, Jennings GS, Strauss M, Sandig V. Reactivation of the previously silenced cytomegalovirus major immediate-early promoter in the mouse liver: involvement of NFkappaB. *J Virol* 1998;72:180–190.

11. Heise C, Sampson-Johannes A, Williams A, McCormick F, Von Hoff DD, Kirn DH. ONYX-015, an E1B gene-attenuated adenovirus, causes tumor-specific cytolysis and antitumoral efficacy that can be augmented by standard chemotherapeutic agents. *Nat Med* 1997;3:639–645.

12. Post DE, Khuri FR, Simons JW, Van Meir EG. Replicative oncolytic adenoviruses in multimodal cancer regimens. *Hum Gene Ther* 2003;14:933–946.

13. Heise C, Hermiston T, Johnson L et al. An adenovirus E1A mutant that demonstrates potent and selective systemic anti-tumoral efficacy. *Nat Med* 2000;6:1134–1139.

14. Cascallo M, Capella G, Mazo A, Alemany R. Ras-dependent oncolysis with an adenovirus VAI mutant. *Cancer Res* 2003;63:5544–5550.

15. McCart JA, Wang ZH, Xu H et al. Development of a melanoma-specific adenovirus. *Mol Ther* 2002;6:471–480.

16. Rodriguez R, Schuur ER, Lim HY, Henderson GA, Simons JW, Henderson DR. Prostate attenuated replication competent adenovirus (ARCA) CN706: a selective cytotoxic for prostate-specific antigen-positive prostate cancer cells. *Cancer Res* 1997;57: 2559–2563.

17. Griffin DE. Measles virus. In: Knipe DM, Howely PM, editors. *Fields Virology.* Lippincott, Williams and Wilkins, Philadelphia; 2001: 1401–1442.

18. Nakamura T, Russell SJ. Oncolytic measles viruses for cancer therapy. *Expert Opin Biol Ther* 2004;4:1685–1692.

19. Ong HT, Timm MM, Greipp PR et al. Oncolytic measles virus targets high CD46 expression on multiple myeloma cells. *Exp Hematol* 2006;34:713–720.

20. Radecke F, Spielhofer P, Schneider H et al. Rescue of measles viruses from cloned DNA. *EMBO J* 1995;14:5773–5784.

21. Dingli D, Peng KW, Harvey ME et al. Image-guided radiovirotherapy for multiple myeloma using a recombinant measles virus expressing the thyroidal sodium iodide symporter. *Blood* 2004;103:1641–1646.

22. Rose JK, Whitt MA. Rhabdoviridae: the viruses and their replication. In: Knipe DM, Howely PM, editors. *Fields Virology*. Lippincott, Williams and Wilkins, Philadelphia; 2001: 1221–1243.

23. Stojdl DF, Lichty BD, tenOever BR et al. VSV strains with defects in their ability to shutdown innate immunity are potent systemic anti-cancer agents. *Cancer Cell* 2003;4:263–275.

24. Balachandran S, Porosnicu M, Barber GN. Oncolytic activity of vesicular stomatitis virus is effective against tumors exhibiting aberrant p53, Ras, or myc function and involves the induction of apoptosis. *J Virol* 2001;75:3474–3479.

25. Lichty BD, Power AT, Stojdl DF, Bell JC. Vesicular stomatitis virus: re-inventing the bullet. *Trends Mol Med* 2004;10:210–216.

26. Goel A, Carlson SK, Classic KL et al. Radioiodide imaging and radiovirotherapy of multiple myeloma using VSV(Delta51)-NIS, an attenuated vesicular stomatitis virus encoding the sodium iodide symporter gene. *Blood* 2007;110:2342–2350.

27. Moss B., Poxviridae: the viruses and their replication. In: Knipe DM, Howely PM, editors. *Fields Virology*. Lippincott, Williams and Wilkins, Philadelphia; 2001: 2849–2884.

28. Shen Y, Nemunaitis J. Fighting cancer with vaccinia virus: teaching new tricks to an old dog. *Mol Ther* 2005;11:180–195.

29. McCart JA, Ward JM, Lee J et al. Systemic cancer therapy with a tumor-selective vaccinia virus mutant lacking thymidine kinase and vaccinia growth factor genes. *Cancer Res* 2001;61:8751–8757.

30. Moss B, Earl PL. Expression of proteins in mammalian cells using vaccinia viral vectors. *Curr Protoc Mol Biol* 1998;43:16.15.1–16.19.11.

31. Smith GL, Moss B. Infectious poxvirus vectors have capacity for at least 25000 base pairs of foreign DNA. *Gene* 1983;25:21–28.

32. McCart JA, Mehta N, Scollard D et al. Oncolytic vaccinia virus expressing the human somatostatin receptor SSTR2: molecular imaging after systemic delivery using [111]In--pentetreotide. *Mol Ther* 2004;10:553–561.

33. Rohrer L, Raulf F, Bruns C, Buettner R, Hofstaedter F, Schule R. Cloning and characterization of a fourth human somatostatin receptor. *Proc Natl Acad Sci USA* 1993;90:4196–4200.

34. Yamada Y, Post SR, Wang K, Tager HS, Bell GI, Seino S. Cloning and functional characterization of a family of human and mouse somatostatin receptors expressed in brain, gastrointestinal tract, and kidney. *Proc Natl Acad Sci USA* 1992;89:251–255.

35. Yamada Y, Reisine T, Law SF et al. Somatostatin receptors, an expanding gene family: cloning and functional characterization of human SSTR3, a protein coupled to adenylyl cyclase. *Mol Endocrinol* 1992;6:2136–2142.

36. Panetta R, Greenwood MT, Warszynska A et al. Molecular cloning, functional charac-terization, and chromosomal localization of a human somatostatin receptor (somatostatin receptor type 5) with preferential affinity for somatostatin-28. *Mol Pharmacol* 1994;45: 417–427.

37. Susini C, Buscail L. Rationale for the use of somatostatin analogs as antitumor agents. *Ann Oncol* 2006;17:1733–1742.

38. Brazeau P, Vale W, Burgus R et al. Hypothalamic polypeptide that inhibits the secretion of immunoreactive pituitary growth hormone. *Science* 1973;179:77–79.

39. Pradayrol L, Jornvall H, Mutt V, Ribet A. N-terminally extended somatostatin: the primary structure of somatostatin-28. *FEBS Lett* 1980;109:55–58.

40. Olias G, Viollet C, Kusserow H, Epelbaum J, Meyerhof W. Regulation and function of somatostatin receptors. *J Neurochem* 2004;89:1057–1091.

41. Selmer I, Schindler M, Allen JP, Humphrey PP, Emson PC. Advances in understanding neuronal somatostatin receptors. *Regul Pept* 2000;90:1–18.

42. Patel YC. Somatostatin and its receptor family. *Front Neuroendocrinol* 1999;20:157–198.

43. Grimberg A. Somatostatin and cancer: applying endocrinology to oncology. *Cancer Biol Ther* 2004;3:731–733.

44. Reubi JC, Laissue J, Krenning E, Lamberts SW. Somatostatin receptors in human cancer: incidence, characteristics, functional correlates and clinical implications. *J Steroid Biochem Mol Biol* 1992;43:27–35.

45. Volante M, Rosas R, Allia E et al. Somatostatin, cortistatin and their receptors in tumors. *Mol Cell Endocrinol* 2008;286:219–229.

46. Prasad V, Fetscher S, Baum RP. Changing role of somatostatin receptor targeted drugs in NET: nuclear medicine's view. *J Pharm Pharm Sci* 2007;10:321s–337s.

47. Florio T, Morini M, Villa V et al. Somatostatin inhibits tumor angiogenesis and growth via somatostatin receptor-3-mediated regulation of endothelial nitric oxide synthase and mitogen-activated protein kinase activities. *Endocrinology* 2003;144: 1574–1584.

48. Lamberts SW, de Herder WW, Hofland LJ. Somatostatin analogs in the diagnosis and treatment of cancer. *Trends Endocrinol Metab* 2002;13:451–457.

49. Buscail L, Saint-Laurent N, Chastre E et al. Loss of sst2 somatostatin receptor gene expression in human pancreatic and colorectal cancer. *Cancer Res* 1996;56: 1823–1827.

50. Hu C, Yi C, Hao Z et al. The effect of somatostatin and SSTR3 on proliferation and apoptosis of gastric cancer cells. *Cancer Biol Ther* 2004;3:726–730.

51. Fisher WE, Wu Y, Amaya F, Berger DH. Somatostatin receptor subtype 2 gene therapy inhibits pancreatic cancer *in vitro*. *J Surg Res* 2002;105:58–64.

52. Vernejoul F, Faure P, Benali N et al. Antitumor effect of *in vivo* somatostatin receptor subtype 2 gene transfer in primary and metastatic pancreatic cancer models. *Cancer Res* 2002;62:6124–6131.

53. Celinski SA, Fisher WE, Amaya F et al. Somatostatin receptor gene transfer inhibits established pancreatic cancer xenografts. *J Surg Res* 2003;115:41–47.

54. Ugur O, Kothari PJ, Finn RD et al. Ga-66 labeled somatostatin analogue DOTA-DPhe[1]-Tyr[3]-octreotide as a potential agent for positron emission tomography imaging and receptor mediated internal radiotherapy of somatostatin receptor positive tumors. *Nucl Med Biol* 2002;29:147–157.

55. Miederer M, Henriksen G, Alke A et al. Preclinical evaluation of the α-particle generator nuclide [225]Ac for somatostatin receptor radiotherapy of neuroendocrine tumors. *Clin Cancer Res* 2008;14:3555–3561.

56. Tiensuu JE, Eriksson B, Oberg K et al. Treatment with high dose [^{111}In-DTPA-D-PHE^1]-octreotide in patients with neuroendocrine tumors: evaluation of therapeutic and toxic effects. *Acta Oncol* 1999;38:373–377.

57. Zinn KR, Chaudhuri TR, Krasnykh VN et al. Gamma camera dual imaging with a somatostatin receptor and thymidine kinase after gene transfer with a bicistronic adenovirus in mice. *Radiology* 2002;223:417–425.

58. Hemminki A, Belousova N, Zinn KR et al. An adenovirus with enhanced infectivity mediates molecular chemotherapy of ovarian cancer cells and allows imaging of gene expression. *Mol Ther* 2001;4:223–231.

59. Verwijnen SM, Sillevis Smith PA, Hoeben RC et al. Molecular imaging and treatment of malignant gliomas following adenoviral transfer of the herpes simplex virus-thymidine kinase gene and the somatostatin receptor subtype 2 gene. *Cancer Biother Radiopharm* 2004;19:111–120.

60. Hemminki A, Zinn KR, Liu B et al. *In vivo* molecular chemotherapy and noninvasive imaging with an infectivity-enhanced adenovirus. *J Natl Cancer Inst* 2002;94:741–749.

61. Rogers BE, Chaudhuri TR, Reynolds PN, Della MD, Zinn KR. Non-invasive gamma camera imaging of gene transfer using an adenoviral vector encoding an epitope-tagged receptor as a reporter. *Gene Ther* 2003;10:105–114.

62. Zinn KR, Chaudhuri TR, Buchsbaum DJ, Mountz JM, Rogers BE. Simultaneous evaluation of dual gene transfer to adherent cells by gamma-ray imaging. *Nucl Med Biol* 2001;28:135–144.

63. Wickham TJ, Tzeng E, Shears LL et al. Increased *in vitro* and *in vivo* gene transfer by adenovirus vectors containing chimeric fiber proteins. *J Virol* 1997;71:8221–8229.

64. Zinn KR, Chaudhuri TR, Buchsbaum DJ, Mountz JM, Rogers BE. Detection and measurement of *in vitro* gene transfer by gamma camera imaging. *Gene Ther* 2001;8:291–299.

65. Zinn KR, Buchsbaum DJ, Chaudhuri TR, Mountz JM, Grizzle WE, Rogers BE. Noninvasive monitoring of gene transfer using a reporter receptor imaged with a high-affinity peptide radiolabeled with ^{99m}Tc or ^{188}Re. *J Nucl Med* 2000;41:887–895.

66. Rogers BE, Zinn KR, Lin CY, Chaudhuri TR, Buchsbaum DJ. Targeted radiotherapy with [^{90}Y]-SMT 487 in mice bearing human nonsmall cell lung tumor xenografts induced to express human somatostatin receptor subtype 2 with an adenoviral vector. *Cancer* 2002;94 (Suppl.): 1298–1305.

67. Buchsbaum DJ, Rogers BE, Khazaeli MB et al. Targeting strategies for cancer radio-therapy. *Clin Cancer Res* 1999;5 (Suppl.): 3048s–3055s.

68. Chaudhuri TR, Rogers BE, Buchsbaum DJ, Mountz JM, Zinn KR. A noninvasive reporter system to image adenoviral-mediated gene transfer to ovarian cancer xenografts. *Gynecol Oncol* 2001;83:432–438.

69. Rogers BE, Parry JJ, Andrews R, Cordopatis P, Nock BA, Maina T. MicroPET imaging of gene transfer with a somatostatin receptor-based reporter gene and ^{94m}Tc-Demotate 1. *J Nucl Med* 2005;46:1889–1897.

70. Bernard BF, Krenning EP, Breeman WA et al. D-lysine reduction of indium-111 octreotide and yttrium-90 octreotide renal uptake. *J Nucl Med* 1997;38:1929–1933.

71. de JM, Breeman WA, Valkema R, Bernard BF, Krenning EP. Combination radionuclide therapy using ^{177}Lu- and ^{90}Y-labeled somatostatin analogs. *J Nucl Med* 2005;46 (Suppl.): 13S–17S.

72. Rogers BE, McLean SF, Kirkman RL et al. *In vivo* localization of [^{111}In]-DTPA-D-Phe^{1}-octreotide to human ovarian tumor xenografts induced to express the somatostatin receptor subtype 2 using an adenoviral vector. *Clin Cancer Res* 1999;5:383–393.

73. Dai G, Levy O, Carrasco N. Cloning and characterization of the thyroid iodide transporter. *Nature* 1996;379:458–460.

74. Smanik PA, Liu Q, Furminger TL et al. Cloning of the human sodium iodide symporter. *Biochem Biophys Res Commun* 1996;226:339–345.

75. Spitzweg C, Joba W, Eisenmenger W, Heufelder AE. Analysis of human sodium iodide symporter gene expression in extrathyroidal tissues and cloning of its complementary deoxyribonucleic acids from salivary gland, mammary gland, and gastric mucosa. *J Clin Endocrinol Metab* 1998;83:1746–1751.

76. Hertz S, Roberts A, Salter WT. Radioactive iodine as an indicator in thyroid physiology. IV. The metabolism of iodine in Graves' disease. *J Clin Invest* 1942;21:25–29.

77. Mazzaferri EL, Kloos RT. Clinical review 128: current approaches to primary therapy for papillary and follicular thyroid cancer. *J Clin Endocrinol Metab* 2001;86:1447–1463.

78. Shen DH, Kloos RT, Mazzaferri EL, Jhian SM. Sodium iodide symporter in health and disease. *Thyroid* 2001;11:415–425.

79. Durante C, Haddy N, Baudin E et al. Long-term outcome of 444 patients with distant metastases from papillary and follicular thyroid carcinoma: benefits and limits of radioiodine therapy. *J Clin Endocrinol Metab* 2006;91:2892–2899.

80. Tsang RW, Brierley JD, Simpson WJ, Panzarella T, Gospodarowicz MK, Sutcliffe SB. The effects of surgery, radioiodine, and external radiation therapy on the clinical outcome of patients with differentiated thyroid carcinoma. *Cancer* 1998;82:375–388.

81. Wapnir IL, van de RM, Nowels K et al. Immunohistochemical profile of the sodium/iodide symporter in thyroid, breast, and other carcinomas using high density tissue microarrays and conventional sections. *J Clin Endocrinol Metab* 2003;88:1880–1888.

82. Tazebay UH, Wapnir IL, Levy O et al. The mammary gland iodide transporter is expressed during lactation and in breast cancer. *Nat Med* 2000;6:871–878.

83. Moon DH, Lee SJ, Park KY et al. Correlation between ^{99m}Tc-pertechnetate uptakes and expressions of human sodium iodide symporter gene in breast tumor tissues. *Nucl Med Biol* 2001;28:829–834.

84. Wapnir IL, Goris M, Yudd A et al. The Na^{+}/I^{-} symporter mediates iodide uptake in breast cancer metastases and can be selectively down-regulated in the thyroid. *Clin Cancer Res* 2004;10:4294–4302.

85. Spitzweg C, Joba W, Morris JC, Heufelder AE. Regulation of sodium iodide symporter gene expression in FRTL-5 rat thyroid cells. *Thyroid* 1999;9:821–830.

86. Simon D, Korber C, Krausch M et al. Clinical impact of retinoids in redifferentiation therapy of advanced thyroid cancer: final results of a pilot study. *Eur J Nucl Med Mol Imaging* 2002;29:775–782.

87. Kitazono M, Robey R, Zhan Z et al. Low concentrations of the histone deacetylase inhibitor, depsipeptide (FR901228), increase expression of the Na^{+}/I^{-} symporter and iodine accumulation in poorly differentiated thyroid carcinoma cells. *J Clin Endocrinol Metab* 2001;86:3430–3435.

88. Furuya F, Shimura H, Suzuki H et al. Histone deacetylase inhibitors restore radioiodide uptake and retention in poorly differentiated and anaplastic thyroid cancer cells by

expression of the sodium/iodide symporter thyroperoxidase and thyroglobulin. *Endocrinology* 2004;145:2865–2875.

89. Fortunati N, Catalano MG, Arena K, Brignardello E, Piovesan A, Boccuzzi G. Valproic acid induces the expression of the Na^+/I^- symporter and iodine uptake in poorly differentiated thyroid cancer cells. *J Clin Endocrinol Metab* 2004;89:1006–1009.

90. Beyer SJ, Jimenez RE, Shapiro CL, Cho JY, Jhiang SM. Do cell surface trafficking impairments account for variable cell surface sodium iodide symporter levels in breast cancer? *Breast Cancer Res Treat* 2008;115:205–212.

91. Dohan O, Baloch Z, Banrevi Z, Livolsi V, Carrasco N. Rapid communication: predominant intracellular overexpression of the Na^+/I^- symporter (NIS) in a large sampling of thyroid cancer cases. *J Clin Endocrinol Metab* 2001;86:2697–2700.

92. Chen RF, Li ZH, Pan QH et al. *In vivo* radioiodide imaging and treatment of pancreatic cancer xenografts after MUC1 promoter-driven expression of the human sodium-iodide symporter. *Pancreatology* 2007;7:505–513.

93. Spitzweg C, Baker CH, Bergert ER, O'Connor MK, Morris JC. Image-guided radioiodide therapy of medullary thyroid cancer after carcinoembryonic antigen promoter-targeted sodium iodide symporter gene expression. *Hum Gene Ther* 2007;18:916–924.

94. Faivre J, Clerc J, Gerolami R et al. Long-term radioiodine retention and regression of liver cancer after sodium iodide symporter gene transfer in wistar rats. *Cancer Res* 2004;64:8045–8051.

95. Kakinuma H, Bergert ER, Spitzweg C, Cheville JC, Lieber MM, Morris JC. Probasin promoter (ARR_2PB)-driven, prostate-specific expression of the human sodium iodide symporter (h-NIS) for targeted radioiodine therapy of prostate cancer. *Cancer Res* 2003;63:7840–7844.

96. Van SJ, Massart C, Beauwens R et al. Anion selectivity by the sodium iodide symporter. *Endocrinology* 2003;144:247–252.

97. Carlin S, Mairs RJ, Welsh P, Zalutsky MR. Sodium-iodide symporter (NIS)-mediated accumulation of [(211)At]astatide in NIS-transfected human cancer cells. *Nucl Med Biol* 2002;29:729–739.

98. Gaut AW, Niu G, Krager KJ, Graham MM, Trask DK, Domann FE. Genetically targeted radiotherapy of head and neck squamous cell carcinoma using the sodium-iodide symporter (NIS). *Head Neck* 2004;26:265–271.

99. Dwyer RM, Bergert ER, O'Connor MK, Gendler SJ, Morris JC. Adenovirus-mediated and targeted expression of the sodium-iodide symporter permits *in vivo* radioiodide imaging and therapy of pancreatic tumors. *Hum Gene Ther* 2006;17:661–668.

100. Dwyer RM, Bergert ER, O'Connor MK, Gendler SJ, Morris JC. Sodium iodide symporter-mediated radioiodide imaging and therapy of ovarian tumor xenografts in mice. *Gene Ther* 2006;13:60–66.

101. Mitrofanova E, Unfer R, Vahanian N, Link C. Rat sodium iodide symporter allows using lower dose of ^{131}I for cancer therapy. *Gene Ther* 2006;13(13): 1052–6.

102. Cho JY, Shen DH, Yang W et al. *In vivo* imaging and radioiodine therapy following sodium iodide symporter gene transfer in animal model of intracerebral gliomas. *Gene Ther* 2002;9:1139–1145.

103. Rahn JJ, Dabbagh L, Pasdar M, Hugh JC. The importance of MUC1 cellular localization in patients with breast carcinoma: an immunohistologic study of 71 patients and review of the literature. *Cancer* 2001;91:1973–1982.

104. Friedman EL, Hayes DF, Kufe DW. Reactivity of monoclonal antibody DF3 with a high molecular weight antigen expressed in human ovarian carcinomas. *Cancer Res* 1986;46:5189–5194.

105. Wang L, Ma J, Liu F et al. Expression of MUC1 in primary and metastatic human epithelial ovarian cancer and its therapeutic significance. *Gynecol Oncol* 2007; 105:695–702.

106. Burdick MD, Harris A, Reid CJ, Iwamura T, Hollingsworth MA. Oligosaccharides expressed on MUC1 produced by pancreatic and colon tumor cell lines. *J Biol Chem* 1997;272:24198–24202.

107. Qu CF, Li Y, Song YJ et al. MUC1 expression in primary and metastatic pancreatic cancer cells for *in vitro* treatment by [213]Bi-C595 radioimmunoconjugate. *Br J Cancer* 2004;91:2086–2093.

108. Baldus SE, Monig SP, Huxel S et al. MUC1 and nuclear beta-catenin are coexpressed at the invasion front of colorectal carcinomas and are both correlated with tumor prognosis. *Clin Cancer Res* 2004;10:2790–2796.

109. Chen L, Liu Q, Qin R et al. Amplification and functional characterization of MUC1 promoter and gene-virotherapy via a targeting adenoviral vector expressing hSSTR2 gene in MUC1-positive Panc-1 pancreatic cancer cells *in vitro*. *Int J Mol Med* 2005;15:617–626.

110. Dwyer RM, Bergert ER, O'Connor MK, Gendler SJ, Morris JC. *In vivo* radioiodide imaging and treatment of breast cancer xenografts after MUC1-driven expression of the sodium iodide symporter. *Clin Cancer Res* 2005;11:1483–1489.

111. Gupta VK, Park JO, Kurihara T et al. Selective gene expression using a DF3/MUC1 promoter in a human esophageal adenocarcinoma model. *Gene Ther* 2003;10:206–212.

112. Abe M, Kufe D. Characterization of *cis*-acting elements regulating transcription of the human DF3 breast carcinoma-associated antigen (MUC1) gene. *Proc Natl Acad Sci USA* 1993;90:282–286.

113. Mitrofanova E, Hagan C, Qi J, Seregina T, Link C, Jr. Sodium iodide symporter/ radioactive iodine system has more efficient antitumor effect in three-dimensional spheroids. *Anticancer Res* 2003;23:2397–2404.

114. Barton KN, Tyson D, Stricker H et al. GENIS: gene expression of sodium iodide symporter for noninvasive imaging of gene therapy vectors and quantification of gene expression *in vivo*. *Mol Ther* 2003;8:508–518.

115. Barton KN, Xia X, Yan H et al. A quantitative method for measuring gene expression magnitude and volume delivered by gene therapy vectors. *Mol Ther* 2004;9:625–631.

116. Siddiqui F, Barton KN, Stricker HJ et al. Design considerations for incorporating sodium iodide symporter reporter gene imaging into prostate cancer gene therapy trials. *Hum Gene Ther* 2007;18:312–322.

117. Grunwald F, Menzel C, Bender H et al. Redifferentiation therapy-induced radioiodine uptake in thyroid cancer. *J Nucl Med* 1998;39:1903–1906.

118. Kogai T, Schultz JJ, Johnson LS, Huang M, Brent GA. Retinoic acid induces sodium/ iodide symporter gene expression and radioiodide uptake in the MCF-7 breast cancer cell line. *Proc Natl Acad Sci USA* 2000;97:8519–8524.

119. Lim SJ, Paeng LC, Kim SJ, Kim SY, Lee H, Moon DH. Enhanced expression of adenovirus-mediated sodium iodide symporter gene in MCF-7 breast cancer cells with retinoic acid treatment. *J Nucl Med* 2007;48:398–404.

120. Das KC, White CW. Activation of NF-kappaB by antineoplastic agents. Role of protein kinase C. *J Biol Chem* 1997;272:14914–14920.

121. Sambucetti LC, Cherrington JM, Wilkinson GW, Mocarski ES. NF-kappa B activation of the cytomegalovirus enhancer is mediated by a viral transactivator and by T cell stimulation. *EMBO J* 1989;8:4251–4258.

122. Kim KI, Kang JH, Chung JK et al. Doxorubicin enhances the expression of transgene under control of the CMV promoter in anaplastic thyroid carcinoma cells. *J Nucl Med* 2007;48:1553–1561.

123. Myers RM, Greiner SM, Harvey ME et al. Preclinical pharmacology and toxicology of intravenous MV-NIS, an oncolytic measles virus administered with or without cyclophosphamide. *Clin Pharmacol Ther* 2007;82:700–710.

124. Cho JY, Xing S, Liu X et al. Expression and activity of human Na^+/I^- symporter in human glioma cells by adenovirus-mediated gene delivery. *Gene Ther* 2000;7:740–749.

125. Heltemes LM, Hagan CR, Mitrofanova EE, Panchal RG, Guo J, Link CJ. The rat sodium iodide symporter gene permits more effective radioisotope concentration than the human sodium iodide symporter gene in human and rodent cancer cells. *Cancer Gene Ther* 2003;10:14–22.

126. Adusumilli PS, Stiles BM, Chan MK et al. Radiation therapy potentiates effective oncolytic viral therapy in the treatment of lung cancer. *Ann Thorac Surg* 2005;80:409–416.

127. Mezhir JJ, Advani SJ, Smith KD et al. Ionizing radiation activates late herpes simplex virus 1 promoters via the p38 pathway in tumors treated with oncolytic viruses. *Cancer Res* 2005;65:9479–9484.

128. Wang YC, Hsu MT. Inhibition of initiation of simian virus 40 DNA replication during acute response of cells irradiated by ultraviolet light. *Nucleic Acids Res* 1996;24:3149–3157.

129. Dingli D, Peng KW, Harvey ME et al. Interaction of measles virus vectors with Auger electron emitting radioisotopes. *Biochem Biophy Res Commun* 2005;337:22–29.

130. Niu G, Gaut AW, Ponto LL et al. Multimodality noninvasive imaging of gene transfer using the human sodium iodide symporter. *J Nucl Med* 2004;45:445–449.

131. Groot-Wassink T, Aboagye EO, Wang Y, Lemoine NR, Reader AJ, Vassaux G. Quantitative imaging of Na/I symporter transgene expression using positron emission tomography in the living animal. *Mol Ther* 2004;9:436–442.

132. Groot-Wassink T, Aboagye EO, Glaser M, Lemoine NR, Vassaux G. Adenovirus biodistribution and noninvasive imaging of gene expression *in vivo* by positron emission tomography using human sodium/iodide symporter as reporter gene. *Hum Gene Ther* 2002;13:1723–1735.

133. Merron A, Peerlinck I, Martin-Duque P et al. SPECT/CT imaging of oncolytic adenovirus propagation in tumors *in vivo* using the Na/I symporter as a reporter gene. *Gene Ther* 2007;14:1731–1738.

134. Hasegawa K, Pham L, O'Connor MK, Federspiel MJ, Russell SJ, Peng KW. Dual therapy of ovarian cancer using measles viruses expressing carcinoembryonic antigen and sodium iodide symporter. *Clin Cancer Res* 2006;12:1868–1875.

135. Yang HS, Lee H, Kim SJ et al. Imaging of human sodium-iodide symporter gene expression mediated by recombinant adenovirus in skeletal muscle of living rats. *Eur J Nucl Med Mol Imaging* 2004;31:1304–1311.

136. Haberkorn U, Beuter P, Kubler W et al. Iodide kinetics and dosimetry *in vivo* after transfer of the human sodium iodide symporter gene in rat thyroid carcinoma cells. *J Nucl Med* 2004;45:827–833.

137. Shen DH, Marsee DK, Schaap J et al. Effects of dose, intervention time, and radionuclide on sodium iodide symporter (NIS)-targeted radionuclide therapy. *Gene Ther* 2004;11:161–169.

138. Jeon YH, Choi Y, Kim HJ et al. Human sodium iodide symporter gene adjunctive radiotherapy to enhance the preventive effect of hMUC1 DNA vaccine. *Int J Cancer* 2007;12:1593–1599.

139. Dwyer RM, Schatz SM, Bergert ER et al. A preclinical large animal model of adenovirus-mediated expression of the sodium-iodide symporter for radioiodide imaging and therapy of locally recurrent prostate cancer. *Mol Ther* 2005;12:835–841.

140. Emami B, Lyman J, Brown A et al. Tolerance of normal tissue to therapeutic irradiation. *Int J Radiat Oncol Biol Phys* 1991;21:109–122.

141. Battey JF, Way JM, Corjay MH et al. Molecular cloning of the bombesin/gastrin-releasing peptide receptor from Swiss 3T3 cells. *Proc Natl Acad Sci USA* 1991;88:395–399.

142. Xiao D, Wang J, Hampton LL, Weber HC. The human gastrin-releasing peptide receptor gene structure, its tissue expression and promoter. *Gene* 2001;264:95–103.

143. Battey J, Wada E. Two distinct receptor subtypes for mammalian bombesin-like peptides. *Trends Neurosci* 1991;14:524–528.

144. Jensen RT, Battey JF, Spindel ER, Benya RV. International Union of Pharmacology. LXVIII. Mammalian bombesin receptors: nomenclature, distribution, pharmacology, signaling, and functions in normal and disease states. *Pharmacol Rev* 2008;60:1–42.

145. Mercer DW, Cross JM, Chang L, Lichtenberger LM. Bombesin prevents gastric injury in the rat: role of gastrin. *Dig Dis Sci* 1998;43:826–833.

146. Presti-Torres J, de Lima MN, Scalco FS et al. Impairments of social behavior and memory after neonatal gastrin-releasing peptide receptor blockade in rats: implications for an animal model of neurodevelopmental disorders. *Neuropharmacology* 2007;52:724–732.

147. Moody TW, Merali Z. Bombesin-like peptides and associated receptors within the brain: distribution and behavioral implications. *Peptides* 2004;25:511–520.

148. Del RM, De la FM. Stimulation of natural killer (NK) and antibody-dependent cellular cytotoxicity (ADCC) activities in murine leukocytes by bombesin-related peptides requires the presence of adherent cells. *Regul Pept* 1995;60:159–166.

149. De la FM, Del RM, Ferrandez MD, Hernanz A. Modulation of phagocytic function in murine peritoneal macrophages by bombesin, gastrin-releasing peptide and neuromedin C. *Immunology* 1991;73:205–211.

150. Li K, Nagalla SR, Spindel ER. A rhesus monkey model to characterize the role of gastrin-releasing peptide (GRP) in lung development. Evidence for stimulation of airway growth. *J Clin Invest.* 1994;94:1605–1615.

151. Lehy T, Puccio F. Influence of bombesin on gastrointestinal and pancreatic cell growth in adult and suckling animals. *Ann N Y Acad Sci* 1988;547:255–267.

152. Roelle S, Grosse R, Buech T, Chubanov V, Gudermann T. Essential role of Pyk2 and Src kinase activation in neuropeptide-induced proliferation of small cell lung cancer cells. *Oncogene* 2008;27:1737–1748.

153. Kang J, Ishola TA, Baregamian N et al. Bombesin induces angiogenesis and neuroblastoma growth. *Cancer Lett* 2007;253:273–281.

154. Zhang Q, Thomas SM, Xi S et al. SRC family kinases mediate epidermal growth factor receptor ligand cleavage, proliferation, and invasion of head and neck cancer cells. *Cancer Res* 2004;64:6166–6173.

155. Zheng R, Iwase A, Shen R et al. Neuropeptide-stimulated cell migration in prostate cancer cells is mediated by RhoA kinase signaling and inhibited by neutral endopeptidase. *Oncogene* 2006;25:5942–5952.

156. Sun B, Schally AV, Halmos G. The presence of receptors for bombesin/GRP and mRNA for three receptor subtypes in human ovarian epithelial cancers. *Regul Pept* 2000;90:77–84.

157. Toi-Scott M, Jones CL, Kane MA. Clinical correlates of bombesin-like peptide receptor subtype expression in human lung cancer cells. *Lung Cancer* 1996;15:341–354.

158. Sun B, Halmos G, Schally AV, Wang X, Martinez M. Presence of receptors for bombesin/gastrin-releasing peptide and mRNA for three receptor subtypes in human prostate cancers. *Prostate* 2000;42:295–303.

159. Bartholdi MF, Wu JM, Pu H, Troncoso P, Eden PA, Feldman RI. *In situ* hybridization for gastrin-releasing peptide receptor (GRP receptor) expression in prostatic carcinoma. *Int J Cancer* 1998;79:82–90.

160. Gugger M, Reubi JC. Gastrin-releasing peptide receptors in non-neoplastic and neoplastic human breast. *Am J Pathol* 1999;155:2067–2076.

161. Baumann CD, Meurer L, Roesler R, Schwartsmann G. Gastrin-releasing peptide receptor expression in cervical cancer. *Oncology* 2007;73:340–345.

162. Tang C, Biemond I, Offerhaus GJ, Verspaget W, Lamers CB. Expression of receptors for gut peptides in human pancreatic adenocarcinoma and tumor-free pancreas. *Br J Cancer* 1997;75:1467–1473.

163. Preston SR, Woodhouse LF, Jones-Blackett S, Miller GV, Primrose JN. High-affinity binding sites for gastrin-releasing peptide on human colorectal cancer tissue but not uninvolved mucosa. *Br J Cancer* 1995;71:1087–1089.

164. Scott N, Millward E, Cartwright EJ, Preston SR, Coletta PL. Gastrin releasing peptide and gastrin releasing peptide receptor expression in gastrointestinal carcinoid tumors. *J Clin Pathol* 2004;57:189–192.

165. Carroll RE, Matkowskyj KA, Chakrabarti S, McDonald TJ, Benya RV. Aberrant expression of gastrin-releasing peptide and its receptor by well-differentiated colon cancers in humans. *Am J Physiol* 1999;276:G655–G665.

166. Carroll RE, Carroll R, Benya RV. Characterization of gastrin-releasing peptide receptors aberrantly expressed by non-antral gastric adenocarcinomas. *Peptides* 1999;20:229–237.

167. Faintuch BL, Teodoro R, Duatti A, Muramoto E, Faintuch S, Smith CJ. Radiolabeled bombesin analogs for prostate cancer diagnosis: preclinical studies. *Nucl Med Biol* 2008;35:401–411.

168. Zhang X, Cai W, Cao F et al. ^{18}F-labeled bombesin analogs for targeting GRP receptor-expressing prostate cancer. *J Nucl Med* 2006;47:492–501.

169. Garcia GE, Schweinsberg C, Maes V et al. New [99mTc]bombesin analogues with improved biodistribution for targeting gastrin releasing-peptide receptor-positive tumors. *Q J Nucl Med Mol Imaging* 2007;51:42–50.

170. Yang YS, Zhang X, Xiong Z, Chen X. Comparative *in vitro* and *in vivo* evaluation of two ^{64}Cu-labeled bombesin analogs in a mouse model of human prostate adenocarcinoma. *Nucl Med Biol* 2006;33:371–380.

171. Van de WC, Phonteyne P, Pauwels P et al. Gastrin-releasing peptide receptor imaging in human breast carcinoma versus immunohistochemistry. *J Nucl Med* 2008;49: 260–264.

172. De VG, Scopinaro F, Varvarigou A et al. Phase I trial of technetium [Leu^{13}] bombesin as cancer seeking agent: possible scintigraphic guide for surgery? *Tumori* 2002;88: S28–S30.

173. Scopinaro F, Di Santo GP, Tofani A et al. Fast cancer uptake of ^{99m}Tc-labelled bombesin (^{99m}Tc BN1). *In Vivo* 2005;19:1071–1076.

174. Rogers BE, Rosenfeld ME, Khazaeli MB et al. Localization of iodine-125-mIP-Des-Met^{14}-bombesin(7-13)NH$_2$ in ovarian carcinoma induced to express the gastrin releasing peptide receptor by adenoviral vector-mediated gene transfer. *J Nucl Med* 1997;38: 1221–1229.

175. Rosenfeld ME, Rogers BE, Khazaeli MB et al. Adenoviral-mediated delivery of gastrin-releasing peptide receptor results in specific tumor localization of a bombesin analogue *in vivo*. *Clin Cancer Res* 1997;3:1187–1194.

176. Rogers BE, Curiel DT, Mayo MS, Laffoon KK, Bright SJ, Buchsbaum DJ. Tumor localization of a radiolabeled bombesin analogue in mice bearing human ovarian tumors induced to express the gastrin-releasing peptide receptor by an adenoviral vector. *Cancer* 1997;80 (Suppl.): 2419–2424.

177. Stackhouse MA, Buchsbaum DJ, Kancharla SR et al. Specific membrane receptor gene expression targeted with radiolabeled peptide employing the erbB-2 and DF3 promoter elements in adenoviral vectors. *Cancer Gene Ther* 1999;6:209–219.

178. Deshane J, Siegal GP, Alvarez RD et al. Targeted tumor killing via an intracellular antibody against erbB-2. *J Clin Invest* 1995;96:2980–2989.

179. Cescato R, Maina T, Nock B et al. Bombesin receptor antagonists may be preferable to agonists for tumor targeting. *J Nucl Med* 2008;49:318–326.

180. Bruss M, Hammermann R, Brimijoin S, Bonisch H. Antipeptide antibodies confirm the topology of the human norepinephrine transporter. *J Biol Chem* 1995;270:9197–9201.

181. Pacholczyk T, Blakely RD, Amara SG. Expression cloning of a cocaine- and antidepressant-sensitive human noradrenaline transporter. *Nature* 1991;350:350–354.

182. Donnan GA, Kaczmarczyk SJ, Paxinos G et al. Distribution of catecholamine uptake sites in human brain as determined by quantitative [^{3}H] mazindol autoradiography. *J Comp Neurol* 1991;304:419–434.

183. Mash DC, Ouyang Q, Qin Y, Pablo J. Norepinephrine transporter immunoblotting and radioligand binding in cocaine abusers. *J Neurosci Methods* 2005;143:79–85.

184. Wehrwein EA, Parker LM, Wright AA et al. Cardiac norepinephrine transporter (NET) protein expression is inversely correlated to chamber norepinephrine content. *Am J Physiol Regul Integr Comp Physiol* 2008;295:R857–R863.

185. Tseng YT, Padbury JF. Expression of a pulmonary endothelial norepinephrine transporter. *J Neural Transm* 1998;105:1187–1191.

186. Bottalico B, Larsson I, Brodszki J et al. Norepinephrine transporter (NET), serotonin transporter (SERT), vesicular monoamine transporter (VMAT2) and organic cation transporters (OCT1, 2 and EMT) in human placenta from pre-eclamptic and normotensive pregnancies. *Placenta* 2004;25:518–529.

187. Bonisch H, Bruss M. The norepinephrine transporter in physiology and disease. *Handbook Exp Pharmacol* 2006;175:485–524.

188. Eisenhofer G. The role of neuronal and extraneuronal plasma membrane transporters in the inactivation of peripheral catecholamines. *Pharmacol Ther* 2001;91:35–62.

189. Hoefnagel CA, Voute PA, de KJ, Marcuse HR. Radionuclide diagnosis and therapy of neural crest tumors using iodine-131 metaiodobenzylguanidine. *J Nucl Med* 1987;28:308–314.

190. Binderup T, Knigge U, Mellon MA, Palnaes HC, Kjaer A. Quantitative gene expression of somatostatin receptors and noradrenaline transporter underlying scintigraphic results in patients with neuroendocrine tumors. *Neuroendocrinology* 2008;87:223–232.

191. Hopfner M, Sutter AP, Beck NI et al. Meta-iodobenzylguanidine induces growth inhibition and apoptosis of neuroendocrine gastrointestinal tumor cells. *Int J Cancer* 2002;101:210–216.

192. Hopfner M, Sutter AP, Huether A, hnert-Hilger G, Scherubl H. A novel approach in the treatment of neuroendocrine gastrointestinal tumors: additive antiproliferative effects of interferon-gamma and meta-iodobenzylguanidine. *BMC Cancer* 2004;4:23.

193. Anton M, Wagner B, Haubner R et al. Use of the norepinephrine transporter as a reporter gene for non-invasive imaging of genetically modified cells. *J Gene Med* 2004;6:119–126.

194. Moroz MA, Serganova I, Zanzonico P et al. Imaging hNET reporter gene expression with ^{124}I-MIBG. *J Nucl Med* 2007;48:827–836.

195. Buursma AR, Beerens AM, de Vries EF et al. The human norepinephrine transporter in combination with ^{11}C-*m*-hydroxyephedrine as a reporter gene/reporter probe for PET of gene therapy. *J Nucl Med* 2005;46:2068–2075.

196. Sealfon SC, Olanow CW. Dopamine receptors: from structure to behavior. *Trends Neurosci* 2000;23 (Suppl.): S34–S40.

197. Baik JH, Picetti R, Saiardi A et al. Parkinsonian-like locomotor impairment in mice lacking dopamine D2 receptors. *Nature* 1995;377:424–428.

198. Ricci A, Mignini F, Tomassoni D, Amenta F. Dopamine receptor subtypes in the human pulmonary arterial tree. *Auton Autacoid Pharmacol* 2006;26:361–369.

199. Hussain T, Lokhandwala MF. Renal dopamine receptors and hypertension. *Exp Biol Med* 2003;228:134–142.

200. Liang Q, Gotts J, Satyamurthy N et al. Noninvasive, repetitive, quantitative measurement of gene expression from a bicistronic message by positron emission tomography, following gene transfer with adenovirus. *Mol Ther* 2002;6:73–82.

201. Yaghoubi SS, Wu L, Liang Q et al. Direct correlation between positron emission tomographic images of two reporter genes delivered by two distinct adenoviral vectors. *Gene Ther* 2001;8:1072–1080.

202. Gambhir SS, Bauer E, Black ME et al. A mutant herpes simplex virus type 1 thymidine kinase reporter gene shows improved sensitivity for imaging reporter gene expression with positron emission tomography. *Proc Natl Acad Sci USA* 2000;97:2785–2790.

203. Ikari H, Zhang L, Chernak JM et al. Adenovirus-mediated gene transfer of dopamine D2 receptor cDNA into rat striatum. *Brain Res* Mol. *Brain Res* 1995;34:315–320.

204. Ogawa O, Umegaki H, Ishiwata K et al. *In vivo* imaging of adenovirus-mediated over-expression of dopamine D2 receptors in rat striatum by positron emission tomography. *Neuroreport* 2000;11:743–748.

205. Umegaki H, Ishiwata K, Ogawa O et al. *In vivo* assessment of adenoviral vector-mediated gene expression of dopamine D(2) receptors in the rat striatum by positron emission tomography. *Synapse* 2002;43:195–200.

206. Umegaki H, Ishiwata K, Ogawa O et al. Longitudinal follow-up study of adenoviral vector-mediated gene transfer of dopamine D2 receptors in the striatum in young, middle-aged, and aged rats: a positron emission tomography study. *Neuroscience* 2003;121:479–486.

207. Auricchio A, Acton PD, Hildinger M et al. *In vivo* quantitative noninvasive imaging of gene transfer by single-photon emission computerized tomography. *Hum Gene Ther* 2003;14:255–261.

208. Lamfers ML, Idema S, Bosscher L et al. Differential effects of combined Ad5- delta 24RGD and radiation therapy in *in vitro* versus *in vivo* models of malignant glioma. *Clin Cancer Res* 2007;13:7451–7458.

209. Geoerger B, Grill J, Opolon P et al. Potentiation of radiation therapy by the oncolytic adenovirus dl1520 (ONYX-015) in human malignant glioma xenografts. *Br J Cancer* 2003;89:577–584.

210. Chen Y, DeWeese T, Dilley J et al. CV706, a prostate cancer-specific adenovirus variant, in combination with radiotherapy produces synergistic antitumor efficacy without increasing toxicity. *Cancer Res* 2001;61:5453–5460.

211. Bieler A, Mantwill K, Holzmuller R et al. Impact of radiation therapy on the oncolytic adenovirus dl520: implications on the treatment of glioblastoma. *Radiother Oncol* 2008;86:419–427.

212. Idema S, Lamfers ML, van BV et al. AdDelta24 and the p53-expressing variant AdDelta24-p53 achieve potent anti-tumor activity in glioma when combined with radiotherapy. *J Gene Med* 2007;9:1046–1056.

213. Adusumilli PS, Chan MK, Hezel M et al. Radiation-induced cellular DNA damage repair response enhances viral gene therapy efficacy in the treatment of malignant pleural mesothelioma. *Ann Surg Oncol* 2007;14:258–269.

214. Lamfers ML, Grill J, Dirven CM et al. Potential of the conditionally replicative adenovirus Ad5-delta24RGD in the treatment of malignant gliomas and its enhanced effect with radiotherapy. *Cancer Res* 2002;62:5736–5742.

215. Maxon HR, Thomas SR, Hertzberg Vs et al. Relation between effective radiation dose and outcome of radioiodine therapy for thyroid cancer. *N Engl J Med* 1983;309:937–41.

CHAPTER 11

Preclinical Cell and Tumor Models for Evaluating Radiopharmaceuticals in Oncology

ANN F. CHAMBERS, EVA A. TURLEY, JOHN LEWIS, AND LEONARD G. LUYT

11.1 INTRODUCTION

In the United States in 2008, it was estimated that 1.4 million new cases of cancer would be diagnosed, and there would be nearly 600,000 deaths due to the disease (1). For many cancer types, the majority of these deaths will be due to metastases—the spread of cancer to distant, vital organs—rather than to the effects of the primary tumor. Patients whose cancers are detected early, when the tumor is localized, have a much better prognosis than patients whose cancers are detected with regional spread, and both do much better than patients whose cancers are diagnosed with distant metastases (1). For example, women with invasive breast cancer that has not spread to the lymph nodes have a 5-year relative survival rate of 98%. However, if the cancer has spread to regional lymph nodes, this rate drops to 84%, and if the disease has metastasized to distant sites the 5-year survival rate is only 27% (1). Similar trends are seen in other cancer types. Thus, early detection must be a key component of improving survival from cancer.

Fortunately, recent advances in cancer detection, coupled with increased availability of cancer screening, are leading to a shift toward earlier detection for many cancer types. For example, recent improvements in survival from breast cancer are thought to be attributable to a combination of detection of earlier, smaller tumors, through organized breast screening programs, coupled with improvements in treatment (2). However, many tumors are still detected only after metastasis has occurred, when survivability is poor, and these rates vary among tumor types (1). As an example, while only ~6% of breast cancers are diagnosed after metastatic spread, diagnosis after the tumor has already spread to distant organs occurs much more frequently for

Monoclonal Antibody and Peptide-Targeted Radiotherapy of Cancer, Edited by Raymond M. Reilly
Copyright © 2010 John Wiley & Sons, Inc.

cancers of the lung, ovary, and pancreas (41%, 68%, and 52%, respectively) (1). In order to improve cancer survival rates, there continues to be a pressing clinical need for early detection of smaller primary tumors for many cancer types, to enable increasingly effective treatments to be used on more treatable cancers. For patients whose tumors have metastasized to distant organs, improved treatment options that are more effective (as well as less toxic) are urgently needed.

Considerable experimental work has clarified many aspects of the process of metastasis (3–7). Some of this work has led to an understanding that cancer cells in secondary sites may coexist in different physiological states: actively growing, vascularized metastases, preangiogenic micrometastases that are active but "dormant" due to a balance between cell division and apoptosis (8), and dormant cells that are resistant to chemotherapy that targets actively dividing cells (3, 4). These distinct cellular states represent therapeutic targets (3, 9–12). For many cancer patients, metastases can arise years (or even decades) after apparently successful treatment of the primary cancer (10, 13). Clinical trials are suggesting that long-term therapy, for example in hormone-responsive breast cancer, may have clinical benefit (14), consistent with the presence of persistent subclinical cancer capable of leading to late recurrences. Minimal residual disease can be monitored by detection of micrometastases present in bone marrow, which may represent a poor prognostic indicator, although the clinical significance of cancer cells detected in bone marrow is still controversial (15, 16). Similarly, while lymph node positivity is a known prognostic indicator, the clinical significance of micrometastatic disease detected in lymph nodes at cancer surgery remains uncertain (17, 18). In addition, these assessments require invasive procedures. Targeted radiopharmaceuticals may offer a less invasive way to detect and potentially treat micrometastatic tumor burden. Once micrometastatic disease is detected, treatment dilemmas may arise in that the significance of the presence of minimal disease may not be known, and if it is deemed to signify poor prognosis, there may be relatively few effective treatment options. Detection of metastases at increasingly earlier stages, when tumors are very small, thus must be accompanied by improvements in understanding the clinical significance of this disease, coupled with improvements in treatment options for early stage metastatic disease. Development of targeted radiopharmaceuticals for the detection of small, hidden lesions, coupled with targeted treatment including targeted radiotherapy, will provide the tools necessary for the clinical assessment of the significance of these lesions.

One key component required for the development of effective targeted diagnostic or therapeutic radiopharmaceutical agents is the need for preclinical testing. In clinical drug development, compounds are often tested against primary tumors established in mice, by for example subcutaneous (ectopic and perhaps inappropriate, for most tumor types) or orthotopic tumor cell inoculation (e.g., mammary fat pad for breast tumors) (19). However, the location in which a tumor grows has been shown to dramatically affect its behavior, malignancy, and even drug responsiveness (20–23). Thus, for therapies to be effective in targeting metastases, it is appropriate that they be assessed preclinically in metastasis models (3, 5, 11, 24). Here we will consider some current approaches to the development and use of targeted radiotherapies in oncology and discuss a variety of approaches using mouse models, which may be useful in the preclinical assessment of therapeutic radiopharmaceuticals.

11.2 TRADITIONAL APPROACHES TO PRECLINICAL EVALUATION OF RADIOTHERAPEUTICS

The primary specification in creating a radiotherapeutic entity is the ability to selectively deliver the radionuclide to the tumor cell. Of nearly equal importance is limiting the uptake of the radionuclide in normal tissues, thereby minimizing toxicity. The preclinical criteria used to evaluate the effectiveness of a novel radio-therapeutic include high tumor uptake, clearance from the blood and minimal uptake in other organs. Of significant concern is the radiation dose to bone marrow, where the tolerated dose is the least among normal tissues, and marrow toxicity is typically dose-limiting for radiolabeled antibodies (25, 26). For smaller peptide derived radiotherapeutics, the radiosensitive kidney is often dose limiting due to tubular reabsorption of peptides that contain positively charged residues (27).

In this chapter, we will be focusing on preclinical animal models for evaluating cancer radiotherapeutics following optimal radiochemical design. Prior to preclinical testing however, there are a number of other biological evaluation criteria that need to be met. The designed peptide or antibody-based radiotherapeutic must demonstrate specificity for the intended target, whether that be an epitope, growth factor receptor, metabolic target, or other. A new radiotherapeutic entity should also be tested for stability in serum to demonstrate that it will have a sufficient biological half-life to allow the intact molecule to reach the tumor target *in vivo*. It is also crucial to ensure that the radionuclide remains stably attached to the targeting molecule and this mandates the wise choice of strong chelators when using radiometals (28–31) or stable radioiodination chemistry (32) (see Chapter 2). Should the radionuclide be released from the targeting molecule *in vivo*, then rapid clearance from the body is imperative to prevent redistribution and toxicity to normal tissues.

Another useful approach to evaluate a radiotherapeutic is to determine the effect of the agent at the cellular level, which is especially prudent if using a radiopharmaceu-tical that emits ultrashort range radiation such as Auger electrons and therefore requires delivery of the therapeutic to the cell nucleus in order to be effective (see Chapter 9). For example, Reilly and coworkers investigated the cytotoxicity of ^{111}In-DTPA-hEGF (human epidermal growth factor) for a number of breast cancer cell lines and correlated the radiation sensitivity of the cells to their EGF receptor expression (33). Once the capability of the radiotherapeutic to destroy tumor cells is demonstrated, one can then proceed with confidence to *in vivo* evaluation in preclinical models for determination of normal tissue distribution and pharmacokinetics. When ^{111}In-DTPA-hEGF was evaluated in such models, it was discovered that blood clearance was fast and although initial liver uptake was high at 41% ID after 1 h, clearance did occur over 72 h. It was also noted that red and white blood cell counts as well as other cells remained within the normal range at high administered doses (34). Further preclinical evaluation was then carried out in mice with breast cancer xenografts, demonstrating strong antitumor effects (35). Determining the toxicology profile of a radiotherapeutic is critical prior to proceeding to human clinical trials and is a requirement for regulatory approval for these trials.

For the treatment of metastasis, there is currently no effective targeted therapeutic with generalized applicability, which is in contrast to diagnostic oncological imaging, where for example ^{18}F-FDG is a generalized approach. Targeted radiotherapeutics in theory have the capacity to reach sites of metastasis and could be especially effective for small tumor sizes. The development of radiotherapeutics for metastatic disease however will likely be dependent upon the location of the metastatic disease as well as upon the specific antigen/receptors present on the tumor cells. For effective preclinical evaluation of radiotherapeutics, it would therefore be best to utilize animal models that properly mimic metastatic disease. This is in contrast to many published preclinical evaluations, where subcutaneous tumor models are used. Exceptions include a report by Breeman et al. who evaluated ^{177}Lu-DOTA-octreotate in a rat model of micro-metastatic CA20948 liver cancer and found a significant increase in survival compared to untreated animals (36).

A standard approach to evaluating the selectivity of a radiopharmaceutical for tumor versus normal tissues is using biodistribution analysis. In this manner, the percent injected dose of the agent per gram of tissue is calculated and a comparison between uptake in the tumor and normal tissues can be made. Although this may be accomplished for an α- or β-particle-emitting radionuclide by liquid scintillation counting, it is often more convenient to evaluate a new radiotherapeutic by preparing a surrogate compound. For *in vitro* evaluation, the surrogate could be the identical compound labeled with a nonradioactive metal, for example, using natural $^{185/187}$Re in place of ^{188}Re or using stable ^{89}Y in place of ^{90}Y. These nonradioactive entities are most useful for evaluation of target binding affinity, but are not typically relevant for *in vivo* evaluation. For *in vivo* evaluation, the surrogate compound could be made with the particle-emitting isotope being replaced by the corresponding γ-emitter, in order to facilitate *in vivo* measurement by γ-scintillation counting or by external imaging (e.g., SPECT or if using a positron-emitting radionuclide, PET). For example, the γ-emitter, ^{99m}Tc could be used in place of the β-emitter, ^{188}Re. Although this is replacing the rhenium with a technetium element, it is documented that the *in vivo* behavior of these coordinated metals is quite similar (37, 38). Another example is to use the γ-emitter, ^{111}In in place of the therapeutic radionuclides ^{90}Y or ^{177}Lu (29, 39–41), although differences in the stability of the ^{111}In and ^{90}Y analogues may confound results (see Chapter 2). Imaging with the corresponding γ-emitter allows estimation of dosimetry data prior to administration of the therapeutic analogue and is also useful for preclinical evaluation of biodistribution by means of small animal imaging.

To allow for evaluation of targeted radiotherapy of both primary and metastatic disease, the survival of treated versus untreated tumor-bearing animals can be compared. Dramatic increases in survival for animals with subcutaneous tumors have been well documented in preclinical studies (42–45). Also notable is that some reports indicate a maximal tolerated dose (MTD), after which animal survival for the treated group is actually lower than that of the control. For example, in one study, an

TABLE 11.1   The Effect of ^{177}Lu-DOTA-Octreotate on the Establishment of CA20948 Liver Metastases Indicating the Number of Animals Having the Given Range of Tumor Lesions

	No. of Tumor Colonies			
	0	1–50	51–100	>100
Untreated	0	0	0	6
185 MBq[a]	1	2	3	0
370 MBq	2	4	0	0

From Ref. (36).

[a] ^{177}Lu-DOTA-Octreotate administered i.v. as a single dose 8 days after direct injection of CA20948 tumor cells into the portal vein; each group consists of six rats with tumor colonies counted postsacrifice at day 21.

^{90}Y-labeled ChL6 antibody targeting prostate cancer was evaluated in mice bearing PC-3 xenografts (46). A radiotherapeutic dose of 150 µCi gave 100% survival over the 84 day study period, while the untreated animals had only 18% survival over this same period. In contrast, a higher dose of the radioimmunotherapeutic (250 µCi) resulted in early toxicity with 40% of the animals surviving only up to day 21, while 100% of those in all other groups still remained alive at that time point.

For a metastatic model where the metastases are localized to a specific organ, one is able to dissect the tissue and quantify the number of metastases. For example, in the instance of the previously described ^{177}Lu-DOTA octreotide, a significantly reduced number of metastases was found in the liver of rats treated with the radiotherapeutic (36). As depicted in Table 11.1, the untreated control animals all had greater than 100 tumor colonies with over half of the liver affected. In contrast, the treated animals showed less than 100 colonies at both 185 and 370 MBq administered doses.

While extending the life span of a tumor-bearing animal demonstrates the potential of a novel radiotherapeutic, one should question how well such data will translate to the human clinical situation. A number of direct preclinical–clinical comparisons of radiation dose estimates have been reported. The ^{90}Y-labeled antibody, ibritumomab tiuxetan (Zevalin) was evaluated in athymic mice bearing CD20+ human tumors and from the biodistribution data, the expected human radiation absorbed dose to key organs was estimated (47). When compared to human clinical trial data, it was determined that the dose estimated based on the preclinical study was equal or less than that determined clinically. However, the mouse biodistribution data resulted in a significant overestimation of the red marrow and kidney dose determined clinically. Another issue with translating preclinical dosimetry calculations to the clinical situation is that the influence of the emitted γ-photons will be different in a human than in experimental animal models, depending on the energy. Calculations have indicated that the absorbed dose rate between different species will be similar for the pure β-emitter, ^{90}Y and the β- and γ-emitter, ^{177}Lu, while large differences in photon irradiation are expected between species for ^{111}In, ^{125}I, and ^{67}Ga (48). The range of the emitted particles will influence the deposition of radiation in organs and tissues, hence using a small animal for evaluating a relatively long range β-particle emission (e.g., from ^{90}Y or ^{188}Re) will be less likely to extrapolate accurately to humans than radionuclides emitting shorter range particles (e.g., ^{177}Lu or ^{131}I) (see Chapter 13).

Recently, several clinical trials have evaluated the efficacy of simultaneous chemotherapy combined with a targeted radiotherapeutic. For example, Zinzani et al. reported on combined fludarabine and mitoxantrone in conjunction with ^{90}Y-labeled ibritumomab tiuxetan for treatment of patients with non-Hodgkin lymphoma achieving complete remission in all 20 patients evaluated (49). This combination of nontargeted chemotherapy with the tumor targeting radiotherapeutic appears to be a very promising treatment approach. As researchers move forward in evaluating novel peptide and antibody-based radiotherapeutics, it may prove valuable to utilize a preclinical model that is appropriate for this comparison. O'Donnell et al. reported on a preclinical evaluation of ^{90}Y-DOTA-ChL6 antibody targeting prostate cancer combined with taxanes and demonstrated a decrease in palpable tumors compared to the control animals (50). A ^{177}Lu-labeled bombesin derivative was investigated by Johnson and coworkers alone or in combination with docetaxel and/or estramustine in PC-3 xenograft bearing mice, with the combined treatment providing the greatest tumor growth suppression in this prostate cancer model (51). The synergistic effects and toxicity of such combination therapy, requires evaluation in animal models prior to clinical trials. A more effective measure of tumor response and associated normal tissue toxicity from combined therapy at the preclinical stage would be of great benefit to those engaged in the development of radiotherapeutic agents. While the literature describes preclinical evaluation of radiotherapeutics and demonstrates encouraging results, there still remains a need for more accurate animal testing methodology.

11.3 MODELS OF CANCER

A key factor in preclinical screening for effective imaging and therapeutic agents in the treatment of cancer, including metastases, is the selection of an appropriate model. Currently, models include the use of human tumor cell lines maintained in two-dimensional (2D) culture (i.e., cells grown in monolayer in tissue culture), 3D culture (i.e., cells grown as spheroids in suspension or embedded in extracellular matrix (ECM) such as collagen type I gels or more complex matrix gels including Matrigel™), or *in vivo* as orthotopic xenografts (52–56). In addition, a large number of genetically modified mouse models are available that are susceptible to neoplastic transformation (52, 57, 58). None of these cell lines or transgenic mouse models replicates all aspects of human cancers. Nonetheless and as noted above, both cell culture models and animal models have been valuable for initial high throughput screening to identify potential therapeutic or imaging agents for further development and for testing the effectiveness of these agents *in vivo*. In the following discussion, breast cancer will be used as an example of a complex and diverse neoplastic process, to examine the availability of cell and animal models that reflect the heterogeneity and subtype complexity of the disease.

11.3.1   Cancer as a Complex Collection of Neoplastic Diseases

Pathologists have long used histological criteria to sort breast tumors and other tumors into grades ranging from benign to aggressive disease. More recently, research has permitted further sorting of cancers based upon identified markers that predict disease outcome and response to targeted therapy. For example, in breast cancer, estrogen/progesterone (ER/PR) and Her2/neu receptor status are not only associated with clinical outcome but also predict response to treatment, for example, estrogen agonists (tamoxifen) and Her2/neu antagonists (Herceptin) (59, 60). Microarray-based analysis of gene expression (61) has more recently provided a molecular blueprint for the complexity of breast cancer as a group of diseases. This and other transcriptome analyses (for reviews see Refs 62–65) confirm that breast cancer can be divided into two broad groups of estrogen receptor positive and estrogen receptor negative tumors. However, these analyses reveal an additional complexity, permitting these two broad groups to be further subdivided into six major molecular subtypes: basal-like (Basal A), Basal B, luminal A, luminal B, Her2+/ER− and normal breast-like (Table 11.2). The molecular differences in these groups are associated not only with distinct clinical outcomes but are also differentially responsive to treatment. In addition to the classification of breast tumors into the above molecular subtypes, there is some evidence to suggest that breast cancers can arise from progenitor cells characterized by a specific phenotype (CD44+/CD24−/ESA+) (66–69). The origin of these "tumor progenitor" cells within breast tissue has not yet been identified, but a gene signature expressed by a subset of CD44+/CD24− primary breast tumor cells separated from other primary tumor cells by fluorescence activated cell sorting predicts poor clinical outcome (68, 70).

Clearly breast cancer is a complex grouping of neoplastic diseases. Identifying and treating these subtypes is a major challenge facing clinicians today while identifying or developing experimental models that replicate this complexity is a significant challenge for scientists to resolve. Since early diagnosis optimizes treatment outcome, developing imaging and therapeutic agents that are specific to transcriptome subtypes or to breast tumor initiating/progenitor cells would be an enormous step forward in the

TABLE 11.2   Human Breast Cancer Cell Lines and Transgenic Mice Grouped According to Tumor Molecular Subtypes

Tumor Subtype (Gene Cluster)	Tumor Cell Line[a]	Transgenic Mice
Nontumorigenic BaB	S1, MCF10A	−
Lu	MCF7; MDA-MB-361	
ERBB2	SKBR3; HCC1569	neu/her2
BaA	HCC1569; SUM190PT	neu/her2[b]
BaB	MDA-MB-231; HCC38	−

BaA: basal subtype A; BaB: basal subtype B; Lu: luminal subtype A; ERBB2: ERBB2 overexpressing; NB: normal breast like.

[a] Ref. 71.

[b] Ref. 114.

treatment of breast cancer as well as in other malignancies. In this section, the extent to which available models capture the diversity of breast cancer subtypes will be discussed.

11.3.2 Human Breast Cancer Cell Culture Models

A bank of human breast cancer cell lines are available that can be classified into each of the different subtypes of clinical disease based upon their resemblance to the aggressiveness, chemoresistance or molecular signatures of these subtypes (Table 11.2) (58, 71–74). To date, most studies using human breast cancer cell lines rely upon two-dimensional cell culture. Several recent reports have attempted to compare the molecular phenotype of these cell lines with those of primary breast tumors using genomic, transcriptome, and proteomic profiling (71, 72, 74, 75). Most of the commonly used human breast cancer cell lines can be grouped into the six molecular subtypes based upon the similarity of their transcriptome to primary human breast tumor subtypes (58, 71). Furthermore, genomic changes including amplification, high frequency of low level gains and losses observed in specific primary breast tumor subtypes are also retained in these breast cancer cell lines (74). Perhaps most importantly, in one study, the molecular signature of these cell lines can be used to successfully predict a response to targeted therapies for breast cancer such as trastuzumab (Herceptin) (71).

Breast cancer cell lines have also been characterized with progenitor properties and progenitor signatures. For example, recently developed MCF-15, HMT348, and EM-G3 breast epithelial cells possess bipotential characteristics since they are able to differentiate into myoepithelial and luminal phenotypes (58). Breast cancer cell lines characterized by molecular profiling, such as MDA-MB-231 cells, also exhibit a progenitor-like status based both on surface phenotyping and the ability to undergo differentiation into a luminal-like epithelium (76–79). Emerging studies also suggest that the transcriptome of the peritumor stroma also affects tumor progression (80–86). Currently, there are no therapeutic or imaging agents developed to target this tumor compartment but this may be an important therapeutic target for the future and will require development of peritumor stromal cell line banks (87). Collectively, these results indicate that currently available human breast cancer cell lines largely mirror the complexity of the clinical disease and that standard culture of these reasonably recapitulates subtype primary tumors both in their genetic makeup and their susceptibility to targeted therapy. However, virtually all of these studies noted that the transcriptomes of tumor cell lines do not perfectly match their primary breast tumor subtype counterpart. These differences are in part due to the restriction of tissue architecture that can be achieved in 2D cultures since the transcriptomes of tumor cells maintained in 3D culture more closely resemble those of primary tumors (54, 88, 89).

A number of studies have demonstrated that signals from the extracellular matrix and 3D architecture of tissues are necessary for the ability of mammary epithelial and other cell types to undergo morphogenesis. Since ECM and tissue architecture cues also affect cell transformation, development of 3D culture assays in matrices such as basement membranes may provide a more faithful replication of cellular responses *in vivo* than is observed in 2D culture (54, 89–92).

3D culture affects cellular responses to signaling inhibitors (93). For example, MDA-MB-231 cells grown in 3D reconstituted basement membrane extracellular matrix undergo phenotypic reversion when treated with β1-integrin and PI3kinase or ERK1/2 kinase inhibitors in this 3D but not in 2D cultures (79). Expression microarray analyses of such tumor cell lines maintained in 3D versus 2D culture have shown that signal transduction pathways are also integrated in a different manner. This was recently confirmed in large scale analysis of gene expression patterns in 25 breast cancer cell lines (94). Since targeted therapies are often directed at specific signaling pathways including peptide growth factor receptors, these results suggest that screening for targeted radiotherapeutics using 3D culture assays may be useful. Consistent with this, studies have shown that radiotherapeutics cause cell stress or death by different mechanisms in monolayer versus spheroid cultures (95, 96). Although most studies have not documented differential targeting in monolayer and spheroid cultures, in some cases (97–99) differential kill by radiotherapeutics has been noted (100). For example, ^{131}I-meta-iodobenzylguanidine (^{131}I-mIBG) more effectively kills noradrenaline transporter expressing glioblastoma cells grown as spheroids than as monolayers. This effect was attributed to more effective radiation cross fire (radiological bystander effect) in the 3D spheroid cultures (99). However, other studies have not observed a differential effect of radiotherapeutics on cell death in 2D versus 3D cultures (101). Even if kill rates are similar in these two culture modalities, cellular responses to radiotherapeutics are clearly different. Future studies using spheroid and other 3D culture methods will allow investigation of the more fundamental questions of how tissue architecture affects radiation-induced apoptotic cell death, cell-cycle events, cell–cell interactions, and cell adhesion phenomena (97). In addition, the use of spheroid cultures may more closely reproduce some of the delivery barriers imposed by tumors *in vivo* for the effective use of targeted radiotherapeutics (e.g., tumor penetration as well as the finite range of the emitted α- or β-particles in irradiating more distant cells).

11.3.3 Animal Models

Two broad groups of animal tumor models are currently used for preclinical assessment of imaging or radiotherapeutic agents: xenograft models and transgenic mouse models of disease susceptibility. Xenograft models offer the advantage of using human cell lines that have been well characterized with respect to their invasive/metastatic properties, subtype and molecular/signaling phenotype (58, 91), and are amenable for use in screening for both radiotherapeutics or imaging agents (53, 56, 102). Most importantly, as noted above, cell lines that resemble each of the molecularly defined subtypes of breast tumors are available for xenograft assays. Indeed several studies have identified gene signatures using human cancer cell lines (e.g., MDA-MB-231) grown as xenografts that predict poor outcome in the human disease (103–105). Furthermore, methods for growing primary tumor xenografts are now available (70). Xenografts can be either grown at subcutaneous or orthotopic sites depending on

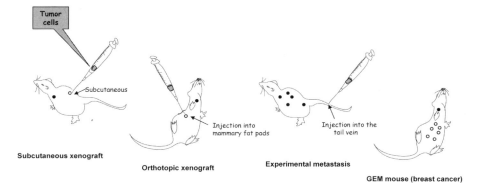

FIGURE 11.1   Preclinical mouse models of breast cancer. Commonly used preclinical models of breast cancer in mice include xenografts of human breast cancer cells grown as subcutaneous implants or orthotopically in immune-compromised mice. With subcutaneous and orthotopic xenografts, primary tumors may be produced, and cells may spontaneously metastasize to internal organs. With experimental metastasis assays, cells are delivered via the bloodstream to internal organs, where metastases may form in various organs. In contrast, in GEMs, primary tumors (often multifocal) arise spontaneously, and may (or may not) metastasize to various distant organs. Various GEM models of breast cancer susceptibility are also available, which can be chosen to match breast cancer molecular subtypes. See Refs 3 and 9 for further details. Open circles depict primary tumors while closed circles depict metastatic lesions.

requirements (19). Subcutaneous tumors are most often the easiest to image with radiopharmaceuticals and are easily measured for quantifying therapeutic response. However, orthotopic xenografts (e.g., breast tumor cells implanted into the mammary fat pads of mice) are more representative of primary tumors since they are growing in the appropriate tissue microenvironment (58). In addition, xenograft models have been developed that are able to metastasize thus offering the opportunity to study this process with human cells as well as evaluate the effectiveness of new targeted therapies including radiotherapeutics (Fig. 11.1). Furthermore, assays designed to quantify tissue colonization, which mimic (but are not an exact replica of) the metastatic process are possible using xenograft models (3, 19, 54, 106–108). However, human cells must be grown in immune-compromised mice and convincing evidence that immune cells affect tumor progression should be considered when using these models (86, 109, 110). In addition, the use of immune-compromised mice does not allow an examination of the immune response to targeted radiotherapeutics (i.e., to murine or humanized monoclonal antibodies; see Chapter 1) or to novel targeting vectors such as cytolytic viruses (see Chapter 10).

The development of transgenic mice that exhibit susceptibility to breast cancer and that resemble important aspects of the human disease (e.g., oncogenic mutations, tumor invasion, and metastasis) have become an important part of preclinical assessments of imaging and therapeutic agents (52, 58, 111–114). In particular, genetic engineering techniques permitting the induction/targeting of genetic modifications to specific tissues have greatly advanced the development of animal models of cancer susceptibility. A large number of transgenic mice now exist and many of these include on the one hand loss of tumor suppressor genes that are known to have significant roles in human breast cancer (e.g., Brca1, Trp53, Pten) and gain of function oncogenes such as Erbb2, Myc, Ccnd1, PyMT, and Hras (115).

Comparative analysis of mammary cancer in transgenic mice with human breast cancer reveals a number of differences. Overall, the histology of most tumors from mice does not resemble the common types of human breast cancers and most mouse tumors metastasize only to the lungs, contain less fibrosis and inflammation and are nearly all hormone independent in comparison to human breast tumors in which about half are hormone-dependent (54, 58). Nonetheless, some transgenic models demonstrate features of the human neoplastic disease, such as metastasis (116). However, in general most tumors from genetically modified mice do not resemble the subtypes of breast cancer. Despite these properties, genetically modified mice have provided a wealth of data about the molecular pathways that are involved in breast cancer. In particular the development of breast cancer in neu/erbb2 transgenic mice closely resembles neoplastic disease of the human ERBB2 subtype (114). In the last 20 years, a great deal of effort has been directed toward characterizing the similarities and differences between genetically modified mouse models of cancer and human cancers in which the oncogenic event is common to human breast cancers (e.g., BRCA1 loss, TP53 mutations, ERBB2 amplification). Analyses of gene expression suggests a significant number of genes are commonly de-regulated when these genetic modifications are present in mouse models or in human breast cancers. Indeed, recent transcriptome analyses of a number of transgenic mouse models were compared to the molecular profiles of human breast cancer subtypes (57, 58). Results from these studies confirmed earlier analyses that mouse tumors are more similar to each other than to human tumors. However, several models shared similarities with human breast cancer including representations of luminal (e.g., MMTV-Neu, MMTV-PyMT, Wap-mYc, and Wap-Int-3) and basal tumors (Wap-Tag and Brca1-deficient) (57, 58, 117). The combination of the histological profiling of transgenic mouse mammary tumors with the growing molecular characterization of the transcriptomes of these mice will provide a basis for choosing the best transgenic mouse model to match the subtype of human tumor to which the imaging or radiotherapeutic agents are targeted.

A general conclusion is that no single animal model of breast cancer susceptibility or human breast cancer cell line matches the complexity of breast cancer as a disease. This is also true for other cancers and therefore a number of models should ideally be used when screening or testing new imaging or radiotherapeutic agents.

11.4   ANIMAL MODELS FOR EVALUATING RADIOPHARMACEUTICALS: UNRESOLVED ISSUES AND CHALLENGES FOR TRANSLATION

Animal models used for the preclinical evaluation of radiopharmaceuticals, including their effectiveness in targeting and treating tumors as well as toxicity, have consisted of a variety of species such as mice, rats, rabbits, dogs, and monkeys. In addition to the experimental factors outlined above, practical factors must be considered when choosing an appropriate animal model. In order to obtain regulatory approval, the determination of efficacy is only part of the scientific workup that must be completed ahead of human clinical trials. In the case of a therapeutic radiopharmaceutical, the agent may be given several times to a patient, sometimes at high radioactivity doses. The efficacy and toxicity of the agent, with and without incorporation of the radionuclide, must be evaluated comprehensively at the preclinical stage (see Chapter 17). In a first step, the subacute toxicity of the unlabeled molecule should be determined, and the duration of treatment in these animal studies needs to be taken into account in relation to the expected maximum duration of treatment in humans. For a single administration of the radiopharmaceutical in humans, only an acute toxicity study with a 2-week follow up is usually needed, but this often needs to be done in a rodent as well as a nonrodent species. In a second step, the toxicological properties using the radiolabeled molecule should be performed, using a scheduled program of treatment adapted from the program of treatment proposed in humans to study potential normal organ radiation toxicity.

An area of increasing concern is being raised by reports that describe differences in radiopharmaceutical bioavailability between the various animal models described above, and more importantly, between these animal models and humans (118–121). For example, the extent to which a radiotherapeutic binds to serum proteins is a key factor, since protein binding may directly limit the rate, and magnitude, of radiopharmaceutical extravasation from blood into the tissues of interest. Interspecies variation in plasma protein binding may impact the reliability of animal models to predict radiopharmaceutical performance in humans. Protein binding issues are most likely to affect small radiolabeled molecules and peptides, rather than antibodies. It was recently reported that copper bis(thiosemicarbazone) complexes can exhibit significant, species-dependent, binding to serum albumin (118, 119). This phenomenon highlights the need both for evaluation and confirmation of similarity in biodistribution between several animal models and for establishing a rigorous and well-informed approach to translating animal data to human clinical trials.

In addition to carefully selecting one or more animal species for preclinical evaluation, the choice between xenograft or genetically engineered models (GEM) of cancer is an important one. With subcutaneous xenografts, the progression of a large number of synchronized, easily observable, and measurable tumors can be followed, such that initiation of treatment can begin when the tumors are of an optimal or desired size. Furthermore, xenografts have a relatively high degree of predictability and rapidity of tumor formation, and this translates to smaller standard errors and group sizes when performing efficacy studies (122). As detailed above, the primary shortcoming of xenografts is the absence of an immune system and the extent to which the tumor cell lines represent the tumor of origin (Table 11.2). This is manifested by qualitative differences in the resultant physical characteristics, genetics, and histology (123). GEMs often more closely resemble the genetic characteristics of the corresponding human cancer. From a practical standpoint, however, they have the disadvantages of heterogeneity with regard to frequency of initiation, latency, and growth rate (124, 125) that impacts the design and implementation of preclinical studies. To accommodate these challenges, new anatomic and molecular *in vivo* imaging techniques such as micro-CT, ultrasound (126–128), optical (129), MRI (107, 130, 131), and bioluminescence imaging (132) are being utilized to enable tumor growth to be followed over time. Given that GEMs provide some characteristics that are arguably more accurate models of cancer, they may be better suited to make correlations between the effectiveness of targeted radiotherapeutics and tumor biology, and are a promising alternative to traditional preclinical tumor xenograft models. However, mixed results have slowed the acceptance of these models as preclinical tools (124, 125). There are some additional factors that have the potential to impede clinical translation using GEM models. In 1988, the "OncoMouse" became the first animal approved for patent protection when a US patent was granted to Harvard University geneticist Philip Leder and Timothy Stewart of the University of California, San Francisco. The ruling was sufficiently broad in scope to include "not simply a transgenic mouse with an activated MYC gene; it is any transgenic mammal, excluding human beings, that contains in all its cells an activated oncogene that had been introduced into it or an ancestor at an embryonic stage" (133). DuPont subsequently licensed this patent and a sublicense must be obtained from DuPont in order to use transgenic mice in biomedical research. The European Patent Office restricted the DuPont patent to "transgenic mice" only, while the Supreme Court of Canada ruled that the OncoMouse cannot be patented. The NIH has reached an agreement with DuPont permitting use of transgenic mice for nonprofit research. Nonetheless, these patents have the potential to impede collaborative studies with industry (134–136).

Ultimately, a successful preclinical strategy for evaluating targeted radiotherapeutics and other radiopharmaceuticals will come down to its predictive utility. How effective is a particular model at selecting efficacious agents? Can it reliably predict on-target and off-target toxicities? What is the probability of failure or unpredictable toxicity when administered to humans for those agents that succeed in preclinical assays? A broad analysis of *in vitro* models and tumor xenografts done at the US National Cancer Institute found poor overall correlations between preclinical testing and therapeutic activity in Phase II clinical trials and generally concluded that only compounds that are successful in a large number of different animal models are likely to be effective in the clinic (125). In 1999, the US NIH Breast Cancer Think Tank prepared a report comparing the pathology of 39 mammary cancer GEMs and human breast cancers. The principle conclusion was that the histology of most GEM tumors did not resemble the histology of the common types of human breast cancer (58). Nevertheless, certain commonalities were identified by the think tank including evidence that tumor formation resulted from multiple genetic mutations, and that genes associated with human cancer similarly cause cancer in mice. These observations and the development of more sophisticated genetically engineered animal models suggest that correlations between GEM and human disease should improve. For the moment, however, an inescapable conclusion is that the use of multiple cell culture and animal models during the preclinical phase increases the probability of successful translation of radiopharmaceuticals to the clinic.

REFERENCES

1. Jemal A, Siegel R, Ward E, et al. Cancer statistics, 2008. *CA Cancer J Clin* 2008; 58:71–96.

2. Berry DA, Cronin KA, Plevritis SK, et al. Effect of screening and adjuvant therapy on mortality from breast cancer. *N Engl J Med* 2005;353:1784–1792.

3. Chambers AF, Groom AC, MacDonald IC. Dissemination and growth of cancer cells in metastatic sites. *Nat Rev Cancer* 2002;2:563–572.

4. Naumov GN, MacDonald IC, Chambers AF, Groom AC. Solitary cancer cells as a possible source of tumour dormancy? *Semin Cancer Biol* 2001;11:271–276.

5. Steeg PS, Theodorescu D. Metastasis: a therapeutic target for cancer. *Nat Clin Pract Oncol* 2008;5:206–219.

6. Duffy MJ, McGowan PM, Gallagher WM. Cancer invasion and metastasis: changing views. *J Pathol* 2008;214:283–293.

7. Taylor J, Hickson J, Lotan T, Yamada DS, Rinker-Schaeffer C. Using metastasis suppressor proteins to dissect interactions among cancer cells and their microenvironment. *Cancer Metastasis Rev* 2008;27:67–73.

8. Holmgren L, O'Reilly MS, Folkman J. Dormancy of micrometastases: balanced proliferation and apoptosis in the presence of angiogenesis suppression. *Nat Med* 1995;1:149–153.

9. Barkan D, Kleinman H, Simmons JL, et al. Inhibition of metastatic outgrowth from single dormant tumor cells by targeting the cytoskeleton. *Cancer Res* 2008;68:6241–6250.

10. Brackstone M, Townson JL, Chambers AF. Tumour dormancy in breast cancer: an update. *Breast Cancer Res* 2007;9:208.

11. Hedley BD, Winquist E, Chambers AF. Therapeutic targets for antimetastatic therapy. *Expert Opin Ther Targets* 2004;8:527–536.

12. Naumov GN, Townson JL, MacDonald IC, et al. Ineffectiveness of doxorubicin treatment on solitary dormant mammary carcinoma cells or late-developing metastases. *Breast Cancer Res Treat* 2003;82:199–206.

13. Townson JL, Chambers AF. Dormancy of solitary metastatic cells. *Cell Cycle* 2006;5:1744–1750.

14. Gligorov J, Pritchard K, Goss P. Adjuvant and extended adjuvant use of aromatase inhibitors: reducing the risk of recurrence and distant metastasis. *Breast* 2007;16 (Suppl. 3): S1–S9.

15. Pantel K, Brakenhoff RH, Brandt B. Detection, clinical relevance and specific biological properties of disseminating tumour cells. *Nat Rev Cancer* 2008;8:329–340.

16. Riethdorf S, Wikman H, Pantel K. Review: biological relevance of disseminated tumor cells in cancer patients. *Int J Cancer* 2008;123:1991–2006.

17. Grabau D, Breast cancer patients with micrometastases only: is a basis provided for tailored treatment? *Surg Oncol* 2008;17:211–217.

18. Kahn HJ, Hanna WM, Chapman JA, et al. Biological significance of occult micrometastases in histologically negative axillary lymph nodes in breast cancer patients using the recent American Joint Committee on Cancer breast cancer staging system. *Breast J* 2006;12:294–301.

19. Welch DR. Technical considerations for studying cancer metastasis *in vivo*. *Clin Exp Metastasis* 1997;15:272–306.

20. Killion JJ, Radinsky R, Fidler IJ. Orthotopic models are necessary to predict therapy of transplantable tumors in mice. *Cancer Metastasis Rev* 1998;17:279–284.

21. Dong Z, Radinsky R, Fan D, et al. Organ-specific modulation of steady-state mdr gene expression and drug resistance in murine colon cancer cells. *J Natl Cancer Inst* 1994; 86:913–920.

22. Fidler IJ, Wilmanns C, Staroselsky A, Radinsky R, Dong Z, Fan D. Modulation of tumor cell response to chemotherapy by the organ environment. *Cancer Metastasis Rev* 1994; 13:209–222.

23. Wilmanns C, Fan D, O'Brian CA, Bucana CD, Fidler IJ. Orthotopic and ectopic organ environments differentially influence the sensitivity of murine colon carcinoma cells to doxorubicin and 5-fluorouracil. *Int J Cancer* 1992;52:98–104.

24. Chambers AF, Naumov GN, Vantyghem SA, Tuck AB. Molecular biology of breast cancer metastasis. Clinical implications of experimental studies on metastatic inefficiency. *Breast Cancer Res* 2000;2:400–407.

25. Sgouros G. Bone marrow dosimetry for radioimmunotherapy: theoretical considerations. *J Nucl Med* 1993;34:689–694.

26. Langmuir VK, Fowler JF, Knox SJ, Wessels BW, Sutherland RM, Wong JY. Radiobiology of radiolabeled antibody therapy as applied to tumor dosimetry. *Med Phys* 1993; 20:601–610.

27. Behr TM, Goldenberg DM, Becker W. Reducing the renal uptake of radiolabeled antibody fragments and peptides for diagnosis and therapy: present status, future prospects and limitations. *Eur J Nucl Med* 1998;25:201–212.

28. Stimmel JB, Stockstill ME, Kull FC, Jr. Yttrium-90 chelation properties of tetraazatetraacetic acid macrocycles, diethylenetriaminepentaacetic acid analogues, and a novel terpyridine acyclic chelator. *Bioconjug Chem* 1995;6:219–225.

29. Liu S. The role of coordination chemistry in the development of target-specific radiopharmaceuticals. *Chem Soc Rev* 2004;33:445–461.

30. Liu S, Edwards DS. Bifunctional chelators for therapeutic lanthanide radiopharmaceuticals. *Bioconjug Chem* 2001;12:7–34.

31. Li WP, Ma DS, Higginbotham C, et al. Development of an *in vitro* model for assessing the *in vivo* stability of lanthanide chelates. *Nucl Med Biol* 2001;28:145–154.

32. Wilbur DS. Radiohalogenation of proteins: an overview of radionuclides, labeling methods, and reagents for conjugate labeling. *Bioconjug Chem* 1992;3:433–470.

33. Cai Z, Chen Z, Bailey KE, Scollard DA, Reilly RM, Vallis KA. Relationship between induction of phosphorylated H2AX and survival in breast cancer cells exposed to ^{111}In-DTPA-hEGF. *J Nucl Med* 2008;49:1353–1361.

34. Reilly RM, Chen P, Wang J, Scollard D, Cameron R, Vallis KA. Preclinical pharmacokinetic, biodistribution, toxicology, and dosimetry studies of ^{111}In-DTPA-human epidermal growth factor: an auger electron-emitting radiotherapeutic agent for epidermal growth factor receptor-positive breast cancer. *J Nucl Med* 2006; 47:1023–1031.

35. Chen P, Cameron R, Wang J, Vallis KA, Reilly RM. Antitumor effects and normal tissue toxicity of ^{111}In-labeled epidermal growth factor administered to athymic mice bearing

epidermal growth factor receptor-positive human breast cancer xenografts. *J Nucl Med* 2003;44:1469–1478.

36. Breeman WA, Mearadji A, Capello A, et al. Anti-tumor effect and increased survival after treatment with [^{177}Lu-DOTA0,Tyr3]octreotate in a rat liver micrometastases model. *Int J Cancer* 2003;104:376–379.

37. Deutsch E, Libson K, Vanderheyden JL, Ketring AR, Maxon HR. The chemistry of rhenium and technetium as related to the use of isotopes of these elements in therapeutic and diagnostic nuclear medicine. *Int J Rad Appl Instrum B* 1986;13:465–477.

38. Mindt T, Struthers H, Garcia-Garayoa E, Desbouis D, Schibli R. Strategies for the development of novel tumor targeting technetium and rhenium radiopharmaceuticals. *CHIMIA* 2007;61:725–731.

39. Vallabhajosula S, Kuji I, Hamacher KA, et al. Pharmacokinetics and biodistribution of ^{111}In- and ^{177}Lu-labeled J591 antibody specific for prostate-specific membrane antigen: prediction of ^{90}Y-J591 radiation dosimetry based on ^{111}In or ^{177}Lu? *J Nucl Med* 2005; 46:634–641.

40. Zhang H, Chen J, Waldherr C, et al. Synthesis and evaluation of bombesin derivatives on the basis of pan-bombesin peptides labeled with indium-111, lutetium-177, and yttrium-90 for targeting bombesin receptor-expressing tumors. *Cancer Res* 2004; 64:6707–6715.

41. Onthank DC, Liu S, Silva PJ, et al. ^{90}Y and ^{111}In complexes of a DOTA-conjugated integrin alpha v beta 3 receptor antagonist: different but biologically equivalent. *Bioconjug Chem* 2004;15:235–241.

42. de Jong M, Breeman WA, Bernard BF, et al. [^{177}Lu-DOTA^{0},Tyr^{3}] octreotate for somatostatin receptor-targeted radionuclide therapy. *Int J Cancer* 2001;92:628–633.

43. Schmitt A, Bernhardt P, Nilsson O, et al. Radiation therapy of small cell lung cancer with ^{177}Lu-DOTA-Tyr3-octreotate in an animal model. *J Nucl Med* 2004;45: 1542–1548.

44. Lantry LE, Cappelletti E, Maddalena ME, et al. ^{177}Lu-AMBA: synthesis and characterization of a selective ^{177}Lu-labeled GRP-R agonist for systemic radiotherapy of prostate cancer. *J Nucl Med* 2006;47:1144–1152.

45. Miao Y, Shelton T, Quinn TP. Therapeutic efficacy of a ^{177}Lu-labeled DOTA conjugated alpha-melanocyte-stimulating hormone peptide in a murine melanoma-bearing mouse model. *Cancer Biother Radiopharm* 2007;22:333–341.

46. O'Donnell RT, DeNardo SJ, DeNardo GL, et al. Efficacy and toxicity of radioimmunotherapy with ^{90}Y-DOTA-peptide-ChL6 for PC3-tumored mice. *Prostate* 2000; 44:187–192.

47. Chinn PC, Leonard JE, Rosenberg J, Hanna N, Anderson DR. Preclinical evaluation of ^{90}Y-labeled anti-CD20 monoclonal antibody for treatment of non-Hodgkin's lymphoma. *Int J Oncol* 1999;15:1017–1025.

48. Uusijarvi H, Bernhardt P, Forssell-Aronsson E. Translation of dosimetric results of preclinical radionuclide therapy to clinical situations: influence of photon irradiation. *Cancer Biother Radiopharm* 2007;22:268–274.

49. Zinzani PL, Tani M, Fanti S, et al. A phase 2 trial of fludarabine and mitoxantrone chemotherapy followed by yttrium-90 ibritumomab tiuxetan for patients with previously untreated, indolent, nonfollicular, non-Hodgkin lymphoma. *Cancer* 2008; 112:856–862.

50. O'Donnell RT, DeNardo SJ, Miers LA, et al. Combined modality radioimmunotherapy for human prostate cancer xenografts with taxanes and ^{90}yttrium-DOTA-peptide-ChL6. *Prostate* 2002;50:27–37.

51. Johnson CV, Shelton T, Smith CJ, et al. Evaluation of combined ^{177}Lu-DOTA-8-AOC-BBN (7-14)NH$_2$ GRP receptor-targeted radiotherapy and chemotherapy in PC-3 human prostate tumor cell xenografted SCID mice. *Cancer Biother Radiopharm* 2006; 21:155–166.

52. Jonkers J, Derksen PW. Modeling metastatic breast cancer in mice. *J Mammary Gland Biol Neoplasia* 2007;12:191–203.

53. Ottewell PD, Coleman RE, Holen I. From genetic abnormality to metastases: murine models of breast cancer and their use in the development of anticancer therapies. *Breast Cancer Res Treat* 2006;96:101–113.

54. Bissell MJ. Modelling molecular mechanisms of breast cancer and invasion: lessons from the normal gland. *Biochem Soc Trans* 2007;35:18–22.

55. MacDonald IC, Chambers AF. Breast cancer metastasis progression as revealed by intravital videomicroscopy. *Expert Rev Anticancer Ther* 2006;6:1271–1279.

56. Man S, Munoz R, Kerbel RS. On the development of models in mice of advanced visceral metastatic disease for anti-cancer drug testing. *Cancer Metastasis Rev* 2007;26:737–747.

57. Shoushtari AN, Michalowska AM, Green JE. Comparing genetically engineered mouse mammary cancer models with human breast cancer by expression profiling. *Breast Dis* 2007;28:39–51.

58. Vargo-Gogola T, Rosen JM. Modelling breast cancer: one size does not fit all. *Nat Rev Cancer* 2007;7:659–672.

59. Rastelli F, Crispino S. Factors predictive of response to hormone therapy in breast cancer. *Tumori* 2008;94:370–383.

60. Prowell TM, Armstrong DK. Selecting endocrine therapy for breast cancer: what role does HER-2/neu status play? *Semin Oncol* 2006;33:681–687.

61. Weigelt B, Hu Z, He X, et al. Molecular portraits and 70-gene prognosis signature are preserved throughout the metastatic process of breast cancer. *Cancer Res* 2005; 65:9155–9158.

62. Peppercorn J, Perou CM, Carey LA. Molecular subtypes in breast cancer evaluation and management: divide and conquer. *Cancer Invest* 2008;26:1–10.

63. Stadler ZK, Come SE. Review of gene-expression profiling and its clinical use in breast cancer. *Crit Rev Oncol Hematol* 2009;69:1–11.

64. Tan DS, Reis-Filho JS. Comparative genomic hybridisation arrays: high-throughput tools to determine targeted therapy in breast cancer. *Pathobiology* 2008;75:63–74.

65. Yulug IG, Gur-Dedeoglu B. Functional genomics in translational cancer research: focus on breast cancer. *Brief Funct Genomic Proteomic* 2008;7:1–7.

66. Charafe-Jauffret E, Monville F, Ginestier C, Dontu G, Birnbaum D, Wicha MS. Cancer stem cells in breast: current opinion and future challenges. *Pathobiology* 2008; 75:75–84.

67. Eyler CE, Rich JN. Survival of the fittest: cancer stem cells in therapeutic resistance and angiogenesis. *J Clin Oncol* 2008;26:2839–2845.

68. Ponti D, Zaffaroni N, Capelli C, Daidone MG. Breast cancer stem cells: an overview. *Eur J Cancer* 2006;42:1219–1224.

69. Shipitsin M, Polyak K. The cancer stem cell hypothesis: in search of definitions, markers, and relevance. *Lab Invest* 2008;88:459–463.

70. Shipitsin M, Campbell LL, Argani P, et al. Molecular definition of breast tumor heterogeneity. *Cancer Cell* 2007;11:259–273.

71. Neve RM, Chin K, Fridlyand J, et al. A collection of breast cancer cell lines for the study of functionally distinct cancer subtypes. *Cancer Cell* 2006;10:515–527.

72. Goncalves A, Charafe-Jauffret E, Bertucci F, et al. Protein profiling of human breast tumor cells identifies novel biomarkers associated with molecular subtypes. *Mol Cell Proteomics* 2008;7:1420–1433.

73. Hughes L, Malone C, Chumsri S, Burger AM, McDonnell S. Characterisation of breast cancer cell lines and establishment of a novel isogenic subclone to study migration, invasion and tumourigenicity. *Clin Exp Metastasis* 2008;25:549–557.

74. Jonsson G, Staaf J, Olsson E, et al. High-resolution genomic profiles of breast cancer cell lines assessed by tiling BAC array comparative genomic hybridization. *Genes Chromosomes Cancer* 2007;46:543–558.

75. Herschkowitz JI, Simin K, Weigman VJ, et al. Identification of conserved gene expression features between murine mammary carcinoma models and human breast tumors. *Genome Biol* 2007;8:R76.

76. Sheridan C, Kishimoto H, Fuchs RK, et al. CD44+/CD24− breast cancer cells exhibit enhanced invasive properties: an early step necessary for metastasis. *Breast Cancer Res* 2006;8:R59.

77. Phillips TM, McBride WH, Pajonk F. The response of CD24(-/low)/CD44+ breast cancer-initiating cells to radiation. *J Natl Cancer Inst* 2006;98:1777–1785.

78. Croker AK, Goodale D, Chu J, et al. High aldehyde dehydrogenase and expression of cancer stem cell markers selects for breast cancer cells with enhanced malignant and metastatic ability. *J Cell Mol Med* 2009;13(8B): 2236–2252.

79. Wang F, Hansen RK, Radisky D, et al. Phenotypic reversion or death of cancer cells by altering signaling pathways in three-dimensional contexts. *J Natl Cancer Inst* 2002; 94:1494–1503.

80. Finak G, Bertos N, Pepin F, et al. Stromal gene expression predicts clinical outcome in breast cancer. *Nat Med* 2008;14:518–527.

81. Beck AH, Espinosa I, Gilks CB, van de Rijn M, West RB. The fibromatosis signature defines a robust stromal response in breast carcinoma. *Lab Invest* 2008;88:591–601.

82. Morris DS, Tomlins SA, Rhodes DR, Mehra R, Shah RB, Chinnaiyan AM. Integrating biomedical knowledge to model pathways of prostate cancer progression. *Cell Cycle* 2007;6:1177–1187.

83. Bacac M, Provero P, Mayran N, Stehle JC, Fusco C, Stamenkovic I. A mouse stromal response to tumor invasion predicts prostate and breast cancer patient survival. *PLoS ONE* 2006;1:e32.

84. Casey T, Bond J, Tighe S, et al. Molecular signatures suggest a major role for stromal cells in development of invasive breast cancer. *Breast Cancer Res Treat* 2009; 114: 47–62.

85. Tammi RH, Kultti A, Kosma VM, Pirinen R, Auvinen P, Tammi MI. Hyaluronan in human tumors: pathobiological and prognostic messages from cell-associated and stromal hyaluronan. *Semin Cancer Biol* 2008;18:288–295.

86. Radisky ES, Radisky DC. Stromal induction of breast cancer: inflammation and invasion. *Rev Endocr Metab Disord* 2007;8:279–287.

87. Kenny PA, Lee GY, Bissell MJ. Targeting the tumor microenvironment. *Front Biosci* 2007;12:3468–3474.

88. Fournier MV, Martin KJ. Transcriptome profiling in clinical breast cancer: from 3D culture models to prognostic signatures. *J Cell Physiol* 2006; 209: 625–630.

89. Hebner C, Weaver VM, Debnath J. Modeling morphogenesis and oncogenesis in three-dimensional breast epithelial cultures. *Annu Rev Pathol* 2008;3:313–339.

90. Jager R. Targeting the death machinery in mammary epithelial cells: Implications for breast cancer from transgenic and tissue culture experiments. *Crit Rev Oncol Hematol* 2007;63:231–240.

91. Dimri G, Band H, Band V. Mammary epithelial cell transformation: insights from cell culture and mouse models. *Breast Cancer Res* 2005;7:171–179.

92. Kim JB, Stein R, O'Hare MJ. Three-dimensional *in vitro* tissue culture models of breast cancer—a review. *Breast Cancer Res Treat* 2004;85:281–291.

93. Kenny PA. Three-dimensional extracellular matrix culture models of EGFR signalling and drug response. *Biochem Soc Trans* 2007;35:665–668.

94. Kenny PA, Lee GY, Myers CA, et al. The morphologies of breast cancer cell lines in three-dimensional assays correlate with their profiles of gene expression. *Mol Oncol* 2007; 1:84–96.

95. Khaitan D, Chandna S, Arya MB, Dwarakanath BS. Differential mechanisms of radiosensitization by 2-deoxy-D-glucose in the monolayers and multicellular spheroids of a human glioma cell line. *Cancer Biol Ther* 2006;5:1142–1151.

96. Santini MT, Romano R, Rainaldi G, et al. ^{1}H-NMR evidence for a different response to the same dose (2 Gy) of ionizing radiation of MG-63 human osteosarcoma cells and three-dimensional spheroids. *Anticancer Res* 2006;26:267–281.

97. Santini MT, Rainaldi G, Indovina PL. Multicellular tumour spheroids in radiation biology. *Int J Radiat Biol* 1999;75:787–799.

98. Mitrofanova E, Hagan C, Qi J, Seregina T, Link C, Jr. Sodium iodide symporter/ radioactive iodine system has more efficient antitumor effect in three-dimensional spheroids. *Anticancer Res* 2003;23:2397–2404.

99. Boyd M, Cunningham SH, Brown MM, Mairs RJ, Wheldon TE. Noradrenaline transporter gene transfer for radiation cell kill by ^{131}I meta-iodobenzylguanidine. *Gene Ther* 1999;6:1147–1152.

100. Genc M, Castro Kreder N, Barten-van Rijbroek A, Stalpers LJ, Haveman J. Enhancement of effects of irradiation by gemcitabine in a glioblastoma cell line and cell line spheroids. *J Cancer Res Clin Oncol* 2004;130:45–51.

101. Lamfers ML, Idema S, Bosscher L, et al. Differential effects of combined Ad5-delta 24RGD and radiation therapy in *in vitro* versus *in vivo* models of malignant glioma. *Clin Cancer Res* 2007;13:7451–7458.

102. Burris HA, 3rd. Preclinical investigations with epothilones in breast cancer models. *Semin Oncol* 2008;35 (Suppl. 2): S15–S21.

103. Minn AJ, Gupta GP, Siegel PM, et al. Genes that mediate breast cancer metastasis to lung. *Nature* 2005;436:518–524.

104. Kluger HM, Chelouche Lev D, Kluger Y, et al. Using a xenograft model of human breast cancer metastasis to find genes associated with clinically aggressive disease. *Cancer Res* 2005;65:5578–5587.

105. Montel V, Huang TY, Mose E, Pestonjamasp K, Tarin D. Expression profiling of primary tumors and matched lymphatic and lung metastases in a xenogeneic breast cancer model. *Am J Pathol* 2005;166:1565–1579.

106. Agrawal D, Chen T, Irby R, et al. Osteopontin identified as colon cancer tumor progression marker. *C R Biol* 2003;326:1041–1043.

107. Heyn C, Ronald JA, Ramadan SS, et al. *In vivo* MRI of cancer cell fate at the single-cell level in a mouse model of breast cancer metastasis to the brain. *Magn Reson Med* 2006;56:1001–1010.

108. MacDonald IC, Groom AC, Chambers AF. Cancer spread and micrometastasis development: quantitative approaches for *in vivo* models. *Bioessays* 2002;24:885–893.

109. Hojilla CV, Wood GA, Khokha R. Inflammation and breast cancer: metalloproteinases as common effectors of inflammation and extracellular matrix breakdown in breast cancer. *Breast Cancer Res* 2008;10(2):205.

110. DeNardo DG, Johansson M, Coussens LM. Immune cells as mediators of solid tumor metastasis. *Cancer Metastasis Rev* 2008;27:11–18.

111. Shen Q, Brown PH. Transgenic mouse models for the prevention of breast cancer. *Mutat Res* 2005;576:93–110.

112. Cardiff RD, Moghanaki D, Jensen RA. Genetically engineered mouse models of mammary intraepithelial neoplasia. *J Mammary Gland Biol Neoplasia* 2000;5:421–437.

113. Borowsky A. Special considerations in mouse models of breast cancer. *Breast Dis* 2007;28:29–38.

114. Hutchinson JN, Muller WJ. Transgenic mouse models of human breast cancer. *Oncogene* 2000;19:6130–6137.

115. Hanahan D, Wagner EF, Palmiter RD. The origins of oncomice: a history of the first transgenic mice genetically engineered to develop cancer. *Genes Dev* 2007;21:2258–2270.

116. Derksen PW, Liu X, Saridin F, et al. Somatic inactivation of E-cadherin and p53 in mice leads to metastatic lobular mammary carcinoma through induction of anoikis resistance and angiogenesis. *Cancer Cell* 2006;10:437–449.

117. Marcotte R, Muller WJ. Signal transduction in transgenic mouse models of human breast cancer—implications for human breast cancer. *J Mammary Gland Biol Neoplasia* 2008;13:323–335.

118. Mathias CJ, Bergmann SR, Green MA. Species-dependent binding of copper(II) bis (thiosemicarbazone) radiopharmaceuticals to serum albumin. *J Nucl Med* 1995;36:1451–1455.

119. Basken NE, Mathias CJ, Lipka AE, Green MA. Species dependence of [^{64}Cu]Cu-bis (thiosemicarbazone) radiopharmaceutical binding to serum albumins. *Nucl Med Biol* 2008;35:281–286.

120. Mukai T, Arano Y, Nishida K, et al. Species difference in radioactivity elimination from liver parenchymal cells after injection of radiolabeled proteins. *Nucl Med Biol* 1999;26:281–289.

121. Knight LC, Romano JE, Bright LT, Agelan A, Kantor S, Maurer AH. Platelet binding and biodistribution of $[^{99m}Tc]$rBitistatin in animal species and humans. *Nucl Med Biol* 2007;34:855–863.

122. Sausville EA, Burger AM. Contributions of human tumor xenografts to anticancer drug development. *Cancer Res* 2006;66:3351–3354.

123. Becher OJ, Holland EC. Genetically engineered models have advantages over xenografts for preclinical studies. *Cancer Res* 2006;66:3355–3358.

124. Olive KP, Tuveson DA. The use of targeted mouse models for preclinical testing of novel cancer therapeutics. *Clin Cancer Res* 2006;12:5277–5287.

125. Johnson JI, Decker S, Zaharevitz D, et al. Relationships between drug activity in NCI preclinical *in vitro* and *in vivo* models and early clinical trials. *Br J Cancer* 2001; 84:1424–1431.

126. Xuan JW, Bygrave M, Jiang H, et al. Functional neoangiogenesis imaging of genetically engineered mouse prostate cancer using three-dimensional power Doppler ultrasound. *Cancer Res* 2007;67:2830–2839.

127. Graham KC, Wirtzfeld LA, MacKenzie LT, et al. Three-dimensional high-frequency ultrasound imaging for longitudinal evaluation of liver metastases in preclinical models. *Cancer Res* 2005;65:5231–5237.

128. Wirtzfeld LA, Graham KC, Groom AC, et al. Volume measurement variability in three-dimensional high-frequency ultrasound images of murine liver metastases. *Phys Med Biol* 2006;51:2367–2381.

129. Henriquez NV, van Overveld PG, Que I, et al. Advances in optical imaging and novel model systems for cancer metastasis research. *Clin Exp Metastasis* 2007;24:699–705.

130. Raman V, Pathak AP, Glunde K, Artemov D, Bhujwalla ZM. Magnetic resonance imaging and spectroscopy of transgenic models of cancer. *NMR Biomed* 2007;20:186–199.

131. Heyn C, Ronald JA, Mackenzie LT, et al. *In vivo* magnetic resonance imaging of single cells in mouse brain with optical validation. *Magn Reson Med* 2006;55:23–29.

132. Kang JH, Chung JK. Molecular-genetic imaging based on reporter gene expression. *J Nucl Med* 2008;49 (Suppl. 2): 164S–179S.

133. Ledar P, Stewart TA. Transgenic non-human mammals. US patent # 4,736,866, 1988.

134. Wadman M. DuPont opens up access to genetics tool. *Nature* 1998;394:819.

135. Marshall E. NIH, DuPont declare truce in mouse war. *Science* 1998;281:1261–1262.

136. Marshall E. Intellectual property. DuPont ups ante on use of Harvard's OncoMouse. *Science* 2002;296:1212.

CHAPTER 12

Radiation Biology of Targeted Radiotherapy

DAVID MURRAY AND MICHAEL WEINFELD

12.1 INTRODUCTION

In this chapter, the radiation biology of targeted radiotherapy (TRT) will be discussed. In particular, the chapter will focus on how our improved understanding of the cellular DNA damage surveillance–response network, especially with regards to the response to DNA double-strand breaks, and cell-death mechanisms, has changed our thinking about the various biological effects of the low dose rate (LDR) exposures typically administered in TRT in comparison to the effectiveness of locally-focused radiotherapy with external irradiation. Radiobiological modeling of TRT has largely depended on extrapolation of data obtained using homogeneous exposures to single or fractionated doses of radiation delivered at high dose rate (HDR). The underlying assumption is that such exposures have biological equivalence to the declining LDR exposures. In recent years, it has become apparent that the cellular and molecular mechanisms by which mammalian cells respond to low dose and/or LDR radiation exposures can be quite different from those occurring at HDR. Many of these low dose/LDR findings were controversial on their initial appearance, often because such effects were not universal. Understanding the mechanisms of these various effects will enable us to move forward to the point where we will be able to better use this information to guide translational and clinical advances. Of major importance in this regard are the phenomena of low-dose hyperradiosensitivity-increased radioresistance, inverse dose-rate effects, adaptive responses, and the bystander effect (which is quite distinct from the "cross-fire" effect that is also operative in TRT). Each of these areas requires a major reconsideration of existing models for radiation action and an understanding of how this knowledge will integrate into the evolution of clinical TRT practice. Validation of a role *in vivo*

Monoclonal Antibody and Peptide-Targeted Radiotherapy of Cancer, Edited by Raymond M. Reilly
Copyright © 2010 John Wiley & Sons, Inc.

for any or all of these effects in TRT would greatly impact the way we would assess therapeutic response to TRT, the design of clinical trials of novel TRT radio-pharmaceuticals, and risk estimates for both therapeutic and diagnostic radiophar-maceuticals. The current state of research in LDR effects therefore offers a major opportunity to critically address the basic science behind clinical TRT practice, to use this new knowledge to expand the use and roles of TRT, and to facilitate the introduction of new radiopharmaceuticals.

12.2   TARGETED RADIONUCLIDE THERAPY: CONCEPTS

The biological principles that underlie the use of ionizing radiation to treat cancer have largely derived from studies using doses and dose rates typical of those employed in external-beam radiotherapy (XRT). In XRT, the dose is typically delivered at a constant high dose rate, usually between 1 and 5 Gy/min. A typical XRT schedule would involve giving a total tumor dose of between 60 and 70 Gy in daily fractions of ~2 Gy, five fractions per week. The overall treatment time would thus be 6–8 weeks. The therapeutic index of XRT is largely achieved by the use of sophisticated computational treatment planning algorithms that result in the locoregional delivery of a relatively homogeneous dose to the tumor while minimizing the dose delivered to critical normal-tissue structures.

In targeted radiotherapy, the objective is the *systemic* delivery of radiation dose to the tumor using radiolabeled receptor-directed metabolic precursors or other ligands, or by monoclonal antibodies such as anti-CD20 antibodies for lymphoma, including Bexxar and Zevalin (1, 2), that is, the so-called "radioimmunotherapy" approach (see Chapter 6). TRT may also be achieved using radiolabeled peptides (see Chapters 3 and 4). The therapeutic index in TRT derives from the expectation that the targeted molecule will preferentially localize at the tumor site where it will deposit most of the dose, while depositing little dose in normal tissues (3). In contrast to XRT, the radiation dose in TRT is delivered at a continuous but exponentially declining low dose rate that will depend on the localization of the therapeutic agent and the half-life of the radionuclide. The typical average dose rate to the tumor for TRT is 2–8 Gy/day (~10–40 cGy/h), with the maximum absorbed dose being up to 50 Gy delivered over a period of many days (4).

In addition to the fundamental differences in the way that XRT and TRT are administered, there are also important radiobiological differences between these approaches that could impact on treatment outcome. Indeed, emerging evidence suggests that the mechanisms by which cells respond to LDR versus HDR radiation exposures are quite different. Clinically, it is also apparent that the patterns of therapeutic response to TRT are not always consistent with classical XRT-derived radiobiological models. In particular, TRT is sometimes effective in palliation and tumor control even when calculated absorbed doses to the tumor are lower than those delivered by XRT (5–11). Despite such caveats, biological models of TRT have been mainly derived by extrapolation of data obtained following homogeneous acute exposures to single or fractionated doses of radiation and have assumed that the doses delivered by XRT and TRT are biologically equivalent (12, 13). This chapter will review several recent research findings that may help to explain some of the discrepancies in the TRT literature and guide future translational studies and clinical practice in TRT.

12.3 RADIATION-INDUCED DNA DAMAGE

Early models of radiation effects assumed that a cell must be directly "hit" by an ionizing-particle track for a biological effect to be manifested in that cell. It was generally held that genomic DNA was the critical intracellular "target" that must be ionized by the passage of the particle in order for a cell to be killed. Exposing mammalian cells to a 1-Gy dose of a low linear energy transfer (LET) beam such as X-rays or γ-rays results in the induction of several thousand DNA lesions of several distinct types (Table 12.1). Isolated lesions such as damaged bases and sugars and single-strand breaks (SSBs) involve only one strand of the DNA duplex, and are therefore relatively easy for a cell to repair by using the undamaged complementary strand as a template for the repair DNA-polymerase activity. In addition to these individual lesions, a "hallmark" of ionizing radiation is the production of complex "clustered lesions" or "multiply damaged sites" that arise because of the microheter-ogeneity of ionizing-particle tracks; along such tracks, large amounts of energy are occasionally deposited in discrete regions, and when these events coincide with the cellular DNA they can cause multiple ionizations within a short span of the DNA molecule (14, 15). Such clustered lesions, which include frank double-strand breaks (DSBs), are widely thought to be the main cause of radiation-induced cell death, probably because they can give rise to additional DSBs during the course of repair or replication (16, 17). For the remainder of this chapter, the term "DSB" will be used to refer to all types of clustered lesions regardless of their complexity. Because DSBs involve both strands of the DNA duplex they are considered to be more difficult to repair than isolated lesions.

TABLE 12.1   Approximate Initial Yields of the Various Classes of Cellular DNA Lesions for a 1 Gy X-Ray or γ-Ray Exposure[a]

Lesion Class	Lesions Induced/Cell/Gy
DSB	$\sim$40
SSB	500–1000
Base damage	1000–2000
Sugar damage	800–1600
DNA–DNA cross-link	$\sim$30
DNA–protein cross-link	$\sim$150

[a] After Murray et al. (18).

12.4   CELLULAR DNA DAMAGE SURVEILLANCE–RESPONSE NETWORKS

Radiation-induced DNA lesions such as DSBs must be rapidly detected and repaired if a cell is to retain its genomic integrity and survive. Mammalian cells respond to a radiation insult by activating a highly orchestrated network of signaling events that results in alteration of the expression or activity of numerous genes/proteins involved in processes such as DNA repair, cell-cycle arrest, and cell death (mechanisms will be discussed later) (19, 20). The purpose of the DNA damage surveillance–response is to (a) detect and remove DNA lesions and stabilize the genome, (b) suppress mutagenesis and promote the survival of cells that have properly processed such DNA lesions, and (c) eliminate heavily damaged cells from the population/tissue. Objectives (a) and (b) are achieved in part by transiently interrupting the progression of damaged cells through the cell cycle via the activation of cell-cycle checkpoints that operate in G_1, S, and G_2 phases, thereby allowing DNA repair to proceed without the complication of ongoing critical DNA transactions such as DNA replication and mitosis.

The cellular response to DNA damage, which is summarized in Fig. 12.1, is carried out by four classes of proteins: sensors, mediators, transducers, and effectors. Sensor (damage-recognition) proteins recognize and bind to DNA lesions and in turn recruit and/or activate the mediator and transducer proteins. Mediator (also known as "adaptor") proteins are important for signaling-complex assembly. The major proteins that transduce DNA damage signals are members of a family of serine/threonine kinases known as the phosphatidylinositol 3-kinase-related kinases (PIKKs). These include ATM (mutated in ataxia telangiectasia), ATR (ATM and RAD3 related), and DNA-PK$_{cs}$ (DNA-dependent protein kinase, catalytic subunit) (20–22). Other serine/threonine kinase transducers transmit and amplify signals from the activated PIKKs to downstream effectors (e.g., p53) that implement the various functional outcomes of the damage surveillance–response network.

Because of their prominent status, the cellular response to DSBs has been studied in some detail. Following a radiation insult, several proteins rapidly relocalize to the site of a DSB, generating a "repair focus." These include MRE11, RAD50, and NBS1, which together constitute the "MRN" complex, as well as mediators such as BRCA1 and BRCA2 (both of which are commonly mutated in human cancers such as those of the breast), 53BP1 (p53-binding protein 1), MDC1 (mediator of DNA damage checkpoint 1), and SMC1 (structural maintenance of chromosomes 1). The MRN complex exhibits a number of distinct damage sensor and mediator activities (20, 23) that indicate roles in cell-cycle checkpoint activation, signal transduction and DNA repair (e.g., removal of excess DNA at 3′-flaps, endonuclease, exonuclease, and helicase activities and possibly resection of 5′-DNA ends). Another complex involving the RAD9, HUS1, and RAD1 proteins (the so-called "9-1-1" complex) appears to be involved in sensing some types of bulky DNA lesions, although its role in sensing radiation-induced DSBs remains to be defined.

The major PIKK involved in the DNA damage-sensing and -signaling response to a radiation-induced DSB is the ATM protein (24). Activation of the ATM kinase function by a DSB depends on autophosphorylation and relocation of ATM to protein complexes at the site of the DSB, these events possibly being stimulated by changes in the chromatin structure of the cell (25). The ability of non-DSB lesions to activate ATM remains uncertain (20). ATM relocation/activation probably involves the MRN complex and perhaps MDC1 (20, 26, 27). Once recruited, activated ATM can

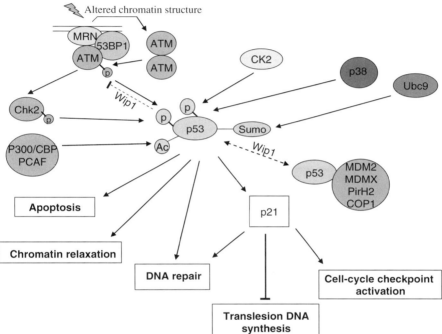

FIGURE 12.1   Cellular DNA damage surveillance–response networks activated by ionizing radiation. This response involves four classes of proteins: sensors, mediators, transducers, and effectors. Sensors sense and bind to the various types of DNA damage, illustrated here for a DSB induced by ionizing radiation, and activate transducers such as the ATM (mutated in ataxia telangiectasia) serine/threonine kinase that amplify and transmit these signals to the effector proteins, such as p53, which execute the various functional outcomes of the damage-response network, such as the activation of cell-cycle checkpoints, DNA repair, and apoptosis. In response to DNA damage p53 can undergo various posttranslational modifications—phosphorylation, acetylation, and sumoylation—which in turn can increase the level of p53 by disrupting the normal regulation of p53, mediated by ubiquitin ligases such as MDM2, PirH2, and COP1. Upregulation of p53 can lead to several consequences including DNA repair or cell death, depending on the level of DNA damage. The result of this response is the triggering of a cascade of signaling events that modifies the expression or posttranslational modification/activity of many downstream genes and proteins which are involved in prosurvival or prodeath mechanisms. Modified from Ref. 35.

phosphorylate its various substrates, which are currently known to number many hundreds and include sensors, mediators, cell-cycle checkpoint activators and DNA repair-related proteins such as the CHK2 checkpoint kinase, BRCA1, BLM (the helicase associated with Bloom's syndrome), NBS1, RAD9, MDC1, 53BP1, SMC1, and c-ABL (which can phosphorylate RAD51, a critical DSB-repair factor) (20, 28–30). ATM phosphorylates its targets either directly or indirectly via CHK2 (24, 31).

The variant histone, H2AX, is another important downstream target of activated ATM. Phosphorylation of H2AX by ATM postirradiation precedes DSB rejoining, and may serve to recruit other mediators and repair factors to the DSB. Typically, there are $\sim$2000 phosphorylated H2AX (γ-H2AX) molecules per DSB, which results in a number of discrete and readily observable "repair foci" when the cells are stained with an anti-γ-H2AX antibody (32). This observation has provided an important reagent for the study of low-dose/LDR radiation effects on cells because it provides a very sensitive, albeit indirect, biomarker for DSBs.

Important from the perspective of low-dose/LDR exposures is the finding that ATM activation (specifically, the step involving autophosphorylation of ATM on serine-1981) in human fibroblasts occurs very rapidly. Indeed, this response was apparent immediately after a 0.5 Gy exposure (25). This phosphorylation event was apparent after relatively low doses ($\sim$0.1 Gy), and persisted for at least 24 h postirradiation (25).

Another key ATM-kinase target is the p53 tumor-suppressor protein, which is commonly mutated in human cancers (33). p53 is a critical effector of several steps in the DNA damage surveillance–response network (34, 35), a role that earned it the title of "guardian of the genome" (36). Following a radiation exposure, the wild-type p53 protein undergoes a complex pattern of posttranslational modification (including phosphorylation, acetylation, and sumoylation) by enzymes such as activated ATM and CHK2 (37, 38) that leads to stabilization and increased levels of p53 as well as to p53 activation. Activated p53 can mediate downstream responses either through direct interactions with other effector proteins or indirectly through the transcriptional regulation of target genes. In general, the former events are rapid and transient, whereas the slower events that require transcription relate to response "maintenance." A large number of target genes are transcriptionally regulated by p53 (e.g., (39–41)). Examples of genes that are transactivated by p53 include the cyclin-dependent kinase inhibitor, $p21^{WAF1}$ (see below), proapoptotic genes, 14-3-3σ (whose encoded protein is important for maintaining the G_2/M arrest), and several DNA-repair genes (35, 42). Genes trans-*repressed* by p53 include antiapoptotic genes such as *BCL-2*. In addition, a negative feedback loop leading to down-regulation of ATM by p53 through activation of the p53-regulated WIP1 phosphatase has recently been described (43, 44).

Many critical cellular decisions that follow a low-dose radiation exposure are coordinated by p53, including prosurvival responses (e.g., activation of cell-cycle checkpoints, DNA-repair) and prodeath responses (e.g., apoptosis). The p53 protein appears to regulate DNA repair by interacting with either DNA-repair proteins or DNA lesions themselves, as well as through transcription-dependent mechanisms (for reviews, see Refs 35 and 42). Regulation of the G_1 checkpoint by p53 mainly involves the transcriptional transactivation of the $p21^{WAF1}$ gene that encodes the p21^{WAF1} (CDKN1A) cyclin-dependent kinase inhibitor. Activation of the "conventional" G_2 checkpoint involves both p53-dependent and -independent mechanisms; transcriptional transactivation by p53 seems to be required to maintain the initially p53-independent G_2/M arrest (45). A second G_2/M phase checkpoint, which may be especially relevant for LDR radiation exposures, was described by Xu and colleagues (46); although activation of this checkpoint is dependent on ATM, the role of p53 therein is uncertain.

An anticipated feature of the DNA damage surveillance–response network is the existence of feedback mechanisms at the level of mediators/transducers such that the initiating signal will not be amplified until the DNA damage exceeds some preset threshold level. Such a threshold is expected in order to ensure that a cell will not "over-react" to a low-level injury, such as that caused by a low dose/LDR radiation exposure. Signals and mechanisms are also necessary to reset the machinery following successful repair and to deactivate the checkpoints so the cell can resume proliferation (47). If repair is ongoing, feedback signals from the sensors presumably direct the maintenance of the checkpoint, such that a moderately damaged cell will optimize its chances of repair and survival. Alternatively, in heavily or persistently damaged cells, genes will be activated that evoke cell-death responses such that, in a multicellular system/ tissue, these cells will be eliminated from the population. To discriminate these various biological outcomes, a cell population must exhibit a graded or stepped dose–response to radiation injury.

Another important early event in the cellular response to DSBs is the remodeling of chromatin structure, first to permit the proper access and assembly of repair complexes, then for chromatin restoration when repair is completed. Posttranslational histone modifications such as acetylation, involving proteins such as the NuA4 histone acetyltransferase (HAT) complex, play an important role in both the early and the late stages of DSB repair (48, 49). The acetylation of ATM in response to DNA damage also depends on another HAT, Tip60 (50).

Although most of the literature concerning the generation of signals in response to a radiation insult have focused on the genome, other cellular structures such as the cell membrane could also be important in this regard (51). Indeed, membrane-derived signals can be a major contributor to the activation of apoptotic pathways.

12.5   MAMMALIAN DNA-REPAIR PATHWAYS

Mammalian cells repair damage to their genome by mobilizing one or more of the error-free or error-prone enzymatic DNA-repair pathways to which they have access (52). In the specific case of an ionizing radiation exposure, the major pathways invoked are base excision repair (BER), which removes oxidized or missing bases, single-strand break repair (SSBR), which processes and rejoins strand-break termini and can be considered a subpathway of BER, nonhomologous end joining (NHEJ) which removes DSBs, and homology-directed recombination repair (HRR) which removes DSBs as well as some types of DNA cross-links. Although cells are endowed with other repair pathways, including nucleotide excision repair (NER), mismatch repair (MMR), and lesion bypass, these mechanisms are not commonly associated with the repair of radiation-induced lesions and will not be considered further in this chapter.

12.5.1 The BER Pathway

BER removes many types of radiation-induced DNA lesions, including a broad range of base damages, base loss and SSBs, from the genome (28, 53). BER is a coordinated sequential multiprotein pathway initiated by a damage-specific DNA glycosylase that removes a damaged base by cutting the N-glycosidic bond to generate an apurinic/apyrimidinic (AP) site. Important human glycosylases include NEIL1, NEIL2, and NEIL3 (the homologs of *E. coli* endonuclease VIII), hNTH1 (endonuclease III homolog), and OGG1 (8-oxoguanine glycosylase 1). There is some degeneracy among the glycosylases; for example, another 8-oxoguanine glycosylase, OGG2, can also act on 8-oxopurine lesions (54). Either an AP lyase or AP endonuclease cleaves the ribose-phosphate chain 3' or 5', respectively, to the AP site generated by the glycosylase activity. In fact, some glycosylases have an associated AP-lyase activity. The main human AP endonuclease is HAP1 (also called APE1). DNA synthesis to replace missing nucleotides requires clean 3'-OH termini, and sealing of the break by DNA ligases also requires clean 3'-OH/5'-phosphate termini. Thus, aberrant termini must first be processed by various exonuclease or deoxyribophosphodiesterase activities. An important enzyme in this regard is polynucleotide kinasephosphatase (PNKP), which has both 5'-DNA kinase and 3'-DNA phosphatase activities (55, 56). The HAP1 enzyme displays both diesterase and weak phosphatase activities (57). Recently another enzyme, aprataxin, has been shown to be capable of removing 3'-phosphate and 3'-phosphoglycolate termini (58).

Cleaved AP sites are processed by either the short- or the long-patch subpathway of BER (59). Short-patch BER involves DNA polymerase β (POLβ) replacing a single nucleotide, with DNA ligase III (LIG3) sealing the gap. In long-patch BER, POLβ is displaced from the lesion, followed by the removal and resynthesis of up to 15 nucleotides in a process involving proliferating cell nuclear antigen (PCNA), replication factor C (RFC), flap endonuclease 1 (FEN1), and POLδ or POLε. LIG1, which interacts with both PCNA and POLβ, seals the resulting gap. Not surprisingly, considering the many coordinated/sequential activities involved in BER, a number of interactions among the various BER proteins have been characterized. For example, LIG3 interacts with POLβ, poly(ADP-ribose) polymerase 1 (PARP-1), and XRCC1. Indeed, complex formation during BER is required for protein stability as noncomplexed proteins are targeted for ubiquitylation and degradation by the 26-S proteosome (60). The XRCC1 protein exhibits no intrinsic catalytic activity in BER but seems to play a role in the detection and coordination of SSB processing by acting as a scaffold for the assembly of these multiprotein complexes (61, 62). XRCC1 also interacts with PNKP and stimulates its end-processing activity (63, 64). Disagreement remains as to the exact role of PARP-1. In one model, it recruits repair proteins to strand-break termini at the site of damage (65), while others have argued that its main function is to protect the termini from nuclease digestion (66). More recently,

Woodhouse et al. (67) have provided evidence that PARP-1 inhibits nuclease-mediated conversion of SSB to DSB, thus allowing more time for completion of BER.

12.5.2 DSB-Repair Pathways

In view of the considerable evidence implicating DSBs as the most cytotoxic class of radiation-induced damage, there has been much interest in how these lesions are repaired by mammalian cells. DSBs in fact are substrates for several pathways, two of which, NHEJ and HRR, are particularly important. These pathways, which may represent either competing or sequential processes (68, 69), appear to share some components. For example, the early recognition and signaling of a DSB by the MRN complex seems to be common to both pathways (70, 71). How pathway choice is dictated is a key question. Cell-cycle phase is clearly a major factor in this respect. HRR, unlike NHEJ, depends on extensive DNA sequence homology, and is therefore preferred in late-S and G_2-phase cells, presumably because of efficient recombination between the sister chromatids; axiomatically, NHEJ is preferred in G_1/G_0-phase cells (71–79). MRN and p53 interface with NHEJ and HRR at various points, and may be important regulators of DSB-processing pathway selection (35, 42). The consensus is that NHEJ is the preferred option for repairing radiation-induced DSBs in mammalian cells but that the coordinated activity of NHEJ and HRR may be required for correct repair (19, 71, 80–82). A detailed discussion of these pathways can be found in the excellent volume by Friedberg and colleagues (52).

12.5.2.1 *The NHEJ Pathway*   NHEJ proteins promote the direct rejoining of DSB ends using mechanisms that do not involve strand exchanges and that require no or as little as one base pair of sequence homology (83–85). Consequently, error-prone or "illegitimate" rejoining events can occur by deletion of nucleotides. Some deletions may be tolerated because they occur in noncoding regions of the genome. NHEJ is mediated by the DNA-PK holoenzyme complex that consists of three proteins: a catalytic PIKK serine/threonine kinase subunit, DNA-PK$_{cs}$ (also known as XRCC7), and a regulatory subunit, Ku, which is a heterodimer of the Ku70 (or XRCC6) and Ku80 (or XRCC5) proteins. NHEJ is initiated by Ku binding to a DSB. Ku binding serves as a scaffolding/alignment factor to tether the broken DNA ends together and to protect these ends from degradation or further reaction. The Ku complex can then translocate from the DNA end in an ATP-independent manner, such that several Ku molecules can bind to each DSB. Ku then recruits DNA-PK$_{cs}$ to the DSB, activating its kinase function. Activated DNA-PK$_{cs}$ can then phosphorylate its various targets, such as p53. DNA-PK$_{cs}$ is also inhibited by autophosphorylation (86). Other proteins involved in NHEJ are LIG4 and XRCC4, which carry out tail-removal and gap-filling/ligation reactions (80), and possibly POLμ and POLλ (52). As with BER, ligation requires the cleaning up of "dirty" termini by proteins such as PNKP (87), FEN1, MRN, and Artemis (88). PARP-1 may also be involved in this process (89).

In view of the scaffold/structural roles of some of these proteins, it is perhaps not surprising that several interactions have been reported among the various NHEJ proteins. For example, Ku interacts with MRE11 and LIG4 (90), while XRCC4 interacts with DNA-PK$_{cs}$ (91, 92) and LIG4, increasing the ability of the latter to bind to and ligate broken DNA ends (93, 94).

12.5.2.2 The HRR Pathway   HRR, in distinction to NHEJ, involves the pairing and exchange of homologous DNA sequences to rejoin a DSB by taking advantage of the undamaged homologous sequence as a template for repairing the damaged duplex, so it is a relatively error-free transaction (95). The MRN complex is probably involved in the early steps of HRR, such as end processing and single-strand degradation. A critical enzyme for HRR is RAD51, a DNA-dependent ATPase that promotes exchanges between homologous DNA strands in the presence of replication protein A (RPA) and forms helical polymeric filaments with single- and double-stranded DNA. RAD51 also interacts with and is regulated by several other proteins, including p53, BRCA2, RAD52, and RAD54. RAD52 enhances homologous pairing and strand exchange by promoting RAD51 polymerization. Phosphorylation of RAD52 postirradiation, which appears to require both the ATM and the c-ABL kinase activities, is important for the formation of DSB-repair foci and for RAD51-RAD52 interaction (96). RAD54, a SNF2/SWI2 DNA-dependent helicase-ATPase family member, can bind to RAD51 and may stabilize protein–DNA complexes, promoting homology searching and strand invasion. It may also cooperate with RAD51 in chromatin remodeling. HRR is completed by DNA polymerases that resynthesize the deleted DNA sequences using the intact homologous sequence for a template, and by DNA ligases that join the newly synthesized fragments. Endonucleases called "resolvases" resolve the resulting Holliday junctions. BRCA1 and BRCA2 are also involved directly or indirectly in the HRR response to DSBs, as is FEN1, which may remove 5′-flaps from HRR intermediates (e.g., see Refs 20, 71, 97, and 98). Five paralogs of RAD51, named RAD51B, RAD51C, RAD51D, XRCC2, and XRCC3, facilitate assembly of the RAD51 nucleoprotein filament. RAD51 and its five paralogs may collectively mediate homology searching and strand pairing/exchanges.

12.6   MODES OF CELL DEATH FOLLOWING RADIATION EXPOSURE

Following exposure to ionizing radiation and other DNA-damaging agents, a cell can lose its reproductive potential through one of several mechanisms depending on factors such as the type of cell, its genotype and local environment (e.g., extent of cell–cell and/or cell–matrix interactions), and the nature and dose of the genotoxic agent (e.g., see Refs 11 and 99–101). From the current perspective, it is important to recognize that these various modes of cell death will probably have their own unique dependency on dose and time, and potentially on dose rate, in different genetic backgrounds. Five such mechanisms will be considered here.

12.6.1   Apoptosis

Apoptosis is a genetically regulated, energy dependent, form of "programmed" cell death that has been reported by some authors to occur preferentially after lower doses of ionizing radiation in some model systems. It involves the activation of proteolytic enzymes called caspases that function as cell executioners, and is characterized by shrinkage of the cytoplasm, condensation of the chromatin and nucleus, nonrandom degradation ("laddering") of DNA, membrane blebbing, and fragmentation of the cell, generating "apoptotic bodies." These apoptotic bodies are phagocytosed by macrophages *in vivo*, so apoptosis of tumor cells postirradiation is generally believed to avoid the triggering of strong inflammatory responses, although this may not be a universal principle (102). All types of mammalian cells, both normal and malignant, appear to include the inherent machinery required for activating the apoptotic program.

Radiation-induced apoptosis can be mediated by either of two pathways: intrinsic or extrinsic. Activation of the cascade of caspases is common to both pathways. Targets for cleavage by caspases include proteins involved in DNA repair, DNA replication and cytoskeleton regulation, as well as the nuclease that causes DNA fragmentation. The intrinsic pathway is triggered by signals derived in the nucleus (e.g., via the activated ATM-kinase) or cell membrane (e.g., via ceramide) (103) that activates the proapoptotic protein, Bax. This event causes the mitochondria to release cytochrome c into the cytosol, where it can bind to apoptotic protease-activating factor-1 (APAF-1) to form an "apoptosome." The apoptosome in turn activates caspase 9, which then activates caspases 3, 6, 7, and 8, leading to apoptosis. Whereas proapoptotic proteins such as Bax permit cytochrome c to exit the mitochondrion, antiapoptotic proteins such as Bcl-2 and Bcl-x_L block this exit. The extrinsic apoptosis pathway is triggered by death receptors that are activated by ligands such as Fas ligand, tumor necrosis factor-α (TNF-α) and tumor necrosis factor-related apoptosis-inducing ligand (TRAIL) (104, 105). Following exposure of some cell types to ionizing radiation, TRAIL can trigger apoptosis by promoting the clustering of DR4 and DR5 death receptors in the cell membrane, leading to the formation of the death-inducing signaling complex (DISC). Recruitment of the Fas-associated protein with death domain (FADD) causes the activation of caspases 8 and 10, which leads to caspase 3-mediated activation of death-effector proteins. It is important to note that apoptosis is regulated both positively and negatively by a number of genes/gene products, many of which (e.g., p53, Bcl-2) are commonly altered in human tumors (e.g., see Ref. 106). The role of p53 in activating apoptosis involves both transcriptional transactivation-dependent and -independent mechanisms (e.g., see Ref. 107).

12.6.2 Necrosis

Necrosis is triggered in cells that enter mitosis with heavily damaged genomes. It is usually described as a generalized, nonspecific, or passive response to injury, and is typically not a genetically regulated process. Necrotic death is characterized by progressive cell swelling, random fragmentation of DNA, denaturation and coagulation of cytoplasmic proteins, disintegration of subcellular organelles and the cell membrane, swelling of mitochondria, and the release of cytotoxic cell components. Unlike apoptosis, necrosis generally occurs after higher doses of radiation and does invoke inflammatory responses *in vivo* (108).

12.6.3 Accelerated Senescence

Accelerated senescence has been variously referred to in the earlier literature as permanent, terminal, or irreversible growth arrest (e.g., see Refs 109 and 110) but is now recognized as a genetically regulated active response to DNA injury (111, 112). The cellular phenotype of accelerated senescence is characterized by enlarged and flattened morphology, granularity, positive staining for senescence-associated β-galactosidase activity, and a long-term hiatus from the cell division cycle even though the cells retain viability. Radiation-induced accelerated senescence was originally described in fibroblasts, where it was apparent after acute exposure to doses as low as 0.1 Gy and seen to exhibit features in common with cells undergoing telomere erosion-driven "replicative" senescence (e.g., see Refs 113 and 114). However, it has now been reported to occur in many cell types, including some human tumor cell lines (115, 116). An important molecular mediator of the accelerated senescence program postirradiation is the p53-regulated radiation-inducible p21^{WAF1} (CDKN1A) protein, which not only inhibits various cyclin-dependent kinases (CDKs) (see above) but also is a major transcriptional regulator (110, 111). Other potential contributors are the p16^{INK4a} CDK inhibitor and the p105^{Rb} retinoblastoma protein (111). Accelerated senescence is generally regarded as being p53-dependent, such that tumor cells that either lack or have mutant p53 do not strongly activate this response following irradiation (101, 110). However, exposing cultures of a p53-knockout derivative of the HCT116 colon carcinoma cell line to γ-rays did cause manifestations of accelerated senescence, including upregulation of p21^{WAF1}, in a subset of the cells, so we presume that there is also a (minor) p53-independent mechanism for this response (R. Mirzayans and D. Murray, unpublished data).

12.6.4 Autophagy

Autophagy is another genetically regulated form of programmed cell death in which the cell essentially undergoes self-digestion. It is best defined as a conserved response to nutrient deprivation where the cell exits the cell division cycle, autodigests proteins and damaged organelles, shrinks, and recycles fatty acids and amino acids for synthesis or ATP production (117). The cellular phenotype includes the formation of prominent vacuoles in the cytoplasm that sequester organelles such as mitochondria and ribosomes. Autophagy is regulated by pathways that signal through another PIKK family member, the mammalian target of rapamycin (mTOR) protein, and occurs in a variety of human tumor cell lines following exposure to ionizing radiation, at least in the dose range 2–10 Gy (e.g., see Refs 118–120), although the effect of lower doses remains to be determined. Autophagy is independent of both p53 and caspases (101).

12.6.5 Mitotic Catastrophe

Mitotic catastrophe is usually defined as the failure of a cell to undergo proper mitosis after DNA injury. It is probably caused by chromosome missegregation and cell fusion. The cells often fuse, enlarge, and form multinucleated/polyploid or giant cells containing decondensed chromatin, spindle abnormalities, and micronuclei. Whether mitotic catastrophe is a mode of cell death or represents a step on the way toward death is unclear. Some reports suggest that cell death may occur secondarily to mitotic catastrophe through an apoptotic or accelerated senescence response (112, 121), and such events may contribute significantly to the loss of clonogenic potential following a radiation insult (122–124). Mitotic catastrophe could, under some circumstances, represent a survival mechanism for tumor cells, and clonogenic malignant cells might emerge from irradiated cultures that could contribute to resistance to therapy (125, 126).

12.7 CONVENTIONAL MODELS FOR CELL SURVIVAL CURVES, FRACTIONATION, AND DOSE-RATE EFFECTS

Based on a number of publications that employed clonogenic (colony-forming) assays for assessing cell death following acute single exposures to low-LET radiations, it was apparent that survival curves for cultured mammalian cell lines often display a low-dose "shoulder" that reflects their relative radioresistance at doses below ~0.5 Gy. For many years, such survival curves were modeled and interpreted on the basis of "target theory" in which a number of cellular targets had to be "hit" by an ionizing event in order for the cell to be reproductively inactivated. This observation led to the widespread use of the multitarget model and an adaptation thereof, the "multitarget with initial slope" model, which accounted for the nonzero initial slope of most survival curves (127). Subsequently, the linear quadratic model found increasing favor for fitting clonogenic cell survival data and for predicting the effect of different dose-fractionation protocols or dose rates on biological/clinical effects (12). In this model, the surviving fraction (SF) of cells and the dose (D) are related by the equation: $-\ln[\text{SF}] = \alpha D + \beta D^2$. The basic principles of this model are illustrated in Fig. 12.2a. In the linear quadratic model there are two components of cell killing. The linear (αD) component represents single-hit lethal events that are induced in direct proportion to D. The quadratic (βD^2) component represents two-hit lethal events that result from an interaction between two sublesions, each of which is produced in proportion to D.

Because it requires an interaction in space and time between two separate sublesions, the β-component of cytotoxicity reflects underlying types of damage whose contribution to cell killing can be minimized by fractionating the dose. In other words, the contribution of β-type events to cell killing depends on the time taken to administer the dose. The resulting sparing effect presumably reflects repair of the individual β-type "sublethal" lesions that can occur in the interval between the dose fractions; if repaired, then these sublesions can no longer contribute to β-type cell killing, and the overall extent of cell killing decreases as the dose is fractionated over time (Fig. 12.2b). The half-time for the repair of the individual sublethal lesions is generally reported to be on the order of 1 h for cultured human cells. The linear α-type lethal events, on the other hand, do not depend on time and fractionation. For large numbers of small fractions, the survival curves will approximate the initial slope (α) of the HDR curve (Fig. 12.2b).

The dose rates typically used in clinical XRT and in laboratory investigations are in the range of 1–5 Gy/min. Delivery of a dose of 2 Gy therefore takes less than 2 min, which means that biological responses to DNA damage do not occur to a significant

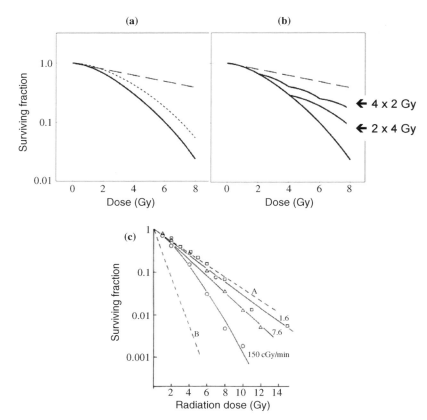

FIGURE 12.2 (a) The solid line represents the idealized linear quadratic survival curve for a hypothetical human tumor cell line following exposure to γ- or X-rays. The individual curves representing the α (single-hit: ——) and β (two-hit: - - -) components of the linear quadratic model are shown by the dashed lines. (b) Effect of fractionation on cell survival for this hypothetical cell line, based on the linear quadratic equation, for either 2 × 4 Gy fractions or 4 × 2 Gy fractions, assuming complete repair. Dashed line; α-component, redrawn from (a). Reproduced with permission from Ref. 11. (c) Dose-rate effect for the killing of a human melanoma cell line as the dose rate is lowered from 150 to 1.6 cGy/min. The dashed curves represent the best fit to the data set obtained using the lethal-potentially lethal model assuming either full repair (A) or no repair (B), respectively. The estimated half-time for repair was ~10 min. Reproduced with permission from Ref. 129.

degree during the irradiation period. At lower dose rates, however, the time required to deliver the dose will be longer, thus enabling biological responses to occur during the irradiation and thus to modify the observed response (128). As with fractionation, the conventional linear quadratic model predicts that lowering the dose rate from 1 Gy/min down to 1 cGy/min should be associated with a gradual decrease in radiosensitivity, with the LDR survival curve becoming progressively straighter and eventually extrapolating the initial slope (α) of the HDR curve at low dose rates

(Fig. 12.2c). By analogy to dose fractionation, this sparing effect is ascribed to the effective repair of sublethal/β-type lesions by the cells during the protracted exposure. Even lower dose rates below ~1 cGy/min allow cell proliferation to occur during the irradiation period, and this will lead to the repopulation of clonogenic cells and to a further, albeit "artifactual," increase in radioresistance (128).

The conventional wisdom is therefore that a given dose of TRT (i.e., a protracted LDR exposure) will be less clinically effective than the same dose of XRT (i.e., an acute HDR exposure). However, as was noted previously, this prediction is inconsistent with some clinical observations, suggesting that there may be some biological differences in the cellular response to LDR versus HDR exposures, a point to which we will return below. Of additional interest in the context of the potential therapeutic application of TRT, it has been noted that the LDR survival curves tend to exaggerate differences in the intrinsic radiosensitivity of human tumor cells (128).

Given that repair processes mediate this sparing effect, it is perhaps not surprising that the dose rate effect for cell killing is highly dependent on the cellular DNA-repair status; thus, DSB repair/Ku-deficient cells such as xrs5 and xrs6 (130), DNA-PK_{cs}-deficient irs-20 cells (131), and fibroblasts from AT patients (132) show a much reduced or absent dose-rate effect for cell killing.

12.8   LOW-DOSE HYPERRADIOSENSITIVITY-INCREASED RADIORESISTANCE

In recent years there have been several unexpected findings in fundamental cellular radiobiology that have raised major questions concerning the use of conventional models to describe the biological and clinical effects of low dose/LDR ionizing radiation exposures. These include the identification and mechanistic description of low-dose hyperradiosensitivity (HRS)-increased radioresistance (IRR), inverse dose-rate effects, the radiobiological bystander effect, and the adaptive response. The following sections will summarize the current understanding of these effects. The potential implications for these and other recently identified low dose/LDR radiobiological phenomena will be discussed later in this chapter.

12.8.1   Phenomenology of HRS-IRR

Early radiobiological models of cell killing relied on measuring the surviving subpopulation of cells within an irradiated culture, rather than identifying those cells that have lost their reproductive potential, and were therefore relatively insensitive to changes occurring at low doses where the surviving fraction of cells is high. With the emergence of computerized imaging or flow-cytometric methods for discriminating clonogenically "dead" cells within an *in vitro*-irradiated population, many mammalian cell lines were found to display an intriguing structure in their survival curves after low single/acute doses. This behavior has now been designated HRS-IRR, and is illustrated in Fig. 12.3a and b for human fibroblasts and for three human tumor cell lines, respectively. Most of the survival curves are characterized by an initial hypersensitive

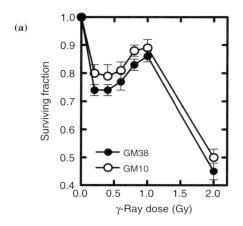

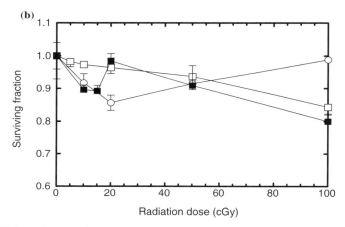

FIGURE 12.3 γ-Ray survival curves for (a) the normal human fibroblast strains GM38 (●) and GM10 (○) (reproduced from Ref. 18) and (b) the human A549 lung adenocarcinoma (■), T98G glioma (○), and MCF7 breast carcinoma (□) tumor cell lines (data taken from Ref. 162). Note the lack of a significant low-dose HRS response in MCF7 cells.

response at low doses (typically below ~0.25 Gy), followed by a region (typically up to ~0.5 Gy) in which the cells display increasing resistance to a single-dose exposure (133, 134); an exception is the MCF7 breast cancer cell line, which did not exhibit an HRS-IRR response. At doses above ~1 Gy the survival curve again turns downward and closely follows the typical linear quadratic response. For cell types and conditions where an HRS-IRR response is in evidence, the conventional linear quadratic model underestimates the level of cell killing occurring after an acute low-dose exposure. Accordingly, a modified linear quadratic model incorporating the concept of an "induced repair threshold" was developed that has two parameters for single-hit/α-type cell killing: α_s (sensitive) defines the low-dose response, whereas α_r (resistant) defines the high-dose response (134, 135). For tumor cell lines, α_s/α_r is typically greater than 1, that is, HRS responses are relatively commonplace. Malignant cells also appear to exhibit more pronounced HRS responses than normal cells (136).

12.8.2 Mechanistic Basis of HRS-IRR

The HRS-IRR response presumably reflects alterations in the way that cells process DNA-damaging events as the dose increases. The overall pattern of cellular responses to radiation-induced DNA injury was outlined above and is summarized in Fig. 12.1. Clearly, alterations in damage processing could occur at many steps within this DNA damage surveillance–response network, including at the levels characterized by the four major classes of damage-response proteins, that is, the sensors, mediators, transducers, and effectors. Based on this model, cells would exhibit hypersensitivity to low doses of radiation that do not produce sufficient DNA damage to trigger critical cytoprotective processes; rather, these processes would only be activated by higher doses. Some possible scenarios will now be discussed.

12.8.2.1 Is There a Cellular Dose Threshold for Sensing and Responding to DNA Damage?   The suspicion here is that cells might be hypersensitive to low doses below ~0.3 Gy because such doses do not activate the cytoprotective DNA damage surveillance–response network as efficiently as higher doses. A number of experimental studies have directly addressed this question. As was noted earlier, the activation of ATM (the major protein kinase responsible for p53 and H2AX phosphorylation) is an important early step in the response to radiation-induced DSBs. The dose–response for the autophosphorylation of ATM at serine-1981, which is believed to be important for activation of the ATM kinase activity postirradiation, was examined in human fibroblasts and found to exhibit a graded increase over the dose range 0.1 to 0.5 Gy; autophosphorylation at 15 min postirradiation was detectable after only 0.1 Gy and increased thereafter, becoming maximal at ~0.4 Gy and being relatively insensitive to further increases in dose above this point (25). Clearly the dose–response curve for ATM autophosphorylation in this cell type exhibits some similarity to the HRS-IRR response insofar as ATM is not fully activated until the same general range of doses that cellular cytoprotective mechanisms appear to fully engage. These widely cited data have been interpreted to suggest that activated ATM may regulate the induction of at least some of the cytoprotective responses that mediate the IRR process, and axiomatically that HRS is a result of the suboptimal activation of the damage-response network at lower doses (137). However, in another study, 10B635 human fibroblasts showed a quite different pattern, with a slight *decrease* in ATM serine-1981 phosphorylation at 15 min postirradiation for doses below 0.2 Gy, followed by a progressive increase in activity at doses above 0.5 Gy (138). In contrast, T98G human glioma cells showed a response similar to that seen for fibroblasts in the Bakkenist and Kastan (25) study, with serine-1981 phosphorylation increasing progressively between 0.1 and 0.4 Gy and being relatively nonresponsive above this dose (138). MR4 and 3.7 rat fibroblasts showed a different response again, with serine-1981 phosphorylation increasing at doses between ~0.3 and 0.6 Gy before declining again at doses up to 1 Gy (138). In EBV-immortalized LCL-N human lymphoblastoid

cells, ATM phosphorylation at serine-1981 as well as *in vitro* kinase activity, measured at 30 min after γ-irradiation, increased progressively with dose over the range 0.25–2 Gy (139). Interestingly, both the increased ATM autophosphorylation and kinase activity were transient after doses below 0.5 Gy but were longer lived above this dose (139).

A key question arises as to the nature of the dose-dependency for the phosphorylation of downstream targets of ATM. With respect to H2AX phosphorylation, the initial yield of γ-H2AX foci in primary human fibroblasts was found to be a linear function of dose all the way from 1 mGy to 100 Gy (140). This is obviously a very different scenario from the above-mentioned dose–responses for ATM autophosphorylation in human fibroblasts. While these collective observations suggest that ATM phosphorylation at serine-1981 may not be rate limiting for H2AX phosphorylation, it should be noted that the phosphorylation of H2AX in irradiated cells is not totally dependent on ATM. Both DNA-PK$_{cs}$ and ATR have been shown to phosphorylate H2AX (141–143). A linear no-threshold dose–response curve was also seen for γ-H2AX, measured 30 min after exposure to doses of 0.05–2 Gy, in two human glioma cell lines, T98G and U373, that do and do not exhibit an HRS response, respectively (144). A linear response with no threshold was also seen for phosphorylation of the CHK1 and CHK2 checkpoint kinases in that study over the range 0.2–2 Gy (144). In LCL-N lymphoblastoid cells the initial phosphorylation of CHK2 at threonine-68 was readily apparent after 0.25 Gy and further increased with dose, indicating no threshold; however, in spite of this low-dose phosphorylation at threonine-68, the full activation of CHK2 kinase (indicated by *in vitro* enzyme protein mobility shift and autophosphorylation at threonine-387) exhibited a pronounced threshold, requiring doses in excess of ∼1 Gy (139). The authors note that these observations underscore the complex multistep nature of the activation of CHK2, which undergoes a cascade of phosphorylation events (145), and suggest that the later phosphorylation steps may involve kinases such as the polo-like kinases 1 and 2 that may require higher doses than ATM for activation (139).

A noteworthy recent addition to the literature monitored responses to DNA damage in live proliferating cell cultures and confirmed that the cellular response to γ-ray-induced DSBs in HT1080 human fibrosarcoma cells was a linear function of dose over the extended range of 5 mGy to 1 Gy (146); this study looked at the formation of 53BP1 foci (in addition to the γ-H2AX foci measured in primary human fibroblasts in the above-mentioned earlier study by Rothkamm and Löbrich (140)), a response that depends on the phosphorylation of 53BP1 by PIKKs such as ATM (147).

Another important downstream target of the ATM kinase activity is p53. Defining the shape of the dose–response for p53 activation at low doses is, however, confounded by the fact that both the stability and the activity of p53 are impacted by a variety of concomitant posttranslational modifications, and many of these combinations of modifications are of unknown significance at the present time. One of the most extensive studies of this type was performed by Offer and colleagues (148) using selected cell types, including the p53 wild-type 70Z/3 murine pre-B-cell leukemia cell line. Their findings, summarized in Fig. 12.4, show that p53 protein accumulation in this cell line was already significant after a 0.5 Gy γ-ray exposure and further increased

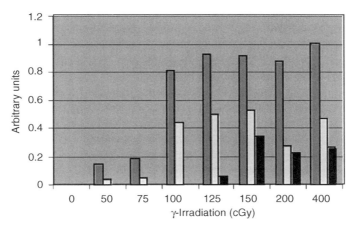

FIGURE 12.4   p53 protein levels and posttranslational modifications in p53 wild-type 70Z/3 murine pre-B-cell leukemia cells exposed to various doses of γ-rays. Data are from Western blot analyses illustrating the relative levels of (i) total p53 protein (gray bars), (ii) dephosphorylation of p53 at serine-376 (open bars), and (iii) phosphorylation of p53 at serine-15 (dark bars). Reproduced with permission from (148).

at higher doses. In contrast, dephosphorylation of p53 at serine-376 (which appears to be related to the BER activity of the cells; see below) did not occur until a dose of 1 Gy, and there was no further change above this dose. Phosphorylation of p53 at serine-15 was evident only after doses in excess of 1.5 Gy and again was not dose-dependent above this point. Furthermore, the ability of p53 from these cells to specifically bind to its DNA consensus sequence was only apparent after doses of 1.5 Gy and higher. If generally applicable, such dose-dependent effects on the cellular levels, pattern of posttranslational modification and functional activity of p53 would be expected to impact on its ability to crosstalk with the various downstream events in the DNA damage surveillance–response network. Indeed, depending on how it is posttranslationally modified, p53 appears to selectively bind to different promoters (38, 149, 150). Phosphorylation of p53 at serine-15 at 30 min postexposure in the T98G (HRS-positive) glioma cell line indicated a linear dose–response with no threshold over the range 0.05–2 Gy; a similar response was seen for the U373 (HRS-negative) glioma cell line (144). In LCL-N lymphoblastoid cells the phosphorylation of p53 at serine-15 was modest after doses of 0.25 and 0.5 Gy, but increased strongly above 1 Gy (139). In a more complex model—organ cultured human skin—the dose–response for total p53 and for p53 serine-15 phosphorylation in epidermal cells at 4 h postirradiation showed linear to logarithmic increases over the dose ranges 0.05–1.0 Gy and 0.10–1.0 Gy, respectively (151). This response was presumed to reflect primarily keratinocytes. In the same study, dermal cells exhibited a weak p53 phosphorylation at serine-15 that was more transient and, at 4 h after irradiation, exhibited little dose responsiveness; this result may reflect differing responses in different cell types, but may also have been an artifact of the stress caused by the tissue processing procedures (151).

A number of studies have addressed the question of whether there is a threshold dose for the transcriptional transactivation of specific genes following exposure to ionizing radiation. Not surprisingly, many of these responses are mediated by p53. In ML-1 human myeloid tumor cells there was an approximately linear dose–response for the induction of $p21^{WAF1}$ and $GADD45$ transcripts by γ rays up to 0.5 Gy, measured at 2–3 h postirradiation; both genes were significantly induced by as little as 0.2 Gy (152). At doses above $\sim$2 Gy a saturation of induction was apparent. For three other genes—$MDM2$, $ATF3$, and BAX—the dose–response was linear below $\sim$0.1 Gy and then began to flatten out. Thus, none of the five genes studied showed any threshold for transactivation. However, this may not be a general response in all cell types as the accumulation of p21^{WAF1} protein in LCL-N lymphoblastoid cells at 3 h postirradiation, as well as inhibition of DNA replication, was only seen after doses above 1 Gy (139).

In summary, it is apparent that many of the published studies for various cell types do not support the existence of a low-dose threshold for the activation of aspects of the cellular DNA damage surveillance–response network. Most of the existing dose–response data are either linear or linear at low doses and then reach a plateau at higher doses, indicating that there is either an equal response per induced DSB or an optimal response per DSB at lower doses. However, there are some notable exceptions to this behavior. For example, in the study of Buscemi et al. (139) the CHK2 protein, unlike ATM, was enzymatically activated only by doses in excess of $\sim$1 Gy even though doses below this threshold did cause significant ATM-dependent phosphorylation of CHK2 at threonine-68. This study highlights the importance of assaying the protein kinase activity *per se* in addition to individual posttranslational events such as the phosphorylation of specific serine and threonine residues. Indeed, for proteins that are multiply posttranslationally modified, such as CHK2 and p53, it may be misleading to study a single posttranslational modification of uncertain biological significance.

12.8.2.2 Is There a Cellular Dose Threshold for Repairing DNA Damage?

It has been widely proposed that, for those cell types that exhibit an HRS-IRR response, the transition from HRS to IRR may be a consequence of the triggering of some type of DNA-repair activity once a critical threshold level of DNA damage is exceeded. Thus, for the specific fibroblast lines shown in Fig. 12.3a, doses below $\sim$0.4 Gy would be predicted to be below the activation threshold. Either the relatively low levels of DNA lesions in these cells would escape detection by the sensor proteins and thus evade repair, or they might be detected properly but the signal does not fully trigger a repair response. In contrast, doses exceeding the IRR threshold (i.e., above $\sim$0.4 Gy) would be anticipated to fully activate these DNA-repair pathways.

That DNA-repair mechanisms may be involved in the IRR response was suggested by observations such as the attenuated IRR response of some repair-deficient cell lines, especially those with defects in NHEJ (e.g., see Refs 153 and 154), and of cells treated with inhibitors of various DNA-repair pathways (133, 155, 156). Human tumor lines with a pronounced HRS response were also reported to exhibit decreased DNA-PK activity when assayed 2 h after a 0.2 Gy γ-ray exposure, whereas cell lines lacking an HRS response displayed *increased* DNA-PK activity (157).

Because of their presumed critical role in initiating cell-death responses following radiation exposure, DSBs and their repair are a logical starting point for this discussion. The expectation that DSB-repair efficiency might be dose-dependent (at least for some cell types) is already implied from the observations, discussed above, that (i) there appears to be a nadir for ATM activation at ~0.4 Gy in human fibroblasts, (ii) p53 accumulation and activation occurs at relatively low doses, (iii) posttranslational modifications of p53 change with dose, (iv) both ATM and p53 positively influence the rate of DSB rejoining in mammalian cells (158, 159), and (v) p53 in general plays multiple direct and indirect roles in cellular DSB-repair pathways (35, 42).

So are there differences in the repair response to DSBs at different dose ranges? In one study, the extent of DSB rejoining in confluent primary human fibroblast cultures following a single acute X-ray exposure was seen to decrease with decreasing dose (140). These authors actually monitored the disappearance of DSB-related γ-H2AX foci as a surrogate biomarker for DSB rejoining and found that, after doses between 0.5 cGy and 2 Gy, extensive repair of DSBs occurred. However, after doses below 0.5 cGy, DSB repair was considerably slower, and after 1.2 mGy, essentially no removal of DSB-related foci was apparent even after 24 h. Although these observations could have important consequences for LDR effects and for TRT in general, the suggestion that they are consistent with a role for activation of DNA repair at the HRS to IRR transition (137, 140) is questionable considering that the doses at which the major changes occurred were not in the same range that HRS-IRR responses are typically observed.

In our laboratory, we addressed this question using the GM38 normal human fibroblast cell line that we had found (Fig. 12.3a) to exhibit significant HRS and IRR responses (18). Because the γ-H2AX assay represents an indirect biomarker for cellular DSBs that are detectable by the damage sensors, and because there has been some suggestion that lesions other than DSBs may contribute to the output of this assay, our study also assessed DSB rejoining using the neutral (pH 7) comet assay. Whereas exposing proliferating GM38 cultures to 2 Gy of γ rays was followed by extensive rejoining of DSBs as indicated by either the γ-H2AX assay (Fig. 12.5) or the comet assay (Fig. 12.6), minimal repair of DSBs was apparent in these cells after a low-dose (e.g., 0.2 Gy) exposure, suggesting that repair in this cell type is indeed accelerated at doses above the IRR threshold. Recent unpublished data from the Weinfeld laboratory similarly indicate that, for the human A549 nonsmall cell lung adenocarcinoma (Fig. 12.7) and HCT116 colon cancer cell lines (data not shown), the rate of removal of γ-H2AX foci was more rapid after a dose of 25 cGy than after a 10 cGy exposure.

In contrast, Wykes et al. (161) did *not* observe a dose dependency of the kinetics of γ-H2AX foci removal in two pairs of isogenic human cell lines with differing DNA repair status: M059K/M059J (DNA-PK$_{cs}$⁺/DNA-PK$_{cs}$⁻) and EBS7/EBSYZ5 (ATM⁻/ATM⁺). In general, there was no major difference in the rate of removal of γ-H2AX foci up to 4 h after exposure to doses of 0.2 Gy versus 2 Gy. The rate of removal of γ-H2AX foci between 30 min and 4 h postirradiation in the T98G (HRS-positive) and U373 (HRS-negative) human glioma cell lines was also equivalent following exposures in the dose range 0–0.5 Gy versus 0.5–2 Gy, although

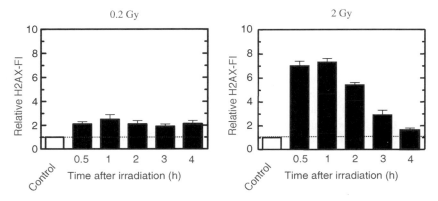

FIGURE 12.5 Kinetics of the formation and removal of radiation-induced γ-H2AX (the phosphorylated form of the variant histone, H2AX) after exposure of human cells to doses above and below the threshold for IRR. GM38 fibroblasts were exposed to either 0.2 or 2 Gy of γ-rays, and γ-H2AX was determined at various times thereafter. The data are normalized to the γ-H2AX levels for the corresponding nonirradiated control samples. Reproduced with permission from Ref. 18.

measurements of RAD51 foci suggest that DSB rejoining may shift toward the HR-dependent pathway at lower doses (144).

Offer and colleagues (148) examined whether there might be a radiation dose–response for the activation of BER activity using an assay in which nuclear extracts from selected cell types were incubated with plasmids containing an AP site, and repair resynthesis was quantified based on ^{32}P-dGTP incorporation. Whereas low

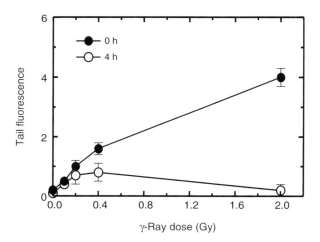

FIGURE 12.6 Effect of dose on the extent of repair of radiation-induced DSBs in GM38 human normal fibroblasts. DSBs were evaluated by the neutral comet assay either immediately after (●) or 4 h after (○) exposure to different doses of γ-rays. Reproduced with permission from Ref. 18.

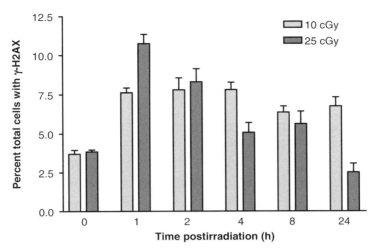

FIGURE 12.7   Induction and removal of radiation-induced γ-H2AX in human A549 non-small cell lung adenocarcinoma cells following either a 10 or a 25 cGy exposure to ^{137}Cs γ-rays. Phosphorylated histone H2AX (γ-H2AX) was measured by flow cytometry following the method of Kurose et al. (160).

doses of γ-rays resulted in increased p53-dependent BER activity in M1/2 mouse myeloid cells expressing wild-type p53, with a peak enhancement at ∼0.5 Gy, higher doses caused a rapid inhibition of BER activity. A similar behavior was seen in the p53 wild-type 70Z/3 pre-B-cell line and in normal human NL 5J3 lymphoblastoid cells. These changes in BER paralleled the observed posttranslational modifications of p53; specifically, the enhancement of BER activity after low doses was associated with dephosphorylation of p53 at serine-376 and was ATM-dependent, whereas the decrease in BER after higher doses correlated with p53 serine-15 phosphorylation. It is of obvious interest in the present context that the dose-dependency of p53-dependent BER activation and inhibition in that study (148) exhibits some resemblance to the dose–response for HRS-IRR. Note also that the BER activity of p53-deficient cells did not show this biphasic pattern, but rather continuously increased with dose.

There is clearly a need for rigorous comparative studies in several different cell types and genetic backgrounds before it will be possible to make any general conclusions with respect to the existence (or not) of cellular thresholds for the repair of DNA damage and their actual relationship to the HRS-IRR phenotype. Technically, we should also note that the question of whether IRR requires efficient DNA repair or whether the signal to induce IRR might initiate efficient repair cannot be discriminated at the present time.

Finally, a recent report examined the removal of 53BP1 foci (rather than γ-H2AX foci) in living cells using 53BP1 fused to yellow fluorescent protein (YFP-53BP1) as a surrogate marker for DSBs, rather than incubating the cells with labeled antibodies after fixation (146). These authors observed that the half-time for the repair of DSBs was very similar (about 3 h) following exposure to doses of γ-radiation all the way from 5 mGy to 1 Gy. These observations suggest that the rate of repair of DSBs is essentially independent of dose and are therefore quite different from the data reported for γ-H2AX foci removal at low doses by Rothkamm and Löbrich (140). The authors (146) suggest that the main difference between these studies lies in the use of live-cell imaging, which improves the background signal-to-noise ratio. As in the earlier study (140), this later study (146) is not particularly informative for events in the dose region of the HRS to IRR transition, although these findings may prove to be especially relevant to the TRT community, even if they appear at this point in time to confuse our current knowledge relating to the dose dependency of cellular responses to DSBs.

12.8.2.3   A Possible Role for P53 in the Dose-Dependency of the Activation of the DNA Damage Surveillance–Response Network?   Given the dose–responsive nature of the accumulation and activation of the wild-type p53 protein in irradiated cells, and given the extensive crosstalk between p53 and the DNA-repair machinery of cells and indeed with other effector responses, it is somewhat intuitive to suspect that p53 will be an important determinant of how different aspects of the DNA damage surveillance–response network are engaged as a function of dose. Sengupta and Harris (35) coined the term "cellular rheostat" to describe the role of wild-type p53 in the context of how cells respond to different radiation dose levels. They hypothesized that, after a low-dose exposure, p53 may interact with the DNA-repair machinery and facilitate repair; in contrast, after a high-dose exposure (when DNA damage cannot easily be managed by p53 alone), p53 may undergo stabilization and activation as a transcription factor.

The modest literature relating to the impact of p53 on the BER pathway does support this model. Specifically, the shape of the dose–response for the stimulation of BER by radiation in both mouse and human cells in the above-mentioned studies by Offer and colleagues (148) was greatly dependent on p53. As noted above, in the p53-wild-type background, doses below $\sim$0.5 Gy stimulated BER activity, whereas higher doses inhibited BER and invoked p53-dependent apoptosis (148). In contrast, for p53-deficient cells, BER activity increased progressively with dose. In that study, it was also apparent from using the apoptosis inhibitor z-VAD-fmk that the decision of p53-proficient cells to disable their BER activity at higher doses was independent of their decision to activate the apoptotic cascade.

As regards the impact of p53 status on DSB rejoining in mammalian cells, we (158) showed that wild-type p53 in both the human fibroblast and the HCT116 human colon carcinoma background positively influenced the rate of DSB rejoining after doses below 10 Gy; given that both the stabilization and the activation (by posttranslational modification) of p53 is highly dose dependent (e.g., see Ref. 148), such changes could easily be anticipated to contribute to the differences in the rate of DSB rejoining with dose observed in some studies and cell types. It is perhaps also not too surprising to find that some studies indicate that the HRS-IRR phenotype is itself highly dependent on cellular p53 status. Thus, some cell lines with abrogated p53 (notably A549 lung cancer and T98G glioma cells treated with the p53 inhibitor pifithrin, the

HCT116-p53 knockout line 379.2, and the p53-insufficient Li-Fraumeni Syndrome human fibroblast strain 2800T) did not display an HRS-IRR response, in marked contrast to their p53-proficient counterparts (162). Interestingly the T98G tumor line, which has a p53 mutation at methionine-237 that causes a severe deficit in transcriptional transactivation, showed a normal HRS-IRR response; thus, p53-dependent transcription is probably not important for this activity. A recent study by Krueger and colleagues also indicated a role for p53-dependent apoptosis in HRS; however, these authors noted that the role of p53 in HRS-IRR is complex and remains incompletely understood (163). As noted elsewhere (164), it would be of great interest to define the p53 response after low doses of radiation in cell lines with differing HRS-IRR phenotypes.

12.8.2.4 Is There a Threshold for Activating the Critical Cell-Cycle Checkpoint?
As was discussed earlier, the key checkpoint implicated in the HRS-IRR transition is an early G_2 checkpoint initially described by Xu et al. (46). Unlike the "classical" G_2/M checkpoint (165), this early G_2 checkpoint, although ATM dependent, is dose independent at doses above $\sim$1 Gy, but is activated progressively at lower doses (138). In some cell lines, for example, MR4 rat fibroblasts, there appears to be a threshold at $\sim$0.3 Gy for activation of this early checkpoint, while in others, for example, 3.7 rat fibroblasts, there is no evidence for such a threshold dose (138). The dose-dependency range for the induction of the checkpoint is similar to that for activation of ATM via serine-1981 autophosphorylation (138). Furthermore, AT cells, which by definition lack active ATM, show an HRS response, but no IRR response (138). On the other hand, ATM-expressing cells in which ATM has been activated by treatment with chloroquine prior to irradiation no longer display an HRS response (138), suggesting that the IRR response may override HRS.

The phase of the cell cycle in which cells are irradiated is also a critical factor (138, 166). Whereas irradiation of an asynchronous cell population typically leads to 10–15% cell death at the optimal HRS dose, enrichment of the cells in G_2 phase by flow cytometry prior to irradiation greatly increased the percentage of cells displaying the HRS response (30–40%), but did not completely abolish the IRR response, in comparison to an asynchronous population.

Based on these and other observations, Marples and coworkers have put forward a dose and cell-cycle dependent model to explain the HRS-IRR phenomenon (summarized in Refs 136 and 164). In cell lines that exhibit HRS, a fraction of cells exposed to doses below $\sim$0.3 Gy fail to activate the ATM-dependent early G_2 checkpoint, and the cells enter mitosis with unrepaired DNA damage and die by apoptosis (138). Because activation of the checkpoint increases with dose, a balance point is reached above which a greater number of cells survive, thus giving rise to the observed IRR response.

12.8.2.5 Are There Different Thresholds for Activating Various Cell-Death Mechanisms?
Several investigators have suggested that low doses of ionizing radiation tend to trigger cell death in some cell types by activating apoptosis, whereas high doses invoke necrosis (e.g., see Refs 7 and 167). Indeed, low doses of

XRT were found to induce extensive apoptosis in tumor cells, with the dose–response curve for apoptosis reaching a plateau at ~7.5 Gy (168). In the same study, multiple small fractions of XRT produced a higher level of apoptosis than a large single dose (168).

Based on such observations, it was logical to ask whether the HRS-IRR phenomenon might reflect the shape of the underlying dose–response curve for radiation-induced apoptosis. However, no clear relationship was apparent between the α_s/α_r ratio and apoptosis (134, 136). Similarly, no apoptosis was seen in BMG-1 human glioma cells at low doses where these cells exhibited pronounced HRS (169), nor was HRS obviously related to p53 status. In contrast, a significant correlation was apparent between the HRS-IRR and the extent of low-dose apoptosis (as measured by Annexin V binding or caspase 3 activation) in a panel of human cell lines (162); the caspase 3 activation data from that study are shown in Fig. 12.8. The corresponding survival curve data for these cell lines are shown in Fig. 12.3b; consideration of these data indicates an essentially mirror image relationship between the HRS-IRR response and the extent of apoptosis over the 0–1 Gy dose range (162). In the study by Enns et al. (162), both Annexin V binding and HRS in irradiated p53 wild-type cells were blocked by the p53 inhibitor pifithrin, as were these responses in p53-deficient cell lines, suggesting a relationship between p53-dependent apoptosis and HRS. A recent report by Krueger et al. (163) also suggested a correlation between the apoptosis (caspase 3 activation) and the incidence of HRS in a panel of mammalian cell lines.

Although the balance of data support a role for apoptosis in the HRS-IRR phenotype, there are clearly some disparate findings in this area, the reasons for which need to be established if such information is to be useful in the development of

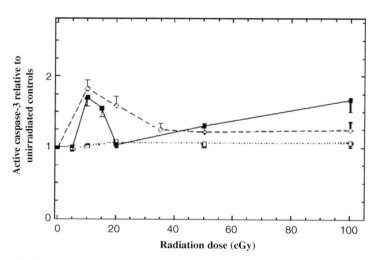

FIGURE 12.8 γ-Ray dose–response curves for apoptosis in human tumor cell lines that either do (A549 (■) and T98G (○)) or do not (MCF7 (□)) exhibit a low-dose HRS-IRR response. Apoptosis was measured by caspase 3 activation at 24 h postirradiation. The dose rate was 22 cGy/min. Reproduced with permission from Ref. 162.

TRT. Enns and colleagues (162) note that the timing at which apoptosis assays are performed relative to irradiation may contribute to interstudy variations and suggest that these discrepant observations on the role of apoptosis in HRS-IRR might be reconcilable if the above-mentioned early G_2 checkpoint hypothesized to mediate the transition from HRS to IRR (137) was to regulate p53-dependent apoptosis. The role of p53 in the HRS-IRR response is also in need of clarification.

Although a number of other mechanisms of cell death, such as autophagy, mitotic catastrophe and accelerated senescence, were outlined earlier, it is unfortunate that little is known about the activation of these other cytotoxic responses at low doses/dose rates which may be common in TRT. This is clearly a major void in the literature, especially considering that different types of cells preferentially follow different death pathways. Such effects could have a major impact on the role of HRS-IRR in the human body and may also help to explain some of the above-mentioned discrepancies in the existing literature.

12.8.3 Ultrafractionation

One strategy for cancer therapy that might exploit HRS by avoiding triggering the IRR response is the use of "ultrafractionation" by delivering multiple acute low-dose fractions of radiation of a size that should not trigger an IRR response. Several groups have reported that human tumor model systems that are known to display HRS exhibit significant hyperradiosensitive responses after ultrafractionated XRT regimens (e.g., see Refs 170–172). In the first such study, three radioresistant human glioma cell lines—T98G, A7, and U87—that exhibited HRS were found to display a greater radiosensitivity following exposure to three consecutive 0.4 Gy fractions given at 2–4 h intervals compared with a single 1.2 Gy exposure (170). The data for T98G cells are reproduced in Fig. 12.9. The T98G cell line also exhibited greater radiosensitivity following fifteen 0.4 Gy fractions given three times a day for 5 days than following exposure to five 1.2 Gy fractions once a day for 5 days (170). This effect was not seen in a radioresistant astrocytoma cell line, U373, which does not show HRS. These authors (170) suggested that the lack of a positive ultrafractionation response in an earlier study (173) could indicate that the rodent cell lines used in that study may not exhibit an HRS response. Dey and colleagues (171) also reported a hyperradiosensitive ultrafractionation response with two human cell lines—SCC-61 and SQ-20B—derived from moderately differentiated squamous cell carcinomas of the head and neck. For both lines, giving two fractions of 1 Gy or a single fraction of 2 Gy resulted in a similar level of cell survival; however, for four fractions of 0.5 Gy there was a slight increase in cytotoxicity. Recent reports are beginning to suggest that ultrafractionated irradiation protocols might indeed be clinically effective in tumor control or palliation in some cases (e.g., see Refs. 174 and 175).

Hypersensitivity or "reverse fractionation" effects have also been reported in normal human skin (176–178) and salivary gland (P. Lambin, personal communication cited by Joiner and colleagues (134)). A pattern consistent with an HRS response was also reported for the persistence of γ-H2AX foci in epidermal skin cells in patient biopsy samples when measured at 30 min after exposure to doses below 0.3 Gy (179).

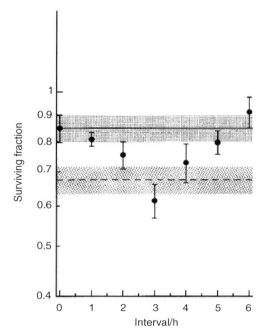

FIGURE 12.9 Survival data for T98G human glioma cells receiving three consecutive 0.4-Gy fractions of X-rays at various interfraction intervals between 1 and 6 h. The solid line and associated shading represents cell survival following a single 1.2 Gy exposure; the dashed line and associated shading represents the maximum low-dose HRS response predicted on the basis of the single-dose survival curve. Reproduced with permission from Ref. 170.

Clinical studies on metastatic tumor nodules in skin are also consistent with a low-dose HRS response (180). These studies raise the possibility that there may be a potential use for ultrafractionated XRT in the clinic. Although not all studies suggest an increase in cell killing after ultrafractionation (e.g., see Ref. 181), it should be noted that the doses used in such studies are not always consistent.

An interesting feature of the study by Dey et al. (171) is the effect of these different fractionation schemes on the activity of the NF-κB transcription factor, which is involved in cellular responses to various stress stimuli, and on various pro- and antiapoptotic mediators. Whereas exposing SCC-61 tumor cells to 2 Gy of X-rays caused an increase in their NF-κB activity as well as induction of the NF-κB-dependent antiapoptotic Bcl-2 protein, repeated low doses of radiation (0.5 Gy) in the HRS region of the survival curve did not activate NF-κB but instead induced the proapoptotic Bax protein. On this basis, the authors suggested that this dose-dependent apoptosis/survival signaling may underlie the HRS-IRR phenotype.

Whereas the ultrafractionation approach is an attempt to exploit HRS in tumors for therapeutic advantage, there has also been some concern expressed that the operation of HRS (as well as bystander effects discussed later in this chapter) in some normal tissues may be a problem in intensity-modulated radiation therapy (IMRT), where a large volume of normal tissue can receive a significant "low" dose of radiation (182, 183). An obvious question is "might such effects also play a role in the normal-tissue toxicity of TRT?" If so, then it might be cogent to reconsider the use of radioprotective agents such as amifostine, especially during the early delivery phase of TRT (184), as a means of ameliorating such effects. Whether such effects might contribute to the antitumor efficacy of TRT delivered as multiple doses will be considered later in this chapter.

12.9   INVERSE DOSE-RATE EFFECTS

From the perspective of the biology of TRT, how cells respond to radiation delivered at low dose rates is of greater relevance than their response to acute low doses. As outlined earlier, the conventional description of the dose-rate effect predicts a progressive increase in cell survival as the dose rate is reduced, until the point is reached where there is no further sparing and the survival is defined by $e^{-\alpha D}$. This "sparing" of cells is equated to the repair of the individual "sublethal" DNA lesions that contribute to the 2-hit/β-/quadratic-type cell death in the linear quadratic model. The range of dose rates typically spanned in such studies is between 1 and 5 Gy/min (acute exposures) all the way down to 1 cGy/h, under which conditions it would take 200 h (and thus several cell-cycle times) to deliver a 2 Gy dose. However, not all studies have observed such "conventional" behavior over the entire range of dose rates. Thus, R-1 rat rhabdomyosarcoma cells *in vitro* and tumors *in vivo* exhibited atypical responses to decreasing dose rates; for example, the survival of cells irradiated *in vivo* showed no obvious dependency on the dose rate between 150 and 75 cGy/h (185). In some cell lines under specific conditions, the extent of cell killing has actually been seen to *increase* as the dose rate is reduced over a particular range. Jim Mitchell and his colleagues (186, 187) originally observed such "inverse dose-rate effects" with a number of mammalian cell lines, including HeLa human cervix carcinoma cells, L-P69 mouse fibroblasts and PK-15 pig kidney cells. For example, HeLa cells irradiated at a dose rate of 37 cGy/h showed a *lower* survival for a given total dose than following a 154 cGy/h exposure. These authors suggested that such inverse dose-rate effects were caused by certain continuous low dose rates having the ability to hold the cells in G_2 (which is generally a radiosensitive phase of the cell cycle) by persistently activating the G_2/M checkpoint (186, 187). Several other laboratories subsequently reported similar inverse dose-rate effects and associated G_2-synchronization in a variety of cell types, including human cervix carcinoma cells at dose rates less than 86 cGy/h (188), human glioma cells below 49 cGy/h (189), and astrocytic tumor lines at $\sim$37 cGy/h (190). It is well established that LDR exposures in the range of 10–300 cGy/h can partially synchronize some tumor cell lines in G_2/M phase (10). In addition to rendering LDR radiation exposures more biologically effective, a prolonged G_2/M checkpoint arrest could also positively impact on LDR therapies by abrogating the effects of tumor-cell proliferation (191) and by activating apoptotic pathways (192). Although the literature with respect to this mechanism is somewhat contradictory, G_2 synchronization may be an important factor in the cellular response to LDR exposures such as those typical of TRT, and has been suggested to underlie the good tumor control achieved with protracted LDR radioimmunotherapy in the treatment of patients with lymphoma and other cancers (192).

A correlation between sensitivity to LDR radiation exposures (assessed by the regrowth delay assay) and the extent of G_2/M arrest was also seen in five human tumor cell lines—HeLa, HeLa S3, and NHIK-3025 (adenocarcinoma of the uterine cervix), Me180 (squamous cell carcinoma of the cervix) and A431 (squamous cell carcinoma of the vulva)—growing and irradiated *in vivo* as xenografts (193). *In vivo* studies by Knox and colleagues using 38C13 murine B-cell lymphoma cells as well as xenografts of the human HT29 colon cancer and the SNB75 glioblastoma also support the idea that the arrest of cells in G_2 plays a major role in tumor radioresponsiveness following LDR exposures (194).

As noted above, the generality of this G_2-synchronization mechanism is uncertain. For example, Cao et al. (195) suggested that the inverse dose-rate effect exhibited by mouse Bp8 ascites sarcoma cells *in vivo* was probably not caused by G_2/M-synchronization but rather might be related to a failure to induce repair processes at the lower dose rates. An inverse dose-rate effect was also seen in several human prostate cancer cell lines even though these cells did not accumulate appreciably in G_2 (196). One mechanism other than G_2 synchronization that could contribute to such inverse dose-rate effects is the HRS-IRR phenomenon described earlier (197, 198). In this case, an inverse dose-rate effect would be anticipated if a particular LDR exposure failed to trigger the cytoprotective IRR response, analogous to the situation for single HDR exposures below $\sim$0.3 Gy. If so, then it would be expected that cell types that exhibit a pronounced HRS response would be the most likely to exhibit an inverse dose-rate effect. Indeed, asynchronous cultures of three human tumor cell lines—T98G (glioma), A7 (glioma), and PC-3 (prostate cancer)—that exhibited a clear HRS response after acute low-dose exposures were also found to exhibit an inverse dose-rate effect for LDR exposures at dose rates below 100 cGy/h, with radiosensitivity increasing by $\sim$4 fold when the dose rate was lowered from 100 cGy/h down to 2–5 cGy/h (197, 198). An HRS-negative glioma cell line, U373MG, did not show such an effect. At 5 cGy/h there was no accumulation of either T98G or U373MG cells in G_2 at any dose, suggesting that the inverse dose-rate effect observed in T98G cells was not due to a synchrony effect. As shown in Fig. 12.10, a marked inverse dose-rate effect was also apparent with confluent T98G cultures for which cell-cycle progression/synchrony effects should be further minimized (from Ref. 198). These collective observations are consistent with the idea that inverse dose-rate effects are somehow related to the HRS phenomenon (137).

The next obvious question is whether chronic LDR exposures that invoke inverse dose-rate effects might fail to induce the typical cytoprotective DNA damage-response pathways seen after acute doses of ionizing radiation and described earlier. As noted previously, DNA damage caused by a low single HDR exposure (i.e., a dose below the threshold for inducing the IRR response) does appear to be relatively inefficient for triggering some aspects of the early damage-sensing/signaling responses. So what occurs with LDR exposures that would be more typical of TRT; might low levels of DSBs associated with such exposures similarly evade these

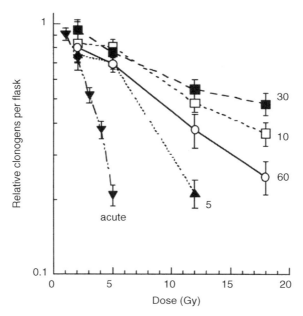

FIGURE 12.10   Survival curves for confluent cultures of the T98G cell line exposed to γ-radiation at different dose rates. At higher doses, an inverse dose-rate effect was apparent at dose rates between 30 and 5 cGy/h. Data reproduced with permission from Ref. 198.

responses? The autophosphorylation of ATM at serine-1981 in human tumor cell lines was indeed reported by Collis and colleagues (199) to be greatly reduced following an LDR exposure compared with an HDR exposure. These authors compared the response of cells to an equivalent dose of radiation delivered at either acute HDR (45 Gy/h, which produces ~1800 DSBs/h and takes ~3 min to deliver 2 Gy) or continuous LDR (9.4 cGy/h, which produces only ~4 DSB/h and takes ~20 h to deliver 2 Gy). Their findings can be summarized as follows: (i) all four human tumor cell lines studied—HCT116 and RKO (colon cancer), DU145 and PC-3 (prostate cancer)—exhibited an unusually dramatic inverse dose-rate effect for cell killing insofar as cytotoxicity was actually *greater* for LDR than HDR exposures per unit dose, although this only reached statistical significance for the RKO and DU145 lines; (ii) activation of the DNA damage surveillance–response network was diminished at LDR as evidenced by greatly attenuated levels of phosphorylated ATM (serine-1981) and NBS1 (serine-343) after LDR versus HDR exposures; and (iii) the phosphorylation of γ-H2AX (an important downstream target of the ATM kinase and a signal of *detected* DSBs) was diminished after exposure at LDR compared with HDR exposures.

This study therefore suggests that low levels of DSBs produced at LDR can evade detection to some degree, with the increased cytotoxicity possibly representing a protective mechanism for a multicellular organism after sustaining low-level DNA damage (199). These authors also stress the fact that inverse dose-rate effects and abrogated activation of the DNA damage surveillance–response network do not occur in all cell types. Findings similar to those of Collis and coworkers (199) have been reported for human fibroblasts immortalized with the human telomerase reverse transcriptase (hTERT) gene (200). Thus, acute irradiation of confluent cultures of these cells at HDR (1.8 Gy/min) resulted in significant levels of γ-H2AX foci, whereas exposure at chronic LDR (0.3 mGy/min) induced few of these foci. Phosphorylation of p53 at serine-15 (which is largely mediated by ATM) was also much less pronounced for LDR versus HDR exposures (200). Sugihara and colleagues (201) used mouse NIH/PG13Luc cells stably transfected with a p53-dependent luciferase reporter plasmid to study the transcriptional activity of p53 in response to γ-radiation at dose rates between 0.1 and 10 cGy/h. At the lowest dose rate studied, 0.1 cGy/h, p53 responses were detected after doses as low as 0.2 cGy. There appeared to be an inverse dose-rate effect for transactivation between 1 and 0.1 cGy/h. These authors (202) subsequently examined the phosphorylation of p53 in these mouse cells after LDR exposures. p53 phosphoserine-15/18 levels did not increase after continuous LDR exposures at 1.5 or 9 cGy/h, presumably because the activated protein is rapidly degraded during the protracted (72 h) irradiation period. In contrast, levels of the p53-dependent p21^{WAF1} protein were significantly elevated at dose rates above 1.5 cGy/h, as was $p21^{WAF1}$ mRNA expression, possibly reflecting differences in stability of these species; however, it should also be noted that the sustained induction of $p21^{WAF1}$ mRNA/p21^{WAF1} protein is characteristic of the activation of the accelerated senescence phenotype, as discussed previously.

In view of the critical role attributed to gene induction in the cellular and tissue response to ionizing radiation, a number of investigators have attempted to define relationships between such events and dose/dose rate. In one such study, a number of genes were evaluated for their inducibility in ML-1 human myeloid tumor cells (p53 wild type) after protracted LDR exposures spanning three orders of magnitude between 0.28 and 290 cGy/min (203). These included the $p21^{WAF1}$, *GADD45*, and *MDM2* genes that had been shown earlier (152) to be dose-responsive over the low dose range. All transcripts were quantified at 2 h after the completion of the irradiation (total dose 2–50 cGy). Whereas $p21^{WAF1}$ and *GADD45* displayed a "conventional" dose-rate effect, that is, induction generally decreased with decreasing dose rate, *MDM2* expression was independent of dose rate. A subsequent microarray/hierarchical clustering analysis of a broader group of genes after a 50 cGy exposure indicated three basic types of behavior: (i) *MDM2*-like, where induction was strong but independent of dose rate, (ii) $p21^{WAF1}$-like where induction decreased with decreasing dose rate, and (iii) genes that only appeared to be strongly induced at HDR. Interestingly, these studies of ML-1 cells suggested a possible functional consequence for these differing patterns of gene induction insofar as many of the dose rate-dependent genes were involved in apoptosis, and for ML-1 cells there was a clear dose-rate effect on the induction of apoptosis, the extent of apoptosis at 48 h postexposure being much less at the lowest dose rate; in contrast, many dose rate-independent genes were involved in regulation of the cell cycle, and for ML-1 cells there was no significant dose-rate effect on cell-cycle progression. An exception to this correspondence between gene function and dose-rate effect was apparent for the cell-cycle regulator $p21^{WAF1}$, which showed a clear dose-rate protective effect;

however, the p21^{WAF1} protein is involved in many activities in addition to its CDK-inhibitory function, including being a major transcriptional regulator with a potential role in mediating radiation-induced accelerated senescence. It would be of obvious interest in the present context to know if the ML-1 cell line exhibits an HRS-IRR and/or inverse dose-rate effect for cell survival.

The overwhelming conclusion from the study of Amundson et al. (203) is that most genes are either "conventionally" dose responsive or not responsive to changes in dose rate; there was only a hint that the induction of some genes, for example, *PHLDA3*, might exhibit an inverse dose-rate effect for specific dose rates, although this was not statistically significant. Although these data suggest that inverse dose-rate effects for gene induction are unlikely to be a major determinant of the above-mentioned inverse dose-rate effects for cell killing seen in some cell types, it is important to consider that little is known about the stability of the transcripts and the potential effect of different doses/dose rates on their degradation, which could obscure the true nature of these obviously highly dynamic molecular responses. Indeed, *any* study of the mechanisms underlying LDR/inverse dose-rate effects will encounter considerable logistical issues because of the protracted periods during which the cells are being irradiated, a time during which the cells can respond to DNA injury as well as divide to differing extents (depending in part on the magnitude of the radiation-induced cell-cycle arrest) before they are subjected to molecular or clonogenic-survival assay. Thus, mechanistic interpretation of such studies is enormously complicated. This quandary was clearly in evidence in studies by Sugihara and colleagues (201, 202) who did observe an apparent inverse dose-rate effect for the activation of some genes, notably those encoding extracellular matrix-related proteins such as collagen-2, tenascin-C, and fibulin-5, in mouse cells. However, detailed studies by this group (202) revealed that this inverse dose-rate effect was actually a result of the fact that these genes exhibit a delayed induction response, such that they are not strongly activated soon after acute/HDR exposures. Thus, the nature of the dose-rate effect depends greatly on the kinetics of induction of the particular gene, and presumably of its reversal. This problem is compounded by the fact that inverse dose-rate effects are not universal, being restricted to certain cell types for which they may occur within a narrow window of dose-rates. In fact, for this reason they would be quite difficult to exploit for clinical advantage (191).

A major unanswered question is how inverse dose-rate effects might depend on cellular p53 status, which is critically important to cancer-therapeutic applications such as TRT. Furthermore, as noted above, some studies indicate a requirement for an HRS phenotype for cell lines to exhibit an inverse dose-rate effect (197, 198). Another interesting question is therefore whether there is any relationship between inverse dose-rate effects and activation of the early G_2 checkpoint, which might be addressed through the use of specific inhibitors of ATM.

12.10 CROSS FIRE

The "cross-fire" effect refers to the fact that an ionizing particle originating from a radionuclide localized in one cell can deposit much of its energy in a neighboring or distant cell by virtue of its long range in tissue (which is typically on the order of millimeters for β-particles) relative to cellular dimensions (10). Cross fire is therefore inherent to and plays a major role in the efficacy of any TRT protocol using a long range β-emitting isotope such as ^{131}I. For example, in a study of UVW glioblastoma cells transfected with the noradrenaline transporter, ^{131}I-labeled metaiodobenzylguanidine (^{131}I-mIBG) was considerably more cytotoxic to multicellular spheroids than to monolayer cultures, supporting the idea that cross fire from the β-particles contributes additional cell killing in the three-dimensional system (204, 205). Cross-fire effects with β-emitters should contribute positively to TRT/radioimmunotherapy outcomes by overcoming limitations caused by the heterogeneous distribution of the radiopharmaceutical, especially in larger solid tumors, but this also has the downside of increasing the dose to adjacent normal tissues, a point to which we will return later in this chapter.

12.11   THE RADIOBIOLOGICAL BYSTANDER EFFECT

The radiobiological "bystander" effect and its possible role in TRT responses are discussed by Mothersill and Seymour in Chapter 14, and will not be covered in any detail in this chapter. Bystander effects are invoked to explain experimental observations in which manifestations of injury, such as DNA damage, mutation, or cytotoxicity, can be detected in cells within irradiated cultures but which have not themselves been traversed or "hit" by an ionizing-particle track. However, the terminology is also used to describe the similar events that are observed when cell-free medium from irradiated cultures is transferred to cultures of nonirradiated cells. It should be emphasized that the bystander effect in the context of TRT using radiopharmaceuticals is quite distinct from the cross-fire effect in that the former does not involve cross-irradiation of cells; rather, it appears to reflect the generation of a "damage signal" by an irradiated cell that is communicated to a nonirradiated cell through a variety of signaling mechanisms, some of which clearly involve secreted diffusible factors (see Chapter 14). The importance of this phenomenon from the TRT perspective is that, at low doses/dose rates, a cytotoxic response in which a bystander effect is operative will be greater than that predicted on the basis of dosimetric estimates and conventional models (see Chapter 13).

Some investigators (206) have remarked that the bystander effect is, by definition, a low dose and/or LDR phenomenon because, at higher doses, every cell in the target population will be "hit" by an ionizing-particle track. Based on this thinking, an increased efficiency of cell killing with respect to absorbed dose should only be observed with lower doses. Axiomatically, it has been suggested that the HRS response could be caused by a bystander effect occurring in the low-dose region of the survival curve, although data showing an inverse relationship between the bystander effect and the HRS suggest otherwise (206).

Considering the suggestion that PIKKs other than ATM may contribute to the ionizing radiation-induced phosphorylation of H2AX described earlier, it is interesting to note that ATR appears to be the kinase primarily responsible for the phosphorylation of H2AX observed in cells that are subject to bystander effects (207).

12.12 THE ADAPTIVE RESPONSE

Exposing cells to a low "priming" dose of radiation can sometimes result in resistance to a subsequent high-dose exposure delivered some time later. Such "adaptive responses" were originally observed with human lymphocytes that had been radiolabeled with tritiated thymidine, and which were found to be more resistant than nonlabeled cells to the induction of chromosome aberrations following a subsequent exposure to X-rays (208). Adaptive responses in several human cell types, but most notably in lymphocytes, were then described for various combinations of priming treatments and biological end points (e.g., see Refs 209–211). Some discrepancies and controversies in the literature are possibly caused in part by the fact that adaptive responses may not be universal and occur over a narrow dose range of priming doses (~0.5–20 cGy) (212).

Adaptive responses, like the HRS-IRR response, could be mediated in part or whole by inducible cytoprotective mechanisms such as DNA-repair pathways that are activated by exposure to the priming dose. Naturally, the question of whether cellular DNA-repair processes are stimulated following a low-dose priming exposure has been the subject of some interest. Priming treatments have indeed been found to stimulate the repair of some types of radiation-induced DNA lesions. Thus, exposure of A549 human lung cancer cells to a priming dose of γ radiation (25 cGy) stimulated the repair of thymine glycol lesions, a simple type of base damage that would be processed by the BER pathway, when the cultures were exposed 4 h later to a higher "test" dose (2 Gy) of γ rays (213). In contrast, no induction of the genes encoding the BER enzymes NTH1, OGG1, APE1, POLβ, and NEIL1/2/3, or the accessory factors XPG, LIG3, and XRCC1, was apparent in TK6 human lymphoblastoid cells within 24 h after exposure to γ-ray doses between 1 cGy and 2 Gy, nor was there an increase in BER enzyme activity or in the levels of the APE1 or NTH1 proteins in extracts of these cells after doses of 0.5–2.0 Gy (214).

The literature regarding adaptation and DSBs is also perplexing insofar as priming exposures have been reported to accelerate the repair of DSBs in some studies but not others. In V79 cultures, priming with 5 cGy of γ-rays resulted in the more rapid rejoining of DSBs (as measured using the neutral comet assay) following exposure to a higher "test" dose of radiation delivered 4 h later (215). Similarly, an enhancement of DSB-repair rate by a low dose (2 cGy) exposure to X-rays was apparent in extracts of mouse m5S cells that were subsequently exposed to 3 Gy of X-rays 5 h later and incubated with a DSB-containing plasmid substrate (216). Other studies have failed to demonstrate an adaptive response for DSB repair. For example, we saw no significant change in the rate of removal of DSB-related γ-H2AX foci when human fibroblasts were given a priming dose of 10 cGy of γ rays 4 h prior to a "test" dose of 2 Gy (Fig. 12.11) (18). It should be noted, however, that as in many of the molecular studies cited in this section, the presence or absence of an adaptive response for cell killing under these same conditions was not established.

Among other possible explanations for these differing findings are the ranges of cell types, DNA-damage assays, priming treatments and times of observation used in different studies. The fact that adaptive responses occur only within a narrow dose

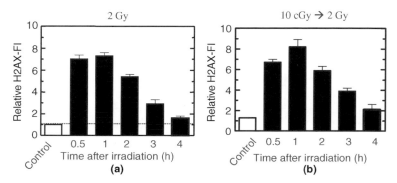

FIGURE 12.11 Kinetics of the disappearance of γ-H2AX (the phosphorylated form of the variant histone, H2AX) from human GM38 fibroblasts following a 2 Gy exposure to γ-rays either without (a) or with (b) a 10 cGy priming dose of γ-rays given 4 h prior to the "test" dose of 2 Gy. Data are normalized to the relative γ-H2AX levels in nonirradiated controls. Reproduced with permission from Ref. 18.

window may also be an issue. Understanding the roles of such variables will require a controlled evaluation of each of these factors in defined cell types and after multiple time point and dose combinations.

At least some adaptive responses to radiation, such as the formation of chromosomal abnormalities, require wild-type p53, which may channel DSBs into an adaptive "legitimate" repair pathway (217). However, at the level of BER activity, we (M. Weinfeld et al., unpublished data) have observed that, in the human fibroblast background, the repair of thymine glycol was stimulated in both p53-wild-type and p53-deficient cells receiving a priming dose (25 cGy) of γ-rays, although the total level of thymine glycol removed was much lower in the p53-deficient than wild-type cells.

Both the HRS-IRR and the adaptive responses have been suggested to involve the activation of cellular DNA repair mechanisms following stress. It is therefore of interest to note that, at least in some cases, priming cells with DNA-damaging agents such as low-dose X-rays or ³H-thymidine can eliminate the HRS response by preinducing IRR (133). Another interesting observation is that there can be a significant adaptive response in the bystander effect; thus, in mouse C3H 10T(¹/₂) cells, a low-dose exposure to 2 cGy of γ-rays was cytoprotective when given 6 h before exposure to microbeam-delivered α-particles (218). Short and colleagues (170) have also suggested that adaptive responses might play a protective role in ultrafractionation responses if the interfraction interval is too short (1–2 h) to permit an HRS response to develop.

One protein linked to both the bystander and the adaptive responses is clusterin (219). This protein exists in two forms: secreted (sCLU) and nuclear (nCLU). sCLU is an extracellular chaperone that binds to hydrophobic regions of partially unfolded proteins, thereby inhibiting their stress-induced precipitation, and is considered to be cytoprotective. It is transcriptionally upregulated in a p53-dependent manner at relatively low doses (0.5 Gy) (220) and is secreted into the medium of cultured cells following a dose as low as 2 cGy (unpublished data cited in Ref. 219). nCLU is activated by posttranslational modification and binds to the Ku70 NHEJ protein, possibly inhibiting DNA repair, and is considered to be a prodeath protein. It remains unclear if both forms of the protein are involved in the bystander effect and adaptive response and how either form of CLU exerts its effect in response to low-dose radiation exposures.

12.13   A POSSIBLE CONTRIBUTION FROM LOW-DOSE RADIOBIOLOGICAL MECHANISMS TO TRT TUMOR RESPONSES?

As was noted earlier in this chapter, TRT can sometimes be clinically effective even when calculated absorbed doses to the tumor are lower than those delivered by XRT. Similar observations have been reported in a variety of *in vivo* animal model systems (e.g., see Ref. 194 and references cited therein). Whether the observed clinical efficacy of TRT might, at least in part, be related to the above-mentioned "nonconventional" effects of ionizing radiation that are now fairly well established for low dose/LDR exposures remains to be answered. An obvious question that needs to be addressed here is, does the HRS-IRR phenomenon have any relevance to LDR therapeutics? Specifically, might the efficacy of LDR therapies such as TRT be related to their avoidance of triggering the cytoprotective IRR response in tumor cells?

A similar question may be posed about the role of apoptosis in tumor responses to protracted LDR exposures/TRT. Data obtained with cultured cells, summarized earlier, indicate that some cell lines are extremely prone to apoptosis after an acute low-dose radiation exposure. That apoptosis may similarly contribute to the efficacy of LDR therapies, that is, to TRT, has been previously proposed in the literature (7, 10, 168, 191). Indeed, it has been noted that TRT appears to kill cells primarily through apoptosis, whereas responses to conventional XRT are characterized by cell necrosis (7). Several studies suggest that LDR exposures do induce apoptosis as their primary mechanism of cytotoxicity. For example, LDR exposure of human adenocarcinoma cells to γ-rays at doses as low as 2 Gy efficiently induced apoptosis; furthermore, apoptosis exhibited an inverse dose-rate effect (8). In another study, LDR exposure of HL60 human leukemia cells to 10 Gy of β radiation (^{188}Re) over a period of 24 h caused more apoptosis than when the same dose was delivered at HDR over 0.5, 1, or 3 h (221). Again, therefore, apoptosis exhibited an inverse dose-rate effect in this model. Inevitably, such behavior is not seen in all cell types. Thus, ML-1 human myeloid tumor cells displayed a "conventional" dose-rate effect for apoptosis, the extent of apoptosis measured at 48 h postirradiation decreasing progressively as the dose rate was lowered from 290 to 0.28 cGy/min (203).

It therefore appears that apoptosis may indeed be the preferred mode of cell death following LDR radiation exposures for some cell types. Some types of tumor cells (e.g., lymphomas) also appear to have greater propensity than normal cells to undergo radiation-induced apoptosis (7). As noted earlier, for single acute low-dose exposures, there is virtually no information about the dose–response for accelerated senescence at the low dose rates typical of TRT even *in vitro*, and this is clearly one area in need of further investigation in animal models. There is, however, some indirect clinical evidence that might support a role for accelerated senescence with LDR therapies. The fact that mIBG and radiopeptide treatments often result in stable disease over a period of many months and cycles of treatment (222) and that delayed reductions in tumor volume may be seen even after several months might suggest that alternative cell-death mechanisms such as accelerated senescence are occurring (11).

The potential role of adaptive responses in TRT is uncertain. However, any LDR irradiation protocol such as TRT might be regarded as a series of priming doses briefly separated in time (191), in which case adaptive responses (*if* they are activated under these conditions) could exert a negative influence on the efficacy of LDR therapies. Furthermore, adaptive responses could potentially be invoked in some cases when a low diagnostic dose of a radiopharmaceutical is administered at some time prior to the subsequent delivery of the higher therapeutic dose of a TRT agent, for example, as is done in some centers with Zevalin where ^{111}In-labeled antibody is given first for dosimetry, followed by ^{90}Y-labeled antibody for therapy (e.g., see Ref. 223); however, there is no definitive information in this regard.

The implications of the bystander effect for TRT, while considerable, have not been considered here as they are discussed in Chapter 14. However, we should note that the operation of bystander effects *in vivo* would have implications for patient risk estimates for diagnostic radiopharmaceuticals and for assessing therapeutic responses to TRT (224). Several observations suggest that bystander effects do occur *in vivo* with radiopharmaceuticals (e.g., see Refs 224 and 225) but there is no available clinical information in this regard. As noted above for the cross-fire effect, bystander effects generated in TRT may be advantageous because they may minimize the negative impact on clinical outcome related to the expected heterogeneous tumor distribution of the radiopharmaceutical.

12.14   USE OF RADIONUCLIDES OTHER THAN β-PARTICLE EMITTERS

Although radionuclides such as the β-particle emitting ^{131}I or ^{90}Y are the most widely used in current TRT protocols, another β-emitting isotope, ^{177}Lu, has attracted some interest (226). Because these β-particles typically have a maximum range of several millimeters in tissue with most of the energy deposited at the end of the β-particle track, much of the dose deposited in the tumor is through cross-fire events, rather than targeting individual cells. This scenario might be advantageous insofar as it may help to overcome any disadvantages caused by the heterogeneity of radiopharmaceutical distribution (and thus of dose) in large tumors. Unfortunately, cross-fire events can also deliver considerable dose to the adjacent normal tissues. Partly for this reason, the use of α-particle emitters such as [212]Bi and [211]At in TRT is an area of active research. Unlike β-particles, α-particles have a short path length in tissue of only several cell diameters, so they could be useful for treating microscopic/disseminated cancers (227–229). Of interest in this context is the observation that the α-emitting radiopharmaceutical meta-[[211]At]-astatobenzylguanidine was $\sim 10^3$ fold more toxic toward human neuroblastoma cells than its β-emitting counterpart, [^{131}I]-mIBG (227).

Auger electron-emitting isotopes such as ^{125}I and ^{111}In (see Chapter 9) are also of interest in TRT because these low energy ($\sim$1 keV) electrons have a very short range in tissue of only several nanometers to at most a few micrometers. Auger-emitting radiopharmaceuticals therefore deposit most of their dose close to their site of localization, which may be exploited by intracellular targeting using, for example, internalizing antibodies for radioimmunotherapy in some cases modified with nuclear localizing peptide sequences (see Chapter 9) or ^{125}I-deoxyuridine which incorporates into cellular DNA (230, 231). These radionuclides exhibit relatively little nonspecific toxicity, presumably because of the short range (and thus minimal contribution to absorbed dose in normal tissue from cross-fire events) of the Auger electrons (230). Auger electrons may, however, generate significant bystander effects that might contribute to their therapeutic activity (224, 225).

12.15   ROLE OF TUMOR HYPOXIA AND FRACTIONATION EFFECTS

It has been known for many decades that tumors may develop regions of hypoxia (cells with low oxygenation status) within their mass. It is also apparent that hypoxia represents a significant barrier to tumor control by XRT. For one thing, transiently hypoxic cells are highly resistant to single HDR radiation exposures. This resistance is quantitated by the parameter "oxygen enhancement ratio" (OER). The OER is usually in the range of 2.5–3 for cultured mammalian cells (232). Chronically hypoxic tumors may also become more clinically aggressive or metastatic (233), so the early elimination of hypoxic cells may be critical for therapeutic success.

Conventional dose fractionation of XRT partially overcomes the negative effect of tumor-cell hypoxia by allowing for the reoxygenation of some hypoxic cells in the $\sim$24 h interval between consecutive dose fractions (234). It is not unreasonable to assume that the same would be true for protracted LDR therapies such as TRT (10). There are also some experimental reports to suggest that the OER is decreased at lower dose rates (235). Unfortunately, there is no clinical experience to draw on in this regard, although if the radiobiology of TRT reduces the importance of hypoxia in poor therapeutic outcomes there are significant implications for the role of TRT in cancer therapy.

In general, for acute high-dose exposures, the OER also decreases with increasing LET, such that the negative impact of hypoxia on tumor control probability with conventional XRT might be less important for high-LET beams (232). On this basis, there is some expectation that the use of radiopharmaceuticals that emit high-LET α-particles or Auger electrons should be more effective than β-emitters for treating hypoxic solid tumors (10). This effect of diminished OER appears to be most relevant for Auger electron emitters that incorporate directly into DNA, such as ^{125}I-deoxyuridine, as opposed to those located at extranuclear sites (236, 237).

Fractionated TRT is designed mostly to compensate for the anticipated heterogeneity of TRT dose distribution, especially for large poorly vascularized tumors that contain regions of hypoxia (7). Tumor control is therefore compromised because of regions within the tumor to which the radionuclide has limited or no access. Under some conditions, fractionated delivery of radiolabeled antibodies and peptides has shown efficacy while also causing less toxicity than a single administration. In general, both preclinical and clinical evidence suggest that TRT fractionation can provide a beneficial effect and a more uniform radiation dose distribution (7). This has proven to be an effective clinical strategy with ^{131}I-mIBG, and fractionated treatment with radiopeptides may also be more efficacious (238). Whether or not effects analogous to the "ultrafractionation" phenomenon seen with multiple small doses of radiation could contribute to TRT tumor responses in general, or to the relative efficacy of TRT delivered as multiple fractionated doses versus a single large maximally tolerated dose, is not known at this time.

12.16   SUMMARY AND FUTURE DIRECTIONS

In this chapter, the current knowledge relating to low-dose/LDR radiobiological mechanisms that might impact on the efficacy of TRT has been summarized. These include the increasingly characterized phenomena of HRS-IRR, inverse dose-rate effects, bystander effects and adaptive responses. Where possible, speculation as to the possible relevance of these findings to the clinical practice of TRT has been provided. Although our understanding of these low-dose/LDR phenomena has increased dramatically in the last few years, especially in the context of activation of the cellular DNA damage surveillance–response networks, it is important to recognize that many of these phenomena need to be validated in the clinical setting as not all of these findings will extrapolate directly to more complex models, let alone cancer patients. For example, cell-contact effects (which are largely ignored in most *in vitro* studies) are clearly important in the bystander effect, and may also be relevant to the HRS response, which seems to be suppressed in some three-dimensional model systems that invoke cell–cell contact (169). This is clearly a time where improved technologies are enabling major research advances into low-dose/LDR effects on biological systems, and these advances represent a huge opportunity for understanding the basic science underlying clinical TRT practice.

ACKNOWLEDGMENTS

This work was supported by a grant from the Canadian Breast Cancer Foundation, Alberta Chapter (to David Murray) and by grant MOP 15385 from the Canadian Institutes for Health Research (to Michael Weinfeld). We are grateful to Dr. Raymond Reilly for his insightful comments and suggestions.

REFERENCES

1. Wahl RL. Tositumomab and ^{131}I therapy in non-Hodgkin's lymphoma. *J Nucl Med* 2005; 46(Suppl 1):128S–140S.

2. Borghaei H, Wallace SG, Schilder RJ. Factors associated with toxicity and response to yttrium 90-labeled ibritumomab tiuxetan in patients with indolent non-Hodgkin's lymphoma. *Clin Lymphoma* 2004;5(Suppl 1):S16–S21.

3. Larson SM, Krenning EP. A pragmatic perspective on molecular targeted radionuclide therapy. *J Nucl Med* 2005;46(Suppl 1):1S–3S.

4. Flower MA, Fielding SL. Radiation dosimetry for ^{131}I-mIBG therapy of neuroblastoma. *Phys Med Biol* 1996;41:1933–1940.

5. Blake GM, Zivanovic MA, Blaquiere RM, et al. Strontium-89 therapy: measurement of absorbed dose to skeletal metastases. *J Nucl Med* 1988;29:549–557.

6. Goldenberg DM. Radioimmunotherapy. In: Freeman LM, editor. *Nuclear Medicine Annual*. Lippincott Williams & Wilkins, Philadelphia, PA, 2001, pp. 169–206.

7. DeNardo GL, Schlom J, Buchsbaum DJ, et al. Rationales, evidence, and design considerations for fractionated radioimmunotherapy. *Cancer* 2002;94(Suppl 4): 1332–1348.

8. Mirzaie-Joniani H, Eriksson D, Sheikholvaezin A, et al. Apoptosis induced by low-dose and low-dose-rate radiation. *Cancer* 2002;94(Suppl 4):1210–1214.

9. Koral KF, Francis IR, Kroll S, et al. Volume reduction versus radiation dose for tumors in previously untreated lymphoma patients who received iodine-131 tositumomab therapy. Conjugate views compared with a hybrid method. *Cancer* 2002;94(Suppl 4):1258–1263.

10. Dixon KL. The radiation biology of radioimmunotherapy. *Nucl Med Commun* 2003; 24:951–957.

11. Murray D, McEwan AJ. Radiobiology of systemic radiation therapy. *Cancer Biother Radiopharm* 2007;22:1–23.

12. Dale R, Carabe-Fernandez A. The radiobiology of conventional radiotherapy and its application to radionuclide therapy. *Cancer Biother Radiopharm* 2005;20:47–51.

13. Kassis AI, Adelstein SJ. Radiobiologic principles in radionuclide therapy. *J Nucl Med* 2005;46(Suppl 1):4S–12S.

14. Ward JF. Complexity of damage produced by ionizing radiation. *Cold Spring Harb Symp Quant Biol* 2000;65:377–382.

15. Goodhead DT. Initial events in the cellular effects of ionizing radiations: clustered damage in DNA. *Int J Radiat Biol* 1994;65:7–17.

16. Weinfeld M, Rasouli-Nia A, Chaudhry MA, Britten RA. Response of base excision repair enzymes to complex DNA lesions. *Radiat Res* 2001;156:584–589.

17. Georgakilas AG. Processing of DNA damage clusters in human cells: current status of knowledge. *Mol Biosyst* 2008;4:30–35.

18. Murray D, Wang JYJ, Mirzayans R. DNA repair after low doses of ionizing radiation. *Int J Low Radiat* 2006;3:255–272.

19. Jackson SP. Sensing and repairing DNA double-strand breaks. *Carcinogenesis* 2002;23:687–696.

20. Kurz EU, Lees-Miller SP. DNA damage induced activation of ATM and ATM-dependent signaling pathways. *DNA Repair* 2004;3:889–900.

21. Kastan MB, Lim DS. The many substrates and functions of ATM. *Nat Rev Mol Cell Biol* 2000;1:179–186.

22. Shiloh Y. ATM and related protein kinases: safeguarding genome integrity. *Nat Rev Cancer* 2003;3:155–168.

23. Petrini JH, Stracker TH. The cellular response to DNA double-strand breaks: defining the sensors and mediators. *Trends Cell Biol.* 2003;13:458–462.

24. Shiloh Y. ATM: ready, set, go. *Cell Cycle* 2003;2:116–117.

25. Bakkenist CJ, Kastan MB. DNA damage activates ATM through intermolecular autophosphorylation and dimer dissociation. *Nature* 2003;421:499–506.

26. Uziel T, Lerenthal Y, Moyal L, et al. Requirement of the MRN complex for ATM activation by DNA damage. *EMBO J* 2003;22:5612–5621.

27. Lee JH, Paull TT. Direct activation of the ATM protein kinase by the Mre11/Rad50/Nbs1 complex. *Science* 2004;304:93–96.

28. Christmann M, Tomicic MT, Roos WP, Kaina B. Mechanisms of human DNA repair: an update. *Toxicology* 2003;193:3–34.

29. Niida H, Nakanishi M. DNA damage checkpoints in mammals. *Mutagenesis* 2006; 21:3–9.

30. Matsuoka S, Ballif BA, Smogorzewska A, et al. ATM and ATR substrate analysis reveals extensive protein networks responsive to DNA damage. *Science* 2007;316(5828): 1160–1166.

31. Bartek J, Lukas J. Chk1 and Chk2 kinases in checkpoint control and cancer. *Cancer Cell* 2003;3:421–429.

32. Pilch DR, Sedelnikova OA, Redon C, et al. Characteristics of gamma-H2AX foci at DNA double-strand breaks sites. *Biochem Cell Biol* 2003;81:123–129.

33. Olivier M, Hussain SP, Caron de Fromentel C, et al. TP53 mutation spectra and load: a tool for generating hypotheses on the etiology of cancer. *IARC Sci Publ* 2004; 157:247–270.

34. Kastan MB, Lim DS, Kim ST, Yang D. ATM: a key determinant of multiple cellular responses to irradiation. *Acta Oncol* 2001;40:686–688.

35. Sengupta S, Harris CC. p53: traffic cop at the crossroads of DNA repair and recombination. *Nat Rev Mol Cell Biol* 2005;6:44–55.

36. Lane DP. p53, guardian of the genome. *Nature* 1992;358:15–16.

37. Appella E, Anderson CW. Post-translational modifications and activation of p53 by genotoxic stresses. *Eur J Biochem* 2001;268:2764–2772.

38. Brooks CL, Gu W. Ubiquitination, phosphorylation and acetylation: the molecular basis for p53 regulation. *Curr Opin Cell Biol* 2003;15:164–171.

39. Zhao R, Gish K, Murphy M, et al. Analysis of p53-regulated gene expression patterns using oligonucleotide arrays. *Genes Dev* 2000;14:981–993.

40. Vogelstein B, Lane D, Levine AJ. Surfing the p53 network. *Nature* 2000;408:307–310.

41. Mirza A, Wu Q, Wang L, et al. Global transcriptional program of p53 target genes during the process of apoptosis and cell cycle progression. *Oncogene* 2003;22:3645–3654.

42. Murray D, Mirzayans R. Role of p53 in the repair of ionizing radiation-induced DNA damage. In: Landseer BR, editor. *New Research on DNA Repair*, Nova Science Publishers, Hauppauge, NY, 2007, pp. 325–373.

43. Batchelor E, Mock CS, Bhan I, et al. Recurrent initiation: a mechanism for triggering p53 pulses in response to DNA damage. *Mol Cell* 2008;30:277–289.

44. Lu X, Nguyen TA, Moon SH, et al. The type 2C phosphatase Wip1: an oncogenic regulator of tumor suppressor and DNA damage response pathways. *Cancer Metastasis Rev* 2008;27:123–135.

45. Taylor WR, Stark GR. Regulation of the G2/M transition by p53. *Oncogene* 2001; 20:1803–1815.

46. Xu B, Kim ST, Lim DS, Kastan MB. Two molecularly distinct G(2)/M checkpoints are induced by ionizing irradiation. *Mol Cell Biol* 2002;22:1049–1059.

47. Wang JYJ, Cho SK. Coordination of repair, checkpoint, and cell death responses to DNA damage. *Adv Protein Chem* 2004;69:101–135.

48. Downs JA, Allard S, Jobin-Robitaille O, et al. Binding of chromatin-modifying activities to phosphorylated histone H2A at DNA damage sites. *Mol Cell* 2004; 16:979–990.

49. Utley RT, Lacoste N, Jobin-Robitaille O, et al. Regulation of NuA4 histone acetyltransferase activity in transcription and DNA repair by phosphorylation of histone H4. *Mol Cell Biol* 2005;25:8179–8190.

50. Sun Y, Jiang X, Chen S, et al. A role for the Tip60 histone acetyltransferase in the acetylation and activation of ATM. *Proc Natl Acad Sci USA* 2005;102:13182–13187.

51. Averbeck D, Testard I, Boucher D. Changing views on ionising radiation-induced cellular effects. *Int J Low Radiat* 2006;3:117–134.

52. Friedberg EC, Walker GC, Siede W, et al. *DNA Repair and Mutagenesis*, 2nd edition. ASM Press, Washington, DC, 2006.

53. Almeida KH, Sobol RW. A unified view of base excision repair: lesion-dependent protein complexes regulated by post-translational modification. *DNA Repair (Amst)* 2007; 6:695–711.

54. Boiteux S, Le Page F. Repair of 8-oxoguanine and Ogg1-incised apurinic sites in a CHO cell line. *Progr Nucleic Acid Res Mol Biol* 2001;68:95–105.

55. Jilani A, Ramotar D, Slack C, et al. Molecular cloning of the human gene, PNKP, encoding a polynucleotide kinase 3′-phosphatase and evidence for its role in repair of DNA strand breaks caused by oxidative damage. *J Biol Chem* 1999;274: 24176–24186.

56. Karimi-Busheri F, Daly G, Robins P, et al. Molecular characterization of a human DNA kinase. *J Biol Chem* 1999;274:24187–24194.

57. Izumi T, Hazra TK, Boldogh I, et al. Requirement for human AP endonuclease 1 for repair of 3′-blocking damage at DNA single-strand breaks induced by reactive oxygen species. *Carcinogenesis* 2000;21:1329–1334.

58. Takahashi T, Tada M, Igarashi S, et al. Aprataxin, causative gene product for EAOH/AOA1, repairs DNA single-strand breaks with damaged 3′-phosphate and 3′-phosphoglycolate ends. *Nucleic Acids Res* 2007;35:3797–3809.

59. Dianov GL, Sleeth KM, Dianova II, Allinson SL. Repair of abasic sites in DNA. *Mutat Res* 2003;531:157–163.

60. Parsons JL, Tait PS, Finch D, et al. CHIP-mediated degradation and DNA damage-dependent stabilization regulate base excision repair proteins. *Mol Cell* 2008; 29:477–487.

61. Masson M, Niedergang C, Schreiber V, et al. XRCC1 is specifically associated with poly (ADP-ribose) polymerase and negatively regulates its activity following DNA damage. *Mol Cell Biol* 1998;18:3563–3571.

62. Tomkinson AE, Chen L, Dong Z, et al. Completion of base excision repair by mammalian DNA ligases. *Progr Nucleic Acid Res Mol Biol* 2001;68:151–164.

63. Whitehouse CJ, Taylor RM, Thistlethwaite A, et al. XRCC1 stimulates human polynucleotide kinase activity at damaged DNA termini and accelerates DNA single-strand break repair. *Cell* 2001;104:107–117.

64. Mani RS, Fanta M, Karimi-Busheri F, et al. XRCC1 stimulates polynucleotide kinase by enhancing its damage discrimination and displacement from DNA repair intermediates. *J Biol Chem* 2007;282:28004–28013.

65. Okano S, Lan L, Caldecott KW, et al. Spatial and temporal cellular responses to single-strand breaks in human cells. *Mol Cell Biol* 2003;23:3974–3981.

66. Satoh MS, Lindahl T. Role of poly(ADP-ribose) formation in DNA repair. *Nature* 1992;356:356–358.

67. Woodhouse BC, Dianova II, Parsons JL, Dianov GL. Poly(ADP-ribose) polymerase-1 modulates DNA repair capacity and prevents formation of DNA double strand breaks. *DNA Repair (Amst)* 2008;7:932–940.

68. Pierce AJ, Hu P, Han M, et al. Ku DNA end-binding protein modulates homologous repair of double-strand breaks in mammalian cells. *Genes Dev* 2001;15:3237–3242.

69. Delacote F, Han M, Stamato TD, et al. An xrcc4 defect or Wortmannin stimulates homologous recombination specifically induced by double-strand breaks in mammalian cells. *Nucleic Acids Res* 2002;30:3454–3463.

70. Kuschel B, Auranen A, McBride S, et al. Variants in DNA double-strand break repair genes and breast cancer susceptibility. *Hum Mol Genet* 2002;11:1399–1407.

71. Valerie K, Povirk LF. Regulation and mechanisms of mammalian double-strand break repair. *Oncogene* 2003;22:5792–5812.

72. Richardson C, Moynahan ME, Jasin M. Double-strand break repair by interchromosomal recombination: suppression of chromosomal translocations. *Genes Dev* 1998; 12:3831–3842.

73. Takata M, Sasaki MS, Sonoda E, et al. Homologous recombination and non-homologous end-joining pathways of DNA double-strand break repair have overlapping roles in the maintenance of chromosomal integrity in vertebrate cells. *EMBO J* 1998;17:5497–5508.

74. Johnson RD, Jasin M. Sister chromatid gene conversion is a prominent double-strand break repair pathway in mammalian cells. *EMBO J* 2000;19:3398–3407.

75. Thompson LH, Schild D. Homologous recombinational repair of DNA ensures mammalian chromosome stability. *Mutat Res* 2001;477:131–153.

76. Rothkamm K, Kruger I, Thompson LH, Löbrich M. Pathways of DNA double-strand break repair during the mammalian cell cycle. *Mol Cell Biol* 2003;23:5706–5715.

77. Hinz JM, Yamada NA, Salazar EP, et al. Influence of double-strand-break repair pathways on radiosensitivity throughout the cell cycle in CHO cells. *DNA Repair (Amst)* 2005;4:782–792.

78. Cann KL, Hicks GG. Regulation of the cellular DNA double-strand break response. *Biochem Cell Biol* 2007;85:663–674.

79. Shrivastav M, De Haro LP, Nickoloff JA. Regulation of DNA double-strand break repair pathway choice. *Cell Res* 2008;18:134–147.

80. Jeggo PA. Identification of genes involved in repair of DNA double-strand breaks in mammalian cells. *Radiat Res* 1998;150(Suppl 5):S80–S91.

81. Haber JE. Partners and pathways repairing a double-strand break. *Trends Genet* 2000;16:259–264.

82. Cromie GA, Connelly JC, Leach DR. Recombination at double-strand breaks and DNA ends: conserved mechanisms from phage to humans. *Mol Cell* 2001;8:1163–1174.

83. Liang F, Han M, Romanienko PJ, Jasin M. Homology-directed repair is a major double-strand break repair pathway in mammalian cells. *Proc Natl Acad Sci USA* 1998; 95:5172–5177.

84. Lieber MR. The mechanism of human nonhomologous DNA end joining. *J Biol Chem* 2008;283:1–5.

85. Weterings E, Chen DJ. The endless tale of non-homologous end-joining. *Cell Res* 2008;18:114–124.

86. Chan DW, Lees-Miller SP. The DNA-dependent protein kinase is inactivated by autophosphorylation of the catalytic subunit. *J Biol Chem* 1996;271:8936–8941.

87. Koch CA, Agyei R, Galicia S, et al. Xrcc4 physically links DNA end processing by polynucleotide kinase to DNA ligation by DNA ligase IV. *EMBO J* 2004;23:3874–3885.

88. Jeggo PA, Lobrich M. Artemis links ATM to double strand break rejoining. *Cell Cycle* 2005;4:359–362.

89. Ruscetti T, Lehnert BE, Halbrook J, et al. Stimulation of the DNA-dependent protein kinase by poly(ADP-ribose) polymerase. *J Biol Chem* 1998;273:14461–14467.

90. Goedecke W, Eijpe M, Offenberg HH, et al. Mre11 and Ku70 interact in somatic cells, but are differentially expressed in early meiosis. *Nature Genet* 1999;23:194–198.

91. Hsu HL, Yannone SM, Chen DJ. Defining interactions between DNA-PK and ligase IV/XRCC4. *DNA Repair (Amst)* 2002;1:225–235.

92. Calsou P, Delteil C, Frit P, et al. Coordinated assembly of Ku and p460 subunits of the DNA-dependent protein kinase on DNA ends is necessary for XRCC4-ligase IV recruitment. *J Mol Biol* 2003;326:93–103.

93. Critchlow SE, Bowater RP, Jackson SP. Mammalian DNA double-strand break repair protein XRCC4 interacts with DNA ligase IV. *Curr Biol* 1997;7:588–598.

94. Lee KJ, Jovanovic M, Udayakumar D, et al. Identification of DNA-PKcs phosphorylation sites in XRCC4 and effects of mutations at these sites on DNA end joining in a cell-free system. *DNA Repair (Amst)* 2004;3:267–276.

95. Li X, Heyer WD. Homologous recombination in DNA repair and DNA damage tolerance. *Cell Res* 2008;18:99–113.

96. Chen G, Yuan SS, Liu W, et al. Radiation-induced assembly of Rad51 and Rad52 recombination complex requires ATM and c-Abl. *J Biol Chem* 1999; 274:12748–12752.

97. Zhang J, Powell SN. The role of the BRCA1 tumor suppressor in DNA double-strand break repair. *Mol Cancer Res* 2005;3:531–539.

98. Zhuang J, Zhang J, Willers H, et al. Checkpoint kinase 2-mediated phosphorylation of BRCA1 regulates the fidelity of nonhomologous end-joining. *Cancer Res* 2006; 66:1401–1408.

99. Abend M. Reasons to reconsider the significance of apoptosis for cancer therapy. *Int J Radiat Biol* 2003;79:927–941.

100. Okada H, Mak TW. Pathways of apoptotic and non-apoptotic death in tumour cells. *Nat Rev Cancer* 2004;4:592–603.

101. Brown JM, Attardi LD. The role of apoptosis in cancer development and treatment response. *Nat Rev Cancer* 2005;5:231–237.

102. McBride WH, Chiang CS, Olson JL, et al. A sense of danger from radiation. *Radiat Res* 2004;162:1–19.

103. Kolesnick R, Fuks Z. Radiation and ceramide-induced apoptosis. *Oncogene* 2003; 22:5897–5906.

104. Lawen A. Apoptosis-an introduction. *Bioessays* 2003;25:888–896.

105. Shankar S, Singh TR, Srivastava RK. Ionizing radiation enhances the therapeutic potential of TRAIL in prostate cancer *in vitro* and *in vivo*: intracellular mechanisms. *Prostate* 2004;61:35–49.

106. McGill G, Fisher DE. Apoptosis in tumorigenesis and cancer therapy. *Front Biosci* 1997;2:d353–d379.

107. Sionov RV, Haupt Y. The cellular response to p53: the decision between life and death. *Oncogene* 1999;18:6145–6157.

108. Hatfield P, Merrick A, Harrington K, et al. Radiation-induced cell death and dendritic cells: potential for cancer immunotherapy?. *Clin Oncol (R Coll Radiol)* 2005;17:1–11.

109. Waldman T, Zhang Y, Dillehay L, et al. Cell-cycle arrest versus cell death in cancer therapy. *Nat Med* 1997;3:1034–1036.

110. Mirzayans R, Murray D. Cellular Senescence: implications for cancer therapy. In: Garvey RB, editor. *New Research on Cell Aging*. Nova Science Publishers, Hauppauge, NY, 2007, pp. 1–64.

111. Roninson IB. Tumor cell senescence in cancer treatment. *Cancer Res* 2003; 63:2705–2715.

112. Shay JW, Roninson IB. Hallmarks of senescence in carcinogenesis and cancer therapy. *Oncogene* 2004;23:2919–2933.

113. Di Leonardo A, Linke SP, Clarkin K, Wahl GM. DNA damage triggers a prolonged p53-dependent G1 arrest and long-term induction of Cip1 in normal human fibroblasts. *Genes Dev.* 1994;8:2540–2551.

114. Linke SP, Clarkin KC, Wahl GM. p53 mediates permanent arrest over multiple cell cycles in response to gamma-irradiation. *Cancer Res* 1996;57:1171–1179.

115. Chang BD, Broude EV, Dokmanovic M, et al. A senescence-like phenotype distinguishes tumor cells that undergo terminal proliferation arrest after exposure to anticancer agents. *Cancer Res* 1999;59:3761–3767.

116. Mirzayans R, Scott A, Cameron M, Murray D. Induction of accelerated senescence by gamma radiation in human solid tumor-derived cell lines expressing wild-type TP53. *Radiat Res* 2005;163:53–62.

117. Hait WN, Jin S, Yang JM. A matter of life or death (or both): understanding autophagy in cancer. *Clin Cancer Res* 2006;12:1961–1965.

118. Ito H, Daido S, Kanzawa T, et al. Radiation-induced autophagy is associated with LC3 and its inhibition sensitizes malignant glioma cells. *Int J Oncol* 2005;26:1401–1410.

119. Daido S, Yamamoto A, Fujiwara K, et al. Inhibition of the DNA-dependent protein kinase catalytic subunit radiosensitizes malignant glioma cells by inducing autophagy. *Cancer Res.* 2005;65:4368–4375.

120. Paglin S, Yahalom J. Pathways that regulate autophagy and their role in mediating tumor response to treatment. *Autophagy* 2006;2:291–293.

121. Chu K, Teele N, Dewey MW, et al. Computerized video time lapse study of cell cycle delay and arrest, mitotic catastrophe, apoptosis and clonogenic survival in

irradiated 14-3-3sigma and CDKN1A (p21) knockout cell lines. *Radiat Res* 2004;162:270–286.

122. Jonathan EC, Bernhard EJ, McKenna WG. How does radiation kill cells? *Curr Opin Chem Biol* 1999;3:77–83.

123. Brown JM, Wouters BG. Apoptosis: mediator or mode of cell killing by anticancer agents? *Drug Resist Updat* 2001;4:135–135.

124. Ianzini F, Bertoldo A, Kosmacek EA, et al. Lack of p53 function promotes radiation-induced mitotic catastrophe in mouse embryonic fibroblast cells. *Cancer Cell Int* 2006; 6:11.

125. Erenpreisa J, Cragg MS. Mitotic death: a mechanism of survival? A review. *Cancer Cell Int* 2001;1:1.

126. Sundaram M, Guernsey DL, Rajaraman MM, Rajaraman R. Neosis: a novel type of cell division in cancer. *Cancer Biol Ther* 2004;3:207–218.

127. Alper T, *Cellular Radiobiology*. Cambridge University Press, Cambridge, UK, 1979.

128. Steel GG. The dose rate effect: brachytherapy and targeted radiotherapy. In: Steel GG, editor. *Basic Clinical Radiobiology*, 3rd edition. Hodder Arnold, London, UK, 2002, pp. 192–204.

129. Steel GG, Deacon JM, Duchesne GM, et al. The dose-rate effect in human tumour cells. *Radiother Oncol* 1987;9:299–310.

130. Nagasawa H, Chen DJ, Strniste GF. Response of X-ray-sensitive CHO mutant cells to gamma radiation. I. Effects of low dose rates and the process of repair of potentially lethal damage in G1 phase. *Radiat Res* 1989;118:559–567.

131. Stackhouse MA, Bedford JS. An ionizing radiation-sensitive mutant of CHO cells: irs-20. II. Dose-rate effects and cellular recovery processes. *Radiat Res* 1993; 136:250–254.

132. Nagasawa H, Little JB, Tsang NM, et al. Effect of dose rate on the survival of irradiated human skin fibroblasts. *Radiat Res* 1992;132:375–379.

133. Marples B, Lambin P, Skov KA, Joiner MC. Low dose hyper-radiosensitivity and increased radioresistance in mammalian cells. *Int J Radiat Biol* 1997;71: 721–735.

134. Joiner MC, Marples B, Lambin P, et al. Low-dose hypersensitivity: current status and possible mechanisms. *Int J Radiat Oncol Biol Phys* 2001;49:379–389.

135. Marples B, Joiner MC. The response of Chinese hamster V79 cells to low radiation doses: evidence of enhanced sensitivity of the whole cell population. *Radiat Res* 1993; 133:41–51.

136. Marples B, Wouters BG, Collis SJ, et al. Low-dose hyper-radiosensitivity: a consequence of ineffective cell cycle arrest of radiation-damaged G2-phase cells. *Radiat Res* 2004;161:247–255.

137. Marples B. Is low-dose hyper-radiosensitivity a measure of G2-phase cell radiosensitivity? *Cancer Met Rev* 2004;23:197–207.

138. Krueger SA, Collis SJ, Joiner MC, et al. Transition in survival from low-dose hyper-radiosensitivity to increased radioresistance is independent of activation of ATM Ser1981 activity. *Int J Radiat Oncol Biol Phys* 2007;69:1262–1271.

139. Buscemi G, Perego P, Carenini N, et al. Activation of ATM and Chk2 kinases in relation to the amount of DNA strand breaks. *Oncogene.* 2004;23:7691–7700.

140. Rothkamm K, Löbrich M. Evidence for a lack of DNA double-strand break repair in human cells exposed to very low x-ray doses. *Proc Natl Acad Sci USA* 2003; 100:5057–5062.

141. Stiff T, O'Driscoll M, Rief N, et al. ATM and DNA-PK function redundantly to phosphorylate H2AX after exposure to ionizing radiation. *Cancer Res* 2004; 64:2390–2396.

142. Wang H, Wang M, Bocker W, Iliakis G. Complex H2AX phosphorylation patterns by multiple kinases including ATM and DNA-PK in human cells exposed to ionizing radiation and treated with kinase inhibitors. *J Cell Physiol* 2005; 202:492–502.

143. Ward IM, Chen J. Histone H2AX is phosphorylated in an ATR-dependent manner in response to replicational stress. *J Biol Chem* 2001;276:47759–47762.

144. Short SC, Bourne S, Martindale C, et al. DNA damage responses at low radiation doses. *Radiat Res* 2005;164:292–302.

145. Schwarz JK, Lovly CM, Piwnica-Worms H. Regulation of the Chk2 protein kinase by oligomerization-mediated *cis*- and *trans*-phosphorylation. *Mol Cancer Res* 2003; 1:598–609.

146. Asaithamby A, Chen DJ. Cellular responses to DNA double-strand breaks after low-dose γ-irradiation. *Nucleic Acids Res* 2009;37: 3912–3923.

147. Lee H, Kwak HJ, Cho IT, et al. S1219 residue of 53BP1 is phosphorylated by ATM kinase upon DNA damage and required for proper execution of DNA damage response. *Biochem Biophys Res Commun* 2009;378:32–36.

148. Offer H, Erez N, Zurer I, et al. The onset of p53-dependent DNA repair or apoptosis is determined by the level of accumulated damaged DNA. *Carcinogenesis* 2002; 23:1025–1032.

149. Giaccia AJ, Kastan MB. The complexity of p53 modulation: emerging patterns from divergent signals. *Genes Dev* 1998;12:2973–2983.

150. Oren M. Decision making by p53: life, death and cancer. *Cell Death Differ* 2003; 10:431–442.

151. Pond CD, Leachman SA, Warters RL. Accumulation, activation and interindividual variation of the epidermal TP53 protein in response to ionizing radiation in organ cultured human skin. *Radiat Res* 2004;161:739–745.

152. Amundson SA, Do KT, Fornace AJ, Jr. Induction of stress genes by low doses of gamma rays. *Radiat Res* 1999;152:225–231.

153. Skov K, Marples B, Matthews JB, et al. A preliminary investigation into the extent of increased radioresistance or hyper-radiosensitivity in cells of hamster cell lines known to be deficient in DNA repair. *Radiat Res* 1994;138:S126–S129.

154. Marples B, Cann NE, Mitchell CR, et al. Evidence for the involvement of DNA-dependent protein kinase in the phenomena of low dose hyper-radiosensitivity and increased radioresistance. *Int J Radiat Biol* 2002;78:1139–1147.

155. Marples B, Joiner MC. Modification of survival by DNA repair modifiers: a probable explanation for the phenomenon of increased radioresistance. *Int J Radiat Biol* 2000;76:305–312.

156. Chalmers A, Johnston P, Woodcock M, et al. PARP-1, PARP-2, and the cellular response to low doses of ionizing radiation. *Int J Radiat Oncol Biol Phys* 2004;58:410–419.

157. Vaganay-Juery S, Muller C, Marangoni E, et al. Decreased DNA-PK activity in human cancer cells exhibiting hypersensitivity to low-dose irradiation. *Br J Cancer* 2000; 83:514–518.

158. Mirzayans R, Severin D, Murray D. Relationship between DNA double strand break rejoining and cell survival following exposure to ionizing radiation in human fibroblast strains with differing ATM/p53 status: implications for the evaluation of clinical radiosensitivity. *Int J Radiat Oncol Biol Phys* 2006;66:1498–1505.

159. Kato TA, Nagasawa H, Weil MM, et al. Gamma-H2AX foci after low-dose-rate irradiation reveal ATM haploinsufficiency in mice. *Radiat Res* 2006; 166:47–54.

160. Kurose A, Tanaka T, Huang X, et al. Assessment of ATM phosphorylation on Ser-1981 induced by DNA topoisomerase I and II inhibitors in relation to Ser-139-histone H2AX phosphorylation, cell cycle phase, and apoptosis. *Cytometry A* 2005;68:1–9.

161. Wykes SM, Piasentin E, Joiner MC, et al. Low-dose hyper-radiosensitivity is not caused by a failure to recognize DNA double-strand breaks. *Radiat Res* 2006;165: 516–524.

162. Enns L, Bogen KT, Wizniak J, et al. Low-dose radiation hypersensitivity is associated with p53-dependent apoptosis. *Mol Cancer Res* 2004;2:557–566.

163. Krueger SA, Joiner MC, Weinfeld M, et al. Role of apoptosis in low-dose hyper-radiosensitivity. *Radiat Res* 2007;167:260–267.

164. Marples B, Collis SJ. Low-dose hyper-radiosensitivity: past, present, and future. *Int J Radiat Oncol Biol Phys* 2008;70:1310–1318.

165. Sinclair WK. Cyclic X-ray responses in mammalian cells *in vitro*. *Radiat Res* 1968; 33:620–643.

166. Short SC, Woodcock M, Marples B, Joiner MC. Effects of cell cycle phase on low-dose hyper-radiosensitivity. *Int J Radiat Biol* 2003;79:99–105.

167. Lennon SV, Martin SJ, Cotter TG. Dose-dependent induction of apoptosis in human tumour cell lines by widely diverging stimuli. *Cell Prolif* 1991;24:203–214.

168. Meyn RE. Apoptosis and response to radiation: implications for radiation therapy. *Oncology (Williston Park)* 1997;11:349–356.

169. Chandna S, Dwarakanath BS, Khaitan D, et al. Low-dose radiation hypersensitivity in human tumor cell lines: effects of cell–cell contact and nutritional deprivation. *Radiat Res* 2002;157:516–525.

170. Short SC, Kelly J, Mayes CR, et al. Low-dose hypersensitivity after fractionated low-dose irradiation *in vitro*. *Int J Radiat Biol* 2001;77:655–664.

171. Dey S, Spring PM, Arnold S, et al. Low-dose fractionated radiation potentiates the effects of Paclitaxel in wild-type and mutant p53 head and neck tumor cell lines. *Clin Cancer Res* 2003;9:1557–1565.

172. Spring PM, Arnold SM, Shajahan S, et al. Low dose fractionated radiation potentiates the effects of taxotere in nude mice xenografts of squamous cell carcinoma of head and neck. *Cell Cycle* 2004;3:479–485.

173. Smith LG, Miller RC, Richards M, et al. Investigation of hypersensitivity to fractionated low-dose radiation exposure. *Int J Radiat Oncol Biol Phys* 1999;45:187–191.

174. Pulkkanen K, Lahtinen T, Lehtimäki A, et al. Effective palliation without normal tissue toxicity using low-dose ultrafractionated re-irradiation for tumor recurrence after radical or adjuvant radiotherapy. *Acta Oncol* 2007;46:1037–1041.

175. Arnold SM, Regine WF, Ahmed MM, et al. Low-dose fractionated radiation as a chemopotentiator of neoadjuvant paclitaxel and carboplatin for locally advanced squamous cell carcinoma of the head and neck: results of a new treatment paradigm. *Int J Radiat Oncol Biol Phys* 2004;58:1411–1417.

176. Turesson I, Joiner MC. Clinical evidence of hypersensitivity to low doses in radiotherapy. *Radiother Oncol* 1996;40:1–3.

177. Hamilton CS, Denham JW, O'Brien M, et al. Underprediction of human skin erythema at low doses per fraction by the linear quadratic model. *Radiother Oncol* 1996; 40:23–30.

178. Harney J, Shah N, Short S, et al. The evaluation of low dose hyper-radiosensitivity in normal human skin. *Radiother Oncol* 2004;70:319–329.

179. Simonsson M, Qvarnström F, Nyman J, et al. Low-dose hypersensitive gammaH2AX response and infrequent apoptosis in epidermis from radiotherapy patients. *Radiother Oncol* 2008;88:388–397.

180. Harney J, Short SC, Shah N, et al. Low dose hyper-radiosensitivity in metastatic tumors. *Int J Radiat Oncol Biol Phys* 2004;59:1190–1195.

181. Krause M, Prager J, Wohlfarth J, et al. Low-dose hyperradiosensitivity of human glioblastoma cell lines *in vitro* does not translate into improved outcome of ultrafractionated radiotherapy *in vivo*. *Int J Radiat Biol* 2005;81:751–758.

182. Honoré HB, Bentzen SM. A modelling study of the potential influence of low dose hypersensitivity on radiation treatment planning. *Radiother Oncol* 2006;79:115–121.

183. Welsh JS, Limmer JP, Howard SP, et al. Precautions in the use of intensity-modulated radiation therapy. *Technol Cancer Res Treat* 2005;4:203–210.

184. Badger CC, Rasey J, Nourigat C, et al. WR2721 protection of bone marrow in ^{131}I-labeled antibody therapy. *Radiat Res* 1991;128:320–324.

185. Kal HB, Barendsen GW, Hauwe RB, Roelse H. Increased radiosensitivity of rat rhabdomyosarcoma cells induced by protracted irradiation. *Radiat Res* 1975;63: 521–530.

186. Mitchell JB, Bedford JS, Bailey SM. Dose-rate effects on the cell cycle and survival of S3 HeLa and V79 cells. *Radiat Res* 1979;79:520–536.

187. Mitchell JB, Bedford JS, Bailey SM. Dose-rate effects in mammalian cells in culture. III. Comparison of cell killing and cell proliferation during continuous irradiation for six different cell lines. *Radiat Res* 1979;79:537–551.

188. Furre T, Koritzinsky M, Olsen DR, Pettersen EO. Inverse dose-rate effect due to premitotic accumulation during continuous low dose-rate irradiation of cervix carcinoma cells. *Int J Radiat Biol* 1999;75:699–707.

189. Marin LA, Smith CE, Langston MY, et al. Response of glioblastoma cell lines to low dose rate irradiation. *Int J Radiat Oncol Biol Phys* 1991;21:397–402.

190. Schultz CJ, Geard CR. Radioresponse of human astrocytic tumors across grade as a function of acute and chronic irradiation. *Int J Radiat Oncol Biol Phys* 1990; 19:1397–1403.

191. Murtha AD. Review of low-dose-rate radiobiology for clinicians. *Semin Radiat Oncol* 2000;10:133–138.

192. Ning S, Knox SJ. G2/M-phase arrest and death by apoptosis of HL60 cells irradiated with exponentially decreasing low-dose-rate gamma radiation. *Radiat Res* 1999;151:659–669.

193. van Oostrum IE, Erkens-Schulze S, Petterson M, et al. The relationship between radiosensitivity and cell kinetic effects after low- and high-dose-rate irradiation in five human tumors in nude mice. *Radiat Res* 1990;122:252–261.

194. Knox SJ, Sutherland W, Goris ML. Correlation of tumor sensitivity to low-dose-rate irradiation with G_2/M-phase block and other radiobiological parameters. *Radiat Res* 1993;135:24–31.

195. Cao S, Skog S, Tribukait B. Comparison between protracted and conventional dose rates of irradiation on the growth of the Bp8 mouse ascites sarcoma. *Acta Radiol Oncol* 1983;22:35–47.

196. DeWeese TL, Shipman JM, Dillehay LE, Nelson WG. Sensitivity of human prostatic carcinoma cell lines to low dose rate radiation exposure. *J Urol* 1998;159:591–598.

197. Mitchell CR, Joiner MC. Effect of subsequent acute-dose irradiation on cell survival *in vitro* following low dose-rate exposures. *Int J Radiat Biol*. 2002;78:981–990.

198. Mitchell CR, Folkard M, Joiner MC. Effects of exposure to low-dose-rate ^{60}Co gamma rays on human tumor cells *in vitro*. *Radiat Res* 2002;158:311–318.

199. Collis SJ, Schwaninger JM, Ntambi AJ, et al. Evasion of early cellular response mechanisms following low level radiation-induced DNA damage. *J Biol Chem* 2004; 279:49624–49632.

200. Ishizaki K, Hayashi Y, Nakamura H, et al. No induction of p53 phosphorylation and few focus formation of phosphorylated H2AX suggest efficient repair of DNA damage during chronic low-dose-rate irradiation in human cells. *J Radiat Res (Tokyo)* 2004;45:521–525.

201. Sugihara T, Magae J, Wadhwa R, et al. Dose and dose-rate effects of low-dose ionizing radiation on activation of Trp53 in immortalized murine cells. *Radiat Res* 2004; 162:296–307.

202. Sugihara T, Murano H, Tanaka K, Oghiso Y. Inverse dose-rate-effects on the expressions of extra-cellular matrix-related genes in low-dose-rate gamma-ray irradiated murine cells. *J Radiat Res (Tokyo)* 2008;49:231–240.

203. Amundson SA, Lee RA, Koch-Paiz CA, et al. Differential responses of stress genes to low dose-rate gamma irradiation. *Mol Cancer Res* 2003;1:445–452.

204. Boyd M, Cunningham SH, Brown MM, et al. Noradrenaline transporter gene transfer for radiation cell kill by ^{131}I meta-iodobenzylguanidine. *Gene Ther* 1999;6:1147–1152.

205. Boyd M, Mairs RJ, Cunningham SH, et al. A gene therapy/targeted radiotherapy strategy for radiation cell kill by [^{131}I]meta-iodobenzylguanidine. *J Gene Med* 2001;3: 165–172.

206. Mothersill C, Seymour CB, Joiner MC. Relationship between radiation-induced low-dose hypersensitivity and the bystander effect. *Radiat Res* 2002;157:526–532.

207. Burdak-Rothkamm S, Short SC, Folkard M, et al. ATR-dependent radiation-induced gamma H2AX foci in bystander primary human astrocytes and glioma cells. *Oncogene* 2007;26:993–1002.

208. Olivieri G, Bodycote J, Wolff S. Adaptive response of human lymphocytes to low concentrations of radioactive thymidine. *Science* 1984;223:594–597.

209. Ikushima T. Chromosomal responses to ionizing radiation reminiscent of an adaptive response in cultured Chinese hamster cells. *Mutat Res* 1987;180:215–221.

210. Wolff S. The adaptive response in radiobiology: evolving insights and implications. *Environ Health Perspect* 1998;106(Suppl 1):277–283.

211. Upton AC. Radiation hormesis: data and interpretations. *Crit Rev Toxicol* 2001; 31:681–695.

212. Preston RJ. Radiation biology: concepts for radiation protection. *Health Phys* 2004; 87:3–14.

213. Le XC, Xing JZ, Lee J, et al. Inducible repair of thymine glycol detected by an ultrasensitive assay for DNA damage. *Science* 1998;280:1066–1069.

214. Inoue M, Shen GP, Chaudhry MA, et al. Expression of the oxidative base excision repair enzymes is not induced in TK6 human lymphoblastoid cells after low doses of ionizing radiation. *Radiat Res* 2004;161:409–417.

215. Ikushima T, Aritomi H, Morisita J. Radioadaptive response: efficient repair of radiation-induced DNA damage in adapted cells. *Mutat Res* 1996;358:193–198.

216. Tachibana A. Genetic and physiological regulation of non-homologous end-joining in mammalian cells. *Adv Biophys* 2004;38:21–44.

217. Sasaki MS, Ejima Y, Tachibana A, et al. DNA damage response pathway in radioadaptive response. *Mutat Res* 2002;504:101–118.

218. Sawant SG, Randers-Pehrson G, Metting NF, Hall EJ. Adaptive response and the bystander effect induced by radiation in C3H 10T(1/2) cells in culture. *Radiat Res* 2001;156:177–180.

219. Klokov D, Criswell T, Leskov KS, et al. IR-inducible clusterin gene expression: a protein with potential roles in ionizing radiation-induced adaptive responses, genomic instability, and bystander effects. *Mutat Res* 2004;568:97–110.

220. Criswell T, Klokov D, Beman M, et al. Repression of IR-inducible clusterin expression by the p53 tumor suppressor protein. *Cancer Biol Ther* 2003;2:372–380.

221. Friesen C, Lubatschofski A, Kotzerke J, et al. Beta-irradiation used for systemic radioimmunotherapy induces apoptosis and activates apoptosis pathways in leukaemia cells. *Eur J Nucl Med Mol Imaging* 2003;30:1251–1261.

222. Buscombe JR, Cwikla JB, Caplin ME, Hilson AJ. Long-term efficacy of low activity meta-[^{131}I]iodobenzylguanidine therapy in patients with disseminated neuroendocrine tumours depends on initial response. *Nucl Med Commun* 2005;26:969–976.

223. Otte A. Diagnostic imaging prior to ^{90}Y-ibritumomab tiuxetan (Zevalin) treatment in follicular non-Hodgkin's lymphoma. *Hell J Nucl Med* 2008;11:12–15.

224. Xue LY, Butler NJ, Makrigiorgos GM, et al. Bystander effect produced by radiolabeled tumor cells *in vivo*. *Proc Natl Acad Sci USA* 2002;99:13765–13770.

225. Boyd M, Ross SC, Dorrens J, et al. Radiation-induced biologic bystander effect elicited *in vitro* by targeted radiopharmaceuticals labeled with alpha-, beta-, and auger electron-emitting radionuclides. *J Nucl Med* 2006;47:1007–1015.

226. Kwekkeboom DJ, Bakker WH, Kooij PP, et al. [^{177}Lu-DOTAOTyr^3]octreotate: comparison with [^{111}In-DTPA^0]octreotide in patients. *Eur J Nucl Med* 2001; 28:1319–1325.

227. Vaidyanathan G, Zalutsky MR. Targeted therapy using alpha emitters. *Phys Med Biol* 1996;41:1915–1931.

228. Zalutsky MR, Bigner DD. Radioimmunotherapy with alpha-particle emitting radio-immunoconjugates. *Acta Oncol* 1996;35:373–379.

229. Zalutsky MR. Targeted alpha-particle therapy of microscopic disease: providing a further rationale for clinical investigation. *J Nucl Med* 2006;47:1238–1240.

230. Behr TM, Behe M, Lohr M, et al. Therapeutic advantages of Auger electron- over beta-emitting radiometals or radioiodine when conjugated to internalizing antibodies. *Eur J Nucl Med* 2000;27:753–765.

231. Capello A, Krenning EP, Breeman WA, et al. Peptide receptor radionuclide therapy *in vitro* using [^{111}In-DTPA0]octreotide. *J Nucl Med* 2003;44:98–104.

232. Hall EJ, *Radiobiology for the Radiologist*, 5th edition. Lippincott, Williams & Wilkins, Philadelphia, PA, 2000.

233. Chan DA, Giaccia AJ. Hypoxia, gene expression, and metastasis. *Cancer Metastasis Rev* 2007;26:333–339.

234. Cooper RA, Carrington BM, Loncaster JA, et al. Tumour oxygenation levels correlate with dynamic contrast-enhanced magnetic resonance imaging parameters in carcinoma of the cervix. *Radiother Oncol* 2000;57:53–59.

235. Ling CC, Spiro IJ, Mitchell J, Stickler R. The variation of OER with dose rate. *Int J Radiat Oncol Biol Phys* 1985;11:1367–1373.

236. Hofer KG. Biophysical aspects of Auger processes. *Acta Oncol* 2000;39:651–657.

237. Koch CJ, Burki HJ. The oxygen-enhancement ratio for reproductive death induced by 3H or 125I damage in mammalian cells. *Int J Radiat Biol Relat Stud Phys Chem Med* 1975;28:417–425.

238. Teunissen JJ, Kwekkeboom DJ, de Jong M, et al. Endocrine tumours of the gastrointestinal tract. Peptide receptor radionuclide therapy. *Best Pract Res Clin Gastroenterol* 2005;19:595–616.

CHAPTER 13

Dosimetry for Targeted Radiotherapy

SUI SHEN AND JOHN B. FIVEASH

13.1 INTRODUCTION

Information on radiation doses from radiolabeled antibodies or peptides distributed in the patient body are important to clinical studies for radionuclide therapy. For example, radiation dose estimates (cGy or rads) obtained from an imaging study can in theory be used to determine the therapeutic injection dose (mCi or GBq) (Fig. 13.1) based on critical organ dose and target tumor dose estimates. This chapter provides an introductory description of the recent developments in radiation dosimetry for radionuclide cancer therapy with antibodies and peptides. In the brief descriptions of the recent developments, it is not possible to cover all literature on radiation dosimetry of radiolabeled antibodies or peptides. Nonetheless, a preparatory description on practical aspects of the methodologies and clarification on specific roles of radiation dosimetry in some preclinical or clinical applications are provided. Depending on the context, radiation dosimetry can have different levels of desirable "accuracy."

There are two different roles for internal radiation dosimetry: (1) its role in radiation safety protection for patients receiving radiolabeled-antibodies/peptides for diagnostic procedures; (2) its role in treatment planning for patients receiving these radiolabeled molecules for therapy, especially for the treatment of cancer.

For diagnostic procedures, radiation dose estimates are used to assess the risk of radiation-induced genetic effects or cancer development versus the benefit of diagnostic information. Because the benefit of diagnostic information may not be directly related to life-threatening disease at the time of the diagnostic procedure, the amount of radiation dose from these diagnostic procedures is typically limited to a level not significantly greater than natural background radiation, or a minimal radiation absorbed dose acceptable for the procedure. At these low radiation dose levels, the responses (possible hazards) are usually extrapolated, and the required accuracy in dose estimation is not high. Therefore, a population-averaged dose estimation is usually adequate for the purpose of radiation safety protection.

Monoclonal Antibody and Peptide-Targeted Radiotherapy of Cancer, Edited by Raymond M. Reilly
Copyright © 2010 John Wiley & Sons, Inc.

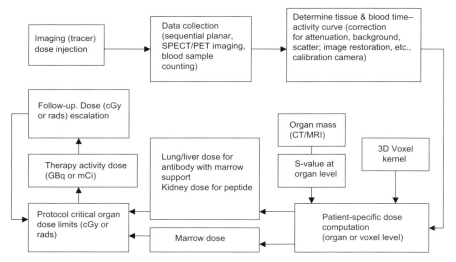

FIGURE 13.1   Imaging and dosimetry in clinical trials of targeted radionuclide therapy. Radiation dosimetry from a tracer dose study can be used to assess the dose to the critical organs (such as red marrow) and determine the therapy radioactivity dose for individual patients. Doses can be escalated in cGy for critical organs, instead of mCi/m^2 for body surface or mCi/kg for body weight.

For therapeutic procedures, especially for the treatment of terminal cancers, the radiation dose estimates are often used to assess the risk of radiation-induced normal tissue toxicity and predict the benefit of tumor control and increased time of survival. In these potentially life-saving therapeutic procedures, the primary concern is the risk of radiation-induced normal tissue toxicity. The risk of radiation-induced genetic effects or secondary cancer development should also be considered, especially if the cancer is likely to be controlled and patient survival time is expected to be long. Although, there is a known uncertainty in tissue dose–response relationship among patients, the magnitude of uncertainty in tissue dose–response is much smaller compared to those uncertainties in dose–response in radiation-induced genetic effects. Therefore, population-averaged dose estimates are usually not adequate for the purpose of assessing the risk of radiation-induced normal tissue toxicity or for the prediction of a benefit of tumor control. Both patient-specific information on the kinetics of biodistribution of the administered radiolabeled antibody/peptide and reliable anatomic data are needed to generate patient-specific dosimetry.

Radiation dosimetry generally has three roles in radionuclide therapy using peptides or monoclonal antibodies. Radiation dosimetry data are used (1) to provide feedback information for developing improved radiolabeled antibody/peptide-based agents for preclinical and clinical trials; (2) to design a dosing scheme for clinical trials; and (3) to maximize the injected dose of radioactivity for individual patients without causing significant normal tissue toxicity.

In preclinical studies, radiation dosimetry for small animals provides critical information on the newly developed radiolabeled antibody/peptide, and this information is used to evaluate the therapeutic advantages or disadvantages of radiopharmaceuticals. Although, these dosimetric advantages or disadvantages of a radiolabeled antibody/peptide should be eventually confirmed by an efficacy/toxicity study, radiation doses to tumors and normal tissues from a biodistribution/dosimetry study are often used as surrogates to predict potential efficacy and toxicity. In clinical trials, dosimetry data are required for all phases of studies. Preclinical dosimetry data are often required by regulatory agencies (see Chapter 18) to design phase I therapeutic or imaging/dosimetry studies. Clinical dosimetry data are required for designing a dosing scheme (including radioactivity dose escalation) and to anticipate any expected normal tissue toxicity. For each patient treatment, radiation dosimetry can be helpful to maximize tumor dose by prescribing the injected radioactivity dose at maximum tolerable level for the dose-limiting organ.

13.2 BASIC CONCEPTS OF MIRD DOSIMETRY

13.2.1 MIRD Equations

Living tissues can be damaged by ionization radiation emitted from radionuclides in the body. When living tissues are exposed to ionizing radiation, the effects of biological damage are often related to the amount of energy absorbed in the tissue. The *mean absorbed dose*, or simply *dose*, is defined as the mean energy, dE, imparted by ionizing radiation to material of mass dm,

$$\text{Dose} = \frac{dE}{dm} \tag{13.1}$$

The standard international (SI) unit of radiation dose is *gray* (Gy), defined as 1 J of energy absorbed per kg mass. The traditional unit for radiation dose is *rad,* defined as 100 ergs energy absorbed per g mass. 1 rad = 1 cGy.

At present, the most common method for absorbed dose calculation was developed from a formalism, MIRD (medical internal radiation dose) schema, proposed by Loevinger and Berman (1, 2). The MIRD schema provides a convenient and simplified method to calculate absorbed radiation dose for radiopharmaceuticals distributed inside the body. The basic equation for absorbed dose to the target region t for radioactivity in the source region s is

$$D(t \leftarrow s) = \tilde{A}(s)\, S(t \leftarrow s) = \tilde{A}(s) \sum_i \Delta_i \phi_i(t \leftarrow s)/m \tag{13.2}$$

where, $\tilde{A}(s)$ is the time integral of radioactivity in the source volume; Δ_i is the "equilibrium dose constant," representing mean energy emitted per unit cumulated activity for i-type radiation; $\phi_i(t \leftarrow s)$ is the absorbed fraction for i-type energy from radioactivity located in source region s deposited in target region t; and m is the mass of the target volume. While $\tilde{A}(s)$ includes the biological parameters for the dose estimation, $S(t \leftarrow s)$ includes radiation energy transport and anatomic parameters for dose estimation:

$$S(t \leftarrow s) = \sum_i \Delta_i \phi_i(t \leftarrow s)/m \qquad (13.3)$$

Based on Equation (13.2), the MIRD schema for dose calculation is a three-step process:

(1) Determine the time integral of radioactivity or cumulated radioactivity in the source volume. This involves a quantitative imaging or biodistribution study based on tissue counting (the former is more commonly used in patients while the latter is often used in preclinical studies in animals).

(2) Obtain mean energy emitted per unit cumulated activity for i-type radiation, Δ_i. This usually involves the use of look-up tables of published data for various radionuclides or use of specific dosimetry software.

(3) Determine the fraction of energy absorbed within the target region. This depends on the specific type of radiation (penetrating or nonpenetrating), and the anatomic/geometric relationship between source and target region. More detailed discussion on basic concepts and examples of basic dosimetry calculations can be found in an excellent introductory book entitled "MIRD Primer for Absorbed Dose Calculation" (3).

13.2.2 Cumulated Radioactivity and Residence Time

The amount of radioactivity and time duration that this remains in the source region depend on a unique interaction between the tissue of interest and the radiopharmaceuticals. The time integral of radioactivity in the source region is often referred to as the cumulated radioactivity or activity $\tilde{A}$ in the MIRD schema. For convenience in MIRD dose computation, the residence time τ is defined as the average time that the administered radioactivity spends in the source region. By definition, $\tilde{A} = A_0 \times \tau$, where A_0 is the administered activity. It is convenient to determine radiation dose per unit administered radioactivity (rads/mCi or Gy/MBq) using the residence time τ. In practice, the exact analytical form of activity as a function of time is unknown and can only be estimated. Cumulated activity is thus commonly determined by curve fitting and extrapolating several activity measurements at certain predefined time intervals.

13.2.3 Radionuclide Data

The radionuclide properties related to radiation dose estimates are $T_{1/2}$ (half-life), radiation type ($\alpha, \beta, \gamma, \ldots$), n_i (particles/transition), E_i (energy/particle), and Δ_i (energy/transition). This information can be found in decay tables in MIRD Pamphlet No. 10 (4) and in a recent update of the book entitled "MIRD: Radionuclide Data and

Decay Schemes" (5, 6). The decay tables can also be found online, such as on the RADAR (RAdiation Dose Assessment Resource) web site (7). In practice, Δ (energy/transition) is directly used in radiation dose calculation (Eq. 13.2).

13.2.4 Penetrating and Nonpenetrating Radiation and the Absorbed Fraction

Another simplification for dose calculation is to classify radiation into two types: penetrating and nonpenetrating radiation. When radiation is nearly fully absorbed within a source volume or energies that escape from the source volume can be neglected in estimating radiation dose to that source volume, that type of radiation is considered nonpenetrating. The classification of nonpenetrating radiation depends on the energy absorption and definition of volume of interest. For nonpenetrating radiation, dose computation is simplified because, by definition, $\Phi_{np} = 1/m$, where m = mass, and Δ_{np} can be easily calculated by simple summation. This can be very helpful if it is desired to have a quick, rough approximation of the radiation dose. The calculation of nonpenetrating dose can provide a convenient way to estimate the total radiation dose, as the penetrating radiation for therapeutic radionuclides, such as the β-emitter, ^{131}I, is typically less than 10% of the total radiation as uptake in normal organs or tumor is higher than their surrounding tissues. Computation of the exact absorbed fraction for penetrating radiation can be very challenging for a user in a small clinic. Typically, simplification is needed to select a phantom model to represent the subject: adult, female, 15 years old, and so on (8, 9). The values of absorbed fraction for penetrating radiation from source organ to target organ in various body phantoms have been calculated and stored in dosimetry software programs. Readers are encouraged to read an excellent chapter by Stabin in a recent book entitled "Radionuclide Peptide Cancer Therapy" (10). In addition to clear explanations on important MIRD concepts, literature and software resources for performing radiation dose calculations have been described in detail.

13.3 PRECLINICAL DOSIMETRY

In preclinical studies, the three-step process for dose estimation often involves determination of cumulated activity from biodistribution studies based on tissue sample counting and determination of the fraction of energy absorbed in the target organ based on existing data using population-averaged animal anatomic models or animal-specific computations based on imaging of individual animals.

13.3.1 Data Collection

Tumor xenografts are often implanted subcutaneously or orthotopically inside the organ in which the human tumor arises (e.g., implantation of breast cancer xenografts in the mammary fat pad of mice; see Chapter 11). Subcutaneous implantation has the advantages of easy implantation and measurement of tumor size for monitoring therapeutic response. Sometime, subcutaneous implantation may not be adequate, however, to represent a true tumor environment. For example, to evaluate uptake of a radiolabeled-antibody/peptide in glioma, tumor xenografts may need to be implanted in the brain, not subcutaneously, as the extent of blood–brain barrier transport is important (11). Animals in each study group are sacrificed at several hours and days postinjection of the radiopharmaceuticals. Selecting the appropriate time points postinjection at which to sample tissues is important for determining accurate time–activity curves. This can be difficult for the first-time investigation of a newly developed radiolabeled-antibody/peptide as the pharmacokinetics are not known. A first-order estimation of the total body elimination rate could be helpful. Total body elimination of a radiolabeled-antibody/peptide in a mouse can be determined by measuring whole-body radioactivity by placing the animal in a radioisotope dose calibrator or positioning a thyroid uptake probe at a consistent distance from the animal. Since there is minimal photon attenuation by the mass of a small animal, the changes in photon attenuation can generally be neglected in estimating the amounts of radioactivity in the animal body over hours and days.

To obtain a more accurate estimate of the radioactivity distribution in various tissues over time, immediately after the animal is sacrificed, organs such as the heart, stomach, small intestine, spleen, kidneys, liver, and thyroid are dissected and weighed. Blood samples are also collected. The amount of radioactivity in each tissue is determined by γ-counting using a NaI(Tl) well counter. The background-corrected counts for tissue samples are converted to radioactivity amounts by comparing them with the counts from a calibrated counting standard of the radionuclide. Concentrations of radiolabeled-antibody/peptide in the tumor and organs at each time point are expressed as the percentage of injected dose per gram of sample weight (%ID/g). It should be noted that error bars for %ID/g are often quite large in many studies published in the literature. These large variations are partly due to large variations in tissue uptake among different animals from pooled data sets. These large variations could also be due to uncertainties in sample handling and counting as these experimental results are often error sensitive because of the small mass of tissue samples. It is very important that the investigator understands the experimental procedures and is aware of pitfalls in handling and counting samples. Some of the common pitfalls include aliquoting of nonhomogeneous sample solutions, evaporation of ^{131}I into the air column inside a sealed test tube, and tissue contamination. These errors, however, are generally small compared to interanimal variability in biodistribution of the radiopharmaceutical, thus emphasizing the importance of sufficient numbers of animals within each group (at least $n = 5$).

Recently, with the development of dedicated high-resolution and high-sensitivity small animal imaging (12–15), large academic centers can now determine cumulated radioactivity in tumors and organs using microSPECT or microPET tomographs. The counts in volumes of interest can be converted to radioactivity by imaging a calibrated source phantom with the identical scan and reconstruction parameters. Accuracy of image quantification can be further improved when microSPECT or microPET are coregistered with microCT images that aids in the delineation of the volumes of interest (i.e. organs). A significant advantage in the image-based approach is that the time–activity curve can be determined for an individual mouse, in contrast to pooled data from different mice in traditional biodistribution studies.

13.3.2 Preclinical Macrodosimetry

13.3.2.1 *Mouse Dose Computation Using Nonpenetrating Radiation*
The purpose of dose estimation for preclinical studies can be generally divided into two categories: (1) to obtain preliminary information for projecting dosimetry in human studies; and (2) to explain tumor response and normal tissue toxicities observed in animals. For the purpose of the second category, radiation dose estimates for each tissue of the mouse need to be as accurate as reasonably achievable. For most studies, the purpose falls into the first category and accurate dosimetry considering cross-radiation from adjacent organs may not be helpful for projecting dosimetry for human study due to substantial differences in distances between organs in mice versus humans that affect the absorbed fraction estimates. For example, in a ^{90}Y-antibody/peptide study, the high-energy beta particles of ^{90}Y ($E_\beta = 2.2$ MeV) that travel for up to 10 mm in tissues should be considered as penetrating radiation for the mouse, but not in a human. The total radiation dose to the mouse kidneys would be contributed from ^{90}Y in the kidneys, as well as in the spleen, stomach, pancreas, and liver (16). It is therefore important to include cross-organ radiation dose for evaluating kidney toxicity in the mouse because only 46% of the beta energy from ^{90}Y in the mouse kidney will be absorbed within the kidney. However, this complete radiation dose estimate for mouse kidneys is not helpful for projecting the radiation dose to patient kidneys because cross-organ ^{90}Y radiation from the spleen or stomach is negligible in the human model because of the much greater distances between these organs and the kidneys (17). When a novice reader evaluates a dosimetry report from a preclinical study using a novel radiolabeled antibody/peptide, he or she often quickly divides the tumor dose by the normal organ dose to obtain a ball park "therapeutic ratio" for patients. In the previous example of a ^{90}Y-labeled antibody/peptide, it would be more appropriate to compare cumulated activity in the mouse tumor and kidneys, instead of radiation dose to the mouse tumor and kidneys, which avoids the problem of the different anatomies affecting the absorbed fractions.

For the above reason, radiation dose calculations can be quite simple for some therapeutic radionuclides. For example, ^{131}I is predominated by nonpenetrating radiation and the contribution from penetrating radiation is fairly small because of the high photon energy of ^{131}I and the relative small size of a mouse. Ninety-nine percent of the photon (or penetrating) radiation will escape from the liver and 97% of the photon radiation will escape from the mouse body (18). For the purpose of projecting dosimetry for humans, penetrating radiation from ^{131}I in the target organs themselves can be ignored and the radiation dose can be calculated by

$$\text{Dose} = \sum \Delta_{\text{np}} \tilde{A}/m \qquad (13.4)$$

The 90-percentile distance (x_{90}) for nonpenetrating radiation from ^{131}I is 0.8 mm. Although, nonpenetrating radiation of ^{131}I may not be technically 100% absorbed in a mouse organ or tumor, complete absorption can be used in the dose calculation for nonpenetrating emissions in the mouse if the purpose is simply projecting dosimetry in patients.

13.3.2.2 3-D Mouse Dose Computations Although tissue responses to radiation in mice are quite different from those in patients, accurate radiation dose estimation is required when there is a need to explain tumor response and organ toxicities observed in mice. In order to calculate the absorbed fraction for a mouse, mouse geometry needs to be modeled. Geometric models initially used are mathematical equations to describe organs in the form of simple geometries. Based on organ sizes measured from 10 nude (athymic) mice of approximately 25 g, Hui et al. modeled the organs as ellipsoids, except for bone and marrow (16). To further refine the mouse bone marrow model, Muthuswamy et al. used slab, cylinder, and spheres to represent marrow at various bone regions (19). To address heterogeneous uptake of radiopharmaceuticals in the kidney, Flynn et al. modeled the cortex and medulla region using ellipsoids and noted differences in radioluminograph measurements (using phosphor plates) of kidney sections for ^{131}I and [90]Y (20). Hindorf et al. modeled organs as ellipsoids partly based on images in anatomic atlases of mice (21) to assess the impact of mass and the shape of organs and their relative locations on mouse S-values (22). More realistic geometric models have been obtained by digital imaging. While Kolbert et al. delineated left and right kidneys, liver, and spleen using magnetic resonance imaging (23), Stabin et al. delineated skeleton, lungs, heart, liver, kidneys, stomach small intestine, spleen, and testes using micro-CT images (24). Using the Moby phantom developed by Segars et al. (25) for 35 delineated regions of C57BL/6 mouse, Larsson et al. calculated the absorbed fractions and S-values with EGS4 and MCNPX 2.6a Monte Carlo method (Monte Carlo method refers to a class of computational algorithms that rely on repeated random sampling to compute their results) (26). Bitar et al. delineated 13 source organs and 25 target regions based on high-resolution digital photographs of frozen thin slices of a nude mouse (27). In these studies, 3D doses were computed using either dose point kernels from Monte Carlo code or full Monte Carlo simulations (dose point kernel refers to a function describing energy deposited in concentric spherical shells around a point source). Dose estimates calculated using a single-dose point kernel can overestimate or underestimate the dose in the lung and bone region when tissue density is not considered. Even if tissue densities are considered, it might be still less accurate at heterogeneous tissue boundaries (such as the bone-to-soft tissue and lung-to-soft tissue boundary) compared to full Monte Carlo simulations. However, the variations in mass and the shape of organs and their relative locations among the mice introduce larger uncertainty in dose estimates compared to these uncertainties at heterogeneous tissue boundaries for a mouse.

13.3.3 Nonuniform Distribution and Multicellular Dosimetry

It has been long recognized that radiolabeled antibodies/peptides are distributed nonuniformly in tissues at the organ, suborgan, multicellular, cellular, and subcellular levels. While conventional clinical radiation dosimetry is performed at the organ level, the MIRD schema has no limitation in target/source size. When time–activity curves are determined at the organ level, *organ dosimetry* is performed. When the time–activity curve can be determined at the multicellular or single cellular level, the MIRD schema can also be applied for *multicellular dosimetry* calculation. Research studies may sometime utilize autoradiography to quantify the distribution of radioactivity in histological sections of tissues of interest in small animals (28–32) or in patients (if surgical samples can be obtained) (33). In the case of tumors, because malignant cells are nonuniformly distributed within the gross tumor volume and radiolabeled antibodies/peptides are nonuniformly distributed to and within malignant cells, the actual radiation dose to the cell nuclei can be substantially higher or lower compared to dose estimations that assume uniform distribution (31, 34). Depending on the spatial distribution of radioactivity relative to the cell nuclei, substantial overestimation or underestimation can occur, especially for short-range emitters such as α-emitters or Auger electron-emitting radionuclides (see Chapter 9).

13.4 CLINICAL DOSIMETRY METHODS

In clinical studies, cumulated radioactivity in the source volume is most typically obtained by sequential quantitative imaging and in some cases by tissue sample counting. Methods for image and sample data collection have been summarized in several reports including MIRD Pamphlet No. 16 (35–37). Imaging methods can generally be categorized into 2D planar methods and 3D SPECT/planar hybrid methods.

For a radiolabeled antibody/peptide to be approved by a regulatory agency, such as the Food and Drug Administration (FDA) in the United States of America, relatively large numbers of patients need to be enrolled in clinical trials, especially in phase 2 and 3 trials. The imaging and dosimetry substudy of these trials can often be a bottleneck for patient recruitment. Sequential imaging over 1 week requires a patient to have multiple trips to hospital and live/stay not too far from the hospital. Sometimes a clinical trial site that has a relatively high patient recruitment may not have a corresponding high level of expertise and resources in nuclear medicine imaging to accommodate this recruitment. Therefore, for a large-scale multicenter clinical study, a complicated image acquisition protocol may not always yield high-quality imaging data. For these reasons, most clinical dosimetry data have been derived from simpler 2D planar imaging methods.

13.4.1 Planar Conjugate View Imaging

The most widely used method for planar gamma camera image quantification was developed by Thomas et al. and corrects for source thickness and attenuation (38). While the original method provided mathematical formulae for $g(\alpha)$ and $k(\gamma)$ as function of α and γ (activity ratios for various regions), for correcting radioactivity in overlapping structures and background radioactivity (38), the widely used geometric-mean equation is a simplified version (Eq. 13.5) (39, 40), which corresponds to no background radioactivity in the original report (38).

$$A = (I_A \, I_P)^{1/2} \, e^{(\mu_e T/2)} \frac{f}{c} \qquad (13.5)$$

This form is widely used because the radioactivity concentration and the size of the overlapping structure or background volume are unknown in practice, without obtaining additional 3D imaging, such as SPECT or PET. Anterior and posterior view count rates (I_A and I_P) in Equation (13.5) should represent count rates from source volume only and counts from background volumes should be subtracted. The self-attenuation correction factor equals $(\mu t/2)/\sinh(\mu t/2)$, for the source region attenuation coefficient μ and source thickness t. The attenuation factor $\exp(\mu_e T/2)$ may be obtained directly from the patient by taking transmission measurements across the region of interest (ROS) (38). It should be noted that this geometric-mean quantification was derived from point source geometry and experimental data were collected using a 5% energy window to minimize scatter effects (38). In clinical imaging for radiolabeled antibodies/peptides, energy windows are typically 15–20% and transmission images are acquired using a large flood/sheet source of radioactivity with a broad-beam geometry.

The difference in the estimation of μ_e determined by measuring the transmission fraction with a small hot source and with a cold source using a flood/sheet source can be quite substantial. Eary et al. suggested that to reduce the large fraction of scattered photons, the flood source should be collimated and an asymmetric energy window should be used (40). While that report has been widely referenced in many studies using geometric-mean image quantification, most investigators have used Equation (13.5) without collimating the flood/sheet source. One method for reducing the broad beam from a rectangular sheet source is the use of a rod source for the transmission scan as described by DeNardo et al. and Macey et al. (41, 42). In order to incorporate the transmission scan and conjugate static and whole-body image quantification, Macey et al. and Erwin et al. developed one of the earliest treatment planning systems that employs the Siemens MicroDelta computer platform (42–44). The attenuation corrections derived from rod source transmission scans have been used only for large organs such as the liver or lungs, and attenuation corrections for kidneys and normal-size spleens are determined by measuring the transmission fraction using a 150 ml phantom to avoid underestimation (45, 46). In our experience, the difference in transmission fraction data using a 10 mL source or 150 mL source is small, but the difference using a 150 mL source versus a 1500–1800 mL source can be substantial (47). Often a long-lived ^{57}Co source is used for transmission measurements and corrected using a calibration factor for the energy difference between this source and the radionuclide administered to the patient. This difference in transmission fraction due to energy differences between the ^{57}Co sheet source and the administered radionuclide should be measured for each γ-camera model used especially when the crystal thickness and collimator are different (47). A simple liver phantom measurement can reveal difficulties in using Equation (13.5). One solution is to experimentally determine the values for $\mu_{\text{Co-57,liver}}$ and $\mu_{\text{In-111,liver}}$ using the large-volume phantom; experimentally determine the value of f (instead of calculation using $\mu_s t / \sinh(\mu_s t/2)$) and then adjust the volume effect if the calibration factor c is determined from a small vial source (48). Similarly, we found that the error in estimating the liver background count rate was substantial if conventional integration of the transmission is performed using a narrow beam μ_e value:

$$C_{\text{BGC}} = \int_{d}^{d+t} (C_{\text{bg}}/T) \exp(-\mu_e x) dx \qquad (13.6)$$

where C_{BGC} is the apparent background concentration.

One great advantage of the geometric mean is the ability to quantify source radioactivity without depth information. This was important in the 1970s and 1980s when CT/MRI images were not readily available. This advantage has become less important now as CT/MRI images are typically available for source depth information. The geometric-mean method should not be used for a source organ that cannot be clearly visualized on both conjugate views. Typically, when a source is not clearly visualized, more than 95% of the counts are contributed from the background noise. Therefore, net counts (background subtracted) will be small and statistically unreliable. If these counts are included in a geometric-mean calculation, more error would be introduced.

13.4.2 Accounting for Scatter Effects in SPECT and Planar Quantification

There are many methods that have been developed to account for scatter effects. Although many of the following methods were originally developed for single-photon emission computed tomography (SPECT) quantification, they can also be applied to planar images.

One may use a narrower or asymmetric energy window to exclude scattered photons. However, the narrow energy window methods often suffer from inadequate counting statistics in clinical dosimetry images of radiolabeled antibodies/peptides because of the limited injection dose of radioactivity used for safety considerations. Asymmetric energy window imaging can suffer from a similar problem of inadequate counting statistics. In addition, asymmetric energy window imaging can be very sensitive to error in daily energy window setup and possible window drift during acquisition.

One widely used scatter correction method is subtraction of counts collected in a subphotopeak energy window. Assuming a spatial correlation between scatter counts in the photopeak window and its subphotopeak energy window, Jaszczak et al. proposed to subtract a k fraction of counts collected in the subphotopeak energy window from the counts in photopeak window for SPECT images (49). The scaling factor k can be estimated experimentally for line sources and cold spheres/cylinders inside a phantom with moderate background radioactivity and subtraction can then be performed in projection images or reconstructed SPECT images (49, 50). The scaling factor k-value can vary with the object being imaged, and a single k-value may not be applied to every source in the field of view (51). Koral et al. developed a scatter correction method by analyzing the energy spectrum (52). In this method, 32 energy channels were used to acquire the γ-photon energy spectrum and the scatter spectrum at each energy bin was determined by least squares fitting (52). Based on the observation that scatter counts contribute more to the lower side of the photopeak than the higher side, King et al. developed a dual-photopeak window scatter correction method by determining a regression relationship for the counts ratio between the two windows and scatter fraction (53). This method allows the estimation of the scatter distribution within the photopeak window without using a k scaling factor (53). Assuming a linear "spill down" distribution of scattered photons from high-energy photons, Ogawa et al. developed a triple-energy window method without using a k scaling factor (54). Because the method is relatively simple to implement, it has been used in many clinical studies. A more rigorous triple-energy window method for ^{131}I was developed by Macey et al. using ^{137}Cs and ^{51}Cr sources to estimate k scaling factors for the fraction of scattered photons in three windows for septal penetrating 637/723 keV photons and primary 364 keV photons of ^{131}I.

One common limitation of these scatter subtraction methods is poor image counting statistics after subtraction, and these techniques are often sensitive to the quality of daily energy window set up by the imaging technologists. We have applied subphotopeak energy window subtraction techniques (49, 50, 54) for ^{67}Cu (55), ^{111}In, ^{166}Ho (56, 57), and ^{131}I in ongoing clinical trials (58–60). We noticed that the counts in the upper part of the subphotopeak energy window are more likely to be "bad" counts, that is, "spill down" from higher energy photons far from the photopeak being imaged for ^{166}Ho or ^{131}I (56, 57, 59, 60). However, the image counts collected at the lower part of the subphotopeak energy window (just below the photopeak) are likely to have some "good" counts for ^{67}Cu, ^{111}In, ^{166}Ho, or ^{131}I, thus subtracting these counts degrades the image quality and even could introduce additional errors for ^{166}Ho (56, 57). For relatively good imaging radionuclides such as ^{67}Cu and ^{111}In, the impact of scatter subtraction on planar image quantification seems not very clear using standard subtraction techniques. This could be due to the fact that point source geometry may still not be applicable to the images after subtraction. The impact become less clear when the source size-dependent attenuation and calibration are applied to images after subtraction. In addition, radiolabeled antibody images often have significant radioactivity in the background volume, which contributes more uncertainty to image quantification compared to scattered photons for ^{111}In and ^{67}Cu. In our ongoing clinical trials using ^{131}I (58–60), we noticed that Ogawa's scatter subtraction technique is sensitive to the accuracy in daily energy window setup for patient image acquisition.

Some investigators believe that scattered photons are "good counts gone bad," and that scattered photons should be repositioned, instead of being permanently subtracted (61). Assuming SPECT projection images to be a convolution of the original photon component with an averaged scatter response function, deconvolution is performed using a digital filter during image reconstruction (62). The scatter response or energy-response function is usually Gaussian in shape (62, 63). When the scatter effect and resolution degradation are estimated by measuring the modulation transfer function (64), the deconvolution filter can be formulated as a modified inverse filter to achieve a practical balance between restoration of image degradation and noise suppression (65–67). Deconvolution methods require accurate estimation of scatter distribution, and an averaged scatter response or modulation transfer function can be optimized only for one selected source geometry, thus undercorrecting or over-correcting for other source geometries. The basic problem is that the scatter response function varies as function of both depth inside the scatter medium and distance from the edge of the scatter medium (68, 69).

One effective method for planar image quantification based on detailed calibration was developed by Wu et al. using depth-dependant buildup factors (70). Subsequently, Siegel et al. introduced a variation, depth-independent, buildup factor method (71, 72). The effectiveness of the depth-independent buildup factor method has been verified by Van Rensburg et al. for ^{111}In and Kojima et al. for ^{99m}Tc (73, 74). Kojima et al. used a thin rectangular source of $7.5 \times 6.0 \times 0.03$ cm to determine transmission fraction data and verified the method for 2 cm and 4 cm thick sources simulating the kidney (74). We also verified the depth-independent buildup factor method for 14–65 mL ^{131}I sources in an abdominal phantom filled with background concentrations of ^{131}I. The source depths were varied from 4 to 17.5 cm and images were acquired using a 20% energy window centered at 364 keV. A buildup factor $B(\infty)$ of 1.225 ± 0.084 was determined and the quantification error was 0.4–3.6%. These results illustrate that a better quantification accuracy can be achieved based on a detailed calibration process. Similarly, a simple version of the calibration process can be applied to geometric-mean quantification to achieve reasonable accuracy when appropriate attenuation correction and camera sensitivity are determined experimentally (47, 48, 75), considering the image quantification as a process of detailed calibration.

In external beam radiotherapy, detailed calibration has been a standard method for radiation dose calculation for treatment planning for many years. Scatter effects versus field sizes are measured in field size ranging from 5×5, 10×10, $15 \times 15, \ldots$ to 40×40 cm for various depths and beam energies. All theoretical calculations used for dose calculations are calibrated based on a large measured data set for each machine. Similar processes can be applied to gamma camera image quantification to provide improved confidence in the radioactivity measurements.

13.4.3   3D CT/Spect/Planar Hybrid Methods

Planar images are intrinsically limited when sources are overlapped. SPECT imaging provides information on 3D radioactivity distribution, thus making it possible to determine patient-specific 3D dose distribution. The disadvantages of obtaining SPECT information are (1) lengthy image acquisition times; (2) poor spatial resolution compared to planar images; (3) and more significant impact from scattered photons. The scattered counts after reconstruction typically account for 20–40% of the recorded counts in a SPECT image (37, 69).

A patient-specific 3D treatment planning system for radiolabeled antibodies/peptides was first described by Sgouros et al. (76, 77), and followed by reports from Giap et al. and Akabani et al. (78, 79). In these 3D planning systems, 3D radioactivity distributions were determined by quantitative SPECT and 3D radioactivity distributions were converted to a 3D dose rate distribution, using convolution of a point dose kernel. Dose volume histograms can be determined at the SPECT voxel level. In the comprehensive treatment planning system developed by Sgouros et al., image registration of CT/MRI/SPECT was built user-friendly into the planning system and structure segmentation on CT/MRI images and 3D contour display were all menu driven and semi-automatic (76, 77, 80). Koral et al. have reported CT-SPECT fusion using fiducial markers on the patient skin (81). Restricted by image acquisition time, Koral et al. have advocated a 3D-SPECT/planar hybrid method to determine cumulated radioactivity (81). In their protocol, planar conjugate view images were acquired at multiple time points and SPECT was acquired at a single time point. It was assumed that the daily planar conjugate views provided the shape of the time-activity curve and the amplitude of the curve was normalized by the quantitative SPECT measurement (81). Although CT was used for structure segmentation, uniform distribution was assumed for organ and tumor dose estimation. In some clinical studies, five sequential SPECT images over 1 week must be performed because radioactivity accumulation can be determined only by 3D imaging. For example, in a recent ^{131}I-TM601 study for glioma, ^{131}I-TM601 was injected into a surgically created cavity to treat residual tumor or microscopic tumor cells outside the surgically created cavity (82, 83). It is desirable to determine ^{131}I concentration in 2-cm margins outside the cavity because most residual tumor cells reside in the 2-cm margin. Planar imaging is incapable of determining ^{131}I concentration in the 2-cm margins given the high concentration of ^{131}I inside the cavity.

One recent advance toward a more accurate SPECT quantification has been Monte Carlo simulation for photon scatter, attenuation, and depth-dependent image degradation (84–89). Some of the results (image quality after correction) were quite impressive. Because Monte Carlo simulation provides a more accurate estimation of the image degradation process, it may provide a better solution in the future for routine image processing for individual patients.

One significant development in planar image quantification is CT-assisted quantification developed by Liu et al. (90). Liu coregistered CT image data to planar conjugate views to provide depth information on structures overlapped in the anterior and posterior direction. Assuming uniform distribution along a line of pixels inside the volume of interest, radioactivity distribution in the body was determined using a matrix inversion method (90). Based on this technique, Liu et al. have developed a treatment planning system that performs semiautomatic planar image quantification and dose volume histogram (DVH) computation after organs and tumors are contoured on CT images. Tang et al. developed a similar method to project 3D organs obtained by segmentation of CT images to the planar images (91). Attenuation and collimator–detector response function were modeled in the projection process. Subsequently, Sjogreen et al. have applied a registered CT scan and a ^{57}Co transmission scan to the planar conjugate view images (92). The scatter-penetration was compensated with

Wiener filtering, and attenuation was determined from the transmission scan and CT in a narrow beam geometry, and correction for background and overlapping organs (92). Recently, He and Frey have developed a QSPECT method based on maximum likelihood estimation of organ radioactivities using 3D organ volumes of interest and a model-based projector that models image-degrading effects including attenuation, scatter, and the full collimator–detector response (93). They have shown improved accuracy using this method with both simulated and physical phantom experiments.

13.4.4 3D PET

PET provides a better quantitative accuracy compared to SPECT with better spatial resolution and more accurate attenuation. ^{124}I-PET has been used to guide ^{131}I therapy in patients with differentiated thyroid carcinoma (94–97). These ^{124}I-PET studies were mainly focused on tumor dosimetry. The dose-limiting factor is determined based on radiation dose to the red marrow from ^{131}I in the circulation blood. Dosimetric analysis of the ^{124}I-PET data could identify those patients likely to benefit from radioiodine therapy. ^{90}Y has been used in many radionuclide therapy studies because its long-range beta energy provides a uniform dose coverage for tumor. While its lack of photon emission is considered an advantage for radiation safety to the public, a surrogate is required for the biodistribution and dosimetry study. ^{86}Y, a positron emitter with a 14.7 h half-life, is naturally a better surrogate (isotope) compared to ^{111}In. The biodistribution and dosimetry of ^{90}Y-pharmaceuticals have been evaluated using ^{86}Y-pharmaceuticals and PET imaging for citrate (98), EDTMP (99), and SMT487 (100). Although ^{67}Cu, a beta emitter with a 61.9 h half-life, showed a 50% higher tumor to marrow dose ratio compared to ^{64}Cu, a positron emitter with a 12.7 h half-life, when used to radiolabel the Lym-1 antibody (101), production of ^{67}Cu has been challenging. ^{64}Cu can be produced with high specific activity and ^{64}Cu can be used for PET with good image quality. Anderson et al. have developed method for radiolabeling with ^{64}Cu with high specific activity and good stability and ^{64}Cu has been used for various clinical/preclinical studies including labeling 1A3 antibodies for colon cancer (102), octreotide for neuroendocrine tumors (103), and cetuximab for cervical cancer (104). The limitation of PET imaging-based dosimetry for a large clinical trial will be the supply of the positron-emitter as most of the radionuclides mentioned above have short half-lives, except for ^{124}I.

13.5 DOSIMETRY FOR DOSE-LIMITING ORGANS AND TUMORS

In principle, there is no difference in the methodology for dosimetry applied to antibodies or peptides. In this section, discussions on clinical dosimetry are therefore mainly focused on methodologies. Clinical dosimetry data for normal organs and tumors for various individual radiolabeled antibodies are not discussed. Clinical dosimetry data and relevant methods for peptide radionuclide receptor therapy have been summarized in an excellent chapter by Cremonesi in a recent book entitled "Radionuclide Peptide Cancer Therapy" (105).

13.5.1 Marrow Dosimetry

13.5.1.1 Without Bone Marrow Reconstitution   Radiation-induced myelotoxicity is dose limiting for most radionuclide therapies that do not involve bone marrow reconstitution. For radiolabeled antibodies/peptides that do not bind to marrow cells or bone, the radiation dose to the marrow can be considered solely from circulating radioactivity in the blood as described by Siegel et al. (106), who suggested a red marrow to blood ratio (RMBLR) of 0.2–0.4. By assuming rapid equilibration of radiolabeled antibodies in the plasma and extracellular fluid of the red marrow, Sgouros developed a method to estimate RMBLR using patient-specific hematocrit values (107), and his method has been widely used. In murine studies, a range of 0.3–0.4 for RMBLR has been confirmed experimentally for intact, 150-kDa antibodies but not for lower molecular weight fragments that yield a red marrow to blood concentration ratio closer to 1.0 as reported by Behr et al. (108). For marrow dosimetry based on radioactivity per gram of blood, an explicit estimate of marrow mass may not be required since the majority of therapeutic radiation (i.e., nonpenetrating) dose is directly related to the radioactivity per gram of tissue (109). In addition, the fraction of electron energy absorbed in the marrow cavity depends primarily on the size and distribution of marrow cavity in the trabecular bone and much less on the total grams of the red marrow (110–113). Ideally, patient-specific doses to the marrow from penetrating radiation in other source organs in the body other than the blood should be determined using 3D data sets (114) incorporating multiple imaging time points. Because these data are typically unavailable, there have been efforts to estimate the photon dose contribution from all source organs in the body without using a 3D data set (109, 115–118). These practical calculations used various assumptions and the magnitudes of errors associated with these assumptions are expected to be determined soon (119–121).

For radiolabeled antibodies/peptides where the marrow or skeleton are clearly visualized in sequential gamma camera images, an imaging-based method is recommended because the antibodies/peptides may directly bind to marrow or bone components as noted by Sgouros et al. (122). Sacral (123, 124) and lumbar vertebral (125–127) regions have been used to represent the whole skeleton. While the radioactivity in the bladder can affect radioactivity quantification in the sacral region, radioactivity in the major blood vessels (aorta and vena cava) can also affect quantification in lumbar vertebral region. The background radioactivity in the major blood vessels can be corrected using blood concentrations measured in a gamma counter and vessel dimensions from CT as described by Meredith et al. (128). The marrow mass in the region of interest is required to calculate a patient-specific dose. Although high-resolution MRI or MRI combined with spectroscopy (MRS) has potential to determine marrow mass, this method is difficult to use routinely. One practical approach is to use trabecular bone volume in the ROI to scale from a reference marrow mass (127, 129). One of the underlying assumptions of using any regional imaging method is that regional radioactivity concentrations represent the mean radioactivity concentration in the total marrow. This assumption can be invalid if marrow heterogeneity is compounded by marrow involvement with cancer or is altered by prior radiation (122). Substantial regional variation in marrow uptake has been observed in patients with leukemia (130). Further research is needed to resolve this problem as quantification of all marrow regions in the body is impractical and inaccurate when patient data are acquired by whole-body planar imaging or a single-field SPECT image.

13.5.1.2 *With Bone Marrow Reconstitution*   Myelosuppresion can be ameliorated with autologous bone marrow transplantation or peripheral blood stem cell (PBSC) infusion, as illustrated first by Press et al. for radioimmunotherapy of lymphoma (131) and later by others in the treatment of breast cancer (132–138), colon cancer (139), and lymphoma (140). These studies suggested that a higher tumor response rate was correlated with a higher radioactivity injection dose followed by bone marrow reconstitution.

In treatments incorporating PBSC support, the time interval between the radioactivity dose injection and PBSC infusion is an important parameter for optimal patient management. While PBSCs can be harmed by irradiation from radiopharmaceuticals in the body if they are transfused too early, patients can also develop serious health complications from low neutrophils or platelets if PBSCs are transfused too late (i.e., infections or bleeding episodes). Initially, the time interval for PBSC infusion was based on the residual ^{131}I concentration level in the blood (below 1 μCi/mL) in breast cancer patients treated with ^{131}I-chimeric L6 antibody (132). In limited observations of patients receiving high-dose ^{131}I-chimeric L6, this PBSC infusion time based on ^{131}I concentrations in the blood worked reasonably well. However, in patients receiving high-dose ^{90}Y-2IT-BAD-m170, results were unsatisfactory when the time for PBSC infusion was based on ^{90}Y concentration in the blood. The ^{90}Y concentration threshold level was determined for PBSC *in vitro*. Recovery of blood counts was delayed even when the ^{90}Y concentration in the blood was well below this activity concentration threshold.

^{90}Y concentrations in the bone and marrow can be directly determined from bone marrow biopsy. Wong et al. used bone marrow biopsy to estimate the ^{90}Y radiation dose to autologous stem cells (136). Initially, they transfused 25% of stem cells at 5 days post-^{90}Y injection (0.56 GBq/m^2 (15.1 mCi/m^2)) regardless of the variation in ^{90}Y distribution among the patients. The remaining 75% of stem cells were transfused at the time point at which the estimated remaining marrow dose was $\leq$5 cGy, as determined by patient-specific bone marrow biopsy. Their protocol was subsequently modified for a dose level of 0.83 GBq/m^2 (22.4 mCi/m^2), at which 25% of the stem cells were transfused when the remaining marrow dose was $\leq$5 cGy, and the remaining 75% were transfused when the absolute granulocyte count was <1000/μL. All patients demonstrated hematopoietic recovery after stem cell infusion (136).

Radioactivity concentrations in the liver, spleen, and kidneys can be much higher than those in the blood, so that PBSCs can be damaged during their circulation through these organs. PBSCs can be further damaged after homing (a process that describes the migration of PBSCs from the peripheral blood to the bone marrow) if the marrow or bone has radioactivity uptake (141). Therefore, the radiation dose from radioactivity in the blood is only one part of the total radiation dose PBSCs are exposed to. Shen et al. proposed to determine the optimal PBSC infusion time based on time-varying radiation dose rates to PBSCs from radioactivity in the blood and other source organs; time-varying PBSC distribution in the body during the homing process, and radiation dose to PBSCs from radioactivity in the bone and marrow after PBSC homing (141). In that analysis, it was found that the remainder of the blood (excluding blood volume in the liver, spleen, lungs, kidneys, bones, and marrow), which has 74% of the blood volume, contributes 12% of the total dose to PBSCs. The majority of the radiation dose to PBSCs was from ^{90}Y in the marrow and bone matrix as the PBSCs accumulate in the marrow over time (141). As ^{111}In was used as a surrogate for ^{90}Y for dosimetry imaging, the difference between calculated ^{111}In amounts in the marrow and ^{90}Y amounts in the bone was considered using data reported for patient core biopsies; these studies showed that the mean difference between ^{90}Y and ^{111}In concentration in bone and marrow was 0.003%ID/g with the MX-DTPA chelator conjugated to the antibodies (142) (see Chapter 2). This infusion time interval worked well as all three patients demonstrated evidence of hematological recovery for a single-dose injection of 38–62 mCi ^{90}Y (137, 141).

Based on quantitative planar and SPECT/CT scans for ^{111}In-labeled antibodies specific for non-Hodgkin's lymphoma (NHL), Flinn et al. determined that the time for stem cell infusion based on the estimated marrow dose rate was <1 cGy/h (143). Despite the fact that the liver is the organ that receives the highest absorbed dose of radiation, no significant hepatotoxicity has been seen. They found that the most common treatment-related toxicities were hematologic requiring stem cell support (143).

13.5.1.3 *Correlation with Myelotoxicity*

The ability to predict the hematologic toxicity after radionuclide therapy is essential for planning the tolerable radioactivity injection dose and for maximizing the radiation dose to tumors. Conceptually, the marrow radiation dose should be a natural choice for planning the radioactivity injection dose for individual patients. However, in practice, the marrow radiation dose has not been commonly used for planning the injection dose. The most widely used methods for dose planning are based on patient body surface area (mCi/m^2) or body weight (mCi/kg), similar to the approach taken for dose prescription of chemotherapy. Recently, patient-specific lean body dose has been used for planning the injection dose for one radiolabeled antibody, assuming lean body dose can serve as an indicator for marrow dose (144).

The usefulness of any radiation dose estimate relies on the establishment of a dose–response correlation. While Wiseman et al. and Erwin et al. found poor or no correlation between marrow dose and myelotoxicity for ^{90}Y-anti-CD20 monoclonal antibody for NHL (145, 146), other investigators found positive marrow dose–toxicity correlations for various antibodies. Correlations with myelotoxicity include body dose of ^{131}I-antibody for treatment of gastrointestinal (GI) cancer (147), marrow dose based on sacral imaging with ^{131}I-antibody for NHL (124), a better correlation with lumbar imaging with ^{131}I-antibody for NHL (126), marrow dose better than body dose with ^{131}I-antibody for GI cancer (148); similar correlation with marrow dose, body dose, or mCi administered with ^{186}Re-antibody for solid tumor (149); better correlation using marrow dose with pretargeted ^{90}Y-antibody (150); better correlation using lumbar imaging and patient-specific trabecular bone volume with ^{90}Y-antibody for lung cancer (127); better correlation using marrow dose with ^{131}I-antibodies/fragments for various cancers (151); better correlation using marrow dose with ^{131}I-antibody for prostate cancer (152); and better correlation using marrow dose for ^{177}Lu-antibody compared to ^{90}Y-antibody for prostate cancer (153). In these reports, toxicity was most often expressed as the percentage decrease in blood cell counts from their baseline values.

Marrow dosimetry for treatment planning may be practical for patients without prior myelosuppressive chemotherapy or radiation. For such a patient population, marrow-based radiation dose escalation may be used in clinical trials instead of mCi dose escalation. In patients with prior myelosuppressive chemotherapies, prediction of myelotoxicity solely based on radiation dosimetry of the radiotherapeutic agent becomes inadequate. Even with precise radiation dosimetry, it would not predict toxicity since biological factors influence toxicity. For chemotherapies with short-term myelosuppression, prediction of myelotoxicity can be relatively simple. However, many myelosuppressive chemotherapies induce long-term damage leading to poor self-renewal capability of the progenitor cells (154). This long-term damage is considered a result of damage to the stromal microenvironment (155–157). In these patients, the recovery of peripheral blood cell counts after radionuclide therapy depends on the condition of the stromal cell population and their production of cytokines in regulating homeostasis. Direct assessment of the condition of the stromal microenvironment is difficult. However, high levels of stimulatory cytokines may be indicative of "excess toxicity" (158). In patients with solid tumors treated with various chemotherapy drugs (doxorubicin, methotrexate, topotecan, lomustine, and mitomy-cin), Blumenthal et al. showed that the plasma FLT3-L (FMS-related tyrosine kinase 3 ligand) level predicted excess platelet toxicity in 13 of the 16 patients and resulted in a false-positive prediction in only 3 of the 27 other patients (158). Subsequently, in patients without marrow or bone involvement, Siegel et al. demonstrated a significantly improved correlation of red marrow dose with 1/(platelet nadir) (from $r = 0.20$ to $r = 0.86$) with FLT3-L adjustment (159). These results suggest that combined patient-specific marrow dosimetry and plasma cytokine levels may become useful in future treatment planning and further investigations are desirable.

13.5.2 Other Normal Organ Toxicity

Radiation-associated renal toxicity has been observed in several clinical studies using radiolabeled peptides (e.g., ^{90}Y-DOTATOC; see Chapter 4). Often, these renal toxicities in the early studies were unexpected because radiation dose estimates for kidneys derived from the traditional MIRD calculation method, without accounting for high dose-rate effect or nonuniform distribution, were significantly lower than 23 Gy, a reported tolerated dose with 5% risk of complication within 5 years of external beam radiotherapy (160). To address the nonuniform uptake in the kidney, Bouchet et al. developed a multiregion kidney model incorporating the outer shape of the kidney, medullary pyramid, medullary papillae, and renal pelvis in MIRD 19 (161). Depending on the subregion and radionuclide used, the traditional uniform distribution model calculations can overestimate doses by 50% or underestimate them by 80% (161). The significant variation in the dose in subregions has been illustrated by Konijnenberg et al. using autoradiogram data of kidney section samples from three patients (33). Using the Lea-Catcheside time–dose factor G as a function of the repair constant μ and the dose rate $D(t)$, Konijnenberg addressed the dose rate effect by incorporating the linear-quadratic model in determining the biologically effective dose to the kidney for peptide receptor radionuclide therapy with ^{90}Y-DOTA^{-}Tyr^{3}-octreotide (^{90}Y-DOTATOC) (162). These findings have been summarized and expanded in a recent MIRD Pamphlet by Wessels et al. (163). The MIRD Pamphlet 20 also provided many examples illustrating how to apply dose rate effects and the multiregion model in clinical settings (163).

In high dose ^{131}I-labeled antibody therapy for B cell lymphoma with autologous bone marrow support, Press et al. escalated the injected dose of radioactivity until a single, grade 3 or 4, nonhematopoietic toxicity was observed (131, 164). In 17 of the 19 cases, lung was the normal organ receiving dose-limiting radiation exposure (164). Pulmonary grade 3 toxicity was found in one patient, and cardiac grade 3 toxicity was found in another. Gastrointestinal, renal, and hepatic grade 2 toxicity was also observed (164).

As mentioned previously, the liver would be predicted to be a dose-limiting organ because it receives the highest ^{90}Y dose per injected amount of radioactivity in lymphoma studies with bone marrow reconstitution (143). The results suggesting that the liver radiation dose was not radioactivity dose limiting were also found with high dose ^{90}Y-labeled antibodies for treatment of breast cancer and prostate cancer (132, 134, 137, 138). As the TD5/5 (tolerance dose with the probability of 5% complication within 5 years) for external beam radiotherapy is 30 Gy (160) and the projected ^{90}Y radiation dose to the liver was expected to reach hepatic toxicity, these protocols were designed for dose escalation assuming liver as the dose-limiting organ (138, 143). While no significant hepatic toxicity was observed, significant hematologic toxicities were found with single high-dose injections of 20–143 mCi of ^{90}Y-Zevalin for NHL (143) and 12–22 mCi/m^{2} ^{90}Y-mAb 170 for breast and prostate cancers (137, 138). These hematologic toxicities were not expected at the level of the calculated radiation dose to the marrow microenvironment based on experience from total body irradiation. This may reflect a more significant long-term damage to the marrow microenvironment from ^{90}Y compared to that from external beam total body irradiation (for a discussion of the differences in radiobiological effects of radionuclide versus external radiation therapy, see Chapter 12).

13.5.3 Tumor Dosimetry

13.5.3.1 Clinical Tumor Dosimetry The MIRD three-step process also applies to tumor dose estimation. Nevertheless, because the tumor has a nonstandard volume and is located in a nonstandard geometry, it creates a problem for dose calculation using MIRD pamphlet tables or MIRDOSE/OLINDA software based on the reference man phantom (165). However, for the dose contribution from radioactivity in the tumor itself (i.e., where the tumor is both the source and the target), tumor penetrating and nonpenetrating radiation dose can be obtained using OLINDA software or MIRD Pamphlets. OLINDA has a sphere model to calculate these dose estimates assuming that the tumor is a sphere composed of unit density material for a chosen radionuclide with sphere masses ranging from 0.01 to 6000 g (9). Using MIRD Pamphlets, photon-absorbed fraction data can be obtained from MIRD Pamphlets No. 3 and No. 8 for spheres and ellipsoids (18, 166). Nonpenetrating radiation absorbed fractions are considered to be 1.0 for most therapeutic radionuclides in patient tumors, except for high-energy beta emitters in small-sized tumors. For example, the absorbed fraction for ^{90}Y is less than 0.89 for tumor mass less than 20 g (167). The OLINDA software or MIRD data tables list only estimates for selected discrete masses. Linear interpolation between sphere masses may not produce a correct result for intermediate sphere sizes as the absorbed fractions do not change linearly with mass. Stabin suggested to fit a simple function through the results as a function of mass and use this function to estimate any intermediate values (9).

Tumor dose contribution from other source organs in the body cannot be simply obtained from MIRD Pamphlets or OLINDA software. Johnson has developed a Monte Carlo computer code to address this shortcoming by simulating photon transport in all source organs identified in the body (168, 169). Tumors are approximated as spheres with their centers identified by means of a mouse-driven cursor on a graphical representation of the Reference Man (a model of the human body used for dosimetry calculations). Patient-specific tumor dose can be determined using a 3D approach based on SPECT imaging (77–79). The penetrating radiation dose from radioactivity distributed in 3D voxels outside the tumor can be determined by dose point kernel or voxel S-values (77–79, 170, 171). Using CT image data and whole-body planar images, radioactivity distribution in the body can be determined using a matrix inversion method, assuming uniform distribution along a pixel line inside the volume of interest (90). The radioactivity distribution in voxels can then be used to compute penetrating radiation to tumor using the voxel kernel (172). Typical dose contribution to the tumor from penetrating radiation outside the tumor is less than 10% for ^{90}Y, ^{131}I, and ^{67}Cu, except for a tumor in a host organ with high radioactivity uptake.

For a clinical site that does not have the tools mentioned above, penetrating radiation from other source organs may be estimated on the basis of simple assumptions. For example, when the tumor is not close to source organs with high uptake (such as the liver), the photon dose of the total body dose (OLINDA output lists contributions from alpha, beta, photon) may be used as a first-order estimate for penetrating radiation from other source organs. Using cumulated radioactivities assigned to the liver, spleen, whole body, and tumor in Table I of Johnson and Colby (168), the calculated photon dose from the total body dose represented 8% of the tumor to tumor self-dose. When added to the tumor to tumor self-dose, the estimated tumor doses were 0–11% underestimated compared to the results obtained using MABDOSE that does not take this into account (168).

13.5.3.2 Tumor Response with Calculated Tumor Dose   Ideally, tumor and normal organ dosimetry should be used in treatment planning. Radioactivity dose can be prescribed to deliver a tumor dose at a level expected to achieve significant tumor growth control while further tumor dose increases would not significantly improve control. Furthermore, the prescribed radioactivity dose is expected to have a tolerable normal tissue toxicity before reaching a steep dose-toxicity response region in the curve. However, in clinical practice, tumor dosimetry has not been used for treatment planning in antibody- or peptide-targeted radiotherapy. This is somewhat against our natural instincts, or the "principles of radionuclide therapy," to predict the therapeutic dose to the tumor by detailed dosimetry (173). In theory, failure to demonstrate uptake of the radiotherapeutic agent in a tumor in gamma camera images should help exclude from unnecessary radiation treatment for those patients whose tumors are without avidity for the agent, although the spatial resolution limitations of the gamma camera in revealing such uptake must also be considered (173).

With autologous bone marrow support, Press et al. achieved impressive remission rates of 84% complete response (CR), 11% partial response (PR), and 5% minor response using high-dose radiolabeled antibody therapy (131), demonstrating that higher response rates were associated with higher injection doses of radioactivity. Using ^{131}I- and ^{67}Cu-labeled antibody (Lym-1), Lamborn et al., however, found no correlation between response and tumor radiation dose, and a weak correlation ($p = 0.09$) between CR and peak tumor uptake (%ID/g) for tumors receiving the maximum tumor dose (174). Using ^{131}I-labeled antibody (tositumomab), Kaminski et al. found higher response rates and more durable CRs were seen with increasing total body radiation dose (175). The response rate to a total body dose under 65 cGy was 57% compared to 86% to doses of 65 cGy or higher. Using ^{90}Y-labeled antibody (Zevalin), Wiseman et al. found no correlation between tumor response and blood clearance $T_{1/2}$ or blood cumulated activity (176). Subsequently, Wiseman et al. found an overall response rate of 80% for Zevalin compared to 44% for rituximab (nonradiolabeled CD20 antibody) (177). The most positive correlations between tumor dose and response were reported by Koral et al. (178, 179). For a selected subset of data (tumors with an initial mass less than or equal to 10 g, tumors from patients only having PR, excluding axillary tumors, using hybrid SPECT/planar imaging), Koral et al. found a statistically significant correlation between tumor dose and response in a patient population in which many were previously untreated (178, 179). It was noted that while the excluded axillary tumor doses were substantially lower than tumors at other locations, all patients with axillary tumor had a CR (178, 179). For the same ^{131}I-labeled antibody (tositumomab), Sgouros et al. did not find a statistically significant correlation between tumor response and tumor dose mean, maximum, minimum, or uniformity for tumors in 15 patients using hybrid SPECT/planar imaging (180). Using ^{90}Y-epratuzumab, Sharkey et al. found no correlation between response and tumor dose or tumor visibility on images (181). In a considerable number of instances, tumors could not be discerned by either planar or SPECT imaging but could be discerned by CT, and were nonetheless found to respond to the treatment (181). Similar findings were reported in a study by Iagaru et al. in which a higher rate of complete response after ^{90}Y Zevalin treatment was seen in patients with negative pretherapy imaging findings, whereas a higher rate of disease progression despite therapy was noted in patients with positive pretherapy imaging findings (182).

Various explanations have been discussed for the above paradoxical clinical results. One possible explanation that has not been receiving much attention is that a tumor uptake identified by planar or SPECT imaging may not reflect the concentration of radioactivity uptake per number of active clonogenic tumor cells. Because only a proportion of cells in a gross tumor volume (identified by planar, SPECT, CT, or MRI) are active clonogenic cells, poor image identification (by planar/SPECT) could be due to lower active clonogenic cell population in the gross tumor volume. Although a bulky tumor usually will be clearly visualized by planar/SPECT images, it may not have a true high radiation dose per number of active clonogenic tumor cells. As discussed earlier, depending on the location of the target (cell surface versus internalized) and radiation type, the actual radiation dose to the cell nucleus can substantially vary. In addition, biological response is much more complicated than a single parameter of tumor radiation dose, even if we have an accurate tumor dosimetry to account for various problems. It is common to find a weak or no statistically significant correlation between tumor local control and tumor dose in studies of external beam therapy (183) where accuracy of tumor dose is typically within $\pm 5\%$ error (routine verification standard for each patient prior treatment is within $\pm 3\%$). Furthermore, in targeted radiotherapy, the targeting molecule itself often has growth inhibitory properties. For example, in the Zevalin protocol, the rituximab component may produce objective tumor response, in addition to the ^{90}Y-ibritumomab that may have antiproliferative effects from both the radionuclide and the antibody. The same would be the case for ^{90}Y-DOTATOC, as octreotide is a therapeutic agent in its own right without the radionuclide. Therefore, in these circumstances it may be difficult to obtain a correlation with radiation-absorbed dose alone.

13.6 CONCLUSIONS

Great progress has been made in the past decade in radiation dosimetry for targeted radionuclide cancer therapy with antibodies and peptides. In the dose computation aspect, more and more accurate dosimetry models and realistic anthropomorphic phantoms have been developed. These developments in models and phantoms have provided the desirable level of accuracy for dose computation in clinical studies. In the quantification of radioactivity distribution aspect, substantial progress has been achieved in clinical studies mainly because 3D anatomic information (CT/MRI) is now routinely available. Standardizing planar image quantification techniques with detailed calibration procedures for individual imaging systems can improve quantification accuracy in large-scale clinical trials. More accurate 3D SPECT imaging methods have been developed and these can also play a more important role in clinical trials. Nonetheless some critically important challenges (such as nonuniform radioactivity distribution) have been identified preclinically using autoradiography or microSPECT/PET studies and these challenges remain for the foreseeable future in clinical studies because the poor spatial resolution of planar gamma camera imaging or SPECT/PET has not yet been sufficiently improved to aid in accurate dosimetry calculations.

REFERENCES

1. Loevinger R, Berman M. A formalism for calculation of absorbed dose from radionuclides. *Phys Med Biol* 1968;13:205–217.
2. Loevinger R, Bermam M.A Revised Schema for Calculating the Absorbed Dose from Biologically Distributed Radionuclides. MIRD Phamplet No. 1, Society of Nuclwear Medicine, New York, 1976.
3. Loevinger R, Budinger TF, Watson EE. *MIRD Primer for Absorbed Dose Calculations.* Society of Nuclwear Medicine, New York, 1988.
4. Dillman L, Von der Lage F. Radionuclide Decay Schemes and Nuclear Parameters for Use in Radiation-Dose Estimation. MIRD Pamphlet No. 10. Society of Nuclwear Medicine, New York, 1975.
5. Weber DA, Eckerman KF, Dillman LT, Ryman JC. *MIRD: Radionuclide Data and Decay Schemes.* Society of Nuclear Medicine, Reston, VA, 1989.
6. Eckerman KF, Endo A. *MIRD: Radionuclide Data and Decay Schemes*, 2nd edition. Society of Nuclear Medicine, Reston, VA, 2008.
7. RADAR: the RAdiation Dose Assessment Resource (http://www.doseinfo-radar.com/ RADARDecay.html). http://www-ndsi aeaorg/nsdd html, and http://www nndc bnl gov/ mird/. 2008.
8. Stabin MG. MIRDOSE: personal computer software for internal dose assessment in nuclear medicine *J Nucl Med* 1996;37:538–546.
9. Stabin MG, Sparks RB, Crowe E. OLINDA/EXM: the second-generation personal computer software for internal dose assessment in nuclear medicine. *J Nucl Med* 2005;46:1023–1027.
10. Stabin MG. Radiation dosimetry methods for therapy. In: Chinol M, Paganelli G, editor. *Radionuclide Peptide Cancer Therapy*, Informa Healthcare, 2006; 239–261.
11. Shen S, Khazaeli MB, Gillespie GY, Alvarez VL. Radiation dosimetry of ^{131}I-chlorotoxin for targeted radiotherapy in glioma-bearing mice. *J Neurooncol* 2005;71:113–119.
12. Strand SE, Ivanovic M, Erlandsson K, et al. Small animal imaging with pinhole single-photon emission computed tomography. *Cancer* 1994;73:981–984.
13. Weber DA, Ivanovic M, Franceschi D, et al. Pinhole SPECT: an approach to *in vivo* high resolution SPECT imaging in small laboratory animals. *J Nucl Med* 1994;35:342–348.
14. Funk T, Sun M, Hasegawa BH. Radiation dose estimate in small animal SPECT and PET. *Med Phys* 2004;31:2680–2686.
15. Deroose CM, De A, Loening AM, et al. Multimodality imaging of tumor xenografts and metastases in mice with combined small-animal PET, small-animal CT, and bioluminescence imaging. *J Nucl Med* 2007;48:295–303.

16. Hui TE, Fisher DR, Kuhn JA, et al. A mouse model for calculating cross-organ beta doses from yttrium-90-labeled immunoconjugates. *Cancer* 1994;73:951–957.

17. Snyder WS, Ford MR, Warner GG, Watson SB. 'S', Absorbed Dose Per Unit Cumulated Activity for Selected Radionuclides and Organs. MIRD Pamphlet No. 11. Society of Nuclear Medicine, New York, 1975.

18. Ellett WH, Humes RM. Absorbed fractions for small volumes containing photon-emitting radioactivity. MIRD Pamphlet No. 8. *J Nucl Med* 1971;12 (Suppl. 5): 25–32.

19. Muthuswamy MS, Roberson PL, Buchsbaum DJ. A mouse bone marrow dosimetry model. *J Nucl Med* 1998;39:1243–1247.

20. Flynn AA, Green AJ, Pedley RB, Boxer GM, Boden R, Begent RH. A mouse model for calculating the absorbed beta-particle dose from ^{131}I- and ^{90}Y-labeled immunoconjugates, including a method for dealing with heterogeneity in kidney and tumor. *Radiat Res* 2001;156:28–35.

21. Cook MJ. *The Anatomy of the Laboratory Mouse*. Academic Press, London, 1965.

22. Hindorf C, Ljungberg M, Strand SE. Evaluation of parameters influencing S values in mouse dosimetry. *J Nucl Med* 2004;45:1960–1965.

23. Kolbert KS, Watson T, Matei C, Xu S, Koutcher JA, Sgouros G. Murine S factors for liver, spleen, and kidney. *J Nucl Med* 2003;44:784–791.

24. Stabin MG, Peterson TE, Holburn GE, Emmons MA. Voxel-based mouse and rat models for internal dose calculations. *J Nucl Med* 2006;47:655–659.

25. Segars WP, Tsui BM, Frey EC, Johnson GA, Berr SS. Development of a 4-D digital mouse phantom for molecular imaging research. *Mol Imaging Biol* 2004; 6:149–159.

26. Larsson E, Strand SE, Ljungberg M, Jonsson BA. Mouse S-factors based on Monte Carlo simulations in the anatomical realistic Moby phantom for internal dosimetry. *Cancer Biother Radiopharm* 2007;22:438–442.

27. Bitar A, Lisbona A, Thedrez P, et al. A voxel-based mouse for internal dose calculations using Monte Carlo simulations (MCNP). *Phys Med Biol* 2007;52:1013–1025.

28. Soremark R, Hunt VR. Autoradiographic studies of the distribution of polonium-210 in mice after a single intravenous injection. *Int J Radiat Biol Relat Stud Phys Chem Med* 1966;11:43–50.

29. Jonsson BA, Strand SE, Larsson BS. A quantitative autoradiographic study of the heterogeneous activity distribution of different indium-111-labeled radiopharmaceuticals in rat tissues. *J Nucl Med* 1992;33:1825–1833.

30. Roberson PL, Buchsbaum DJ, Heidorn DB, Ten Haken RK. Three-dimensional tumor dosimetry for radioimmunotherapy using serial autoradiography. *Int J Radiat Oncol Biol Phys* 1992;24:329–334.

31. Humm JL, Macklis RM, Bump K, Cobb LM, Chin LM. Internal dosimetry using data derived from autoradiographs. *J Nucl Med* 1993;34:1811–1817.

32. Akabani G, Kennel SJ, Zalutsky MR. Microdosimetric analysis of alpha-particle-emitting targeted radiotherapeutics using histological images. *J Nucl Med* 2003; 44:792–805.

33. Konijnenberg M, Melis M, Valkema R, Krenning E, de Jong M. Radiation dose distribution in human kidneys by octreotides in peptide receptor radionuclide therapy. *J Nucl Med* 2007;48:134–142.

34. Humm JL. Dosimetric aspects of radiolabeled antibodies for tumor therapy. *J Nucl Med* 1986;27:1490–1497.

35. Leichner PK, Koral KF, Jaszczak RJ, Green AJ, Chen GT, Roeske JC. An overview of imaging techniques and physical aspects of treatment planning in radioimmunotherapy. *Med Phys* 1993;20:569–577.

36. Macey DJ, Williams LE, Breitz HB, Liu A, Johnson TK, Zanzonico PB. *A Primer for Radioimmunothearpy and Radionuclide Therapy.* AAPM Report No. 71. Medical Physics Publishing, Madison, WI, 2001.

37. Siegel JA, Thomas SR, Stubbs JB, et al. MIRD pamphlet no. 16: techniques for quantitative radiopharmaceutical biodistribution data acquisition and analysis for use in human radiation dose estimates. *J Nucl Med* 1999;40:37S–61S.

38. Thomas SR, Maxon HR, Kereiakes JG. *In vivo* quantitation of lesion radioactivity using external counting methods. *Med Phys* 1976;03:253–255.

39. Hammond ND, Moldofsky PJ, Beardsley MR, Mulhern CB, Jr. External imaging techniques for quantitation of distribution of I-131 F(ab′)$_2$ fragments of monoclonal antibody in humans. *Med Phys* 1984;11:778–783.

40. Eary JF, Appelbaum FL, Durack L, Brown P. Preliminary validation of the opposing view method for quantitative gamma camera imaging. *Med Phys* 1989;16: 382–387.

41. DeNardo GL, DeNardo SJ, Macey DJ, Mills SL. Quantitative pharmacokinetics of radiolabeled monoclonal antibodies for imaging and therapy in patients. In: *Radiolabeled Monocolonal Antibodies for Imaging and Therapy,* Srivastava SC, editor. Plenum, New York, NY, 1988; 293–310.

42. Macey DJ, DeNardo GL, DeNardo SJ. A treatment planning program for radioimmunotherapy. In: *The Present and Future Role of Monoclonal Antibodies in the Management of Cancer. Frontiers of Radiation Therapy and Oncology,* Vaeth JM, Meyer JL, editors. Basel, Karger, San Francisco, California, 1990; 123–131.

43. DeNardo GL, Raventos A, Hines HH, et al. Requirements for a treatment planning system for radioimmunotherapy. *Int J Radiat Oncol Biol Phys* 1985;11:335–348.

44. Erwin WD, Groch MW, Macey DJ, DeNardo GL, DeNardo SJ, Shen S. A radioimmunoimaging and MIRD dosimetry treatment planning program for radioimmunotherapy. *Nucl Med Biol* 1996;23:525–532.

45. Shen S, DeNardo GL, DeNardo SJ. Quantitative bremsstrahlung imaging of yttrium-90 using a Wiener filter. *Med Phys* 1994;21:1409–1417.

46. Shen S, DeNardo GL, DeNardo SJ, Yuan A, DeNardo DA, Lamborn KR. Reproducibility of operator processing for radiation dosimetry. *Nucl Med Biol* 1997;24:77–83.

47. Shen S, Forero A, LoBuglio AF, et al. Patient-specific dosimetry of pretargeted radioimmunotherapy using CC49 fusion protein in patients with gastrointestinal malignancies. *J Nucl Med* 2005;46:642–651.

48. Shen S, Forero A, Meredith R, et al. Impact of rituximab treatment on ^{90}Y-ibritumomab dosimetry for patients with non-Hodgkin's lymphoma. *J Nucl Med* 2010;51:150–157.

49. Jaszczak RJ, Greer KL, Floyd CE, Jr., Harris CC, Coleman RE. Improved SPECT quantification using compensation for scattered photons. *J Nucl Med* 1984; 25:893–900.

50. Gilland DR, Jaszczak RJ, Turkington TG, Greer KL, Coleman RE. Quantitative SPECT imaging with indium-111. *IEEE Trans Nucl Sci* 1991;38:761–766.

51. Floyd CE, Jr., Jaszczak RJ, Coleman RE. Scatter detection in SPECT imaging: dependence on source depth, energy, and energy window. *Phys Med Biol* 1988; 33:1075–1081.

52. Koral KF, Wang XQ, Rogers WL, Clinthorne NH, Wang XH. SPECT Compton-scattering correction by analysis of energy spectra. *J Nucl Med* 1988;29:195–202.

53. King MA, Hademenos GJ, Glick SJ. A dual-photopeak window method for scatter correction. *J Nucl Med* 1992;33:605–612.

54. Ogawa K, Harata Y, Ichihara T, Kubo A, Hashimoto S. A practical method for position-dependent Compton-scatter correction in single photon emission CT. *IEEE Trans Med Imaging* 1991;10:408–412.

55. Shen S, DeNardo GL, DeNardo SJ. Quantitative SPECT imaging with Cu-67 in an Alderson phantom. *J Nucl. Med.* 1993;34:189–189.

56. Breitz HB, Wendt RE, III, Stabin MS, et al. ^{166}Ho-DOTMP radiation-absorbed dose estimation for skeletal targeted radiotherapy. *J Nucl Med* 2006;47:534–542.

57. Shen S, Duan J, Ye S, Breitz H. Combined scatter subtraction and digital restoration of Ho-166 images for quantification. *Med Physics* 2005;32:2054–2054.

58. Fiveash JB, Conry RM, Shen S. Tumor-specific targeting of intravenous ^{131}I-chlorotoxin (^{131}I-TM-601) in patients with metastatic melanoma. *J Clin Oncol* 2008;26:20003. (abstract).

59. Fiveash JB, Nabors LB, Raizer JJ, Avgeropoulos N, Modarresifar H, Shen S. Tumor specific targeting of intravenous I-131-chlorotoxin (TM-601) in patients with recurrent glioma. *IJROBP* 2007;69:S257–S258.

60. Shen S, Lustig R, Judy KD, Fiveash JB, Shan JS. Open-label, dose confirmation and dosimetry study of interstitial 131 ll-ch1TNT-1/B MAb for the treatment of recurrent glioblastoma multiforme (GBM). *J Clin Oncol* 2008;26:2072 (abstract).

61. Links JM. Scattered photons as "good counts gone bad": are they reformable or should they be permanently removed from society? *J Nucl Med* 1995;36:130–132.

62. Floyd CE, Jr., Jaszczak RJ, Greer KL, Coleman RE. Deconvolution of Compton scatter in SPECT. *J Nucl Med* 1985;26:403–408.

63. Wang X, Koral KF. A regularized deconvolution-fitting method for Compton-scatter correction in SPECT. *IEEE Trans Med Imaging* 1992;11:351–360.

64. Metz CE, Doi K. Transfer function analysis of radiographic imaging systems. *Phys Med Biol* 1979;24:1079–1106.

65. King MA, Doherty PW, Schwinger RB, Penney BC. A Wiener filter for nuclear medicine images. *Med Phys* 1983;10:876–880.

66. King MA, Penney BC, Glick SJ. An image-dependent Metz filter for nuclear medicine images. *J Nucl Med* 1988;29:1980–1989.

67. King MA, Schwinger RB, Penney BC, Doherty PW, Bianco JA. Digital restoration of indium-111 and iodine-123 SPECT images with optimized Metz filters. *J Nucl Med* 1986;27:1327–1336.

68. Msaki P, Axelsson B, Larsson SA. Some physical factors influencing the accuracy of convolution scatter correction in SPECT. *Phys Med Biol* 1989;34:283–298.

69. Tsui BM, Zhao X, Frey EC, McCartney WH. Quantitative single-photon emission computed tomography: basics and clinical considerations. *Semin Nucl Med* 1994;24:38–65.

70. Wu RK, Siegel JA. Absolute quantitation of radioactivity using the buildup factor. *Med Phys* 1984;11:189–192.

71. Siegel JA, Wu RK, Maurer AH. The buildup factor: effect of scatter on absolute volume determination. *J Nucl Med* 1985;26:390–394.

72. Siegel JA. The effect of source size on the buildup factor calculation of absolute volume. *J Nucl Med* 1985;26:1319–1322.

73. van Rensburg AJ, Lotter MG, Heyns AD, Minnaar PC. An evaluation of four methods of 111In planar image quantification. *Med Phys* 1988;15:853–861.

74. Kojima A, Takaki Y, Matsumoto M, et al. A preliminary phantom study on a proposed model for quantification of renal planar scintigraphy. *Med Phys* 1993;20:33–37.

75. Fisher DR, Shen S, Meredith RF, MIRD Dose Estimate Report No. 20: radiation absorbed-dose estimates for ^{111}In- and ^{90}Y-ibritumomab tiuxetan. *J Nucl Med* 2009; 50:644–652.

76. Sgouros G, Barest G, Thekkumthala J, et al. Treatment planning for internal radionuclide therapy: three-dimensional dosimetry for nonuniformly distributed radionuclides. *J Nucl Med* 1990;31:1884–1891.

77. Sgouros G, Chiu S, Pentlow KS, et al. Three-dimensional dosimetry for radioimmunotherapy treatment planning. *J Nucl Med* 1993;34:1595–1601.

78. Giap HB, Macey DJ, Podoloff DA. Development of a SPECT-based three-dimensional treatment planning system for radioimmunotherapy. *J Nucl Med* 1995;36:1885–1894.

79. Akabani G, Hawkins WG, Eckblade MB, Leichner PK. Patient-specific dosimetry using quantitative SPECT imaging and three-dimensional discrete Fourier transform convolution. *J Nucl Med* 1997;38:308–314.

80. Kolbert KS, Sgouros G, Scott AM, et al. Implementation and evaluation of patient-specific three-dimensional internal dosimetry. *J Nucl Med* 1997;38:301–308.

81. Koral KF, Zasadny KR, Kessler ML, et al. CT-SPECT fusion plus conjugate views for determining dosimetry in iodine-131-monoclonal antibody therapy of lymphoma patients. *J Nucl Med* 1994;35:1714–1720.

82. Shen S, Mamelak A, Rosenfeld S, et al. Penetration and retention of intracavitary administered I-131-TM-601 peptide in patients with recurrent high-grade glioma. *IJROBP* 2006;66:S560–S561.

83. Mamelak AN, Rosenfeld S, Bucholz R, et al. Phase I single-dose study of intracavitary-administered iodine-131-TM-601 in adults with recurrent high-grade glioma. *J Clin Oncol* 2006;24:3644–3650.

84. Ljungberg M, Strand SE. A Monte Carlo program for the simulation of scintillation camera characteristics. *Comput Methods Programs Biomed* 1989;29:257–272.

85. Ljungberg M, Strand SE. Scatter and attenuation correction in SPECT using density maps and Monte Carlo simulated scatter functions. *J Nucl Med* 1990;31:1560–1567.

86. Zaidi H. Relevance of accurate Monte Carlo modeling in nuclear medical imaging. *Med Phys* 1999;26:574–608.

87. Dewaraja YK, Ljungberg M, Koral KF. Characterization of scatter and penetration using Monte Carlo simulation in ^{131}I imaging. *J Nucl Med* 2000;41:123–130.

88. Autret D, Bitar A, Ferrer L, Lisbona A, Bardies M. Monte Carlo modeling of gamma cameras for I-131 imaging in targeted radiotherapy. *Cancer Biother Radiopharm* 2005;20:77–84.

89. King MA, Xia W, deVries DJ, et al. A Monte Carlo investigation of artifacts caused by liver uptake in single-photon emission computed tomography perfusion imaging with technetium 99m-labeled agents. *J Nucl Cardiol* 1996;3:18–29.

90. Liu A, Williams LE, Raubitschek AA. A CT assisted method for absolute quantitation of internal radioactivity. *Med Phys* 1996;23:1919–1928.

91. Tang HR, Brown JK, Da Silva AJ, et al. Implementation of a combined X-ray CT-scintillation camera imaging system for localizing and measuring radionuclide uptake: experiments in phantoms and patients. *IEEE Trans Nucl Sci* 1999;46:551–557.

92. Sjogreen K, Ljungberg M, Strand SE. An activity quantification method based on registration of CT and whole-body scintillation camera images, with application to [131]I. *J Nucl Med* 2002;43:972–982.

93. He B, Frey EC. Comparison of conventional, model-based quantitative planar, and quantitative SPECT image processing methods for organ activity estimation using In-111 agents. *Phys Med Biol* 2006;51:3967–3981.

94. Erdi YE, Macapinlac H, Larson SM, et al. Radiation dose assessment for I-131 therapy of thyroid cancer using I-124 PET imaging. *Clin Posit. Imaging.* 1999;2:41–46.

95. Eschmann SM, Reischl G, Bilger K, et al. Evaluation of dosimetry of radioiodine therapy in benign and malignant thyroid disorders by means of iodine-124 and PET. *Eur J Nucl Med Mol Imaging* 2002;29:760–767.

96. Jentzen W, Weise R, Kupferschlager J, et al. Iodine-124 PET dosimetry in differentiated thyroid cancer: recovery coefficient in 2D and 3D modes for PET(/CT) systems. *Eur J Nucl Med Mol Imaging* 2008;35:611–623.

97. Jentzen W, Freudenberg L, Eising EG, Sonnenschein W, Knust J, Bockisch A. Optimized [124]I PET dosimetry protocol for radioiodine therapy of differentiated thyroid cancer. *J Nucl Med* 2008;49:1017–1023.

98. Herzog H, Rosch F, Stocklin G, Lueders C, Qaim SM, Feinendegen LE. Measurement of pharmacokinetics of yttrium-86 radiopharmaceuticals with PET and radiation dose calculation of analogous yttrium-90 radiotherapeutics. *J Nucl Med* 1993; 34:2222–2226.

99. Rosch F, Herzog H, Plag C, et al. Radiation doses of yttrium-90 citrate and yttrium-90 EDTMP as determined via analogous yttrium-86 complexes and positron emission tomography. *Eur J Nucl Med* 1996;23:958–966.

100. Jamar F, Barone R, Mathieu I, et al. [86]Y-DOTA^0-D-Phe^1-Tyr^3-octreotide (SMT487)—a phase 1 clinical study: pharmacokinetics, biodistribution and renal protective effect of different regimens of amino acid co-infusion. *Eur J Nucl Med Mol Imaging* 2003;30:510–518.

101. Shen S, DeNardo GL, DeNardo SJ, et al. Dosimetric evaluation of copper-64 in copper-67-2IT-BAT-Lym-1 for radioimmunotherapy. *J Nucl Med* 1996;37:146–150.

102. Cutler PD, Schwarz SW, Anderson CJ, et al. Dosimetry of copper-64-labeled monoclonal antibody 1A3 as determined by PET imaging of the torso. *J Nucl Med* 1995; 36:2363–2371.

103. Anderson CJ, Dehdashti F, Cutler PD, et al. ^{64}Cu-TETA-octreotide as a PET imaging agent for patients with neuroendocrine tumors. *J Nucl Med* 2001;42:213–221.

104. Eiblmaier M, Meyer LA, Watson MA, Fracasso PM, Pike LJ, Anderson CJ. Correlating EGFR expression with receptor-binding properties and internalization of 64Cu-DOTA-cetuximab in 5 cervical cancer cell lines. *J Nucl Med* 2008;49:1472–1479.

105. Cremonesi M. Dosimetry applied to peptide radionuclide receptor therapy. In: Radionuclide Peptide Cancer Therapy, Chinol M, Paganelli G, editors. Informa Healthcare, 2006; 263–299.

106. Siegel JA, Wessels BW, Waston EE, et al. Bone marrow dosimetry and toxicity for radioimmunotherapy. *Antibody Immunoconj Radiopharmacol* 1990;3:213–233.

107. Sgouros G. Bone marrow dosimetry for radioimmunotherapy: theoretical considerations. *J Nucl Med* 1993;34:689–694.

108. Behr TM, Behe M, Sgouros G. Correlation of red marrow radiation dosimetry with myelotoxicity: empirical factors influencing the radiation-induced myelotoxicity of radiolabeled antibodies, fragments and peptides in pre-clinical and clinical settings. *Cancer Biother Radiopharm* 2002;17:445–464.

109. Shen S, DeNardo GL, Sgouros G, O'Donnell RT, DeNardo SJ. Practical determination of patient-specific marrow dose using radioactivity concentration in blood and body. *J Nucl Med* 1999;40:2102–2106.

110. Eckerman KF, Stabin MG. Electron absorbed fractions and dose conversion factors for marrow and bone by skeletal regions. *Health Phys* 2000;78:199–214.

111. Bouchet LG, Bolch WE, Howell RW, Rao DV. S values for radionuclides localized within the skeleton. *J Nucl Med* 2000;41:189–212.

112. Stabin MG, Eckerman KF, Bolch WE, Bouchet LG, Patton PW. Evolution and status of bone and marrow dose models. *Cancer Biother Radiopharm* 2002;17:427–433.

113. Shah AP, Bolch WE, Rajon DA, Patton PW, Jokisch DW. A paired-image radiation transport model for skeletal dosimetry. *J Nucl Med* 2005;46:344–353.

114. Boucek JA, Turner JH. Validation of prospective whole-body bone marrow dosimetry by SPECT/CT multimodality imaging in ^{131}I-anti-CD20 rituximab radioimmunotherapy of non-Hodgkin's lymphoma. *Eur J Nucl Med Mol Imaging* 2005;32: 458–469.

115. de Keizer B, Hoekstra A, Konijnenberg MW, et al. Bone marrow dosimetry and safety of high ^{131}I activities given after recombinant human thyroid-stimulating hormone to treat metastatic differentiated thyroid cancer. *J Nucl Med* 2004;45:1549–1554.

116. Wessels BW, Bolch WE, Bouchet LG, et al. Bone marrow dosimetry using blood-based models for radiolabeled antibody therapy: a multiinstitutional comparison. *J Nucl Med* 2004;45:1725–1733.

117. Hindorf C, Linden O, Tennvall J, Wingardh K, Strand SE. Evaluation of methods for red marrow dosimetry based on patients undergoing radioimmunotherapy. *Acta Oncol* 2005;44:579–588.

118. Siegel JA. Establishing a clinically meaningful predictive model of hematologic toxicity in nonmyeloablative targeted radiotherapy: practical aspects and limitations of red marrow dosimetry. *Cancer Biother Radiopharm* 2005;20:126–140.

119. Siegel JA, Stabin MG. Mass scaling of S values for blood-based estimation of red marrow absorbed dose: the quest for an appropriate method. *J Nucl Med* 2007;48:253–256.

120. Petoussi-Henss N, Bolch WE, Zankl M, Sgouros G, Wessels B. Patient-specific scaling of reference S-values for cross-organ radionuclide S-values: what is appropriate? *Radiat Prot Dosimetry* 2007;127:192–196.

121. Traino AC, Ferrari M, Cremonesi M, Stabin MG. Influence of total-body mass on the scaling of S-factors for patient-specific, blood-based red-marrow dosimetry. *Phys Med Biol* 2007;52:5231–5248.

122. Sgouros G, Stabin M, Erdi Y, et al. Red marrow dosimetry for radiolabeled antibodies that bind to marrow, bone, or blood components. *Med Phys* 2000;27:2150–2164.

123. Siegel JA, Lee RE, Pawlyk DA, Horowitz JA, Sharkey RM, Goldenberg DM. Sacral scintigraphy for bone marrow dosimetry in radioimmunotherapy. *Int J Rad Appl Instrum B* 1989;16:553–559.

124. Juweid M, Sharkey RM, Siegel JA, Behr T, Goldenberg DM. Estimates of red marrow dose by sacral scintigraphy in radioimmunotherapy patients having non-Hodgkin's lymphoma and diffuse bone marrow uptake. *Cancer Res* 1995;55:5827s–5831s.

125. Macey DJ, DeNardo SJ, DeNardo GL, DeNardo DA, Shen S. Estimation of radiation absorbed doses to the red marrow in radioimmunotherapy. *Clin Nucl Med* 1995; 20:117–125.

126. Lim SM, DeNardo GL, DeNardo DA, et al. Prediction of myelotoxicity using radiation doses to marrow from body, blood and marrow sources. *J Nucl Med* 1997;38:1374–1378.

127. Shen S, Meredith RF, Duan J, et al. Improved prediction of myelotoxicity using a patient-specific imaging dose estimate for non-marrow targeting ^{90}Y-antibody therapy. *J Nucl Med* 2002;43:1245–1253.

128. Meredith RF, Shen S, Forero A, LoBuglio A. A method to correct for radioactivity in large vessels that overlap the spine in imaging-based marrow dosimetry of lumbar vertebrae. *J Nucl Med* 2008;49:279–284.

129. Bolch WE, Patton RW, Shah AR, Rajon DA, Jokisch DW. Considerations of anthropometric, tissue volume, and tissue mass scaling for improved patient specificity of skeletal S values. *Med Phys* 2002;29:1054–1070.

130. Sgouros G, Jureidini IM, Scott AM, Graham MC, Larson SM, Scheinberg DA. Bone marrow dosimetry: regional variability of marrow-localizing antibody. *J Nucl Med* 1996;37:695–698.

131. Press OW, Eary JF, Appelbaum FR, et al. Radiolabeled-antibody therapy of B-cell lymphoma with autologous bone marrow support. *N Engl J Med* 1993;329:1219–1224.

132. Richman CM, DeNardo SJ, O'Grady LF, DeNardo GL. Radioimmunotherapy for breast cancer using escalating fractionated doses of ^{131}I-labeled chimeric L6 antibody with peripheral blood progenitor cell transfusions. *Cancer Res* 1995;55:5916s–5920s.

133. DeNardo SJ, Richman CM, Goldstein DS, et al. Yttrium-90/indium-111-DOTA-peptide-chimeric L6: pharmacokinetics, dosimetry and initial results in patients with incurable breast cancer. *Anticancer Res* 1997;17:1735–1744.

134. Richman CM, Schuermann TC, Wun T, et al. Peripheral blood stem cell mobilization for hematopoietic support of radioimmunotherapy in patients with breast carcinoma. *Cancer* 1997;80:2728–2732.

135. Cagnoni PJ, Ceriani RL, Cole W, et al. Phase 1 study of high-dose radioimmunotherapy with Y-90-hu-Bre-3 followed by autologous stem cell support (ASCS) in patients with metastatic breast cancer. *Cancer Biother Radiopharm* 1998;13:328 (abstract).

136. Wong JY, Somlo G, Odom-Maryon T, et al. Initial clinical experience evaluating yttrium-90-chimeric T84.66 anticarcinoembryonic antigen antibody and autologous hematopoietic stem cell support in patients with carcinoembryonic antigen-producing metastatic breast cancer. *Clin Cancer Res* 1999;5:3224s–3231s.

137. Richman CM, DeNardo SJ, O'Donnell RT, et al. High-dose radioimmunotherapy combined with fixed, low-dose paclitaxel in metastatic prostate and breast cancer by using a MUC-1 monoclonal antibody, m170, linked to indium-111/yttrium-90 via

a cathepsin cleavable linker with cyclosporine to prevent human anti-mouse antibody. *Clin Cancer Res* 2005;11:5920–5927.

138. Richman CM, DeNardo SJ, O'Donnell RT, et al. Dosimetry-based therapy in metastatic breast cancer patients using ^{90}Y monoclonal antibody 170H.82 with autologous stem cell support and cyclosporin A. *Clin Cancer Res* 1999;5:3243s–3248s.

139. Tempero M, Leichner P, Dalrymple G, et al. High-dose therapy with iodine-131-labeled monoclonal antibody CC49 in patients with gastrointestinal cancers: a phase I trial. *J Clin Oncol* 1997;15:1518–1528.

140. Knox SJ, Goris ML, Trisler K, et al. Yttrium-90-labeled anti-CD20 monoclonal antibody therapy of recurrent B-cell lymphoma. *Clin Cancer Res* 1996;2:457–470.

141. Shen S, DeNardo SJ, Richman CM, et al. Planning time for peripheral blood stem cell infusion after high-dose targeted radionuclide therapy using dosimetry. *J Nucl Med* 2005;46:1034–1041.

142. Carrasquillo JA, White JD, Paik CH, et al. Similarities and differences in ^{111}In- and ^{90}Y-labeled 1B4M-DTPA antiTac monoclonal antibody distribution. *J Nucl Med* 1999;40:268–276.

143. Flinn IW, Kahl BS, Frey EC, et al. Dose finding trial of yttrium 90 ibritumomab tiuxetan ((YIT)-Y-90) with autologous stem cell transplantation (ASCT) in patients with relapsed or refractory B-cell non-Hodgkin's lymphoma (NHL). *J Clin Onco* 2006;24:430S–430S.

144. Wahl RL, Kroll S, Zasadny KR. Patient-specific whole-body dosimetry: principles and a simplified method for clinical implementation. *J Nucl Med* 1998;39:14S–20S.

145. Wiseman GA, Kornmehl E, Leigh B, et al. Radiation dosimetry results and safety correlations from ^{90}Y-ibritumomab tiuxetan radioimmunotherapy for relapsed or refractory non-Hodgkin's lymphoma: combined data from 4 clinical trials. *J Nucl Med* 2003; 44:465–474.

146. Erwin WD, Spies SM, Kelly ME, et al. Correlation of marrow dose estimates based on serial pretreatment radiopharmaceutical imaging and blood data with actual marrow toxicity in anti-CD20 yttrium-90 monoclonal antibody radioimmunotherapy of non-Hodgkin's B-cell lymphoma. *Nucl Med Commun* 2001;22:247–255.

147. Meredith RF, Khazaeli MB, Liu T, et al. Dose fractionation of radiolabeled antibodies in patients with metastatic colon cancer. *J Nucl Med* 1992;33:1648–1653.

148. Liu T, Meredith RF, Saleh MN, et al. Correlation of toxicity with treatment parameters for ^{131}I-CC49 radioimmunotherapy in three phase II clinical trials. *Cancer Biother Radiopharm* 1997;12:79–87.

149. Breitz HB, Fisher DR, Wessels BW. Marrow toxicity and radiation absorbed dose estimates from rhenium-186- labeled monoclonal antibody. *J Nucl Med* 1998; 39:1746–1751.

150. Breitz HB, Fisher DR, Goris ML, et al. Radiation absorbed dose estimation for 90Y-DOTA-biotin with pretargeted NR-LU-10/streptavidin. *Cancer Biother Radiopharm* 1999;14:381–395.

151. O'Donoghue JA, Baidoo N, Deland D, Welt S, Divgi CR, Sgouros G. Hematologic toxicity in radioimmunotherapy: dose–response relationships for I-131 labeled antibody therapy. *Cancer Biother Radiopharm* 2002;17:435–443.

152. Shen S, Meredith RF, Duan J, Brezovich I, Khazaeli MB, LoBuglio AF. Comparison of methods for predicting myelotoxicity for non-marrow targeting I-131-antibody therapy. *Cancer Biother Radiopharm* 2003;18:209–215.

153. Vallabhajosula S, Goldsmith SJ, Hamacher KA, et al. Prediction of myelotoxicity based on bone marrow radiation-absorbed dose: radioimmunotherapy studies using ^{90}Y- and ^{177}Lu-labeled J591 antibodies specific for prostate-specific membrane antigen. *J Nucl Med* 2005;46:850–858.

154. Testa NG, Hendry JH, Molineux G. Long-term bone marrow damage in experimental systems and in patients after radiation or chemotherapy. *Anticancer Res* 1985; 5:101–110.

155. Jones TD, Morris MD, Young RW. A mathematical model for radiation-induced myelopoiesis. *Radiat Res* 1991;128:258–266.

156. Shen S, DeNardo GL, Jones TD, Wilder RB, O'Donnell RT, DeNardo SJ. A preliminary cell kinetics model of thrombocytopenia after radioimmunotherapy. *J Nucl Med* 1998;39:1223–1229.

157. Jones TD, Morris MD, Young RW. Mathematical models of marrow cell kinetics: differential effects of protracted irradiations on stromal and stem cells in mice. *Int J Radiat Oncol Biol Phys* 1993;26:817–830.

158. Blumenthal RD, Lew W, Juweid M, Alisauskas R, Ying Z, Goldenberg DM. Plasma FLT3-L levels predict bone marrow recovery from myelosuppressive therapy. *Cancer* 2000;88:333–343.

159. Siegel JA, Yeldell D, Goldenberg DM, et al. Red marrow radiation dose adjustment using plasma FLT3-L cytokine levels: improved correlations between hematologic toxicity and bone marrow dose for radioimmunotherapy patients. *J Nucl Med* 2003;44:67–76.

160. Emami B, Lyman J, Brown A, et al. Tolerance of normal tissue to therapeutic irradiation. *Int J Radiat Oncol Biol Phys* 1991;21:109–122.

161. Bouchet LG, Bolch WE, Blanco HP, et al. MIRD pamphlet no. 19: absorbed fractions and radionuclide S values for six age-dependent multiregion models of the kidney. *J Nucl Med* 2003;44:1113–1147.

162. Konijnenberg MW. Is the renal dosimetry for [90Y-DOTA0,Tyr3]octreotide accurate enough to predict thresholds for individual patients? *Cancer Biother Radiopharm* 2003; 18:619–625.

163. Wessels BW, Konijnenberg MW, Dale RG, et al. MIRD pamphlet no. 20: the effect of model assumptions on kidney dosimetry and response—implications for radionuclide therapy. *J Nucl Med* 2008;49:1884–1899.

164. Press OW, Eary JF, Appelbaum FR, Bernstein ID. Treatment of relapsed B-cell lymphomas with high-dose radioimmunotherapy and bone marrow transplantation. In: *Cancer Therapy with Radiolabeled Antibodies* Goldenberg DM, editor. CRC Press, Boca Raton, 1995; 229–237.

165. Meredith RF, Johnson TK, Plott G, et al. Dosimetry of solid tumors. *Med Phys* 1993;20:583–592.

166. Brownell GL, Ellett WH, Reddy AR. Absorbed Fractions for Photon Dosimetry. MIRD Pamphlet No. 3. Society of Nuclear Medicine, New York, 1968.

167. Siegel JA, Stabin MG. Absorbed fractions for electrons and beta particles in spheres of various sizes. *J Nucl Med* 1994;35:152–156.

168. Johnson TK, Colby SB. Photon contribution to tumor dose from considerations of ^{131}I radiolabeled antibody uptake in liver, spleen, and whole body. *Med Phys* 1993; 20:1667–1674.

169. Johnson TK, McClure D, McCourt S. MABDOSE. I: characterization of a general purpose dose estimation code. *Med Phys* 1999;26:1389–1395.

170. Erdi AK, Yorke ED, Loew MH, Erdi YE, Sarfaraz M, Wessels BW. Use of the fast Hartley transform for three-dimensional dose calculation in radionuclide therapy. *Med Phys* 1998;25:2226–2233.

171. Bolch WE, Bouchet LG, Robertson JS, et al. MIRD Pamphlet No. 17: the Dosimetry of Nonuniform Activity Distributions—Radionuclide S Values at the Voxel Level. Medical Internal Radiation Dose Committee. *J Nucl Med* 1999;40:11S–36S.

172. Liu A, Williams LE, Wong JY, Raubitschek AA. Monte Carlo-assisted voxel source kernel method (MAVSK) for internal beta dosimetry. *Nucl Med Biol* 1998;25:423–433.

173. Britton KE. Radioimmunotherapy of Non-Hodgkin's lymphoma. *J Nucl Med* 2004;45:924–925.

174. Lamborn KR, DeNardo GL, DeNardo SJ, et al. Treatment-related parameters predicting efficacy of Lym-1 radioimmunotherapy in patients with B-lymphocytic malignancies. *Clin Cancer Res* 1997;3:1253–1260.

175. Kaminski MS, Estes J, Zasadny KR, et al. Radioimmunotherapy with iodine (131)I tositumomab for relapsed or refractory B-cell non-Hodgkin lymphoma: updated results and long-term follow-up of the University of Michigan experience. *Blood* 2000; 96:1259–1266.

176. Wiseman GA, White CA, Stabin M, et al. Phase I/II ^{90}Y-Zevalin (yttrium-90 ibritumomab tiuxetan, IDEC-Y2B8) radioimmunotherapy dosimetry results in relapsed or refractory non- Hodgkin's lymphoma. *Eur J Nucl Med* 2000;27:766–777.

177. Wiseman GA, White CA, Sparks RB, et al. Biodistribution and dosimetry results from a phase III prospectively randomized controlled trial of Zevalin radioimmunotherapy for low-grade, follicular, or transformed B-cell non-Hodgkin's lymphoma. *Crit Rev Oncol Hematol* 2001;39:181–194.

178. Koral KF, Kaminski MS, Wahl RL. Correlation of tumor radiation-absorbed dose with response is easier to find in previously untreated patients. *J Nucl Med* 2003; 44:1541–1543.

179. Koral KF, Dewaraja Y, Li J, et al. Update on hybrid conjugate-view SPECT tumor dosimetry and response in ^{131}I-tositumomab therapy of previously untreated lymphoma patients. *J Nucl Med* 2003;44:457–464.

180. Sgouros G, Squeri S, Ballangrud AM, et al. Patient-specific. 3-dimensional dosimetry in non-Hodgkin's lymphoma patients treated with ^{131}I-anti-B1 antibody: assessment of tumor dose-response. *J Nucl Med.* 2003;44:260–268.

181. Sharkey RM, Brenner A, Burton J, et al. Radioimmunotherapy of non-Hodgkin's lymphoma with ^{90}Y-DOTA humanized anti-CD22 IgG (^{90}Y-Epratuzumab): do tumor targeting and dosimetry predict therapeutic response? *J Nucl Med* 2003;44:2000–2018.

182. Iagaru A, Gambhir SS, Goris ML. ^{90}Y-ibritumomab therapy in refractory non-Hodgkin's lymphoma: observations from ^{111}In-ibritumomab pretreatment imaging. *J Nucl Med* 2008;49:1809–1812.

183. Fiveash JB, Hanks G, Roach M, et al. 3D conformal radiation therapy (3DCRT) for high grade prostate cancer: a multi-institutional review. *Int J Radiat Oncol Biol Phys* 2000; 47:335–342.

CHAPTER 14

The Bystander Effect in Targeted Radiotherapy

CARMEL MOTHERSILL AND COLIN SEYMOUR

14.1 INTRODUCTION

The bystander effect in the context of cancer therapy describes the cytotoxicity experienced by otherwise viable, nontargeted neighboring cancer cells from mediators released by cells that have been targeted and killed by cancer therapeutics (see schematic in Fig. 14.1). The net effect of the bystander phenomenon can be to enhance the effectiveness of cancer treatments beyond that would normally be expected based on the delivery of the agent to tumor cells. It is a complex effect, however, and can affect distant or proximal normal cells causing tissue damage and systemic radiation effects. The bystander effect has been studied previously for tumors treated with cytotoxic agents and external radiation, but has only very rarely been investigated in the case of targeted radiotherapy. This chapter will deal with the historical emergence and general characteristics of bystander effects as a new phenomenon that is actually leading to a paradigm shift in radiobiology impacting our understanding of both low- and high-dose radiobiology. While we usually think of radiotherapy as requiring high doses, the bystander effect means that low, scatter doses and systemic effects including doses to normal tissues in the vicinity of targeted tumors can result in significant triggering of bystander responses. It is essential therefore that we understand the mechanisms involved in these "collateral" effects so we can build them in to models seeking to determine optimal outcome in therapeutic situations. This is particularly relevant for targeted therapy approaches because these seek to deliver a cytotoxic dose to precise locations in the body where the tumor or metastatic cells reside. Following this discussion, this chapter will focus on the mechanisms that could be harnessed to provide better therapeutic targeting of tumors for radiotherapy. Finally, some of the many possible directions forward in the field will be discussed.

Monoclonal Antibody and Peptide-Targeted Radiotherapy of Cancer, Edited by Raymond M. Reilly
Copyright © 2010 John Wiley & Sons, Inc.

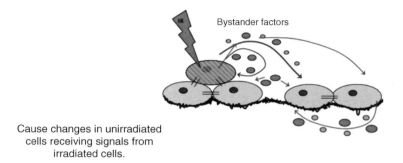

Bystander factors

Cause changes in unirradiated
cells receiving signals from
irradiated cells.

**"Good" and "bad" changes
observed**

FIGURE 14.1   Schematic showing the radiation-induced bystander effect operating through a gap junction intercellular communication mechanism or by diffusible factors.

14.2   HISTORICAL REVIEW OF BYSTANDER EFFECTS IN THE CONTEXT OF RADIATION DAMAGE TO CELLS

Historically, radiation has been thought of in two different ways. Radiative energy may be transferred either through a wave-like diffusion or through discrete units of energy (quanta). This is known as the Copenhagen paradox, or the dual particle theory of matter. This paradox is still unresolved, but has been thought to be irrelevant to radiotherapy. In radiotherapy, target theory has always been conceptually important (1–3). The aim of radiotherapy has been to target the radiation to the tumor area, while sparing the adjacent or proximal normal tissue to give an optimal therapeutic ratio, and within the tumor each cell had a radiation-sensitive target (thought to be DNA) that had to be inactivated. Targeted radiation therapy and radiation target theory were conceptually linked in a common goal (2, 4). The bystander effect, which has become thought of as a nontargeted effect, has challenged cellular target theory (5–8). Through analogy, it could be argued that it also provides both opportunities and potential problems in the targeted radiotherapy field. The bystander effect works through cell-to-cell communication (Fig. 14.1). This may be through direct gap junction communication between cells (9, 10), or may be through a liquid (or gaseous) intermediary (11, 12). The importance of the liquid intermediary is that signals will travel around the entire body very rapidly. The old paradigm of radiation biology that is still widely held and underpins radiotherapy basically holds that there is a linear or log linear relationship between radiation dose and biological effect. It holds that DNA is the critical "target" for radiation damage and that the DNA double strand break is the critical lesion. The number of double strand breaks can be directly related to the dose. Arising from this DNA damage, chromosome aberrations can occur due to changes in the sequence of DNA bases (code sequences). It should be noted that the old paradigm held that low-dose chronic irradiation does not necessarily have as great an impact as a brief higher dose exposure—a division factor of 2 was applied to the "dose" if this was accumulated over a long period (13) but critically, the concept of adaptive or epigenetic influences (i.e., effects due to processes or mechanisms not involving mutation or direct DNA damage) were not considered and are still not integrated into fractionated radiotherapy models (14). The direct relationship between dose and DNA damage lent weight to the linear nonthreshold (LNT) model that was supported by high-dose epidemiological data from the Japanese atomic bomb survivors who had an increasing rate of cancer incidence (biological effect) as the dose received increased (15–18). To determine effects at low doses, the high-dose data were extrapolated to zero dose where there was assumed to be a zero effect. Of course, in the low-dose region, it was not easy to assign causation to radiation exposure due to the high background incidence of cancer and other diseases associated with radiation but the model is used even though flawed to relate dose and effect both in radiation protection and in radiotherapy (19, 20).

14.3 NEW KNOWLEDGE AND THE PILLARS OF THE DEVELOPING NEW PARADIGM

Within conventional radiobiology as accepted in the 1950s continuing through to the late 1990s, there was no room for epigenetic effects because the traditional concept of radiobiology was based on target theory. In order to work, radiation had to hit defined targets within the cell, assumed to be DNA. Assumptions about the number of targets hit could then be made from measurements of dose and dose rate. The evolution of nontargeted radiobiology meant that the previous assumptions could no longer hold particularly in the low-dose region or where targeted radiotherapy meant that the whole field was not equally exposed. This meant that radiation effects no longer had to be genome based (due to DNA mutation or reproductive death), but that radiation could cause or acerbate systemic disease and could kill tumor cells without necessarily depositing lethal amounts of energy in the cell itself (for reviews, see Refs (21–24)). These concepts, although largely accepted theoretically by the radiobiology community, have been difficult to prove in the population or clinic because of enormous confounding variables (smoking, drinking, age, sex, concurrent, past, or future exposures to the same or a different toxic agent). It has also been argued that the radiation might actually boost the immune system and be good for you—and there is a history of using radium baths in Europe and the United States as health spas. The key point, however, is that there will be huge individual variation due to involvement of epigenetic factors in the response. At any one time, we are as unique epigenetically as we are genetically. Epigenetic differences are linked to sex and lifestyle. In theory, therefore a dose of radiation could cause any number of effects at any hierarchical level ranging from beneficial to death-inducing damage. Understanding the underlying mechanisms and harnessing these variables is essential if we are to optimize treatments using targeted radiotherapy approaches.

Another key pillar of the changing paradigm is that we now realize that radiation effects even at the level of the cell are not simply due to DNA double strand breaks. Cell communication, microenvironment, tissue infrastructure, and a whole host of systemic variables influence outcome from a cellular track of ionizing radiation. These points are discussed in the various reviews cited above. The key milestones in the development of this new paradigm are listed in the following time line.

14.3.1 Key Points and Historical Time Line

1954: First report of persistent "clastogenic activity" in the plasma of children who received irradiation to the spleen. Clastogenic activity refers to the presence of factors in the blood that could cause chromosome damage (25).

1962: Souto reported that plasma of irradiated rats could induce tumors in unirradiated rats at a greater rate than plasma form unexposed rats. This was published in *Nature* (26). They also called this "clastogenic activity."

1967–1972: Several experimental studies using animals and human radiotherapy, patients or victims of radiation exposure in the Marshall Islands and in Hiroshima show the presence of chromosome damaging agents circulating in the blood many years after exposure. Papers highlighted in a review by Mothersill and Seymour 2001 (27).

1972–1985: During this period, no major work reported in this area. The data on clastogenic factors were forgotten or shelved because there was no mechanism known by which these factors could act and because there was no epidemiological link to increased tumors at the population level.

1986: Report of "lethal mutations" in the distant progeny of irradiated surviving cells (28) started a renewed interest in the persistence of radiation damage and the mechanisms by which it could be perpetuated.

1986–1991: Several reports in the radiobiology literature of delayed effects both mutagenic and lethal in "normal progeny" cells surviving radiation (29–32).

1992: Kadhim et al. (33) publish a paper in *Nature* showing chromosomal aberrations in the progeny of bone marrow cells exposed to low-fluence alpha particles. This paper was very widely cited as it suggested a mechanism for radiation-induced leukemia. It also led to the term "genomic instability" being widely known and defined as "nonclonal aberrations appearing in clonal progeny of irradiated cells." Definitive proof that the cells were unstable was facilitated because bone marrow cell lineages were very well worked out and descendants were known to be clonal. Therefore, finding a high rate of new chromosome defects in descendants of irradiated stem cells meant a process was occurring upstream that made the occurrence of genetic mutations more probable in the clonal descendants. This was demonstrated in human bone marrow by the same group in 1994 (34) and also showed big differences between patients, a clear indication that individual variation was occurring. Implicit in these chromosomal studies and the lethal mutation studies was the concept that more cells were demonstrating damage than could have been directly targeted by the radiation.

1992: Little et al. (35) report that more cells than could have been hit by a low fluence of alpha particles, show chromosome damage. This report could not be accommodated by target theory because a cell did not need to be hit by radiation to show an effect. This together with the earlier discoveries of genomic instability and lethal mutations set the scene for the "paradigm shift" in radiobiology.

These papers were very controversial at the time and were regarded as "artifacts," "irrelevant to radiation protection," and "just plain wrong" by the International Commission on Radiological Protection (ICRP) and the U.K. National Radiological Protection Board (NRPB). European Union funding was denied to groups proposing to work in this area and there was much bitterness continuing into the late 90s despite mounting evidence of a real effect, in terms of chromosomal aberrations.

1996: A paper by Clutton et al. (36) links genomic instability to oxidative stress in the cell population. Papers reported later on by Limoli (37), Murphy (38), Prise, (39), Hei (40), and others confirm the link and suggest that cell and organ stress leads to production of bystander factors that in turn drive the persistent instability and thus increase the probability of mutations.

1996: First report in recent times of a soluble factor produced by irradiated cells that can reduce the survival of distant unirradiated cells (41).

1997: Genotype dependence of genomic instability outcomes reported in a widely read and cited paper by Ponniaya et al. (42). The experiments demonstrated clearly in mice that depending on the genetic strain, the cells would favor a "cell death pathway" or a "survival carrying genetic damage" pathway. The latter strain developed cancers as a late effect of radiation exposure while the former did not. Watson et al. (43) in the same year injected gender mismatch normal bone marrow cells into mice that had their bone marrow ablated by a whole body radiation dose. The group showed strain-dependent induction of genomic instability in the normal injected marrow cells. This had to be due to soluble factors from the host affecting the newly injected cells.

1997–2001: Many reports in the literature (reviewed in Ref. (27)) suggesting so called "bystander effects," that is, chromosome damage, death, DNA damage, and an assortment of other effects in cells receiving signals from irradiated cells. The key papers that were considered relevant to human carcinogenesis and cancer therapy were by the group at Columbia University in New York, who showed mutagenic and carcinogenic effects in cells receiving signals from irradiated cells (reviewed in Ref. (44)) and a paper by Weber et al., (45) confirming that intercellular signals could induce transformation (an *in vitro* state where cells acquire characteristics similar to those seen in malignant cells and if injected into animals, form metastatic tumors). Another key paper was by Seymour and Mothersill in 1997 (46) showing in human skin cells that both lethal mutations and genomic instability were induced by bystander signals. Lorimore et al. in 1998 (47) then showed using alpha particle irradiation that the population of cells that received the signals but were not actually targeted were in fact the population that went on to develop genomic instability. This led to suggestions of wider targets for radiation effects than DNA and the field of "nontargeted effects of radiation" was born.

2001: First report (48) written in Russian of an allelopathic (i.e., a communicated signal inducing a damaging response in another organism) effect of urine from irradiated mice (4 Gy) on the bone marrow and immune response of unexposed animals sharing the same cage. Allelopathic effects are those communicated by chemicals from one organism to another. They are well known in plants, less so in animals.

2006: First report (49) of communicated effects of radiation exposure from irradiated fish (0.5 Gy) to unirradiated fish.

2007: Demonstration by Smith et al. (50) of a unique protein profile induced in fish receiving signals from irradiated fish. The significance of this and Surinov's earlier finding in mice is that they demonstrated the *in vivo* relevance of nontargeted effects and revealed in two widely different vertebrate animals—fish and mice—in which signals transmitted from one animal that has been irradiated can cause induction of proteins in another animal (unirradiated) that received the signal. This is unequivocal proof that bystander effects/responses occur *in vivo* and provides a way of dissecting signal production from bystander response that is not possible in models using shielding or microbeam approaches because of the confounding problems of scatter dose and blood circulation. While it may be argued that the latter models are more relevant to the therapeutic situation, because the reality is that the dose is delivered to part of a whole organism and thus "bystander effects" are going to be a result of direct effects due to the actual dose, the scatter dose and the dose to circulating blood, the communicated signal model is essential if we are to understand the unique mechanisms involved in bystander responses and harness them for the benefit of patients.

2007: Wide acceptance that the "bystander effect" is actually akin to a stress response inducible in susceptible individuals, which due to the link between cellular stress, oxidative stress, and DNA damage, provides a mechanism by which low or nonuniform doses of radiation can have profound and unexpectedly widespread effects in susceptible individuals. While repair mechanisms may take care of the actual damage, these can be compromised by immune deficiency due to infectious disease and other illnesses, nutritional insufficiency, age, and a host of other factors including emotional "stress."

2007: Demonstration by Liu et al. (51) and Prise et al. (52) of bystander effects occurring in cells at doses in the region of 2–3 mGy. Prise's group also showed a "binary response" at very low doses where some cells responded to bystander signals but others did not. They regarded the effect as a stochastic or random event. Demonstration of adaptive responses has been documented also at doses as little as 10 µGy. The point here is not whether the effect is "good" or "bad," but that an effect occurs at all. Given what we know about individual variation, due to both genetic and epigenetic factors, it is likely that the same dose of radiation could have beneficial effects in one person and adverse effects in another. Averaged out over the population, this would not be apparent in the epidemiology but is clearly important to understand when planning radiotherapy, especially if this is targeted because this aims to leave large areas of the body untargeted and thus available for bystander mechanisms to dominate.

2008: Reports of the demonstration of a "radiation signature" in the form of a discriminatory biomarker that can be induced by radiation but not by a toxic heavy metal by Nakamori et al. (53). This offers the possibility to test for radiation-induced effects as opposed to chemotherapy or chemical carrier effects in targeted therapy and is a "holy grail" of radiobiology.

14.4   CONCEPT OF HIERARCHICAL LEVELS OF ASSESSMENT OF TARGETED RADIATION EFFECTS

Enormous confusion in the radiobiology field arises from a lack of consideration of this concept. Most of the arguments about whether radiation is "good" or "bad " in a given situation fail due to lack of consideration of the level at which the effect is taking place. For example, cell death is seen as a "bad" effect in radiation protection but if it removes a cancer cell or a metastatic cell from the population of cells in a tissue it could prevent the cancer from spreading or a metastatic focus from starting and could be seen as "good." Similarly, in nonhuman populations, for example, a competing rodent species—death of radiosensitive population members that cannot adapt to the changed (now radioactive) environment, could be "good" for the population although "bad" for the individual. It is only by considering responses in context that any conclusions can be drawn about risk or benefit of a proposed targeted therapy.

14.5   THE NEW MEANING OF THE LNT MODEL

Given all the new uncertainties and the emerging mechanistic understanding of bystander effects, the LNT model cannot be called an LNT hypothesis anymore. It is clearly not correct to say a linear extrapolation describes low or nonuniform radiation dose effects. The new paradigm contains complexity and unpredictability. There are arguments and data to support any relationship between dose and effect after low and nonuniform doses but the reality is that any outcome can happen to an individual and there are ample data showing *effects* in these situations. The purpose of the LNT model that plots a linear relationship between dose of radiation and carcinogenic risk based on the A-bomb survivors is now to provide a tool for regulation in an environment of uncertainty. The problem with the LNT model is the lack of actual data at low doses—the model used extrapolation to make a relationship showing zero risk at zero dose and without any threshold. This is hugely controversial with both pro- and antiradiation groups arguing that the relationship is either hypo or hyper linear. On scientific analysis, the LNT dose effect relationship that has been used for regulation of human exposure has been rejected by various radiological organizations or committees asked to consider the evidence, such as the CERRIE (Committee Examining Radiation Risks of Internal Emitters) minority and majority reports of 2003/2004 and the French Academy of Sciences (54–56). The cause of the uncertainty is simply that the simple DNA damage paradigm does not hold at low doses such as those experienced in nontargeted regions of the body and therefore dose and effect cannot always be linearly related. Which way the curve will go depends on other factors—including genetic background and environmental conditions.

14.5.1   Relating Dose to Effect, Harm, and Risk

This is the key issue. It is always controversial and in dose range senarios or dose distributions where epidemiology is a weak tool, it is usually difficult to assess whether a dose produced a specific consequence in an individual. The reverse relationship (that, for example, an adverse health effect is caused by a dose) is also difficult to assess. The gold standard is, of course, chromosome aberrations as these are evidence of fixed genetic change in dividing cells and are relevant to both cancer and hereditary effects but induction of cancer associated proteomes, stress pro-teomes, or genomic changes in cancer associated genes are also important espe-cially for monitoring therapeutic outcomes even though these are not necessarily fixed and transmissible.

14.6   TECHNIQUES FOR STUDYING BYSTANDER EFFECTS

The new paradigm emerged initially as a result of reexamination of firmly held beliefs and some odd results in the laboratory that did not fit the existing theory. Proof of the new hypotheses required the application of techniques such as cellular molecular imaging, M-FISH (multiple fluorescence *in situ* hybridization), SKY (spectral karyotyping) as well as the development of culture techniques for human *normal and tumor* tissues which permitted functional studies to be done (57). Older studies tended to be performed on a limited number of cell lines that grew well in the laboratory but were radioresistant and very unrepresentative of either normal or tumor tissues in the body. However, the major driver was the entry into radiobiology of biologists with cell and molecular biology approaches, experience in tissue and organ physiology and no preconceived ideas.

14.6.1   Emerging Biomarkers of Nontargeted Radiation Effects

The new techniques are of great power in the study of disease processes or therapeutic outcomes where frank changes in tissues have occurred and where normal tissue samples along with the tumor biopsies are available for comparative purposes. Most of the new techniques other than those based on live cell imaging and cytogenetics are collectively grouped as "omics" technologies, for example, genomics, proteomics, and so on. All "omics" studies rely on comparison of altered patterns against normal patterns preferably from the same patient or mouse strain, or at least from a large bank of normal or diseased tissues. The question posed in this chapter is not whether the techniques are useful but whether they can address the specific issue of nonuniform dose effects in targeted radiotherapy situations.

14.7 BYSTANDER PHENOMENA IN TARGETED AND CONVENTIONAL RADIOTHERAPY

These may be considered from several aspects—What is the difference in the type of signal produced by normal versus tumor cells? What is the impact of signals produced by targeted cells on surrounding or systemically affected normal cells? Do different tissues (normal or malignant) give rise to different bystander signals? Does one tissue respond to signals sent by another tissue whether normal or malignant? What is the impact of imaging techniques and treatment planning? Traditional radiotherapy used two or three fields to cover the tumor area. Doses to the tumor were limited by damage to the normal surrounding tissue. Recently, several articles have appeared exploiting bystander effects to increase tumor cell kill. These studies have shown that indirect effects of ionizing radiation may contribute significantly to the effectiveness of radiotherapy by sterilizing malignant cells that are not directly hit by the radiation. Reports by Kassis (58), Bodei et al. (59), Marples et al. (60), and Mothersill et al. (14) define or discuss the importance of bystander effects *in vivo* in clinical situations. More recently, GRID using a lead grid placed over the field to be irradiated and synchrotron beam approaches (synchrotrons can be used to generate parallel microbeams of high-energy electrons) both aim to increase the total dose to the irradiated region by sparing areas of the tumor target and appear to involve bystander processes operating in areas where mimimized dose (i.e., valley dose) occurs (61, 62). The mechanism underlying the clinical success of these techniques has not been elucidated but the techniques raise very important questions about bystander signal effects occurring in targeted cells as well as in nontargeted cells—a concept that is only now beginning to be considered (63).

Targeted radionuclide therapy aims to increase accuracy, range of delivery and dose by increasing the number of fields, specifically targeting the dose to tumor cells wherever they may be in the body (64–66). This should decrease damage to the near surrounding tissue and increase dose to the targeted tumor cells, but also increases the whole body scatter dose—that is, the dose reaching areas not in the planned radiation field. The bystander effect is the dominant effect at these low scatter doses, but is thought to saturate at higher doses; however, this is not to say it is irrelevant at high doses and Fig. 14.2 shows that it contributes about 60% of the cell kill after a 2 Gy conventional radiotherapy dose. If it is important in targeted therapy, logic would suggest there would be a difference between conventional and targeted radiotherapy in effects on targeted and nontargeted tissues that may be attributable to the bystander effect. This may explain the effectiveness of targeted approaches such as those employed by Boyd et al. (64–70). Boyd et al. have optimized several aspects of targeted radiotherapy/gene therapy strategies using noradrenaline transporter (NAT)

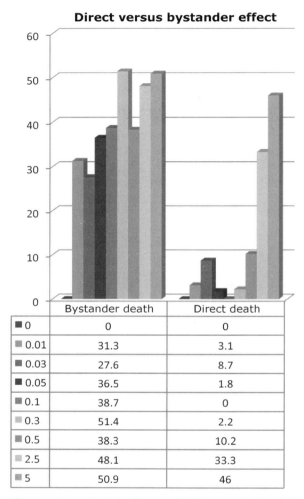

	Bystander death	Direct death
■ 0	0	0
■ 0.01	31.3	3.1
■ 0.03	27.6	8.7
■ 0.05	36.5	1.8
■ 0.1	38.7	0
■ 0.3	51.4	2.2
■ 0.5	38.3	10.2
■ 2.5	48.1	33.3
■ 5	50.9	46

FIGURE 14.2 Clonogenic survival of cells treated with bystander signals derived from cells exposed to radiation (left panel) and cells directly irradiated (right panel) where the bystander contribution to the total result has been subtracted leaving the effect attributable to the direct irradiation only.

gene-transfected tumor cells to achieve tumor-specific transcriptional regulation of therapeutic genes and to maximize collateral cell damage via cross-fire irradiation between cells, thereby overcoming the problem of heterogeneity of transgene expression. The efficacy of these ploys has been demonstrated in their unique transfectant mosaic spheroid model. Recognizing that, in addition to the physical bystander effect (cross fire), there is a more subtle biologic bystander effect associated with targeted radionuclide therapy, they embarked on a study of the characterization of this phenomenon. They employ an adaptation of the media transfer procedure developed by Mothersill and Seymour (12) to compare the induction of bystander effects by external beam cobalt 60 gamma radiation with those generated by radiolabeled metaiodobenzylguanidine (MIBG), a radiotherapeutic agent for neuro-blastoma, incorporating radionuclides (^{131}I, ^{211}At, or ^{123}I) emitting β-particles, α-particles, or Auger electrons, respectively. This is a good example of the use of gene transfection to construct a radiotherapy model, inasmuch as it allowed the creation of an excellent control—that is, non-(noradrenaline transporter (NAT) gene) transfected cells that were incapable of active uptake of radiolabeled MIBG. For a more detailed discussion of gene-transfection approaches for targeted radiotherapy, see Chapter 10. Boyd et al. (64–70) refrain from commenting on the relationship between absorbed dose to the cell and radioactivity concentration. To do so, they would need more complete information concerning uptake and washout dynamics and transfer constants between media, cell surface, and intracellular and nuclear compart-ments. Instead, the investigators estimated effective dose by comparing the clonogenic cell kill achieved by external beam radiation or radiopharmaceutical treatment. The results of this study indicated that intracellularly accumulated radionuclides power-fully stimulated the production of bystander effects. Active cellular accumulation was necessary for the induction of bystander effects. Those cells that had not been transfected with the NAT gene produced no toxin. Ultimately, however, serious dose modeling will be needed to carry these techniques forward.

In terms of treatment planning techniques and the use of IMRT (intensity modulated radiotherapy) to target tumors, bystander effects are theoretically relevant but difficult to quantify. There is as yet little evidence for or against IMRT being advantageous. This is because it is a relatively new technology and is expensive in terms of time and machine utilization. Large-scale comparative studies have not yet been done.

If the only important bystander effect in determining outcome is that produced by the tumor, then the shift from conventional to IMRT treatment will have no detectable effect. If the bystander factor(s) produced by the normal tissue is(are) important, then the shift from conventional to IMRT treatments may have significant effects. If all cells produce bystander factor, then the total volume of irradiated cells will be important (especially with a threshold as there will be effectively no dose-dependent gradients). If tumor cells and normal cells produce different bystander factors then respective volumes of irradiated tissue will also be important. In both of these instances, it can be predicted that the increase in whole body scatter dose produced by IMRT will be beneficial in tumor suppression. This makes the assumption that the bystander effect we observe *in vitro* happens *in vivo*, and there is some evidence for this. While it is not possible to predict exact effects, the argument can be made that the bystander effect would be different between conventional and IMRT treatments, and may be sufficient to produce different treatment outcomes. These questions are only beginning to be addressed.

In relation to the specific topic of this book that is about targeted radiotherapy using molecular tools rather than physical techniques to focus the dose to the tumor numerous clinical trials of radioimmunotherapy have shown positive treatment out-comes (partial and complete remissions) in some but not all patients, and these have occurred at very modest radiation absorbed doses, that is, less than 2000 cGy, sometimes at only a few hundred centigray units (see Chapter 13). This suggests that there could be a substantial bystander effect at play. In addition, in the area of Auger electron radiotherapy, the dogma has been that these subcellular types of radiations are only effective if the radionuclides are targeted and internalized into tumor cells; however, Kassis et al. have shown that there is a bystander effect from these types of radiations that can kill nontargeted cells, thus potentially greatly amplifying their effects (see Chapter 9).

14.8 MECHANISMS UNDERLYING BYSTANDER EFFECTS AND DETECTION TECHNIQUES

The mechanisms underlying the bystander effect have been extensively reviewed (21–24, 27, 71–73). Basically the consensus now is that bystander signals are products of stress pathways in cells responding to environmental insults. As such, roles for ROS (reactive oxygen species), NOS (reactive nitrogen species), p53, cytokine, and MAPK (mitogen-activated protein kinase) damaging sensing pathways have all been implicated. At the more "macro" level, immune and inflammatory pathways have been implicated (74). Ion channels are thought to mediate the transformation of radiative energy into chemical signals (75, 76). The majority of papers in this field are concerned with carcinogenesis rather than therapy for cancer but this is changing. Recent advances in the fields of targeted radiotherapy and synchrotron therapy, suggest that bystander mechanisms if harnessed properly, could be a major therapeutic tool in the fight against cancer. This raises the important question of how to measure bystander effects in a clinical or preclinical setting. Our group developed an explant assay to experimentally examine the bystander effect that can be done if surgery precedes therapy (77–79). The assay involves culturing small tissue fragments from an irradiated host (either an animal or a human patient undergoing radiotherapy) *in vitro* and harvesting medium from the culture (Fig. 14.3). The medium is added to reporter cells known to produce a response if bystander signals are present. Extensive controls including sham irradiated samples and unhandled samples ensure the signals are due to actual irradiation of the tissues. For the purposes of radiotherapy related bystander assays, the end point is a change in the cloning efficiency of the cell line using the Puck and Marcus assay (80). Decreased cloning efficiency is indicative of a reduced reproductive survival due to the cells receiving bystander signals while increased cloning efficiency is indicative of increased reproductive survival—all relative to the control cloning efficiency. The assay was validated using mice irradiated *in vivo* that were already known to known to produce (C57 BlJ6) or not produce bystander CBA H) signals. The big advantage of the explant assay is that normal and tumor tissue from the same organ can be sampled—thus addressing the question of therapeutic ratio. Because tissue biopsies are not always available, our laboratory also tried to develop a blood assay. Samples were obtained before, during and after conventional radiotherapy courses. Serum was harvested by centrifugation and the serum samples were used in lieu of the fetal calf serum normally added to culture

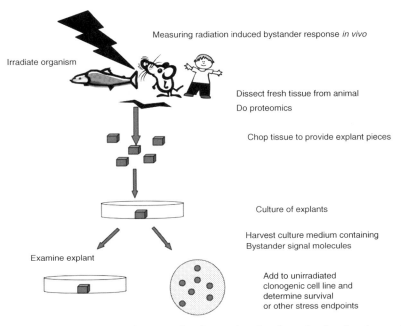

FIGURE 14.3 The protocol for assessing bystander signal production by tissues from exposed organisms or tissues.

medium. The assay results were published as a pilot study (81) but sufficient patients and clinical follow-up were not obtained in the study to validate this method. Perhaps, it is now timely to revisit this approach especially in the case of targeted radiotherapy patients because it would indicate a systemic bystander response (or not) in patients undergoing treatment and could perhaps be used to monitor response. There are very few data concerning tumor versus normal tissue bystander signals. In our hands, tumor cells *per se* do not produce toxic (apoptosis inducing) bystander signals. We suspect that the apoptotic signals are coming from normal cells in the tumor and that the beneficial effects of protocols such as targeted radiotherapy may be due to bystander signals resulting from low collateral doses to normal tissues in the tumor such as vascular or stromal elements. This are needs considerable research. One of the limitations of our own reporter assay is that it is limited to detection of toxic signals. There is no reason to believe that there are not other types of signals that could do other things. To date, virtually all bystander research has focused on low-dose radiation protection issues as discussed previously but the major benefits of bystander research are most likely to come form studies in the therapy field. To our knowledge at the time of writing, no clinical or preclinical studies have been published which actually harness bystander responses for radiotherapy.

Apart from the practical clinical need to measure bystander responses, there is also a major need to answer mechanistic questions of specific relevance to radiotherapy. As mentioned earlier, most bystander research is concerned with the relevance to low-dose radiogenic cancers. In the specific case of targeted radiotherapy, the key mechanistic questions include differences in response between normal (e.g., tumor stromal bed and vasculature) and tumor cells, tumor death evasion mechanisms, long-term transmission of genomic instability and clastogenic effects due to systemic signaling effects. One of the long-term effects of targeted radiotherapy, for example, treatment of lymphoma with Zevalin, or treatment of neuroendocrine malignancies with ^{90}Y-DOTATOC and ^{111}In-pentetreotide has been the induction of myelodys-plastic syndrome (MDS), a precursor to acute myeloid leukemia (AML) (see Chapters 7 and 9). Presumably, this is due to direct effects of the radiation on bone marrow stem cells leading to genomic instability, but it could also be due to bystander effects at low-radiation fluxes. In fact, the current ideas in low-dose radiobiology suggest that the genomic instability is in fact "driven" by bystander mechanisms although how is far from clear.

The mechanisms of "scatter" and abscopal effects (82–85) also are very relevant because a key mechanistic question in targeted radiotherapy is actually: Does the estimated radiation absorbed dose correlate with the therapeutic outcome, or is it possible to have a greater therapeutic outcome than expected because of the bystander effect? These questions should be central for funding agencies in the cancer radio-therapy field because the evidence now (mainly anecdotal) cannot be ignored any more and strongly suggests major therapeutic benefits of whole body or targeted low-dose exposure with consequent induction of apoptotic bystander signals that optimize tumor response. The specific role of biogenic amines and cytokines in targeted radiotherapy approaches, which offer adjuvant therapeutics is also important to study (86–88) because these signaling agents can interfere with the planned radio-therapeutic outcome and if inhibited or stimulated using inhibitors or activators could optimize the desired outcome.

14.9   THE FUTURE

The possible exploitation of bystander mechanisms as novel targets for targeted radiotherapy needs to be explored. In particular, there is a case for manipulating the differences between normal and tumor tissue and the differences in response between different individuals to create "designer therapies," which in the case of targeted radiotherapy, could optimize the cell killing effect using inhibitors or activators of bystander signals. Several drugs are also now known to interfere with or enhance bystander signaling cascades. Among these are L-deprenyl and ondansetron—already used in external radiotherapy for other reasons (86, 87). Ondansetron is an inhibitor of serotonin known to be important in the bystander mechanism (86, 87) and it is already known to improve tolerance of radiotherapy but the link to the bystander mechanism was not known. Clearly, in a targeted therapy situation where bystander mechanisms may be even more important—this drug or in fact its antagonist could be very valuable. Similarly, L-deprenyl, used to treat Parkinson's disease, is a potent inhibitor of bystander effects by inducing the antiapoptotic protein—bcl 2—in normal tissues (it is usually already induced in tumors), thus optimizing the therapeutic ratio. The rational inclusion of these and other drugs in targeted radiotherapy situations needs to be explored. Also the idea of preconditioning patients with low-dose radiation to turn on the bystander signal cascades prior to targeted radiotherapy needs to be critically evaluated.

REFERENCES

1. Lea DE. *The Action of Radiation on Living Cells*. Cambridge University Press, London, 1946.

2. Timofeeff-Ressovsky NW, Zimmer KG. *Das Trefferprinzip in der Biologie*. S. Hirzel Verlag, Leipzig, 1947.

3. Alper T. *Cellular Radiobiology*. Cambridge University Press, Cambridge, 1979.

4. Elkind MM, Whitmore GF. *The Radiobiology of Cultured Cells*. Gordon and Breach Science Publishers, New York, 1967.

5. Matsumoto H, Hamada N, Takahashi A, Kobayashi Y, Ohnishi T. Vanguards of paradigm shift in radiation biology: radiation-induced adaptive and bystander responses. *J Radiat Res* 2007;48:97–106.

6. Schwartz JL. Abandon hope all ye target theory modelers: on the effects of low dose exposures to ionizing radiation and other carcinogens. *Mutat Res* 2004;568:3–4.

7. Liu Z, Prestwich WV, Stewart RD, et al. Effective target size for the induction of bystander effects in medium transfer experiments. *Radiat Res* 2007;168:627–630.

8. Morgan WF. Will radiation-induced bystander effects or adaptive responses impact on the shape of the dose response relationships at low doses of ionizing radiation? *Dose Response* 2006;4:257–262.

9. Lumniczky K, Sáfrány G. Cancer gene therapy: combination with radiation therapy and the role of bystander cell killing in the anti-tumor effect. *Pathol Oncol Res* 2006; 12: 118–124.

10. Suzuki M, Tsuruoka C. Heavy charged particles produce a bystander effect via cell-cell junctions. *Biol Sci Space* 2004;18:241–246.

11. Mothersill C, Lyng F, Seymour C, Maguire P, Lorimore S, Wright E. Genetic factors influencing bystander signaling in murine bladder epithelium after low-dose irradiation *in vivo*. *Radiat Res* 2005;163:391–399.

12. Mothersill C, Seymour CB. Cell-cell contact during gamma irradiation is not required to induce a bystander effect in normal human keratinocytes: evidence for release during irradiation of a signal controlling survival into the medium. *Radiat Res* 1998; 149:256–262.

13. Schimmerling W, Cucinotta FA. Dose and dose rate effectiveness of space radiation. *Radiat Prot Dosimetry* 2006;122:349–353.

14. Mothersill CE, Moriarty MJ, Seymour CB. Radiotherapy and the potential exploitation of bystander effects. *Int J Radiat Oncol Biol Phys* 2004;58:575–579.

15. Kellerer AM. Risk estimates for radiation-induced cancer: the epidemiological evidence. *Radiat Environ Biophys* 2000;39:17–24.

16. Sinclair WK. The linear no-threshold response: why not linearity? *Med Phys.* 1998; 25:285–300. Erratum in: *Med Phys.* 1998;25:794.

17. Little MP. Leukaemia following childhood radiation exposure in the Japanese atomic bomb survivors and in medically exposed groups. *Radiat Prot Dosimetry* 2008; 132:156–165.

18. Schöllnberger H, Mitchel RE, Redpath JL, Crawford-Brown DJ, Hofmann W. Detrimental and protective bystander effects: a model approach. *Radiat Res* 2007;168:614–626; Erratum in *Radiat Res* 2008;169:481.

19. Preston RJ. Update on linear non-threshold dose–response model and implications for diagnostic radiology procedures. *Health Phys* 2008;95:541–546.

20. Hall EJ. Radiation, the two-edged sword: cancer risks at high and low doses. *Cancer J* 2000;6:343–350.

21. Hei TK, Zhou H, Ivanov VN, et al. Mechanism of radiation-induced bystander effects: a unifying model. *J Pharm Pharmacol* 2008;60:943–950.

22. Kovalchuk O, Baulch JE. Epigenetic changes and nontargeted radiation effects: is there a link? *Environ Mol Mutagen* 2008;49:16–25.

23. Wright EG. Microenvironmental and genetic factors in haemopoietic radiation responses. *Int J Radiat Biol* 2007;83:813–818.

24. Mothersill C, Seymour C. Radiation-induced bystander and other non-targeted effects: novel intervention points in cancer therapy? *Curr Cancer Drug Targets* 2006;6:447–454.

25. Parsons WB Jr, Watkins CH, Pease GL, Childs DS Jr. Changes in sternal marrow following roentgen-ray therapy to the spleen in chronic granulocytic leukemia. *Cancer* 1954; 7:179–189.

26. Souto J. Tumour development in the rat induced by blood of irradiated animals. *Nature* 1962;195:1317–1318.

27. Mothersill C, Seymour C. Radiation-induced bystander effects: past history and future directions. *Radiat Res.* 2001;155:759–767.

28. Seymour CB, Mothersill C, Alper T. High yields of lethal mutations in somatic mammalian cells that survive ionizing radiation. *Int J Radiat Biol Relat Stud Phys Chem Med* 1986;50:167–179.

29. Gorgojo L, Little JB. Expression of lethal mutations in progeny of irradiated mammalian cells. *Int J Radiat Biol* 1989;55:619–630.

30. Born R, Trott KR. Clonogenicity of the progeny of surviving cells after irradiation. *Int J Radiat Biol Relat Stud Phys Chem Med* 1988;53:319–330.

31. Mendonca MS, Kurohara W, Antoniono R, Redpath JL. Plating efficiency as a function of time postirradiation: evidence for the delayed expression of lethal mutations. *Radiat Res* 1989;119:387–393.

32. Alper T, Mothersill C, Seymour CB. Lethal mutations attributable to misrepair of Q-lesions. *Int J Radiat Biol* 1988;54:525–530.

33. Kadhim MA, Macdonald DA, Goodhead DT, Lorimore SA, Marsden SJ, Wright EG. Transmission of chromosomal instability after plutonium alpha-particle irradiation. *Nature* 1992;355:738–740.

34. Kadhim MA, Lorimore SA, Hepburn MD, Goodhead DT, Buckle VJ, Wright EG. Alpha-particle-induced chromosomal instability in human bone marrow cells. *Lancet* 1994; 344:987–988.

35. Nagasawa H, Little JB. Induction of sister chromatid exchanges by extremely low doses of alpha-particles. *Cancer Res* 1992;52:6394–6396.

36. Clutton SM, Townsend KM, Walker C, Ansell JD, Wright EG. Radiation-induced genomic instability and persisting oxidative stress in primary bone marrow cultures. *Carcinogenesis* 1996;17:1633–1639.

37. Limoli CL, Giedzinski E. Induction of chromosomal instability by chronic oxidative stress. *Neoplasia* 2003;5:339–346.

38. Murphy JE, Nugent S, Seymour C, Mothersill C. Mitochondrial DNA point mutations and a novel deletion induced by direct low-LET radiation and by medium from irradiated cells. *Mutat Res* 2005;585:127–136.

39. Tartier L, Gilchrist S, Burdak-Rothkamm S, Folkard M, Prise KM. Cytoplasmic irradiation induces mitochondrial-dependent 53BP1 protein relocalization in irradiated and bystander cells. *Cancer Res* 2007;67:5872–5879.

40. Chen S, Zhao Y, Han W, et al. Mitochondria-dependent signalling pathway are involved in the early process of radiation-induced bystander effects. *Br J Cancer* 2008; 98:1839–1844.

41. Mothersill C, Seymour C. Medium from irradiated human epithelial cells but not human fibroblasts reduces the clonogenic survival of unirradiated cells. *Int J Radiat Biol* 1997; 71:421–427.

42. Ponnaiya B, Cornforth MN, Ullrich RL. Radiation-induced chromosomal instability in BALB/c and C57BL/6 mice: the difference is as clear as black and white. *Radiat Res* 1997; 147:121–125.

43. Watson GE, Lorimore SA, Macdonald DA, Wright EG. Chromosomal instability in unirradiated cells induced *in vivo* by a bystander effect of ionizing radiation. *Cancer Res* 2000;60:5608–5611.

44. Hei TK, Persaud R, Zhou H, Suzuki M. Genotoxicity in the eyes of bystander cells. *Mutat Res* 2004;568:111–120.

45. Weber TJ, Siegel RW, Markillie LM, Chrisler WB, Lei XC, Colburn NH. A paracrine signal mediates the cell transformation response to low dose gamma radiation in JB6 cells. *Mol Carcinog* 2005;43:31–37.

46. Seymour CB, Mothersill C. Delayed expression of lethal mutations and genomic instability in the progeny of human epithelial cells that survived in a bystander-killing environment. *Radiat Oncol Investig* 1997;5:106–110.

47. Lorimore SA, Kadhim MA, Pocock DA, et al. Chromosomal instability in the descendants of unirradiated surviving cells after alpha-particle irradiation. *Proc Natl Acad Sci USA* 1998;95:5730–5733.

48. Surinov BP, Isaeva VG, Tokarev OIu. Allelopathic activity of volatile secretions in irradiated animals. *Radiats Biol Radioecol* 2001;41:645–649.

49. Mothersill C, Bucking C, Smith RW, et al. Communication of radiation-induced stress or bystander signals between fish *in vivo*. *Environ Sci Technol* 2006;40:6859–6864.

50. Smith RW, Wang J, Bucking CP, Mothersill CE, Seymour CB. Evidence for a protective response by the gill proteome of rainbow trout exposed to X-ray induced bystander signals. *Proteomics* 2007;7:4171–4180.

51. Liu Z, Mothersill CE, McNeill FE, et al. A dose threshold for a medium transfer bystander effect for a human skin cell line. *Radiat Res* 2006;166:19–23.

52. Schettino G, Folkard M, Michael BD, Prise KM. Low-dose binary behavior of bystander cell killing after microbeam irradiation of a single cell with focused c(k) X rays. *Radiat Res* 2005;163:332–336.

53. Nakamori T, Fujimori A, Kinoshita K, Ban-Nai T, Kubota Y, Yoshida S. Application of HiCEP to screening of radiation stress-responsive genes in the soil microarthropod *Folsomia candida* (Collembola). *Environ Sci Technol* 2008;42:6997–7002.

54. U.K. CHERRIE Report. www.bandepleteduranium.org/en/a/59.html.

55. U.K. CHERRIE Minority Report. www.llrc.org/wobblyscience/subtopic/cerrie.htm.

56. Report of the French Academy of Science on the problems associated with low and non-uniform doses of ionizing radiation. www.iop.org/EJ/abstract/0952-4746/18/4/002.

57. Wouters BG. Proteomics: methodologies and applications in oncology. *Semin Radiat Oncol* 2008;18:115–125.

58. Kassis AI. *In vivo* validation of the bystander effect. *Hum Exp Toxicol* 2004;23:71.

59. Bodei L, Kassis AI, Adelstein SJ, Mariani G. Radionuclide therapy with iodine-125 and other Auger-electron-emitting radionuclides: experimental models and clinical applications. *Cancer Biother Radiopharm* 2003;18:861–877.

60. Marples B, Greco O, Joiner MC, Scott SD. Molecular approaches to chemo-radiotherapy. *Eur J Cancer* 2002;38:231–239.

61. Kennedy C, Dilmanian A. Prospects for microbeam radiation therapy of brain tumours in children. *Dev Med Child Neurol* 2007;49:566.

62. Dilmanian FA, Qu Y, Feinendegen LE, et al. Tissue-sparing effect of X-ray microplanar beams particularly in the CNS: is a bystander effect involved? *Exp Hematol* 2007; 35:69–77.

63. Mackonis EC, Suchowerska N, Zhang M, Ebert M, McKenzie DR, Jackson M. Cellular response to modulated radiation fields. *Phys Med Biol* 2007;52:5469–5482.

64. Mairs RJ, Fullerton NE, Zalutsky MR, Boyd M. Targeted radiotherapy: microgray doses and the bystander effect. *Dose Response* 2007;5:204–213.

65. Boyd M, Sorensen A, McCluskey AG, Mairs RJ. Radiation quality-dependent bystander effects elicited by targeted radionuclides. *J Pharm Pharmacol* 2008;60:951–958.

66. McCluskey AG, Boyd M, Ross SC, et al. [^{131}I]Meta iodobenzylguanidine and topotecan combination treatment of tumors expressing the noradrenaline transporter. *Clin Cancer Res* 2005;11:7929–7937.

67. Boyd M, Mairs RJ, Mairs SC, et al. Expression in UVW glioma cells of the noradrenaline transporter gene, driven by the telomerase RNA promoter, induces active uptake of [^{131}I] MIBG and clonogenic cell kill. *Oncogene* 2001;20:7804–7808.

68. Boyd M, Mairs SC, Stevenson K, et al. Transfectant mosaic spheroids: a new model for the evaluation of bystander effects in experimental gene therapy. *J Gene Med* 2002; 4:567–579.

69. Boyd M, Mairs RJ, Keith WN, et al. An efficient targeted radiotherapy/gene therapy strategy utilising human telomerase promoters and radioastatine and harnessing radiation-mediated bystander effects. *J Gene Med* 2004;6:937–945.

70. Boyd M, Spenning HS, Mairs RJ. Radiation and gene therapy: rays of hope for the new millennium? *Curr Gene Ther* 2003;3:319–339.

71. Hei TK, Zhou H, Ivanov VN, et al. Mechanism of radiation-induced bystander effects: a unifying model. *J Pharm Pharmacol* 2008;60:943–950.

72. Morgan WF. Is there a common mechanism underlying genomic instability, bystander effects and other nontargeted effects of exposure to ionizing radiation? *Oncogene* 2003; 22:7094–7099.

73. Chen S, Zhao Y, Han W, et al. Mitochondria-dependent signalling pathway are involved in the early process of radiation-induced bystander effects. *Br J Cancer* 2008; 98:1839–1844.

74. Lorimore SA, Wright EG. Radiation-induced genomic instability and bystander effects: related inflammatory-type responses to radiation-induced stress and injury? A review. *Int J Radiat Biol* 2003;79:15–25.

75. Lyng FM, Maguire P, McClean B, Seymour C, Mothersill C. The involvement of calcium and MAP kinase signaling pathways in the production of radiation-induced bystander effects. *Radiat Res* 2006;165:400–409.

76. Hamada N, Matsumoto H, Hara T, Kobayashi Y. Intercellular and intracellular signaling pathways mediating ionizing radiation-induced bystander effects. *J Radiat Res (Tokyo)* 2007;48:87–95.

77. Mothersill C, Cusack A, Seymour CB. Radiation-induced outgrowth inhibition in explant cultures from surgical specimens of five human organs. *Br J Radiol* 1988;61:226–230.

78. Mothersill C, O'Malley K, Seymour CB. Characterisation of a bystander effect induced in human tissue explant cultures by low let radiation. *Radiat Prot Dosimetry* 2002; 99:163–167.

79. Belyakov OV, Folkard M, Mothersill C, Prise KM, Michael BD. A proliferation-dependent bystander effect in primary porcine and human urothelial explants in response to targeted irradiation. *Br J Cancer* 2003;88:767–774.

80. Puck TT, Marcus PI. Action of X-rays on mammalian cells. *J Exp Med* 1956;103:653–666.

81. Seymour C, Mothersill C. Development of an *in vivo* assay for detection of non-targeted radiation effects. *Dose Response* 2006;4:277–282.

82. Mancuso M, Pasquali E, Leonardi S, et al. Oncogenic bystander radiation effects in patched heterozygous mouse cerebellum. *Proc Natl Acad Sci USA* 2008; 105:12445–12450.

83. Tamminga J, Koturbash I, Baker M, et al. Paternal cranial irradiation induces distant bystander DNA damage in the germline and leads to epigenetic alterations in the offspring. *Cell Cycle* 2008;7:1238–1245.

84. Kaminski JM, Shinohara E, Summers JB, Niermann KJ, Morimoto A, Brousal J. The controversial abscopal effect. *Cancer Treat Rev* 2005;31:159–172.

85. Koturbash I, Loree J, Kutanzi K, Koganow C, Pogribny I, Kovalchuk O. *In vivo* bystander effect: cranial X-irradiation leads to elevated DNA damage, altered cellular proliferation and apoptosis, and increased p53 levels in shielded spleen. *Int J Radiat Oncol Biol Phys* 2008;70:554–562.

86. Seymour CB, Mothersill C, Mooney R, Moriarty M, Tipton KF. Monoamine oxidase inhibitors L-deprenyl and clorgyline protect nonmalignant human cells from ionising radiation and chemotherapy toxicity. *Br J Cancer* 2003;89:1979–1986.

87. Poon RC, Agnihotri N, Seymour C, Mothersill C. Bystander effects of ionizing radiation can be modulated by signaling amines. *Environ Res* 2007;105:200–211.

88. Shao C, Folkard M, Prise KM. Role of TGF-beta1 and nitric oxide in the bystander response of irradiated glioma cells. *Oncogene* 2008;27:434–440.

CHAPTER 15

The Role of Molecular Imaging in Evaluating Tumor Response to Targeted Radiotherapy

NORBERT AVRIL

15.1 INTRODUCTION

Molecular imaging combines molecular biology and *in vivo* imaging and enables the visualization of cellular processes and functions. It is different from traditional anatomical imaging as specific biomarkers are used for particular targets or defined cellular pathways. Using radiolabelled ligands to noninvasively derive functional information *in vivo* has been established in nuclear medicine over several decades. Positron emission tomography (PET) has become one of the most important molecular imaging modalities in clinical oncology. This success is primarily based on the ability of PET to measure and to visualize the increased glucose metabolism of cancer tissue by using the radiolabeled glucose analogue ^{18}F-fluorodeoxyglucose (FDG). Numerous studies have shown that FDG-PET is useful in a variety of clinical situations (1). These include the characterization of suspicious lesions, such as pulmonary nodules, for example, distinguishing between radiation necrosis and tumor recurrence, and determining the extent and resectability of recurrent tumor. Staging of cancer patients has evolved as the most important clinical application by providing an accurate assessment of the localization, extent and spread of disease. FDG-PET can assess tissue glucose utilization with high reproducibility. Following therapy, the decrease of tumor glucose utilization correlates with the reduction of viable tumor cells. Thus, FDG-PET allows for assessment of therapy response by determining the viability of residual masses after completion of treatment.

Monoclonal Antibody and Peptide-Targeted Radiotherapy of Cancer, Edited by Raymond M. Reilly
Copyright © 2010 John Wiley & Sons, Inc.

15.2   POSITRON EMISSION TOMOGRAPHY

15.2.1   Background and Basic Principles

Positron emission tomography is a functional imaging technology that enables the visualization, characterization, and quantification of biological processes *in vivo*. By using positron-emitting radiotracers, PET provides unique information about the molecular and metabolic changes associated with disease. The technology was developed more than 50 years ago and after its use as a research tool, primarily in neurology and cardiology, it has gained clinical acceptance over the last 10–15 years, particularly in oncology. PET measures the distribution of positron-emitting radio-tracers in the body and converts the data into cross-sectional images. Dedicated PET scanners have been developed and refined over the past decade, which now enable rapid, reliable, and reproducible imaging in humans (Fig. 15.1). These devices generally comprise an array of scintillation crystals forming a ring around the patient. The crystals convert the two 511 keV gamma rays created when the positrons emitted by PET radionuclides are annihilated by interaction with electrons in tissues, into light, and attached photomultipliers subsequently produce electrical signals, which are further processed in a proportional manner. PET is highly sensitive, with the capacity to detect subnanomolar concentrations of radiotracer and provides superior image resolution to conventional gamma camera imaging (i.e., single photon emission computerized tomography (SPECT)).

The first application of positron emitters in humans took place in 1951 by employing a simple probe system with coincidence detectors to localize tumors in the brain (2). Almost 25 years later, the first images were acquired using a ring tomograph for studying oxygen metabolism with ^{15}O-oxygen, glucose metabolism

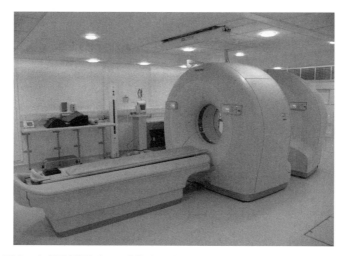

FIGURE 15.1   A GEMINI time-of-flight PET/CT. The PET/CT scanner consists of two devices; the first is a 64-slice CT and then the scanner table with the patient is moved to the second device, a time-of-flight PET.

with ^{11}C-glucose, bone metabolism with ^{18}F-fluoride as well as blood pool and perfusion of the heart and brain with ^{13}NH$_3$ and ^{11}CO-hemoglobin (carboxyhemoglobin) (3, 4). Significant advances in PET technology were the introduction of bismuth–germanium–oxide (BGO) as a scintillator material for PET scanners (5) and the successful synthesis of ^{18}F-2-fluoro-2-deoxyglucose (FDG) by Wolf and Fowler in 1978 (6).

The accomplishments of PET in oncology can be primarily attributed to the fact that depending on the radiopharmaceutical used, PET has the capability and potential to visualize various biological processes of tumors, such as glucose metabolism, cell proliferation, receptor expression, angiogenesis, and hypoxia. The amounts of PET radiotracers administered are extremely small, generally in the pico- and nanomolar range and have essentially no pharmacologic effect.

15.2.2 PET Radiotracers

Glucose metabolism is often upregulated in malignant tumors resulting in increased cellular uptake of the glucose analogue FDG. The uptake mechanism and biochemical pathway of FDG have been extensively studied *in vitro* and *in vivo* (Fig. 15.2). Active transport of the radiotracer through the cell membrane via glucose transport proteins (GLUT) and subsequent intracellular phosphorylation by hexokinase (HK) are the key steps for cellular accumulation (7). As FDG-6-phosphate is a poor substrate for glucose-6-phosphate isomerase, and levels of glucose-6-phosphatase are generally low in tumors, FDG-6-phosphate accumulates within cells proportional to the level of exogenous glucose consumption and can be visualized by PET.

^{18}F-fluoro-3'-deoxy-3'-L-fluorothymidine (FLT) (Fig. 15.3) has been developed as a cell proliferation tracer for PET (8). Cell proliferation as a biological target is particularly attractive in cancer imaging. Knowledge of tumor cell proliferation could be used in the evaluation of tumor growth and may provide a prognostic indicator of

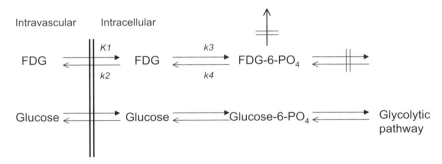

FIGURE 15.2 Tumor uptake mechanism of FDG. FDG crosses the cell membrane via glucose transport proteins (GLUT) followed by intracellular phosphorylation to FDG-6-phosphate. As FDG-6-phosphate is a poor substrate for glucose-6-phosphate isomerase, and levels of glucose-6-phosphatase are generally low in tumors, FDG-6-phosphate accumulates within cells proportional to the level of exogenous glucose consumption. Glucose 6-phosphate continues in the glycolytic pathway.

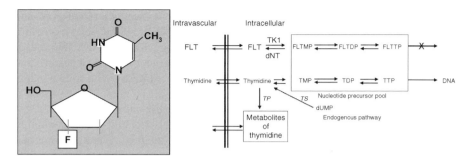

FIGURE 15.3 Chemical structure of FLT. FLT is phosphorylated by TK1 to form FLT-monophosphate. FLT-monophosphate may continue to be phosphorylated into FLT-diphosphate and FLT-triphosphate; however, it is not incorporated into DNA and is trapped in the cytosol.

tumor aggressiveness. Recent developments in the introduction of anticancer drugs that inhibit cell proliferation could possibly be monitored specifically by noninvasive imaging of proliferation. Cell proliferation is required for tumor growth and thymidine is utilized by proliferating cells for DNA replication. Therefore, thymidine and its analogues have been radiolabeled to prove cellular proliferation rates by PET (8). The cellular uptake of FLT is mainly not only via equilibrative nucleoside transporters but also via Na^+-dependent active nucleoside transporters. The concentrative transporters are mostly located in normal tissues while the equilibrative transporters have been found in tumor cells (9).

FLT is phosphorylated by thymidine kinase 1 (TK1), a principal enzyme in the salvage pathway of DNA synthesis, to form FLT-monophosphate (10). FLT is a selective substrate for TK1; in contrast, thymidine also reacts with TK2, the unregulated isozyme, which is used for mitochondrial DNA replication and repair. In quiescent cells, TK1 activity is virtually absent but in proliferating and malignant cells it is increased particularly in the S-phase of the cell cycle. FLT-monophosphate may continue to be phosphorylated into FLT-diphosphate and FLT-triphosphate. It is at this point in the pathway to DNA synthesis that the FLT phosphorylation process differs from thymidine. Due to the 3'-fluorine substitution, FLT-triphosphate is not incorporated into DNA and is trapped in the cytosol (10). The rate-limiting step for cellular FLT retention is the initial phosphorylation by TK1. The phosphorylated FLT can be dephosphorylated by 5'-deoxynucleotidase but this occurs at a slow rate relative to TK1 activity (8). Since TK1 is a control point in the salvage pathway of DNA synthesis, assessing the uptake and retention of FLT through TK1 enzyme activity provides a direct assessment of cellular proliferation. The protein Ki-67 identified by MIB-1 antibody staining is a histopathological measure of proliferation that can be correlated to FLT uptake measured by PET (11, 12).

15.2.3 PET and PET/CT Imaging of Biologic Features of Cancer

15.2.3.1 *Glucose Consumption* Imaging the metabolic activity of tumors provides both more sensitive and more specific information about the extent of disease compared with anatomical (e.g., CT or MRI) imaging alone (13). FDG-PET has become a standard imaging procedure for staging many types of cancer. An important limitation of FDG-PET alone, however, is the precise localization of abnormalities due to the lack of reliable anatomical landmarks and the limited spatial resolution of current PET tomographs (4–5 mm). Clinical interpretation of metabolic FDG-PET imaging is particularly challenging in the neck, abdomen and pelvis due to variable physiologic FDG uptake in lymphatic, bowel, and muscle tissue as well as by renal excretion of the radiotracer, which can confound image interpretation. Combined positron emission tomography and computed tomography (PET/CT) is a new imaging technology, which acquires PET and CT images that are concurrent and coregistered, merging the functional information from PET with the anatomical information from CT (14). PET/CT is unique because it provides tissue characterization as well as assessment of the exact location and the extent of tumor tissue. The use of FDG-PET/CT has been shown to improve diagnostic accuracy compared to either imaging procedure alone by localizing areas of increased FDG uptake with improved anatomic specificity and by providing better characterization of suspicious morphological abnormalities (15).

15.2.3.2 FDG-PET Procedures and Image Analysis

To ensure a standardized metabolic state, especially low plasma glucose and insulin levels, it is necessary that patients for oncologic PET imaging have fasted for at least 4 to 6 h prior to administration of FDG. The blood glucose level should be tested prior to tracer injection and should not exceed 8.5 mmol/L (150 mg/100 mL) (16). It is important to note that most published studies have excluded diabetic patients and the diagnostic performance of FDG-PET is generally lower in patients with elevated blood glucose levels. Intravenous administration of about 300–400 MBq ($\sim$10 mCi) of ^{18}F-FDG is used in most centers, although some inject more than 800 MBq ($\sim$20 mCi) particularly in larger patients. Most FDG is taken up by tissues within 1 h after injection and PET data acquisition is initiated after 60–90 min. Some studies found increasing target-to-background ratios over time suggesting benefits to longer waiting periods between radiotracer injection and data acquisition (17). However, lower image quality, due to radionuclide decay (^{18}F has a half-life of 2 h), has to be taken into account.

Correction for photon attenuation is required for quantification of FDG uptake; this can be determined by measuring tissue attenuation using longer-lived radioactive positron rod sources (e.g., germanium-68) that are rotated around the patient or by coregistered CT data (automatically adjusted for the differences in attenuation of the energies of the low energy X-ray CT beam and that of the higher energy 511 keV gamma photons). Calculating standardized uptake values (SUVs) by normalization of tissue FDG uptake to the dose of injected radioactivity and body weight is the most common method for tumor uptake quantification (18). Dynamic data acquisition allows calculation of the radiotracer influx constant, although, this procedure is more complex and has not been shown to significantly increase the diagnostic accuracy of PET imaging. Quantification of tumor radiotracer uptake allows for comparison between PET studies performed in the same patient at different times but also among different patients. In the clinical setting, the maximum SUV within the region of interest (ROI), which represents the highest radioactivity concentration in one voxel within the tumor, is often used. Previous studies have demonstrated that SUVs provide highly reproducible parameters of tumor glucose utilization in patients studied twice without any cancer treatment in between (19, 20). Although SUVs have been shown to be important for monitoring treatment effects, visual image analysis is generally sufficient for staging and restaging purposes.

Visual PET interpretation should include analysis of transaxial, coronal, and sagittal views. Cancer typically presents with focally increased FDG uptake, whereas most benign tumors are negative in FDG-PET. Normal increased FDG uptake or FDG retention is found within the brain cortex, the myocardium, and the urinary tract. Low to moderate normal uptake is often seen in the base of the tongue, the salivary glands, thyroid, liver, spleen, gastrointestinal tract, bone marrow, musculature, and reproductive organs. There are inherent limitations of metabolic FDG-PET imaging that can result in false-negative as well as false-positive findings. The most common normal cause of misinterpretation is related to muscle activity. Muscle tension may lead to increased FDG uptake and physical activity immediately before or after tracer injection can lead to spurious muscle activity. False-positive findings are most commonly associated with uptake of FDG in infectious or inflammatory tissue. In different situations, FDG-PET can actually be helpful in imaging infections by visualizing the increased metabolic activity of activated granulocytes and mononuclear cells that accumulate the radiotracer. Another cause of false-positive results is related to cancer treatment. FDG uptake can be seen in tissue after radiation therapy and it is therefore recommended to wait at least 8 weeks after external beam radiation to evaluate the irradiated area for residual disease (21). For chemotherapy, a waiting period of at least 4 weeks after the last cycle is recommended to avoid false-negative PET results due to metabolic stunning of potential residual tumor tissue (22). In addition, the bone marrow frequently exhibits increased metabolic activity following chemotherapy, which makes it more difficult to identify bone metastases. False-positive findings can occur in benign conditions such as Paget's and Graves disease, thyroid, adrenal and villous adenomas, healing fractures, and a few benign tumors. Other pitfalls include increased FDG uptake in normal ovaries, for example, during ovulation as well as normal physiologic activity in bowel, endometrium, and blood vessels and focal retained activity in ureters, bladder diverticula, and pelvic kidneys (23). Whole-body imaging can be improved by intravenous injection of furosemide (20–40 mg) to reduce radiotracer retention in the urinary system and by administration of n-butyl-scopolamine (20–40 mg) to reduce FDG uptake in the bowel (24). Weaknesses of FDG-PET for cancer imaging include its limited spatial resolution of 4–5 mm currently achievable in commercial systems as previously mentioned. Thus, it is important to note that a negative FDG-PET scan cannot exclude the presence of small tumor deposits or microscopic tissue involvement.

15.2.3.3 Cellular Proliferation
^{18}F-FLT uptake has been shown in feasibility studies in a variety of tumors including breast cancer, gliomas, lymphoma, esophageal cancer, lung cancer, and sarcomas (25–29). Physiological uptake of FLT occurs in the bone marrow and liver, which may hamper assessment of liver and bone metastases.

There is generally low background radioactivity distributed throughout the body, particularly in fat, skeletal muscle and myocardium. FLT is excreted via the kidneys and the urinary tract and a significant portion of the injected dose is found in the urinary bladder within the first hour after radiotracer injection. The uptake of FLT in the brain is very low since FLT does not cross the blood brain barrier which suggests that FLT is a promising radiotracer for imaging brain tumors by providing a low background signal (in contrast to FDG that is normally accumulated in the brain).

In general, FLT tumor uptake tends to be lower compared to FDG that limits its sensitivity for cancer staging. The low FLT uptake may be due to thymidine competing with FLT for the active site of the trapping enzyme TK1 and a higher affinity of TK1 for thymidine leading to its preferential phosphorylation. In addition, a study utilizing compartmental pharmacokinetic analysis to assess FLT uptake in nonsmall cell lung cancer patients demonstrated that the dephosphorylation step (k4) is significant (30). This is important as it may lead to errors in response estimation if FLT uptake is evaluated using semiquantitative measures such as SUV depending on the time interval between radiotracer injection and PET imaging. Nonetheless, several groups have reported that FLT uptake reflects tumor proliferation as defined by correlative histopathologic evaluation of Ki-67 expression in nonsmall cell lung cancer, colorectal cancer, lymphoma and gliomas. Buck and coworkers in a prospective study of 26 patients with solitary pulmonary nodules demonstrated FLT to be specific for malignancy and to correlate with Ki-67 (11). They suggested that FLT-PET may be able to differentiate between benign and malignant pulmonary nodules; however, there were three false-negative cases in slowly proliferating tumors.

A prospective study in 34 lymphoma patients demonstrated that FLT uptake was significantly higher in aggressive compared to indolent lymphomas (31). This led to a suggestion that the progression of indolent to aggressive lymphoma might be assessed by FLT-PET. Of note, no correlation between FLT uptake and cell proliferation measured by Ki-67 was identified in breast cancer and esophageal cancer. The reason for a lack of correlation in some tumors is unclear but may be related to differences in tumor growth patterns and biological heterogeneity between different tumor types.

15.3 RESPONSE TO CANCER TREATMENT INCLUDING TARGETED RADIOTHERAPY

15.3.1 Conventional Methods Used to Evaluate Treatment Response

Patient cure is the ultimate goal of cancer treatment. However, with solid tumors this is generally only possible at an early stage, mainly by surgical resection followed, if applicable, by adjuvant chemotherapy and/or radiotherapy. In some tumors, primary radiotherapy or chemoradiotherapy has also been shown to be effective. In the metastatic setting, cure is generally not possible. However, there is an array of cancer treatment options available, which result in various degrees of tumor response documented by the disappearance of macroscopic tumor or significant shrinkage of the tumor burden. Nevertheless, in solid tumors, there is a very wide range of response

(10–80%) to standard cancer treatments depending on the type of tumor, stage of disease, and patient-specific factors, and therefore, a significant number of patients undergo toxic therapy without benefit. The current end points for assessing response to therapy in solid tumors should be disease-free and overall survival. The most frequently used surrogate end point for these effects is change in tumor size (32). Anatomical imaging modalities, predominantly computed tomography (CT), magnetic resonance imaging (MRI), and ultrasound (US) are used to obtain unidimensional or bidimensional measurements of reference tumor lesions from pretreatment scans relative to follow-up. Several criteria to define tumor response have been developed. Based on the WHO-criteria, a tumor is classified as responding when the product of two perpendicular diameters of a mass has decreased by at least 50%. In cases with multiple lesions, the summation of the products should decrease by more than 50% (33). The Response Evaluation Criteria in Solid Tumors (RECISTs) define tumor response as a decrease of the maximum tumor diameter by at least 30% (32). Despite wide acceptance of these criteria, it is important to note that a 50% or 30% decrease of tumor size is a more or less arbitrary convention, which is not based on outcome studies. Therefore, tumor response as defined by morphological imaging techniques may be less relevant as a surrogate end point in the evaluation of new anticancer drugs, particularly those that interfere with biological processes and may be cytostatic (34). Furthermore, full or partial resolution of a tumor mass is the final step in a complex cascade of cellular and sub-cellular changes in response to treatment. Frequently, several cycles of treatment need to be given before treatment response can be reliably assessed by current anatomical imaging modalities. Even the evaluation of response after completion of treatment can pose a challenge. If a residual mass is present, it is difficult to differentiate viable tumor tissue from posttreatment changes such as scarring and fibrosis.

15.3.2 Molecular Imaging for Monitoring Treatment Response

15.3.2.1 *Assessment of Treatment Response to Conventional Therapies*   Restaging after completion of a course of treatment is essential to verify response and determine the need for subsequent additional therapy. Conventional anatomical imaging modalities frequently reveal residual masses where cancer was present and it is often difficult to determine whether this represents viable tumor or fibrotic scar tissue. Even biopsies can be misleading, because residual masses may contain a mixture of viable tumor cells and scar tissue, which can lead to false-negative results. The ability to accurately monitor response to treatment is crucial in order to select patients who need more intensive or salvage treatment. Many studies have shown that the metabolic information from FDG-PET is important for assessment of treatment response.

Several studies in Hodgkin's or high-grade non-Hodgkin's lymphoma have shown that if both, CT and FDG-PET are negative at the end of treatment, patients have a low likelihood of relapse. In contrast, patients with residual masses where FDG uptake is increased have a high likelihood to relapse. In an early study of 44 patients with Hodgkin's disease or aggressive non-Hodgkin's lymphoma, the 2-year relapse-free survival rate was 95% for those with negative FDG-PET compared with 0% for FDG-PET positive patients (35). In another study of 54 lymphoma patients, a positive FDG-PET scan was highly predictive of residual disease and predicted early disease progression (36). Others have reported similar results (37–39). In 93 patients with non-Hodgkin's lymphoma, persistent abnormal FDG uptake was observed in 26 patients, all of whom relapsed with a median progression-free survival of only 73 days compared to a median progression-free survival of 404 days for patients with negative FDG-PET scans (40). In a recent study, 90 patients with newly diagnosed aggressive and mainly diffuse large B-cell lymphoma underwent FDG-PET before and after chemotherapy (41). After completion of treatment, 83% of FDG-PET negative patients achieved complete remission compared with only 58% of FDG-PET positive patients. Outcome differed significantly between FDG-PET negative and FDG-PET positive groups; the 2-year estimates of disease-free survival were 82% and 43%, respectively ($p < 0.001$), and the 2-year estimates of overall survival were 90% and 61%, respectively ($p = 0.006$).

A systematic literature review of FDG-PET in evaluating first-line therapy in Hodgkin's disease and (aggressive) non-Hodgkin's lymphoma included 15 studies and a total of 705 patients (42). Pooled sensitivity and specificity for detecting residual disease in Hodgkin's lymphoma were 84% and 90%, respectively. For non-Hodgkin's lymphoma, sensitivity and specificity were 72% and 100%, respectively. In 54 patients with aggressive non-Hodgkin's lymphoma, integration of FDG-PET into the International Workshop Criteria (IWC) provided a more accurate response assessment than IWC alone (43). Only the IWC combined with FDG-PET were a statistically significant independent predictor for progression-free survival. There is now sufficient evidence in the literature that persistent increased FDG uptake in residual masses after treatment of lymphoma is highly predictive for residual disease.

Comparable results were observed in solid tumors. In patients with esophageal and gastric cancer, residual FDG-uptake after completion of chemoradiotherapy was a specific marker for viable residual tumor tissue and was associated with a poor prognosis (44). In 27 patients with locally advanced esophageal squamous cell carcinoma, therapy induced reduction of tumor FDG uptake was 72% for histopathologic responders compared to 42% for nonresponders. Other studies found similar results, although differing criteria were used to define metabolic response (45). These included patients with osteosarcoma or Ewing sarcoma (46, 47), rectal cancer (48, 49) and germ cell tumors (50, 51). In 73 lung cancer patients, prospectively evaluated for response to chemoradiotherapy by CT and FDG-PET, metabolic FDG-PET responses predicted survival more accurately than CT responses (52).

There have been some concerns regarding the diagnostic value of FDG-PET due to inflammatory changes after external radiotherapy. A recent study evaluated FDG-PET/CT in detecting residual disease after definitive radiochemotherapy in 28 patients with head and neck cancer (21). Regarding the detection of residual disease, the overall accuracy of FDG-PET/CT was 85.7%, compared with 67.9% for CT alone. All three false-negative and one false-positive FDG-PET/CT results occurred between 4 and 8 weeks after treatment. At 8 weeks or later after treatment, the specificity of CT was 28%, compared with 100% for FDG-PET/CT. The authors concluded that the metabolic–anatomic information from FDG-PET/CT provided the most accurate assessment for treatment response when performed later than 8 weeks after the conclusion of radiation therapy.

The clinical data on the assessment of treatment response with FLT-PET to date is limited. Several studies have reported an early reduction in cellular ^{18}F-FLT uptake after treatment with external radiotherapy (53, 54) or chemotherapy (55–57). Several groups have demonstrated a significant correlation between the reduction in tumor FLT accumulation posttherapy and histopathological expression of proliferating cell nuclear antigen (PCNA) a nuclear polypeptide marker of cell proliferation, thus implying the reduction in tumor FLT uptake is due to a reduction in cell proliferation. Wieder and coworkers assessed response to adjuvant chemoradiotherapy in 10 patients with locally advanced rectal cancer (51). There was a significant reduction in tumor FLT uptake 4–6 weeks postchemoradiotherapy in both responders and nonresponders. They concluded FLT-PET does not seem to be a promising method for assessment of tumor response to neoadjuvant chemoradiotherapy in rectal cancer.

15.3.2.2 *Monitoring Response to Targeted Radiotherapy*   Radioimmunotherapy (RIT) and peptide-directed radiotherapy (PDRT) describe an approach in which radiation from radionuclides is delivered more selectively to tumor cells by using antibodies or peptides recognizing tumor-associated antigens or receptors. RIT has been used with success clinically in lymphoma patients employing nonmyeloablative doses of ^{131}I-labeled murine anti-CD20 tositumomab antibodies (Bexxar®, GlaxoSmithKline, Philadelphia, PA, USA) (see Chapter 7), which is FDA approved in the United States. Alternatively, ^{90}Y-labeled ibritumomab tiuxetan (Zevalin®, Biogen IDEC Pharmaceuticals, San Diego, CA, USA) (see Chapter 7) antibodies can be administered, which is approved in Europe and in the United States for treatment of patients with relapsed or refractory follicular lymphoma. Zevalin consists of the anti-CD20 monoclonal antibody ibritumomab covalently linked to the chelator, tiuxetan for complexing ^{90}Y. RIT combines the benefits of both external radiotherapy and biologically targeted immunotherapy, enabling multiple sites of disseminated disease to be treated simultaneously and effectively, with the goal of minimizing toxicity to normal tissues.

To date, there is only very limited data published regarding the use of FDG-PET to monitor response to RIT (Figs. 15.4 and 15.5). A small pilot trial included five men and five women with relapsed or refractory non-Hodgkin's lymphoma who underwent FDG-PET/CT 14–27 days before treatment with Zevalin and 4–6 months after treatment (58). Response after treatment was measured with CT imaging defined as complete response (CR), partial response (PR), stable disease (SD), or progressive disease (PD), according to published criteria from a National Cancer Institute-sponsored international workshop. Response after treatment was similarly measured with FDG-PET as CR, PR, SD or PD, as defined according to published criteria of the European Organization for Research and Treatment of Cancer (EORTC). Interpretation of CT images alone resulted in classification of eight of ten patients as responders, with two patients classified as having a CR. Analyzing fused FDG-PET/CT images, two patients had residual lesions at CT that did not show increased FDG uptake. These

(a)

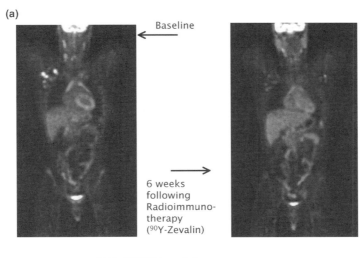

Baseline

6 weeks
following
Radioimmuno-
therapy
(^{90}Y-Zevalin)

(b)   **FDG-PET/CT partial response**

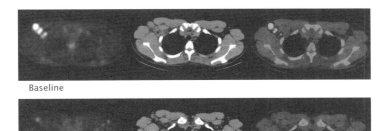

Baseline

6 weeks following radioimmunotherapy (^{90}Y-Zevalin)

FIGURE 15.4   (a) Coronal FDG-PET images of a 58-year-old male patient presenting with increased FDG uptake in right axillary lymph nodes as well as right neck lymph nodes at baseline but much reduced uptake at 6 weeks following radioimmunotherapy with ^{90}Y-Zevalin. (See insert for color representation of the figure.) (b) Axial FDG-PET, CT, and fused FDG-PET/CT images of lymphoma involved axillary lymph nodes in this patient. There is still mild residual metabolic activity noted consistent with a partial response. (See insert for color representation of the figure.)

two patients, classified as PR according to CT criteria alone, were classified as CR by FDG-PET/CT. Both of these patients were free of evident disease at 18 or more months of follow-up and therefore the FDG-PET/CT results were true negative.

In a recent study, 27 patients with relapsed non-Hodgkin Lymphoma underwent FDG-PET after fractionated RIT using anti-CD22 ^{90}Y-labeled epratuzumab antibodies (see Chapter 7) (59). Patients received one or two courses of fractionated RIT. Each course consisted of two or three infusions of 92.5–370 MBq/m^2

FDG-PET/CT complete response

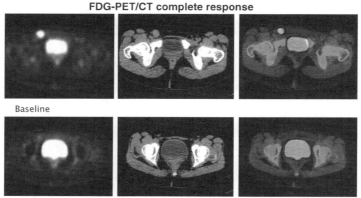

FIGURE 15.5 Patient with lymphoma demonstrating increased FDG uptake in a right inguinal lymph node at baseline that completely resolved 8 weeks after radioimmunotherapy with ^{90}Y-Zevalin. Axial FDG-PET, CT, and fused FDG-PET/CT. (See insert for color representation of the figure.)

(2.5–10 mCi/m^2) of ^{90}Y-epratuzumab, resulting in a cumulative administered activity of 185–1110 MBq/m^2 (5–30 mCi/m^2). The radiolabeled antibody infusions were administered one week apart. FDG-PET was performed at baseline and at 6 weeks after each course of RIT. Treatment response was compared with conventional imaging (CT and MRI) using the International Workshop Response Criteria; CR, unconfirmed CR, PR, SD, or PD. FDG-PET was classified as CR, PR, or PD with histology and follow-up as the reference standard. Sensitivity, specificity, positive predictive value, negative predictive value, and accuracy of FDG-PET 6 weeks after RIT were 86%, 62%, 80%, 71%, and 77% respectively ($p < 0.01$), compared with 36%, 87%, 83%, 44%, and 55% respectively, for conventional imaging. Positive FDG uptake predicted earlier relapse compared to negative FDG-PET results. The mean time-to-progression was 15.6 months when FDG-PET was negative compared to 5.4 months when FDG-PET was positive ($p = 0.008$). There are some theoretical concerns regarding the influence of inflammatory changes following RIT, which could result in false-positive FDG-PET results as mentioned earlier. Nevertheless, in more than 90% of cases, metabolic response could be accurately defined as soon as six weeks after treatment. The results of this pilot study indicate that responders to RIT may be identified with FDG-PET more accurately than with CT. In addition, a benefit was found using FDG-PET to detect residual disease as soon as six weeks post-RIT.

Although the results of FDG-PET for assessment of treatment response to RIT are promising, they rely on small series and there is currently no generally accepted consensus regarding its use. This includes the optimal timing of FDG-PET or PET/CT after treatment as well as defined criteria for PET image analysis. The most important benefit of metabolic imaging to date is likely the identification of treatment failure by persistent markedly increased FDG uptake in lymphoma. In cases with a significant reduction in FDG uptake between baseline and posttreatment FDG-PET, careful and close follow-up is recommended as residual viable tumor cannot be excluded.

Finally, it is important to note that as a result of the increasing evidence regarding the use of FDG-PET in response assessment, revised response criteria developed by an International Harmonization Project (IHP) have just been published together with guidelines for performing and interpreting FDG-PET imaging (22, 60). These guidelines support the use of FDG-PET for end of therapy response assessment in diffuse large B-cell lymphoma (DLBCL) and Hodgkin's lymphoma. However, they state that the use of FDG-PET for response assessment of "aggressive" NHL subtypes other than DLBCL and "indolent" lymphomas is less clear, and FDG-PET should be used only if overall response rate and complete remission rate are important end points of the study.

REFERENCES

1. Fletcher JW, Djulbegovic B, Soares HP, et al. Recommendations on the use of ^{18}F-FDG PET in oncology. *J Nucl Med* 2008;49:480–508.

2. Sweet WH. The uses of nuclear disintegration in the diagnosis and treatment of brain tumor. *N Engl J Med* 1951;245:875–878.

3. Phelps ME, Hoffman EJ, Mullani NA, Ter-Pogossian MM. Application of annihilation coincidence detection to transaxial reconstruction tomography. *J Nucl Med* 1975;16: 210–224.

4. Phelps ME, Hoffman EJ, Coleman RE, et al. Tomographic images of blood pool and perfusion in brain and heart. *J Nucl Med* 1976;17:603–612.

5. Cho ZH, Farukhi MR. Bismuth germanate as a potential scintillation detector in positron cameras. *J Nucl Med* 1977;18:840–844.

6. Fowler JS, Ido T. Initial and subsequent approach for the synthesis of [18]FDG. *Semin Nucl Med* 2002;32:6–12.

7. Avril N. GLUT1 expression in tissue and ^{18}F-FDG uptake. *J Nucl Med* 2004;45:930–932.

8. Bading JR, Shields AF. Imaging of cell proliferation: status and prospects. *J Nucl Med* 2008;49(Suppl 2):64S–80S.

9. Belt JA, Marina NM, Phelps DA, Crawford CR. Nucleoside transport in normal and neoplastic cells. *Adv Enzyme Regul* 1993;33:235–252.

10. Shields AF, Lawhorn-Crews JM, Briston DA, et al. Analysis and reproducibility of 3'-deoxy-3'-^{18}F]fluorothymidine positron emission tomography imaging in patients with non-small cell lung cancer. *Clin Cancer Res* 2008;14:4463–4468.

11. Buck AK, Schirrmeister H, Hetzel M, et al. 3-deoxy-3-^{18}F]fluorothymidine-positron emission tomography for noninvasive assessment of proliferation in pulmonary nodules. *Cancer Res* 2002;62:3331–3334.

12. Vesselle H, Grierson J, Muzi M, et al. *In vivo* validation of 3'deoxy-3'-^{18}F]fluorothymidine (^{18}F]FLT) as a proliferation imaging tracer in humans: correlation of ^{18}F]FLT uptake by positron emission tomography with Ki-67 immunohistochemistry and flow cytometry in human lung tumors. *Clin Cancer Res* 2002;8:3315–3323.

13. Rohren EM, Turkington TG, Coleman RE. Clinical applications of PET in oncology. *Radiology* 2004;231:305–332.

14. Townsend DW, Carney JP, Yap JT, Hall NC. PET/CT today and tomorrow. *J Nucl Med* 2004;45:4S–14S.

15. Beyer T, Townsend DW, Brun T, et al. A combined PET/CT scanner for clinical oncology. *J Nucl Med* 2000;41:1369–1379.

16. Weber WA, Schwaiger M, Avril N. Quantitative assessment of tumor metabolism using FDG-PET imaging. *Nucl Med Biol* 2000;27:683–687.

17. Boerner AR, Weckesser M, Herzog H, et al. Optimal scan time for fluorine-18 fluorodeoxyglucose positron emission tomography in breast cancer. *Eur J Nucl Med* 1999; 26:226–230.

18. Zasadny KR, Wahl RL. Standardized uptake values of normal tissues at PET with 2-[fluorine-18]-fluoro-2-deoxy-D-glucose: variations with body weight and a method for correction. *Radiology* 1993;189:847–850.

19. Minn H, Zasadny KR, Quint LE, Wahl RL. Lung cancer: reproducibility of quantitative measurements for evaluating 2-[F-18]-fluoro-2-deoxy-D-glucose uptake at PET. *Radiology* 1995;196:167–173.

20. Weber WA, Ziegler SI, Thodtmann R, Hanauske AR, Schwaiger M. Reproducibility of metabolic measurements in malignant tumors using FDG PET. *J Nucl Med* 1999;40: 1771–1777.

21. Andrade RS, Heron DE, Degirmenci B, et al. Posttreatment assessment of response using FDG-PET/CT for patients treated with definitive radiation therapy for head and neck cancers. *Int J Radiat Oncol Biol Phys* 2006;65:1315–1322.

22. Juweid ME, Stroobants S, Hoekstra OS, et al. Use of positron emission tomography for response assessment of lymphoma: consensus of the Imaging Subcommittee of International Harmonization Project in Lymphoma. *J Clin Oncol* 2007;25:571–578.

23. Short S, Hoskin P, Wong W. Ovulation and increased FDG uptake on PET: potential for a false-positive result. *Clin Nucl Med* 2005;30:707.

24. Stahl A, Weber WA, Avril N, Schwaiger M. Effect of N-butylscopolamine on intestinal uptake of fluorine-18-fluorodeoxyglucose in PET imaging of the abdomen. *Nuklearmedizin* 2000;39:241–245.

25. van Westreenen HL, Cobben DC, Jager PL, et al. Comparison of ^{18}F-FLT PET and ^{18}F-FDG PET in esophageal cancer. *J Nucl Med* 2005;46:400–404.

26. Chen W, Cloughesy T, Kamdar N, et al. Imaging proliferation in brain tumors with ^{18}F-FLT PET: comparison with ^{18}F-FDG. *J Nucl Med* 2005;46:945–952.

27. Jacobs AH, Thomas A, Kracht LW, et al. ^{18}F-fluoro-L-thymidine and ^{11}C-methylmethionine as markers of increased transport and proliferation in brain tumors. *J Nucl Med* 2005;46:1948–1958.

28. Smyczek-Gargya B, Fersis N, Dittmann H, et al. PET with [^{18}F]fluorothymidine for imaging of primary breast cancer: a pilot study. *Eur J Nucl Med Mol Imaging* 2004;31: 720–724.

29. Buck AK, Bommer M, Juweid ME, et al. First demonstration of leukemia imaging with the proliferation marker ^{18}F-fluorodeoxythymidine. *J Nucl Med* 2008;49:1756–1762.

30. Muzi M, Vesselle H, Grierson JR, et al. Kinetic analysis of 3′-deoxy-3′-fluorothymidine PET studies: validation studies in patients with lung cancer. *J Nucl Med* 2005;46:274–282.

31. Buck AK, Bommer M, Stilgenbauer S, et al. Molecular imaging of proliferation in malignant lymphoma. *Cancer Res* 2006;66:11055–11061.

32. Therasse P, Arbuck SG, Eisenhauer EA, et al. New guidelines to evaluate the response to treatment in solid tumors. European Organization for Research and Treatment of Cancer, National Cancer Institute of the United States, National Cancer Institute of Canada. *J Natl Cancer Inst* 2000;92:205–216.

33. Miller AB, Hoogstraten B, Staquet M, Winkler A. Reporting results of cancer treatment. *Cancer* 1981;47:207–214.

34. Buyse M, Thirion P, Carlson RW, Burzykowski T, Molenberghs G, Piedbois P. Relation between tumour response to first-line chemotherapy and survival in advanced colorectal cancer: a meta-analysis. Meta-Analysis Group in Cancer. *Lancet* 2000;356:373–378.

35. Zinzani PL, Magagnoli M, Chierichetti F, et al. The role of positron emission tomography (PET) in the management of lymphoma patients. *Ann Oncol* 1999;10:1181–1184.

36. Jerusalem G, Beguin Y, Fassotte MF, et al. Whole-body positron emission tomography using ^{18}F-fluorodeoxyglucose for posttreatment evaluation in Hodgkin's disease and non-Hodgkin's lymphoma has higher diagnostic and prognostic value than classical computed tomography scan imaging. *Blood* 1999;94:429–433.

37. de Wit M, Bohuslavizki KH, Buchert R, Bumann D, Clausen M, Hossfeld DK. [18]FDG-PET following treatment as valid predictor for disease-free survival in Hodgkin's lymphoma. *Ann Oncol* 2001;12:29–37.

38. Weihrauch MR, Re D, Scheidhauer K, et al. Thoracic positron emission tomography using ^{18}F-fluorodeoxyglucose for the evaluation of residual mediastinal Hodgkin disease. *Blood* 2001;98:2930–2934.

39. Mikhaeel NG, Timothy AR, O'Doherty MJ, Hain S, Maisey MN. [18]FDG-PET as a prognostic indicator in the treatment of aggressive non-Hodgkin's lymphoma-comparison with CT.u *Leuk Lymphoma* 2000;39:543–553.

40. Spaepen K, Stroobants S, Dupont P, et al. Prognostic value of positron emission tomography (PET) with fluorine-18 fluorodeoxyglucose (^{18}F]FDG) after first-line chemotherapy in non-Hodgkin's lymphoma: is ^{18}F]FDG-PET a valid alternative to conventional diagnostic methods? *J Clin Oncol* 2001;19:414–419.

41. Haioun C, Itti E, Rahmouni A, et al. [18]F]fluoro-2-deoxy-D-glucose positron emission tomography (FDG-PET) in aggressive lymphoma: an early prognostic tool for predicting patient outcome. *Blood* 2005;106:1376–13781.

42. Zijlstra JM, Lindauer-van der Werf G, Hoekstra OS, Hooft L, Riphagen II, Huijgens PC. ^{18}F-fluoro-deoxyglucose positron emission tomography for post-treatment evaluation of malignant lymphoma: a systematic review. *Haematologica* 2006;91:522–529.

43. Juweid ME, Wiseman GA, Vose JM, et al. Response assessment of aggressive non-Hodgkin's lymphoma by integrated International Workshop Criteria and fluorine-18-fluorodeoxyglucose positron emission tomography. *J Clin Oncol* 2005;23:4652–4661.

44. Brucher B, Weber W, Bauer M, et al. Neoadjuvant therapy of esophageal squamous cell carcinoma: Response evaluation by positron emission tomography. *Ann Surg* 2001;233:300–309.

45. Downey RJ, Akhurst T, Ilson D, et al. Whole body [18]FDG-PET and the response of esophageal cancer to induction therapy: results of a prospective trial. *J Clin Oncol* 2003;21:428–432.

46. Hawkins DS, Rajendran JG, Conrad EU, 3rd, Bruckner JD, Eary JF. Evaluation of chemotherapy response in pediatric bone sarcomas by [F-18]-fluorodeoxy-D-glucose positron emission tomography. *Cancer* 2002;94:3277–3284.

47. Schulte M, Brecht-Krauss D, Werner M, et al. Evaluation of neoadjuvant therapy response of osteogenic sarcoma using FDG PET. *J Nucl Med* 1999;40:1637–1643.

48. Amthauer H, Denecke T, Rau B, et al. Response prediction by FDG-PET after neoadjuvant radiochemotherapy and combined regional hyperthermia of rectal cancer: correlation with endorectal ultrasound and histopathology. *Eur J Nucl Med Mol Imaging* 2004;31:811–919.

49. Calvo FA, Domper M, Matute R, et al. ^{18}F-FDG positron emission tomography staging and restaging in rectal cancer treated with preoperative chemoradiation. *Int J Radiat Oncol Biol Phys* 2004;58:528–535.

50. Bokemeyer C, Kollmannsberger C, Oechsle K, et al. Early prediction of treatment response to high-dose salvage chemotherapy in patients with relapsed germ cell cancer using [^{18}F] FDG PET. *Br J Cancer* 2002;86:506–511.

51. De Santis M, Becherer A, Bokemeyer C, et al. 2-[18]fluoro-deoxy-D-glucose positron emission tomography is a reliable predictor for viable tumor in postchemotherapy seminoma: an update of the prospective multicentric SEMPET trial. *J Clin Oncol* 2004;22:1034–1039.

52. Mac Manus MP, Hicks RJ, Matthews JP, et al. Positron emission tomography is superior to computed tomography scanning for response-assessment after radical radiotherapy or chemoradiotherapy in patients with non-small-cell lung cancer. *J Clin Oncol* 2003;21: 1285–1292.

53. Yang YJ, Ryu JS, Kim SY, et al. Use of 3'-deoxy-3'-[^{18}F]fluorothymidine PET to monitor early responses to radiation therapy in murine SCCVII tumors. *Eur J Nucl Med Mol Imaging* 2006;33:412–419.

54. Sugiyama M, Sakahara H, Sato K, et al. Evaluation of 3'-deoxy-3'-^{18}F-fluorothymidine for monitoring tumor response to radiotherapy and photodynamic therapy in mice. *J Nucl Med* 2004;45:1754–1758.

55. Barthel H, Cleij MC, Collingridge DR, et al. 3'-deoxy-3'-[^{18}F]fluorothymidine as a new marker for monitoring tumor response to antiproliferative therapy *in vivo* with positron emission tomography. *Cancer Res* 2003;63:3791–3798.

56. Buck AK, Kratochwil C, Glatting G, et al. Early assessment of therapy response in malignant lymphoma with the thymidine analogue [^{18}F]FLT. *Eur J Nucl Med Mol Imaging* 2007;34:1775–1782.

57. Dittmann H, Dohmen BM, Kehlbach R, et al. Early changes in [^{18}F]FLT uptake after chemotherapy: an experimental study. *Eur J Nucl Med Mol Imaging* 2002;29:1462–1469.

58. Ulaner GA, Colletti PM, Conti PS. B-cell non-Hodgkin lymphoma: PET/CT evaluation after ^{90}Y-ibritumomab tiuxetan radioimmunotherapy-initial experience. *Radiology* 2008;246:895–902.

59. Bodet-Milin C, Kraeber-Bodere F, Dupas B, et al. Evaluation of response to fractionated radioimmunotherapy with ^{90}Y-epratuzumab in non-Hodgkin's lymphoma by ^{18}F-fluorodeoxyglucose positron emission tomography. *Haematologica* 2008;93:390–397.

60. Cheson BD, Pfistner B, Juweid ME, et al. Revised response criteria for malignant lymphoma. *J Clin Oncol.* 2007;25:579–586.

CHAPTER 16

The Economic Attractiveness of Targeted Radiotherapy: Value for Money?

JEFFREY S. HOCH

> Waiter: What can I bring you?
> Customer: I would like a pizza.
> Waiter: Can I cut that into 8 slices for you?
> Customer: No, I'm feeling hungry today. Cut it into 16 slices.

16.1 INTRODUCTION

Health economics is the art of applying economics principles to healthcare. These principles include the following: (a) scarcity exists in healthcare; (b) scarcity forces choice; and (c) smart choices involving not paying more for something than it is worth. While these simple ideas seem trivial, they have profound consequences, especially in cancer control. Often, health economics is viewed in a less than favorable light by patients and physicians. This may be an inevitable consequence of the tension between economic reasoning designed to maximize society's welfare and medical techniques designed to maximize an individual's health outcome. The controversy about how limited resources should be spent (or not spent) is captured frequently by the media.

A recent example from the Canadian press was an article entitled "Ontario won't cover all costs of new cancer drugs" (1). The article juxtaposed views expressed by cancer patients and views expressed by the health minister of Ontario.[1] One anecdote

---

[1] Although there is a focus on the Canadian health care system and Ontario in particular, the resource allocation concepts apply to most publicly and privately funded healthcare systems where scarce health resources force difficult choices.

Monoclonal Antibody and Peptide-Targeted Radiotherapy of Cancer, Edited by Raymond M. Reilly
Copyright © 2010 John Wiley & Sons, Inc.

was about a woman who was diagnosed with late-stage cancer and wanted to know "why the government felt her life was not worth the $18,000 she was billed [for her cancer drug]."

> It became clear to me that early on the drugs I needed to fight my cancer were not being provided by a universal healthcare system that I, as a Canadian, have been taught to be so proud of ... I'm a Canadian first and foremost—I happen to live in Ontario. Who would have thought that this would affect the type of treatment that would be available to me? ... How does the government of Ontario have the audacity to make the choices that deny their citizens the recommended standard of care that is offered in other G8 countries?

These concerns—"what is a life worth?", "what are needed cancer drugs?", and "why all healthcare payers do not cover the same cancer drugs?"—we consider later in this chapter. The response from Ontario's health minister was also reported:

> ... there's no public or private health insurance plan in the country that could afford to pay for all of the latest cancer drugs ... Ontario has more than doubled spending on new cancer drugs, but it would be impossible to cover every new medication that's developed ... I can't imagine an environment, and I can't imagine leadership under any political party ... that could ... offer a solution that said 'every time there's a new cancer drug available on the market that a public system could pay for it.'

This response introduces themes like "increased spending on drugs" and "scarcity" (e.g., not being able to pay for all cancer drugs) that we touch upon later in this chapter.

Throughout this chapter, it is important that the reader not lose sight of the fact that cancer is a devastating disease that costs a lot to treat. Cancer is the leading cause of death in most developed countries, causing more than 25% of all deaths (2), and cancer care accounts for 2.9% of all healthcare direct costs and 8.9% of indirect costs in Canada (3). While the consequence of the application of economics to cancer healthcare may be the denial of treatment, this is not the purpose. Economics offers a framework for an organized consideration of treatment options as we seek to balance the imperatives of treating a very bad disease and not paying more than we can (or should) to do it. The next section lays out the case for applying economic thinking to healthcare. After rehearsing the arguments for why this makes sense in theory, the chapter next explores how economic evaluation is used in practice. Finally, comments on the example of Zevalin® are provided and concluding remarks offered.

16.2   APPLYING ECONOMICS IN THEORY

Most microeconomic problems are constrained optimization problems. These types of problems, in their simplest forms, have two parts: a constraint and an objective. An example of a constraint is a fixed budget (i.e., the amount of money that can be spent is limited). An example of an objective is to maximize how long people live, and in oncology, perhaps maximizing their "quality adjusted" years of life. When considering which healthcare treatments to reimburse, a healthcare payer might face the following problem. Assume a decision maker's (e.g., the government's) objective is to maximize quality-adjusted life years (QALYs) for its population. At the same time, a fixed budget constrains the government's spending. What should the government pay for (and what should it not pay for) to get the most QALYs given its limited budget? Economic evaluation techniques are designed to answer questions like these. Next, we offer a detailed example based on the work of Karlsson and Johannesson (4) to illustrate this claim.

16.2.1 A Simple Constrained Optimization Problem

Assume the Ministry of Health (MOH) must decide which treatments to fund. To simplify matters, assume there are only three types of diseases: breast cancer, prostate cancer, and lymphoma, and there are 1000 patients suffering from each disease. The costs and health effects (QALYs) *per patient* treated by different cancer treatments are presented in Table 16.1 with costs reported in the C column and QALYs reported in the E column. To make the math easier, often it is assumed that the average costs and health effects are independent of the number of patients treated (this is sometimes referred to as constant returns to scale). The constant returns to scale assumption is usually coupled with the idea that treatments are perfectly divisible, so that if 50% of people were treated with treatment F and 50% were treated with treatment G, the average cost would be $300 (i.e., $50\% \times \$200 + 50\% \times \$400 = \$100 + \$200 = \$300$) and the average effect would be 14 QALYs (i.e., $50\% \times 12$ QALYs $+ 50\% \times 16$ QALYs $= 6$ QALYs $+ 8$ QALYs $= 14$ QALYs). Finally, assume if no money is spent on the patients in a disease group, Cost $= \$0$ and Effect $= 0$ QALYs.

Given the information in Table 16.1, which of the mutually exclusive treatments should be funded if the MOH has a budget of $900,000? This question is identical to

TABLE 16.1 Average Cost and Average Effect (QALYs) of New Treatments for Three Conditions

Breast Cancer			Prostate Cancer			Lymphoma		
TX	C	E	C/E	TX	C	E	C/E	TX	C	E	C/E
O	0	0	–	O	0	0	–	O	0	0	–
Z_1	110	4	27.5	F	200	12	17	Z_5	100	1	100
Z_2	100	5	20	G	400	16	25	Z_6	200	4	50
A	100	10	10	Z_4	700	17	41.18	K	100	5	20
B	200	14	14	H	550	18	31	L	200	8	25
C	300	16	19	Z_3	600	18	33.3	M	300	12	25
D	400	19	21								
E	500	20	25								

Note: TX is the treatment option; C is the cost per person; E is the QALY per person; and C/E is the ratio of average cost to average effect. Treatment O is a treatment that has no cost and no effect. Since 0/0 is not defined, the C/E column has a "–" for TX = O.

the constrained optimization problem introduced above, "What should the government pay for to get the most QALYs given its limited budget (in this case, $900,000)?"

16.2.1.1 Funding the Most Effective Treatments

If one were to focus solely on maximizing QALYs, one would recommend funding treatments E, H (or Z_3), and M. This would produce 50,000 total QALYs:

$$1000 \text{ people} \times 20 \text{ QALYs} + 1000 \text{ people} \times 18 \text{ QALYs}$$
$$+ 1000 \text{ people} \times 12 \text{ QALYs} =$$
$$20,000 \text{ QALYs} + 18,000 \text{ QALYs} + 12,000 \text{ QALYs}$$
$$= 50,000 \text{ total QALYs}.$$

However, this would be an impossible thing to do because with only $900,000 to spend, the MOH could not spend the necessary amount:

$$1000 \text{ people} \times \$500 + 1000 \text{ people} \times \$550 + 1000 \text{ people} \times \$300$$
$$= \$500,000 + \$550,000 + \$300,000 = \$1,350,000.$$

If two-thirds of the patients received the new treatments (E, H, and M) and one-third received no treatment, then it would cost $900,000 (i.e., 2/3 × $1,350,000) to produce 33,333.33 QALYs (i.e., 2/3 × 50,000 QALYs). While it may not be politically feasible to give two-thirds of the patients cutting-edge treatment while giving one-third of the patients no treatment at all, the key consideration in constrained optimization is maximizing the objective without using more of the constrained resource than one has. In contrast to this proposed solution of funding only the most effective treatments, the potential of using economics in this setting is that it will suggest a solution that cannot be bettered; that is, no other use of resources will produce more QALYs for a given budget.

16.2.1.2 Funding Treatments with the Highest E/C Ratios (and Most Effect)

It is a common misperception that "using economics" means examining the health outcome per unit cost. In other words, if one were to compute the ratio of effect to cost (or *E/C*), one could simply choose the treatments that have the highest ratio (most bang for the buck) until the budget was exhausted. Table 16.2 shows the results of sorting the treatment options by *E/C* from highest to lowest.

Based on this logic, the first treatment that should be funded is treatment A since it has the highest *E/C* ratio at 0.100 QALYs per dollar spent. The next treatment to be funded should be treatment B (replacing treatment A) because it has the next highest *E/C* ratio at 0.070. At this point, $200,000 has been spent (i.e., 1000 people × $200 = $200,000). Note that since all treatments for the same disease are mutually exclusive, once B is covered for breast cancer, there is no need to pay for A (this is why the total cost is $200,000 and not $300,000). The next treatments that are considered for funding are F, C, K, Z_2, and D. At this point, the MOH is covering

TABLE 16.2   Table 16.1 Sorted by the Ratio of Average Effect (*E*) to Average Cost (*C*) for New Treatments of Three Conditions

	Breast Cancer				Prostate Cancer				Lymphoma		
TX	C	E	E/C	TX	C	E	E/C	TX	C	E	E/C
O	0	0	–	O	0	0	–	O	0	0	–
A	100	10	0.100	F	200	12	0.060	K	100	5	0.050
B	200	14	0.070	G	400	16	0.040	L	200	8	0.040
C	300	16	0.053	H	550	18	0.033	M	300	12	0.040
Z_2	100	5	0.050	Z_3	600	18	0.030	Z_5	100	1	0.020
D	400	19	0.048	Z_4	700	17	0.024	Z_6	200	4	0.010
E	500	20	0.040								
Z_1	110	4	0.036								

Note: TX is the treatment option; C is the cost per person; E is the QALY per person; and E/C is the ratio of average effect to average cost. Treatment O is a treatment that has no cost and no effect. Since 0/0 is not defined, the E/C column has a "–" for TX = O.

treatment D for breast cancer, F for prostate cancer, and K for lymphoma at a cost of $700,000 since

$$1000\ \text{people} \times \$400 + 1000\ \text{people} \times \$200 + 1000\ \text{people} \times \$100$$
$$= \$400,000 + \$200,000 + \$100,000 = \$700,000.$$

There is $200,000 more to spend (since the full budget is $900,000) and four treatments E, G, L, and M have the next highest value for *E/C* of 0.040. Following the theme that treatments with "bigger effects (*E*) are better investments," one might argue that the MOH should fund treatments E (with an effect of 20 QALYs) and then G (with an effect of 16 QALYs). A decision by the MOH to fund treatments E for breast cancer, G for prostate cancer, and K for lymphoma will provide 41,000 QALYs, as

$$1000\ \text{people} \times 20\ \text{QALYs} + 1000\ \text{people} \times 16\ \text{QALYs} + 1000\ \text{people}$$
$$\times 5\ \text{QALYs} = 20,000\ \text{QALYs} + 16,000\ \text{QALYs} + 5000\ \text{QALYs}$$
$$= 41,000\ \text{total QALYs}.$$

However, this treatment combination will cost $1,000,000 because

$$1000\ \text{people} \times \$500 + 1000\ \text{people} \times \$400 + 1000\ \text{people} \times \$100$$
$$= \$500,000 + \$400,000 + \$100,000 = \$1,000,000.$$

To reduce the $1,000,000 expenditure to fit the $900,000 budget, half the patients receiving treatment G will need to receive treatment F. Thus, the cost for prostate cancer treatment will be $300,000 since

$$500\ \text{people} \times \$200 + 500\ \text{people} \times \$400 = \$100,000 + \$200,000 = \$300,000$$

which is $100,000 less than before, and the QALYs gained from this treatment strategy for prostate cancer will be 14,000 QALYs since

$$500 \text{ people} \times 12 \text{ QALYs} + 500 \text{ people} \times 16 \text{ people}$$
$$= 6000 \text{ QALYs} + 8000 \text{ QALYs} = 14,000 \text{ QALYs}$$

which is 2000 QALYs less than before. Therefore, this E/C treatment selection strategy yields 39,000 QALYs for the $900,000 that was spent.

The two strategies that we have reviewed—"fund the most effective treatments" and "fund treatments with the highest E/C ratios"—spend $900,000 to produce 33,333.33 and 39,000 QALYs, respectively. The reason economic evaluation techniques are recommended is that they can identify treatment strategies that produce the most QALYs given a fixed budget. We now explain how economic evaluation techniques can be used to produce 43,000 QALYs for $900,000. The main tool for identifying good investments using economic evaluation is not the ratio of effect to cost (or as it is sometimes reported C/E), but rather the ratio of *extra* cost to *extra* effect (5, 6). This ratio estimates the trade-off of what one spends for what one gets.

16.2.1.3 Using Economic Evaluation

The concept of the trade-off is at the heart of economic evaluation: what is the extra cost for the extra patient benefit? The cost-effectiveness trade-off is calculated as the ratio of the extra cost (ΔC) to the extra effect (ΔE) and is called the incremental cost-effectiveness ratio (ICER). A treatment's extra cost and extra effect are calculated relative to an alternative treatment for that disease (often the treatment options are sorted from least effective to most effective and ICERs are calculated between one treatment and the next most effective option as in the example that follows). Before the ICER is calculated for each treatment strategy under consideration, (a) mutually exclusive treatments must be sorted from least to most effective and (b) inefficient ("dominated") treatment strategies should be removed. Table 16.1 presents the treatment strategies sorted from least to most effective. Table 16.3 reproduces Table 16.1 with dominated strategies crossed out. A *strongly* dominated treatment costs more and provides less health outcome (a bad ΔC and a bad ΔE trade-off). A *weakly* dominated treatment either costs more and provides the same health outcomes (a bad ΔC with $\Delta E = 0$) or costs the same and provides less health outcome (a bad ΔE with $\Delta C = 0$).

For example, in breast cancer treatment Z_1 is strongly dominated by treatment A; it costs $10 more and provides six less QALYs (i.e., $\Delta C > 0$ and $\Delta E < 0$). Treatment Z_2 is weakly dominated by treatment A; it costs the same and provides five less QALYs (i.e., $\Delta C = 0$ and $\Delta E < 0$). In prostate cancer, Z_4 is strongly dominated by H, and Z_3 is weakly dominated by H (same effectiveness but Z_3 costs more). In lymphoma, Z_5 is weakly dominated by K, and Z_6 is strongly dominated by K. The next treatments to weed out are those that are dominated by *combinations* of other treatments. This situation is called *extended* dominance.

If treatment provision can be "resized" without affecting the relationship between costs and effects, it is essential to check for treatments that could be ruled out because of extended dominance. Extended dominance can be detected by using the $\Delta C/\Delta E$

TABLE 16.3   Table 16.1 with Strongly and Weakly Dominated Strategies Crossed Out for New Treatments of Three Conditions

Breast Cancer				Prostate Cancer				Lymphoma			
TX	C	E	$\Delta C/\Delta E$	TX	C	E	$\Delta C/\Delta E$	TX	C	E	$\Delta C/\Delta E$
O	0	0	–	O	0	0	–	O	0	0	–
~~Z_1~~	~~110~~	4	=	F	200	12	17	~~Z_5~~	~~100~~	1	=
~~Z_2~~	~~100~~	5	=	G	400	16	50	~~Z_6~~	~~200~~	4	=
A	100	10	10	~~Z_4~~	~~700~~	~~17~~	=	K	100	5	20
B	200	14	25	H	550	18	75	L	200	8	33
C	300	16	50	~~Z_3~~	~~600~~	~~18~~	=	M	300	12	25
D	400	19	33								
E	500	20	100								

Note: TX is the treatment option; C is the cost per person; E is the QALY per person; and $\Delta C/\Delta E$ is the extra cost per extra effect also known as the incremental cost-effectiveness ratio. Treatment O is a treatment that has no comparator that is less effective, so $\Delta C/\Delta E$ is not defined and the $\Delta C/\Delta E$ column has a "–" for TX = O.

ratios, after removing strongly and weakly dominated treatments. If a less effective treatment has a larger $\Delta C/\Delta E$ ratio than the next more effective treatment, this indicates extended dominance in a set of mutually exclusive treatments. For example, in Table 16.3, the $\Delta C/\Delta E$ ratio for treatment C is 50 and the next $\Delta C/\Delta E$ ratio (for treatment D) is 33. There is extended dominance since the less effective treatment has a higher ICER ($50 per QALY) than the next more effective treatment ($33 per QALY). This means that a combination of treatment B and treatment D (e.g., some patients would get B and some would get D) would be better than treating all patients using treatment C. Specifically, if D were used to treat 50% of breast cancer patients and B to treat the other 50%, this combination would have an average cost of $300 (i.e., $400 × 1/2 + $200 × 1/2 = $300) and would yield on average 16.5 QALYs (i.e., 19 QALYs × 1/2 + 14 QALYs × 1/2 = 16.5 QALYs). Hence, the average cost is $300 for both the combined strategy (using B and D) and for treatment strategy C, so $\Delta C = 0$. However, the combined strategy option of treating the breast cancer population with B or D is more effective than the current treatment of C ($\Delta E = 16.5 - 16 = 0.50 > 0$). A similar extended dominance argument could be made for the removal of treatment option L from additional consideration.

All treatments that are dominated—whether strongly, weakly, or by extension—should be removed because they represent inefficient spending (we can get more health for each dollar spent by not investing in dominated treatments). After removing the inefficient dominated treatments, the $\Delta C/\Delta E$ ratios then must be recalculated (see Table 16.4). The recalculation is necessary since removal of dominated treatments has the potential to change the ΔC and ΔE for those that remain. In this case, the $\Delta C/\Delta E$ ratios for treatments D and M have both increased; they were 33 and 25, and afterward they are 40 and 29, respectively. This is because now treatments D and M are being compared to more efficient treatments; consequently, D and M do not appear as economically attractive.

TABLE 16.4   Table 16.3 with All Dominated Strategies Removed for New Treatments of Three Conditions

Breast Cancer				Prostate Cancer				Lymphoma			
TX	C	E	$\Delta C/\Delta E$	TX	C	E	$\Delta C/\Delta E$	TX	C	E	$\Delta C/\Delta E$
O	0	0	–	O	0	0	–	O	0	0	–
A	100	10	10	F	200	12	17	K	100	5	20
B	200	14	25	G	400	16	50	M	300	12	29
D	400	19	40	H	550	18	75				
E	500	20	100								

Note: TX is the treatment option; C is the cost per person; E is the QALY per person; and $\Delta C/\Delta E$ is the extra cost per extra effect also known as the incremental cost-effectiveness ratio. Treatment O is a treatment that has no comparator that is less effective, so $\Delta C/\Delta E$ is not defined and the $\Delta C/\Delta E$ column has a "–" for TX = O.

At this point, treatments should be considered for funding in order of their $\Delta C/\Delta E$ ratio. Treatments are selected in order of least to most expensive in terms of the extra cost per additional unit of effect. First, treatment A instead of no treatment for breast cancer is selected (ICER = 10), then F instead of no treatment for prostate cancer (ICER = 17), then K instead of no treatment for lymphoma (ICER = 20), then B instead of A (ICER = 25), then M instead of K (ICER = 29), and finally D instead of B (ICER = 40). The ICER estimates ($\Delta C/\Delta E$) indicate the order in which to fund treatments; the MOH should purchase treatments that produce QALYs at the cheapest rate (lowest cost per additional unit of health outcome). In the end, this strategy recommends paying for treatments D, F, and M. This option costs $900,000 as

$$1000 \text{ people} \times \$400 + 1000 \text{ people} \times \$200 + 1000 \text{ people} \times \$300$$
$$= \$400,000 + \$200,000 + \$300,000 = \$900,000$$

and produces 43,000 QALYs since

$$1000 \text{ people} \times 19 \text{ QALYs} + 1000 \text{ people} \times 12 \text{ QALYs} + 1000 \text{ people}$$
$$\times 12 \text{ QALYs} = 19,000 \text{ QALYs} + 12,000 \text{ QALYs} + 12,000 \text{ QALYs}$$
$$= 43,000 \text{ total QALYs}.$$

Using the ICER, it was possible to recommend to the MOH which treatments to fund in such a way that no other treatment funding strategy would produce more QALYs. Using the $\Delta C/\Delta E$ ratio suggested funding D, F, and M producing at least 4000 more QALYs than funding only the most effective treatments or funding the treatments with the highest E/C. For our hypothetical population of 3000 people, it is as if we have given each person at least 16 additional months of good health, simply by allocating our resources in an efficient manner. Funding any other set of treatments incurs an opportunity cost equal to the number of QALYs we lose while pursuing this alternative arrangement. Funding the treatments using the ICER produces the most QALYs with the fixed budget. When researchers report the ICER for a treatment (e.g., Zevalin for the treatment of lymphoma), they show the value of spending on that treatment in the context of other ways the money could be used to purchase health.

The techniques used have been known to health economists for over 35 years (7), but their application can lead to stories in the media of the type reviewed earlier. Furthermore, there is still some debate between economists about how healthcare decision makers should use the ICER (8–12). Recognizing how the ICER should be used to solve resource allocation problems in theory helps one appreciate the consequences of deviations in practice. Before discussing how economics is applied in practice, the theoretical role of the ICER will be reviewed.

16.2.1.4 *The Theoretical Role of the Incremental Cost-Effectiveness ratio*   The ICER is crucial to selecting treatments using economic thinking. It helps identify treatments that yield the most health outcome for a fixed budget in many ways. Parts of the ICER are used to determine strong and weak dominance (i.e., strong dominance is $\Delta C > 0$ and $\Delta E < 0$; weak dominance is $\Delta C = 0$ and $\Delta E < 0$ or $\Delta C > 0$ and $\Delta E = 0$). The $\Delta C/\Delta E$ ratio is used to detect extended dominance, signalling the need to remove certain "inefficient" treatments from further consideration. Afterward, the ICER is recalculated for the remaining treatment options. These subsequent ICER calculations indicate the order in which treatments should be funded (until funding runs out).

Because the ICER is not introduced to clinical audiences as the workhorse of a constrained optimization problem, many clinicians are unaware that the healthcare budget affects which is the last treatment funded, and therefore, the ICER of the last treatment chosen for reimbursement. In this case, the last treatment funded is treatment D with an ICER of $40 per additional QALY. Some may see this as a clear indication that the MOH is willing to pay $40 per QALY, and any treatment with an ICER of $40 per QALY should be funded. The problem with this thinking is that it fails to acknowledge the fact that either the healthcare budget affects the ICER of the last treatment funded or the ICER of the last treatment funded affects the healthcare budget.

For example, if drug I for Lung Cancer is submitted for reimbursement consideration, and it has an ICER of $40 per QALY, advocates might argue that drug I should be covered by the MOH since it is as cost-effective as treatment D (with an ICER of 40). If the healthcare budget is still $900,000, the MOH may be able to afford treatment D or I but not both. To stay within budget, the MOH must take funding from one treatment and then direct it toward another treatment. From an economics perspective, the key issue is the fixed healthcare budget and not whether drug I is standard of care somewhere else. The political cost of saying "No" to cancer patients may be greater than the political cost of overspending the healthcare budget. In this case, the healthcare budget may continue to expand to cover the necessary spending. For example, in 1999/2000 the province of Ontario's MOH had experienced annual rates of increase for total drug expenditures during the previous 3 years of 10.6%, 9.9%, and 10.1%; in 2000/01, the increase in expenditure was 15% (13). The money being allocated to new treatments is coming from other government programs (e.g., education, social services, or other areas of healthcare). Money spent in one area cannot be spent in another area. Governments must try to invest in areas that the public feels will provide the most value.

Sometimes it is argued that society should be willing to pay (WTP) a fixed amount for an additional year of life (e.g., $50,000 per QALY). The way that the ICER is used in this paradigm is quite simple. If the ICER is less than the WTP, the MOH should pay for the new treatment. Along these lines, there has been much interest and speculation in inferring from past decisions what a decision maker's WTP is or should be. Early research suggested that ICERs between CDN$20,000 (US$15,300) and CDN$100,000 (US$76,300) were economically attractive price tags for an additional QALY (in 1992 monetary units) (14); however, later studies and different methods suggested other potential WTP values (US$24,777–161,305) (13, 15). Outside of North America, there is some evidence of WTP thresholds for a QALY of US$10,000 in Denmark (16), US$40,400 in Australia (17), and £30,000 (US$55,000) in the United Kingdom (18). In 2008, a US study inferred lower and upper bounds for WTP of US$183,000 and US$264,000 per life year, respectively, and concluded that it was "very unlikely that $50,000 per QALY is consistent with societal preferences in the United States" (19).

In the example above, if the MOH followed a rule that society's WTP was $50 per QALY, all treatments with an ICER < $50 should be funded. The MOH would cover treatments D, G, and K, spending $1,100,000 to produce 47,000 QALYs. A different WTP would lead to a different amount spent (e.g., if society's WTP were $40, the MOH would spend $900,000 as discussed earlier). Likewise, a different budget leads to a different threshold ICER (or WTP). With a healthcare budget of $900,000, the MOH appears willing to pay $40 per QALY. With a WTP of $50, the healthcare budget must be $1,100,000. Either the budget determines the WTP or vice versa. In theory, the process can happen in either direction, but what happens in practice?

16.3   APPLYING ECONOMICS IN PRACTICE

16.3.1   How are Economic Evaluations Used to Make Decisions in Practice?

A simple test to ascertain if it is the WTP determining the healthcare budget or vice versa is to look at funding decisions in comparison to a drug's ICER. If drugs with ICERs below a certain threshold are usually funded and drugs with ICERs above a certain threshold are not, this is evidence of a fixed WTP. However, such a strict pattern is at odds with the evidence provided by funding decisions throughout the world. A distinct possibility is that there is not an exact cost-effectiveness threshold to which decision makers subscribe. Reviews of decisions made in Australia, Canada, and the United Kingdom suggest that the ICER is not the only factor associated with the reimbursement determination (17, 20, 21). In addition to the ICER estimate, the burden of the disease and uncertainty about the ICER estimate are important contributors in their own right, and together all three factors suggest a threshold range in the United Kingdom somewhat higher than the commonly accepted £20,000–30,000 range (US$35,700–53,600) (20). In sum, the evidence seems consistent with treatments having a better chance of being funded if they have lower ICERs; however, frequently there are extenuating circumstances such that treatments with higher ICERs are funded (e.g., the United Kingdom's decision to fund sunitinib for advanced kidney cancer).

Basically, these findings refute the idea that the healthcare budget strictly determines the threshold ICER or WTP since some treatments with higher ICERs get funded while other treatments with lower ICERs do not. In fact, some suggest that decision makers are able to determine what WTP they feel is reasonable (say λ) and then fund treatments based on this (e.g., if the ICER $<\lambda$ for that particular disease) (22). This idea has been sharply contested on theoretical grounds (23):

> Hoch, Briggs, and Willan (2002) note the crucial role of λ in determining solutions to the constrained maximization problem facing decision-makers but suggest that this may be overcome 'if the decision maker can be assumed to know λ.' But how do decision-makers determine λ? Have they developed or discovered a scientific approach that does not require information on the incremental costs and benefits of all programs? Do they have a solution to the problems of indivisibilities and nonconstant returns to scale in programs? Or is this simply a convenient (albeit invalid) way for analysts to deal with the problem they are unable to solve for themselves?

While decision makers may not be blindly adhering to the solution algorithm proposed for constrained optimization problems when deciding whether to fund new drugs, it is indisputable that many are (a) requiring economic evaluations to support funding decisions (24–26) and (b) deciding whether new treatments will be funded. It is possible that the ICERs required for (a) play some part in (b), regardless of how well the theory fits practice or vice versa.

Economists may argue about whether ICERs are being used correctly (according to theory), but there is no argument that ICERs are being used. Some cancer drugs are not funded by healthcare payers because of poor evidence, but some cancer drugs are not funded because healthcare payers are not willing to buy the drug's benefits (i.e., the extra patient outcomes are not worth the extra cost). To make this decision, healthcare payers must know what they are willing to pay (e.g., what represents good value for money). It may be unknown how decision makers come up with their WTP values, but however this happens, decision makers use them to make decisions. The exact amount decision makers are using when deciding whether a new drug represents good value for money may never be known in theory; however, decision makers' actual decisions reveal their preferences about which treatments seem to be good uses of scarce resources. Decision makers know their WTP, λ, or threshold ICER, even if academics do not; however, the WTP value does appear to vary.

There does not appear to be a universally endorsed WTP value and the ICER threshold appears to be context specific, with different decision makers using different values in different contexts. The inevitable consequence is heterogeneous drug coverage policies throughout the world. Those seeking a drug to be covered may use examples of other countries where the drug is covered. It is an open question whether the same drugs should be covered in all first-world countries given that the context and WTP values may be different due to societal differences. Somewhat more awkward is when funding decisions for the same treatment vary within a country. For example, for a time in Canada, bevacizumab (Avastin®) for colon cancer was covered by public payers in British Columbia, Quebec, Newfoundland, and Labrador but not by public payers in Alberta, Saskatchewan, Manitoba, or Ontario. In 2007, Ontario's MOH changed policies and decided to cover Avastin, while the National Institute for Health and Clinical Excellence (NICE) in the United Kingdom did not find Avastin to be a cost-effective use of resources (27). In theory, reasonable people should be able to agree on an estimate for $\Delta C/\Delta E$. Whether the ICER estimate is below society's WTP and represents a good value is a matter of opinion. An economic evaluation can only provide the $\Delta C/\Delta E$ estimate; value judgements are necessary to decide whether the drug is cost-effective (so a cost-effectiveness analysis based on cost and effect data will not generally be able to conclude if a treatment is cost-effective).

It is important to remember that a cost-effective new treatment that is more costly and more effective is not cost saving from the healthcare payer's point of view. Paying $45,000 per extra year of life, when extra years are "worth" $50,000, costs the healthcare payer $45,000 per extra life year. This must be kept in mind when considering previous funding decisions. Caution is needed when applying reasoning like "if decision makers usually fund new treatments that are approximately $30,000 per additional life year then perhaps they should fund drug X which also has an ICER of $30,000." While this type of reasoning may help provide context, some have argued that decisions made in this way will lead to an uncontrolled growth in expenditures (28, 29). Ultimately, healthcare decision makers do not have enough money to purchase every treatment that is cost-effective (i.e., the selected WTP may require a budget that cannot be met). This is an especially important point if decision makers decide to use the ICER without regard for the budget. Any way the ICER is used, either as a guide for maximizing health outcomes while staying within budget or as a way of identifying efficiently produced health outcomes, the ICER plays a central role in applying economics in healthcare. However, there are some challenges and concerns about how cost-effectiveness is assessed in practice.

16.3.2   Challenges and Concerns

Naturally, there are some challenges that accompany trying to estimate an ICER in practice for cancer treatments. A recent review of cancer-specific issues focused on the difficulty of calculating the relevant costs and health effects (30). All economic evaluations face the twin challenges of which cost perspective to use (e.g., only the costs accruing to the MOH or all costs accruing to society) and which time horizon to choose (e.g., 1 year versus a patient lifetime). Even once these issues have been settled—perhaps looking at an MOH perspective over a patient's lifetime—other challenges remain. For example, the appropriate representation of costs and health outcomes associated with unplanned treatments for metastatic disease administered beyond disease progression, the appropriate extrapolation of long-term outcomes and resources from clinical trials, modeling assumptions concerning survival beyond the duration of the trial, and relationships between surrogate outcomes and final outcomes (30). Because clinical trials can provide cost (e.g., units of service times unit costs) and health outcome data for a short time period (e.g., 6 months to 2 years) on a special population (e.g., those who fit the study entrance criteria), mathematical modeling is almost always required. In fact, none of the economic evaluations of cancer treatments undertaken by the NICE has relied solely on direct trial-based economic evidence; rather, all these economic assessments have been augmented with some element of mathematical modeling (30). Some of the challenges that this introduces are described next.

An initial challenge one faces when modeling is determining which treatment regimen to model. Tappenden et al. (30) report an example where UK and French trials of a cancer treatment were considered for use in an economic model to estimate an ICER:

> The French trial . . . observed a substantially better overall survival rate than the UK . . . trial, yet a statistically significant improvement between treatment groups was not observed within either study. Consequently, it was unclear whether these survival differences, which led to a clear cost-effectiveness advantage for the trial arms included in the French trial, were due to the more intensive use of the cytotoxic agents under consideration (i.e. the effectiveness of the treatment sequence), differences between UK and French healthcare systems, inherent differences in the patient groups, some other bias, or a combination of all of the above.

Clearly, the ICER estimated from the economic model will depend on "potential heterogeneities in terms of patient populations specified within the inclusion criteria and observed across clinical trials, inconsistencies in the administration of the current standard treatment, as well as other potential differences between health service delivery systems" (30). Sensitivity analysis can illustrate the ICER estimate's sensitivity to various assumptions. When a model's output differs depending on which assumption is made, and when it is not clear what the right assumption is, sensitivity analysis can be used to illustrate the uncertainty. In addition to uncertainty about what number to use (e.g., for the probability of recurrence at 15 years), there is often uncertainty about what is the right health outcome to use in an economic evaluation.

This is of critical importance because the defining aspect of an economic evaluation has nothing to do with costs; it is how patient outcomes are treated that makes all the difference, distinguishing one economic evaluation method from another. There are many different types of economic evaluations, among them cost–benefit analysis, cost-effectiveness analysis, cost–utility analysis, and cost-minimization analysis (31). An easy way to distinguish one from the other is by examining the analyst's choice of patient outcome and asking "How is it measured?" In cost–benefit analyses, there are typically many outcomes and all outcomes are valued in dollars; this type of analysis is not very prevalent in healthcare. In cost-effectiveness analysis (CEA), a single outcome is analyzed. Commonly the outcome is measured in clinical units such as adverse events avoided or progression-free survival (PFS). A second form of cost-effectiveness analysis is cost–utility analysis, which values outcomes in QALYs, equal to the number of life years remaining multiplied by a factor reflecting quality of life. In cost-minimization analysis, only costs are compared since patient outcomes are assumed to be identical. It is clear from the names of the different types of economic evaluations that "cost" plays a prominent role; however, it is the choice of outcome that defines the type of economic analysis. But which outcome should be used in a CEA of a cancer treatment?

Typically, health outcome measures employed within economic evaluations of cancer therapies include overall survival, quality-adjusted survival, progression-free survival, tumor response, and adverse events avoided. Intermediate outcomes such as tumor response and adverse events avoided are considered less informative in economic evaluations of cancer treatments (30). There are important advantages and disadvantages for each "survival" measure. The "overall survival" outcome is clearly understood and easily measured. However, when the majority of patients live longer than the study period, their survival times are "censored." Patients alive at the end of a 6-month clinical trial have lived at least 6 months, but it is unknown how much longer they will live without making some assumptions. This means to estimate the exact average survival time on a new treatment, the analyst must make assumptions about the survival function (e.g., exponential, Weibull, Gompertz, etc.). Another limitation of using overall survival as the health outcome in economic evaluations of cancer treatments is that it does not adjust for quality of life.

Quality-adjusted life years are a popular way to incorporate important facets of health that are not captured in a measure such as length of life. Another advantage is that an ICER for ibritumomab tiuxetan (Zevalin) versus standard care using a QALY as the outcome measure can be compared to an ICER for a treatment for another type of cancer and to ICERs for treatments for entirely different diseases (e.g., heart disease, diabetes, mental illness, etc.). In the example presented earlier in the chapter, QALYs gained from the treatment of breast cancer were compared to QALYs gained from the treatment of prostate cancer and lymphoma. If the health outcomes in Table 16.1 had differed for each of the three diseases, it would be difficult to use the ICER to make decisions. Drug B (for breast cancer) might have an ICER of $30,000 for an extra year of life, drug G (for prostate cancer) might have an ICER of $10,000 for an extra year of progression-free survival, and drug M (for lymphoma) might have an ICER of $40,000 for an extra QALY. It is difficult to determine the drug providing the best value when what one is buying is not comparable. Typically, healthcare payers must make decisions about drugs for a variety of diseases, so a QALY potentially can serve as a universal currency.

However, some feel the benefits of using the QALY are outweighed by its limitations (32). There are various methods of calculating QALYs, and there is evidence showing important differences in estimates depending on the methods used (33–35). Furthermore, one can get different QALY estimates using the same method depending upon the population that is asked about their values (36, 37). Thus, the advantage of doing a CEA using QALYs (i.e., a cost–utility analysis) is that the cost per QALY for a new treatment can be compared with other healthcare interventions' cost per QALY price tags. However, some have argued that to enjoy this advantage, one must risk using an erroneous measure to elicit incorrect values from the wrong people (38, 39), and others have argued that while quality adjustment has the potential in cancer to make an important difference, it usually does not change decisions based on the ICER (40). Regardless, since QALYs are the product of a quality of life weight and a length of life estimate, they share a limitation with the "overall survival" measure; many patients' length of life estimates have to be guessed at since many patients are not dead when the study ends.

Progression-free survival is a commonly reported event that is likely to occur during a clinical trial. In addition, if a clinical trial allows patients to crossover to a different treatment following disease progression, PFS will be an uncontaminated outcome in contrast to overall survival or quality-adjusted survival (30). However, PFS values are recorded at checkups; therefore, PFS may be related to checkup schedules (30). In addition, PFS may not be related to longer life. A treatment could alter the distribution of time in various stages of disease without affecting a patient's years of life. Finally, ICERs made using PFS as the outcome may not be easy for decision makers to use. For example, it may be difficult for decision makers to determine an acceptable range of cost-effectiveness for cancer treatments valued in terms of cost per progression-free life year avoided (30).

There are additional areas of uncertainty related to the stochastic nature of sample data (41) or the desire to blend the model's results with the analyst's prior beliefs (42). While these important issues are beyond the scope of this introductory chapter, detailed treatments are given in more advanced texts about economic evaluation (43, 44). Before concluding this section, some contentious challenges related to the source of the economic analysis are reviewed.

16.3.3   How Does the Model Affect the Results?

Most of the time, an economic evaluation designed to inform a funding decision for a cancer treatment will need to fill in the variables in the following statement: In A years, it will cost $B to get one more unit of C when using D instead of E in patients of type F in context G. Frequently, pharmaceutical manufacturers choose different values for A–G when creating economic models for their product compared to independent reviewers constructing independent models. Different choices for A–G create different cost-effectiveness estimates. An early evaluation of conflict of interest in economic analyses of new drugs used in oncology found that "pharmaceutical company sponsorship of economic analyses is associated with reduced likelihood of reporting unfavorable results" (45). This finding does not appear unique to cancer as a review of economic studies of antidepressants found that "among industry studies, modelling studies are more favourable to the sponsor than administrative studies, and . . . studies sponsored by industry are significantly more favourable to industry" (46). In fact, in their analysis of English language studies measuring health outcomes in QALYs, Bell and colleagues found that about half the published ICERs were below $20,000 per QALY, and studies funded by industry were more likely to report lower ICER estimates (47). They also found that studies of higher quality and those conducted in Europe or the United States were less likely to report ICERs below $20,000 per QALY (47).

So what if manufacturers publish economic evaluations with lower ICERs? Maybe they are studying drugs with greater potential than those studied by others. A review of the literature published in *Current Oncology Reports* (48) found that

> ... there are some causes for concern, given the fact that most pharmacoeconomic studies report positive findings for the sponsor's drug. However, a more detailed analysis suggests that, although the methodologic quality of some published studies may be poor, the main reason for positive results is that companies only sponsor economic studies where a positive outcome is likely.

A better comparison would be to look at pharmacoeconomic analyses of the same drug conducted by different analysts. Recently, studies have done just that (49, 50) finding, in general, that ICER estimates by drug manufacturers were lower than those submitted by independent assessment groups. Moreover, in one study over 80% of manufacturers' estimates (21 of 25) were less than the independent assessment groups' estimates ($p < 0.001$). A crucial consideration to keep in mind is that no one knows what the true ICER is. Often, the ICER is being estimated over a time period for which there are no data available (e.g., 30 years after the clinical trial ended). It may be more productive to avoid ascribing labels such as "right" or "wrong" to the different ICER estimates and instead explore the reasons for the different estimates.

A fascinating study by Chauhan and coworkers (51) examined economic models presented to NICE by pharmaceutical manufacturers and independent academic groups and found independent groups tended to estimate larger differences in cost (ΔC) and smaller differences in effectiveness (ΔE) compared to manufacturers. Since, $\Delta C_{academics} > \Delta C_{manufacturers}$, *and* $\Delta E_{academics} < \Delta E_{manufacturers}$, this explains why $\Delta C/\Delta E$ estimated by independent academic groups is a larger number than $\Delta C/\Delta E$ estimated by pharmaceutical manufacturers. There were two factors driving the results. Academic groups appeared to use higher estimates for the average cost of a new cancer treatment and higher estimates for the average effectiveness of usual care ($p \leq 0.010$) (51). Using the hypothetical scenario from earlier in the chapter, a manufacturer's analysis might compare treatment K (cost = $100 and effect = 5 QALYs) for lymphoma to no treatment at all (cost = $0 and effect = 0 QALYs). The ICER estimate would equal $\Delta C/\Delta E = (\$100 - 0)/(5 - 0) = \$20$ per QALY. An academic group might feel the real average cost of the new drug would be $150 and that standard care would provide at least 2 QALYs. With a higher average cost for the new drug and a higher average effectiveness for standard care, the academic ICER would be $(\$150 - 0)/(5 - 2) = \$50$ per QALY.

If Chauhan et al.'s findings hold in other jurisdictions where ICERs submitted by pharmaceutical manufacturers are independently assessed, these findings suggest a way forward for getting the various and sundry parties to agree on a single ICER estimate. Decision makers and pharmaceutical manufacturers could work together either to use the data that exist or collect new data (e.g., coverage with evidence development) to get estimates of the average cost of the new cancer treatment and the average effectiveness of standard care. Even if such an arrangement is impractical, it seems possible that data could be collected on patients currently receiving standard care, and these data could be used to inform estimates on how the average patient does on standard care. In this setting, at least in theory, both decision makers and pharmaceutical manufacturers could agree on three of the four parts of the ICER. With some additional negotiation, it seems possible to fashion a mutually agreeable $\Delta C/\Delta E$ estimate.

Using the concepts developed throughout this chapter, the cost-effectiveness of ibritumomab tiuxetan (Zevalin) as a new targeted radiotherapeutic agent for lymphoma is discussed next.

16.4   THE ECONOMIC ATTRACTIVENESS OF TARGETED RADIOTHERAPY: THE CASE OF ^{90}Y-IBRITUMOMAB TIUXETAN (ZEVALIN)

The use of monoclonal antibodies or peptides conjugated to radionuclides for treatment of malignancies is an exciting development. One of the most well-known and commercially available examples at the moment is Zevalin, a ^{90}Y-conjugated monoclonal antibody against CD20 that has received approval for treatment of non-Hodgkin's B-cell lymphomas. Zevalin, used as a targeted radiotherapeutic agent, has been demonstrated to be more effective for treatment of lymphoma than rituximab (52, 53):

> 80% of the patients receiving radiolabelled Zevalin responded, compared with 56% for rituximab. However, the time taken for the disease to get worse after treatment was the same in both groups (about 10 months). In the additional study, radiolabelled Zevalin brought about a response in about half of the patients.

Next we consider the reasons for Zevalin's relatively poor adoption by the oncologic community, despite its therapeutic efficacy. These thoughts offer a perspective on the application of health economics analyses to the introduction of other targeted radiotherapeutics that may be developed in the future.

16.4.1   What has Been Published about the Cost-Effectiveness of Zevalin?

In 2008, searching PubMed (www.pubmed.gov) using the search terms *cost effectiveness of ibritumomab tiuxetan* produced two "hits." One was an article written in the Czech language (54) and the other (55) was a letter to the editor in the quarterly journal of AΩA, an honour medical society, taking issue with an editorial in a previous edition of the same journal (56). From the Czech article's translated title and abstract, the article appears to be a review of the therapeutic results of rituximab and other approved monoclonal antibodies (e.g., ibritumomab tiuxetan ^{90}Y) with a discussion of cost-effectiveness. The debate in the honour medical society's quarterly journal is about whether Zevalin is a good use of money. The initial editorial claimed that when Americans develop cancer they want "as much surgery, as much radiation, as much chemotherapy as possible" since they are "unlikely to ever accept less than the most advanced medical care" (56). The editorial goes on to claim that "the case for arbitrary rationing of medical care has gained little traction in the United States. Quite the opposite. Look at recent examples of the standard of care: Zevalin and Bexxar, new radioimmunotherapy anti-cancer drugs, at the cost of $25,000 per treatment" (56). This drew a response from Dr. Singer, chief medical officer of Cell Therapeutics Inc., which was in the process of purchasing the rights to market and further develop Zevalin. He wrote,

> I strongly take issue with the appropriateness of one of the examples used to argue that the cost of new therapies is too high—that of Zevalin and Bexxar for relapsed or refractory indolent lymphoma. Although the cost of a single administration of these radio-immunotherapeutics is approximately $25,000, that single dose frequently produces durable multiyear remissions in patients with relapsed or refractory indolent lymphomas, whose other options are at least equally costly, more toxic, and probably less effective . . . Although cost must be part of the overall health care equation, benefit must be considered as well, particularly for therapies for which there is compelling evidence of clinical benefit. It is of note, that both Bexxar and Zevalin are approved for use in the European Union despite their review process which includes a cost/benefit analysis.

It is not uncommon for countries around the world to approve for use a drug (as safe) that subsequently is not reimbursed by healthcare payers (since it is not deemed to provide good value for money). Is Zevalin a good example? What are the results of economic evaluations published in the scientific literature?

In 2008, the Scopus™ database had 25 hits related to *cost effectiveness of ibritumomab tiuxetan*. Among them was an editorial in *Nuclear Medicine Communications* by A. Otte (who received a consultancy grant from Schering AG) and S.L. Thompson (who works for Schering AG) about how the practical and clinical benefits of radioimmunotherapy lead to advantages in cost-effectiveness in the treatment of patients with non-Hodgkin's lymphoma (57). The article covers topics including radiation exposure risk, hospitalization, and time commitments for medical staff and patients, practical safety, interdisciplinary cooperation, reimbursement, and cost-effectiveness. With regard to cost-effectiveness, Otte and Thompson make the interesting claim that

> . . . it is unacceptable that an approved and efficacious radioimmunotherapy in follicular non-Hodgkin's lymphoma is sometimes reimbursed at only a small fraction of the treatment cost, as is currently the case in . . . Germany and Italy. Growing pressures on health care budgets worldwide have led to an increasing interest in the use of health economics data to support the added value of new therapies in terms of both outcomes and cost and such data is [sic] available for ^{90}Y-ibritumomab tiuxetan demonstrating cost-effectiveness relative to rituximab.

It is ironic that the concern that a drug's reimbursement cost is too low is mentioned in the same paragraph as the concern about growing pressures on healthcare budgets. What are the studies being marshalled as evidence of cost-effectiveness of Zevalin?

Both studies are poster abstracts published in 2005 (58, 59). Three years later, neither study has been published in the scientific literature. The abstract by Gabriel and coauthors (58) estimated the cost-effectiveness of ^{90}Y ibritumomab tiuxetan versus rituximab (four-dose scheme) for outpatient treatment in Germany (based on costs from 2004) in patients with relapsed or refractory follicular non-Hodgkin's lymphoma. Otte and Thompson summarize that

> ... cost-effectiveness was determined as cost per year in remission by relating costs to the overall response rate and duration of response; cost per disease-free year was based on complete response rate and duration of response of complete response patients. The conclusion of the analysis was that although the total cost of the ^{90}Y-ibritumomab tiuxetan regimen was higher (€19,567 vs. €9,756), the cost per year in remission (€14,862 vs. €16,967) and in particular cost per disease-free year (€22,235 vs. €80,077) were clearly in favour of ^{90}Y-ibritumomab tiuxetan. This conclusion was driven largely by the superior response rates and in particular complete response rates for ^{90}Y-ibritumomab tiuxetan over rituximab monotherapy.

The results are reported in the form of less useful average cost-effectiveness ratios (i.e., C/E) and not more useful ICERs (i.e., $\Delta C/\Delta E$); as noted previously (5, 6), the C/E ratio provides misleading information. To calculate the appropriate cost-effectiveness estimate, the ICER, for Zevalin, one needs an estimate for ΔC and ΔE. Based on the results, $\Delta C = €9811$ (i.e., €19,567 − €9756 = €9811). Using the numbers reported in the abstract, Zevalin is likely to produce 0.74 more years in remission and 0.76 more disease-free years. Therefore, the ICER estimates for Zevalin based on these numbers are €13,258 per additional year in remission (€9811/0.74 more years in remission = €13,258) and €12,909 per additional disease-free year (€9811/0.76 more disease-free years = €12,909). This is approximately US$18,694 per additional year in remission and US$18,202 per additional disease-free year (using an exchange rate of €1 = US$1.41). Although these preliminary results are encouraging, Gabriel et al. note that "due to the relatively short follow-up time (median 44 months), survival could not be evaluated yet. Further research is needed to compare both clinical efficacy and actual costs, especially with prolonged rituximab immunotherapy ('maintenance')" (58).

The poster abstract by Thompson and van Agthoven (59) compared ^{90}Y-ibritumomab tiuxetan with four-dose or eight-dose schemes of rituximab for treatment in the Netherlands (based on costs from 2001). Otte and Thompson report that

> ... the mean total costs were estimated as follows: ^{90}Y-ibritumomab tiuxetan €16,345, rituximab 4-dose scheme €9,510 and rituximab 8-dose scheme €19,020. The expected number of months in remission per patient treated were 14.4 months for ^{90}Y-ibritumomab tiuxetan, 11.4 months for the rituximab 8-dose scheme and 6.2 months for the rituximab 4-dose scheme, resulting in a mean cost per month in remission for ^{90}Y-ibritumomab tiuxetan of €1,138, followed by €1,544 for the rituximab 4-dose scheme and €1,674 for the rituximab 8-dose scheme.

Again it appears that the ICERs have not been calculated. From a theoretical perspective, the interesting trade-off is between Zevalin and the four-dose scheme for rituximab (since the eight-dose scheme can be ruled out on the basis of extended dominance). In their conclusion, Thompson and Agthoven (59) emphasize this, "where ^{90}Y-Zevalin is used rather than four doses of rituximab, the additional cost to the payer would be, on average, €6,835. For this additional cost, the benefit to the patient would be an average 8.2 additional disease-free months, over and above what would have been gained with 4-dose rituximab therapy . . . when the costs and benefits of ^{90}Y-Zevalin are compared with the 8-dose rituximab regimen, ^{90}Y-Zevalin is the more cost-effective strategy." This interpretation is consistent with an ICER of €10,002 per extra additional disease-free year (€6,835/8.2 extra months/ 12 months/1 year = €10,002). This is approximately US\$14,103 per additional disease-free year (using an exchange rate of €1 = US\$1.41). Thus, both poster abstracts provide similar ICER estimates for the extra cost of an additional disease-free year. Whether the extra cost is worth it involves a value judgment.

Based on the two poster abstracts, Otte and Thompson conclude that "the cost-effectiveness data for ^{90}Y-ibritumomab tiuxetan . . . provide convincing evidence in favour of the added value of ^{90}Y-ibritumomab tiuxetan in terms of cost per month in remission or cost per disease-free month despite higher initial product acquisition costs." This assessment is repeated by Otte in a recent letter to the editor (60): "Cost-effectiveness data for Zevalin already provide convincing evidence in favor of the added value of Zevalin in terms of cost per month in remission or cost per disease-free month despite higher initial product acquisition costs." Otte supports this claim by referencing his prior work with Thompson (57), and then concludes that it is

> . . . unacceptable that radiolabeled immunotherapy with Zevalin is sometimes reimbursed at only a small fraction of the treatment cost. In fact, there is no definitive reimbursement system in many European countries. Germany, for example, is still awaiting a specific radiolabeled immunotherapy DRG code, and here Zevalin is paid out of the hospital budgets or by privately insured patients. The situation is similar in many other European countries. So far, only Portugal has achieved a DRG code for radiolabeled immunotherapy (60).

Other claims about the cost-effectiveness of Zevalin (61) have been supported by referring to clinically exciting findings (e.g., Zevalin is well tolerated by the older patients and does not require age-related dose adjustments and delays, or patients with relapsed non-Hodgkin's lymphoma are more likely to respond to Zevalin). So why do healthcare payers not cover Zevalin? As a brief case study, we examine the reasons put forward by the Scottish Medicines Consortium (SMC) (62).

In April 2005, following a full submission, the Scottish Medicines Consortium advised that "ibritumomab tiuxetan (Zevalin) is not recommended for use within NHS Scotland for the preparation of a radiopharmaceutical incorporating Yttrium 90 [^{90}Y] for the treatment of adult patients with rituximab-relapsed or refractory CD20 + follicular B-cell non-Hodgkin's lymphoma" (62). The reason, "No economic information was submitted to allow an assessment of its cost effectiveness" (62). Before 2005, there were no cost-effectiveness data comparing the use of ^{90}Y-Zevalin with rituximab in follicular lymphoma (59). This initial rejection by the SMC is a classic example of why submissions for reimbursement must be accompanied by economic evidence: if they are not, they are likely to be rejected. Perhaps the CEA abstracts published in 2005 were then used to support the resubmission in 2007. In 2007, the SMC published its reassessment (62): "... ibritumomab tiuxetan (Zevalin) is not recommended for use within NHS Scotland for the preparation of a radiopharmaceutical incorporating Yttrium 90 [^{90}Y] for the treatment of adult patients with rituximab relapsed or refractory CD20 + follicular B-cell non-Hodgkin's lymphoma ..." The reason this time? "The manufacturer did not present a sufficiently robust economic analysis to gain acceptance by SMC" (61). How could one set of experts conclude that the cost-effectiveness of Zevalin is convincing and another set of experts conclude that the economic analysis is not sufficient? This is explored in the next section.

16.4.2   Why the Scottish Medicines Consortium Said "No" (Again)

The SMC's concerns about the economic model for Zevalin are related to the assumptions made to get data inputs for the model. The model time horizon was 15 years. Individual patient-level data were used where possible, and a model was constructed for patients whose data were censored. Zevalin was compared with a conventional care arm composed of a range of other treatments including chemotherapy, radiotherapy, and stem cell transplant. The model took data from the licensing trial, but since the trial was a phase II single-arm trial, no comparative data on active comparatives were available (62). For this Scottish analysis, the manufacturer used data from a case review of 46 Canadian patients, of whom 17 matched the inclusion criteria of the licensing trial and were used as the basis for the comparison (the proportion of patients receiving comparator treatments was drawn from the Canadian data), and quality-of-life weights were drawn from a survey of 24 medical consultants and specialist oncology nurses (62).

As for the results, the manufacturer estimated that Zevalin on average would cost an additional £8535 per patient over the 15 year time horizon and would result in an additional 0.38 QALYs to give an ICER of £22,445 per QALY; curtailing the time horizon to 5 years slightly raised the estimate of cost effectiveness to £25,589 per QALY (62). The reason the SMC said "No" was *not* based primarily on whether these ICERs were felt to represent good value for money. The main concern appears to be whether the estimates were made using the right data:

> Limited data were presented ... regarding the clinical parameters of the model. Clarification by the manufacturer indicated that the ... model ... derived the transition probabilities from the case note review, and extrapolated using these. This differentiated transition probabilities by treatment arm, the likelihood for moving from progressive disease to palliation being assumed to be considerably worse for the conventional care arm than for the ibritumomab tiuxetan arm. Equalising the transition probabilities

between the two arms of the model increased the cost per QALY to £31000. The quality of life value derived for progressive disease was also very low and additional sensitivity analysis provided by the manufacturer indicated that adjusting this value caused further increases in the cost per QALY. The effect on the QALY of the combined effect of these two weaknesses is not known. Therefore, the manufacturer has not presented a sufficiently robust economic analysis to gain acceptance by SMC.

No one seems to have collected the data that the people who might pay for the cancer treatment wanted to know. This is not an unusual occurrence.

Earlier in the chapter, we discussed the choices of A–G in the statement "In A years, it will cost $B to get one more unit of C when using D instead of E in patients of type F in context G." It is wrong to believe that all ICERs produced (or funded) by pharmaceutical companies are bad and all ICERs produced (or funded) independently are good. However, pharmaceutical manufacturers do seem to choose different values for A–G when creating economic models for their product compared to independent reviewers making independent models. If there is not more discussion earlier between pharmaceutical manufacturers submitting their economic models and decision makers reviewing the models about choices for A–G, we may always remain in the current state of affairs where a promising new cancer treatment is being rejected on cost-effectiveness grounds while at the same time the most current review articles are trumpeting the convincing cost-effectiveness of these products (63). However, even once a convincing economic argument has been made for these new products, other economic aspects (e.g., incentives) may play an important role (64) in treatment acceptance.

16.5   CONCLUSIONS

Scientific journals and popular media teem with the results of clinical trials proclaiming evidence of more effective new treatments or interventions. Without the resources to be able to provide all new treatments that are more effective, how should decision makers choose? Often, clinical enthusiasm is tempered with economic discipline, in the form of an economic evaluation. In many decision-making contexts, this involves estimating the extra cost of an extra unit of a health outcome. Do healthcare decision makers reject paying for new cancer treatments because they are unaware of the clinical evidence? No. In the process of not recommending Zevalin, the Scottish Medicines Consortium referenced key clinical findings (65–67). A recurring theme is that the economic analysis is not answering the question that healthcare decision makers want answered. This problem has a relatively simple solution (e.g., asking healthcare decision makers what they want to see in the economic analysis and then providing it). It is a waste of everyone's resources bringing a cancer treatment to market just to discover it will not be covered because the economic evaluation was an afterthought.

New cancer treatments appear to be getting more expensive. Consequently, healthcare decision makers will continue to use economic evaluation to inform their decisions about how to spend scarce resources. While new cancer treatments have faired well in comparison to other treatments for other diseases (68–70), this may not always be the case (71). Now is the right time to make a commitment to justify the value of new cancer treatments. Economic evaluation offers a framework for an organized consideration of treatment options as we seek to balance the imperatives of treating a very bad disease with not spending more than we should. The discussion in this chapter of the concepts and methods used by health economists provide the tools for evaluating these new therapies, and the resulting analyses can form the basis for decisions by healthcare payers and providers.

REFERENCES

1. Leslie K. Ont. Won't cover all costs of new cancer drugs. *Can Press*. May 2, 2007 (accessed online on May 19, 2007).
2. www40.statcan.ca/l01/cst01/health36.htm (accessed on February 1, 2008).
3. Public Health Agency of Canada. *The Economic Burden of Illness* 1998;102.
4. Karlsson G, Johannesson M. The decision rules of cost-effectiveness analysis. *Pharmacoeconomics* 1996;9:113–120.
5. Briggs A, Fenn P. Trying to do better than average: a commentary on 'statistical inference for cost-effectiveness ratios.' *Health Econ* 1997;6:491–495.
6. Hoch JS, Dewa CS. A clinician's guide to correct cost-effectiveness analysis: think incremental not average. *Can J Psychiatry* 2008;53:267–274.
7. Weinstein M, Zeckhauser R. Critical ratios and efficient allocation. *J Pub Econ* 1973; 2:147–158.
8. Birch S, Gafni A. Cost effectiveness/utility analyses: do current decision rules lead us to where we want to be? *J Health Econ* 1992;1:279–296.
9. Johannesson M, Weinstein MC. On the decision rules of cost-effectiveness analysis. *J Health Econ* 1993;12:459–467.
10. Johannesson M. On the estimation of cost-effectiveness ratios. *Health Policy* 1995; 31:225–229.
11. Garber AM, Phelps CE. Economic foundations of cost-effectiveness analysis. *J Health Econ* 1997;16:1–31.
12. Stinnett AA, Paltiel AD. Mathematical programming for the efficient allocation of health care resources. *J Health Econ* 1996;15:641–653.
13. Laupacis A. Inclusion of drugs in provincial drug benefit programs: who is making these decisions, and are they the right ones? *CMAJ* 2002;166:44–47.
14. Laupacis A, Feeny D, Detsky AS, Tugwell PX. How attractive does a new technology have to be to warrant adoption and utilization: tentative guidelines for using clinical and economic evaluations. *CMAJ* 1992;146:473–481.
15. Hirth RA Chernew ME Miller E, Fendrick AM Weissert WG. Willingness to pay for a quality-adjusted life year: in search of a standard. *Med Decis Making* 2000;20: 332–342.
16. Gyrd-Hansen D. Willingness to pay for a QALY. *Health Econ* 2003;12: 1049–1060.

17. George B, Harris A, Mitchell A. Cost effectiveness analysis and the consistency of decision making: evidence from pharmaceutical reimbursement in Australia (1991 to 1996). *Pharmacoeconomics* 2001;19:1103–1109.

18. Towse A. What is NICE's threshold? An external view. In: *Cost Effectiveness Thresholds: Economic and Ethical Issues*, Towse A, Pritchard C, Devlin N,editors, King's Fund and Office for Health Economics, London, UK, 2002, 25–30.

19. Braithwaite RS, Meltzer DO, King JT Jr., Leslie D, Roberts MS. What does the value of modern medicine say about the $50,000 per quality-adjusted life-year decision rule? *Med Care* 2008;46:349–356.

20. Devlin N, Parkin D. Does NICE have a cost-effectiveness threshold and what other factors influence its decisions? A binary choice analysis. *Health Econ* 2004;13:437–452.

21. Rocchi A, Menon D, Verma S, Miller E. The role of economic evidence in Canadian oncology reimbursement decision-making: to lambda and beyond. *Value Health* 2008; 11:771–783.

22. Hoch JS, Briggs AH, Willan AR. Something old, something new, something borrowed, something blue: a framework for the marriage of health econometrics and cost-effectiveness analysis. *Health Econ* 2002;11:415–430.

23. Gafni A, Birch S. Incremental cost-effectiveness ratios (ICERs): the silence of the lambda. *Soc Sci Med* 2006;62:2091–2100.

24. Morgan SG, McMahon M, Mitton C, Roughead E, Kirk R, Kanavos P, Menon D. Centralized drug review processes in Australia, Canada, New Zealand, and the United Kingdom. *Health Aff* 2006;25:337–347.

25. Sarría-Santamera A, Matchar DB, Westermann-Clark EV, Patwardhan MB. Evidence-based practice center network and health technology assessment in the United States: bridging the cultural gap. *Int J Technol Assess Health Care* 2006;22:33–38.

26. Oliver A, Mossialos E, Robinson R. Health technology assessment and its influence on health-care priority setting. *Int J Technol Assess Health Care* 2004;20:1–10.

27. 118TA Colorectal Cancer (metastatic) — Bevacizumab & Cetuximab: Guidance 24 January 2007. Accessed online on December 25, 2008 at http://www.nice.org.uk/Guidance/TA118/NiceGuidance/pdf/English.

28. Gafni A, Birch S. Guidelines for the adoption of new technologies: a prescription for uncontrolled growth in expenditures and how to avoid the problem. *CMAJ* 1993; 148: 913–917.

29. Gafni A, Birch S. Inclusion of drugs in provincial drug benefit programs: should "reasonable decisions" lead to uncontrolled growth in expenditures? *CMAJ* 2003; 168:849–851.

30. Tappenden P, Chilcott J, Ward S, Eggington S, Hind D, Hummel S. Methodological issues in the economic analysis of cancer treatments. *Eur J Cancer* 2006;42: 2867–2875.

31. Hoch JS, Dewa CS. An introduction to economic evaluation: what's in a name? *Can J Psychiatry* 2005;50:159–166.

32. McGregor M, Caro JJ. QALYs: are they helpful to decision makers? *Pharmacoeconomics* 2006;24:947–952.

33. Gold MR, Stevenson D, Fryback DG. HALYs and QALYs and DALYs, oh my: similarities and differences in summary measures of population health. *Annu Rev Public Health* 2002;23:115–134.

34. Krabbe PFM, Essink-Bot ML, Bonsel GJ. The comparability and reliability of five health-state valuation methods. *Soc Sci Med* 1997;45:1641–1652.

35. Arnesen T, Trommald M. Roughly right or precisely wrong? Systematic review of quality-of-life weights elicited with the time trade-off method. *J Health Serv Res Policy* 2004; 9:43–50.

36. Shumway M. Preference weights for cost-outcome analyses of schizophrenia treatments: comparison of four stakeholder groups. *Schizophr Bull* 2003;29:257–266.

37. Prosser LA, Kuntz KM, Bar-Or A, Weinstein MC. Patient and community preferences for treatments and health states in multiple sclerosis. *Mult Scler* 2003;9:311–319.

38. Hoch J. Cost-effectiveness lessons from disease-modifying drugs in the treatment of multiple sclerosis. *Expert Rev Pharmacoecon Outcomes Res* 2004;4:537–547.

39. Heller JG. Will public health survive QALYs? *Can J Clin Pharmacol* 2002;9:5–6.

40. Tengs TO. Cost-effectiveness versus cost-utility analysis of interventions for cancer: does adjusting for health-related quality of life really matter? *Value Health* 2004; 7:70–78.

41. Hoch JS, Dewa CS. Lessons from trial-based cost-effectiveness analyses of mental health interventions: why uncertainty about the outcome, estimate and willingness to pay matters. *Pharmacoeconomics* 2007;25:807–816.

42. Hazen GB, Huang M. Large-sample Bayesian posterior distributions for probabilistic sensitivity analysis. *Med Decis Making* 2006;26:512–534.

43. Drummond MF, O'Brien BJ, Torrance GW, Stoddart GL. *Methods for the Economic Evaluation of Health Care Programmes*, 3rd edition, Oxford University Press, New York, 2005.

44. Drummond MF, McGuire A. *Economic Evaluation in Health Care: Merging Theory with Practice*, Oxford University Press, New York, 2001.

45. Friedberg M, Saffran B, Stinson TJ, Nelson W, Bennett CL. Evaluation of conflict of interest in economic analyses of new drugs used in oncology. *JAMA* 1999;282:1453–1457.

46. Baker CB, Johnsrud MT, Crismon ML, Rosenheck RA, Woods SW. Quantitative analysis of sponsorship bias in economic studies of antidepressants. *Br J Psychiatry* 2003; 183:498–506.

47. Bell CM, Urbach DR, Ray JG, et al. Bias in published cost effectiveness studies: systematic review. *BMJ* 2006;332:699–703.

48. Barbieri M, Drummond MF. Conflict of interest in industry-sponsored economic evaluations: real or imagined? *Curr Oncol Rep* 2001;3:410–413.

49. Chilcott J, McCabe C, Tappenden P, et al. Modelling the cost effectiveness of interferon beta and glatiramer acetate in the management of multiple sclerosis. Commentary: evaluating disease modifying treatments in multiple sclerosis. *BMJ* 2003;326:522.

50. Miners AH, Garau M, Fidan D, Fischer AJ. Comparing estimates of cost effectiveness submitted to the National Institute for Clinical Excellence (NICE) by different organisations: retrospective study. *BMJ* 2005;330:65.

51. Chauhan D, Miners AH, Fischer AJ. Exploration of the difference in results of economic submissions to the National Institute of Clinical Excellence by manufacturers and assessment groups. *Int J Technol Assess Health Care* 2007;23:96–100.

52. Witzig TE, Gordon LI, Cabanillas F, et al. Randomized controlled trial of yttrium-90-labeled ibritumomab tiuxetan radioimmunotherapy versus rituximab immunotherapy

for patients with relapsed or refractory low-grade, follicular, or transformed B-cell non-Hodgkin's lymphoma. *J Clin Oncol* 2002;20:2453–2463.

53. European Public Assessment Report (EPAR) for Zevalin: EPAR summary for the public. Accessed at www.emea.europa.eu/humandocs/PDFs/EPAR/zevalin/535103en1. pdf (accessed on June 8, 2009).

54. Trnený M, Klener P. Deset let od úispesného zavedení prvé monoklonální protilátky (rituximab) do lécby lymfomu [Ten years since the successful introduction of the first monoclonal antibody (rituximab) into the therapy of lymphomas]. *Cas Lek Cesk* 2007; 146:578–85.

55. Singer JW. Putting a cost on extra days given by drugs. *Pharos Alpha Omega Alpha Honor Med Soc* 2008;71:44. Accessible at http://alphaomegaalpha.org/PDFs/Pharos/Articles/ 2008Spring/Letters.pdf.

56. Harris ED. Editorial: The American culture and health care. *Pharos Alpha Omega Alpha Honor Med Soc* 2007;70:1. Accessible at http://alphaomegaalpha.org/pharos/New/ 2007Autumn.htm.

57. Otte A, Thompson SL. Practical and clinical benefits of radioimmunotherapy lead to advantages in cost-effectiveness in the treatment of patients with non-Hodgkin's lymphoma. *Nucl Med Commun* 2006;27:753–756.

58. Gabriel A, Hänel M, Wehmeyer J, Griesinger F. Advantages in cost effectiveness of Zevalin radioimmunotherapy vs. rituximab immunotherapy in patients with relapsed or refractory follicular non-Hodgkin's lymphoma [abstract]. *Onkologie* 2005;28:730.

59. Thompson S, van Agthoven M. Cost-effectiveness of ^{90}Y-ibritumomab tiuxetan (^{90}Y-Zevalin) versus rituximab monotherapy in patients with relapsed follicular lymphoma [abstract]. *Blood* 2005;106:436.

60. Otte A. Does health economics have an impact on non-Hodgkin's lymphoma patients' options? *Nucl Med Commun* 2008;29:748–750.

61. DeNardo GL. The conundrum of personalized cancer management, drug development, and economics. *Cancer Biother Radiopharm* 2007;22:719–721.

62. Scottish Medicines Consortium document accessed at http://www.scottishmedicines.org. uk/smc/files/ibritumomab%20tiuxetan%20(Zevalin)%20Resubmission%20FINAL% 20June%202007%20for%20website.pdf (accessed on June 8, 2009)

63. Otte A, van de Wiele C, Dierckx RA. Radiolabeled immunotherapy in non-Hodgkin's lymphoma treatment: the next step. *Nucl Med Commun* 2009;30:5–15.

64. Berenson A. Market forces cited in lymphoma drugs' disuse. *New York Times* July 14 2007.

65. Witzig TE, Flinn IW, Gordon LI , et al. Treatment with ibritumomab tiuxetan radio-immunotherapy in patients with rituximab-refractory follicular non-Hodgkin's lymphoma. *J Clin Oncol* 2002;20:3262–3269.

66. Wiseman GA, Gordon LI, Multani PS , et al. Ibritumomab tiuxetan radioimmunotherapy for patients with relapsed or refractory non-Hodgkin's lymphoma and mild thrombocytopenia: a phase II multicenter trial. *Blood* 2002;99:4336–4342.

67. Gordon LI, Witzig T, Molina A , et al. Yttrium 90-labeled ibritumomab tiuxetan radio-immunotherapy produces high response rates and durable remissions in patients with previously treated B-cell lymphoma. *Clin Lymphoma* 2004;5:98–101.

68. Summerhayes M, Catchpole P. Has NICE been nice to cancer? *Eur J Cancer* 2006; 42:2881–2886.

69. DiMasi JA, Grabowski HG. Economics of new oncology drug development. *J Clin Oncol* 2007;25:209–216.

70. Drummond MF, Mason AR. European perspective on the costs and cost-effectiveness of cancer therapies. *J Clin Oncol* 2007;25:191–195.

71. Low E. Many new cancer drugs in the United Kingdom are facing negative NICE rulings. *J Clin Oncol* 2007;25:2635–2636.

███████ **CHAPTER 17**

Selected Regulatory Elements in the Development of Protein and Peptide Targeted Radiotherapeutic Agents

THOMAS R. SYKES AND CONNIE J. SYKES

17.1 INTRODUCTION

The use of radiopharmaceuticals for cancer therapy has a long history. At present, there are a number of approved drugs that are simple radioactive compounds that display enhanced tumor accumulation and continue to be a treatment of choice in many patients (e.g., [89]Sr Chloride, [32]P Phosphate, and ^{131}I metaiodobenzylguanadine). Recent research has focused on utilizing more biologically based targeting molecules with hopefully higher tumor specificity that should enhance their therapeutic potential. With the advent of recombinant genetic engineering capabilities (that have superseded hybridoma technology; see Chapter 1) and large-scale cell culture feasibilities, the reliable supply of significant quantities of large proteins including antibodies as well as recombinant peptides is currently achievable. This has led to a considerable variety of protein-based tumor targeting agents (of which monoclonal antibodies (MAbs) and their derivatives are a main group) that can be prepared with variation in, for example, their size (small single-chain fragments to large extended multimeric constructs and natural immunoglobulins), their species-specific content (murine antibodies to chimeric antibodies to humanized antibodies to human antibodies), their pharmacokinetic (PK) behavior (blood clearance half-lives from weeks to hours), and immunogenicity (high to negligible) (see Chapter 1). The screening of exceptionally large random peptide sequence libraries using cell and receptor-based systems has invigorated the selection of suitable peptides with high affinities that are potentially suitable for radiotherapeutic applications (see Chapter 3). The refinement of peptide synthesis has enabled their large-scale preparation in cost-effective yields with minimal impurities and longer chain length. The incorporation of D-amino acids,

Monoclonal Antibody and Peptide-Targeted Radiotherapy of Cancer, Edited by Raymond M. Reilly
Copyright © 2010 John Wiley & Sons, Inc.

selective side-chain, and end-chain modifications has increased the variety of available synthetic peptides. A number of commercially feasible radionuclide production systems have also recently been reported that should be capable of supplying more optimized radionuclides for radiotherapeutic purposes (see Chapter 2). As yet, these advances have not been reflected in the number of regulatory agency-approved therapeutic radiopharmaceuticals of this type from the current two (^{131}I-tositumomab and ^{90}Y-ibritumomab tiuxetan) but there are a number of promising agents in clinical trials that have potential to achieve this milestone in the near future.

Radioprotein and radiopeptide therapy has been evaluated in numerous formats designed to match the (radio)pharmacological behavior of the agent and the clinical therapeutic risk management profile. The great majority of agents use the simple direct approach involving tumor targeting conjugate radiopharmaceutical administration (usually via intravenous or intracavity injection) as single or multiple doses. Alternate approaches utilize administration of the unlabeled tumor targeting conjugate followed at a specified interval by a unique radioligand (with affinity for the conjugate) administration termed two-step targeting or pretargeting (see Chapter 8). Other pretargeting approaches incorporate specific agent administration (in two or three steps) designed to remove the circulating tumor targeting agent after tumor accumulation (secondary antibody), to clear circulating uncomplexed radioisotope after tumor targeting radiopharmaceutical tumor accumulation (clearing agent) or to minimize nonspecific and undesirable tumor targeting radiopharmaceutical (such as minimizing renal accumulation) or other approaches. This chapter focuses on the simple direct approach but mention is also made of other approaches in some sections. The use of protein and peptide-based radiopharmaceuticals as therapeutic agents may resemble diagnostic radiopharmaceuticals in many ways but the former also have potential for a larger total protein or peptide mass dose, a larger radioactivity dose and different radionuclide radiation emission characteristics that implies some unique regulatory implications regarding their pharmaceutical, nonclinical, and clinical evaluation. While radiolabeled MAb may function by multiple mechanisms, this chapter assumes their primary effect is due to selective tumor targeting with localized tumor radionuclide induced radiotherapy and therefore they are grouped with other protein radiotherapeutic agents (protein RA) and peptide radiotherapeutic agents (peptide RA) within the overall group of protein and peptide radiotherapeutic agents (PPRA). This chapter will also focus on the types of studies and information required by the regulatory agencies in Canada, the European Union, and the United States for the development and eventual approval of these agents. As new science and technology is incorporated into the drug development process so will the regulatory requirements be updated to reflect these realities; thus, it is imperative to be aware of applicable advance notices that are often circulated from regulatory agencies as draft guidance, concept, or policy documents. The exact requirements for any given drug product are best discussed with the specific regulatory agency by appropriate means prior to critical development decisions that involve regulatory compliance expectations. The enormous volume of information applicable to this task dictates that only a limited summary of the selected elements can be presented but with appropriate references to more authoritative documents.

17.2 ADMINISTRATIVE AND ORGANIZATIONAL ELEMENTS

The operational regulation of PPRA development in Canada, the European Union, and the United States is the responsibility of government administrative departments (regulatory agencies) with the necessary expertise to evaluate the scientific and clinical information provided by the sponsor to enable appropriate decisions of product quality and patient risk/benefit for clinical testing and/or approved use. The sponsor's clinical trial application (CTA); (Canada and European Union) or investigational new drug (IND) application (United States) must be filed and approved before the trial can proceed for progressive phases of the clinical assessment of PPRA (with some exceptions). Sponsors are encouraged to have pre-CTA meetings (Canada and European Union) or pre-IND meetings (United States) with the respective regulatory agencies to inform them of the new drug under development and to obtain valuable guidance on the contents of the CTA/IND and on the design and conduct of the associated clinical trial.

A summary overview of the departments within the regulatory agencies that may be involved in the approval of PPRA clinical trials and PPRA new drug submissions for Canada, the European Union, and the United States are presented in Fig. 17.1. In Canada, the CTA and New Drug Submission (NDS) approval process for PPRA is the responsibility of the Centre for Evaluation of Radiopharmaceuticals and Biotherapeutics in the Biologics and Genetic Therapies Directorate with expertise from the other divisions depending upon whether it is a peptide RA (Clinical Trials Division, Cytokines Division or Hormones and Enzymes Division, Therapeutic Products

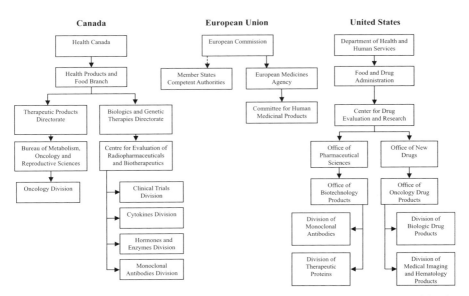

FIGURE 17.1 Overview of selected regulatory agency organizations with potential relevance to protein and peptide radiotherapeutic agents.

Directorate) or protein RA (Clinical Trials Division, Monoclonal Antibodies Division) or from the Oncology Division (Bureau of Metabolism, Oncology and Reproductive Sciences) as required. In the European Union, PPRA CTA are approved by each member state's competent authority (such as the Medicines and Healthcare Products Regulatory Agency (MHRA) in the United Kingdom) but the applications and trial conduct follow the centralized guidance provided by the Committee for Human Medicinal Products of the European Medicines Agency (EMEA, currently unofficially designated as EMA). Marketing authorization (MA) submissions and approval for PPRA must follow the centralized procedure established by EMA (which allows simultaneous adoption by all member states) because PPRA are classified as cancer therapeutics and/or as biotechnology-derived products. In the United States, the clinical trial and New Drug Application (NDA) approval processes for PPRA are under the auspices of the Center for Drug Evaluation and Research (CDER) and depending upon whether it is a protein RA or peptide RA will be evaluated either by the Office of Pharmaceutical Sciences and its Office of Biotechnology Products or by the Office of New Drugs and its Office of Oncology Drug Products.

CTA in Canada (1) follow the Common Technical Document (CTD) format but need only include Modules 1–3. Currently, paper submissions supported by certain electronic documents are the only submission formats being accepted. Module 1 contains the administrative information required (application form, clinical trial protocol, investigator brochure (IB), informed consent form, clinical trial site information, and so on). It is encouraged that the clinical trial protocol, IB and informed consent form are formatted as described by good clinical practice (GCP) guidance (2). Module 2 contains the quality information summary-radiopharmaceuticals (QIS-R) and if the agent is of biological/biotechnological origin, then the appropriate quality information summary-biologicals (QIS-B) Part 1 to Part 4 are also included here. Summary nonclinical and clinical data are included in the IB and are not required in Module 2. Full reports of nonclinical or clinical testing are also not required to be filed in Canada to support the CTA but must be available upon request. Module 3 contains the detailed information on the quality (Chemistry and Manufacturing) of the investigational PPRA to support the summary information supplied in Module 2 but full process validation and full assay validation data is not expected in early phase submissions. For imported PPRA, the receiving site/responsible person must ensure that local radiation safety requirements are met and a copy of the CTA approval (no objection) letter must accompany all shipments of the material into Canada. CTA are reviewed and approved/rejected within 30 days of submission.

In the European Union, CTA also follow the CTD format (3), with each member state's requirements for forms and supporting information to be supplied in addition to the EMA requirements for clinical trial applications, including specifications for paper or electronic submissions. The investigational medicinal product (IMP) dossier is constructed using the CTD Module 2 headings and it is recognized that not all information will be available depending upon the stage of development (particularly full process validation and full assay validation data). An IB cross-reference to the nonclinical and clinical data can be supplied for the appropriate sections of the IMP dossier. For imported PPRA, the qualified person (4) ensures that local radiation safety requirements are met and must ascertain appropriate product quality including product release. Each member state's competent authority sets the review period for approval/rejection of the CTA (e.g., 30 days for the MHRA).

In the United States, clinical trial applications are covered under the IND regulations (5) and are currently not required to be submitted in the CTD format, except for electronic submissions. As PPRA are handled by CDER, electronic submissions are encouraged for IND but paper submissions are also currently still acceptable. Under the current regulations, the IND is divided into 11 sections. Sections 1–6 encompass Form FDA 1571, the table of contents, introductory statement and general investigational plan, the IB and the clinical protocol. Section 7 presents the relevant information regarding the chemistry, manufacturing, and controls (CMC) of the clinical trial material PPRA with specific guidance provided for content requirements for Phase 1 studies (6) and Phase 2 and 3 studies (7). Section 8 requires detailed information regarding the pharmacology and toxicology information available for the PPRA (along with the full draft or final reports of each study) in addition to the information supplied in the IB. Section 9 requires a summation of the previous human experience (again, in addition to that provided in the IB) along with the full (draft or final) reports for each study. Section 10 covers additional information such as radiation dosimetry studies and any other information the sponsor wishes to provide to aid in the review of the IND. Section 11 will provide any other information specifically requested by FDA to be included in the document. Foreign sponsors are also required to have an agent located in the United States to file and communicate officially with FDA. For imported PPRA, the receiving site/responsible person (United States agent) must ensure that local radiation safety requirements are met and the IND number must be clearly marked on all shipments of the material into the United States. INDs are not approved by the FDA but are in force 30 days after receipt by FDA unless a clinical hold notification is issued.

Regardless of the jurisdiction, nonclinical (8, 9) and clinical trial data (10) must be documented in the appropriate report format. Marketing applications (NDS, MA, and NDA) for PPRA approval for sale in all three regions are to be made in the CTD format with all jurisdictions moving to electronic submissions to make review faster and more efficient. Approval times will vary depending upon the type of application and can range from 6–9 months (e.g., priority review for Canada, accelerated assessment for the European Union, fast track review for the United States) to several years.

17.3 PHARMACEUTICAL QUALITY ELEMENTS

The basic requirements for single-step molecularly targeted oncologic radiotherapy agents include a well defined chemobiological (for this chapter, a protein or peptide) entity that is capable of highly specific association or binding to a particular target biomolecule *in vivo*, the target having an established pattern of preferential localization in or on a cancer cell; and a natural or synthetic appendage or inclusion structure capable of appropriately strong and facile attachment to a radionuclide with appropriate therapeutic properties. For this chapter, protein tumor targeting entities can be MAb or other protein molecules derived from recombinant cells engineered specifically for this purpose (rather than from MAb hybridoma approaches) and characterized for safety and quality to meet regulatory requirements. Proteins from this source enable robust manufacture using large-scale cell culture bioreactor systems (rather than *in vivo* ascites production approaches) specifically designed for optimized cell growth and product generation that, along with primarily chromatographic techniques for high-level purification, must be characterized to demonstrate process control, cell-based impurity minimization and product reproducibility using a number of analytical methods. For this chapter, the source of peptide targeting component is via wholly synthetic procedures such that they can be reliably obtained from large-scale solid phase, solution phase, or combination production techniques for chain lengths typically up to 40–60 amino acids with a similar range of chromatographic purification techniques and more traditional chemical-based analysis for quality and purity. Larger peptides may be produced by recombinant DNA (rDNA) technology similar to MAb. The radionuclide binding component is typically specifically designed for easy, controlled incorporation in or on the tumor targeting component with fast, reliable radiolabeling and optimum stability. These will be either complexing/chelating structures or, less frequently, agents incorporating structures for radiohalogenation (see Chapter 2). The selection of therapeutic radionuclide has many criteria (11) but the control of production regarding the radionuclidic, radiochemical, chemical, and safety profiles must be established to meet regulatory requirements preferably by the supplier but definitely in relation to final product quality. Additional components may be present in molecularly targeted oncologic radiotherapy agent products (in the product container or an ancillary container) that address selected aspects of pharmaceutical quality (e.g., lyoprotectants, tonicity and pH adjusters, stabilizers), (radio)chemical stability (antioxidants), radiolabeling efficiency (pH, counter ion and other adjusters), and safety (scavenging chelating agents) while several on-site procedures may also be part of the final radiolabeled product preparation, both of which require regulatory scrutiny and approval to assure control of all ingredients, adequate instructions for use and ultimately safe and efficacious use in patients. For all pharmaceuticals, the quality of the final product is critically dependent on the documented quality of the personnel, facilities/equipment, raw materials, and containers/labels in conjunction with control of the manufacturing and related processes that is supported by acceptable individual batch quality analytical data and an overall quality assurance program. Regulatory agency approval for all these elements is evaluated on a case-by-case basis in each submission but must be sufficient to assure the high regulatory expectations of a safe product prior to any human use. Standards for this good manufacturing practice (GMP) requirement are available for clinical trial material (12–15), for active pharmaceutical ingredients (16) and for final stage/approved products (17–22), and compliance with these regulatory concepts and guidance is critical to a successful PPRA pharmaceutical development strategy. It should be appreciated that GMP extends beyond consideration of the facilities/equipment required for production to all drug components and their quality ranging from raw materials to intermediates to the final product and its packaging. Knowledge of various pharmaceutical quality elements including a high degree of physicobiochemical characterization and a sound potency assay with human relevance are important in overall quality control and in determining whether a high-risk designation is warranted that would impact the development strategy (23).

17.3.1 Tumor Targeting Component

17.3.1.1 Recombinant Protein

Cell Line and Production Bank Starting Materials   The large molecular size and complexity of biological proteins (e.g., MAb), possessing secondary and tertiary structures as well as glycosylation and other posttranslational alterations, make chemical synthesis virtually impossible and therefore the use of biosynthetic processes within living cells are the only realistic source (see Chapter 1). Recombinant DNA technology has enabled genetic alteration of a variety of host cell lines that, as a result, acquire specific protein expression capabilities and become the originating cell source. The research-based, well-established procedures for identification of the DNA sequence of interest need only be described in general with appropriate references but newer approaches such as the use of transgenic mice may require a more detailed safety assessment (24). Typically an expression vector that carries the DNA sequence of the protein of interest (and perhaps other beneficial cellular DNA signals such as enhancer/promoter/antibiotic resistance/purification aid sequences) is constructed and transfected into a suitable host cell line. Several suitable host cell lines include those from bacteria (e.g., *Escherichia coli*), fungi (e.g., *Pichia stipitis*), insect (e.g., *Spodoptera frugiperda*), and mammalian (e.g., Chinese hamster ovary) sources, which possess the necessary genetic, biochemical, and regulatory quality attributes preferred for this purpose. The exact nucleic acid sequence of the chosen construct and the vector (if appropriate) must be verified by established sequencing procedures to assure that the synthesized and final coding sequences match the intended theoretical sequence. The entire vector sequence should be provided showing the appropriate function of each DNA segment. The detailed procedure for transfection and subsequent selection of the final recombinant cell line (including materials, biologicals, and cell lines as well as the procedures and controls) must be described in detail even for wholly research-based generation. For transient expression systems, master vector seed stocks must be established and their quality and stability documented (25, 26). This basic cell line origin information must be provided in the initial regulatory submission for approval prior to any human use of the cell-derived protein.

The origin cell line is cultured under high-quality-controlled conditions until a sufficient quantity of cells is available as a homogeneous culture from which a predetermined number of identical vials (typically 200 vials at $1-10 \times 10^6$ cells/vial) can be prepared and subsequently stored frozen in liquid nitrogen under defined and secure conditions that result in the master cell bank (MCB). A similar scale culture is then prepared from a single MCB vial to generate a working cell bank (WCB), a single vial of which is typically used as the seed stock for each production lot maximizing the probability of (recombinant) protein biological source continuity/homogeneity. The critical importance of cell bank quality for all future production and safety assurance strongly suggests their generation under GMP conditions. Procedures, materials and controls regarding each cell bank preparation should be thoroughly described in appropriate regulatory submissions. In keeping with the general requirements of GMP for material identity verification and more specifically for the potential of originating cell safety considerations, the MCB, WCB, and end of production cell bank (EOPCB) should be extensively characterized. A variety of tests are required to provide evidence of cell bank quality depending on the bank origin (bacteria, mammal), history, genetic alteration and intended utilization including cell identity and genetic integrity (morphology on selected media, DNA sequencing, plasmid retention, copy number and restriction endonuclease mapping, and isoenzyme analysis), product identity (biochemical testing), appropriate contaminants (sterility, bacteriophage, and mycoplasma), culture stability (viability), and virus content. The latter is of particular regulatory concern as the potential presence of virus particles in the cell banks may enable (*trans*-species) infection of human recipients. The extent of virus testing depends primarily on the type of host cells for evaluating endogenous viruses and animal-derived component exposure during generation and cultivation to evaluate adventitious (unintentionally introduced) viruses that may indicate potential for infectious and/or pathogenic virus content risk. Virus tests cover a range of methodologies including electron microscopy, susceptible cell line infectivity, antibody production, specific enzyme determination and virus nucleic acid determination and these specialized assays are normally contracted to analytical service groups operating under good laboratory practice (GLP)/GMP standards (26, 27). Documentation of cell bank virus status and safety assessment as evidenced by the results of all virus (including virus removal/inactivation validation studies) and other quality testing will be an important part of initial regulatory submissions.

Production Activities   The expansion of a vial of WCB cells is conducted in scaled-up stages as required to provide the necessary cell numbers under the appropriate growth/protein expression conditions for a suitable time period to obtain the targeted amount of desired protein in the unprocessed bulk fluid. Numerous process technologies may be used to achieve this goal including many variations of batch culture bioreactor and hollow-fiber systems. Early development lots should be used to characterize and optimize significant process parameters (media/supplement feed rate, pH control, oxygenation control, etc.). Control of all materials used in the clinical production lots is particularly important including all media, supplements and sparging gases as potential sources of viral and microbial contamination with chemical quality typically in line with pharmacopeial requirements. At an appropriate stage of development (typically prior to Phase 3 regulatory submission) at the completion of cell culture production for a representative production run, a sample of cells is removed just prior to batch termination from which an EOPCB is generated. Specific testing needs to be conducted on the EOPCB and unprocessed bulk fluid to verify viral load consistency in the WCB with no aberrant virus presence from large-scale culture induction or contamination, and absence of microbial contamination. The terminal unprocessed bulk fluid is harvested and typically subjected to cell disruption (for intracellular protein products) or cell removal (for secreted protein products) prior to undergoing a variety of sequential downstream processing steps intended to purify the product from the rest of the bulk harvest. Techniques such as centrifugation, ultra/dia/nanofiltration, affinity/ion exchange/hydrophobic interaction/size exclusion chromatography (and specialized viral inactivation/removal steps such as low pH or detergent treatment) are typically employed with precise operating conditions designed to maximize protein product recovery and purity. At the final production scale, all physical downstream processes (e.g., ultrafiltration, centrifugation) and all chromatographic column processes should be validated (such as cleaning, reuse, etc.) and acceptable ranges for all critical operational parameters (such as pressure differentials, loading and elution fluids, pH, ionic strength, capacity, flow rate, etc.) with respect to purity, endotoxin removal, and so on should be determined. Chemical materials must be of pharmacopeial or equivalent quality and attention should be paid to the final formulation solution contents regarding material stability and compatibility with further manufacturing operations (concentration, buffer counter ions, pH, metal ions, etc.) that may be involved to produce the final radiolabeled protein RA. A critical element of downstream process validation is its capability for virus inactivation and removal. Prior to any human use, a risk assessment of the potential for virus presence in the protein product should be conducted taking into consideration the MCB/WCB/unprocessed bulk fluid virus testing results, the use of animal source materials for bank preparation and production (if any), and the scientific literature and any proprietary data for supporting viral inactivation/removal for justifiably comparable downstream platform processes. In general, MCB/WCB/unprocessed bulk fluid containing known pathogenic or unidentified viruses should not be used for human studies. Especially for cell lines of mammalian origin for MAb production, viral validation studies consisting of spiking each evaluable downstream process intermediate input material (such as preultrafiltrates or chromatographic purification loading solutions) with known titers of endogenous or model viruses followed by the downstream process conduct with assessment of (intermediate) output material model virus titer is required, enabling calculation of each downstream process viral clearance/inactivation value (log reduction value) (LRV)). These studies are conducted on appropriately scaled-down versions of each evaluable downstream process for practical reasons. Typically LRV data from 2 model viruses are required for Phase 1 submissions and a further 2 model viruses are required for later development. The properties of the acceptable model viruses available for this testing (27, 28) cover a range of features including genomic constituents (RNA or DNA), encapsulation (enveloped or nonenveloped), size (from $\sim$10 to $\sim$200 nm), families (retrovirus, etc.), and resistance level (to pH or solvent extremes) and thus demonstration of an adequate LRV (typically $>$6 units higher than the initial virus load log value of the bulk harvest for the overall downstream process) provide sufficient regulatory assurance of virus absence (and/or risk minimization) in the protein product. Viral validation studies must be summarized for regulatory agency review and may require provision of all virus validation study reports. It is a regulatory expectation that all materials that contact the cells are obtained from transmissible spongiform encephalitis (TSE)-free sources or if not, that a TSE risk minimization evaluation document can be provided (29).

Quality   The final purified tumor targeting protein component must be adequately analyzed to establish basic and unique physicochemical, biochemical, and biological properties (24). A summary of some useful characterization and testing studies that may be expected by regulatory agencies is shown in Table 17.1 (and can also be selectively applicable to suitable peptide tumor targeting components).

As development progresses, an increasing degree of characterization should be documented leading to an in-depth understanding of the structural and functional properties that complement the requirements of scale-up manufacturing operations

TABLE 17.1   Summary of Potential Regulatory Requirements for Characterization and Testing Studies of the Protein Tumor Targeting Component of Protein Radiotherapeutic Agents

Type	Characterization/Test	Method Examples
Physico biochemical	Peptide map	Enzymatic/chemical degradation with HPLC analysis
	Amino acid composition	Chemical hydrolysis with HPLC analysis
	Amino acid sequence (total, N-terminal, C-terminal)	Selective chemical degradation with HPLC analysis, HPLC-MS/MS
	Disulfide bridge location	Selective chemical degradation with HPLC analysis
	Molecular weight	MS, LC-dynamic light scattering detection, ultracentrifugation, SE-HPLC
	Charge isoform profile	IEF, CIEF
	Electrophoretic molecular weight analysis	SDS-PAGE (R/NR) with silver stain, CGE
	Liquid chromatographic analysis	IE, HI, SE, RP, affinity chromatography
	Spectroscopic analysis	UV, NMR, CD spectroscopy
	Crystalline structure	X-ray diffraction
	Solution properties (pH, solubility, etc.)	Chemical assays
	Oligosaccharide content, sequence, and attachment site	Selective chemical and biochemical degradation with HPLC analysis
Biological (potency)	Target receptor/binding site specificity, affinity/response	*In vitro* cell binding, ELISA, inhibition RIA, Western blot
	Target tissue binding specificity/ intensity	Semiquantitative immunohistochemistry
	Biological response	Animal or isolated organ pharmacological response

CD: circular dichroism; CGE: capillary gel electrophoresis; CIEF: capillary isoelectric focusing; ELISA: enzyme-linked immunosorbent assay; HI: hydrophobic interaction; HPLC: high-performance liquid chromatography; IE: ion exchange; IEF: isoelectric focusing; LC: liquid chromatography; MS: mass spectrometry; NMR: nuclear magnetic resonance; R/NR: reducing/nonreducing; RIA: radioimmunoassay; RP: reverse phase; SDS-PAGE: sodium dodecyl sulfate polyacrylamide gel electrophoresis; SE: size exclusion; UV: ultraviolet and batch heterogeneity, quality control, and stability data interpretation. Several of these characterization studies should lead to scientifically sound (progressing to validated) assay methods capable of verifying critical protein attributes on a lot-to-lot analysis basis as routine quality control tests. In addition, the quality of the overall protein product formulation must be established with regards to the extent and nature of impurities. These include residual bioprocess additives (media, media components, reagents, etc.), those from downstream processing (column matrix leachables, elution reagents, refolding agents, organic solvents, etc.), cellular materials (DNA and host cell proteins) and product related materials (aggregated, truncated, degraded forms, etc.). Given its susceptibility to microbial contamination, specific pharmaceutical testing (sterility and endotoxin) should also be conducted. A summary of typical quality control tests as potential batch analysis release tests is shown in Table 17.2 (and can also be selectively applied to suitable peptide tumor targeting components). It is noteworthy that a reference material lot should be prepared early in development (and updated thereafter) and be well characterized to serve as a baseline for test method development, for subsequent lot results and for associated specification delineation. Stability studies should be initiated at an early stage for the bulk material under conditions that approximate anticipated routine storage conditions (concentration, formulation, container, temperature, etc.) and analyzed for real time degradation with special attention to physical changes such as visible and subvisible particle/covalent and noncovalent aggregate formation (30) as well as potency.

17.3.1.2 Synthetic Peptides

Chemical Starting Materials The initial building blocks for peptide synthesis are specially derivatized amino acids that incorporate the necessary activating and protecting groups to facilitate traditional carboxy to amino group coupling (see Chapter 3). These compounds must be obtained from reliable suppliers with certificates of analysis that include testing results of optical rotation, chemical purity (by HPLC and/or thin layer chromatography (TLC)), melting point, chiral purity, and potentially other characterization tests (NMR, infrared (IR) spectroscopy, and MS) depending on their complexity and criticality. For solid-phase peptide synthesis (SPPS), high-quality resins characterized by their backbone chemical composition, functionalization type and amount (molar ratio), solvent swelling ratios, and general appropriateness should be described (31).

Production Activities For SPPS, sequential processing stages of the synthesis should be described by a schematic flow diagram indicating the initial amino acid derivative coupling to the resin, peptide elongation involving a cycle of deprotection, washing, (neutralization and washing, if required), coupling, washing followed by terminal peptide coupling (and capping, if required). In general the solution compositions (e.g., amino acids, coupling reagents, neutralization reagents, and solvents volumes) and the timing of each reaction type should be reported. The process for peptide cleavage from the resin should be described including all timing and reagent conditions (e.g., cleavage/deprotection, scavenger agent amounts, and solvents), and

TABLE 17.2   Summary of Potential Regulatory Requirements for Batch Analysis Release Tests of Protein Tumor Targeting Component of Protein Radiotherapeutic Agent

Test	Primary Quality	Typical Quality Attribute Acceptance Criteria
Appearance	Description	Color, solid-particle characteristics or solution clarity acceptable
Peptide map	Identity/purity–impurity	Profile comparable to reference
Amino acid composition	Identity	Composition comparable to theoretical
Amino acid sequencing (total, N-terminal, C-terminal)	Identity/purity–impurity	Sequence identical to theoretical
Molecular weight	Identity/purity–impurity	Molecular weight comparable to theoretical
Charge isoform analysis	Identity/purity–impurity	Profile comparable to reference, pI values comparable to reference, purity–impurity levels acceptable
Electrophoretic molecular weight analysis	Identity/purity–impurity	Molecular weight comparable to reference, purity–impurity levels acceptable
Liquid chromatographic analysis	Identity/purity–impurity	Profile comparable to reference, purity–impurity levels acceptable
Protein determination	Quantity	Protein concentration value acceptable
Residual water (dry or lyophilized form)	Additional pharmacopeial	Water content value (%) acceptable
pH (liquid form)	General	Solution pH value acceptable
Bioassay	Potency	Potency value (activity/mass unit) or Reference ratio value acceptable
Bioprocess additives	Purity–impurity	Impurity levels acceptable
Downstream process additives	Purity–impurity	Impurity levels acceptable
Residual solvents	Additional pharmacopeial	Impurity levels acceptable
Host cell proteins	Purity–impurity	Impurity levels acceptable
Host cell DNA	Purity–impurity	Impurity levels acceptable
Sterility	Additional pharmacopeial	Passes test
Pyrogen or endotoxin	Additional pharmacopeial	Pyrogen test result or endotoxin content value acceptable

DNA: deoxyribonucleic acid

isolation/recovery steps. For solution phase synthesis, all reaction conditions, intermediate processing or purification steps and final deprotection steps should be described. For both methods, results of in-process monitoring for reaction completion should be provided (ninhydrin test, TLC, or Ellman's test as appropriate). A key component of peptide synthesis manufacturing involves final purification typically by single or multiple reverse-phase chromatography steps and therefore adequate descriptions of all physical (e.g., column packing type, particle size, and column dimensions) and operational (e.g., column loading and elution solvent composition, solvent flow rates and gradients, temperature, fraction collection method, etc.) parameters are required. Intermediate fraction collection and final product lyophilization or drying procedures should be described (31).

Quality   As chemical entities, peptide quality may be ascertained by a number of traditional analytical methods in addition to those specific for their amino acid content and peptidic nature. Evidence of the chemical structure should be obtained from chemical methods (e.g., elemental analysis), spectroscopic methods (e.g., IR, UV, NMR, CD) or other methods (e.g., MS, X-ray crystallography). Other properties such as solubility (e.g., pH, solvent variations) and physical characteristics (e.g., melting point, pK_a, particle size, crystalline structure) may also provide useful information. Characterization of product-related impurities should progress with development to ascertain the identity and quantity of truncated, deleted, oxidized, deaminated, racemized, and partially deprotected residues as well as other process-related products, residual solvents, and reagents (31). If the radionuclide binding component is added as part of the peptide synthesis to create the tumor targeting conjugate (see Chapter 3), some analytical testing to provide full characterization may be delayed until a later stage or be performed at both stages. Table 17.3 contains a typical list of tests that may be required for routine batch release testing. Potency tests for *in vitro* confirmation of tumor targeting function such as receptor binding assays are usually recommended especially if peptide configuration or formulation is critical to biological activity. If sterility is not required at this stage due to subsequent drug formulation, bulk peptide microbial limit testing may be substituted for the sterility testing. Specific additional pharmacopeial testing may be required on early lots (e.g., heavy metals, residue on ignition) to provide basic impurity information. Stability study data of dry and solubilized peptide should be provided under various conditions (as for protein RA) with special attention to degradation products associated with specific amino acid content, sequence and conformation. A supply of well-characterized reference material should be prepared for ongoing use.

17.3.2   Radionuclide Binding Component

The radionuclide binding portion of a PPRA may take several forms. They may be natural endogenous constituents (e.g., available aromatic or metal complexing amino acids), modified endogenous structures (e.g., thiols from reduction of disulfides), attached ligands (e.g., metal or halogen binding structures) or combinations thereof (e.g., attached ligands using thiols) (see Chapter 2). Tumor targeting components with

TABLE 17.3 Summary of Potential Regulatory Requirements for Batch Analysis Release Tests of a Peptide Tumor Targeting Component of Peptide Radiotherapeutic Agents

Test	Primary Quality Attribute	Description	Method Examples	Typical Quality Attribute Specification
Appearance	Identity/purity–impurity	Visual inspection		Color, solid-particle characteristics or solution clarity acceptable
Molecular weight	Identity/purity–impurity	HPLC-MS		Molecular weight value comparable to theoretical
Optical rotation	Identity/purity–impurity	Polarimetry		$[\alpha]$ 25°C, 589 nm value acceptable
Amino acid composition	Identity/purity–impurity	Chemical hydrolysis with HPLC analysis		Composition comparable to theoretical
Amino acid sequence	Identity/purity–impurity	HPLC-MS/MS		Sequence identical to theoretical
Liquid chromatographic analysis	Identity/purity–impurity	RP-HPLC, IE-HPLC, HI-HPLC, SE-HPLC		Profile comparable to reference, purity–impurity levels acceptable
Electrophoretic analysis	Identity/purity–impurity	SDS-PAGE, IEF		Profile comparable to reference, purity–impurity levels acceptable
Peptide determination	Quantity	Quantitative RP-HPLC (with reference standard)		Peptide concentration value acceptable
Bioassay	Potency	Cell/receptor binding assay (with reference standard)		Potency value (activity/unit mass) acceptable
Peptide content	Purity–impurity	Calculation		Peptide content value (%) acceptable
Moisture content (dry or lyophilized form)	Additional pharmacopeial	Karl Fischer		Water content value (%) acceptable
pH (liquid form)	General	pH electrode		pH value acceptable
Counter-ion content	Purity–impurity	HPLC		Counter-ion content level acceptable
Residual solvents	Additional pharmacopeial	Organic volatile impurities (USP ⟨467⟩)		Individual solvent level acceptable
Specific residual reagents	Purity–impurity	As required		Individual reagent level acceptable
Heavy metals	Additional pharmacopeial	Heavy metals (USP ⟨231⟩)		Heavy metal content value acceptable
Residue on ignition	Additional pharmacopeial	Residue on ignition (USP ⟨281⟩)		Residue on ignition level acceptable
Sterility	Additional pharmacopeial	Sterility test (USP ⟨71⟩)		Passes test
Microbial examination of nonsterile products	Additional pharmacopeial	Microbial enumeration test (USP ⟨61⟩), Tests for specified Microorganisms (USP ⟨62⟩)		Microbial type and levels acceptable
Pyrogens or endotoxin	Additional pharmacopeial	Pyrogen test (USP ⟨151⟩) or bacterial endotoxins test (USP ⟨85⟩)		Pyrogen test result or endotoxin content value acceptable

USP: United States Pharmacopeia; other abbreviations see Table 17.1

584

any of the cited radionuclide binding components will be termed tumor targeting conjugates (TTC) as they are essentially ready for radionuclide binding. Based on the desirable properties of many metallic radionuclides, chelating structures are commonly attached via bifunctional ligands (i.e., those ligands that incorporate two functional groups, a chelate for radionuclide binding and a linker moiety for attachment to the tumor targeting component) hence the term bifunctional chelate (BFC) is often used. These BFC may contain a range of chelating groups (tailored to the metallic radionuclide) and attachment chemistries (tailored to the tumor targeting component) and may include spacer or other groups that optimize subsequent radiolabeling procedures, maintain potency and/or influence biological function (e.g., pH or enzyme sensitivities). For BFC, the characterization of the metal binding (complexation) site with regards to endogenous metal content, metal ion specificity, oxidation state requirements, kinetic and thermodynamic binding properties, and compatibility with proposed radiolabeling procedures (such as the use of intermediate radionuclide binding agents) should be provided in the initial regulatory submission. This is especially important if these attributes may be altered from the unmodified chelate by the presence of the linker functionality or from its proximity to the tumor targeting component in the final conjugated form. Other bifunctional ligands may be attached to alter the pharmacological behavior of the tumor targeting component such as selective glycosylation or polyethylene glycol attachment and their pharmaceutical quality and process use must be similarly documented. All ligands must possess high chemical quality and stability that translates into reproducible conjugation to the tumor targeting component with minimal undesirable side reactions. For manufacturer synthesized ligands, both the physicochemical characterization and the specific quality attributes should be reported (similar to the nonpeptide-specific chemical structure, characterization, and batch release testing described earlier). When supplied as raw materials from qualified suppliers, certificates of analysis containing this type of quality testing results and ideally certification of their manufacture under GMP would be required depending on the development stage. Reagents used in the creation of radionuclide binding sites (such as chemical reduction of disulfides to thiols) should be of similar quality.

17.3.3 Radionuclide Component

The radionuclide component provides the primary therapeutic action of the PPRA by virtue of particulate emissions associated with its radioactive decay (see Chapter 2). There are a limited number of regulatory agency-approved therapeutic radionuclides suitable for use with PPRA (e.g., ^{32}P, ^{131}I, and ^{90}Y) and given their less than ideal characteristics; numerous other radionuclides may be preferable. Therefore, for most PPRA, codevelopment of the radionuclide component is desirable as evidenced by the large number of radionuclides that have been used for clinical trials in conjunction with PPRA approval. The following regulatory expectations for the radionuclide component may differ slightly depending on the intended PPRA development strategy, as a Kit PPRA format requires provision of the separate radionuclide in radiolabeling form whereas for a ready-to-use (RTU) PPRA format the radionuclide is incorporated into the PPRA during manufacturing (see Chapter 2). Radionuclide generator systems will not be specifically discussed but similar considerations for both parent and daughter radionuclide production and quality would apply. In general, the regulatory expectations for radionuclide quality are based on the radionuclidic, radiochemical, chemical, and pharmaceutical attributes as it impacts these parameters in the final product. The manufacture of the radionuclide must demonstrate that the radionuclide is consistently generated to control the potentially undesirable impurities to an acceptable level. In early development, reproducibility of production at a specified level of radioactivity (a least that proposed for clinical use) by a well-defined process is critical to evaluating the radionuclide quality as process deviations significantly impact quality and influence PPRA development. For example, significant amounts of radioisotopic radiochemical impurities (due to a specific radionuclide production process) that are present during radiolabeling would be expected to be incorporated into the product in proportion to the desired radiochemical and may contribute to radiodosimetry and/or radioassay differences when alternate radionuclide production processes are utilized (e.g., ^{124}I produced from the various tellurium isotope targets and nuclear reactions). Unwanted radiochemical forms may not participate in the intended binding reactions to the TTC, thus reducing the (apparent) radiolabeling yield and potentially requiring postlabeling purification (both of which may be acceptable during early development but complicate later stage development/commercialization). Both radionuclidic and radiochemical impurities may alter the stability of the final drug product if they contribute to excessive radiolysis. The presence of chemical impurities such as trace complexing agents and metal or halogen ions may complicate radiolabeling development and, in a worst case, present toxicity concerns. Depending on the final format, a low level of endotoxin and sterility of the radionuclide source solution may be required. A description of the radionuclide production systems consisting of accelerator/reactor design, targetry design and fabrication, irradiation system and control, irradiated target processing, and, if applicable, generator fabrication/characterization with summary flowcharts would be required to be submitted for new radionuclides. Of particular interest to regulators are target material quality testing, especially for enriched nuclide targets where verification of critical stable nuclide composition is normally required, and postirradiation target handling/purification systems where contact material quality and operational conditions directly impact both radiochemical quality and production reproducibility. Unusual radionuclides may require development of a specific dose calibrator assay to quantitate the level of radioactivity and assay development reports including calibration standards, and so on may be required to be submitted to support this activity as assurance of adherence to patient radioactivity dose limitations.

Data on radionuclidic impurities (typically evaluated by high-resolution gamma-ray or beta-spectroscopy), radiochemical impurities (typically by TLC or other chromatographic methods), chemical impurities (typically by inorganic and organic analysis with special emphasis on carrier or potentially interfering substances), and pharmaceutical quality (pH, endotoxin, and bioburden or sterility) should be provided on several lots of radionuclide unless justified and adequate specifications are required to support its clinical use. Stability studies on the radionuclide source solution under actual storage conditions with respect to impurity levels (radionuclidic in-growth of longer lived radionuclides, and radiochemical decomposition due to radiolysis) and radiolabeling efficiency using standardized conditions with reference product should be conducted with special consideration for the time of manufacture rather than the time of receipt or on-site calibration. When supplied as raw materials from qualified suppliers, certificates of analysis containing this type of quality testing results and ideally certification of their manufacture under GMP would be required depending on the development stage. Similar studies of other potential radionuclide source forms (e.g., dilute liquid, concentrated liquid, and dry) and differing radioactivity concentrations would be required when these issues are part of scale-up or later stage radionuclide component development.

17.3.4 Drug Product

The components described earlier (tumor targeting component, radionuclide binding component, and radionuclide component, respectively) constitute the starting materials for the assembly of the PPRA. Protein RA with a tumor targeting component that are produced from biotechnological processes or peptide RA produced from complex peptide synthesis procedures require complete documentation of this component's production and quality using regulatory submission drug substance (DS) content descriptors regardless of further ligand conjugation (e.g., BFC) and radiolabeling. The format of the PPRA dictates how the manufacturing is conducted and there are regulatory implications to each format. A ready-to-use (RTU) PPRA format consists of the components assembled by the manufacturer as a radiolabeled PPRA and supplied to the end user in this final functional form without further processing. A Kit PPRA format has two constituents, one of which is the TTC and the other of which is the radionuclide component, each provided in separate containers for radiolabeling onsite by the end user. Manufacturing and quality regulatory documents distinguish between DS and drug product (DP) that differ from each other where the DP could, for example, have additional formulation ingredients and be subjected to further processing (such as dilution) and a final sterilizing grade filtration into the final DP container (and still further processing such as lyophilization). While the DS may be sufficiently isolated in the manufacturing process and could be subject to specific quality testing, especially for the nonradioactive TTC in the Kit format, we will consider only DP manufacturing and quality in this review. For the RTU format the final dosage form (solution) in its final container/closure is the DP (RTU-DP) and for the Kit format there are two DP, one for the TTC (TTC-DP) and one for the radionuclide (RNC-DP) each in final dosage form in its final container/closure. The final radiolabeled Kit PPRA (KIT-DP) prepared from the TTC-DP and the RNC-DP must be fully characterized by the manufacturer as part of the regulatory submission documentation, even though it will not be supplied to the end user in this format. Additional separate component vials containing, for example, pH adjustment, intermediate radionuclide binding and other reagents may also be required in the Kit format to facilitate on-site step-wise preparation of the KIT-DP but only the optimum two vial kit format will be considered here. Many other formats can be envisioned that would be feasible for early development, such as those necessitating on-site postradiolabeling purification leading to the final DP, but these generally would present more complex commercialization expectations and will not be specifically discussed. Further consideration of the physical forms of the TTC-DP (solution or lyophilized) and RNC-DP (solution or dry) enables specific manufacturing process development for each form.

17.3.4.1 Manufacturing Activities   The unique requirements for this type of DP manufacturing process utilizing biological, chemical, and radiochemical components requires, even during early development, the manufacturing process to be sufficiently characterized to enable a reasonable understanding of all the critical reaction parameters and critical processing steps. While early stage DP may be prepared using defined reproducible conditions and described as such in regulatory submissions, it is advisable that studies using variations of the reaction parameters be conducted (with the help of design-of-experiment statistical methods) to elucidate critical parameters and appropriate specification ranges that establish minimum manufacturing control and assist in future process scale-up and more complete process validation for later stage submissions. While manufacturing changes are anticipated as development proceeds, their potential impact on product quality must always be reviewed and, in concert with actual product data, enable a legitimate product comparability assessment. All manufacturing operations should be completely described stating all materials, equipment, operational conditions as well as in-process testing involved including synthetic pathways and flow diagrams. Demonstrated control of all raw materials and quality of the final container closure system must be provided.

For the conjugation of the radionuclide binding component (e.g., a BFC) to the protein tumor targeting component, numerous reaction parameters should be considered for process characterization such as variations in the reactant mass, concentrations (and ratio), duration, pH, temperature, and organic solvent requirements. Conditions may be limited by the radionuclide binding component or tumor targeting component compatibility/stability/reactivity (such as optimum pH and organic solvents) that limit reaction yield optimization. However, a balance between final conjugation level to preserve tumor targeting function (often presented as a final conjugation ratio versus biological activity graph) and other quality attributes as well as adequate radionuclide radiolabeling yield (purity)/conditions (often presented as final conjugation ratio versus radiolabeling yield graph) must be established. The conjugation site may also be affected by the conjugation reaction conditions as specific peptides and/or side-chain sites can be preferentially activated especially in large proteins with multiple accessible reactive groups and a description of the known or most probable ligand attachment site should be provided (if not defined by structural analysis). For protein RA development, final conjugation ratios (e.g., average number of chelating groups per protein molecule) are often determined by HPLC-MS, colorimetric or radiometric assays (see Chapter 2) and the TTC may consist of a pool of unconjugated, single and multiply conjugated protein species. Purification processes (typically chromatographic and/or ultrafiltration-based) that are employed for the removal of residual (such as unreacted or degraded) ligands and undesirable conjugate types (such as those with excess or limited ligand attachment or polymeric forms) must be well documented and characterized in light of their critical role in preparing a reproducible TTC-DP. Residual chelating ligands and trace metal ions from all reaction constituents can have a significant impact on subsequent metallic radionuclide radiolabeling yield and stability. Thus, careful attention to material and process details for the conjugation reaction as defined by the process characterization is required to achieve acceptable results.

For radionuclide attachment to the TTC, limits on total radioactivity, radiolabeling yield, postradiolabeling stability, and clinical utility (mass dose) require balancing. The RTU-DP format should assure high radiochemical purity by virtue of a postradiolabeling purification process conducted by the manufacturer while any use of the Kit format must ensure a similarly high radiolabeling yield (purity) on-site, both of which are designed to minimize unwanted radiation dose from small amounts of radiochemical impurities at therapeutic DP radioactivity levels. This requirement for high radiochemical purity (>95%) is unique to PPRA relative to agents incorporating imaging radionuclides (>90%) where the tolerance for impurities is normally higher. Even simple reaction parameter (e.g., duration, temperature, TTC mass/concentration, reactant radionuclide and carrier concentration/volume, pH, age, formulation, etc.), variation can alter the radiolabeling reaction rate and yield. Thus multivariable studies lead to characterization of the radiolabeling process and critical parameter identification. Similar to the conjugation reaction, parameter variation may also influence the radiolabeling site especially for (multiple potential) endogenous binding sites on a TTC and in this case may in turn affect biological activity as well as overall radiolabeling efficiency and stability. It is also essential that the radiolabeling conditions are demonstrated to be robust enough to account for foreseeable on-site variations and that explicit instructions for use are provided with the Kit format for regulatory review.

For both the conjugation of the radionuclide binding component and the radionuclide attachment to the TTC, other pharmaceutical manufacturing steps may also be required depending on the format, materials and conditions. Procedures for solution degassing, component endotoxin reduction, headspace gas, lyophilization, final sterilization, and other operational activities need evaluation and control at the appropriate stages of development. At the late development stage, all of these processes and parameters can be fine-tuned to supply the basis for final process validation studies that provide an adequate design space for these parts of the manufacturing process.

Similar to the staged development of components, the early regulatory submissions typically require data from minimal lots of acceptable materials, manufactured at a clinically relevant scale using a suitably defined and controlled operational system. To evaluate DP acceptability, at least three independent batches of RTU-DP and for the Kit format TTC-DP and RNC-DP should be manufactured, taking into consideration the available number, mass (and radioactivity) quantity, and quality of available component batches. More complete delineation of component batch and DP batch relationship to establish a sufficiently characterized and robust manufacturing process should be established as development progresses as should various manufacturing scale and operational issues. These three lots provide DP to enable evidence of manufacturing consistency through product quality testing results, DP for other studies as well as nonclinical and clinical studies. Each stage may be supported by a number of separate DP batches and associated quality testing depending on the manufacturing and clinical development strategy. The development of PPRA analogues incorporating suitable imaging radionuclides whereby a radioisotope (e.g., ^{123}I for ^{131}I) or radionuclide of the same basic radiolabeling chemistry (e.g., ^{111}In for ^{90}Y) is substituted for the radiotherapeutic radionuclide to be used for biodistribution/ radiation dosimetry evaluation or other uses, should be considered during development with all the necessary radionuclide and radiolabeling studies, manufacturing, and quality implications.

17.3.4.2 Quality The final complement of individual component characterization/quality assessments and process control is the evaluation of manufactured DP quality, incorporating many of the same analytical tests used for the components. In addition, the DP needs to be characterized for specific pharmaceutical development requirements that demonstrate adequate control of various content and container/ closure features. These development studies (Table 17.4) need to be considered for their jurisdictional applicability, relevance to the specific PPRA in terms of its ingredients and packaging and, unless otherwise justified, completed prior to regulatory approval (32, 33). As with all quality control tests, their use (or omission) must be justified and a potential panel of batch release tests for a PPRA is provided in Table 17.5. Individual PPRA specifications must be initially established and

TABLE 17.4 Summary of Potential Regulatory Requirements for Pharmaceutical Development Studies for Protein and Peptide Radiotherapeutic Agents

Study Type[a]	Description
Container adsorption	Verification of nature and extent of product loss to contact materials (vial and closure)
Container closure leachables and extractables	Verification of identity and amount of leached and extracted materials from contact materials (vial and closure) under specified conditions in final dosage form
Antioxidant stability	Verification of identity, amount and stability of pharmaceutical anti oxidant in final dosage form
Container closure integrity	Sterility assurance of vial closure system
Critical excipient content	Verification of critical excipients(s) identity and amount in final dosage form
Buffering capacity	Verification of buffer capacity in final dosage form
Nonaqueous solvent content	Verification of pharmaceutical solvent(s) identity and amount in final dosage form
Compatibility	Verification of nature and extent of active/excipients interactions in final dosage form
Administration system compatibility	Verification of nature and extent of radiopharmaceutical interaction with patient administration system components under typical use conditions

[a] Assumes simple nonviscous (reconstituted) solutions, nonplastic containers, no antimicrobial preservatives, no unusual excipients, nonparticulate formulation, and no modified release components are present.

TABLE 17.5 Summary of Potential Regulatory Requirements for DP Batch Analysis Release Tests for Protein and Peptide Radiotherapeutic Agents

Test	Quality Attribute	Method Examples	Typical Quality Attribute Specification
Appearance	Description	Visual	Color, solid-particle characteristics or solution clarity acceptable
Molecular weight	Identity	HPLC-MS, SDS-PAGE (R/NR)	Identity profile of tumor targeting conjugate acceptable
Content assay	Quantity	RP-HPLC, chemical protein assay	Amount of tumor targeting conjugate quantity value acceptable
Biological activity	Potency	Immunoassay, cell binding assay	Potency value (activity/unit mass) of tumor targeting conjugate acceptable
Biochemical purity	Purity–impurity	SDS-PAGE (R/NR), IEF, RP-HPLC, IE-HPLC, HI-HPLC, SE-HPLC	Biochemical purity–impurity profile acceptable
Solution pH	General	pH electrode	Solution pH value acceptable
Specified chemical purity	Purity–impurity	RP-HPLC, IE-HPLC, HI-HPLC, SE-HPLC, Spot test	Chemical purity–impurity profile acceptable
Headspace gas content assay	General	Gas chromatography	Gas concentration (pressure) value acceptable
Radiochemical quality	Identity/purity–impurity	TLC, SE-HPLC, RP-HPLC	Radiochemical identity/purity–impurity profile acceptable
Radionuclidic quality	Identity/purity–impurity	Gamma-spectrometry	Radionuclidic identity/purity–impurity values acceptable
Radioactivity assay	Quantity	Dose calibrator	Radioactivity quantity value acceptable
Osmolality	General	Osmometer	Osmolality value acceptable
Water content (lyophilized, dry components)	Additional pharmacopeial	Karl Fischer	Water content .value (%) acceptable
Particulate matter in injections	Additional pharmacopeial	Particulate matter in injections (USP ⟨788⟩)	Particulate content value acceptable
Pyrogen or endotoxin	Additional pharmacopeial	Bacterial endotoxins test (USP ⟨85⟩) or pyrogen test (USP ⟨151⟩)	Pyrogen test result or endotoxin content value acceptable
Sterility	Additional pharmacopeial	Sterility test (USP ⟨71⟩)	Passes test

TLC: thin layer chromatography; other abbreviations see Table 17.1 and Table 17.3.

progressively tuned based on historical PPRA batch results, manufacturing capabilities, nonclinical (and previous clinical study, if any) results as well as regulatory/pharmaceutical requirements and other considerations. The unique radioactive nature of PPRA necessitates specific quality tests related to this attribute (Table 17.5) but for some pharmaceutical development and quality testing, DP that are nonradioactive but maintain all other features can be employed. A simulated (or mock) radiolabeled DP may be useful as a substitute for the radiolabeled DP when the radioactivity is not central to the test conduct and when the presence of an isotopic (radiolabel) substitute and the radiolabeling process effects should be accounted for in the test for the DP quality attribute. Simulated radiolabeled DP is prepared by substituting comparable levels of stable isotope for the radionuclide component in the same chemical form in an identical formulation for use in the (radio)labeling (and purification) process, or the use of decayed radiolabeled DP may be justified. In addition to previous characterization studies of the radiolabeling process, the manufacturer must also ensure that each batch of TTC-DP is tested (up to the shelf-life limit of the KIT-DP) prior to release using typical radiolabeling procedures to produce the KIT-DP. Additional stress radiolabeling studies using a selected range of critical parameters (e.g., high radioactivity levels, volume or pH extremes) that have been shown during development to have a potential for significant impact on selected KIT-DP quality attributes (e.g., radiochemical yield/purity, biochemical quality, or potency) may also be valuable. Also unique to PPRA is the general requirement for on-site quality testing on an individual container basis for radioactivity assay and radiochemical identity/purity, especially for KIT-DP. Thus, dose calibrator assay methods and simple instant TLC methods are recommended to be developed for use as facile on-site preadministration tests for product quality. The panel of potential tests for batch release (Table 17.5) reflects all necessary DP quality attributes involving identity, purity, potency, quantity and general (parenteral) pharmaceutical tests. Some batch release tests may be conducted retrospectively depending upon the format of the DP and process validation data. Quality control analytical methods, even at the early development stages, must be scientifically sound, conducted with a standardized method using appropriate reference materials, positive and negative controls, where appropriate as well as possess the necessary precision, sensitivity, specificity and accuracy to demonstrate fitness for the measurement purpose. Full analytical method validation data is normally not required until late stage regulatory submissions but initial assay quality and improvement should be a cornerstone of DP development. For synthetic peptide RA DP nearing final development all impurities greater than 0.1% need to be identified but a somewhat greater flexibility in this limit is usually extended to rDNA protein PPRA when appropriate (34). Of particular regulatory concern are potency assays, for which well-defined *in vitro* antigen (for MAbs) or (cellular) receptor binding assays should be developed (rather than *in vivo* potency studies). These assays should be readily amenable to relatively rapid RTU-DP, KIT-DP, or nonradioactive content testing (TTC, TTC-DP) and can be based on specific biointeraction (e.g., affinity or reactive fraction) parameters (see Chapter 2). For small peptide RA that are structurally well characterized and have adequate antigen/receptor binding studies included in radiopharmacological development, the routine application of potency assays may be negated. Radiochemical purity assays should be developed with an important regulatory consideration of complete test sample recovery such that all radioactivity can be accounted for as either the DP or an impurity category.

As part of the assessment of the quality of the PPRA, it is strongly recommended that early in development, a well-characterized reference lot of an appropriate form of the TTC should be prepared and appropriately maintained as they constitute the critical targeting function of the PPRA, and be replaced at the Phase 3 manufacturing stage with a final reference lot prepared for each as representative of final manufactured product. These reference materials serve a critical function in establishing a baseline for early and late product quality testing specifications (and the comparability between the two), and can provide early indications of storage instability.

Stability studies should be conducted on the nonradioactive TTC (RTU format) and TTC-DP (Kit format) in relation to the maintenance of appropriate quality attribute specifications under normal storage conditions supported by intermediate, accelerated and/or stress studies as appropriate with emphasis on biochemical purity, aggregation, potency, and radiolabeling efficiency (consistency). The duration of the real time stability studies is initially dictated by the intended duration of the clinical trial progressing to an achievable and commercially viable period. For the radioactive RNC-DP, the KIT-DP (Kit format) and the RTU-DP (RTU format), stability in relation to the maintenance of appropriate quality attribute specifications under normal storage conditions supported by intermediate, accelerated and/or stress studies as appropriate should also be conducted with special consideration to radiolytic degradation causing radiochemical impurities and nonradiochemical effects (such as aggregation and potency reduction), which necessitates upper limit radioactivity amount/concentration studies. Stability study duration for radioactive DP is dictated by the initial radioactivity level and half-life of the radionuclide in relation to a minimal useable clinical dose. For all shipped DP, transportation stability may need to be demonstrated, especially for sensitive DP solution forms that may under go significant temperature variations (freeze/thaw/heat) and/or with unstabilized formulations and/or agitation sensitive materials (e.g., proteins). Maintenance of sterility, nonpyrogenicity and acceptable particulate level status in all cases should be confirmed. Appropriate liquid formulations should be tested in the upright and inverted positions. Stability indicating assays may need to be developed depending on the extent and nature of the impurities potentially present or observed during stress/accelerated studies (35).

17.4   NONCLINICAL STUDY ELEMENTS

As with all drugs under development, PPRA must undergo varying levels of nonclinical testing to ensure the safety and potential efficacy of the drug both prior to initiating human clinical trials (preclinical) and in later development stages. For drugs developed to treat life-threatening diseases, such as cancer, some relief in the development process is available in order to accelerate the access to market of these potentially life-saving therapeutics and the specific test relevance to this patient population, especially patients with late stage or advanced disease and short life expectancy. Due to the variability of PPRA composition, targets and application, nonclinical testing plans should be reviewed with and approved by the appropriate regulatory jurisdiction prior to execution as some flexibility in the traditional testing plan may be justified.

PPRA can generally be classified as wholly synthetic or as predominantly biological (including the synthetic radionuclide binding component) and overall as therapeutic radiopharmaceuticals (including the radionuclide component) based on their composition and origin for nonclinical regulatory purposes. The synthetic classification (peptide RA) is characteristic of virtually all small-molecule drugs while the biological classification is indicative of large complex proteinaceous biotechnology-derived molecules (protein RA). These two types of PPRA are subject to a set of normally independent nonclinical regulatory guidance in contrast to the overall therapeutic radiopharmaceutical classification that carries some relevance to diagnostic radiopharmaceuticals but there is no encompassing therapeutic radiopharmaceutical category regulatory guidance. Both classifications need to address the types of (additional) studies that are needed to assess the safety/toxicity and supporting (radio)pharmacological properties of the radiotherapeutic form. Thus the regulations governing the development and approval of these agents must consider all facets of each classification that apply, often necessitating the involvement of multiple sets of regulatory agency groups and multiple sets of regulatory guidelines that must be interpreted for compliance. For all safety pharmacology, radiopharmacology and toxicology testing, justification of the species chosen, animal gender, the number of animals, size and the use of negative and/or positive control groups, route of administration, mass (and radioactivity) doses, statistical analysis, the methodological details, and appropriate tabulation of findings/results must be adequately described and discussed in appropriate regulatory submissions. Starting mass doses should be based on pilot studies and all other relevant information, or be based on multiples (typically 10–100) of the scaled (typically via milligram per kilogram or milligram per square meters) human dose estimate. In general, traditional safety pharmacology and toxicology nonclinical studies can be conducted with simulated radiolabeled PPRA to establish these effects independent of the radionuclide and simplify study operations. The requirement for the use of scaled therapeutic radioactivity levels with the (development) DP for these studies should be evaluated on an individual case basis, being more likely for a PPRA with a new radionuclide (and its impurities), for a PPRA where there is significant unusual tissue accumulation, or other situations where the relationship between local radioactivity and biological consequences have not been previously established or cannot be soundly projected from cellular and/or whole body radiodosimetry studies. It is important that the formulation used is as relevant as possible to that of the intended final DP (especially when novel excipients, radiolabeling additives, and so on are used) and for most PPRA where meaningful radiolytic decomposition of the mass dose may occur over the useful life, the application of decayed radiolabeled (development) DP may be suitable as a worst-case composition. These studies are also required to be conducted according to GLP standards (8, 9), which assure the quality and integrity of the safety data used to support a clinical trial or marketing application. It is recognized that all material used for nonclinical testing must be appropriately manufactured and characterized to ensure both an acceptable quality, especially purity and potency, and stability) and verifiable continuity between the material tested in the nonclinical setting with that used in the clinical trial and for marketing approval. The selection of studies is essentially the same in Canada, the European Union or the United States.

17.4.1 (Radio)pharmacology Studies

In general, for either peptides or proteins, regardless of their origin, certain supporting studies are required to be completed in addition to the safety pharmacology and toxicity testing mandated by the regulatory agencies prior to any human clinical trials. Depending on the nature of the components and the proposed pharmacological handling of the PPRA, several studies should be undertaken to characterize specific aspects of its biological behavior as indicators of suitability and quality. *In vitro* studies of stability of the PPRA in normal whole blood and in normal plasma at expected physiological concentrations at various incubation times with appropriate analysis serve to determine potential circulating (radio)pharmaceutical associations, degradation and metabolism. Thus, uptake/binding to (normal) circulating cells could have PK implications leading to altered tissue uptake, metabolism by dehalogenases, endogenous peptidases, or transchelation by certain plasma fraction elements, and could generate significant amounts of radionuclide that are no longer associated with the tumor targeting component. Cell culture studies using receptor (binding site) bearing target cells are valuable to confirm the internalization or exterior retention of the PPRA and/or radionuclide under anticipated conditions that are especially relevant if these features were designed into the structure to confer a radiobiological advantage, for example, intracellular pH or enzyme sensitive conjugate component cleavage. Evaluation of the saturation of the target cell receptors in culture by the native (or reference) ligand in competition with the PPRA has value for clinical dose and specific radioactivity estimation especially in conjunction with biodistribution findings and may serve to reinforce the receptor up/down regulation or nonpharmacological activity under conditions that have a degree of physiological relevance. Various other studies of the tumor targeting capacity may be important from cell culture (or isolated receptor/binding site) studies prior to clinical evaluation such as association/dissociation constant (affinity) determination and estimates of radiopharmaceutical (immuno) reactive fractions (see Chapter 2). Many of these types of studies are useful during development to determine appropriate type, site and levels of conjugation, radiolabeling procedure effects, and so on, that may impact the tumor targeting component activity. It is expected that from at least one of these types of studies a reliable assay of the overall biological function will be developed to serve as the quality control test for potency.

An important regulatory characterization study for large protein conjugates (typically MAb) that has a significant potential for numerous cellular interactions by specific and nonspecific mechanisms is immunohistochemical staining (cross-reactivity testing) of (normal and cancerous) human tissue sections exposed to the protein RA. Binding of the protein RA to the tissue is detected with reagent combinations (typically biotin conjugated secondary reagent/antibodies reactive with an unbound portion of the protein RA and subsequent addition of streptavidin bound dye), which permits a visual semiquantitative indication of a wide panel of potential human tissue interactions. For protein RA for which no suitable secondary reagent is available conduct of this type of study may be interpretable, with an understanding of its limitations, by using a specially prepared biotinylated form of the primary protein or peptide. The results of these *in vitro* immunohistochemical staining studies are particularly relevant to the evaluation of potential pharmacological side effects/toxicity of nonspecific and low level receptor/binding site prevalence tissue interactions as well as confirming specific reactivity of the protein RA with tumor target tissues.

Once the basic *in vitro* characterization has been completed, further studies using *in vivo* systems are required to establish potential efficacy and biodistribution/metabolism/radiodosimetry elements that are used to justify an acceptable risk/benefit profile for human use. Of primary concern to the regulatory agencies is the evidence from *in vivo* animal models that the PPRA shows tumor targeting capability and tumor volume/mass stabilizing or reducing effects at radioactivity (and total PPRA mass) doses that produce acceptable toxicity profiles and are likely scalable to the anticipated human dose and exposure. The total PPRA mass administered affects the specific activity of the radiopharmaceutical as well as the potential for receptor/binding site saturation (and *in vivo* up-/downregulation), for dose-dependent PK evaluation and toxicity implications. Where at all possible, models used for these assessments should consist of tumors that express the specific tumor targeting component receptor/binding site. Animal models that utilize human tumors with this property (often adapted to grow in immunocompromised rodents, such as athymic nu/nu or severe combined immunodeficiency (SCID) mice) provide the most realistic scenario and are to be recommended (see Chapter 11). Each PPRA should be evaluated for the best approach to provide the appropriate scientific evidence of *in vivo* efficacy to support the clinical trial application. Of equal importance to efficacy evaluation is the determination of the biodistribution of the PPRA with subsequent analysis of metabolism, PK, and radiation dosimetry (see Chapter 13). Animal biodistribution studies are typically first carried out in normal or tumor-bearing rodents with analysis of blood, organs/tissues, and excreta to determine how rapidly the PPRA radionuclide is cleared and which tissues have the highest accumulation as a function of time after administration (see Chapter 2). Metabolic and PK studies may also be completed at the time of normal rodent biodistribution or be completed as special studies in larger animals especially if metabolism is expected to be significant within the radionuclide lifetime and the assays (and sampling periods) required to assess these parameters are not conducive to the use of small animals. Analysis of blood/plasma PPRA content with time after administration using appropriate software enables estimates of useful PK parameters such as area under the curve (AUC), half-life(s), clearance (CL), and volume of distribution (V_d). Analysis of both the tumor targeting component (using a suitable bioanalytical method such as ELISA) and the radionuclide (using a suitable radiation detection instrument such as a γ-counter or liquid scintillation counter) serves to highlight radionuclide dissociation. PK results may also be obtained from toxicokinetic (TK) analysis or as part of other nonclinical studies. From the animal biodistribution (and radionuclide emission) data, estimates of the radiation dose exposure for standardized organs/tissues can be tabulated using well-established software programs (see Chapter 13). The animal radiation dosimetry estimates must be scaled for overall human exposure and specific identification of critical organ(s) that provide a critical basis for regulatory acceptability for clinical use.

Other specialized studies to evaluate the use of accessory reagents (amount/timing) to enhance clearance/reduce uptake in specified organs and thus reduce nonspecific exposure to the PPRA, may be required to support their application with the PPRA in human clinical trials. For two and three-step pretargeting approaches (see Chapter 8), studies of the dosing levels/ratios of all components and various timing intervals in animal models should be provided with biodistribution/radiodosimetry estimates and toxicity implications that suggest a reasonable starting point for these parameters for proposed human use. The use of PPRA radioisotope/radionuclide analogues to facilitate biodistribution or other radiopharmacological studies, must provide adequate evidence of analogue PPRA quality and behavioral comparability especially if this approach will also be employed clinically. All *in vivo* nonclinical (radio) pharmacology studies should be described in the initial regulatory filing including species/model justification, methodology (including all basic data, calculations, and dosimetry assumptions), results, and discussion of relevant findings. Ideally, all (radio) pharmacological animal studies, but particularly large animal studies, should be conducted according to GLP (8, 9) regulations.

17.4.2 Safety Pharmacology Studies

Safety pharmacology studies are carried out to evaluate effects of the drug on critical organ systems. Cardiovascular studies consist of the measurement of blood pressure, heart rate, and an evaluation of the electrocardiogram. *In vivo*, *in vitro*, and/or *ex vivo* evaluations, including methods for repolarization and conductance abnormalities should also be considered. Central nervous system studies can include motor activity, behavioral changes, coordination, sensory/motor reflex responses, and body temperature and this is most often accomplished by using a functional observational battery or other tests. Respiratory studies can include respiratory rate and other measures of respiratory function (e.g., tidal volume or hemoglobin oxygen saturation) with quantification by appropriate methodologies. Evaluation of these three systems (the cardiovascular, central nervous, and respiratory) constitute the core battery studies with studies of other functions (renal, gastrointestinal, immune, etc.) added as necessary (36, 37). These studies are normally conducted in two relevant species, usually rats and dogs (although justification can be made for other species) and are normally required to be completed prior to initial human use. If concerns with the PPRA are generated during core battery testing, follow-up studies are required to further delineate these effects (38). Safety pharmacology studies may not be required for highly specific tumor targeting protein RA (based on supporting (radio)pharmacology and toxicological evaluation/results) and usually not at all for PPRA that are solely intended for clinical testing and approval in end stage cancer patients. However, on a case-by-case basis for some protein RA, it may be necessary to conduct these studies using *ex vivo* or *in vitro* models, depending upon their cellular targets and the appropriateness of available models for performing these studies. Further, the safety pharmacology testing may be incorporated into toxicity testing which may negate the necessity of doing stand-alone studies, but these combination studies must satisfy the intent and outcome of both types of studies (37). All studies associated with safety pharmacology testing must be conducted according to GLP standards (8, 9), however, it is recognized that some studies employing specialized test systems which are often needed for protein RA may not be able to comply fully with GLP and in this case documentation of the GLP shortfalls is required.

Special considerations would be applied to PPRA safety pharmacology studies that would be used to support pre-Phase 1 and early Phase 1 clinical investigations. In the European Union, applications for microdose exploratory investigations do not require the conduct of safety pharmacology studies unless toxicological or other information warrants but for subtherapeutic or therapeutic doses in exploratory clinical trials, core battery safety pharmacology is required (36). For the United States exploratory IND submissions, core battery safety pharmacology studies are not required for microdose PK or imaging studies but are required for submissions using dosing at pharmacological levels (39).

17.4.3 Toxicology Studies

Synthetic peptide RA toxicity testing should be comparable to standard small-molecule drug candidates and thus will include acute and repeat dose toxicity, TK/PK analysis, local tolerance, genotoxicity analysis, immunotoxicity analysis, phototoxicity potential, carcinogenicity analysis, and reproductive toxicity analysis and are relatively regulatory jurisdiction independent. A notable exception is the planned requirement by the FDA for late radiation toxicity studies applicable to peptide RA to capture acute and delayed radiation effects in long term studies (40). The studies required to proceed with a first in human safety study (Phase 1) will generally include acute and repeat dose toxicity, TK/PK analysis, local tolerance and genotoxicity analysis with other appropriate studies to be completed prior to drug approval. It is generally recommended that two species (one rodent and the other nonrodent) are used for acute and repeat dose toxicity testing studies. Acute (single administration) and repeat dose (up to 14 daily administrations) toxicity studies are typically designed to incorporate multidose levels using the intended route of administration with an aim to provide justifiable evidence of the maximum tolerated dose (MTD) level or no observable adverse effect level (NOAEL) that serve as a basis for estimating a safe starting human dose (typically 1/50 to 1/100 of the scaled NOAEL), provide indications of specific organ toxicity, toxicity reversibility, and potential clinical monitoring parameters. Assessment usually consists of mortality, clinical hematology and biochemistry, clinical observations, weight changes, and postmortem gross and microscopic evaluation on all major organ systems. Toxicology study options for peptide RA (and protein RA) prior to clinical trials in late stage cancer patients place less emphasis on NOAEL and MTD determinations and more on establishing safety by mimicking the proposed clinical administration schedule. TK/PK analysis (to verify peptide RA exposure in the toxicity studies) is generally conducted from blood sampling after the first and last administration in the repeat dose toxicity study and involves assay of the tumor targeting conjugate only (using a suitably validated bioanalytical assay) if supplemented with radiopharmacology (plasma PK) study support. Small peptide RA may have very short plasma half-lives due to fast translocation from the plasma and extensive intravascular metabolism or organ sequestration (e.g., liver or kidneys) that necessitates an appropriate sampling plan and requires very high bioanalytical assay sensitivity for low mass dose administration. Local tolerance effects can be incorporated into acute and repeat dose studies with macroscopic and microscopic histological evaluation of the injection site/ surrounding area and if a deliberate extravasation/misadministration tolerance study is indicated based on product-/formulation-specific characteristics then the radiolabeled peptide RA (and protein RA) should be employed. Genotoxicity studies assess the potential for DNA damage that could result in the permanent mutation of genes (that may play a role in the multistep process of malignancy). Two testing approaches are available to meet regulatory requirements consisting of a test for gene mutation in bacteria, a cytogenetic test for mammalian chromosomal damage and an *in vivo* test of rodent hematopoietic cell chromosomal damage or a test for gene mutation in bacteria with two *in vivo* tests for chromosomal damage (41). Analogous to traditional drug impurities risk analysis, peptide RA that consist of a relatively small mass dose ($<1.5\,\mu g$) should normally be excluded from genotoxicity testing (42, 43). As discussed previously, since radioactivity (ionizing radiation) is genotoxic and this effect/risk can usually be reliably estimated from radionuclidic data, genotoxicity using radiolabeled PPRA is generally not required. Phototoxicity studies can be excluded if it can be demonstrated that no skin or ocular uptake and retention is evident (from supporting (radio)pharmacology studies) and that *in vitro* photochemical properties are negligible. Immunotoxicity studies include evaluation of direct effects on the immune system (stimulation/inhibition) and need only be performed when a cause for concern is derived from other toxicity studies, relevant pharmacological properties, or structural features of the peptide RA. Immunogenicity studies to evaluate the consequences of antiproduct antibody generation are required as part of repeat dose toxicity studies (36, 44). The extent and nature of nonclinical immunogenicity studies should be evaluated on a risk-based analysis in light of their generally poor predictability to humans. Factors that should be taken into account include the cross-species expectation (e.g., human proteins in animals and other combinations); animal model suitability (immunocompetent versus immunocompromised), peptide RA composition, impurity profile, and molecular size, the peptide RA mechanism of action, peptide RA comparability to native (if any) analog, potential seriousness of clinical response, dose, dose frequency, and interval of peptide RA (and route of administration if not intravenous) as well as potential immunocompromised status of the animals and the intended oncology patient population. Nonclinical immunogenicity study design should include peptide RA PK/pharmacodynamic assessment in the presence of an antigenic response (antipeptide RA antibodies) as well as overall toxicological implications (inhibition, enhancement, or undetermined) that adds to the understanding of peptide RA behavior. Carcinogenicity studies can be postponed or eliminated for cancer therapeutics if genotoxicity is unremarkable and justification based on exposure/frequency of use and lack of implications on the specific cancer population can be substantiated. Reproductive toxicity studies involving preconception, preimplantation, and peripostnatal elements are also generally not required prior to early clinical trials if only men, women not of child-bearing potential and women of child-bearing potential with adequate pregnancy testing and pregnancy prevention are eligible but could be required prior to late stage trials and approval if the intended population is expanded beyond these restrictions.

Protein RA toxicity testing should be comparable to other biological drug candidates and thus could include acute and repeat dose toxicity, TK/PK analysis, local tolerance, genotoxicity analysis, immunotoxicity analysis, carcinogenicity, and reproductive toxicity analysis and are more regulatory jurisdiction dependent. However, the diversity of structure, mechanism of action, and biological properties that typically translates into exaggerated pharmacological actions rather than overt toxicity allows greater flexibility in study selection and conduct in comparison to traditional small-molecule toxicological analysis. The studies that are required to proceed with a first in human traditional Phase 1 clinical trials will generally include acute and repeat dose toxicity, TK/PK analysis, and local tolerance with other appropriate studies to be completed prior to approval. It is generally acceptable that only one relevant species (usually nonrodent) is used for acute toxicity studies. Repeat dose toxicity testing studies are typically of shorter duration covering at least one cycle of the intended clinical dosing schedule. TK analysis is required as for peptide RA, but the bioanalytical method for protein PPRA in plasma determination may be more complex (e.g., an immunoassay to differentiate between the protein RA and endogenous proteins and other interfering substances) and generally less specific (reactive with fragments and intact protein RA) requiring more method development to confirm suitability. Local tolerance testing must be conducted with similar considerations as those for peptide RA. Genotoxicity studies are generally not required for protein RA if no cell nucleus association of the tumor targeting component is supportable and the mutagenic potential of the radionuclide binding component is rationally excluded. Immunotoxicity studies are generally not applicable for protein RA but unlike peptide RA, immunogenicity in humans is expected to be more prevalent for large molecular weight proteins and therefore a review of the evidence available for the immunogenic potential of the protein RA (as for peptide RA) should be conducted to determine whether specific immunogenicity studies are required to be conducted prior to initiating clinical trials. Carcinogenicity and reproductive toxicity testing can be handled in the same fashion as for peptide RA.

Special considerations would be applied to PPRA toxicity studies that would be used to support pre-Phase 1 and early Phase 1 clinical investigations. In the European Union and the United States, applications for microdose PPRA clinical trials require the filing of information from an extended acute single dose level study in one justifiable mammalian species with TK profiles and no other toxicology studies. For an exploratory clinical trial using a single subtherapeutic or therapeutic dose, the European Union requires the conduct of an extended single dose acute toxicity study in both a rodent and nonrodent species with TK profiles. If the clinical trial is designed to use a single dose or if it is a repeat dose (up to 14 days) exploratory therapeutic clinical trial and is not intended to evaluate MTD, a 2-week repeated dose toxicity study in a rodent and nonrodent species are required using appropriate dosing criteria. These clinical investigations also require the conduct of an appropriate genotoxicity study unless it can be otherwise justified as previously discussed. For an exploratory clinical trial on subtherapeutic or therapeutic dose(s), the United States requires the conduct of a repeated dose 14-day toxicity study in a single rodent species with TK (and confirming or dose escalation study in a nonrodent species) and the appropriate genotoxicity study. For mechanism of action-based exploratory clinical trials, the United States will also accept modified pharmacological and toxicological study plans to select clinical starting doses and dose escalation schemes (36, 39).

17.5 CLINICAL STUDY ELEMENTS

Human clinical trials are the most critical element of the regulatory approval process to obtain marketing authorizations in Canada, the European Union, and the United States and generally require the most resources and time to complete. Adherence to GCP (2) is required for all clinical trials that are used to support marketing applications (NDS, MA, and NDA). This multijurisdictional standard, established by the International Conference on Harmonization (ICH) for the design, conduct, monitoring, and reporting of human clinical trials, ensures that subjects' rights and safety are maintained and that the data generated in the trial is reliable. GCP also encompasses specific responsibilities for independent ethics committees (Research Ethics Board (REB), Canada; Ethics Committee (EC), European Union; and Institutional Review Board (IRB), United States), investigators (including medical care of subjects, compliance with protocol, informed consent of subjects, and various reports, in particular, safety reports), and sponsors (including trial quality assurance, trial design, financing, regulatory authority notification/submission, in particular, adverse event reports, and product quality). GCP also provides format and content for the most important clinical trial documents, the clinical protocol, the IB and the informed consent form with cross-jurisdictional acceptance of these documents if constructed according to GCP standards. A further complication for PPRA clinical investigations is the need to adhere to international radiation safety principles (e.g., as low as reasonably achievable (ALARA)) and nuclear safety regulatory agency guidelines for minimization of radiation exposure for all personnel from radioactive dose administration, treated patients and their by-products by providing, for example, specialized radioactivity containment suites and a system of operational control of all radioactive patients and radioactive material handling activities.

Consideration for first in human studies should carefully evaluate the perceived risks associated with the potential PPRA. These would include the mechanism of action (e.g., targets that are ubiquitously expressed in the immune system or biological cascade cytokine release systems such as the blood coagulation system), nature of the human target (e.g., structure, tissue distribution, etc.), and the relevance of any animal species to be used for nonclinical safety evaluation (e.g., comparison of animal to human target through structure, homology, nature of pharmacologic effect, etc.) where the questionable relevance of available species will add to the perceived risk and may directly impact study design, dosing, and monitoring intensity (23). If the PPRA lacks significant detectable mass dose pharmacologic action (as the mechanism of action is tumor-targeted radioactivity and assuming that immunogenic responses are negligible), nonpharmacological levels can be estimated from animal toxicity study results (NOAEL or other parameter) appropriately scaled to humans with the human starting (safe) dose established using a 50–100 reduction factor. The radioactivity dose levels should be based on critical organ values (derived from animal dosimetry studies) with the aim of minimizing patient radiation exposure. In both cases, radiolabeling considerations and the intended study goals such as sufficient radioactivity for an imaging biodistribution study or increasing mass dose for saturation analysis studies must also be considered. The importance of early clinical investigation has led to several regulatory mechanisms that would allow the evaluation of PPRA outside of the traditional Phase 1 development approach. Evaluation at this early stage allows for fine-tuning of certain aspects of the clinical development of the PPRA prior to pursuing and/or complementing traditional Phase 1 testing. Further efficacy and safety studies via the traditional approach (Phase 1/2/3) will normally be required prior to drug approval.

17.5.1 Pre-Phase 1 and Exploratory Phase 1 Clinical Investigations

Canada and the United States have programs for clinical investigation prior to Phase 1 studies that do not require a regulatory submission and are confined to subpharmacologic dose level investigations only. In Canada, the positron-emitting radiopharmaceuticals (PER) basic research program is applicable to nonbiologic (nonprotein RA) PER, where their study for research purposes is allowed without the filing of a CTA (45). Limitations on this use of PER in research include the requirement of an established product safety profile, the use of subpharmacologic doses, whole body radiation dose limits, adult nonpregnant subjects, documented evidence of PER quality and oversight by an REB. These research studies are intended to determine the pharmacokinetics or metabolism of a drug, to obtain information related to normal human biochemistry or physiology, or to assess changes caused by aging, disease or treatment interventions. However, these studies cannot be used to study pharmacodynamics, look specifically for adverse events (although any adverse events occurring during the investigation must be documented), provide immediate diagnostic or therapeutic feedback or assess the safety or efficacy of the drug. This approach would be limited to PER radiolabeled (e.g., ^{124}I or ^{64}Cu) peptide RA studies that meet all the necessary criteria (46). This Health Canada policy is currently under review (March 2010) and regulations based on the policy may be codified with changes that are contradictory to the above discussion and information in Table 17.6. In the United States, radiolabeled drugs (both diagnostic and therapeutic) can be assessed on

TABLE 17.6   Summary of Selected Features of Early and Pre-Phase 1 Clinical Investigations for Protein and Peptide Radiotherapeutic Agents

Designation	Jurisdiction	Major Advantages[a] (Compared to Traditional Phase 1 Clinical trials)	Examples of Potential Applications for PPRA
PER basic research (41, 42)	Canada	No nonclinical testing Site-specific product quality compliance Potential limited product supply requirements Full GCP compliance not expected No CTA submission	Screening (multiple) established safe synthetic peptide RA with PER radionuclide in healthy volunteers or patients to ascertain biodistribution (PK)/radiation dosimetry and tumor localization confirmation
RDRC program (43)	United States	No nonclinical testing Site-specific product quality compliance Potential limited product supply requirements Full GCP compliance not expected No IND submission	Screening (multiple) established safe PPRA in healthy volunteers or patients to ascertain biodistribution (PK)/radiation dosimetry and tumor localization confirmation
Exploratory investigations (35) (microdose, sub therapeutic/therapeutic, multidose studies)	European Union	Reduced nonclinical testing No formal safety evaluation required (e.g., MTD) Potential limited product supply requirements	Screening (multiple) new PPRA in healthy volunteers or patients to ascertain biodistribution (PK)/radiation dosimetry and tumor localization confirmation
Exploratory IND (37) (microdose, subtherapeutic/ therapeutic, mechanistic studies)	United States	Reduced nonclinical testing No formal safety evaluation required (e.g., MTD) Potential limited product supply requirements	Screening (multiple) new PPRA in healthy volunteers or patients to ascertain biodistribution (PK)/radiation dosimetry and tumor localization confirmation

[a] See text for description of limitations and other issues.

603

a research basis in humans at licensed facilities by qualified investigators under an FDA-approved Radioactive Drug Research Committee (RDRC) without filing an IND (47). These types of studies may be applicable when the PPRA tumor targeting component is based on approved or endogenous peptides or proteins (e.g., a radiolabeled form of an antibody already approved for treatment of cancer). The nonradiolabeled drug must be generally recognized as safe and administered at doses that are known to have no clinically detectable pharmacologic effect in humans (immune responses to PPRA component can be considered a pharmacological response), while the radioactive form must meet acceptable radiopharmaceutical quality standards and its use must meet the prescribed radiation dose limits; all based on valid scientific evidence. These studies can be designed to evaluate metabolism (including kinetics, distribution, dosimetry, and localization), determine information regarding human physiology, pathophysiology, or biochemistry but cannot provide immediate therapeutic, diagnostic or similar feedback (e.g., preventive benefit to the study subject from the research). They must also not determine the safety (other than recording adverse events) and effectiveness of a radioactive drug in humans. A scientifically justifiable protocol with limits on subject enrollment number and on selection criteria that is approved by the local IRB is required. Individual jurisdictions in the European Union may have other mechanisms for the clinical evaluation of research PPRA, but those details are beyond the scope of this chapter.

In the United States and the European Union, the use of exploratory studies (36, 39) is being encouraged to assist sponsors in determining, for example, which one of multiple drug candidates to move forward into full clinical testing. These studies cover microdosing, subtherapeutic dosing, and therapeutic dosing components requiring the filing of an IND or CTA but importantly there is some regulatory relief for the type and quantity of nonclinical studies required to proceed as well as a further potential limitation on European Union Investigational Medicinal Product (IMP) GMP expectations governing manufacturing quality systems, especially for nonbiologic drugs. Exploratory studies are generally defined as those clinical studies that are conducted early in Phase 1, involve limited human exposure (subject number, dose number, and time), are not intended to produce a therapeutic or diagnostic outcome and are not intended to assess maximum tolerated dose. Microdose is defined, in general, as less than one-hundredth of the dose calculated to provide a pharmacologic effect (up to a maximum of 100 μg for drugs and up to 30 nmol for proteins).

Other features of these early and pre-Phase 1 studies are presented in Table 17.6 with a summary of potential applications to PPRA development. While PPRA radiation dose (and mass dose) limitations could restrict the use of therapeutic radionuclides in these studies, the use of imaging radionuclide analogue replacements may be especially valuable for these initial human studies. The use of pre-Phase 1 and exploratory Phase 1 studies in selecting a PPRA candidate by verifying adequate biodistribution (PK)/radiation dosimetry and tumor localization (with limited regulatory involvement) provides a basis for further clinical investigation that addresses PPRA safety (and efficacy) at therapeutic dose levels in the context of a traditional Phase 1 study. Potential limitations of the microdosing study data to higher mass doses (e.g., dose-dependent PK) that may be desirable for traditional Phase 1 studies should be considered.

17.5.2 Traditional Phase 1 to Phase 3 Clinical Trials

Clinical studies beyond pre-Phase 1 and exploratory Phase 1 studies tend to follow the traditional Phases 1–3 route with all of the associated regulatory documentation (including any nonclinical studies not conducted previously) required to be filed and approved prior to conduct. Traditional Phase 1 drug studies are usually carried out in healthy volunteers, however, for PPRA used in cancer therapeutic trials this is generally considered unethical based on the subjects unwarranted exposure to therapeutic doses of radioactivity. In addition, healthy volunteers may have altered biodistribution due to the lack of an appropriate tumor target and an obvious inability to assess tumor uptake/response. Therefore, regulatory agencies will allow the study of these agents at the Phase 1 stage in cancer patients. The usual cohort size is 10–50 patients. As discussed previously, it is important that a suitable starting dose for both PPRA mass and radioactivity is determined based on the (appropriately scaled) data from nonclinical animal studies with an acceptable safety tolerance and intended study goal considerations. Traditional Phase 1 studies for PPRA generally examine the safety, dose escalation, PK, metabolism, distribution, major side effects, and human radiation dosimetry (see Chapter 13). Bioanalytical assays for PPRA mass determination in biological fluids may need cross-validation studies from those used for animal PK/TK and provide value in the estimation of PPRA mass PK versus PPRA radiolabel PK that may highlight the nature and extent of radionuclide dissociation in humans. Although determination of individual patient (biodistribution and) dosimetry prior to administration of the PPRA at therapeutic radioactivity levels would be optimal to potentially maximize the radiation dose to the tumor while sparing other sensitive organs, this is rarely practiced due to the potential for induction of PPRA mass dose and radiation dose consequences (such as immunogenic response that alters the biodistribution of subsequent doses), the operational/practical hindrances, the general variability of such estimate values, and issues related to the correlation of tumor radiation dose to effective tumor response (see Chapter 13). Nonetheless, suitable imaging is often required for currently approved PPRA to verify an acceptable biodistribution pattern (48, 49) and to acquire total body counts over several days that enable therapeutic radioactivity dose calculations, based on a set safe total body radiation dose (49). Therefore, dosimetry in Phase 1 clinical trials should aim for target population-based estimates that would be useful for radiation dose estimates in (at least) critical normal organs/tissues as a basis for verifying maximum safe (radioactivity) total dose estimate as clinical development progresses and include tumor-specific dosimetry where possible. The radiation dosimetry estimates would be expected to include contributions from radionuclidic impurities if relevant. The critical organ system is often the bone marrow that may also be compromised by cancer involvement and concurrent myelosuppressive therapy suggesting careful patient inclusion criteria for these studies. The conduct of dosimetry studies requires the detection of whole body radionuclide emissions using external devices (gamma-cameras) at various times after radiolabeled PPRA injection and to be meaningful this data must be regionally quantitative; thus, PET and SPECT imaging techniques are preferred (if possible) but attenuation corrected data from any relevant radionuclide is acceptable (see Chapter 13). It is recommended that well-established software developed specifically for this purpose can be used to convert the time/radioactivity data to organ-specific cumulative uptake and further calculations (using phantom model data) to the final dosimetry estimates (such as MIRDOSE or OLINDA) as this simplifies regulatory agency acceptability. Complete documentation of the instrumental parameters, raw data, calculations, and assumptions as well as complete tabulated input and output results will be required to be submitted to support Phase 2 clinical trials.

As stated earlier, immunogenicity is of regulatory concern for PPRA, especially protein RA. The considerations for the nature and extent of immunogenicity testing as part of a Phase 1 clinical trial are similar to that outlined for nonclinical immunogenicity testing for a risk-based approach but regardless of even minimal risk (i.e., for humanized antibodies), its evaluation in human clinical trials is required. The primary assay methodology should be developed to enable reliable overall anti-PPRA reactive responses and further refinements should test for ability to neutralize the pharmacological effects of the PPRA using *in vitro* receptor/binding site cell assays or the presence of anti-idiotype antibodies and nonspecific anti-isotype antibodies in the case of MAb protein RA and there may be a regulatory expectation to extrapolate the observed titers to a proportion of the mass dose affected. Since immunogenicity may be late developing, clinical trial protocols need to assure sample availability for at least six months. Results should be evaluated for their potential effects on PPRA immune complex mediated adverse events and diminished therapeutic response and full disclosure of methods, findings and implications reported to regulatory agencies prior to further clinical development (44).

Phase 2 studies continue with safety determination as well as looking at efficacy outcomes in a narrower range of cancer types and PPRA dose level/frequency that help define these parameters in usual cohort sizes from 50 to 200 patients. It should be noted that because some Phase 1 PPRA clinical trials involve cancer patients, some facets of efficacy evaluation can be included earlier than Phase 2 that may provide a commercial designation as Phase 1/2 clinical trials. At this stage, efficacy outcome determination for cytotoxic PPRA generally requires determination of objective patient response rates based on well-documented criteria and/or time to tumor progression improvement with or without comparison to the current standard of care. Surrogate markers of efficacy such as an approved diagnostic radiopharmaceutical imaging procedure in the specified indication to estimate tumor size and/or functional aspects (e.g., molecular imaging) would be useful as an adjunct evaluation (see Chapter 15). An unapproved diagnostic radiopharmaceutical should only be used for this purpose when it is approved as part of the PPRA clinical trial submission, or as a separate clinical trial submission with overlapping patient inclusion criteria, and assurance of scientific and clinical validity is provided. In both cases, objective responses will be required to confirm efficacy.

When the PPRA reaches the Phase 3 stage, the intended cancer indication is fully evaluated for final safety and efficacy (usually overall survival or progression/disease free survival) in a clinical trial population in the order of 200–1000 cancer patients to provide statistically valid evidence of a therapeutic advantage with an acceptable safety profile against an appropriate control group. These are typically multicenter trials and while treatment blinding may be difficult due to the radiological aspects, outcome evaluation should maintain blinded status. In general, regulatory agencies require at least two well-controlled studies (usually Phase 3) for approval but on rare occasions a PPRA completing extended Phase 2 clinical trials and demonstrating outstanding efficacy may be eligible for approval as would the completion of one well controlled Phase 3 study but both of these require extensive discussion with the regulatory agency and may come with conditions to be fulfilled after approval.

17.6 SUMMARY

The development of PPRA leading to an approved drug product requires extensive knowledge of both the scientific/technical/clinical operational requirements and the national and international regulatory requirements if efficient commercialization is to be achieved. Government regulatory agencies and various departments are responsible for PPRA submission review in Canada, the United States, and the European Union with requirements for submission contents specific for each jurisdiction. The research involving the design and initial manufacture of the recombinant cell culture-derived protein or synthetic chemistry-derived peptide as the TTC must consider appropriate production controls (stage-specific GMP) and a variety of analytical/safety testing requirements for the starting materials (cell lines and chemicals). Further progressive biochemical and physicochemical characterization and quality considerations that meet regulatory expectations must demonstrate an adequate safety profile as this component is the main constituent of the final drug product. The enablement for radiolabeling resides with the creation of (or endogenous) suitable radionuclide component binding site typically by bifunctional ligand (chelate) attachment to form a TTC via a process that requires considerable operational and reaction parameter understanding to demonstrate adequate scientific and regulatory control as a key manufacturing process. Radionuclide component development requires attention to the method of production and on the required testing of radionuclidic purity, radiochemical purity and assay that continues after attachment to the tumor targeting conjugate. Depending on various factors, the TTC can be directly radiolabeled by the manufacturer as a RTU drug product format, or may be formatted as a TTC Kit DP for on-site radiolabeling with separate radionuclide Kit DP component that, for commercialization in either format, requires all elements of GMP regulatory compliance and pharmaceutical and radiopharmaceutical batch quality testing. The developmental DP should be evaluated for a variety of radiopharmacological activities including tumor localization/reduction capabilities and biodistribution (or radiation dosimetry) studies in suitable animal models and a regulatory mandated nonclinical testing scheme in animals to assess potential safety and toxicity. The extent of safety and toxicological testing studies is primarily dependent on the tumor targeting component origin (recombinant or synthetic), the stage of the clinical investigation (i.e., Phase 1, 2, or 3) with some jurisdictional-specific requirements but normally involves acute and repeat dose toxicity studies in two species (rodent and nonrodent), core battery safety pharmacology studies, and a genotoxicity study. Approval to conduct a traditional clinical trial requires a regulatory agency submission (CTA/IND) containing a compilation of all DP manufacturing and quality information, all safety and toxicological testing study results (and previous clinical testing data/results) as well as proposed clinical testing details and commitment of trial conduct according to GCP regulatory requirements to enable regulatory agency evaluation of safety and potential risk/benefit of the proposed trial. For PPRA, determination of radiation dosimetry and immunogenicity potential should be evaluated early on in clinical development. The drug development regulatory framework enables all applicable jurisdictions to have input into the appropriate development pathway that is based on sponsor basic and clinical research science and tailored to an individual PPRA with the outcome of an NDS/NDA/MA submission/application that has sufficient nature, quality, and quantity of adequately documented data/results to justify agency approval for commercialization of a PPRA DP that advances patient treatment benefit.

DEDICATION

This chapter is dedicated to the memory of Antoine A. Noujaim, Ph.D., D.Sc.(Hon), (1937–2006), a learned professor, respected business entrepreneur, and gracious gentleman to all as well as an inspirational mentor, colleague, and friend to the authors.

REFERENCES

1. Guidance for Clinical Trial Sponsors Clinical Trial Applications. Health Products and Foods Branch, Ottawa; 2003 Jun 25. 39 p. (ISBN 0-662-33127-3).
2. ICH Harmonised Tripartite Guideline: Guideline for Good Clinical Practice. E6(R1). Step 5. Geneva: International Conference on Harmonisation of Technical Requirements for Registration of Pharmaceuticals for Human Use (ICH); 1996 Jun 10. 53 p.
3. Detailed Guidance for the Request for Authorisation of a Clinical Trial on a Medicinal Product for Human Use to the Competent Authorities, Notification of Substantial Amendments and Declaration of the End of the Trial. European Commission, Brussels; 2004 April. 44 p. (ENTR/CT 1).
4. Council Directive No. 2001/83/EC of the European Parliament and of the Council of 6 November 2001 on the "Community code relating to medicinal products for human use" as amended, O.J. L311/67 (2004).
5. 21 C.F.R. 312 Investigational New Drug Application. U.S. Food and Drug Administration; April 2009.
6. Guidance for Industry Content and Format of Investigational New Drug Applications (INDs) for Phase 1 Studies of Drugs, Including Well-characterized, Therapeutic, Biotechnology-Derived Products. Food and Drug Administration, Rockville, MD; 1995 Nov. 14 p.

7. Guidance for Industry INDs for Phase 2 and Phase 3 Studies Chemistry, Manufacturing, and Controls Information. U.S. Food and Drug Administration, Rockville, MD; 2003 May. 24 p.

8. 21 C.F.R. 58 Good Laboratory Practice for Nonclinical Laboratory Studies. U.S. Food and Drug Administration; April 2009.

9. OECD Series on Principles of Good Laboratory Practice and Compliance Monitoring. No. 1. OECD Principles on Good Laboratory Practice (as revised in 1997). Paris: Environment Directorate, Organisation for Economic Co-operation and Development; 1998. 41 p. (ENV/MC/CHEM(98)17).

10. ICH Harmonised Tripartite Guideline: Structure and Content of Clinical Study Reports. E3. Step 5. Geneva: International Conference on Harmonisation of Technical Requirements for Registration of Pharmaceuticals for Human Use (ICH); 1995 Nov 30. 41 p.

11. Reilly RM. The radiopharmaceutical science of monoclonal antibodies and peptides for imaging and targeted *in situ* radiotherapy of malignancies. In: Gad SC, editor. *Handbook of Pharmaceutical Biotechnology.* Wiley Interscience, Hoboken, NJ, 2007, pp. 884–942.

12. Guidance Document Annex 13 to the Current Edition of the Good Manufacturing Practices Guidelines Drugs Used in Clinical Trials GUI-0036. Health Products and Foods Branch Inspectorate, Ottawa; 2009 Dec 1. 25 p.

13. Volume 4 EU Guidelines to Good Manufacturing Practice Medicinal Products for Human and Veterinary Use Annex 13 Investigational Medicinal Products. European Commission, Brussels; 2010 Feb. 19 p. (ENTR/F/2/AM/an D(2010) 3374).

14. Guidance for Industry cGMP for Phase 1 Investigational Drugs. U.S. Food and Drug Administration, Rockville, MD; 2008 Jul. 17 p.

15. Guidance for Industry Preparation of Investigational New Drug Products (Human and Animal). Food and Drug Administration, Rockville, MD; 1991 Mar. 7 p.

16. ICH Harmonised Tripartite Guideline: Good Manufacturing Practice Guide for Active Pharmaceutical Ingredients. Q7. Step 5. International Conference on Harmonisation of Technical Requirements for Registration of Pharmaceuticals for Human Use (ICH), Geneva; 2000 Nov 10. 43 p.

17. Good Manufacturing Practices; C.R.C. Food and Drug Regulations, Canada. 2010, c. 870, s. C.02.

18. Good Manufacturing Practices (GMP) Guidelines- 2009 Edition, Version 2. Health Products and Foods Branch Inspectorate, Ottawa; 2009 May 8. 99 p.

19. Annex to the GMP Guidelines. Good Manufacturing Practices for Schedule C Drugs. Health Products and Foods Branch Inspectorate, Ottawa; 1999 Jun 30. 12 p.

20. Council Directive No. 2003/94/EC of the European Parliament and of the Council of 8 October 2003 on the "Laying down the principles and guidelines of good manufacturing practice in respect of medicinal products for human use and investigational medicinal products for human use", O.J. L262/22 (2003).

21. Good Manufacturing Practices, Vol. 4. Selected Annexes. European Commission, Brussels; 1998–2010.

22. 21 C.F.R. 210 Current Good Manufacturing Practice in Manufacturing, Processing, Packing, or Holding of Drugs and 21 C.F.R. 211 Current Good Manufacturing Practice for Finished Pharmaceuticals. U.S. Food and Drug Administration; April 2009.

23. Guideline on Requirements for First-in-Man Clinical Trials for Potential High-Risk Medicinal Products. Draft. Committee for Medicinal Products for Human Use, European Medicines Agency, London; 2007 May 23. 11 p. (EMEA/CHMP/SWP/28367/2007 Corr.).

24. Guideline on Production and Quality Control of Monoclonal Antibodies and Related Substances. Draft. Committee for Medicinal Products for Human Use, European Medicines Agency, London; 2007 Apr 5. 15 p. (EMEA/CHMP/BWP/157653/2007).

25. ICH Harmonised Tripartite Guideline: Quality of Biotechnology Products: Analysis of the Expression Construct in Cells used for Production of r-DNA Derived Protein Products. Q5B. Step 5. International Conference on Harmonisation of Technical Requirements for Registration of Pharmaceuticals for Human Use (ICH), Geneva; 1996 Feb. 5 p.

26. Points to Consider in the Manufacture and Testing of Monoclonal Antibody Products for Human Use. U.S. Food and Drug Administration, Rockville, MD; 1997 Feb 28. 50 p.

27. ICH Harmonised Tripartite Guideline: Viral Safety Evaluation of Biotechnology Products Derived from Cell Lines of Human or Animal Origin. Q5A(R1). Step 5. International Conference on Harmonisation of Technical Requirements for Registration of Pharmaceuticals for Human Use (ICH), Geneva; 1999 Sep 23. 27 p.

28. Guideline on Virus Safety Evaluation of Biotechnology Investigational Medicinal Products. Committee for Medicinal Products for Human Use, European Medicines Agency, London; 2008 Jul 24. 9 p. (EMEA/CHMP/BWP/398498/2005).

29. Note for Guidance on Minimising the Risk of Transmitting Animal Spongiform Encephalopathy Agents via Human and Veterinary Medical Products. Draft. Committee for Medicinal Products for Human Use, Committee for Medicinal Products for Veterinary Use, European Medicines Agency, London; 2008 Jul 24. 5 p. (EMEA/CHMP/BWP/398498/2005).

30. Carpenter JF, Randolph TW, Jiskoot W, et al. Overlooking subvisible particles in therapeutic protein products: gaps that may compromise product quality. *J Pharm Sci* 2009;98:1201–1205.

31. Guidance for Industry for the Submission of Chemistry, Manufacturing and Controls Information for Synthetic Peptide Substances. U.S. Food and Drug Administration, Rockville, MD; 1994 Nov. 13 p.

32. Guidance for Industry: Pharmaceutical Quality of Aqueous Solutions. Health Products and Foods Branch, Ottawa; 2005 Feb 15. 8 p. (ISBN 0-662-39615-4).

33. Note for Guidance on Development Pharmaceutics. Committee for Medicinal Products for Human Use, European Medicines Agency, London; 1998 Jan 28. 8 p. (EMEA/CPMP/QWP/155/96).

34. ICH Harmonised Tripartite Guideline: Impurities in New Drug Products. Q3B(R2). Step 5. International Conference on Harmonisation of Technical Requirements for Registration of Pharmaceuticals for Human Use (ICH), Geneva; 2006 Jun 02. 12 p.

35. ICH Harmonised Tripartite Guideline: Stability Testing of New Drug Substances and Products. Q1A(R2). Step 5. International Conference on Harmonisation of Technical Requirements for Registration of Pharmaceuticals for Human Use (ICH), Geneva; 2003 Feb 06. 18 p.

36. ICH Harmonised Tripartite Guideline: Guidance on Nonclinical Safety Studies for the Conduct of Human Clinical Trials and Marketing Authorization for Pharmaceuticals. M3

(R2). Step 4. International Conference on Harmonisation of Technical Requirements for Registration of Pharmaceuticals for Human Use (ICH), Geneva; 2009 June 11. 25 p.

37. ICH Harmonised Tripartite Guideline: Safety Pharmacology Studies for Human Pharmaceuticals. S7A. Step 5. International Conference on Harmonisation of Technical Requirements for Registration of Pharmaceuticals for Human Use (ICH), Geneva; 2000 Nov 8. 9 pages.

38. ICH Harmonised Tripartite Guideline: Preclinical Safety Evaluation of Biotechnology-Derived Pharmaceuticals. S6. Step 5. International Conference on Harmonisation of Technical Requirements for Registration of Pharmaceuticals for Human Use (ICH), Geneva; 1997 Jul 16. 28 p.

39. Guidance for Industry, Investigators, and Reviewers: Exploratory IND Studies. U.S. Food and Drug Administration, Rockville, MD; 2006 Jan. 13 p.

40. Guidance for Industry: Nonclinical Evaluation of Late Radiation Toxicity of Therapeutic Radiopharmaceuticals. Draft. U.S. Food and Drug Administration, Rockville, MD; 2005 Jun. 10 p.

41. ICH Harmonised Tripartite Guideline: Guidance on Genotoxicity Testing and Data Interpretation for Pharmaceuticals Intended for Human Use. S2(R1). Step 3. International Conference on Harmonisation of Technical Requirements for Registration of Pharmaceuticals for Human Use (ICH), Geneva; 2008 Mar 06. 28 p.

42. Guideline on the Limits of Genotoxic Impurities. Committee for Medicinal Products for Human Use, European Medicines Agency, London; 2006 Jun 28. 8 p. (EMEA/CPMP/QWP/251344/2006).

43. Guidance for Industry: Genotoxic and Carcinogenic Impurities in Drug Substances and Products: Recommended Approaches. Draft. U.S. Food and Drug Administration, Rockville, MD; 2008 Dec. 13 p.

44. Guideline on Immunogenicity Assessment of Biotechnology-Derived Therapeutic Proteins. Committee for Medicinal Products for Human Use, European Medicines Agency, London; 2007 Dec 13. 18 p. (EMEA/CHMP/BMWP/14327/2006).

45. Guidance Policy: Use of Positron Emitting Radiopharmaceuticals (PER) in Basic Research. Policy 0053. Health Products and Foods Branch Inspectorate, Ottawa; 2006 Feb 24. 4 p.

46. Guidance Document: Factors Considered in the Assessment of Risks Involved in the Use of Positron Emitting Radiopharmaceuticals in Basic Research Involving Humans. Health Products and Foods Branch, Ottawa; 2006 Feb 24. 7 p.

47. 21 C.F.R. 361 Prescription Drugs for Human Use Generally Recognized as Safe and Effective and not Misbranded: Drugs used in Research. U.S. Food and Drug Administration; April 2009.

48. Zevalin® Prescribing Information. Cell Therapeutics, Inc. 2008.

49. Bexxar® Prescribing Information. GlaxoSmithKline. Oct 2005.

INDEX

Monoclonal Antibody and Peptide-Targeted Radiotherapy of Cancer, Edited by Raymond M. Reilly
Copyright © 2010 John Wiley & Sons, Inc.

613